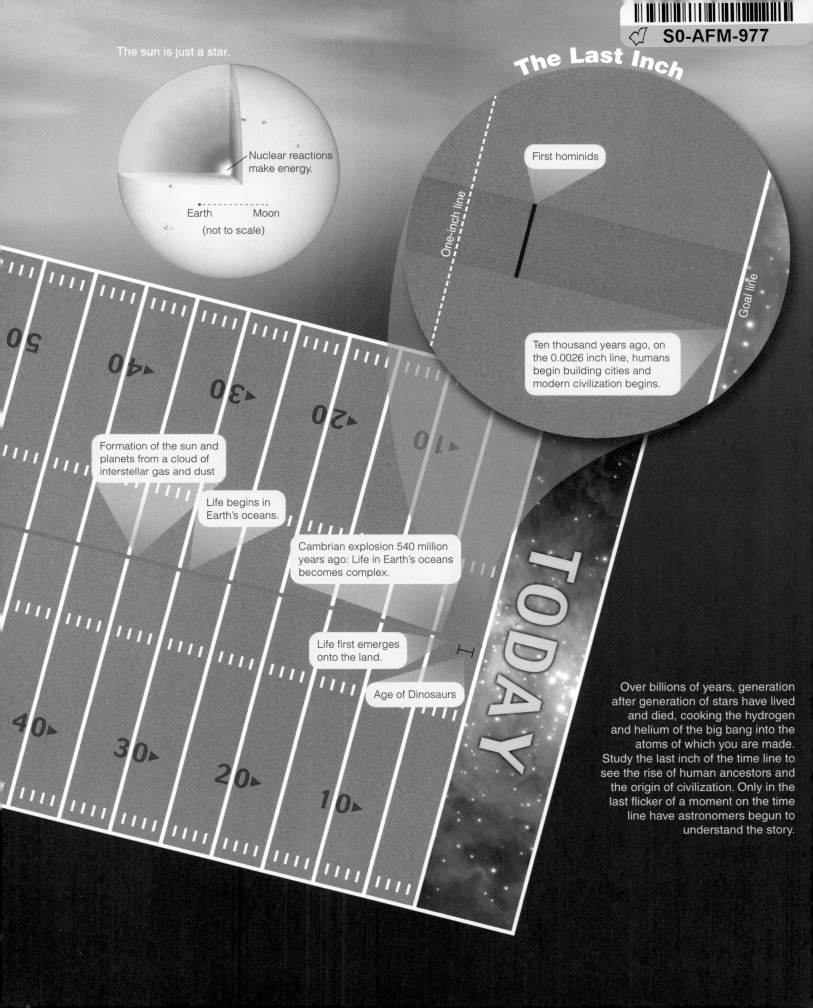

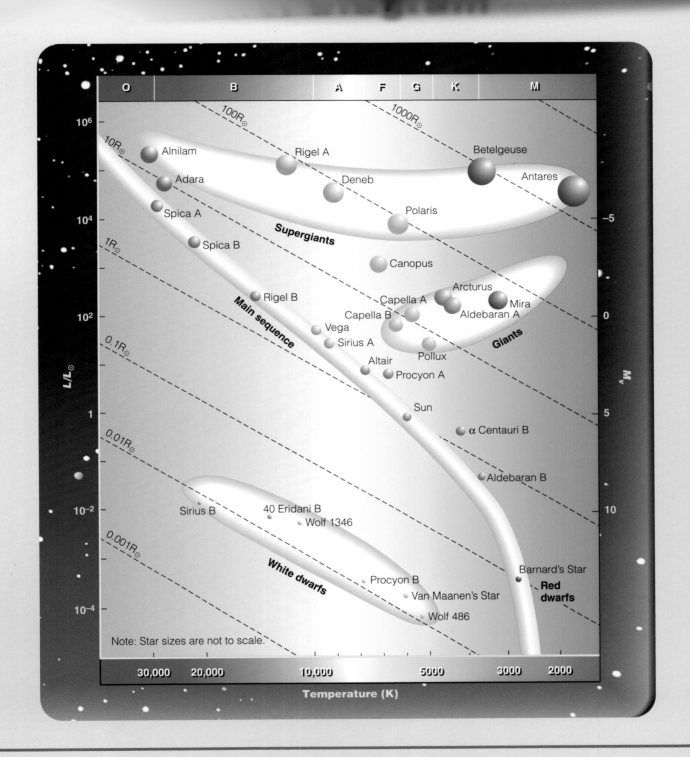

The H–R diagram is the key to understanding stars, their birth, their long lives, and their eventual deaths. Luminosity ($L/L_\odot$) refers to the total amount of energy that a star emits in terms of the sun's luminosity, and the temperature refers to the temperature of its surface. Together, the temperature and luminosity of a star locate it on the H–R diagram and tell astronomers its radius, its family relationships with other stars, and a great deal about its history and fate.

The terrestrial or Earthlike planets lie very close to the sun, and their orbits are hardly visible in a diagram that includes the outer planets.

Mercury, Venus, Earth and its moon, and Mars are small worlds made of rock and metal with little or no atmospheric gases.

The outer worlds of our solar system orbit far from the sun. Jupiter, Saturn, Uranus, and Neptune are Jovian or Jupiter-like planets much bigger than Earth. They contain large amounts of low-density gases.

Pluto is a tiny, icy world orbiting far from the sun, and some astronomers argue that it isn't really a planet at all.

This book is designed to use arrows to alert you to important concepts in diagrams and graphs. Some arrows point things out, but others represent motion, force, or even the flow of light. Look at arrows in the book carefully and use this Flash Reference card to catch all of the arrow clues.

The Terrestrial Worlds

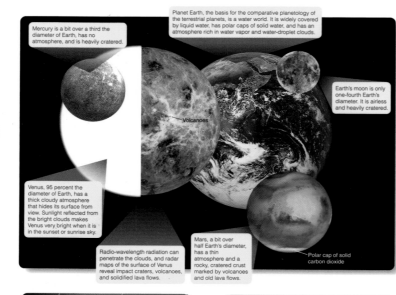

Planet Earth, the basis for the comparative planetology of the terrestrial planets, is a water world. It is widely covered by liquid water, has polar caps of solid water, and has an atmosphere rich in water vapor and water-droplet clouds.

Mercury is a bit over a third the diameter of Earth, has no atmosphere, and is heavily cratered.

Earth's moon is only one-fourth Earth's diameter. It is airless and heavily cratered.

Venus, 95 percent the diameter of Earth, has a thick cloudy atmosphere that hides its surface from view. Sunlight reflected from the bright clouds makes Venus very bright when it is in the sunset or sunrise sky.

Volcanoes

Radio-wavelength radiation can penetrate the clouds, and radar maps of the surface of Venus reveal impact craters, volcanoes, and solidified lava flows.

Mars, a bit over half Earth's diameter, has a thin atmosphere and a rocky, cratered crust marked by volcanoes and old lava flows.

Polar cap of solid carbon dioxide

Planetary Orbits

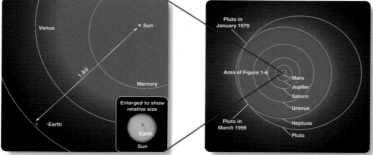

Venus

Sun

1 AU

Mercury

Earth

Enlarged to show relative size

Earth

Sun

Pluto in January 1979

Area of Figure 1-6

Mars
Jupiter
Saturn
Uranus
Neptune
Pluto

Pluto in March 1999

The Outer Worlds

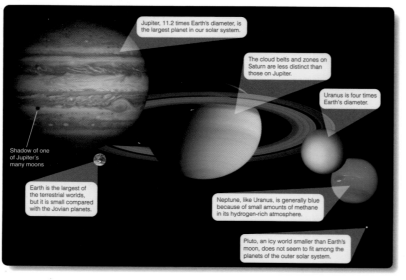

Jupiter, 11.2 times Earth's diameter, is the largest planet in our solar system.

The cloud belts and zones on Saturn are less distinct than those on Jupiter.

Uranus is four times Earth's diameter.

Shadow of one of Jupiter's many moons

Earth is the largest of the terrestrial worlds, but it is small compared with the Jovian planets.

Neptune, like Uranus, is generally blue because of small amounts of methane in its hydrogen-rich atmosphere.

Pluto, an icy world smaller than Earth's moon, does not seem to fit among the planets of the outer solar system.

Point at things:

You are here

Force:

Process flow:

Measurement:

Direction:

Radio waves, infrared, photons:

Motion:

Rotation 2-D

Rotation 3-D

Linear

Light flow:
Updated arrow style

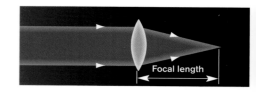

Focal length

See page 478 for the terrestrial planets. See pages 6 and 7 for the two orbital diagrams. See page 555 for the outer worlds.

The Solar System

About the Author

Mike Seeds is Professor of Astronomy at Franklin and Marshall College, where he has taught astronomy since 1970. His research interests have focused on peculiar variable stars and the automation of astronomical telescopes. He has been serving since 1987 as Principal Astronomer in charge of the Phoenix 10, the first fully robotic telescope, located in southern Arizona. In 1989, he received the Christian R. and Mary F. Lindback Award for Distinguished Teaching. In addition to teaching, writing, and research, Mike has published educational tools for use in computer-smart classrooms. His interest in the history of astronomy led him to offer upper-level courses "Archaeoastronomy" and "Changing Concepts of the Universe." He has also published educational software for preliterate toddlers. Mike is Senior Consultant in the creation of the telecourse *Astronomy: Observations and Theories*.

He is the author of *Astronomy: The Solar System and Beyond,* Fourth Edition (2005), and *Horizons: Exploring the Universe,* Ninth Edition (2005), published by Brooks/Cole.

About the Cover

The dome of the 4-meter Blanco Telescope atop Cerro Tololo in Chile is superimposed here on the glowing gases of the Ring Nebula exhaled from an aging star thousands of years ago. The cycle of star death and star birth includes the formation of planets around forming stars just as our solar system formed in the same nebula with the sun about 4.6 billion years ago. (Observatory photograph credit, StockTrek/Getty Images; background photo [digital composite with observatory], Helix Nebula, Hubble Images: NASA, ESA, C. R. O'Dell [Vanderbilt University], M. Melxner and P. I. McCullough [STScI]; small detail photos, left: Jupiter and Io, Royalty Free/Corbis, middle, Saturn and its moons, Sightseeing Archive/Getty Images; right, [digital composite] planet with water-bearing moon, StockTrek/Getty Images)

5

FIFTH EDITION

The Solar System

Michael A. Seeds

Joseph R. Grundy Observatory
Franklin and Marshall College

THOMSON

BROOKS/COLE

Australia • Canada • Mexico • Singapore • Spain
United Kingdom • United States

For Gary and Karen

THOMSON
BROOKS/COLE

The Solar System, **Fifth Edition**
Michael A. Seeds

Astronomy Acquisitions Editor: *Chris Hall*
Development Editor: *Alyssa White*
Assistant Editor: *Brandi Kirksey*
Editorial Assistant: *Jessica Jacobs*
Editorial Intern: *David Marks*
Technology Project Manager: *Samuel Subity*
Publisher/Executive Editor: *David Harris*
Marketing Manager: *Mark Santee*
Marketing Assistant: *Michele Colella*
Managing Marketing Communications Manager: *Bryan Vann*
Project Manager, Editorial Production: *Teri Hyde*
Executive Art Director: *Vernon Boes*
Print and Media Buyer: *Karen Hunt*
Permissions Editor: *Stephanie Lee, Bob Kauser*
Production Service: *Heckman & Pinette*
Text Designer: *Linda Beaupré/Stone House Art*

Art Editor: *Lisa Torri*
Photo Researchers: *Mike Seeds and Kathleen Olson*
Copy Editor: *Margaret Pinette*
Illustrator: *Precision Graphics*
Cover Designer: *Irene Morris*
Cover Image: *Observatory photograph credit, StockTrek/Getty Images; background photo [digital composite with observatory], Helix Nebula, Hubble Images: NASA, ESA, C. R. O'Dell [Vanderbilt University], M. Melxner and P. I. McCullough [STScI]; small detail photos, left: Jupiter and Io, Royalty Free/Corbis, middle, Saturn and its moons, Sightseeing Archive/Getty Images; right, [digital composite] planet with water-bearing moon, StockTrek/Getty Images*
Cover Printer: *Transcontinental-Interglobe*
Compositor: *Thompson Type*
Printer: *Transcontinental-Interglobe*

Printed in Canada
1 2 3 4 5 6 7 10 09 08 07 06

Library of Congress Control Number: 2005936362

ISBN 0-495-01575-X

Thomson Higher Education
10 Davis Drive
Belmont, CA 94002-3098
USA

For more information about our products, contact us at:
Thomson Learning Academic Resource Center
1-800-423-0563

For permission to use material from this text or product, submit a request online at **http://www.thomsonrights.com.**
Any additional questions about permissions can be submitted by e-mail to **thomsonrights@thomson.com.**

Part 1: Exploring the Sky

Part 2: The Stars

Part 4: The Solar System

Part 5: Life

Contents

Part 1: Exploring the Sky

Windows on Science

Concept Art Portfolios

Part 2: The Stars

Part 4: The Solar System

Windows on Science

Concept Art Portfolios

A Note to the Student

From Mike Seeds

Hi,

I'm really glad you are taking an astronomy course. You are going to see some amazing things, from the icy rings of Saturn to powerful eruptions on the sun. Our universe is so beautiful, it is sad to think that not everyone gets to take an astronomy course.

Two Goals

You will learn a lot of new stuff in this course, but there are two things I hope you find especially satisfying. This astronomy course will help you answer two important questions:

- What are we?
- How do we know?

By "What are we?" I mean, "Where do we fit into the history of the universe?" The atoms you are made of had their first birthday in the big bang when the universe began, but those atoms have been cooked and remade inside stars, and now they are inside you. Where will they be in a billion years? Astronomy is the only course on campus than can tell you that story, and it is a story that everyone should know.

By "How do we know?" I mean, "How does science work?" How can anyone know there was a big bang? In today's world, you need to think carefully about the things so-called experts say. You should demand explanations. Scientists have a special way of knowing based on evidence. Scientific knowledge isn't just opinion or policy or marketing or public relations. It is humanity's best understanding of nature. To understand the world around you, you should understand how science works.

These two questions are the focus of astronomy, and they are just for you. You need to know the answers to these questions so you can appreciate how wonderful the universe is and how special you are.

Favorite Stars

Along the way, I want to introduce you to eight of my favorite stars. Maybe you will add a few favorites of your own, but these are eight stars I enjoy looking at in the sky. When I see Betelgeuse in the winter sky, for example, I see more than just a bright red star. In my imagination, I see a star over 800 times larger than the sun—an aging star that will soon die, perhaps in a violent explosion. Here's my list of favorite stars:

Sirius	Brightest star in the sky	Winter
Betelgeuse	Bright red star in Orion	Winter
Rigel	Bright blue star in Orion	Winter
Aldebaran	Red eye of Taurus the Bull	Winter
Polaris	The North Star	Year round
Vega	Bright star overhead	Summer
Spica	Bright southern star	Summer
Alpha Centauri	Nearest star to the sun	Spring, far south

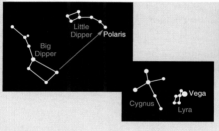

You can see Polaris year round, but Sirius, Betelgeuse, Rigel, and Aldebaran are in the winter sky. Spica is in the summer sky, and Vega is visible evenings in the late summer or fall. Alpha Centauri is a special star, but you will have to travel at least as far south as southern Florida to get a glimpse of it above the southern horizon. Use the charts here along with the star charts in the back of this book to locate these stars.

Every one of these stars has its own secret, and when you find out why they are my favorites, you'll enjoy finding them in the sky just as I do.

Expect to Be Astonished

One reason astronomy is exciting is that astronomers discover new things every day. Astronomers expect to be astonished. You can share in the excitement because I've worked hard to include the newest images, the newest discoveries, and the newest insights that will take you, in an introductory course, to the frontier of human knowledge. You'll see new evidence of ancient oceans and lakes on Mars and a vast ocean on Jupiter's moon Europa. You'll visit the moon Titan, where it rains liquid methane, and Pluto, where most gases freeze solid. You'll see stars born from clouds of gas and dust. You'll see how the life cycle of stars created the atoms you are made of. Huge telescopes in space and on remote mountaintops provide a daily dose of excitement that goes far beyond sensationalism. These new discoveries in astronomy are exciting because they are about us. They tell us more and more about what we are.

As you read this book, notice that it is not organized as lists of facts for you to memorize. That could make even astronomy boring. Rather this book is organized to show you how scientists use evidence and theory to create logical arguments that show how nature works. Look at the list of special features that follows this note. Those features were carefully designed to help you understand astronomy as evidence and theory. Once you see science as logical arguments, you hold the key to the universe.

Do Not Be Humble

As a teacher, my quest is simple. I want you to understand your place in the universe—not just your location in space but your location in the unfolding history of the physical universe. Not only do I want you to know where you are and what you are in the universe, but I want you to understand how scientists know. By the end of this book, I want you to know that the universe is very big, but that it is described by a small set of rules and that we humans have found a way to figure out the rules—a method called *science.*

Do not be humble. Astronomy tells us that the universe is vast and powerful, but it also tells us that we are astonishing creatures. We humans are the parts of the universe that think. You are a small creature, but remember that it is your human brain that is capable of understanding the depth and beauty of the cosmos.

To appreciate your role in this beautiful universe, you must learn more than just the facts of astronomy. You must understand what we are and how we know. Every page of this book reflects that ideal.

Mike Seeds
mike.seeds@fandm.com

Key Content and Pedagogical Changes to the Ninth Edition

- Chapter 8, "The Sun," has been thoroughly rewritten and reorganized. The new presentation is designed to help you see the sun as a physical object in which energy is generated in its core, flows outward through the interior, and generates the activity we see on the sun's surface.

- The Concept Art Portfolios were redesigned to aid you in navigating through these key visual summaries. The flow through the material has been clarified by adding numerical markers and arrows to better guide you to a stronger conceptual understanding.

- The Guideposts and Summaries were redesigned to make them a more effective study tool. The chapter-opening Guideposts contain a short set of Essential Questions focused on the key ideas in each chapter. The questions are answered in the narrative and then are presented as bulleted insights in the Summary.

- All figures have been updated, redesigned, and carefully annotated to help you pick out the critical details related to the discussions.

- A new type of question has been added to the end-of-chapter material. "Learning to Look" questions give you a chance to critically examine images to test your understanding of fundamental concepts.

- Of course, I have updated the newest exciting discoveries in astronomy, including images from the surface of Titan, volcanism on icy moons, the Kuiper belt and the status of Pluto, dark energy, the search for water on Mars, and much more.

Special Features

- **Special two-page Concept Art Portfolios** provide an opportunity for you to create your own understanding and share in the satisfaction that scientists feel as they uncover the secrets of nature.

- **Guided discovery figures** illustrate important ideas visually and guide you to understand relationships and contrasts interactively.

- **Windows on Science** provide a parallel commentary on how all of science works. For example, the Windows point out where astronomers use statistical evidence, where they are reasoning by analogy, and where they are building a scientific model rather than a scientific hypothesis.

- **Focus on Fundamentals** will help you understand concepts from physics that are critical to understanding modern astronomy.

- **Guideposts** on the opening page of each chapter help you see the organization of the book. Cast as a series of Essential Questions that are answered within the text narrative, each Guidepost further connects the chapter with preceding and following chapters to provide an overall organizational guide.

- **Building Scientific Arguments** at the end of each text section are carefully designed questions to help you review and synthesize concepts from the section. A short answer follows to show how scientists construct scientific arguments from observations, evidence, theories, and natural laws that lead to a conclusion. A further question then gives you a chance to construct your own argument on a related issue.

- **End-of-Chapter Review Questions** are designed to help you review and test your understanding of the material.

- **End-of-Chapter Discussion Questions** go beyond the text and invite you to think critically and creatively about scientific questions. You can think about these questions yourself or discuss them in class.

- **Media Clusters** list all of the Web resources you can call on to illustrate, review, and expand the material in each chapter.

- **Critical Inquiries for the Web** challenge you to use the World Wide Web to explore further and to think creatively and analytically about astronomy.

- **Exploring *TheSky*** are experiments you can perform with the software *TheSky* or, by making slight modifications, with any planetarium software. These experiments will allow you to see the sky and celestial bodies in new ways on your own computer screen.

As additional aids, notice that most versions of this book also include access to the following electronic enhancements:

- **Virtual Astronomy Laboratories.** This set of online lab exercises covers 20 of the most important concepts of introductory astronomy in an interactive online environment. You may be asked to run simulations, observe animations, perform calculations, or collect data to help you master these fundamental concepts. The labs cover topics from helioseismology to dark matter and allow you to submit your results electronically to your instructor or print them out to hand in. The Media Cluster at the end of each chapter in this textbook notes which labs correlate to that chapter.

- **AceAstronomy.** Take charge of your learning with the first assessment-centered student learning tool for astronomy. AceAstronomy uses a series of diagnostic tests to provide you with a personalized learning plan, so you can begin to maximize your study time with a host of interactive tutorials and quizzes that help you focus on what you need to learn to master astronomy.

Acknowledgments

I started writing astronomy textbooks in 1973, and over the years I have had the guidance of a great many people who care about astronomy and teaching.

I would like to thank all of the students and teachers who have responded so enthusiastically to *Foundations of Astronomy*. Their comments and suggestions have been very helpful in shaping this book.

I would especially like to thank the reviewers whose careful analysis and thoughtful suggestions have been invaluable in completing this new edition.

Many observatories, research institutes, laboratories, and individual astronomers have supplied figures and diagrams for this edition. Their names are given on the page with the illustrations they provided, and I would like to thank them specifically for their generosity.

Writing about every branch of astronomy is a daunting task, and I could not do it without helpful contributions from experts in various fields. I especially want to think Dana Backman, George Jacoby, Victoria Kaspi, and William Keel for their helpful guidance on technical issues.

I am happy to acknowledge the use of images and data from a number of important programs. In preparing materials for this book I used NASA's Sky View facility located at NASA Goddard Space Flight Center. I have used atlas images and mosaics obtained as part of the Two Micron All Sky Survey (2MASS), a joint project of the University of Massachusetts and the Infrared Processing and Analysis Center/California Institute of Technology, funded by the National Aeronautics and Space Administration and the National Science Foundation. A number of solar images are used by the courtesy of the SOHO consortium, a project of international cooperation between ESA and NASA.

It is always a pleasure to work with the Thomson Learning team at Brooks/Cole and Wadsworth. Special thanks go to all of the people who have contributed to this project, including Teri Hyde, Kathleen Olson, Sam Subity, Lisa Torri, Vernon Boes, Mark Santee, and Brandi Kirksey. I have enjoyed working on production with Margaret Pinette and Bill Heckman of Heckman & Pinette, and I appreciate their understanding and goodwill. I would especially like to thank my editor Chris Hall and my development editor Alyssa White for their help and guidance on this project.

Most of all, I would like to thank my wife, Janet, and my daughter, Kate, for putting up with "the books." They know all too well that textbooks are made of time.

Mike Seeds

Manuscript Reviewers

Steve Desch, Arizona State University
Michael Frey, California State University, Long Beach
Marc Gagné, West Chester University
Paul Hintzen, California State University, Long Beach
Don McCarthy, University of Arizona
Scott Miller, Penn State University
Sharon L. Montgomery, Clarion University of Pennsylvania
Richard Scarborough, University of Nebraska, Lincoln

1 | The Scale of the Cosmos

The longest journey begins with a single step.

CONFUCIUS

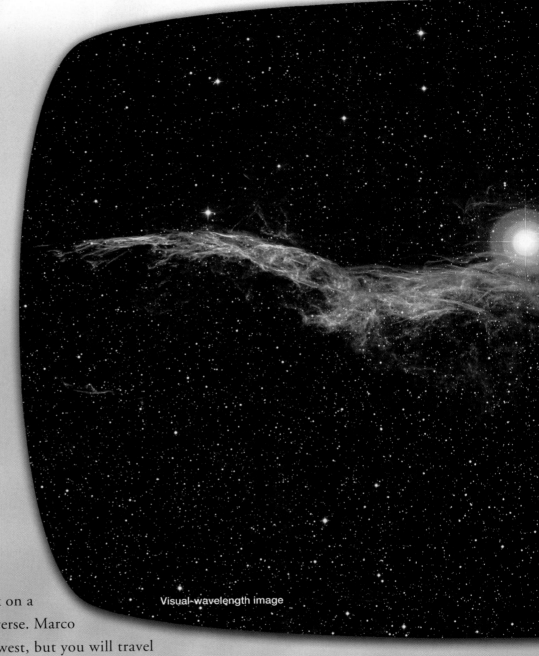

Visual-wavelength image

Y OU ARE ABOUT to embark on a voyage out to the end of the universe. Marco Polo journeyed east, Columbus west, but you will travel outward away from your home on Earth, out past the moon, sun, and other planets, past the stars you see in the evening sky, and past billions more that can be seen only with the aid of the largest telescopes. You will journey through great whirlpools of stars to the most distant galaxies visible from Earth—and then you will continue on, carried only by experience and imagination looking for the structure of the universe itself. Imagination is your key to discovery; it will be your scientific time machine transporting you into the past and into the future. Go back to watch the birth of the universe, the formation of the first stars, and the origin of the sun and Earth. Then rush

Continued on page 4

Guidepost

Getting Started

You are already an expert on astronomy. You have enjoyed sunsets and moonrises. You have admired the stars and may know a few constellations. You have probably read about Mars and eclipses, and you may have read about distant galaxies. That is more than most Earthlings know about astronomy. But you have a right to know more.

You are a planetwalker, and you should understand what it means to live on a planet that whirls around a star drifting through a universe of stars and galaxies. You owe it to yourself to know where you are. That is the first step to knowing what you are.

This Chapter

Here you will take a quick trip, a cosmic zoom, from objects you know up to the largest things in the universe. That quick survey will answer four essential questions:

Where is Earth in relation to the sun, the planets, and the galaxies?

How do astronomers talk about these huge astronomical distances?

Which objects are big, and which are small?

Are there other worlds like Earth?

This chapter will give you a sense of the scale of the universe that will carry you to new discoveries.

Looking Ahead

It is easy to learn a few facts, but it is the relationships among facts that are important. This chapter will give you the sense of scale that you need to understand where you are in the universe. The remaining chapters of this book will fill in the details, cite evidence and theories, and illustrate the wonderful intricacy and beauty of the universe. It is an exciting journey, and it begins here.

into the future to see what will happen when the sun dies and Earth withers.

Although you will discover a beginning to the universe, you will not find an edge in space. No matter how far you voyage, you will not run into a wall or limit beyond which you cannot go. Rather you will discover evidence that our universe may be infinite, that it may extend in all directions without limit. Such vastness may dwarf our earthly dimensions but not our human curiosity and imagination.

Astronomy is more than the study of stars and planets. It is the study of the universe in which we humans exist. You and I live on a small planet circling a small sun drifting through the universe, but astronomy can take us beyond these boundaries and help us not only see where we are in the universe but understand what we are. You have a right to know these things. Perhaps you have a duty to know them.

Do not be humble. Although astronomical sizes and distance may dwarf you, remember that you are an intelligent creature, and you are capable of understanding your universe. It is, after all, yours.

Astronomy will introduce you to sizes, distances, and times far beyond your usual experience on Earth. Your task in this chapter is to grasp the meaning of these unfamiliar sizes, distances, and times. Believe it or not, the solution lies in a single word, *scale*.

In this chapter, you will compare objects of different sizes to grasp the scale of the universe.

Let's begin with something familiar. ■ Figure 1-1 shows a region about 52 feet across occupied by a human being, a sidewalk, and a few trees—all objects whose size you can understand. Each successive picture in this chapter will show you a region of the universe that is 100 times wider than the preceding picture. That is, each step will widen your field of view by a factor of 100.

In ■ Figure 1-2 your field of view widens by a factor of 100, and you can see an area 1 mile in diameter. The arrow points to the scene shown in the preceding photograph. People, trees, and sidewalks have vanished, but now you can see a college campus and the surrounding streets and houses. The dimensions of houses and streets are familiar. This is the world you know, and you can relate such objects to the scale of your body.

You started your adventure using feet and miles, but you should use the metric system of units. Not only is it used by all scientists around the world, but it makes calculations much easier. If you are not already familiar with the metric system, or if you need a review, study Appendix A before reading on.

■ **Figure 1-1**

(Michael A. Seeds)

■**Figure 1-2**

(USGS)

range of "colors," from X rays to radio waves, to reveal sights invisible to unaided human eyes.

At the next step in your journey, you will see our entire planet (■ Figure 1-4), which is 12,756 km in diameter. Earth rotates on its axis once a day, exposing half of its surface to daylight at any particular moment. The photo shows most of the daylight side of the planet. The blurriness at the extreme right is the sunset line. The rotation of Earth carries you eastward, and as you cross the sunset line into darkness, you say the sun has set. It is the rotation of the planet that causes the cycle of day and night.

Earth's interior is made mostly of iron and nickel, and its crust is mostly silicate rocks. Only a thin layer of water makes up

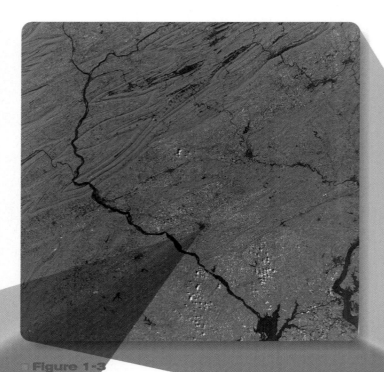

■ Figure 1-3

(NASA infrared photograph)

■ Figure 1-4

(NASA)

The photo in Figure 1-2 is 1 mile in diameter. A mile equals 1.609 kilometers, so you can see in the photo that a kilometer is a bit over two-thirds of a mile—a short walk across a neighborhood.

The view in ■ Figure 1-3 spans 160 kilometers. In this infrared photo, the green foliage shows up as various shades of red. The college campus is now invisible, and the patches of gray are small cities, with the suburbs of Philadelphia visible at the lower right. At this scale, you see the natural features of Earth's surface. The Allegheny Mountains of southern Pennsylvania cross the image in the upper left, and the Susquehanna River flows southeast into Chesapeake Bay. What look like white bumps are a few puffs of clouds.

These features are a reminder that you live on the surface of a changing planet. Forces in Earth's crust pushed the mountain ranges up into parallel folds, like a rug wrinkled on a polished floor. The clouds tell you that Earth's atmosphere is rich in water, which falls as rain and erodes the mountains, washing material down the rivers and into the sea. Mountains and valleys are only temporary features on Earth; they are constantly changing. As you explore the universe, you will come to see that it, like Earth's surface, is always evolving.

Take a closer look at Figure 1-3 and notice the red color. This is an infrared photograph in which healthy green leaves and crops show up as red. Human eyes are sensitive to only a narrow range of colors. As you explore the universe, you will learn to use a wide

the oceans, and the atmosphere is only a few hundred kilometers deep. On the scale of this photograph, the depth of the atmosphere on which life depends is less than the thickness of a piece of thread.

Enlarge your field of view by a factor of 100, and you will see a region 1,600,000 km wide (■ Figure 1-5). Earth is the small blue dot in the center, and the moon, whose diameter is only one-fourth that of Earth, is an even smaller dot along its orbit 380,000 km from Earth.

These numbers are so large that it is inconvenient to write them out. Astronomy is the science of big numbers, and you will use numbers much larger than these to discuss the universe. Rather than writing out these numbers as in the previous paragraph, it is convenient to write them in **scientific notation.** This is nothing more than a simple way to write numbers without writing lots of zeros. In scientific notation, you would write 380,000 as 3.8×10^5. If you are not familiar with scientific notation, read the section on powers of 10 notation in the Appendix. The universe is too big to discuss without using scientific notation.

When you once again enlarge your field of view by a factor of 100 (■ Figure 1-6), Earth, the moon, and the moon's orbit all lie in the small red box at lower left. But now you can see the sun and two other planets that are part of our solar system. Our **solar system** consists of the sun, its family of planets, and some smaller bodies such as moons and comets.

Like Earth, Venus and Mercury are **planets,** small, non-luminous bodies that shine by reflected light. Venus is about the size of Earth, and Mercury is a bit larger than Earth's moon. On this diagram, they are both too small to be seen as anything but tiny dots. The sun is a **star,** a self-luminous ball of hot gas that generates its own energy. The sun is 109 times larger in diameter than Earth (inset), but it too is nothing more than a dot in this diagram.

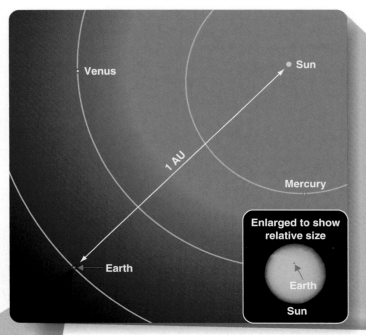

Enlarged to show relative size

Earth

Sun

■ **Figure 1-6**

(NOAO)

■ **Figure 1-5**

(NASA)

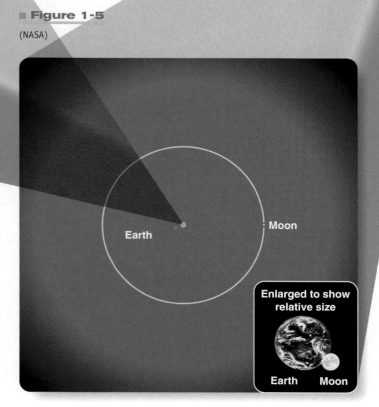

Enlarged to show relative size

Earth   Moon

This diagram has a diameter of 1.6×10^8 km. One way astronomers deal with large numbers is to define new units. The average distance from Earth to the sun is a unit of distance called the **astronomical unit (AU),** a distance of 1.5×10^{11} m. Using this unit, you can say that the average distance from Venus to the sun is about 0.7 AU. The average distance from Mercury to the sun is about 0.39 AU.

The orbits of the planets are not perfect circles, and this is particularly apparent for Mercury. Its orbit carries it as close to the sun as 0.307 AU and as far away as 0.467 AU. You can see this variation in the distance from Mercury to the sun in Figure 1-6. Earth's orbit is more circular, and its distance from the sun varies by only a few percent.

Your first field of view was only 52 feet (about 16 m) in width. After only six steps of enlarging by a factor of 100, you can now see the entire solar system (■ Figure 1-7). Your field of view is 1 trillion (10^{12}) times wider than in your first view. The details of the preceding figure are now lost in the red square at the center of this diagram. You see only the brighter, more widely separated objects as you enlarge your view. The sun, Mercury, Venus, and Earth lie so close together that you cannot separate them at this scale. Mars, the next outward planet, lies only 1.5 AU from the sun. In contrast, Jupiter, Saturn, Uranus, Neptune, and Pluto are so far from the sun that they are easy to place in this diagram. These are cold worlds far from the sun's warmth. Light from the sun reaches Earth in only 8 minutes, but it takes over 4 hours to reach Neptune. Notice that Pluto's orbit is so elliptical that Pluto can come closer to the sun than Neptune does, as Pluto did between 1979 and 1999.

When you again enlarge your field of view by a factor of 100, the solar system vanishes (■ Figure 1-8). The sun is only a point of

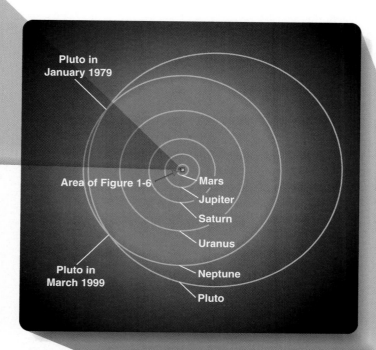

Pluto in
January 1979

Area of Figure 1-6

Mars
Jupiter
Saturn
Uranus

Pluto in
March 1999

Neptune

Pluto

■ Figure 1-7

light, and all the planets and their orbits are now crowded into the small red square at the center. The planets are too small and reflect too little light to be visible so near the brilliance of the sun.

Nor are any stars visible except for the sun. The sun is a fairly typical star, and it seems to be located in a fairly average neighborhood in the universe. Although there are many billions of stars like the sun, none is close enough to be visible in this diagram, which shows an area only 11,000 AU in diameter. The stars are typically separated by distances about 10 times larger than the diameter of this diagram. You will see stars in the next field of view, but, except for the sun at the center, this diagram is empty.

It is difficult to grasp the isolation of the stars. If the sun were represented by a golf ball in New York City, the nearest star would be another golf ball in Chicago. Except for the widely scattered stars and a few atoms of gas drifting between the stars, the universe is nearly empty.

In ■ Figure 1-9, your field of view has expanded to a diameter a bit over 1 million AU. The sun is at the center, and you can see a few of the nearest stars. These stars are so distant that it is not reasonable to give their distances in astronomical units. To express distances so large, astronomers define a new unit of distance, the light-year. One **light-year (ly)** is the distance that light travels in one year, roughly 10^{13} km or 63,000 AU. The diameter of your field of view is 17 ly. The nearest star to the sun, Alpha Centauri, is 4.2 ly from Earth. In other words, light from Alpha Centauri takes 4.2 years to reach Earth.

Do you remember that Alpha Centauri is one of our Favorite Stars? (See p. xi for a list of Favorite Stars.) It is in the southern sky, so it is invisible from all but the southernmost parts of the United States where it occasionally peeks above the southern horizon. If you ever have the chance, you should locate the sun's nearest companion in space. In a later chapter, you will discover that Alpha Centauri is actually three stars orbiting around each other. Our Favorite Stars have interesting secrets.

Although stars are roughly the same size as the sun, they are so far away that you cannot see them as anything but points of light. Even with the largest telescopes on Earth, you would see only points of light when you looked at stars, and any planets that might circle those stars are much too small and too faint to be visible. Using indirect methods, astronomers have found nearly 200 planets orbiting other stars.

■ Figure 1-8

Sun

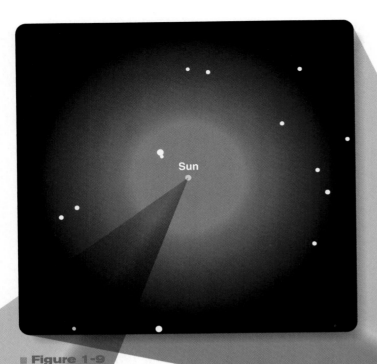

Milky Way, and our galaxy is called the **Milky Way Galaxy.** Of course, no one can journey far enough into space to look back and photograph our home galaxy, so the photo in Figure 1-11 shows a galaxy similar to our own. Our sun would be invisible in such a photo, but if you could see it, you would find it in the disk of the galaxy about two-thirds of the way out from the center.

Our galaxy, like many others, has graceful **spiral arms** winding outward through the disk. You will discover that stars are born in great clouds of gas and dust as they pass through the spiral arms.

■ **Figure 1-10**

This box ■ represents the relative size of the previous frame. (NOAO)

In Figure 1-9, the sizes of the dots represent not the sizes of the stars but their brightness. This is the custom in astronomical diagrams, and it is also how star images are recorded on photographs. Bright stars make larger spots on a photograph than faint stars. The size of a star image in a photograph tells you not how big the star is but only how bright it looks.

In ■ Figure 1-10, you expand your field of view by another factor of 100, and the sun and its neighboring stars vanish into the background of thousands of other stars. The field of view is now 1700 ly in diameter. Of course, no one has ever journeyed thousands of light-years from Earth to look back and photograph the solar neighborhood, so this is a representative photograph of the sky. The sun is a relatively faint star that would not be easily located in a photo at this scale.

What you do not see in this photograph is critically important. You do not see the thin gas that fills the spaces between the stars. Although those clouds of gas are thinner than the best vacuum on Earth, it is those clouds that give birth to new stars. Our sun formed from such a cloud about 5 billion years ago. You will see evidence of star formation in your next field of view.

If you expand your field of view by a factor of 100, you see our galaxy (■ Figure 1-11). A **galaxy** is a great cloud of stars, gas, and dust bound together by the combined gravity of all the matter. Galaxies range from 1500 to over 300,000 ly in diameter and can contain over 100 billion stars. In the night sky, you see our galaxy as a great, cloudy wheel of stars ringing the sky as the

Ours is a fairly large galaxy, roughly 75,000 ly in diameter. Only a century ago astronomers thought it was the entire universe—an island universe of stars in an otherwise empty vastness. Now they know that our galaxy is not unique. Indeed ours is only one of many billions of galaxies scattered throughout the universe.

As you expand your field of view by another factor of 100, our galaxy appears as a tiny luminous speck surrounded by other specks (■ Figure 1-12). This diagram includes a region 17 million ly in diameter, and each of the dots represents a galaxy. Notice that our galaxy is part of a cluster of a few dozen galaxies.

Galaxies are commonly grouped together in such clusters. Some of these galaxies have beautiful spiral patterns like our own galaxy, but others do not. Some are strangely distorted. One of the mysteries of modern astronomy is what produces these differences among the galaxies. Astronomers have found some important clues and have developed a fascinating theory.

If you again expand your field of view, you see that the clusters of galaxies are connected in a vast network (■ Figure 1-13). Clusters are grouped into superclusters—clusters of clusters—and the superclusters are linked to form long filaments and walls outlining voids that seem nearly empty of galaxies. These appear to be the largest structures in the universe. Were you to expand your field of view another time, you would probably see a uniform fog of filaments and voids. When you puzzle over the origin of these structures, you are at the frontier of human knowledge.

■ Figure 1-13

This box ■ represents the relative size of the previous frame. (Detail from galaxy map from M. Seldner, B. L. Siebers, E. J. Groth, and P. J. E. Peebles, *Astronomical Journal 82* [1977])

Milky Way Galaxy

□ **Figure 1-12**

■ Figure 1-11

(© Anglo-Australian Telescope Board)

The first hurdle in studying astronomy is keeping a proper sense of scale. Remember that each of the billions of galaxies contains billions of stars. Later you will see clear evidence that many of those stars have families of planets like our solar system, and on some of those planets liquid-water oceans and protective atmospheres may have sheltered the spark of life. It is possible that at least a few other planets are inhabited by intelligent creatures who share our curiosity and our wonder at the scale of the cosmos.

Summary

Your goal in this chapter is to preview the scale of astronomical objects. To do so, you journeyed outward from a familiar campus scene by expanding your field of view by factors of 100. Only 12 such steps took you to the largest structures in the universe.

Where is Earth in relation to the sun, the planets, and the galaxies?

■ You live on planet Earth, which orbits our star, the sun, once a year. As Earth rotates once a day, you see the sun rise and set.

■ The other planets in our solar system, Mercury, Venus, Mars, Jupiter, Saturn, Uranus, Neptune, and Pluto, orbit the sun in ellipses that are nearly circular.

■ The sun is just one out of the billions of stars that fill our home galaxy, the Milky Way spiral galaxy.

■ Our galaxy is just one of billions of galaxies that fill the universe in great clusters, clouds, filaments, and walls.

How do astronomers talk about these huge astronomical distances?

■ Astronomers use the metric system because it simplifies calculations and scientific notation for very large or very small numbers.

■ The astronomical unit (AU) is the average distance from Earth to the sun. Mars, for example, orbits 1.5 AU from the sun. The light-year (ly) is the distance light can travel in one year. The nearest star is 4.2 ly from the sun.

Which objects are big, and which are small?

■ The moon is only a fourth the diameter of Earth, but the sun is 109 times larger in diameter than Earth—a typical size for a star.

■ The solar system includes the sun at the center and all of the planets that orbit around it.

■ Galaxies contain many billions of stars. Our galaxy is about 75,000 ly in diameter and contains over 100 billion stars.

■ The largest things in the universe are the vast filaments and walls containing many clusters of galaxies.

Are there other worlds like Earth?

■ Many stars seem to have planets, but such small, distant worlds are difficult to detect. Fewer than 200 or so have been found so far, but planets seem to be common, so you can probably trust that there are other worlds like Earth.

New Terms

scientific notation (p. 6)

solar system (p. 6)

planet (p. 6)

star (p. 6)

astronomical unit (AU) (p. 6)

light-year (ly) (p. 7)

galaxy (p. 8)

Milky Way (p. 8)

Milky Way Galaxy (p. 8)

spiral arm (p. 8)

Review Questions

Ace ☯ Astronomy™   Assess your understanding of this chapter's topics with additional quizzing and animations at http://ace .brookscole.com/sf9

1. What is the largest dimension you have personal knowledge of? Have you run a mile? Hiked 10 miles? Run a marathon?

2. In Figure 1-4, the division between daylight and darkness is at the right on the globe of Earth. How do you know this is the sunset line and not the sunrise line?

3. What is the difference between our solar system, our galaxy, and the universe?

4. Look at Figure 1-6. How can you tell that Mercury follows an elliptical orbit? Can you detect the elliptical shape of any other orbits in this figure or the next?

5. Which is the outermost planet in our solar system? Why does that change?

6. Why are light-years more convenient than miles, kilometers, or astronomical units for measuring certain distances?

7. Why is it difficult to detect planets orbiting other stars?

8. What does the size of the star image in a photograph tell you?

9. What is the difference between the Milky Way and the Milky Way Galaxy?

10. What are the largest known structures in the universe?

11. Of the objects listed here, which would be contained inside the object shown in the photograph at the right? Which would contain the object in the photo? Stars, planets, galaxy clusters, filaments, spiral arms

(Bill Schoening/ NOAO/AURA/NSF)

Problems

1. The diameter of Earth is 7928 miles. What is its diameter in inches? In yards? If the diameter of Earth is expressed as 12,756 km, what is its diameter in meters? In centimeters?

2. If a mile equals 1.609 km and the moon is 2160 miles in diameter, what is its diameter in kilometers?

3. One astronomical unit is about 1.5×10^8 km. Explain why this is the same as 150×10^6 km.

4. Venus orbits 0.7 AU from the sun. What is that distance in kilometers?

5. Light from the sun takes 8 minutes to reach Earth. How long does it take to reach Mars?

6. The sun is almost 400 times farther from Earth than is the moon. How long does light from the moon take to reach Earth?

7. If the speed of light is 3×10^5 km/s, how many kilometers are in a light-year? How many meters?

8. How long does it take light to cross the diameter of our Milky Way Galaxy?

9. The nearest galaxy to our own is about 2 million light-years away. How many meters is that?

10. How many galaxies like our own would it take laid edge to edge to reach the nearest galaxy? (*Hint:* See Problem 9.)

Media Cluster

Critical Inquries for the Web

1. Locate photographs of Earth taken from space. What do cities look like? Can you see highways? Is the presence of our civilization detectable from space?

2. Locate photographs of nearby galaxies and compare them with photos of very distant galaxies. What kind of detail is invisible for distant galaxies?

3. One of the biggest clusters of galaxies is the Virgo cluster. Find out how many and what kind of galaxies are in the cluster. Is it nearby or far away?

Exploring *TheSky*

1. Locate and center one example of each of three different types of objects:

   a. A planet, such as Saturn. Find its rising and setting time. Such objects have distances measured in astronomical units (AU).
      How to proceed: Decide on the object you want to locate. Then find and center the object by pressing the **Find** button on the **Object Toolbar.** The second method is to press the **F** key. The third is to click **Edit,** then **Find.** Once you have the **Object Information** window, press the **center** button.

   b. A star. All stars in *TheSky* belong to our Milky Way Galaxy. Give the star's name, its magnitude, and its distance in light-years.
      How to proceed: Click on any star, which brings up an **Object Information** window containing a variety of information about the star.

   c. A galaxy; give its name and/or its designation.
      How to proceed: To show galaxies, click on the **Galaxies** button in the **Object Toolbar,** then click on any galaxy. Distances to galaxies are on the order of millions and billions of light years.

2. Look at the solar system from beyond Pluto by clicking on **View** and then on **3D Solar System Mode.** Tip the solar system edge-on and then face-on. Zoom in to see the inner planets. Under **Tools,** set the **Time Skip Increment** to 1 day and then go **forward in time** to watch the planets move.

3. Identify some of the brightest constellations located along the Milky Way. (*Hint:* See **View, Reference Lines.**)

 Go to the Brooks/Cole Astronomy Resource Center **(http://astronomy.brookscole.com)** for critical thinking exercises, articles, and additional readings from InfoTrac College Edition, Brooks/Cole's online student library.

2 | The Sky

The Southern Cross I saw
every night abeam. The sun
every morning came up astern;
every evening it went down
ahead. I wished for no other
compass to guide me,
for these were true.

CAPTAIN JOSHUA SLOCUM,
SAILING ALONE AROUND THE WORLD

Visual-wavelength image

THE NIGHT SKY is the rest of the universe as seen from our planet. When you look up at the stars, you look out through a layer of air only a few hundred kilometers deep. Beyond that, space is nearly empty, with stars scattered light-years apart. Here you will begin your search for the natural laws that govern the universe by trying to understand what the universe looks like.   As you read this chapter, keep in mind that you live on a planet. The stars are scattered into the void all around you, most very distant and some closer. Our planet rotates on its axis once a day and revolves around the sun in its orbit once a year, and that makes the sky appear to revolve around you in a daily and yearly cycle. Not only does the sun rise and set, but so also do the stars. The entire sky seems to whirl around you because our planet whirls through the universe.

Continued on page 14

Guidepost

Looking Back

Astronomy is about us. As we learn about astronomy, we learn about ourselves. We search for an answer to the question, "What are we?" The cosmic zoom in the previous chapter showed you Earth in relation to other objects in the universe. Even in that quick preview, as you learned about stars and galaxies, you were also learning about yourself.

This Chapter

Now it is time to return to Earth and look closely at the sky. You have noticed stars and may know a few constellations and star names, but now you need to refine your terms and answer a few essential questions:

How do astronomers refer to stars?

How can you compare the brightness of the stars?

How does the sky move as Earth moves?

What causes the seasons?

How do astronomical cycles affect Earth's climate?

Answering these questions will tell you a great deal about yourself and your home on planet Earth.

Looking Ahead

In the next chapter, you will learn more about the beautiful cycles of the sky, and in this chapter and the next it is important to remember that the motions you see in the heavens are produced in part by the motions of Earth. Chapter 4 will show you how hard it was for humanity to understand that we live on a moving planet. In fact, you will discover that modern science was born when people tried to understand the motions in the sky.

ON A DARK NIGHT far from city lights, you can see a few thousand stars in the sky. As you begin your study of the sky, the first step is to organize what you see by naming groups of stars and individual stars and by specifying the brightness of individual stars. That will make the sky familiar territory, and you will be ready to explore further.

Constellations

All around the world, ancient cultures celebrated heroes, gods, and mythical beasts by naming groups of stars—**constellations** (■ Figure 2-1). You should not be surprised that the star patterns do not look like the creatures they represent any more than Columbus, Ohio, looks like Christopher Columbus. The constellations "celebrate" the most important mythical figures in each culture. The constellations named within Western culture originated in Mesopotamia over 5000 years ago, with other constellations added by Babylonian, Egyptian, and Greek astronomers during the classical age. Of these ancient constellations, 48 are still in use today.

Different cultures grouped stars and named constellations differently. The constellation you probably know as Orion was

■ Figure 2-1

The constellations are an ancient heritage handed down for thousands of years as celebrations of great heroes and mythical creatures. Here Sagittarius and Scorpius hang above the southern horizon.

known as Al Jabbar, the giant, to the ancient Syrians, as the White Tiger to the Chinese, and as Prajapati in the form of a stag in ancient India. The Pawnee Indians knew the constellation Scorpius as two groupings. The long tail of the scorpion was the Snake, and the two bright stars at the tip of the scorpion's tail were the Two Swimming Ducks.

Many ancient cultures around the world, including the Greeks, northern Asians, and Native Americans, associated the stars of the Big Dipper with a bear. The concept of the celestial bear may have crossed the land bridge into North America with the first Americans over 10,000 years ago. Some of the constellations you see in the sky may be among the oldest surviving traces of human culture.

To the ancients, a constellation was a loose grouping of stars. Many of the fainter stars were not included in any constellation, and regions of the southern sky, not being visible to the ancient astronomers of northern latitudes, were not organized into constellations. Constellation boundaries, when they were defined at all, were only approximate (■ Figure 2-2a), so a star like Alpheratz could be thought of as part of Pegasus or part of Andromeda. In recent centuries, astronomers have added 40 modern constellations to fill gaps, and in 1928 the International Astronomical Union established 88 official constellations with clearly defined boundaries (Figure 2-2b). Consequently, a constellation now represents not a group of stars but an area of the sky, and any star within the boundaries of the area belongs to the constellation. Because the entire sky is covered by constellations, every star is a member of one and only one constellation.

In addition to the 88 official constellations, the sky contains a number of less formally defined groupings called **asterisms.** The Big Dipper, for example, is a well-known asterism that is part of the constellation Ursa Major (the Great Bear). Another asterism is the Great Square of Pegasus (Figure 2-2b), which includes three stars from Pegasus and one from Andromeda. The star charts at the end of this book will introduce you to the brighter constellations and asterisms.

Although you refer to constellations and asterisms by name, most are made up of stars that are not physically associated with one another. Some stars may be many times farther away than others and moving through space in different directions. The only thing they have in common is that they lie in approximately the same direction from Earth (■ Figure 2-3).

The Names of the Stars

In addition to naming groups of stars, ancient astronomers named the brighter stars,

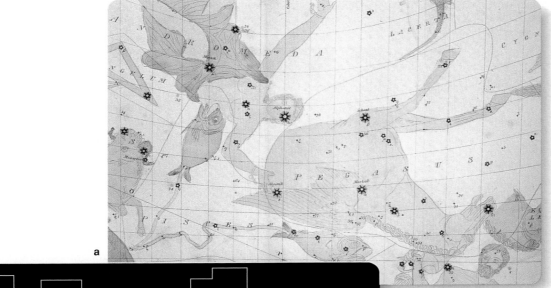

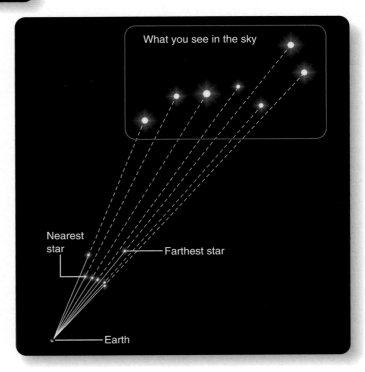

Goat), and Aldebaran (the Follower of the Pleiades) are beautiful additions to the mythology of the sky.

Giving the stars individual names is not very helpful, because you can see thousands of stars, and these names do not help you locate the star in the sky. In which constellation is Antares, for example? In 1603, Bavarian lawyer Johann Bayer published an atlas of the sky called *Uranometria* in which he assigned lowercase Greek letters to the brighter stars of each constellation in approximate order of brightness. Astronomers have used those Greek letters ever since. (See the Appendix table with the Greek alphabet.) In this way, the brightest star is usually designated α (alpha), the second-brightest β (beta), and so on (■ Figure 2-4). To identify a star in this way, give the Greek letter followed by the possessive form of the constellation name, such as α Scorpii (sometimes written alpha Scorpii) for Antares. That designation reveals that Antares is in the constellation Scorpius and that it is probably the brightest star in the constellation.

■ Figure 2-2

(a) In antiquity, constellation boundaries were poorly defined, as shown on this map by the curving dotted lines that separate Pegasus from Andromeda. (From Duncan Bradford, *Wonders of the Heavens,* Boston: John B. Russell, 1837) (b) Modern constellation boundaries are precisely defined by international agreement.

and modern astronomers still use many of those names. The constellation names come from Greek versions translated into Latin—the language of science from the fall of Rome to the 19th century—but most star names come from ancient Arabic, though much altered by the passing centuries. The name of Betelgeuse, the bright red star in Orion, for example, comes from the Arabic *yad al jawza,* meaning "armpit of Jawza [Orion]." Names such as Sirius (the Scorched One), Capella (the Little She

■ Figure 2-3

The stars you see in the Big Dipper—the brighter stars of the constellation Ursa Major, the Great Bear—are not at the same distance from Earth. You see the stars in a group in the sky because they lie in the same general direction as seen from Earth. The size of the star dots in the star chart represents the apparent brightness of the stars.

The brighter stars in a constellation are usually given Greek letters in order of decreasing brightness.

Orion

α Orionis is also known as Betelgeuse.

β Orionis is also known as Rigel.

In Orion β is brighter than α, and κ is brighter than η. Fainter stars do not have Greek letters or names, but if they are located inside the constellation boundaries, they are part of the constellation.

■ **Figure 2-4**

Stars in a constellation can be identified by Greek letters and by names derived from Arabic. The spikes on the star images in the photograph were produced by the optics in the camera. (William Hartmann)

This method of identifying a star's brightness, however, is only approximate. To be precise, you need an accurate way of referring to the brightness of stars, and for that you must consult one of the first great astronomers.

The Brightness of Stars

Astronomers measure the brightness of stars using the **magnitude scale,** a system that first appeared in the writings of the ancient astronomer Claudius Ptolemy about 140 AD. The system may have originated earlier than Ptolemy, and most astronomers attribute it to the Greek astronomer Hipparchus (160–127 BC) (■ Figure 2-5).

The ancient astronomers divided the stars into six classes. The brightest were called first-class stars and those that were fainter, second-class. The scale continued downward to sixth-class stars, the faintest visible to the human eye. Thus, the larger the magnitude number, the fainter the star. This makes sense if you think of the bright stars as first-class stars and the faintest stars visible as sixth-class stars.

Hipparchus is believed to have compiled the first star catalog, and he may have used the magnitude system in that catalog.

Almost 300 years later Ptolemy used the magnitude system in his own catalog, and successive generations of astronomers have continued to use the system. Later astronomers had to revise the ancient magnitude system to include very faint and very bright stars. Telescopes reveal many stars fainter than those the eye can detect, so the magnitude scale was extended to include these faint stars. The Hubble Space Telescope, currently the most sensitive astronomical telescope, can detect stars as faint as 30th magnitude. In contrast, the brightest stars are actually brighter than the first brightness class in the ancient magnitude system. Consequently, modern astronomers extended the magnitude system into negative numbers to account for brighter objects. For instance, Favorite Star Vega (alpha Lyrae) is almost zero magnitude at 0.04, and Sirius, the brightest star in the sky, has a magnitude of -1.42. You can even place the sun on this scale at -26.5 and the moon at -12.5 (■ Figure 2-6).

These numbers are known as **apparent visual magnitudes (m_v),** and they describe how the stars look to human eyes observing from Earth. Although some stars emit large amounts of infrared or ultraviolet light, human eyes can't see it, and it is not included in the apparent visual magnitude. The subscript "v" reminds you that visual magnitudes include only the light human eyes can see. Another problem is the distance to the stars. Very

■ **Figure 2-5**

Hipparchus (2nd century BC) was the first great observational astronomer. He is believed to have compiled the first star catalog and may have invented the magnitude scale. He is honored here on a Greek stamp that also shows one of his observing instruments.

The scale of apparent visual magnitudes extends into negative numbers to represent the brightest objects and to positive numbers larger than 6 to represent objects fainter than the human eye can see.

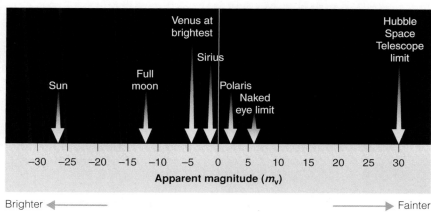

Brighter ◄──────────────── ────────────────► Fainter

distant stars look fainter, and nearby stars look brighter. Apparent visual magnitude ignores the effect of distance and tells you only how bright the star looks as seen from Earth.

Magnitude and Intensity

Nearly every star catalog from today back to the time of the ancients uses the magnitude scale of stellar brightness. However, brightness is subjective, depending on such things as the physiology of the eye and the psychology of perception. To be accurate in measurements and calculations astronomers use the more precise term *intensity*—a measure of the light energy from a star that hits 1 square meter in 1 second. That means astronomers need a way to convert between magnitudes and intensities.

The human eye senses the brightness of objects by comparing the ratios of their intensities. If two stars have intensities I_A and I_B, they can be compared by writing the ratio of their intensities, I_A/I_B. Hipparchus's magnitude system is based on a constant intensity ratio of about 2.5 for each magnitude. That is, if two stars differ by 1 magnitude, then they have an intensity ratio of about 2.5. If they differ by 2 magnitudes, their ratio is 2.5 × 2.5, and so on.

In the 19th century, astronomers began measuring starlight, and they realized they needed to define a mathematical magnitude system that was precise; but for convenience they wanted a system that agreed at least roughly with that of Hipparchus. The stars that ancient astronomers classified as first and sixth differ by 5 magnitudes and have an intensity ratio of almost exactly 100, so the modern system of magnitudes specifies that a magnitude difference of 5 magnitudes corresponds to an intensity ratio of exactly 100. That means that 1 magnitude corresponds to an intensity ratio of 2.512, the fifth root of 100. That is, $100 = (2.512)^5$.

By giving the ancient magnitude scale this precise definition, astronomers can compare the light from two stars. The light from a first-magnitude star is 2.512 times more intense than that from a second-magnitude star. The light from a third-magnitude star is $(2.512)^2$ more intense than the light from a fifth-magnitude star. In general, the intensity ratio equals 2.512 raised to the power of the magnitude difference. That is:

$$\frac{I_A}{I_B} = (2.512)^{(m_B - m_A)}$$

For instance, if two stars differ by 6.32 magnitudes, you can calculate the ratio of their intensities as $2.512^{6.32}$. Your pocket calculator tells you that the ratio is 337.

When you know the intensity ratio and want to find the magnitude difference, you might like to rearrange the equation above and write it as

$$m_A - m_B = 2.5 \, \text{Log} \left(\frac{I_B}{I_A} \right)$$

For example, if you knew that the light from Sirius is 24.2 times more intense than light from Polaris, you could find the magnitude difference easily. It is just 2.5 Log (24.2). Your pocket calculator tells you that the logarithm of 24.2 is 1.38, so the magnitude difference is 2.5 × 1.38, which equals 3.4 magnitudes. Sirius is 3.4 magnitudes brighter than Polaris.

This modern magnitude system has some big advantages. It compresses a tremendous range of intensity into a small range of magnitudes (■ Table 2-1) and makes it possible for modern astronomers to compare their measurements with all the recorded

■ Table 2-1 | Magnitude and Intensity

Magnitude Difference	Intensity Ratio
0	1
1	2.5
2	6.3
3	16
4	40
5	100
6	250
7	630
8	1600
9	4000
10	10,000
⋮	⋮
15	1,000,000
20	100,000,000
25	10,000,000,000
⋮	⋮

Scientific Arguments: The Structure of Science

An argument can be a shouting match, but another definition is "a discourse intended to persuade." Scientists construct arguments as part of the business of science not because they want to persuade others they are right, but because they want to test their ideas. For example, a team of biologists might construct a scientific argument to show that ants communicate by smell. If they do their best and cite all the evidence and all the theories and the argument seems convincing, they gain confidence that they do understand how ants "talk" to each other. If the scientists discover a hole in their argument, they know they have more work to do.

A scientific argument is a logical presentation of evidence and theory with interpretations and explanations that help you understand some aspect of nature. Scientists would be free to include any evidence or theory that helps persuade, but they must observe one fundamental rule of science: They must be totally honest—they must include all of the evidence and

all of the theories. The purpose of a scientific argument is to test understanding, not to win votes or sell soap. Dishonesty in a scientific argument is self-deluding, and scientists consider such dishonesty disgraceful.

In the chapters ahead, you will meet lots of scientific arguments about sophisticated theories and detailed evidence involving everything from the formation of Earth to the fate of the universe, but you can begin using scientific arguments now as a way to review your reading. If you can organize what you know about naming stars, for example, and explain why most star names come from Arabic but most constellation names are Latin, and why Greek letters give you clues to the brightness of stars, you will be building a scientific argument that will help you line up the facts you know and convert them to real understanding.

Scientists think in scientific arguments not to convince others, but to test their own understanding. (Roger Ressmeyer/Corbis)

Each section of this book ends with a review tool called "Building Scientific Arguments." Use these as a model to organize your thoughts as you review each concept in a section, and they will help you understand nature the way scientists do.

measurements of the past, right back to the first star catalogs over 2000 years ago.

It's time for you to review the preceding section on constellations, star names, and magnitudes. Before you begin, read **Window on Science 2-1** and notice how scientists organize information. How you review is critical when you study a science. Memorizing facts won't help you much, but organizing your understanding into scientific arguments will line the facts up in a meaningful way. The review tool that follows, "Building Scientific Arguments," will help you review an important concept from the section, but you should use the same process to review each concept from the beginning of the section to the end.

Building Scientific Arguments

Nonastronomers sometimes complain that the magnitude scale is awkward. Why would they think it is awkward, and how did it get that way?

Two things might make the magnitude scale seem awkward. First, it is backward; the bigger the magnitude number, the fainter the star. Of course, that arose because ancient astronomers were not measuring the brightness of stars but rather classifying them, and first-class stars would be brighter than

second-class stars. The second awkward feature of the magnitude scale is its mathematical relation to intensity. If two stars differ by one magnitude, one is about 2.5 times brighter than the other. But if they differ by two magnitudes, one is 2.5×2.5 times brighter. This mathematical relationship arises because of the way human eyes perceive brightness as ratios of intensity.

Now build your own scientific argument to analyze the following question: **If the magnitude scale is so awkward, why do you suppose astronomers have used it for over two millennia?**

■ ■ ■

Connections: In this section, you have found a way to identify stars by name and by brightness. Now it is time to look at the sky as a whole and notice its motion.

2-2 The Sky and Its Motion

THE SKY ABOVE seems to be a great blue dome in the daytime and a sparkling ceiling at night. Learning to look at the sky requires that you begin thousands of years ago.

Frameworks for Thinking about Nature: Scientific Models

In everyday language, you use the word *model* in various ways—fashion model, model airplane, model student—but scientists use the word in a very specific way. A **scientific model** is a carefully devised mental conception of how something works, a framework that helps scientists think about some aspect of nature. For example, astronomers use the celestial sphere as a way to think about the motions of the sky, sun, moon, and stars.

Although a scientific model is a mental conception, it can take many forms. Some models are quite abstract—the psychologist's model of how the human mind processes visual information into images, for instance. But other models are so specific that they can be expressed as a set of mathematical equations. For example, an astronomer might use a set of equations to describe in detail how gas falls into a black hole. You could refer to such a calculation as a model. Of course, you could use metal and plastic to build a celestial globe, but the thing you built wouldn't really be the model any more than the equations

are a model. A scientific model is a mental conception, an idea that helps you think about nature.

On the other hand, a model is not meant to be a statement of truth. The celestial sphere is not real; you know the stars are scattered through space at various distances, but you can imagine a celestial sphere and use it to help think about the sky. A scientific model does not have to be true to be useful. Chemists, for example, think about the atoms in molecules by visualizing them as balls joined together by rods. This model of a molecule is not fully correct, but it is a helpful way to think about molecules; it gives chemists a framework within which to organize their ideas.

Because scientific models are not meant to be totally true, you must always remember the assumptions on which they are based. If you begin to think a model is true, it can be misleading instead of helpful. The celestial sphere, for instance, can help you think about the sky, but you must remember that it is only

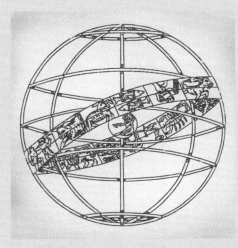

The ancient celestial sphere is a useful model of the sky.

a mental crutch. The universe is much larger and much more interesting than this ancient scientific model of the heavens.

The Celestial Sphere

Ancient astronomers believed the sky was a great sphere surrounding Earth with the stars stuck on the inside like thumbtacks in the ceiling. Modern astronomers know that the stars are scattered through space at different distances, but it is still convenient to think of the sky as a great celestial sphere.

As you study **The Sky around Us** on pages 20–21, notice three important points:

1 The sky appears to rotate westward around Earth each day, but that is a consequence of the eastward rotation of Earth. That produces day and night.

2 Astronomers measure distance across the sky as angles and express them as degrees, minutes, and seconds.

3 What you can see of the sky depends on where you are on Earth. If you lived in Australia, you would see many constellations and asterisms invisible from North America, but you would never see the Big Dipper. Remember our Favorite Star Alpha Centauri? It is in the southern sky and isn't visible from most of the United States. You could just glimpse it above the southern horizon if you were in Miami, but you could see it easily from Australia.

The celestial sphere is an example of a scientific model, a common feature of scientific thought (**Window on Science 2-2**). Notice that a scientific model does not have to be true to be useful. Many scientific models are discussed in the chapters that follow, and you will discover that the most useful models are often quite obviously flawed descriptions of the true facts.

In addition to the daily motion of the sky, Earth's rotation adds a second motion to the sky that can be detected only over centuries.

Precession

Over 2000 years ago, Hipparchus compared a few of his star positions with those made nearly two centuries before and realized that the celestial poles and equator were slowly moving across the sky. Later astronomers understood that this motion is caused by Earth's toplike motion.

If you have ever played with a gyroscope or top, you have seen how the spinning mass resists any change in the direction of its axis of rotation. The more massive the top and the more rapidly it spins, the more difficult it is to change the direction of its axis of rotation. But you probably recall that the axis of even the most

The Sky around Us

1   The eastward rotation of Earth causes the sun, moon,and stars to move westward in the sky as if the **celestial sphere** were rotating westward around Earth. From any location on Earth you see only half of the celestial sphere, the half above the **horizon.** The **zenith** marks the top of the sky above your head, and the **nadir** marks the bottom of the sky directly under your feet. The drawing at right shows the view for an observer in North America. An observer in South America would have a dramatically different horizon, zenith, and nadir.

The apparent pivot points are the **north celestial pole** and the **south celestial pole** located directly above Earth's north and south poles. Halfway between the celestial poles lies the **celestial equator.** Earth's rotation defines the directions you use every day. The **north point** and **south point** are the points on the horizon closest to the celestial poles. The **east point** and the **west point** lie halfway between the north and south points. The celestial equator always touches the horizon at the east and west points.

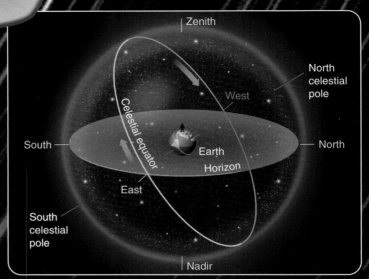

Ace ◉ Astronomy™

Log into AceAstronomy and select this chapter to see Active Figure "Celestial Sphere." Notice how each location on Earth has its unique horizon.

AURA/NOAO/NSF

1a   This time exposure of about 30 minutes shows stars as streaks, called star trails, rising behind an observatory dome. The camera was facing northeast to take this photo. The motion you see in the sky depends on which direction you look, as shown at right. Looking north, you see the star Polaris, the North Star, located near the north celestial pole. As the sky appears to rotate westward, Polaris hardly moves, but other stars circle the celestial pole. Looking south from a location in North America, you can see stars circling the south celestial pole, which is invisible below the southern horizon.

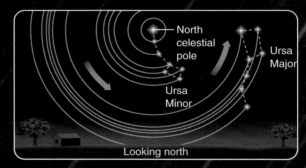

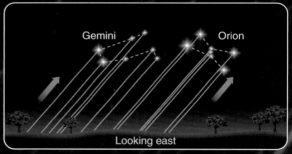

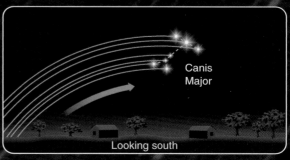

Ace ◉ Astronomy™

Log into AceAstronomy and select this chapter to see Active Figure "Rotation of the Sky." Look in different directions and compare the motions of the stars.

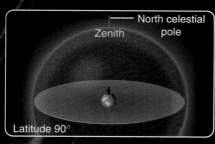

Latitude 90°

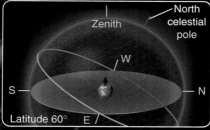

Latitude 60°

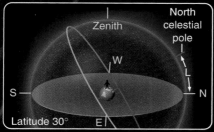

Latitude 30°

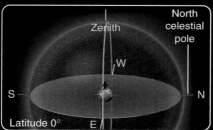

Latitude 0°

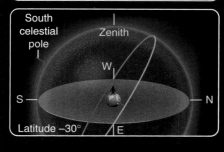

Latitude −30°

Astronomers measure distance across the sky as angles.

Angular distance

2 Astronomers might say, "The star was only 2 degrees from the moon." Of course, the stars are much farther away than the moon, but when you think of the celestial sphere, you can measure distances *on the sky* as **angular distances** in degrees, minutes of arc, and seconds of arc. A **minute of arc** is 1/60th of a degree, and a **second of arc** is 1/60th of a minute of arc. Then the **angular diameter** of an object is the angular distance from one edge to the other. The sun and moon are each about half a degree in diameter, and the bowl of the Big Dipper is about 10° wide.

3 What you see in the sky depends on your latitude as shown at right. Imagine that you begin a journey in the ice and snow at Earth's North Pole with the north celestial pole directly overhead. As you walk southward, the celestial pole moves toward the horizon, and you can see further into the southern sky. The angular distance from the horizon to the north celestial pole always equals your latitude (L)—the basis for celestial navigation. As you cross Earth's equator, the celestial equator would pass through your zenith, and the north celestial pole would sink below your northern horizon.

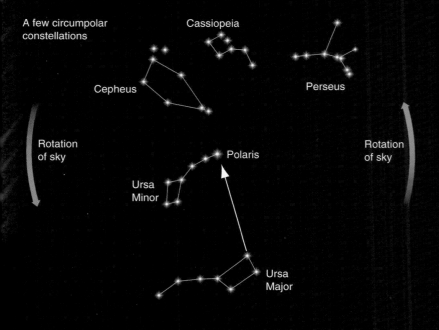

A few circumpolar constellations

Cassiopeia

Cepheus

Perseus

Rotation of sky

Rotation of sky

Polaris

Ursa Minor

Ursa Major

3a **Circumpolar constellations** are those that never rise or set. From mid-northern latitudes, as shown at left, you see a number of familiar constellations circling Polaris and never dipping below the horizon. As the sky rotates, the pointer stars at the front of the Big Dipper always point toward Polaris. Circumpolar constellations near the south celestial pole never rise as seen from mid-northern latitudes. From a high latitude such as Norway, you would have more circumpolar constellations, and from Quito, Ecuador, located on Earth's equator, you would have no circumpolar constellations at all.

rapidly spinning top sweeps around in a conical motion. The weight of the top tends to make it tip over, and this combines with its rapid rotation to make its axis sweep around in a conical motion called **precession** (■ Figure 2-7a).

Earth spins like a giant top, but it does not spin upright in its orbit; it is inclined 23.5° from vertical. Earth's axis of rotation happens to be pointed toward a spot near the star Polaris, the north star, and Earth's large mass and rapid rotation keep its axis fixed in space. The axis would not wander at all if Earth were a perfect sphere. However, Earth, because of its rotation, has a slight bulge around its middle, and the gravity of the sun and moon pull on this bulge, tending to twist Earth upright in its orbit. The combination of these forces and Earth's rotation causes Earth's axis to precess in a conical motion, taking about 26,000 years for one cycle (Figure 2-7b).

Because the celestial poles and equator are defined by Earth's rotational axis, precession moves these reference marks on the sky. You would notice no change at all from night to night or year to year, but precise measurements reveal the precessional motion of the celestial poles and equator.

Over centuries, precession has dramatic effects. Egyptian records show that 4800 years ago the north celestial pole was pointed near the star Thuban (α Draconis).

The pole is now approaching Polaris and will be closest to it in about 2100. In about 12,000 years, the pole will have moved to within 5° of Vega (α Lyrae). Figure 2-7c shows the path followed by the north celestial pole. You will discover in later chapters that precession is common among rotating celestial bodies.

As you study astronomy, notice the special terms used to describe such things as precession and the celestial sphere. You need to know those terms, but science is about understanding nature and not about naming its parts (**Window on Science 2-3**). Science is more than just vocabulary.

Building Scientific Arguments

Does everyone see the same circumpolar constellations?
Here you must use your imagination and build your argument with great care. You can use the celestial sphere as a convenient model of the sky. A circumpolar constellation is one that does not set or rise. Which constellations are circumpolar depends on your latitude. If you live on Earth's equator, you see all the constellations rising and setting, and there are no circumpolar constellations at all. If you live at Earth's North Pole, all the constellations north of the celestial equator never set, and all the constellations south of the celestial equator never rise. In that case, every constellation is circumpolar. At intermediate latitudes, the circumpolar regions are caps on the sky whose angular radius equals the latitude of the observer. If you live in Iceland, the caps are very large, and if you live in Egypt, near the equator, the caps are much smaller.

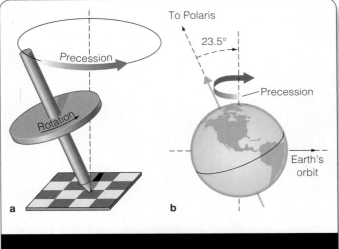

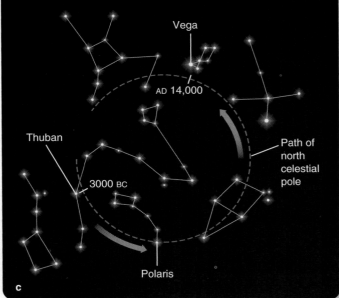

■ **Figure 2-7**

Precession. (a) A spinning top precesses in a conical motion around the perpendicular to the floor because its weight tends to make it fall over. (b) Earth precesses around the perpendicular to its orbit because the gravity of the sun and moon tend to twist it upright. (c) Precession causes the north celestial pole to drift among the stars, completing a circle in 26,000 years.

Locate Ursa Major and Orion on the star charts at the end of this book. For people in Canada, Ursa Major is circumpolar, but people in Mexico see most of this constellation slip below the horizon. From much of the United States, some of the stars of Ursa Major set, and some do not. In contrast, Orion rises and sets as seen from nearly everywhere on Earth. Explorers at Earth's poles, however, never see Orion rise or set.

Now use the argument you have just built. **How would you improve the definition of a circumpolar constellation to clarify the status of Ursa Major? Would your definition help in the case of Orion?**

Understanding versus Naming: The True Goal of Science

One of the fascinations of science is that it can reveal things you normally do not sense, such as a slow drift in the direction of Earth's axis. Furthermore, science can take your imagination into realms beyond your experience, from the inside of an atom to the inside of a star. Because these experiences are so unfamiliar, you need a vocabulary of technical terms just to talk about them, and that leads to a common confusion between naming things and understanding things.

The first step in understanding something is naming the parts, but it is only the beginning. The scientist's true goal is not naming but understanding. Consequently, biologists studying plants go beyond memorizing the names of the flowers' colors. They search for evidence to test their hypothesis that specific plants have developed specific flower colors to attract spe-

cific kinds of insects for efficient pollination. Naming the colors is only the beginning.

Some people complain about jargon in science, but the terminology is unfamiliar only because it describes phenomena you don't notice in everyday life. Scientists don't use this specialized vocabulary because they want to be pretentious or confusing. They use proper terminology so they can discuss nature with precision.

You must know the proper words, but real understanding comes from being able to use that vocabulary to discuss the way nature works. You must be able to tell the stories that scientists call theories and hypotheses. You must be able to use technical terms precisely to cite the evidence that makes us think those stories are true. That goes far beyond merely memorizing terminology; that is real understanding.

Naming the parts is only the first step in understanding how something works. (Royalty-Free/Corbis)

In folklore, naming a thing gives us power over it—recall the story of Rumpelstiltskin. In science, naming things is only the first step toward understanding.

Connections: The motions of the celestial sphere may seem to have little to do with your life, but at the end of this chapter you will see how precession may be one of the causes of the ice ages. Before you can think about ice ages, however, you must consider Earth's orbital motion around the sun and the resulting apparent motion of the sun. That affects you directly through the seasons.

(2-3) The Cycles of the Sun

PERHAPS THE MOST OBVIOUS CYCLES in the sky are those of the sun, but those apparent motions are produced by Earth's rotation and revolution. **Rotation** is the turning of a body about an axis through its center. Thus Earth rotates on its axis once a day. **Revolution** is the motion of a body around a point located outside the body, and Earth revolves around the sun once a year.

The cycle of day and night is caused by the rotation of Earth on its axis. In ■ Figure 2-8 you can see that four people in different places on Earth have different times of day, and that Earth's rotation carries you across the daylight side and then across the night side of Earth, producing the cycle of day and night.

Earth's rotation takes a day and causes day and night. Earth's revolution around the sun takes a year and produces the cycle of the seasons. To understand that motion, you must imagine that you can make the sun fainter.

The Annual Motion of the Sun

Even in the daytime, the sky is filled with stars, but the glare of sunlight fills Earth's atmosphere with scattered light, and you can see only the brilliant sun during the daytime. If the sun were fainter and you could see the stars, you would notice that day by day the sun was moving slowly eastward against the background of stars. That motion is caused by Earth's motion along its orbit as it revolves around the sun.

In ■ Figure 2-9, you can see how the sun appears in front of the stars of the constellation Sagittarius on January 1 and how Earth's motion along its orbit causes the sun to appear to move eastward against the constellations. By March 1 the sun appears to be in Aquarius.

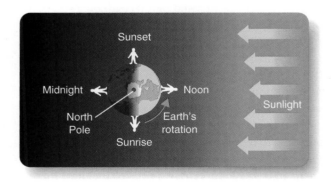

■ Figure 2-8

Looking down on Earth from above the North Pole shows how the time of day or night depends on your location on Earth.

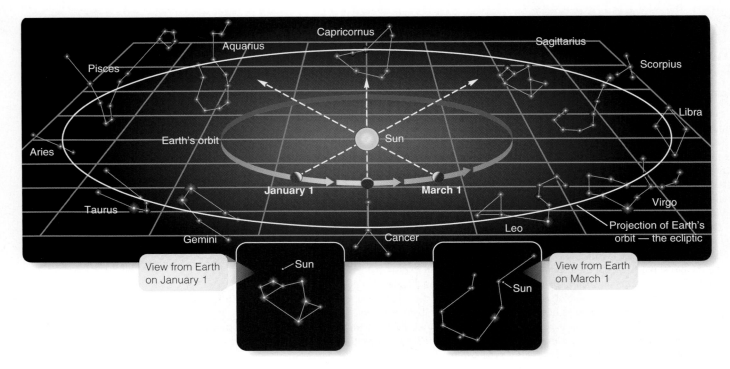

■ Active Figure 2-9

Earth's motion around the sun makes the sun appear to move against the background of the stars. Earth's circular orbit is thus projected on the sky as the circular path of the sun, the ecliptic. If you could see the stars in the daytime, you would notice the sun crossing in front of the distant constellations as Earth moves along its orbit.

Ace◗Astronomy™ Log into AceAstronomy and select this chapter to see the Active Figure called "Constellations in Different Seasons." Notice how the constellations in the morning sky move into the evening sky as the year passes.

Of course, it is not quite correct to say that the sun is "in Aquarius." The sun is only 93 million miles away, and the stars of Aquarius are at least a million times farther away. But in March of each year, you can see the sun against the background of the stars in Aquarius, and thus you could say, "The sun is in Aquarius."

If you continue watching the sun against the background of stars throughout the year, you could plot its path on a star chart. After one full year, you would see the sun begin to retrace its path as it continued its annual cycle of motion around the sky. This line, the apparent path of the sun in its yearly motion around the sky, is called the **ecliptic.** Another way to define the ecliptic is to say it is the projection of Earth's orbit on the sky. If the sky were a great screen illuminated by the sun at the center, then the shadow cast by Earth's orbit would be the ecliptic. Yet a third way to define the ecliptic is to refer to it as the plane of Earth's orbit. These three definitions of the ecliptic are equivalent, and it is worth considering them all because the ecliptic is one of the most important reference lines on the sky.

Earth circles the sun in 365.25 days, and consequently the sun appears to circle the sky in the same period. That means the sun, traveling 360° around the ecliptic in 365.25 days, travels about 1° eastward each day, about twice its angular diameter. You don't notice this motion because you can't see the stars in the daytime, but the motion of the sun has an important consequence that you do notice—the seasons.

The Seasons

The seasons are caused by a simple fact: Earth's axis of rotation is tipped 23.5° from the perpendicular to its orbit (■ Figure 2-10). As you study **The Cycle of the Seasons** on pages 26–27, notice two important principles:

1 The seasons are not caused by any variation in the distance from Earth to the sun. Earth's orbit is nearly circular, so it is always about the same distance from the sun.

2 The seasons are caused by the changes in solar energy that Earth's northern and southern hemispheres receive at different times of the year. Because of circulation patterns in Earth's atmosphere, the northern and southern hemispheres are mostly isolated from each other and exchange little heat.

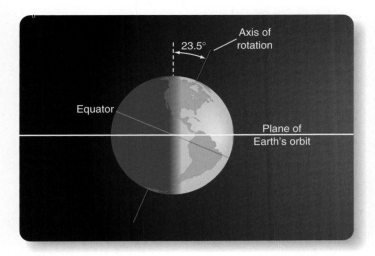

■ Figure 2-10

Earth's axis is inclined 23.5° from the perpendicular to the plane of its orbit. As it moves along its orbit, its axis of rotation remains fixed, pointing at the North Star.

When one hemisphere receives more solar energy than the other, it grows rapidly warmer.

In ancient times, the cycle of the seasons and the solstices and equinoxes were celebrated with rituals and festivals. Shakespeare's play *A Midsummer Night's Dream* describes the enchantment of the summer solstice night. (In Shakespeare's time, the equinoxes and solstices were taken to mark the midpoint of the seasons.) Many North American Indians marked the summer solstice with ceremonies and dances. Early church officials placed Christmas day in late December to coincide with an earlier pagan celebration of the winter solstice.

Building Scientific Arguments

If Earth had a significantly elliptical orbit, how would its seasons be different?

Sometimes, as you review your understanding, it is helpful to build a scientific argument with one factor slightly exaggerated. Suppose Earth had an elliptical orbit so that perihelion occurred in July and aphelion in January. At perihelion, Earth would be closer to the sun, and the entire surface of Earth would be a bit warmer. If that happened in July, it would be summer in the northern hemisphere and winter in the southern hemisphere, and both would be warmer than they are now. It could be a dreadfully hot summer in Canada, and southern Argentina could have a mild winter. Six months later, at aphelion, Earth would be a bit farther from the sun, and if that occurred in January, winter in northern latitudes could be frigid. Argentina, in the southern hemisphere, could be experiencing an unusually cool summer.

Of course, this doesn't happen. Earth's orbit is nearly circular, and the seasons are caused not by a variation in the distance of Earth from the sun but by the inclination of Earth in its orbit.

Nevertheless, Earth's orbit is slightly elliptical. Earth passes perihelion about January 4 and aphelion about July 4. Although Earth's oceans tend to store heat and reduce the importance of this effect, this very slight variation in distance does affect the seasons. Now use your scientific argument to analyze the seasons. **Does the elliptical shape of Earth's orbit make your winters warmer or cooler?**

■　　■　　■

Connections: The sun is not the only body that moves along the ecliptic. The moon and planets maintain constant traffic along the sun's highway.

Ace◐Astronomy™ Log into AceAstronomy and select this chapter to see Astronomy Exercise "Sunrise through the Seasons" to watch the sun moving.

Ace◐Astronomy™ Log into AceAstronomy and select this chapter to see Astronomy Exercise "The Seasons" and see how the sun's altitude changes.

(2-4) The Motion of the Planets

THE ECLIPTIC IS IMPORTANT because it is the path the sun follows around the sky, and that motion gives rise to the seasons. In the next chapter, you will consider the moon's cycle around the ecliptic, but here you should make note of the motion of the planets. Not only are most of the planets visible to the naked eye, but their motion is the basis for one of the longest-living superstitions in human history—that our fate is written in the stars.

The Moving Planets

Most of the planets of our solar system are visible to the unaided eye, though they produce no light of their own. They shine by reflected sunlight. Mercury, Venus, Mars, Jupiter, and Saturn are all visible to the naked eye, but Uranus is usually too faint to be seen (magnitude 5.6 at its brightest), and Neptune is never bright enough. Pluto is even fainter, and you would need a large telescope to find it. Although Uranus, Neptune, and Pluto are too faint to see, their motions are the same as those of the other planets.

All the planets of our solar system move in nearly circular orbits around the sun. If you were looking down on the solar system from the north celestial pole, you would see the planets moving in the same counterclockwise direction around their orbits (Chapter 1). The farther they are from the sun, the more slowly the planets move.

The Cycle of the Seasons

1 You can use the celestial sphere to help you think about the seasons. The celestial equator is the projection of Earth's equator on the sky, and the ecliptic is the projection of Earth's orbit on the sky. Because Earth is tipped in its orbit, the ecliptic and equator are inclined to each other by 23.5° as shown at right. As the sun moves eastward around the sky, it spends half the year in the southern half of the sky and half of the year in the northern half. That causes the seasons.

The sun crosses the celestial equator going northward at the point called the **vernal equinox.** The sun is at its farthest north at the point called the **summer solstice.** It crosses the celestial equator going southward at the **autumnal equinox** and reaches its most southern point at the **winter solstice.**

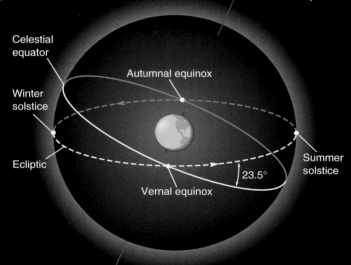

1a The seasons are defined by the dates when the sun crosses these four points, as shown in the table at the right. *Equinox* comes from the word for "equal"; the day of an equinox has equal amounts of daylight and darkness. *Solstice* comes from the words meaning "sun" and "stationary." *Vernal* comes from the word for "green." The "green" equinox marks the beginning of spring.

Event	Date*	Season
Vernal equinox	March 20	Spring begins
Summer solstice	June 22	Summer begins
Autumnal equinox	September 22	Autumn begins
Winter solstice	December 22	Winter begins

* Give or take a day due to leap year and other factors.

Ace ⓢ Astronomy™

Log into AceAstronomy and select this chapter to see Active Figure "Seasons" and watch Earth orbiting the sun.

1b On the day of the summer solstice in late June, Earth's northern hemisphere is inclined toward the sun, and sunlight shines almost straight down at northern latitudes. At southern latitudes, sunlight strikes the ground at an angle and spreads out. North America has warm weather, and South America has cool weather.

Earth's axis of rotation points toward Polaris, and, like a top, the spinning Earth holds its axis fixed as it orbits the sun. On one side of the sun, Earth's northern hemisphere leans toward the sun; on the other side of its orbit, it leans away. However, the direction of the axis of rotation does not change.

23.5°

To Polaris

40° N latitude

Equator

40° S latitude

Sunlight nearly direct on northern latitudes

To sun ⟶

Sunlight spread out on southern latitudes

Earth at summer solstice

Summer solstice light

1c Light striking the ground at a steep angle spreads out less than light striking the ground at a shallow angle. Light from the summer-solstice sun strikes northern latitudes from nearly overhead and is concentrated.

Winter solstice light

Light from the winter-solstice sun strikes northern latitudes at a much steeper angle and spreads out. The same amount of energy is spread over a larger area, so the ground receives less energy from the winter sun.

2 The two causes of the seasons are shown at right for someone in the northern hemisphere. First, the noon summer sun is higher in the sky and the winter sun is lower, as shown by the longer winter shadows. Thus winter sunlight is more spread out. Second, the summer sun rises in the northeast and sets in the northwest, spending more than 12 hours in the sky. The winter sun rises in the southeast and sets in the southwest, spending less than 12 hours in the sky. Both of these effects mean that northern latitudes receive more energy from the summer sun, and summer days are warmer than winter days.

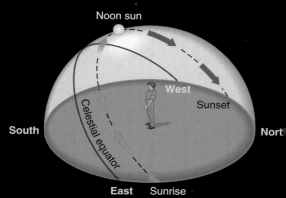

At summer solstice

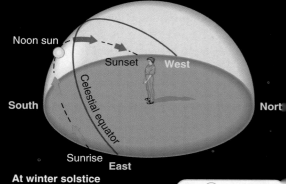

At winter solstice

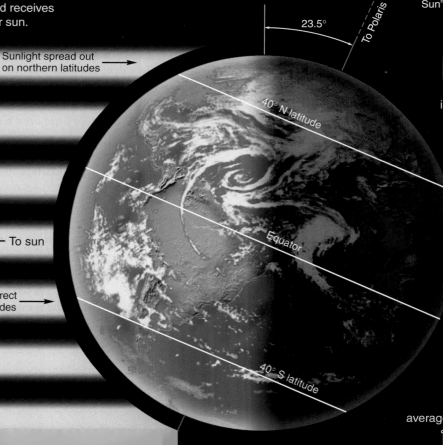

Sunlight spread out on northern latitudes →

← To sun

Sunlight nearly direct on southern latitudes →

23.5°

To Polaris

40° N latitude

Equator

40° S latitude

1d On the day of the winter solstice in late December, Earth's northern hemisphere is inclined away from the sun, and sunlight strikes the ground at an angle and spreads out. At southern latitudes, sunlight shines almost straight down and does not spread out. North America has cool weather and South America has warm weather.

Earth's orbit is only very slightly elliptical. About January 4, Earth is at **perihelion**, its closest point to the sun, when it is only 1.7 percent closer than average. About July 4, Earth is at **aphelion**, its most distant point from the sun, when it is only 1.7 percent farther than average. This small variation does not significantly affect the seasons.

When you look for planets in the sky, you will always find them near the ecliptic, because their orbits lie in nearly the same plane as the orbit of Earth. As they orbit the sun, they appear to move generally eastward along the ecliptic. (In Chapter 4, you will discover an exception to this general eastward motion.) In fact, the word *planet* comes from a Greek word meaning "wanderer." Mars moves completely around the ecliptic in slightly less than two years, but Saturn, being farther from the sun, takes nearly 30 years.

As seen from Earth, Venus and Mercury can never move far from the sun because their orbits are inside Earth's orbit. They sometimes appear near the western horizon just after sunset or near the eastern horizon just before sunrise. Venus is easier to locate because its larger orbit carries it higher above the horizon than Mercury (■ Figure 2-11). Mercury's orbit is so small that it can never get farther than about 28° from the sun. Consequently, it is usually hard to see against the sun's glare and is often hidden in the clouds and haze near the horizon. At certain times when it is farthest from the sun, however, Mercury shines brightly and can be located near the horizon in the evening or morning sky. (See the Appendix for the best times to observe Venus and Mercury.)

By tradition, any planet visible in the evening sky is called an **evening star,** although planets are not stars. Any planet visible in the sky shortly before sunrise is called a **morning star.** Perhaps the most beautiful is Venus, which can become as bright as minus fourth magnitude. As Venus moves around its orbit, it can dominate the western sky each evening for about half a year, but eventually its orbit carries it back toward the sun, and it is lost in the haze near the horizon. In a few weeks it reappears in the dawn sky as a brilliant morning star.

Astrology

Seen from Earth, the planets move gradually eastward along the ecliptic, but they don't follow the ecliptic exactly. Also, each travels at its own pace and seems to speed up and slow down at various times. To the ancients, this complex motion reflected the moods of the sky gods, and astrology was born.

Ancient astrologers defined a **zodiac,** a band 18° wide centered on the ecliptic, as the highway the planets follow. They divided this band into 12 segments named for the constellations along the ecliptic—the signs of the zodiac. A **horoscope** shows the location of the sun, moon, and planets among the zodiacal signs with respect to the horizon at the moment of a person's birth as seen from that longitude and latitude. Even if astrology worked, the generalized horoscopes published in newspapers and tabloids can't have been calculated accurately for individual readers.

Astrology buffs argue that a person's personality, life history, and fate are revealed in his or her horoscope, but the evidence contradicts this belief. Astrology has been tested many times over the centuries, and it just doesn't work. Believers, however, don't give up on it. Thus, astrology is a superstition that depends on blind belief and not a science that depends on evidence (**Window on Science 2-4**).

One reason astronomers find astrology irritating is that it has no link to the physical world. For example, precession has moved the constellations so that they no longer match the zodiacal signs. Whatever sign you were "born under," the sun was probably in the previous zodiacal constellation. In fact, if you were born on or between November 30 and December 17, the sun was passing through a corner of the nonzodiacal constellation Ophiuchus, and you have no official zodiacal sign.* Furthermore, astronomers like to point out, there is no mechanism by which

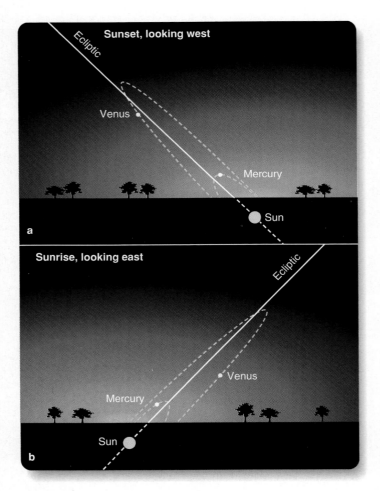

■ **Figure 2-11**

Mercury and Venus follow orbits that keep them near the sun, and they are visible only soon after sunset or before sunrise. Venus takes 584 days to move from morning sky to evening sky and back again, but Mercury zips around in only 116 days.

---

* The author of this book was born on December 14 and thus has no astrological sign. An astronomer friend claims that the author must therefore have no personality.

Astrology and Pseudoscience

Astronomers have a low opinion of astrology, not so much because it is groundless but because it pretends to be a science. It is a pseudoscience, from the prefix *pseudo,* meaning false. There are many examples of pseudoscience, and it is illuminating to consider the difference between pseudoscience and science.

A pseudoscience is a set of beliefs that appear to be based on scientific ideas but that fail to obey the most basic rules of science. For example, some years ago a fad arose in which people placed objects under pyramids made of paper, plastic, wire, and so on. The claim was that the pyramidal shape would focus cosmic forces on anything inside and so preserve fruit, sharpen razor blades, and do other miraculous things. Many books promoted this idea, but simple experiments showed that any shape would protect a piece of fruit from airborne spores and allow it to dry without rotting. Likewise, any shape would allow oxidation to improve the cutting edge of a razor blade. In short, experimental evidence contradicted the claim. Nevertheless, supporters of the theory declined to abandon or revise their claims. Thus, the fad of pyramid power was a pseudoscience.

One characteristic of a pseudoscience is that it appeals to our needs and desires. Thus, some pseudoscientific claims are self-fulfilling. For example, some people bought pyramidal tents to put over their beds and thus improve their rest. While there is no logical mechanism by which such a tent could affect a sleeper, because people wanted and expected the claim to be true they slept more soundly. Many pseudoscientific claims involve medical cures, ranging from magnetic bracelets and crystals to focus spiritual power to astonishingly expensive and illegal treatments for cancer. Logic is a stranger to pseudoscience, but human fears and needs are not.

Astrology is a pseudoscience. Over the centuries, astrology has been tested repeatedly, and no correlation has been found. But it survives, and its supporters disregard any evidence that it doesn't work. Like all pseudosciences, astrology is not open to revision in the face of contradictory evidence. Furthermore, astrology fulfills our human need to believe that there is order and meaning to our lives. It may comfort us to believe that our sweetheart has rejected us because of the motions of the planets rather than to admit that we behaved

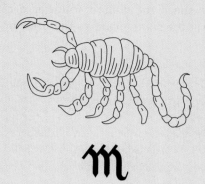

Astrology may be the oldest pseudoscience.

badly on our last date. Comfort aside, astrology is a poor basis for life decisions.

Human nature and human needs probably ensure that pseudoscientific beliefs will continue to plague us, like emotional viruses propagating from person to person. But if we recognize them for what they are we can more easily guide our lives by rational principles and not by giving credit for our successes and blame for our failings to the stars.

---

the planets could influence us. The gravitational influence of a doctor who is delivering a baby is many times more powerful than the gravitational influence of the planets.

The arguments in the preceding paragraph actually miss the point. Astrology is not related to the physical world at all. It does not matter what constellation the sun occupies, because astrology divides the zodiac into equal mathematical sections, sometimes called houses, and it does not matter to the believer in astrology that the constellations don't match. Furthermore, the physical mechanism is beside the point for the true believer. Astrology is not so much an astronomical superstition as it is a mathematical superstition.

Astrology makes sense only when you think of the world as the ancients did. They believed in multiple sky gods whose moods altered events on Earth. They believed in supernatural influences between natural events and human events. The ancients did not understand natural forces such as gravity as the mechanisms that cause things to happen, and thus they did not believe in cause and effect as modern people do. If a house burned, they might conclude that it caught fire because it was cursed and not because someone was careless with an oil lamp.

Modern science left astrology behind centuries ago, but astrology survives as a fascinating part of human history—an early attempt to understand the meaning of the sky.

Building Scientific Arguments

Planets like Mars can sometimes be seen rising in the east as the sun sets. Why is the same not true for Mercury or Venus? In this case, the argument hinges on simple geometry. Mars has an orbit outside the orbit of Earth, and that means it can reach a location opposite the sun in the sky. As the sun sets in the west, Mars rises in the east. But Mercury and Venus follow orbits that are smaller than the orbit of Earth, so they can never reach a point opposite the sun in the sky.

As they follow their orbits around the sun, Earthlings see them gradually swing out on the east side of the sun, reach a maximum distance from the sun, swing back toward the sun, move out to the west side of the sun, reach a maximum distance west, and then move back toward the sun again.

Because they never get very far from the sun in the sky, they can never be seen rising in the east as the sun sets.

The planetary motions in the sky are produced by the orbital motions of the planets around the sun. Build a geometrical argument to analyze the following question: **What would you see if the planets followed orbits that were all in exactly the same plane as the orbit of Earth?**

■   ■   ■

Connections: Modern astronomers have a low tolerance for astrological superstition, yet the motions of the heavenly bodies do affect human lives. The motion of the sun produces the seasons, and, as you will discover in Chapter 5, the moon governs the tides. In addition, small changes in Earth's orbit may partially control global climate.

(2-5) Astronomical Influences on Earth's Climate

WEATHER IS WHAT HAPPENS TODAY; climate is the average of what happens over decades. Earth has gone through past episodes, called ice ages, when the worldwide climate was cooler and dryer and thick layers of ice covered northern latitudes.

The earliest known ice age occurred about 570 million years ago and the next about 280 million years ago. The most recent ice age began only about 3 million years ago and is still going on. You are living in one of the periodic episodes when the glaciers melt and Earth grows slightly warmer. The current warm period began about 20,000 years ago.

Ice ages seem to occur with a period of roughly 250 million years, and cycles of glaciation within ice ages occur with a period of about 40,000 years. These cyclic changes have an astronomical origin.

The Hypothesis

Sometimes a theory or hypothesis is proposed long before scientists can find the critical evidence to test it. That happened in 1920 when Yugoslavian meteorologist Milutin Milankovitch proposed what became known as the **Milankovitch hypothesis**—that changes in the shape of Earth's orbit, in precession, and in inclination affect Earth's climate and trigger ice ages. Let's examine each of these three motions in turn.

First, astronomers know that the elliptical shape of Earth's orbit varies slightly over a period of about 100,000 years. At present, Earth's orbit carries it 1.7 percent closer than average to the sun during northern-hemisphere winters and 1.7 percent farther away in northern-hemisphere summers. This makes the northern climate very slightly less extreme, and that is critical—most of the landmass where ice can accumulate is in the northern hemisphere. If Earth's orbit became more elliptical, northern summers might be too cool to melt all of the snow and ice from the previous winter. That would make glaciers grow larger.

A second factor is also at work. Precession causes Earth's axis to sweep around a cone with a period of about 26,000 years, and that changes the location of the seasons around Earth's orbit. Northern summers now occur when Earth is 1.7 percent farther from the sun, but in 13,000 years northern summers will occur on the other side of Earth's orbit where Earth is slightly closer to the sun. Northern summers will be warmer, which could melt all of the previous winter's snow and ice and prevent the growth of glaciers.

The third factor is the inclination of Earth's equator to its orbit. Currently at 23.5°, this angle varies from 22° to 24° with a period of roughly 41,000 years. When the inclination is greater, seasons are more severe.

In 1920, Milankovitch proposed that these three factors cycled against each other to produce complex periodic variations in Earth's climate and the advance and retreat of glaciers (■ Figure 2-12a). But no evidence was available to test the theory in 1920, and scientists treated it with skepticism. Many thought it was laughable.

The Evidence

By the middle 1970s, Earth scientists could collect the data that Milankovitch needed. Oceanographers could drill deep into the seafloor and collect samples, and geologists could determine the age of the samples from the natural radioactive atoms they contained. From all this, scientists constructed a history of ocean temperatures that convincingly matched the predictions of the Milankovitch hypothesis (Figure 2-12b).

The evidence seemed very strong, and, by the 1980s, the Milankovitch hypothesis was widely discussed as the leading hypothesis. But science follows a mostly unstated set of rules that holds that a hypothesis must be tested over and over against all available evidence (**Window on Science 2-5**). In 1988, scientists discovered contradictory evidence.

A water-filled crack in Nevada called Devil's Hole contains deposits of the mineral calcite. Diving with scuba gear, scientists drilled out samples of the calcite and analyzed the oxygen atoms found there. For 500,000 years, layers of calcite have built up in Devil's Hole, recording in their oxygen atoms the temperature of the atmosphere when rain fell there. Finding the ages of the mineral samples was difficult, but the results seemed to show that the previous ice age ended thousands of years too early to have been caused by Earth's motions.

These contradictory findings are irritating because we naturally prefer certainty, but such circumstances are common in science. The disagreement between ocean floor samples and Devil's Hole samples triggered a scramble to understand the problem. Were the ages of one or the other set of samples wrong? Were the ancient temperatures wrong? Or were scientists misunderstanding the significance of the evidence?

In 1997, a new study of the ages of the samples confirmed that those from the ocean floor are correctly dated. This seems to

The Foundation of Science: Evidence

Science is based on evidence. Every theory and conclusion must be supported by evidence obtained from experiments or from observation. If a theory is supported by many pieces of evidence but is clearly contradicted by a single experiment or observation, scientists quickly abandon it. For a theory to be true, there can be no contradictory evidence.

Of course, scientists argue about the significance of particular evidence and often disagree on the interpretation of evidence. Some observations may seem significant at first glance, but upon closer examination you may find that the procedure was flawed and so

the piece of evidence is not important. Or you might conclude that the observational fact is correct but is being misinterpreted. The observation may not mean what it seems. Some of the most famous disagreements in science, such as those surrounding Galileo and Darwin, have arisen over the interpretation of well-established factual evidence.

Furthermore, scientists are not allowed to be selective in considering evidence. A lawyer in court can call a certain witness and intentionally fail to ask a critical question that would reveal evidence harmful to the lawyer's case. But a scientist may not ignore any known evidence. The difference in their methods is

revealing. The lawyer is attempting to prove only one side of the case and rightly may ignore contradictory evidence. The scientist, however, is searching for the truth and so must test any theory against all available evidence. In a sense, the scientist, in dealing with evidence, must act as both the prosecution and the defense.

As you read about any science, look for the evidence in the form of measurements or observations. Every theory or conclusion should have supporting evidence. If you can find and understand the evidence, the science will make sense. All scientists, from astronomers to zoologists, demand evidence. You should, too.

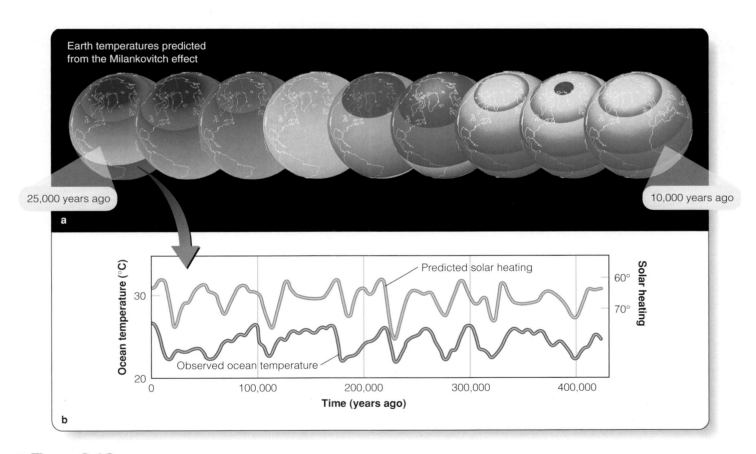

■ Figure 2-12

(a) Mathematical models of the Milankovich effect can be used to predict temperatures on Earth over time. In these Earth globes, cool temperatures are represented by violet and blue and warm temperatures by yellow and red. These globes show the warming that occurred beginning 25,000 years ago, which ended the last ice age. (Courtesy Arizona State University, Computer Science and Geography Departments) (b) Over the last 400,000 years, changes in ocean temperatures measured from fossils found in sediment layers from the seabed match calculated changes in solar heating. (Adapted from Cesare Emiliani)

give scientists renewed confidence in the Milankovitch hypothesis. But the same study found that the ages of the Devil's Hole samples are also correct. Evidently the temperatures at Devil's Hole record local climate changes in the region that became the southwestern United States. The ocean floor samples record global climate changes, and they fit well with the Milankovitch hypothesis. In this way, the Milankovitch hypothesis, though widely accepted today, is still being tested as scientists try to understand the world we live on.

Building Scientific Arguments

How do precession and the shape of Earth's orbit interact to affect Earth's climate?

Here, as in an earlier section of this chapter, exaggeration is a useful analytical tool in your argument. If you exaggerate the variation in the shape of Earth's orbit, you can see dramatically the influence of precession. At present, Earth reaches perihelion during winter in the northern hemisphere and aphelion during summer. The variation in distance is only about 1.7 percent, and that difference doesn't cause much change in the severity of the seasons. But if Earth's orbit were much more elliptical, then winter in the northern hemisphere would be much warmer, and summer would be much cooler.

Now you can see the importance of precession. As Earth's axis precesses, it points gradually in different directions, and the seasons occur at different places in Earth's orbit. In 13,000 years, northern winter will occur at aphelion, and, if Earth's orbit were highly elliptical, northern winter would be terrible. Similarly, summer would occur at perihelion, and the heat would be awful. Such extremes might deposit large amounts of ice in the winter but then melt it away in the hot summer, thus preventing the accumulation of glaciers.

Continue this analysis by exaggeration in your own scientific argument. **What effect would precession have if Earth's orbit were more circular?**

■   ■   ■

Connections: This chapter is about the sky, but it has told you a great deal about Earth. You have discovered that Earth's orbital motion produces the apparent motion of the sun along the ecliptic, causes the seasons, and may even cause the cycle of the ice ages. As is often the case, astronomy helps you understand what you are and where you are. In this case, your study of the sky has helped you understand what it means to live on a planet. However, this discussion has omitted one of the most dramatic objects in the sky—the moon. That omission will be repaired in the next chapter.

Summary

2-1 | The Stars

How do astronomers refer to stars?
- Astronomers divide the sky into 88 constellations. Although the constellations originated in Greek and Middle Eastern mythology, the names are Latin. Even the modern constellations, added to fill in the spaces between the ancient figures, have Latin names.
- The names of stars usually come from ancient Arabic, although modern astronomers often refer to a star by its constellation and a Greek letter assigned according to its brightness within the constellation.

How can you compare the brightness of the stars?
- The magnitude system is the astronomer's brightness scale. First-magnitude stars are brighter than second-magnitude stars, which are brighter than third-magnitude stars, and so on. The magnitude you see when you look at a star in the sky is its apparent visual magnitude, which does not take into account its distance from Earth.
- Apparent visual magnitude, m_v, includes only the light that human eyes can see.

2-2 | The Sky and Its Motion

How does the sky move as Earth moves?
- The celestial sphere is a model of the sky, carrying the celestial objects around Earth. Because Earth rotates eastward, the celestial sphere appears to rotate westward on its axis. The northern and southern celestial poles are the pivots on which the sky appears to rotate.
- The celestial equator, an imaginary line around the sky above Earth's equator, divides the sky in half.
- Astronomers often refer to angles "on" the sky as if the stars, sun, moon, and planets were equivalent to spots painted on a plaster ceiling. These angular distances are unrelated to the true distance between the objects in light-years.
- What you see of the celestial sphere depends on your latitude. Much of the southern hemisphere of the sky is not visible from northern latitudes. To see that part of the sky, you would have to travel southward over Earth's surface.
- The angular distance from the horizon to the north celestial pole always equals your latitude. This is the basis for celestial navigation.
- The gravitational forces of the moon and sun act on the spinning Earth and cause it to precess like a top. Earth's axis of rotation sweeps around in a conical motion with a period of 26,000 years, and consequently the celestial poles and celestial equator move slowly against the background of the stars.

2-3 | The Cycles of the Sun

What causes the seasons?
- Because Earth orbits the sun, the sun appears to move eastward along the ecliptic through the constellations. It circles the entire zodiac in a year.
- Because the ecliptic is tipped 23.5° to the celestial equator, the sun spends half the year in the northern celestial hemisphere and half in the southern celestial hemisphere.
- In the summer, the sun is above the horizon longer and shines more directly down on the ground. Both effects cause warmer weather.

2-4 | The Motion of the Planets
- In addition to the sun, the visible planets also move along the ecliptic, and their positions give rise to the ancient superstition called astrology.
- Mercury and Venus follow orbits inside Earth's orbit and never move far from the sun. They are visible in the east before dawn or in the west after sunset.

2-5 | Astronomical Influences on Earth's Climate

How do astronomical cycles affect Earth's climate?
- Changes in the shape of Earth's orbit, in its precession, and in its axial tilt can alter the planet's heat balance and seem to be at least partly responsible for the ice ages and glacial periods.

New Terms

constellation (p. 14)	second of arc (p. 21)
asterism (p. 14)	angular diameter (p. 21)
magnitude scale (p. 16)	circumpolar constellation (p. 21)
apparent visual magnitude (m_v) (p. 16)	precession (p. 22)
scientific model (p. 19)	rotation (p. 23)
celestial sphere (p. 20)	revolution (p. 23)
horizon (p. 20)	ecliptic (p. 24)
zenith (p. 20)	vernal equinox (p. 26)
nadir (p. 20)	summer solstice (p. 26)
north celestial pole (p. 20)	autumnal equinox (p. 26)
south celestial pole (p. 20)	winter solstice (p. 26)
celestial equator (p. 20)	perihelion (p. 27)
north point (p. 20)	aphelion (p. 27)
south point (p. 20)	evening star (p. 28)
east point (p. 20)	morning star (p. 28)
west point (p. 20)	zodiac (p. 28)
angular distance (p. 21)	horoscope (p. 28)
minute of arc (p. 21)	Milankovitch hypothesis (p. 30)

Review Questions

Ace ⌾ Astronomy™ Assess your understanding of this chapter's topics with additional quizzing and animations at **http://ace.brookscole.com/sf9**

1. Why are most modern constellations composed of faint stars or located in the southern sky?
2. What does a star's Greek-letter designation tell you that its ancient Arabic name does not?
3. From your knowledge of star names and constellations, which of the following stars in each group is the brighter and which is the fainter? Explain your answers. (a) α Ursae Majoris; ε Ursae Majoris (b) ε Scorpii; α Pegasus (c) α Telescopium; α Orionis
4. Give two reasons why the magnitude scale might be confusing.

5. Why do modern astronomers continue to use the celestial sphere when they know that stars are not all at the same distance?

6. Define the celestial poles and the celestial equator.

7. From what locations on Earth is the north celestial pole not visible? the south celestial pole? the celestial equator?

8. If Earth did not turn on its axis, could you still define an ecliptic? Why or why not?

9. Give two reasons why winter days are colder than summer days.

10. How do the seasons in Earth's southern hemisphere differ from those in the northern hemisphere?

11. Why should the eccentricity of Earth's orbit make winter in Earth's northern hemisphere different from winter in the southern hemisphere?

12. The stamp at right shows the constellation Orion. Explain why this looks odd to residents of the northern hemisphere.

Discussion Questions

1. Have you thought of the sky as a ceiling? as a dome overhead? as a sphere around Earth? as a limitless void?

2. How would the seasons be different if Earth were inclined 90° instead of 23.5°? 0° instead of 23.5°?

Problems

1. If one star is 6.3 times brighter than another star, how many magnitudes brighter is it?

2. If one star is 40 times brighter than another star, how many magnitudes brighter is it?

3. If two stars differ by 7 magnitudes, what is their intensity ratio?

4. If two stars differ by 8.6 magnitudes, what is their intensity ratio?

5. If star A is third magnitude and star B is fifth magnitude, which is brighter and by what factor?

6. If star A is magnitude 4 and star B is magnitude 9.6, which is brighter and by what factor?

7. By what factor is the sun brighter than the full moon? (*Hint:* See Figure 2-6.)

8. What is the angular distance from the north celestial pole to the summer solstice? to the winter solstice?

9. As seen from your latitude, what is the angle between the north celestial pole and the northern horizon? between the southern horizon and the noon sun at the summer solstice?

Media Cluster

Ace ⟳ Astronomy™ To access the resources in the Media Cluster, log into AceAstronomy at **http://ace.brookscole.com/sf9** and select Chapter 2.

ACTIVE FIGURES

The Celestial Sphere
Observers on Earth see the sky as a celestial sphere. This animation lets you select your vantage point and visualize that location's half-dome of the sky.

Rotation of the Sky
This animation shows how the stars (including the sun) appear to rotate in the sky every 24 hours, as seen by northern hemisphere observers.

Constellations in Different Seasons
This animation lets you observe how the constellations change with the seasons as Earth moves through its orbit of the sun.

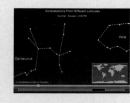

Constellations from Different Latitudes
Use this animation to observe how a change in latitude provides a different view of the constellations.

Seasons
Visit different cities around the world and watch the seasons pass as Earth orbits the sun.

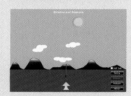

Path of the Sun
Use a shadow stick to observe the altitude of the sun as it crosses the sky in different seasons of the year.

ASTRONOMY EXERCISES

Sunrise through the Seasons
In this animation, you can observe how the position of the sunrise changes throughout the year.

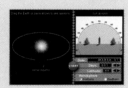

The Seasons
Earlier in this chapter, you saw how the angle of the sun in the sky brings about warmer days in the summer and colder days in the winter for the northern hemisphere. In this animation, you can select the latitude and hemisphere to compare the angle of the sun for various points around the world.

Critical Inquiries for the Web

1. Nearly all cultures have populated the sky with gods, heroes, animals, and objects. What can you learn on the Web about non-Western constellations?

2. Who was Orion? How is he related to the scorpion in the sky?

3. What holidays, rituals, special foods, and beliefs are associated with the winter solstice?

4. What can you find out about Milutin Milankovitch? What is the latest news about the Milankovitch hypothesis?

Exploring *TheSky*

1. As discussed in this chapter, Earth's rotation about its own axis gives the impression that the whole sky rotates around the north celestial pole in a period of one day. This apparent motion of the celestial sphere is difficult to notice because it happens so slowly. However, *TheSky* makes it possible to simulate this motion at a pace that is easy to observe by using a feature called **Time Skip.** Observe and describe the apparent motion of the sky as you see it looking north, east, south, and west.

How to proceed: Set the **Time Skip Increment** (a dropdown menu on the **Time Skip Toolbar**) to 1 minute, and click on the **Go Forward** button to begin the simulation. View the sky from the four cardinal directions, due north, south, east, and west. (You'll find **Time Skip** under the **Tools** menu as well.)

2. In which constellation was the sun located on the date of your birth? (*Hint:* Click **Data** and then **Site Information** to set the location, date, and time of your birth. Turn on constellation figures, constellation boundaries, and labels. Then find the sun.)

3. Set your location to Earth's North Pole and the date to the summer solstice. Turn on the ecliptic and then step forward in time through a day to see what happens to the sun. Repeat for the autumnal equinox and the winter solstice.

4. Repeat Problem 3 above for a location on Earth's equator.

 Go to the Brooks/Cole Astronomy Resource Center **(http://astronomy.brookscole.com)** for critical thinking exercises, articles, and additional readings from InfoTrac College Edition, Brooks/Cole's online student library

3 | Cycles of the Moon

Even a man who is

pure in heart

And says his prayers by night

May become a wolf when

the wolfbane blooms

And the moon shines

full and bright.

PROVERB FROM OLD WOLFMAN MOVIES

Visual-wavelength image

DID ANYONE EVER WARN YOU, "Don't stare at the moon—you'll go crazy"?* For centuries, the superstitious have associated the moon with insanity. The word *lunatic* comes from a time when even doctors thought that the insane were "moonstruck." A "mooncalf" is someone who has been crazy since birth, and the word is probably related to the belief that moonlight can harm unborn children. Everyone "knows" that people act less rationally when the moon is bright, but everyone is

---

* When I was very small, my grandmother told me if I gazed at the moon, I might go crazy. But it was too beautiful, and I ignored her warning. I secretly watched the moon from my window, became fascinated by the sky, and grew up to be an astronomer.

Continued on page 38

Guidepost

Looking Back

In the preceding chapter, you watched Earth rotating on its axis, which makes the sun rise and set. You watched Earth revolving along its orbit around the sun, which produces the cycle of the seasons. These two cycles so completely dominate your life on Earth that you hardly notice them. But there are other cycles in the sky, and now you are ready to study one of the most dramatic and beautiful actors on the celestial stage.

This Chapter

The moon is the brightest object in the night sky, and it moves rapidly against the background of stars, changing its shape and occasionally producing strange events called eclipses. You may feel you know the moon well, but you will probably find some surprises as you answer four essential questions about Earth's satellite:

Why does the moon go through phases?

What causes a lunar eclipse?

What causes a solar eclipse?

How can eclipses be predicted?

This chapter illustrates two powerful ways to analyze certain kinds of problems. Many problems in astronomy depend on seeing how light and shadow move through space in three dimensions. Phases and eclipses are good case studies. Also, if a process repeats, you can analyze it by searching for cycles. The seemingly complex motions of the sun and moon become simple and elegant when you see them as cycles.

Looking Ahead

Once you have a 21st-century understanding of your world and its motion, you will be ready to read the next chapter, where you will see how Renaissance astronomers analyzed what they saw in the sky and came to a revolutionary conclusion—that we live on a planet.

wrong. Careful studies of hospital, school, and police records show that there is no real correlation between the moon and erratic behavior.

The moon is so bright and its cycles through the sky are so dramatic we *expect* it to influence us, and many people are disappointed to learn the moon does not influence us. Some simply refuse to believe the evidence.

The cycles of the moon are indeed dramatic, and the moon is a rich source of myths and traditions. Many different cultures around the world explain eclipses of the sun as invisible monsters devouring the sun (■ Figure 3-1). A Chinese story tells of two astronomers, Hsi and Ho, who were too drunk to predict the solar eclipse of October 22, 2137 BC, or perhaps failed to conduct the proper ceremonies to scare away the dragon snacking on the sun. When the emperor recovered from the terror of the eclipse, he had the two astronomers beheaded. The cycles of the moon provide a wealth of beauty in the sky and an important part of the cultural beliefs of Earth's peoples.

This chapter discusses the lunar cycles of phases and eclipses. Studying these events will help you understand what you see in the night sky. They will also introduce you to some of the most basic concepts in astronomy.

You will discover how light and shadow move through space, how the size and distance of an object affect what you see, and how cyclic events can be analyzed by searching for their patterns.

Most of all, this chapter helps you to begin thinking of your home as a world in space.

3-1 The Changeable Moon

STARTING THIS EVENING, look for the moon in the sky. If the night is cloudy or the moon is in the wrong part of its orbit, you may not see it, but keep trying on successive evenings, and within a week or two you will see the moon. Then watch for the moon on following evenings, and you will see it following its orbit around Earth and cycling through its phases as it has done for billions of years.

The Motion of the Moon

If you watch the moon night after night, you will notice two things about its motion. First, you will see it moving eastward against the background of stars; second, you will notice that the markings on its face don't change. These two observations will help you understand the motion of the moon and the origin of the moon's phases.

The moon moves rapidly among the constellations. If you watch the moon for just an hour, you can see it move eastward against the background of stars by slightly more than its angular diameter. In the previous chapter, you discovered that the moon

■ **Figure 3-1**

(a) A 12th-century Mayan symbol believed to represent a solar eclipse. The black-and-white sun symbol hangs from a rectangular sky symbol, and a voracious serpent approaches from below. (b) This Chinese representation of a solar eclipse shows a dragon flying in front of the sun. (From the collection of Yerkes Observatory) (c) This wall carving from the ruins of a temple at Vijayanagara in southern India symbolizes a solar eclipse as two snakes approaching the disk of the sun. (T. Scott Smith)

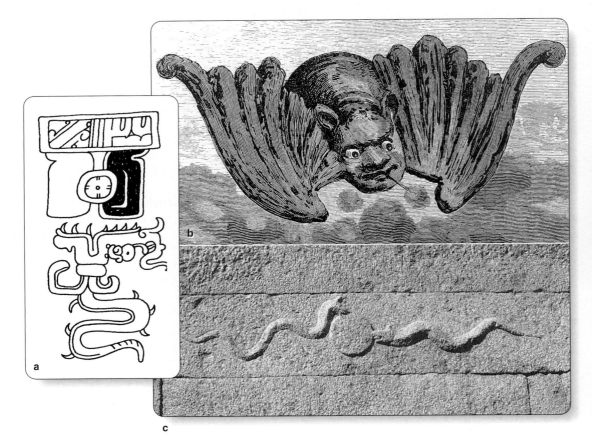

is about 0.5° in angular diameter, so it moves eastward a bit more than 0.5° per hour. In 24 hours, it moves 13°. Each night when you look at the moon, you see it about 13° eastward of its location the night before. This eastward movement is the result of the motion of the moon along its orbit around Earth.

The Cycle of Phases

The changing shape of the moon as it orbits Earth is one of the most easily observed phenomena in astronomy. Everyone has noticed the full moon rising dramatically or a thin crescent moon hanging in the evening sky. Study **The Phases of the Moon** on pages 40–41 and notice three important points:

1 First, the moon always keeps the same side facing Earth, and you never see the far side of the moon. "The man in the moon" is produced by the familiar features on the moon's near side.

2 Second, the changing shape of the moon as it passes through its cycle of phases is produced by sunlight illuminating different parts of the side of the moon you can see.

3 The third thing to notice is the difference between the orbital period of the moon around Earth and the length of the lunar phase cycle. That difference is a good illustration of how your view from Earth is produced by the combined motions of Earth and other heavenly bodies such as the sun and moon.

You can make a moon-phase dial from the middle diagram on page 40 by covering the lower half of the moon's orbit with a sheet of paper, aligning the edge of the paper to pass through the word "Full" at the left and the word "New" at the right. Push a pin through the edge of the paper at Earth's North Pole to make a pivot, and under the word "Full" write on the paper "Eastern Horizon." Under the word "New" write "Western Horizon." The paper now represents the horizon you see facing south.

You can set your moon-phase dial for a given time by rotating the diagram behind the horizon paper. Set the dial to sunset by turning the diagram until the human figure labeled "sunset" is standing at the top of the Earth globe; the dial shows, for example, that the full moon at sunset would be at the eastern horizon.

We Earthlings always see the same side of the moon looking down on us , but the changing shadows make the man in the moon shift his moods as the moon cycles through its phases (■ Figure 3-2). Watch for the moon and enjoy the cycle of phases. It is one of the perks of living on this planet.

Ace★Astronomy™ Log into AceAstronomy and select this chapter to see Astronomy Exercise "Phases of the Moon." You can take control of the moon's phases.

Ace★Astronomy™ Log into AceAstronomy and select this chapter to see Astronomy Exercise "Moon Calendar." Find the phases of the moon for any month.

Building Scientific Arguments

Why is the moon sometimes visible in the daytime?
Lots of people are surprised when they notice the moon in the daytime sky, but you aren't surprised because you can explain it with a simple scientific argument involving the geometry of the moon's motion. The full moon rises at sunset and sets at sunrise, so it is visible in the night sky but never in the daytime sky. At other phases, it is possible to see the moon in the daytime. For example, when the moon is a waxing gibbous moon, it rises a few hours before sunset. If you look in the right spot, you can see it in the late afternoon in the southeast sky. It looks pale and washed out because of sunlight in Earth's atmosphere, but it is quite visible once you notice it. You can also locate the waning gibbous moon in the morning sky. It sets an hour or two after sunrise, so you would look for it in the southwestern sky in the morning.

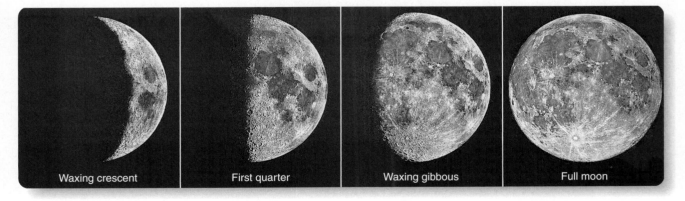

| Waxing crescent | First quarter | Waxing gibbous | Full moon |

■ Figure 3-2

In this sequence of the waxing moon, you see the same face of the moon, the same mountains, craters, and plains, but the changing direction of sunlight produces the lunar phases. (UCO/Lick Observatory)

The Phases of the Moon

1 As the moon orbits Earth, it rotates to keep the same side facing Earth as shown at right. Consequently you always see the same features on the moon, and you never see the far side of the moon. A mountain on the moon that points at Earth will always point at Earth as the moon revolves and rotates.

(Not to scale)

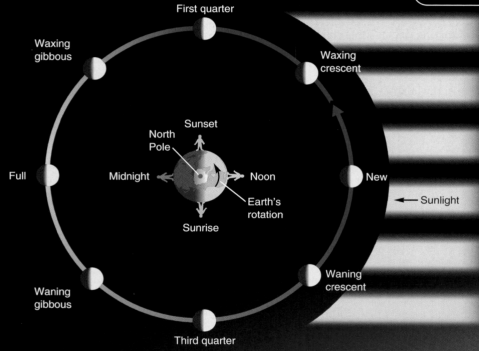

First quarter

Waxing gibbous

Waxing crescent

North Pole
Sunset

Midnight

Noon

Full

New

Earth's rotation

Sunlight

Sunrise

Waning crescent

Waning gibbous

Third quarter

2 As seen at left, sunlight always illuminates half of the moon. Because you see different amounts of this sunlit side, you see the moon cycle through phases. At the phase called "new moon," sunlight illuminates the far side of the moon, and the side you see is in darkness. At new moon you see no moon at all. At full moon, the side you see is fully lit, and the far side is in darkness. How much you see depends on where the moon is in its orbit.

Notice that there is no such thing as the "dark side of the moon." All parts of the moon experience day and night in a month-long cycle.

In the diagram at the left, you see that the new moon is close to the sun in the sky, and the full moon is opposite the sun. The time of day depends on the observer's location on Earth.

2a The first two weeks of the cycle of the moon are shown below by its position at sunset on 14 successive evenings. As the moon grows fatter from new to full, it is said to wax.

Gibbous comes from the Latin word for humpbacked.

The first quarter moon is one week through its 4-week cycle.

Waxing gibbous

Waxing crescent

The full moon is two weeks through its 4-week cycle.

Full moon rises at sunset

THE SKY AT SUNSET

New moon is invisible near the sun

East

South

West

3 The moon orbits eastward around Earth in 27.32 days, its **sidereal period.** This is how long the moon takes to circle the sky once and return to the same position among the stars.

A complete cycle of lunar phases takes 29.53 days, the moon's **synodic period.** (Synodic comes from the Greek words for "together" and "path.")

To see why the synodic period is longer than the sidereal period, study the star charts at the right.

Although you think of the lunar cycle as being about 4 weeks long, it is actually 1.53 days longer than 4 weeks. The calendar divides the year into 30-day periods called months (literally "moonths") in recognition of the 29.53 day synodic cycle of the moon.

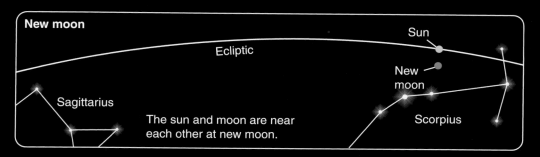

New moon

Ecliptic

Sun

New moon

Sagittarius

The sun and moon are near each other at new moon.

Scorpius

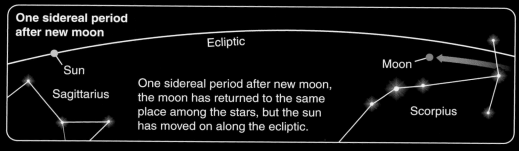

One sidereal period after new moon

Ecliptic

Sun

Sagittarius

Moon

One sidereal period after new moon, the moon has returned to the same place among the stars, but the sun has moved on along the ecliptic.

Scorpius

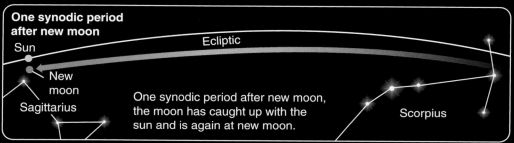

One synodic period after new moon

Sun

Ecliptic

New moon

Sagittarius

One synodic period after new moon, the moon has caught up with the sun and is again at new moon.

Scorpius

You can use the diagram on the opposite page to determine when the moon rises and sets at different phases.

TIMES OF MOONRISE AND MOONSET

Phase	Moonrise	Moonset
New	Dawn	Sunset
First quarter	Noon	Midnight
Full	Sunset	Dawn
Third quarter	Midnight	Noon

2b The last two weeks of the cycle of the moon are shown below by its position at sunrise on 14 successive mornings. As the moon shrinks from full to new, it is said to wane.

The first quarter moon is 3 weeks through its 4-week cycle.

Waning crescent

Waning gibbous

New moon is invisible near the sun

THE SKY AT SUNRISE

Full moon sets at sunrise

East

South

West

A simple scientific argument analyzing the motion of the moon can explain a lot about what you see and what you don't see. **Why is it extremely difficult to see the crescent moon in the daytime?**

■   ■   ■

Connections: The phases of the moon don't affect you directly—you don't act crazier than usual at full moon. The phases are a decoration in the sky, but occasionally, the moon provides a more dramatic treat.

(3-2) Lunar Eclipses

A **LUNAR ECLIPSE** occurs at full moon when the moon moves through Earth's shadow. Because the moon shines only by reflected sunlight, it gradually darkens as it enters the shadow.

Earth's Shadow

Earth's shadow consists of two parts. The **umbra** is the region of total shadow. If you were floating in space in the umbra of Earth's shadow, you would see no portion of the sun; it would be completely hidden behind Earth. However, if you moved into the **penumbra,** you would be in partial shadow and would see part of the sun peeking around Earth's edge. In the penumbra, the sunlight is dimmed but not extinguished.

You can construct a model of Earth's shadow by pressing a map tack into the eraser of a pencil and holding the tack between a lightbulb a few feet away and a white cardboard screen (■ Figure 3-3). The lightbulb represents the sun, and the map tack represents Earth. If you hold the screen close to the tack, you will see that the umbra is nearly as large as the tack and that the penum-

bra is only slightly larger. However, if you move the screen away from the tack, the umbra shrinks and the penumbra expands. Beyond a certain point, the shadow has no dark core at all, indicating that the screen is beyond the end of the umbra.

The umbra of Earth's shadow is about 1.4 million km (860,000 mi) long and points directly away from the sun. A giant screen placed in the shadow at the average distance of the moon would reveal a dark umbra about 9000 km (5700 mi) in diameter, and the faint outer edges of the penumbra would mark a circle about 16,000 km (10,000 mi) in diameter. For comparison, the moon's diameter is only 3476 km (2160 mi). Consequently, when the moon's orbit carries it through the umbra, it has plenty of room to become completely immersed in shadow.

Total Lunar Eclipses

A lunar eclipse occurs when the moon passes through Earth's shadow and grows dark. If the moon passes through the umbra and no part of the moon remains outside the umbra in the partial sunlight of the penumbra, the eclipse is a **total lunar eclipse.**

■ Figure 3-4 illustrates the stages of a total lunar eclipse as a three-dimensional diagram (**Window on Science 3-1**) showing Earth, its shadows, and the path of the moon. As the moon begins to enter the penumbra, it is only slightly dimmed, and a casual observer may not notice anything odd. After an hour, the moon is deeper in the penumbra and dimmer; and, once it begins to enter the umbra, you see a dark bite on the edge of the lunar disk. The moon travels its own diameter in an hour, so it takes about an hour to enter the umbra completely and become totally eclipsed.

Even when the moon is totally eclipsed, it does not disappear completely. Sunlight, bent by Earth's atmosphere, leaks into the umbra and bathes the moon in a faint glow. Because blue

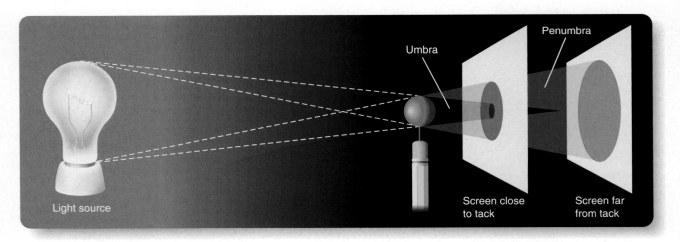

■ Figure 3-3

The shadows cast by a map tack resemble the shadows of Earth and the moon. The umbra is the region of total shadow; the penumbra is the region of partial shadow.

3-D Relationships

Much of science is explained in diagrams; one particular type of diagram can be confusing. When artists draw three-dimensional diagrams on flat sheets of paper, they use lots of artistic clues that have been discovered over the centuries. Perspective, shading, color, and shadows help you see the three-dimensional figure if the drawing is familiar—a house, for example. But when a drawing shows something that is unfamiliar, as is often the case in science, the artist's clues don't work as well. When you look at drawings of a molecule, a nerve cell, layers of rock under the ocean, or Earth and its shadows, you must pay special attention to the three-dimensional nature of the figure.

When you see a three-dimensional diagram in any science book, it helps to decide what the point of view is. For example, if the diagram were a photograph, where was the camera? In the drawing in Figure 3-7, the camera would have to have been in space looking back at Earth and the moon. Of course, astronauts have never been that far from Earth, but you can make such a voyage in your imagination, and that helps you understand the geometry of eclipses. Other three-dimensional diagrams have other points of view, so always be sure to first determine the point of view when you look at any three-dimensional diagram.

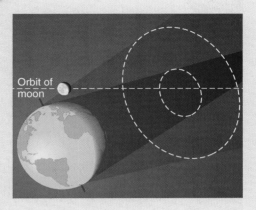

If this were a photograph, where would the photographer have been located?

---

light is scattered more easily than red light, it is red light that penetrates through Earth's atmosphere to illuminate the moon in a coppery glow. If you were on the moon during totality and looked back at Earth, you would not see any part of the sun because it would be entirely hidden behind Earth. However, you would see Earth's atmosphere illuminated from behind by the sun in a spectacular sunset completely ringing Earth. It is the red glow from this sunset that gives the totally eclipsed moon its reddish color.

How dim the totally eclipsed moon becomes depends on a number of things. If Earth's atmosphere is especially cloudy in those regions that must bend light into the umbra, the moon will be darker than usual. An unusual amount of dust in Earth's atmosphere (from volcanic eruptions, for instance) also causes a dark eclipse. Also, total lunar eclipses tend to be darkest when the moon's orbit carries it through the center of the umbra (■ Figure 3-5).

As the moon moves through Earth's umbral shadow, you can see that the shadow is circular. From this the Greek philosopher Aristotle (384–322 BC) concluded that Earth had to be a sphere, because only a sphere could cast a shadow that was always circular.

Depending on the geometry of the eclipse, the moon can take as long as 1 hour 40 minutes to cross the umbra and another hour to emerge into the penumbra. Still another hour passes as it emerges into full sunlight. A total eclipse of the moon, including the penumbral stage, can take almost six hours from start to finish.

■ Figure 3-4

During a total lunar eclipse, the moon passes through Earth's shadow, as shown in this multiple-exposure photograph. A longer exposure was used to record the moon while it was totally eclipsed. The moon's path appears curved in the photo because of photographic effects. (© 1982 Dr. Jack B. Marling)

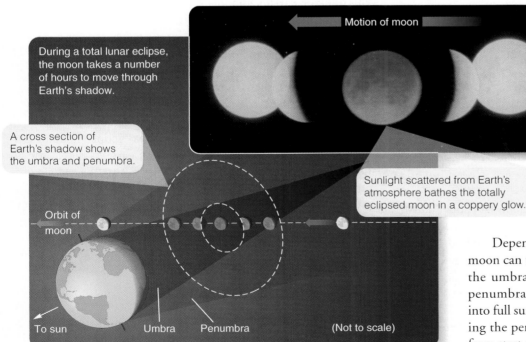

During a total lunar eclipse, the moon takes a number of hours to move through Earth's shadow.

Motion of moon

A cross section of Earth's shadow shows the umbra and penumbra.

Sunlight scattered from Earth's atmosphere bathes the totally eclipsed moon in a coppery glow.

Orbit of moon

To sun   Umbra   Penumbra   (Not to scale)

■ Figure 3-5

During a total lunar eclipse, the moon turns coppery red. In this photo, the moon is darkest toward the lower right, the direction toward the center of the umbra. The edge of the moon at upper left is brighter because it is near the edge of the umbra. (Celestron International)

Year	Time* of Mideclipse (GMT)	Length of Totality (Hr:Min)	Length of Eclipse[†] (Hr:Min)
2006 Sept. 7	18:52	Partial	1:30
2007 Mar. 3	23:22	1:14	3:40
2007 Aug. 28	10:38	1:30	3:32
2008 Feb. 21	3:27	0:50	3:24
2008 Aug. 16	21:11	Partial	3:08
2009 Dec. 31	19:24	Partial	1:00
2010 June 26	11:40	Partial	2:42
2010 Dec. 21	8:18	1:12	3:28
2011 June 15	20:13	1:40	3:38
2011 Dec. 10	14:33	0:50	3:32
2012 June 4	11:03	Partial	2:08

■ Table 3-1 I Total and Partial Eclipses of the Moon, 2006 to 2013

*Times are Greenwich Mean Time. Subtract 5 hours for Eastern Standard Time, 6 hours for Central Standard Time, 7 hours for Mountain Standard Time, and 8 hours for Pacific Standard Time. From your time zone, lunar eclipses that occur between sunset and sunrise will be visible, and those at midnight will be best placed.

[†]Does not include penumbral phase.

Partial and Penumbral Lunar Eclipses

Not all eclipses of the moon are total. Because the moon's orbit is inclined a bit over 5° to the plane of Earth's orbit, the moon might not pass through the center of the umbra.

If the moon's orbit carries the full moon too far north or south of the umbra, the moon may only partially enter the umbra. The resulting **partial lunar eclipse** is usually not as dramatic as a total lunar eclipse. Because part of the moon remains outside the umbra, it receives some direct sunlight and looks bright in contrast with the dark part of the moon inside the umbra. Unless the moon almost completely enters the umbra, the glare from the illuminated part of the moon drowns out the fainter red glow inside the umbra. Partial lunar eclipses are interesting because part of the full moon is darkened, but they are not as beautiful as a total lunar eclipse.

If the orbit of the moon carries the moon far enough north or south of the umbra, the moon may only pass through the penumbra and never reach the umbra. Such **penumbral eclipses** are not dramatic at all. In the partial shadow of the penumbra, the moon is only partially dimmed. Most people glancing at a penumbral eclipse would not notice any difference from a full moon.

Total, partial, or penumbral, lunar eclipses are interesting events in the night sky and are not difficult to observe. When the moon passes through Earth's shadow, the eclipse is visible from anywhere on Earth's dark side. One or two lunar eclipses occur in most years. Consult ■ Table 3-1 to find the next lunar eclipse visible in your part of the world.

Building Scientific Arguments

Why doesn't Earth's shadow on the moon look red during a partial lunar eclipse?

Once again, you need to build an argument based on geometry, but now you must consider the direction of sunlight. During a partial lunar eclipse, part of the moon protrudes from Earth's umbral shadow into sunlight. This part of the moon is very bright compared to the fainter red light inside Earth's shadow, and the glare of the reflected sunlight makes it difficult to see the red glow. If a partial eclipse is almost total, so that only a small sliver of moon extends out of the shadow into sunlight, you can sometimes detect the red glow in the shadow.

Of course, this red glow does not happen for every planet–moon combination in the universe. Adapt your argument for a new situation. **Would a moon orbiting a planet that had no atmosphere glow red during a total eclipse? Why or why not?**

■   ■   ■

Connections: Lunar eclipses are slow and stately. For drama and excitement, there is nothing like a solar eclipse, as you will discover in the next section.

3-3 Solar Eclipses

A SOLAR ECLIPSE OCCURS when the moon moves between Earth and the sun. If the moon covers the disk of the sun completely, the eclipse is a **total solar eclipse.** If the moon covers only part of the sun, the eclipse is a **partial solar eclipse.** During a particular solar eclipse, people in one place on Earth may see a total eclipse, while people only a few hundred kilometers away see a partial eclipse.

These spectacular sights are possible because we on Earth are very lucky. Our moon has the same angular diameter as our sun, so it can cover the sun almost exactly. That lucky coincidence allows you to see total solar eclipses.

The Angular Diameter of the Sun and Moon

You learned about the angular diameter of an object in Chapter 2; now you need to think carefully about how the size and distance of an object like the moon determine its angular diameter. This is the key to understanding solar eclipses.

Linear diameter is simply the distance between an object's opposite sides. You use linear diameter when you order a 16-inch pizza—the pizza is 16 inches in diameter. The linear diameter of the moon is 3476 km. The angular diameter of an object is the angle formed by lines extending toward you from opposite sides of the object and meeting at your eye (■ Figure 3-6). Clearly, the farther away an object is, the smaller its angular diameter.

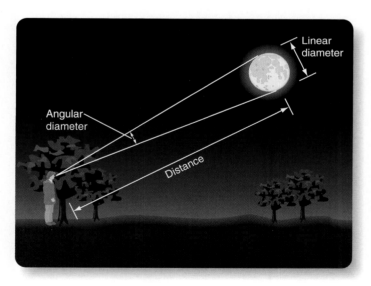

■ Active Figure 3-6

The angular diameter of an object is related to its linear diameter and also to its distance.

Ace ◐Astronomy™ Log into AceAstronomy and select this chapter to see the Active Figure called "Small-Angle Formula." Notice that both distance and linear diameter affect an object's angular diameter.

The **small-angle formula** gives you a way to figure out the angular diameter of any object, whether it is a pizza, the moon, or a galaxy. In the small-angle formula, you should always express angular diameter in seconds of arc* and always use the same units for distance and linear diameter:

$$\frac{\text{angular diameter}}{206{,}265''} = \frac{\text{linear diameter}}{\text{distance}}$$

You can use this formula to find any one of these three quantities if you know the other two; here you are interested in finding the angular diameter of the moon.

The moon has a linear diameter of 3476 km and a distance from Earth of about 384,000 km. What is its angular diameter? The moon's linear diameter and distance are both given in the same units, kilometers, so you can put them directly into the small-angle formula:

$$\frac{\text{angular diameter}}{206{,}265''} = \frac{3476 \text{ km}}{384{,}000 \text{ km}}$$

To solve for angular diameter, you can multiply both sides by 206,265 and find that the angular diameter is 1870 seconds of arc. If you divide by 60, you get 31 minutes of arc or, dividing by 60 again, about 0.5°. The moon's orbit is slightly elliptical, so it can sometimes look a bit larger or smaller, but its angular diameter is always close to 0.5°.

You can repeat this calculation for the angular diameter of the sun. The sun is 1.39×10^6 km in linear diameter and 1.50×10^8 km from Earth. If you put these numbers into the small-angle formula, you will discover that the sun has an angular diameter of 1900 seconds of arc, which is 32 minutes of arc or about 0.5°. Earth's orbit is slightly elliptical, and consequently the sun can sometimes look slightly larger or smaller, but it, like the moon, is always close to 0.5° in angular diameter.

By fantastic good luck, you live on a planet with a moon that is almost exactly the same angular diameter as your sun. When the moon passes in front of the sun, it is almost exactly the right size to block the brilliant surface of the sun. Then you see the most exciting sight in astronomy—a total solar eclipse. There are few other worlds where this can happen, because the angular diameters of the sun and a moon rarely match so closely. To see this beautiful sight, all you have to do is arrange to be in the moon's shadow when the moon crosses in front of the sun. That means you have to chase the moon's shadow wherever it sweeps over Earth's surface.

The Moon's Shadow

Like Earth's shadow, the moon's shadow consists of a central umbra of total shadow and a penumbra of partial shadow. What you see when the moon crosses in front of the sun depends on where you

---

* The number 206,265″ is the number of seconds of arc in a radian. When you divide by 206,265″, you convert the angle from seconds of arc to radians.

are in the moon's shadow. The moon's umbral shadow produces a spot of darkness roughly 269 km (167 mi) in diameter on Earth's surface (■ Figure 3-7). (The exact size of the umbral shadow depends on the location of the moon in its elliptical orbit and the angle at which the shadow strikes Earth.) If you are in this spot of total shadow, you see a total solar eclipse. If you are just outside the umbral shadow but in the penumbra, you see part of the sun peeking around the moon, and the eclipse is partial. Of course, if you are outside the penumbra, you see no eclipse at all.

Because of the orbital motion of the moon, its shadow sweeps across Earth at speeds of at least 1700 km/h (1060 mph). To be sure of seeing a total solar eclipse, you must select an appropriate eclipse (■ Table 3-2), plan far in advance, travel to the right place on Earth, and place yourself in the **path of totality,** the path swept out by the umbral spot.

Total Solar Eclipses

A total solar eclipse begins when you first see the edge of the moon encroaching on the sun's disk. This is the moment when the edge of the penumbra sweeps over your location.

During the partial phase, part of the sun remains visible, and it is hazardous to look at the eclipse without protection. Dense filters and exposed film do not necessarily provide protection, because some filters do not block the invisible heat radiation (infrared) that can burn the retina of your eyes. This has led officials to

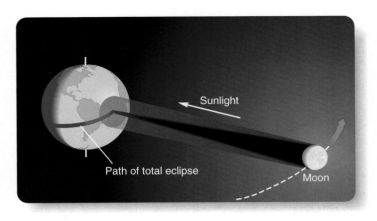

■ **Figure 3-7**

Observers in the path of totality see a total solar eclipse when the umbral shadow sweeps over them. Those in the penumbra see a partial eclipse.

warn the public not to look at solar eclipses and has even frightened some people into locking themselves and their children into windowless rooms during eclipses. In fact, the sun is a bit less dangerous than usual during an eclipse because part of the brilliant surface is covered by the moon. But an eclipse is dangerous in that it can tempt people to look at the sun directly and burn their eyes.

The safest and simplest way to observe the partial phases of a solar eclipse is to use pinhole projection. Poke a small pinhole

	■ Table 3-2	Total and Annular Eclipses of the Sun, 2006 to 2016**			

Date	Total/Annular (T/A)	Time of Mideclipse* (GMT)	Maximum Length of Total or Annular Phase (Min:Sec)	Area of Visibility
2006 Mar. 29	T	10h	4:07	Atlantic, Africa, Turkey
2006 Sept. 22	A	12h	7:09	N.E. of S. America, Atlantic
2008 Feb. 7	A	4h	2:14	S. Pacific, Antarctica
2008 Aug. 1	T	10h	2:28	Canada, Arctic, Siberia
2009 Jan. 26	A	8h	7:56	S. Atlantic, Indian Ocean
2009 July 22	T	3h	6:40	Asia, Pacific
2010 Jan. 15	A	7h	11:10	Africa, Indian Ocean
2010 July 11	T	20h	5:20	Pacific, S. America
2012 May 20	A	23h	5:46	Japan, N. Pacific, W. U.S.
2012 Nov. 13	T	22h	4:02	Australia, S. Pacific
2013 May 10	A	0h	6:04	Australia, Pacific
2013 Nov. 3	AT	13h	1:40	Atlantic, Africa
2015 March 20	T	10h	2:47	N. Atlantic, Arctic
2016 March 9	T	2h	4:10	Borneo, Pacific
2016 Sept. 1	A	9h	3:06	Atlantic, Africa, Indian Oc.

The next major total solar eclipse visible from the United States will occur on August 21, 2017.

*Times are Greenwich Mean Time. Subtract 5 hours for Eastern Standard Time, 6 hours for Central Standard Time, 7 hours for Mountain Standard Time, and 8 hours for Pacific Standard Time.

hhours.

**There are no total or annular eclipses of the sun during 2014.

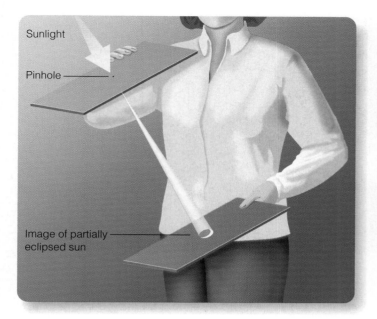

■ Figure 3-8

A safe way to view the partial phases of a solar eclipse. Use a pinhole in a card to project an image of the sun on a second card. The greater the distance between the cards, the larger (and fainter) the image will be.

A Total Solar Eclipse

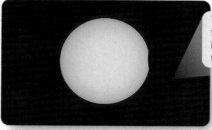

The moon moving from the right just begins to cross in front of the sun.

The disk of the moon gradually covers the disk of the sun.

Sunlight begins to dim as more of the sun's disk is covered.

During totality, pink prominences are often visible.

A longer-exposure photograph during totality shows the fainter corona.

■ Figure 3-9

This sequence of photos shows the first half of a total solar eclipse. (Daniel Good)

in a sheet of cardboard. Hold the sheet with the hole in sunlight and allow light to pass through the hole to a second sheet of cardboard (■ Figure 3-8). On a day when there is no eclipse, the result is a small, round spot of light that is an image of the sun. During the partial phases of a solar eclipse, the image shows the dark silhouette of the moon obscuring part of the sun. These pinhole images of the partially eclipsed sun can also be seen in the shadows under trees as sunlight peeks through the tiny openings between the leaves and branches. This can produce an eerie effect just before totality as the remaining sliver of sun produces thin crescents of light on the ground under trees.

Throughout the partial phases of a solar eclipse, the moon gradually covers the bright disk of the sun (■ Figure 3-9). Totality begins as the last sliver of the sun's bright surface disappears behind the moon. This is the moment when the edge of the umbra sweeps over your location. So long as any of the sun is visible, the countryside is bright; but, as the last of the sun disappears, dark falls in a few seconds. Automatic streetlights come on, car drivers switch on their headlights, and birds go to roost. The darkness of totality depends on a number of factors, including the weather at the observing site, but it is usually dark enough to make it difficult to read the settings on cameras.

The totally eclipsed sun is a spectacular sight. With the moon covering the bright disk of the sun, called the **photosphere,**[*] you

*The photosphere, corona, chromosphere, and prominences will be discussed in detail in Chapter 8. Here the terms are used as the names of features you see during a total solar eclipse.

can see the sun's faint outer atmosphere, the **corona,** glowing with a pale, white light so faint you can safely look at it directly. This corona is made of low-density, hot gas, which is given a wispy appearance by the solar magnetic field, as shown in the last frame of Figure 3-9. Also visible just above the photosphere is a thin

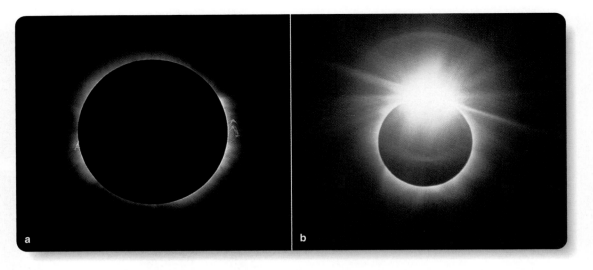

(a) During a total solar eclipse, the moon covers the photosphere, and the white corona and pink prominences are visible. Note the streamers in the corona caused by the sun's magnetic field. (Daniel Good) (b) The diamond ring effect can sometimes occur momentarily at the beginning or end of totality if a small segment of the photosphere peeks out through a valley at the edge of the lunar disk. (National Optical Astronomy Observatory)

layer of bright gas called the **chromosphere.** The chromosphere is often marked by eruptions on the solar surface called **prominences** (■ Figure 3-10a), which glow with a clear, pink color due to the high temperature of the gases involved. The small-angle formula reveals that a large prominence is about 3.5 times the diameter of Earth.

Totality cannot last longer than 7.5 minutes under any circumstances, and the average is only 2 to 3 minutes. Totality ends when the sun's bright surface reappears at the trailing edge of the moon. This corresponds to the moment when the trailing edge of the moon's umbra sweeps over the observer.

Just as totality begins or ends, a small part of the photosphere can peek out from behind the moon through a valley at the edge of the lunar disk. Although it is intensely bright, such a small part of the photosphere does not completely drown out the fainter corona, which forms a silvery ring of light with the brilliant spot of photosphere gleaming like a diamond (Figure 3-10b). This **diamond ring effect** is one of the most spectacular of astronomical sights, but it is not visible during every solar eclipse. Its occurrence depends on the exact orientation and motion of the moon.

Once totality is over, daylight returns quickly, and the corona and chromosphere vanish. Not too many years ago astronomers traveled great distances to forbidding places to get their instruments into the path of totality and study the faint outer corona visible only during the few minutes of a total solar eclipse. Now many of those observations can be made every day by solar telescopes in space, but eclipse fans still journey to exotic places for the thrill of seeing a total solar eclipse.

Not every solar eclipse is total. Sometimes when the moon crosses in front of the sun, it is too small to fully cover the sun. That can happen because the orbit of the moon is elliptical and its distance from Earth varies. When the moon is at **perigee,** its closest point to Earth, the moon is almost 12 percent larger in angular diameter than when it is at **apogee,** its most distant point

from Earth. Because Earth's orbit around the sun is very slightly elliptical, the sun's angular diameter can vary by a total of about 3.4 percent. If a solar eclipse occurs when the moon's angular diameter is too small, you would see an **annular eclipse,** a solar eclipse in which a ring (or annulus) of the photosphere is visible around the disk of the moon (■ Figure 3-11). With a portion of the photosphere visible, the eclipse never becomes total; it never quite gets dark; and you can't see the prominences, chromosphere, and corona. An annular eclipse swept across the United States on May 10, 1994.

Building Scientific Arguments

If you were on Earth watching a total solar eclipse, what would astronauts on the moon see when they looked at Earth?
Building this argument requires that you change your point of view and imagine seeing the geometry from a new direction. Astronauts on the moon could see Earth only if they were on the side that faces Earth. Because solar eclipses always happen at new moon, the near side of the moon would be in darkness, and the far side of the moon would be in full sunlight. The astronauts would be standing in darkness, and they would be looking at the fully illuminated side of Earth. They would see a "full Earth." The moon's shadow would be crossing Earth; and, if the astronauts looked closely, they might be able to see the spot of darkness where the moon's umbral shadow touched Earth. It would take hours for the shadow to cross Earth.

Standing on the moon and watching the moon's umbral shadow sweep across Earth would be a cold, tedious assignment. Perhaps you can imagine a more interesting assignment for the astronauts. **What would astronauts on the moon see while people on Earth were seeing a total lunar eclipse?**

■   ■   ■

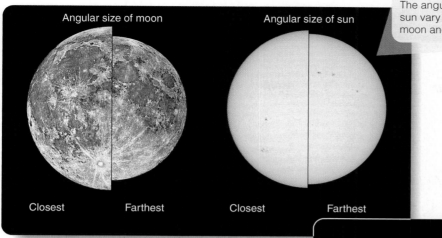

Angular size of moon   Angular size of sun

Closest   Farthest   Closest   Farthest

The angular diameters of the moon and sun vary slightly because the orbits of the moon and Earth are slightly eliptical.

If the moon is too far from Earth during a solar eclipse, the umbra does not reach Earth's surface.

Sunlight

Path of annular eclipse

Moon

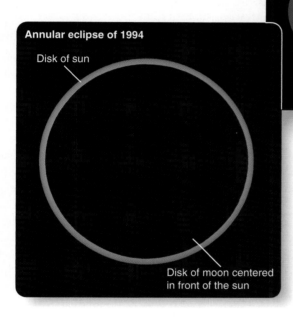

Annular eclipse of 1994

Disk of sun

Disk of moon centered in front of the sun

■ **Figure 3-11**

An annular eclipse occurs when the moon is far enough from Earth that its umbral shadow does not reach Earth's surface. From Earth, you see an annular eclipse because the moon's angular diameter is smaller than the angular diameter of the sun. In the photograph of the annular eclipse of 1994, the dark disk of the moon is almost exactly centered on the bright disk of the sun. (Moon: UCO/Lick Observatory; Sun: Daniel Good)

Connections: Eclipses of the sun and moon are often dramatic and mysterious. Ancient astronomers studied them and found ways to predict the coming of an eclipse. In the section that follows, you will see how eclipses can be predicted from a basic understanding of the motion of the sun and the moon.

3-4 Predicting Eclipses

TO MAKE EXACT ECLIPSE PREDICTIONS, you would need to calculate the precise motions of the sun and moon, and that requires a computer and proper software. Such software is available for desktop computers, but it isn't necessary if you are satisfied with making less exact predictions. In fact, many primitive peoples, such as the builders of Stonehenge and the ancient Maya, are believed to have made eclipse predictions.

You should think about eclipse prediction for three reasons. First, it is an important part of the history of science. Second, it illustrates how apparently complex phenomena can be analyzed in terms of cycles. Third, eclipse prediction will exercise your mental muscles and force you to see Earth, the moon, and the sun as objects moving through space.

Conditions for an Eclipse

You can predict eclipses by understanding the conditions that make them possible. As you begin to think about these conditions, be sure you are aware of your point of view. (See Window on Science 3-1.) Later you will change your point of view, but to begin you must imagine that you can look up into the sky from your home on Earth and see the sun moving along the ecliptic and the moon moving along its orbit.

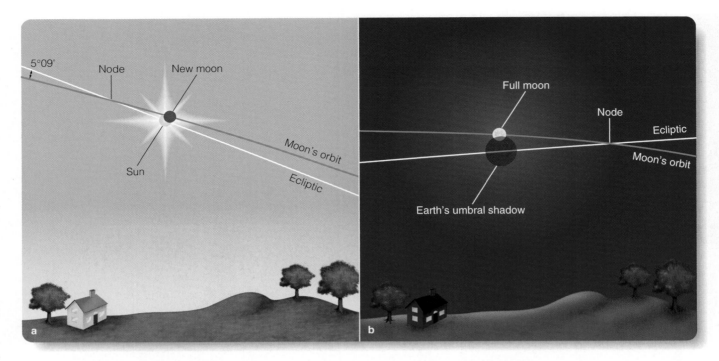

■ Figure 3-12

Eclipses can occur only near the nodes of the moon's orbit. (a) A solar eclipse occurs when the moon meets the sun near a node. (b) A lunar eclipse occurs when the sun and moon are near opposite nodes. Partial eclipses are shown here for clarity.

The orbit of the moon is tipped 5°8'43" to the plane of Earth's orbit, so you see the moon follow a path tipped by that angle to the ecliptic. Each month, the moon crosses the ecliptic at two points called **nodes.** It crosses at one node going southward, and two weeks later it crosses at the other node going northward.

Eclipses can only occur when the sun is near one of the nodes of the moon's orbit. A solar eclipse happens at new moon if the moon passes in front of the sun. Most new moons pass too far north or too far south of the sun to cause an eclipse. Only when the sun is near a node in the moon's orbit can the moon cross in front of the sun, as shown in ■ Figure 3-12a. A lunar eclipse doesn't happen at every full moon because most full moons pass too far north or too far south of the ecliptic and miss Earth's shadow. The moon can enter Earth's shadow only when the shadow is near a node in the moon's orbit, and that means the sun must be near the other node. This is shown in Figure 3-12b.

So there are two conditions for an eclipse: The sun must be crossing a node, and the moon must be crossing either the same node (solar eclipse) or the other node (lunar eclipse). That means, of course, that solar eclipses can occur only when the moon is new, and lunar eclipses can occur only when the moon is full.

Now you know the secret of predicting eclipses. An eclipse can only occur during a period called an **eclipse season** during which the sun is close to a node in the moon's orbit. For solar eclipses, an eclipse season is about 32 days long. Any new moon during this period will produce a solar eclipse. For lunar eclipses, the eclipse season is a bit shorter, about 22 days. Any full moon in this period will encounter Earth's shadow and be eclipsed.

This makes eclipse prediction easy. All you have to do is keep track of where the moon crosses the ecliptic (where the nodes of its orbit are). When the sun is near one of these nodes, you can predict that the nearest new moon will cause a solar eclipse and the nearest full moon will cause a lunar eclipse. This system works fairly well, and ancient astronomers such as the Maya may have used such a system. You could be a very successful ancient Mayan astronomer with what you know about eclipse seasons, but you can do even better if you change your point of view.

The View from Space

Change your point of view and imagine that you are looking at the orbits of Earth and the moon from a point far away in space. You would see the moon's orbit as a small disk tipped at an angle to the larger disk of Earth's orbit. As Earth orbits the sun, the moon's orbit remains fixed in direction. The nodes of the moon's orbit are the points where it passes through the plane of Earth's orbit; an eclipse season occurs each time the line connecting these nodes, the **line of nodes,** points toward the sun. Study ■ Figure 3-13 and notice that the line of nodes does not point at the sun in the example at lower left, and no eclipses are possible. At lower right, the line of nodes points toward the sun, and the shadows produce eclipses.

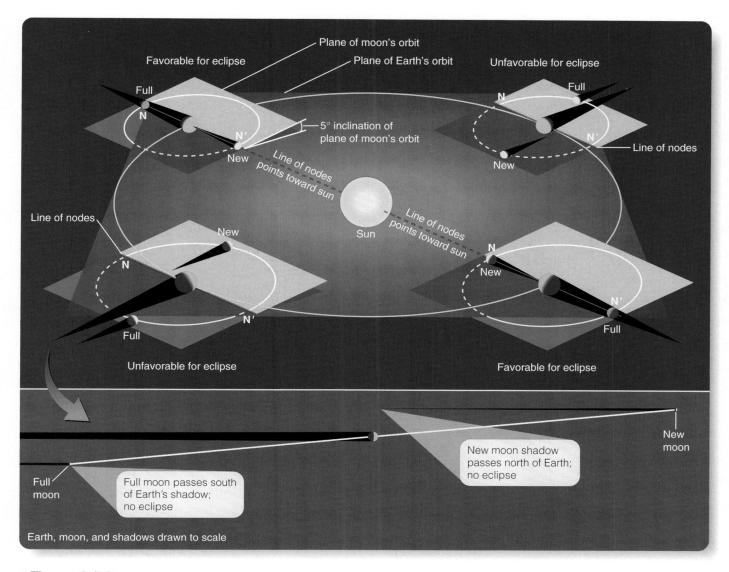

Figure 3-13

The moon's orbit is tipped about 5° to Earth's orbit. The nodes N and N' are the points where the moon passes through the plane of Earth's orbit. If the line of nodes does not point at the sun, the long narrow shadows miss, and there are no eclipses at new moon and full moon. At those parts of Earth's orbit where the line of nodes points toward the sun, eclipses are possible at new moon and full moon.

The shadows of Earth and moon, seen from space, are very long and thin, as shown in the lower part of Figure 3-13. It is easy for them to miss their mark at new moon or full moon and fail to produce an eclipse. Only during an eclipse season, when the line of nodes points toward the sun, do the long, skinny shadows produce eclipses.

If you watched for years from your point of view in space, you would see the orbit of the moon precess like a hubcap spinning on the ground. This precession is caused mostly by the gravitational influence of the sun, and it makes the line of nodes rotate once every 18.6 years. People back on Earth see the nodes slipping westward along the ecliptic 19.4° per year, and the sun takes only 346.62 days (an **eclipse year**) to return to a node. This means

that, according to the calendar, the eclipse seasons begin about 19 days earlier every year (■ Figure 3-14).

The cyclic pattern of eclipses shown in Figure 3-14 should give you another clue how to predict eclipses. Eclipses follow a pattern, and if you were an ancient astronomer and understood the pattern, you could predict eclipses without ever knowing what the moon was or how an orbit works. All you need know is the cycle.

The Saros Cycle

Ancient astronomers could predict eclipses in a crude way using the eclipse seasons, but they could have been much more accurate if they recognized that eclipses occur following certain patterns.

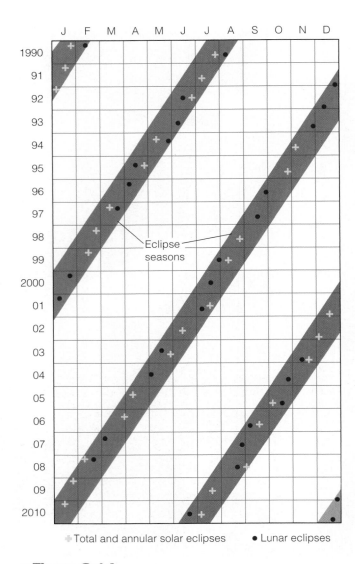

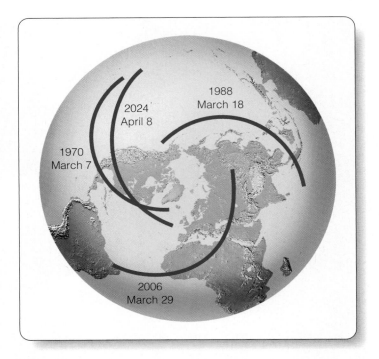

■ **Figure 3-15**

The saros cycle at work. The total solar eclipse of March 7, 1970, recurred after 18 years $11\frac{1}{3}$ days over the Pacific Ocean. After another interval of 18 years $11\frac{1}{3}$ days, the same eclipse was visible from Asia and Africa. After a similar interval, the eclipse will again be visible from the United States.

■ **Figure 3-14**

A calendar of eclipse seasons. Each year the eclipse seasons begin about 19 days earlier. Any new moon or full moon that occurs during an eclipse season results in an eclipse. Only total and annular eclipses are shown here.

The most important of these is the **saros cycle** (sometimes referred to simply as the saros). After one saros cycle of 18 years $11\frac{1}{3}$ days, the pattern of eclipses repeats. In fact, *saros* comes from a Greek word that means "repetition."

The eclipses repeat because, after one saros cycle, the moon and the nodes of its orbit return to the same place with respect to the sun. One saros contains 6585.321 days, which is equal to 223 lunar months. Therefore, after one saros cycle the moon is back to the same phase it had when the cycle began. But one saros is also equal to 19 eclipse years. After one saros cycle, the sun has returned to the same place it occupied with respect to the nodes of the moon's orbit when the cycle began. If an eclipse occurs on a given day, then 18 years $11\frac{1}{3}$ days later the sun, the moon, and the nodes of the moon's orbit return to nearly the same relationship, and the eclipse occurs all over again.

Although the eclipse repeats almost exactly, it is not visible from the same place on Earth. The saros cycle is one-third of a day longer than 18 years 11 days. When the eclipse happens again, Earth will have rotated one-third of a turn farther east, and the eclipse will occur one-third of the way westward around Earth (■ Figure 3-15). That means that after three saros cycles—a period of 54 years 1 month—the same eclipse occurs in the same part of Earth.

One of the most famous predictors of eclipses was the Greek philospher Thales of Miletus (about 640–546 BC), who supposedly learned of the saros cycle from the Chaldeans, who had discovered it. No one knows which eclipse Thales predicted, but some scholars suspect the eclipse of May 28, 585 BC. In any case, the eclipse occurred at the height of a battle between the Lydians and the Medes, and the mysterious darkness in midafternoon so startled the two factions that they concluded a truce.

In fact, many historians doubt that Thales actually predicted the eclipse. It would have been very difficult to gather enough information about past eclipses; total solar eclipses are rare as seen from any one place. If you stay in one city, you will see a total solar eclipse about once in 360 years. Also, 585 BC is very early for the Greeks to have known of the saros cycle. The important point is not that Thales did it, but that he could have done it. If he had had records of past eclipses of the sun visible from the area, he could have discovered that they tended to recur with a period of 54 years 1 month (three saros cycles). Indeed, he could have predicted the eclipse without ever understanding what the sun and moon were or how they moved.

Building Scientific Arguments

Why can't two successive full moons be totally eclipsed?
Most people suppose that eclipses occur at random or in some pattern so complex you need a big computer to make predictions. You know the geometry is fairly simple, so you can build a simple argument to analyze this question. Remember that a total lunar eclipse occurs when the moon passes through Earth's shadow. Earth's shadow always points toward the ecliptic exactly opposite the sun, and most of the time the moon will pass north or south of Earth's shadow, and there will be no eclipse. A lunar eclipse can happen only when the sun is near one node and the moon crosses Earth's shadow at the other node.

Now you can apply what you know about the moon's phases. An eclipse season for a total lunar eclipse is only 22 days long, but the moon takes 29.5 days to go from one full moon to the next. If one full moon is totally eclipsed, the next full moon 29.5 days later will occur long after the end of the eclipse season and there will be no eclipse.

Now use your knowledge of the cycles of the sun and moon to revise your argument. **Why can the sun be eclipsed by two successive new moons?**

■ ■ ■

Connections: Predicting eclipses isn't very hard, and many ancient peoples were probably familiar enough with the cycles of the sun and moon to know when eclipses were likely. Some people call those first predictions early science, but science involves understanding nature, not just predicting events. In the next chapter, you will see how modern science was born when astronomers tried to understand the cycles they saw in the sky.

Study and Review Tools

Summary

3-1 | The Changeable Moon

Why does the moon go through phases?
■ The moon orbits eastward around Earth once a month.

■ The moon rotates on its axis as it orbits around Earth and keeps the same side facing Earth throughout the month.

■ Because you see the moon by reflected sunlight, its shape appears to change as it orbits Earth and sunlight illuminates different amounts of the side facing Earth.

■ The lunar phases wax from new moon to first quarter to full moon and wane from full moon to third quarter to new moon.

■ A complete cycle of lunar phases takes 29.53 days.

3-2 | Lunar Eclipses

What causes a lunar eclipse?
■ If the full moon passes through Earth's shadow, sunlight is cut off, and the moon darkens in a lunar eclipse. If the moon only grazes the shadow, the eclipse is partial or penumbral and not total.

■ The totally eclipsed moon looks copper-red because of sunlight scattered through Earth's atmosphere.

3-3 | Solar Eclipses

What causes a solar eclipse?
■ A total solar eclipse occurs if the new moon passes directly between the sun and Earth and the moon's shadow sweeps over Earth's surface. Observers inside the path of totality see a total eclipse, and those just out-

side the path of totality see a partial eclipse if the penumbra sweeps over their location.

■ During a total eclipse of the sun, the bright photosphere of the sun is covered, and the fainter corona, chromosphere, and prominences become visible.

■ If during a solar eclipse the moon is in the farther part of its orbit, near apogee rather than perigee, its angular diameter is not large enough to cover the entire photosphere, and you see only an annular eclipse.

3-4 | Predicting Eclipses

How can eclipses be predicted?
■ Solar eclipses must occur at new moon, and lunar eclipses must occur at full moon.

■ The moon's orbit crosses the ecliptic at two locations called nodes, and eclipses can occur only when the sun is crossing a node. During these periods, called eclipse seasons, a new moon will cause a solar eclipse, and a full moon can cause a lunar eclipse. Knowing when the eclipse seasons occur would allow you to guess which new moons and full moons could cause eclipses.

■ Eclipses follow a pattern called the saros cycle. After one saros of 18 years $11\frac{1}{3}$ days, the pattern of eclipses repeats. Some ancient astronomers knew of the saros cycle and used it to predict eclipses.

New Terms

sidereal period (p. 41)

synodic period (p. 41)

lunar eclipse (p. 42)

umbra (p. 42)

penumbra (p. 42)

total eclipse (lunar or solar) (pp. 42, 45)

partial eclipse (lunar or solar)
(pp. 44, 45)

penumbral eclipse (p. 44)

small-angle formula (p. 45)

path of totality (p. 46)

photosphere (p. 47)

corona (p. 47)

chromosphere (p. 48)

prominences (p. 48)

diamond ring effect (p. 48)

perigee (p. 48)

apogee (p. 48)

annular eclipse (p. 48)

nodes (p. 50)

eclipse season (p. 50)

line of nodes (p. 50)

eclipse year (p. 51)

saros cycle (p. 52)

Review Questions

Ace ⑤ Astronomy™ Assess your understanding of this chapter's topics with additional quizzing and animations at **http://ace .brookscole.com/sf9**

1. Which lunar phases would be visible in the sky at dawn? at midnight?

2. If you looked back at Earth from the moon, what phase would Earth have when the moon was full? new? a first-quarter moon? a waxing crescent?

3. If a planet has a moon, must that moon go through the same phases that Earth's moon displays?

4. Could a solar-powered spacecraft generate any electricity while passing through Earth's umbral shadow? through Earth's penumbral shadow?

5. Draw the umbral and penumbral shadows onto the diagram in the middle of page 40. Explain why lunar eclipses can occur only at full moon and solar eclipses can occur only at new moon.

6. How did lunar eclipses lead Aristotle to conclude that Earth was a sphere?

7. Why isn't the corona visible during partial or annular solar eclipses?

8. Why can't the moon be eclipsed when it is halfway between the nodes of its orbit?

9. Why aren't solar eclipses separated by one saros cycle visible from the same location on Earth?

10. How could Thales of Miletus have predicted the date of a solar eclipse without observing the location of the moon in the sky?

11. The stamp at right shows a crescent moon. Explain why the moon could never look this way.

12. The photo at right shows the annular eclipse of May 30, 1984. How is it different from the annular eclipse shown in Figure 3-11?

Discussion Questions

1. How would eclipses be different if the moon's orbit were not tipped with respect to the plane of Earth's orbit?

2. Are there other planets in our solar system from whose surface you could see a lunar eclipse? a total solar eclipse?

3. Can you detect the saros cycle in Figure 3-14?

Problems

1. Identify the phases of the moon if on March 21 the moon is located at the point on the ecliptic called (a) the vernal equinox, (b) the autumnal equinox, (c) the summer solstice, (d) the winter solstice.

2. Identify the phases of the moon if at sunset the moon is (a) near the eastern horizon, (b) high in the southern sky, (c) in the southeastern sky, (d) in the southwestern sky.

3. About how many days must elapse between first-quarter moon and third-quarter moon?

4. If on March 1 the full moon is near the star Spica, when will the moon next be full? When will it next be near Spica?

5. How many times larger than the moon is the diameter of Earth's umbral shadow at the moon's distance? (*Hint:* See the photo in Figure 3-4.)

6. Use the small-angle formula to calculate the angular diameter of Earth as seen from the moon.

7. During solar eclipses, large solar prominences are often seen extending 5 minutes of arc from the edge of the sun's disk. How far is this in kilometers? in Earth diameters?

8. If a solar eclipse occurs on October 3: (a) Why can't there be a lunar eclipse on October 13? (b) Why can't there be a solar eclipse on December 28?

9. A total eclipse of the sun was visible from Canada on July 10, 1972. When did this eclipse occur next? From what part of Earth was it total?

10. When will the eclipse described in Problem 9 next be total as seen from Canada?

11. When will the eclipse seasons occur during the current year? What eclipse(s) will occur?

Media Cluster

ACTIVE FIGURES

Ace Astronomy™ To access the resources in the Media Cluster, log into AceAstronomy at http://ace.brookscole.com/sf9 and select Chapter 3.

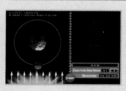

Small-Angle Formula
The small-angle formula animation illustrates how the size and distance of an object affect its angular size and how well this simple animation works.

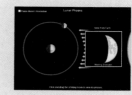

Lunar Phases
Take control of the position of the moon in its orbit and notice how its phases change as seen from Earth.

ASTRONOMY EXERCISES

Phases of the Moon
Sunlight always illuminates half of the moon. Because you see different amounts of this sunlit side, you see the moon cycle through phases. Explore this relationship further with this interactive exercise.

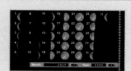

Moon Calendar
This exercise lets you select any month or year in the past, present, or future and see the moon phases for that month.

Critical Inquiries for the Web

1. Search the web for myths about the moon. What common themes do different cultures ascribe to the moon and its cycle of phases?

2. Search for web pages showing photos and observations of a recent total solar eclipse. Can you find web pages that show cultural responses to the eclipse such as celebrations or religious ceremonies?

3. Most people see more total lunar eclipses in a lifetime than total solar eclipses. Why is this so? Compare the regions of visibility for a number of past and upcoming eclipses and determine which future events will be visible from your area.

Exploring *TheSky*

1. How many days must elapse between new moons? Use the **Moon Phase Calendar** under the **Tools** menu to locate new moons.

2. Where would you have to go to see a total solar eclipse in the next year or two? Use the **Eclipse Finder** under the **Tools** menu to select a total solar eclipse. Check **Show Path of Totality** to see a map of Earth.

3. View the total solar eclipse you located in activity 2. In the **Eclipse Finder**, click **View** to see the sun and moon from your present location. Then use **Site Information** under the **Data** menu to change your location to a spot in the path of totality. Set the **Time Step** to 5 minutes and click the **arrows** to move the moon across the sun.

4. Locate and observe an annular eclipse.

5. Locate and observe a total lunar eclipse.

Go to the Brooks/Cole Astronomy Resource Center (**http://astronomy.brookscole.com**) for critical thinking exercises, articles, and additional readings from InfoTrac College Edition, Brooks/Cole's online student library.

4 | The Origin of Modern Astronomy

How you would burst out laughing, my dear Kepler, if you would hear what the greatest philosopher of the Gymnasium told the Grand Duke about me . . .

FROM A LETTER BY GALILEO GALILEI

GALILEO LAUGHED AT JOKES just as you do. We tend to forget that famous scientists in history laughed and joked, felt satisfaction, and experienced frustration day by day. The astronomers you are about to meet loved their work, and they must have laughed often and taken great satisfaction when their observations went well. But there were moments of frustration too. The history of astronomy is the story of great scientific advances made by astronomers working in many lands, but it is also the story of people who loved the sky and felt joy in their work. ▌ As you read, notice how two subplots twist through the story. The first plot involves astronomers trying to understand the place of the Earth. Is it the center of the universe? Does it move? This is not an idle question of philosophy; it led Galileo before the Inquisition.

▌ Continued on page 58 ▌

Astronomers like Galileo Galilei and Johannes Kepler struggled to understand the place of the Earth and the motion of the planets.

Guidepost

Looking Back

The previous chapters gave you a modern view of the heavens. You are now familiar with what you see in the sky, and you also can imagine how the Earth, moon, and sun move through space. Now you are ready to understand one of the most sweeping revolutions in human thought: the realization that we live on a planet.

This Chapter

All cultures have tried to explain their place under the sky, and the ancient Greek philosophers developed an elaborate theory. By the 16th century, however, many astronomers were uncomfortable with the ancient theory that Earth sat unmoving at the center of a spherical universe. In this chapter, you will discover how an astronomer named Copernicus changed the ancient theory, how Galileo Galilei changed the rules of the debate, and how astronomers changed the way they thought about motion in the heavens. Here you will find answers to four essential questions:

How did the ancients describe the place of the Earth?

How did Copernicus change the place of the Earth?

Why was Galileo condemned by the Inquisition?

How did Copernican astronomers solve the puzzle of planetary motion?

This chapter is not just about the history of astronomy. As they struggled to understand Earth and the heavens, the astronomers of the Renaissance invented a new way of understanding nature—a way of thinking that is now called science.

Looking Ahead

This is a book of science, and every chapter that follows will use the methods that were invented when Copernicus tried to repair the ancient theory that Earth was the center of the universe. One of the greatest adventures in that story began when an Englishman wondered why objects fall. That key to the universe is the subject of the next chapter.

The second plot, the puzzle of planetary motion, reveals astronomers observing the sky and trying to understand how the sun, moon, and planets move. That puzzle led eventually to the discovery of gravity, understanding orbital motion, and putting astronauts in space.

This story goes far beyond the birth of astronomy. The quest for the place of the Earth and the puzzle of planetary motion became, in the 16th and 17th centuries, the center of a storm of controversy over the best way to understand our world and ourselves. That conflict is now known as the scientific revolution. As you read about the birth of astronomy, notice that you are also reading about the birth of modern science.

4-1 The Roots of Astronomy

ASTRONOMY HAS ITS ORIGIN in that most noble of all human traits, curiosity. Just as modern children ask their parents what the stars are and why the moon changes, so did ancient peoples ask themselves those same questions. The answers, often couched in mythical or religious terms, reveal a great reverence for the order of the heavens.

Archaeoastronomy

Most of the history of astronomy is lost forever. You can't go to a library or search the Internet to find out what the first astronomers thought about their world because they left no written record. The study of the astronomy of ancient peoples has been called **archaeoastronomy**, and, in spite of the fog of eons gone, it tells you one thing: trying to understand the heavens is part of human nature.

Perhaps the best-known example of archaeoastronomy is also a huge tourist attraction. Stonehenge, standing on Salisbury Plain in southern England, was built in stages from about 3000 BC to about 1800 BC, a period extending from the late Stone Age into the Bronze Age. Though the public is most familiar with the massive stones of Stonehenge, those were added late in its history. In its first stages, Stonehenge consisted of a circular ditch slightly larger in diameter than the length of a football field, with a concentric bank just inside the ditch and a long avenue leading away toward the northeast. A massive stone, the Heelstone, stood then, as it does now, outside the ditch in the opening of the avenue.

As early as AD 1740, the English scholar W. Stukely suggested that the avenue pointed toward the rising sun at the summer solstice, but few accepted the idea. More recently, astronomers have recognized significant astronomical alignments at Stonehenge. Seen from the center of the monument, the summer-solstice sun rises behind the Heelstone. Other sight lines point toward the most northerly and most southerly risings of the moon (■ Figure 4-1).

The significance of these alignments has been debated. Some have claimed that the Stone Age people who built Stonehenge

The central horseshoe of upright stones is only the most obvious part of Stonehenge. The best-known astronomical alignment at Stonehenge is the summer solstice sun rising over the Heelstone. Although a number of astronomical alignments have been found at Stonehenge, experts debate their significance. (Photo: Jamie Backman)

were using it as a device to predict lunar eclipses. After studying eclipse prediction in the previous chapter, you understand that predicting eclipses is easier than most people assume. You could use Stonehenge to predict eclipses, but was that the intention of the people who built it? Some experts doubt that eclipse prediction was the main use of the monument. The truth may never be known. The builders of Stonehenge had no written language and left no records of their intentions. Nevertheless, the presence of solar and lunar alignments at Stonehenge and at many other Stone Age monuments dotting England and continental Europe shows that so-called primitive peoples were paying detailed attention to the sky. The roots of astronomy lie not in sophisticated science and logic but in human curiosity and wonder.

Astronomical alignments in sacred structures are common all around the world. For example, many tombs are oriented toward the rising sun, and Newgrange, a 5000-year-old passage-grave in Ireland (■ Figure 4-2), faces southeast so that, at dawn on the day of the winter solstice, light from the rising sun shines into its long passageway and illuminates the central chamber. No one today knows what the alignment meant to the builders of Newgrange, and many experts doubt that it was actually intended to be a tomb. Whatever its original purpose, Newgrange is clearly a sacred site linked by its alignment to the order and power of the sky. Astronomical alignments in sacred structures are not just an ancient custom. Many cathedrals are carefully oriented to face east.

Building astronomical alignments into structures gives them meaning by connecting them with the heavens. Navajo Indians of the American southwest, for example, have for centuries built their hogans with the door facing east so they can greet the rising sun each morning with prayers and offerings.

Some alignments may have served calendrical purposes. The 2000-year-old Temple of Isis in Dendera, Egypt, was build to align with the rising point of the bright star Sirius. The first appearance of this star in the dawn twilight marked the flooding of the Nile, so it was an important calendar indicator. The symbolism goes further. According to Egyptian mythology, the goddess Isis was associated with the star Sirius, and her husband, Osirus, was linked to the constellation you know as Orion and also to the Nile, the source of Egypt's agricultural fertility.

An American site in New Mexico is known as the Sun Dagger, but there is no surviving mythology to help you understand it. At noon on the day of the summer solstice, a narrow dagger of sunlight shines across the center of a spiral carved on a cliff face high above the desert floor (■ Figure 4-3). The purpose of the

Sunrise on the morning of the summer solstice

Ditch
Bank
Stone
Aubrey Hole

■ Figure 4-2

Newgrange was built on a small hill in Ireland about 3200 BC. A long passageway extends from the entryway back to the center of the mound, and sunlight shines down the passageway into the central chamber at dawn on the day of the winter solstice. Other passage graves have similar alignments, but their purpose is unknown. (bottom, Newgrange: Benelux Press/Index Stock Imagery)

Sun Dagger is open to debate, but similar examples have been found throughout the American southwest. It may have been more a symbolic and ceremonial marker than a precise calendrical indicator. In any case, it is just one of the many astronomical alignments that ancient people built into their structures to link themselves with the sky.

Some scholars are looking not at temples but at small artifacts from thousands of years ago. Scratches on certain bone and stone implements seem to follow a pattern and may be an attempt to keep a record of the phases of the moon (■ Figure 4-4). Some scientists contend that humanity's first attempts at writing were stimulated by a desire to record and predict lunar phases.

Archaeoastronomy is uncovering the earliest roots of astronomy and simultaneously revealing some of the first human efforts at systematic inquiry. The most important lesson of archaeoastronomy is that humans don't have to be technologically sophisticated to admire and study the universe.

One thing about archaeoastronomy is especially sad. Although archaeoastronomy can show how ancient people observed the sky, their thoughts about their universe are in many cases lost. Many had no written language. In other cases, the written record has been lost. Dozens, perhaps hundreds, of beautiful Mayan manuscripts, for instance, were burned by Spanish missionaries who believed that the books were the work of Satan. Only four of these books have survived, and all four contain astronomical references. One contains sophisticated tables that allowed the Maya to predict the motion of Venus and eclipses of the moon. No one will ever know what was burnt.

The fate of the Mayan books illustrates one reason why histories of astronomy usually begin with the Greeks. Some of their writing has survived, and you can discover what they thought about the shape and motion of the heavens.

■ **Figure 4-3**

In the ancient Native American settlement known as Chaco Canyon, New Mexico, sunlight shines between two slabs of stone high on the side of 440-foot-high Fajada Butte to form a dagger of light on the cliff face. About noon on the day of the summer solstice, the dagger of light slices through the center of a spiral pecked into the sandstone. (NPS Chaco Culture National Historic Park)

High on Fajada Butte, the Sun Dagger is off limits to visitors.

The spiral pattern is the size of a dinner plate.

The Astronomy of Greece

Greek astronomy was based on the astronomy of Babylon and Egypt, but these astronomies were heavily influenced by religion and astrology. The astronomers of that time studied the motions of the heavens as a way of worship and divination. The Greek astronomers studied astronomy in an entirely new way—they tried to understand the universe.

This new attitude toward the heavens, a truly scientific attitude, was made possible by two early Greek philosophers. Thales of Miletus (c. 624–547 BC) lived and worked in what is now Turkey. He taught that the universe is rational and that the human

■ **Figure 4-4**

A fragment of a 27,000-year-old mammoth tusk found at Gontzi in Ukraine contains scribe marks on its edge, simplified in this drawing. These markings have been interpreted as a record of four cycles of lunar phases. Although controversial, such finds suggest that some of the first human attempts at recording events in written form were stimulated by astronomical phenomena.

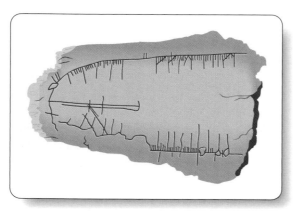

mind can understand why the universe works the way it does. This view contrasts sharply with those of earlier cultures, which believed that the ultimate causes of things are mysteries beyond human understanding. To Thales and his followers, the mysteries of the universe are mysteries because they are unknown, not because they are unknowable.

The second philosopher who made the new scientific attitude possible was Pythagoras (c. 570–500 BC). He and his students noticed that many things in nature seem to be governed by geometrical or mathematical relations. Musical notes, for example, are related in a regular way to the lengths of plucked strings. This led Pythagoras to propose that all nature was underlain by musical principles, by which he meant mathematics. One result of this philosophy was the later belief that the harmony of the celestial movements produced actual music, the music of the spheres. But, at a deeper level, the teachings of Pythagoras made Greek astronomers look at the universe in a new way. Thales said that the universe could be understood, and Pythagoras said that the underlying rules were mathematical.

In trying to understand the universe, Greek astronomers did something that Babylonian astronomers had never done—they tried to construct descriptions based on geometrical forms. Anaximander (c. 611–546 BC) described a universe made up of wheels filled with fire: The sun and moon are holes in the wheels through which the flames can be seen. Philolaus (fifth century BC) argued that Earth moves in a circular path around a central fire (not the sun), which is always hidden behind a counterearth located between the fire and Earth. This was the first theory to suppose that Earth is in motion.

Plato (428–347 BC) was not an astronomer, but his teachings influenced astronomy for 2000 years. Plato argued that the reality humans see is only a distorted shadow of a perfect, ideal form. If human observations are distorted, then observation can be misleading, and the best path to truth is through pure thought on the ideal forms that underlie nature.

Plato also argued that the most perfect geometrical form was the sphere, and therefore, he said, the perfect havens must be made up of spheres rotating at constant rates carrying objects around in circles. Consequently, later astronomers tried to describe the motions of the heavens by imagining multiple, rotating spheres. This became known as the principle of **uniform circular motion.**

■ Figure 4-5

The spheres of Eudoxus explain the motions in the heavens by means of nested spheres rotating about various axes at different rates. Earth is located at the center.

Pythagoras had taught that Earth is a sphere and that the other heavenly bodies were divine, perfect spheres moving in perfect circles. Eudoxus of Cnidus (409–356 BC), a student of Plato, combined a system of 27 nested spheres rotating at different rates about different axes at constant rates to produce a mathematical description of the motions of the universe (■ Figure 4-5).

At the time of the Greek philosophers, it was common to refer to systems such as that of Eudoxus as descriptions of the world, where the word *world* included not only Earth but all of the heavenly spheres. The reality of these spheres was open to debate. Some thought of the spheres as nothing more than mathematical ideas that described motion in the world model, while others began to think of the spheres as real objects made of perfect celestial material. Aristotle, for example, seems to have thought of the spheres as real.

Aristotle and the Nature of Earth

Aristotle (384–322 BC), one of Plato's students, taught and wrote on philosophy, history, politics, ethics, poetry, drama, and so on (■ Figure 4-6). Because of his sensitivity and insight, he became the great authority of antiquity, and astronomers for almost 2000 years cited him as their authority in adopting the Greek model of the universe.

Much of what Aristotle wrote about scientific subjects was wrong, but he was not a scientist. He was trying to understand the universe by creating a system of formal philosophy. Modern science depends on evidence and hypothesis, but scientific meth-

■ Figure 4-6

Aristotle, honored on this Greek stamp, wrote on such a wide variety of subjects and with such deep insight that he became the great authority on all matters of learning. His opinions on the nature of Earth and the sky were widely accepted for almost two millennia.

ods had not been invented at the time of Aristotle. Rather, he attempted to use the most basic observations combined with first principles (ideas that he believed were obviously true) to understand the world. The perfection of the heavens was, for Aristotle, a first principle.

Aristotle believed that the universe was divided into two parts—Earth, corrupt and changeable; and the heavens, perfect and unchanging. Like most of his predecessors, he believed that Earth was the center of the universe, so his model is called a **geocentric universe.** The heavens surrounded Earth, and he devised 55 crystalline spheres turning at different rates and at different angles to carry the sun, moon, and planets across the sky. The lowest sphere, that of the moon, marked the boundary between the changeable imperfect region of Earth and the unchanging perfection of the celestial realm above the moon.

Because he believed Earth to be immobile, Aristotle had to make these spheres whirl westward around Earth to produce day and night and move more slowly with respect to one another to produce the motions of the sun, moon, and planets against the background of the stars. Because his model was geocentric, he taught that Earth could be the only center of motion. All of his whirling spheres had to be centered on Earth. Like most other Greek philosophers, Aristotle viewed the universe as a perfect heavenly machine that was not many times larger than Earth itself.

About a century after Aristotle, the Alexandrian philosopher Aristarchus proposed a theory that Earth rotated on its axis and revolved around the sun. This theory is, of course, correct, but most of the writings of Aristarchus were lost, and his theory was not well known. Later astronomers rejected any suggestion that

Earth could move because it conflicted with the teachings of the great philosopher Aristotle.

Aristotle had taught that Earth had to be a sphere because it always casts a round shadow during lunar eclipses, but he could only estimate its size. About 200 BC, Eratosthenes, working in the great library in Alexandria, found a way to calculate Earth's radius. He learned from travelers that the city of Syene (Aswan) in southern Egypt contained a well into which sunlight shone vertically on the day of the summer solstice. This told him that the sun was at the zenith at Syene; but, on that same day in Alexandria, he noted that the sun was $\frac{1}{50}$ of the circumference of the sky (about 7°) south of the zenith.

Because sunlight comes from such a great distance, its rays arrive at Earth traveling almost parallel. That allowed Eratosthenes to use simple geometry to find that the distance from Alexandria to Syene was $\frac{1}{50}$ of Earth's circumference (■ Figure 4-7).

To find Earth's circumference, Eratosthenes had to know the distance from Alexandria to Syene. Travelers told him it took 50 days to cover the distance, and he knew that a camel can travel about 100 stadia per day. That meant the total distance was about 5000 stadia. If 5000 stadia is $\frac{1}{50}$ of Earth's circumference, then Earth must be 250,000 stadia around, and, dividing by 2π, Eratosthenes found Earth's radius to be 40,000 stadia.

How accurate was Eratosthenes? The stadium (singular of *stadia*) had different lengths in ancient times. If you assume 6 stadia to the kilometer, then Eratosthenes's result was too big by only 4 percent. If he used the Olympic stadium, his result was 14 percent too big. In any case, this was a much better measurement of Earth's radius than Aristotle's estimate, which was much too small, about 40 percent of the true radius.

You might think this is just a disagreement between two ancient philosophers, but it had an interesting consequence. At the time of Christopher Columbus, all educated people knew the world was round and not flat, but they weren't sure how big it was. Columbus, like many others, adopted Aristotle's diameter for Earth, so he thought Earth was small enough that he could sail west and reach Japan and the Spice Islands of the East Indies. If he had accepted Eratosthenes' diameter, Columbus would never have risked the voyage. As it turned out, he and his crew were lucky that North America was in the way. If there had been open ocean all the way, they would have starved to death long before they reached Japan.

Aristotle was a philosopher and, following Plato's recommendation, made few direct observations. Eratosthenes made some measurements to find the diameter of Earth, but the next person you need to meet was a real astronomer who observed the sky in detail. Little is known about him, but he is clearly the first great observer. Hipparchus (see Figure 2-5) lived during the second century BC, about two centuries after Aristotle. He is usually credited with the invention of trigonometry, he compiled the first star catalog, and he discovered precession (Chapter 2). Instead of describing the motion of the sun and moon using nested spheres, as most Greek philosophers did, Hipparchus proposed that the sun and moon traveled around circles with Earth near, but not at, their centers. These off-center circles are now known as **eccentrics.** Hipparchus recognized that he could produce the same motion by having the sun, for instance, travel around a small circle that followed a larger circle around Earth. The compounded circular motion that he devised became the key element in the masterpiece of the last great astronomer of classical times, Ptolemy.

The Ptolemaic Universe

Claudius Ptolemaeus was one of the great astronomer-mathematicians of antiquity. His nationality and birth date are unknown, but he lived and worked in the Greek settlement at

■ **Active Figure 4-7**

On the day of the summer solstice, sunlight fell to the bottom of a well at Syene, but the sun was about $\frac{1}{50}$ of a circle (7°) south of the zenith at Alexandria. From this, Eratosthenes was able to calculate Earth's radius.

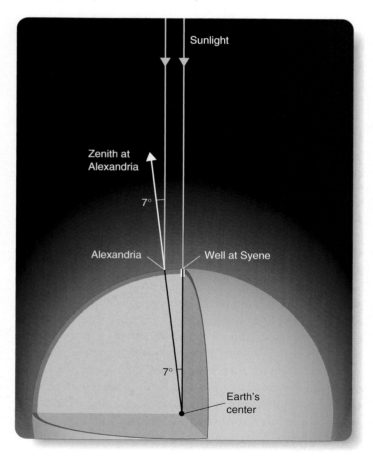

Alexandria in what is now Egypt about AD 140. He ensured the survival of Aristotle's universe by transforming it into a sophisticated mathematical model.

Study **The Ancient Universe** on pages 64–65 and notice three important ideas that show how first principles influenced early descriptions of the universe and its motions:

1 Ancient philosophers and astronomers accepted as first principles that Earth was located at the center of the universe and that the heavens moved in uniform circular motion.

2 Second, notice how the observed motion of the planets, the evidence, did not fit the theory very well. The retrograde loops the planets made were very difficult to explain using geocentrism and uniform circular motion.

3 Finally, notice how Ptolemy attempted to explain the motion of the planets by devising a small circle rotating along the edge of a larger circle, which enclosed a slightly off-center Earth. He even allowed the speed of the planets to vary slightly as they circled Earth. Ptolemy lived roughly five centuries after Aristotle, and although Ptolemy believed in the Aristotelian universe, he was interested in a different problem—the motion of the planets. He was a brilliant mathematician, and he was mainly interested in creating a mathematical description of the motions he saw in the heavens. For him, first principles took second place to mathematical precision.

Aristotle's universe, as embodied in the mathematics of Ptolemy, dominated ancient astronomy, but it was wrong. The planets don't follow circles at uniform speeds. At first the Ptolemaic system predicted the positions of the planets well; but, as centuries passed, errors accumulated. If your watch gains only one second a year, it will keep time well for many years, but the error gradually accumulates. After a century your watch will be 100 seconds fast. So, too, did the errors in the Ptolemaic system gradually accumulate as the centuries passed. Islamic and later European astronomers tried to update the system, computing new constants and adjusting epicycles. In the middle of the 13th century,

a team of astronomers supported by King Alfonso X of Castile studied the *Almagest* for 10 years. Although they did not revise the theory very much, they simplified the calculation of the positions of the planets using the Ptolemaic system and published the result as *The Alfonsine Tables,* the last great attempt to make the Ptolemaic system of practical use.

Ace★Astronomy™ Log into AceAstronomy and select this chapter to see Astronomy Exercise "Parallax I." Notice that the parallax angle depends on the length of the baseline.

Building Scientific Arguments

How did the astronomy of Hipparchus and Ptolemy violate the principles of the early Greek philosophers Plato and Aristotle?

Today, scientific arguments depend on evidence and theory; but, in classical times, they started from first principles. Hipparchus and Ptolemy lived very late in the history of classical astronomy, and they concentrated more on the mathematical problems and less on philosophical principles. They replaced the perfect spheres of Plato with nested circles in the form of epicycles and deferents. Earth was moved slightly away from the center of the deferent, so their models of the universe were not exactly geocentric, and the epicycles moved uniformly only as seen from the equant. The celestial motions were no longer precisely uniform, and the principles of geocentrism and uniform circular motion were weakened.

The work of Hipparchus and Ptolemy led eventually to a new understanding of the heavens, but first astronomers had to abandon uniform circular motion. Construct a scientific argument in the classical style based on first principles to answer the question: **Why did Plato argue for uniform circular motion?**

■   ■   ■

Connections: Ptolemy was the last of the great classical astronomers, and his work dominated astronomical thought for almost 1500 years. The collapse of the Ptolemaic model and the rise of a new model of the universe make an exciting story, but it is not just the story of an astronomical idea. It includes the invention of science as a new way of knowing and understanding what we are and where we are.

4-2 The Copernican Revolution

NICOLAUS COPERNICUS (■ Figure 4-8) triggered an Earthshaking revision in human thought by proposing that the universe is not centered on Earth but rather on the sun. Such a model is called a **heliocentric universe.** That idea eventually brought

■ Figure 4-8

Copernicus proposed that the sun and not Earth was the center of the universe. Notice the heliocentric model on this stamp issued in 1973 to commemorate the 500th anniversary of his birth.

The Ancient Universe

1 For 2000 years, the minds of astronomers were shackled by a pair of ideas. The Greek philosopher Plato argued that the heavens were perfect. Because the only perfect geometrical shape is a circle and the only perfect motion is uniform motion, Plato concluded that all motion in the heavens must be made up of combinations of circles turning at uniform rates. This was called **uniform circular motion.**

Plato's student Aristotle argued that Earth was imperfect and lay at the center of the universe. That is, he argued for a **geocentric universe.** He devised a model universe with 55 spheres turning at different rates and at different angles to carry the seven known planets (the moon, Mercury, Venus, the sun, Mars, Jupiter, and Saturn) across the sky.

Aristotle was known as the greatest philosopher in the ancient world, and his authority, lasting for 2000 years, chained the minds of astronomers and forced them to expect the universe to be geocentric and to move in uniform circular motion. See model at right by Peter Apian.

From *Cosmographica* by Peter Apian (1539).

Seen by left eye

Seen by right eye

1a Ancient astronomers believed that Earth did not move because they saw no **parallax,** the apparent motion of an object because of the motion of the observer. To demonstrate parallax, close one eye and cover a distant object with your thumb held at arm's length. Switch eyes, and your thumb appears to shift position as shown at left. If Earth moves, ancient astronomers reasoned, you should see the sky from different locations at different times of the year, and you should see parallax distorting the shapes of the constellations. They saw no parallax, so they concluded Earth could not move. Actually, the parallax of the stars is too small to see with the unaided eye.

2 Planetary motion was a big problem for ancient astronomers. In fact, the word *planet* comes from the Greek word for "wanderer," referring to the eastward motion of the planets against the background of the fixed stars. The planets did not, however, move at a constant rate, and they could occasionally stop and move westward for a few months before resuming their eastward motion. This backward motion is called **retrograde motion.**

Jan. 29, 2008

East

Nov. 15, 2007

Position of Mars at 5-day intervals

Ecliptic

Dec. 12, 2005

Oct. 3, 2005

Taurus

West

Betelgeuse

Every 2.14 years, Mars passes through a retrograde loop. Two successive loops are shown here. Each loop occurs further east along the ecliptic and has its own shape.

Orion

2a Simple uniform circular motion centered on Earth could not explain retrograde motion, so ancient astronomers combined uniformly rotating circles much like gears in a machine to try to reproduce the motion of the planets.

Rigel

3 Uniformly rotating circles were key elements of ancient astronomy. Claudius Ptolemy created a mathematical model of the Aristotelian universe in which the planet followed a small circle called the **epicycle** that slid around a larger circle called the **deferent.** By adjusting the size and rate of rotation of the circles, he could approximate the retrograde motion of a planet. See illustration at right.

To adjust the speed of the planet, Ptolemy supposed that Earth was slightly off center and that the center of the epicycle moved such that it appeared to move at a constant rate as seen from the point called the **equant.**

To further adjust his model, Ptolemy added small epicycles riding on top of larger epicycles (not shown here), producing a highly complex model.

3a Ptolemy's great book *Mathematical Syntaxis* (c. AD 140) contained the details of his model. Islamic astronomers preserved and studied the book through the Middle Ages, and they called it *Al Magisti* (The Greatest). When the book was found and translated from Arabic to Latin in the 12th century, it became known as *Almagest*.

3b The Ptolemaic model of the universe shown below was geocentric and based on uniform circular motion. Note that Mercury and Venus were treated differently from the rest of the planets. The centers of the epicycles of Mercury and Venus had to remain on the Earth–Sun line as the sun circled Earth through the year.

Equants and smaller epicycles are not shown here. Some versions contained nearly 100 epicycles as generations of astronomers tried to fine-tune the model to better reproduce the motion of the planets.

Notice that this modern illustration shows rings around Saturn and sunlight illuminating the globes of the planets, features that could not be known before the invention of the telescope.

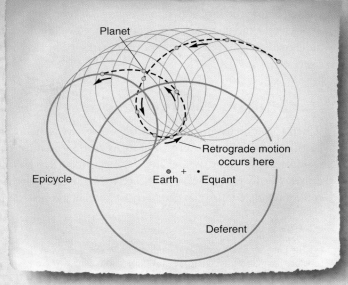

Ace ◐ Astronomy™ Go to AceAstronomy and click Active Figures to see "Epicycles." Notice how the counterclockwise rotation of the epicycle produces retrograde motion.

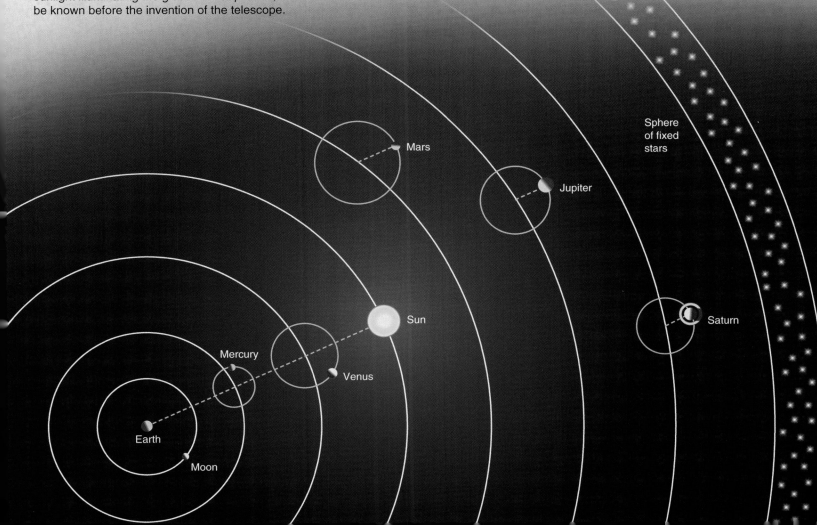

Galileo before the Inquisition and changed forever how we think of our world and ourselves. This Copernican Revolution was much more than an upheaval in astronomy; it marks the birth of modern science.

Copernicus the Revolutionary

When Copernicus proposed that the universe was heliocentric, he risked controversy. According to Aristotle, the most perfect region was the starry sphere, and the most imperfect was Earth's center. In this way the classical geocentric universe matched the commonly held Christian geometry of heaven and hell, and any-one who criticized the geometry of the Aristotelian universe challenged Christian belief and thus risked a charge of heresy.

Throughout his life, Copernicus was associated with the Church. His uncle, by whom he was raised and educated, was an important bishop in Poland; and, after studying canon law and medicine in some of the finest universities in Europe, Copernicus became a canon of the Church at the unusually young age of 24. He was secretary and personal physician to his powerful uncle for 15 years. When his uncle died, Copernicus moved to quarters adjoining the cathedral in Frauenburg, where he was a canon. No doubt his long association with the Church added to his reluctance to publish controversial ideas.

■ **Figure 4-9**

(a) The Copernican universe as reproduced in *De Revolutionibus*. Earth and all the known planets revolve in separate circular orbits about the sun (Sol) at the center. The outermost sphere carries the immobile stars of the celestial sphere. Notice the orbit of the moon around Earth (Terra). (Yerkes Observatory) (b) The model was elegant not only in its arrangement of the planets but also in their motions. Orbital speed (blue arrows) decreased from Mercury, the fastest, to Saturn, the slowest. Compare the elegance of this model with the complexity of the Ptolemaic model on page 65.

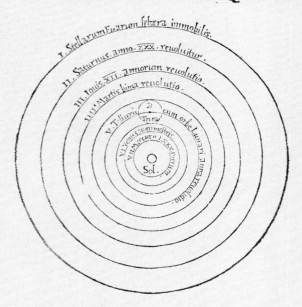

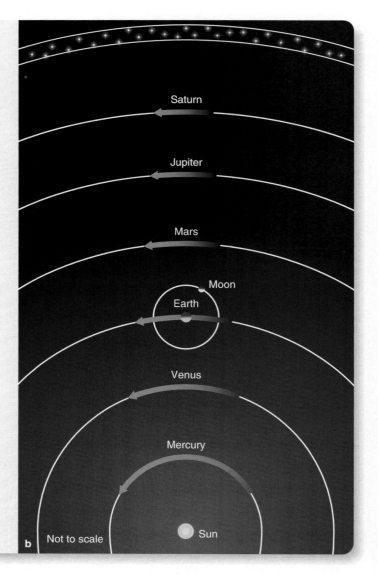

He first wrote about a heliocentric universe sometime before 1514 in a short pamphlet that he distributed in handwritten form and in some cases anonymously. Over the years, Copernicus worked on his book *De Revolutionibus Orbium Coelestium (On the Revolutions of the Celestial Spheres)* (■ Figure 4-9a), but he hesitated to publish it even though other astronomers knew of his theories.

This was a time of rebellion—Martin Luther was speaking harshly about fundamental Church teachings, and others, scholars and scoundrels, questioned the authority of the Church. Even astronomy could stir argument; moving Earth from its central place was a controversial and perhaps heretical idea. Copernicus was concerned about how his ideas would be received, but in 1540 he allowed the visiting astronomer Joachim Rheticus (1514–1576) to publish an account of the Copernican universe in Rheticus's book *Prima Narratio (First Narrative)*. In 1542, Copernicus sent the manuscript for *De Revolutionibus* off to be printed. He died in the spring of 1543 before the printing was completed.

The Copernican Model

You should notice the difference between the Copernican hypothesis and the Copernican model. The Copernican hypothesis was that the universe was heliocentric, and that was correct. The Copernican model was the arrangement and motion of the planets that Copernicus devised to explain the motions in the sky. The model won gradual acceptance, but, as you will discover, it was flawed.

The most important idea in *De Revolutionibus* was the placement of the sun at the center of the universe. That single innovation had an astonishing consequence—the retrograde motion of the planets was immediately explained in a straightforward way without the large epicycles that Ptolemy used. In the Copernican system, Earth moves faster along its orbit than the planets that lie further from the sun. Consequently, Earth periodically overtakes and passes these planets, and they appear to slow and fall behind (■ Figure 4-10). Because the planetary orbits do not lie in precisely the same plane, a planet does not resume its eastward motion in precisely the same path it followed earlier. Consequently, it describes a loop whose shape depends on the angle between the orbital planes.

Copernicus could explain retrograde motion without epicycles, and that was impressive. But he could not do away with epicycles completely. Copernicus, a classical astronomer, had tremendous respect for the old concept of uniform circular motion. In fact, he objected strongly to Ptolemy's use of the equant. It seemed arbitrary to Copernicus, a direct violation of the elegance of Aristotle's philosophy of the heavens. Copernicus called equants "monstrous" in that they violated uniform circular motion. In his model, Copernicus returned to a strong belief in uniform circular motion. Although he did not need epicycles to explain retrograde motion, Copernicus discovered that the sun, moon, and planets suffered small variations in their motions—variations that

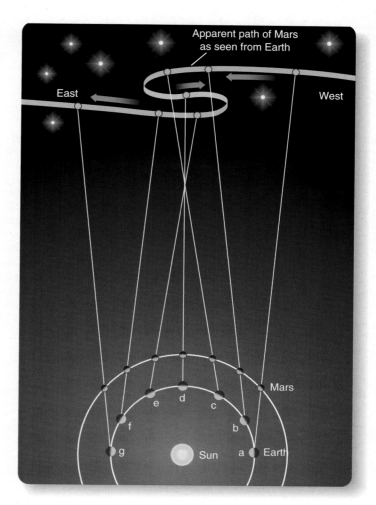

■ Figure 4-10

The Copernican explanation of retrograde motion: As Earth overtakes Mars (a–c), Mars appears to slow its eastward motion. As Earth passes Mars (d), Mars appears to move westward. As Earth draws ahead of Mars (e–g), Mars resumes its eastward motion against the background stars. Compare with the illustration of retrograde motion on page 64. The positions of Earth and Mars are shown at equal intervals of one month.

he could not explain with the concept of uniform circular motion centered on the sun. Today, we recognize those variations as typical of objects following elliptical orbits, but Copernicus held firmly to uniform circular motion, so he had to use small epicycles to produce these minor variations in the motions of the sun, moon, and planets.

Because Copernicus imposed uniform circular motion on his model, it could not accurately predict the motions of the planets. *The Prutenic Tables* (1551) were based on the Copernican model, and they were not significantly more accurate than *The Alfonsine Tables* (1251), which were based on Ptolemy's model. Both could be in error by as much as 2°, four times the angular diameter of the full moon.

The Copernican *model* was inaccurate, but the Copernican *hypothesis* that the universe is heliocentric was correct. There are probably a number of reasons why the hypothesis gradually won

Window on Science | 4-1

Creating New Ways to See Nature: Scientific Revolutions

The Copernican Revolution is often cited as the perfect example of a scientific revolution. Over a few decades, astronomers abandoned a way of thinking about the universe that was almost 2000 years old and adopted a new set of ideas and assumptions. The American philosopher of science Thomas Kuhn has referred to a commonly accepted set of scientific ideas and assumptions as a scientific **paradigm.** The pre-Copernican astronomers had a geocentric paradigm that included uniform circular motion and the perfection of the heavens. That paradigm survived for many centuries until a new generation of astronomers was able to overthrow the old paradigm and establish a new paradigm that included heliocentrism and Earth's motion.

A scientific paradigm is powerful because it shapes your perceptions. It determines what you judge to be important questions and what you judge to be significant evidence. Consequently, it is often difficult to recognize how your paradigms limit what you can understand.

For example, the geocentric paradigm contained problems that seem obvious today, but because astronomers before Copernicus lived and worked inside that paradigm, they had difficulty seeing the problems. Overthrowing an outdated paradigm is not easy, because you must learn to see nature in an entirely new way. Galileo and Kepler saw nature from a new paradigm that would have been almost incomprehensible to astronomers of earlier centuries.

You can find examples of scientific revolutions in many fields, including biology, geology, genetics, and psychology. These scientific revolutions have been difficult and controversial because they have involved the overthrow of accepted paradigms. But that is why scientific revolutions are exciting. They provide not just a new idea or a new theory, but an entirely new insight into how nature works—a new way of seeing the world.

The ancients believed the stars were attached to a starry sphere. (NOAO and Nigel Sharp)

---

acceptance in spite of the inaccuracy of the model's predictions. The most important factor may be the elegance of the idea. Placing the sun at the center of the universe produced a symmetry among the motions of the planets that was pleasing to the eye and to the intellect (Figure 4-9b). All of the planets moved in the same direction at speeds that were simply related to their distance from the sun. In the Ptolemaic model, Venus and Mercury were treated differently from the other planets. In the Copernican model, all planets were treated the same. In this way, the model may have won support not for its accuracy, but for its elegance.

The most astonishing consequence of the Copernican hypothesis was not what it said about the sun, but what it said about Earth. By placing the sun at the center, Copernicus made Earth move along a path around the sun just as the other planets did, and that made Earth a planet. By revealing that Earth is a planet, Copernicus revolutionized humanity's view of our place in the universe and triggered a controversy that would eventually bring Galileo before the Inquisition.

Although astronomers throughout Europe read and admired *De Revolutionibus,* they did not usually accept the Copernican hypothesis. The mathematics was elegant, and the astronomical observations and calculations were of tremendous value. Yet few astronomers believed, at first, that the sun actually was the center of the planetary system and that Earth moved. How the Copernican hypothesis became gradually recognized as correct has been named the Copernican Revolution because it involved not just the adoption of a new idea but a total revolution in the way astronomers thought about the place of the Earth (**Window on Science 4-1**).

Galileo the Defender

Most people know two facts about Galileo, and both facts are wrong. You should begin by getting those facts right: Galileo did not invent the telescope, and he was not condemned by the Inquisition for believing that Earth moved around the sun. Then why is Galileo so important that in 1979, almost 400 years after his trial, the Vatican reopened his case? As you read about Galileo, you will discover that his trial concerned not just the place of the Earth but a new way of finding truth, a new way of knowing about the world, a method known today as science.

Galileo Galilei (■ Figure 4-11) was born in Pisa, a city in what is now Italy, in 1564, and he studied medicine at the university there. His true love, however, was mathematics; and, although he had to leave school early because of financial difficulties, he returned only four years later as a professor of mathematics. Three years later, he became professor of mathematics at the university at Padua. He remained there for 18 years.

During this time, Galileo seems to have adopted the Copernican model, although he admitted in a 1597 letter to the Ger-

Galileo, remembered as the great defender of Copernicanism, also made important discoveries in the physics of motion. He is honored here on an Italian 2000-lira note. The reverse side shows one of his telescopes at lower right and a modern observatory above it.

man astronomer Kepler that he did not support Copernicanism publicly because of the criticism such a declaration would bring. It was the telescope that drove Galileo to publicly defend the heliocentric model.

Galileo did not invent the telescope. It was apparently invented around 1608 by lens makers in Holland. Galileo, hearing descriptions in the fall of 1609, was able to build telescopes in his workshop. Galileo was not the first person to look at the sky through a telescope, but he was the first person to observe the sky systematically and apply his observations to the theoretical problem of the day—the place of the Earth.

What Galileo saw through his telescopes was so amazing he rushed a small book into print. *Sidereus Nuncius (The Sidereal Messenger)* reported three major discoveries. First, the moon was not perfect.

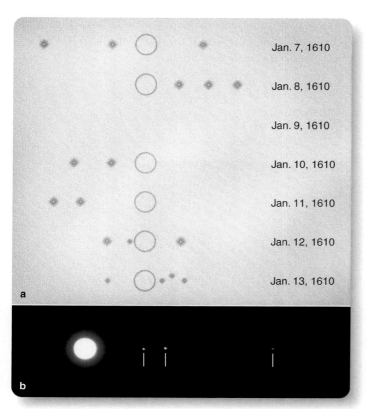

Jan. 7, 1610

Jan. 8, 1610

Jan. 9, 1610

Jan. 10, 1610

Jan. 11, 1610

Jan. 12, 1610

Jan. 13, 1610

a

b

It had mountains and valleys on its surface, and Galileo used the shadows to calculate the height of the mountains. The Ptolemaic model held that the moon was perfect, but Galileo showed that it was not only imperfect, it was a world like Earth.

The second discovery reported in the book was that the Milky Way was made up of a myriad of stars too faint to see with the unaided eye. While intriguing, this discovery could not match the third discovery. Galileo's telescope revealed four new "planets" circling Jupiter, planets known today as the Galilean moons of Jupiter (■ Figure 4-12).

The moons of Jupiter supported the Copernican model over the Ptolemaic model. Critics of Copernicus had said Earth could not move because the moon would be left behind. But Jupiter moved yet kept its satellites, so Galileo's discovery suggested that Earth, too, could move and keep its moon. Also, the Ptolemaic model included the Aristotelian belief that all heavenly motion was centered on Earth at the center of the universe. Galileo showed that Jupiter's moons revolve around Jupiter, so there could be other centers of motion.

■ **Figure 4-12**

(a) On the night of January 7, 1610, Galileo saw three small "stars" near the bright disk of Jupiter and sketched them in his notebook. On subsequent nights (excepting January 9, which was cloudy), he saw that the stars were actually four moons orbiting Jupiter. (b) This photo taken through a modern telescope shows the overexposed disk of Jupiter and three of the four Galilean moons. (Grundy Observatory)

Sometime after *Sidereus Nuncius* was published, Galileo made another discovery that made Jupiter's moons even stronger evidence supporting the Copernican universe. Eventually, he was able to measure the orbital periods of the four moons, and he found that the innermost moon moved fastest and that the moons further from Jupiter moved more slowly. In this way, Jupiter's moons made up a harmonious system ruled by Jupiter, just as the planets in the Copernican universe were a harmonious system ruled by the sun (Figure 4-9b). The similarity isn't proof, but Galileo saw it as an argument that the solar system was sun centered and not Earth centered.

Also soon after *Sidereus Nuncius* was published, Galileo made two additional discoveries. When he observed the sun, he discovered sunspots, raising the suspicion that the sun was less than perfect. Further, by noting the movement of the spots, he concluded that the sun was a sphere and that it rotated on its axis. When he observed Venus, Galileo saw that it was going through phases like those of the moon. In the Ptolemaic model, Venus moves around an epicycle centered on a line between Earth and the sun. In that epicycle, it would always be seen as a crescent (■ Figure 4-13a). But Galileo saw Venus go through a complete set of phases, proving that it did indeed revolve around the sun (Figure 4-13b).

Sidereus Nuncius was very popular and made Galileo famous. He became chief mathematician and philosopher to the Grand Duke of Tuscany in Florence. In 1611, Galileo visited Rome and was treated with great respect. He had long, friendly discussions with the powerful Cardinal Barberini, a man with a deep interest in the new discoveries being made. But Galileo also made enemies. Personally, Galileo was outspoken, forceful, and sometimes tactless. He enjoyed debate, but most of all he enjoyed being right. In lectures, debates, and letters he offended important people who questioned his telescopic discoveries.

By 1616, Galileo was the center of a storm of controversy. Some critics said he was wrong, and others said he was lying. Some refused to look through a telescope lest it mislead them, and others looked and claimed to see nothing (hardly surprising given the awkwardness of those first telescopes) (■ Figure 4-14). Pope Paul V decided to end the disruption, so when Galileo visited Rome in 1616 Cardinal Bellarmine interviewed him privately and ordered him to cease debate. There is some controversy today about the nature of Galileo's instructions, but he did not pursue astronomy for some years after the interview. Books relevant to Copernicanism were banned, including *De Revolutionibus,* although owners were allowed to keep their books if they changed certain phrases to make it clear that the central place of the sun was only a theory and not a fact.

In 1621 Pope Paul V died, and his successor, Pope Gregory XV, died in 1623. The next pope was Galileo's friend Cardinal Barberini, who took the name Urban VIII. Galileo rushed to Rome hoping to have the prohibition of 1616 lifted; and, although the new pope did not revoke the orders, he did encourage Galileo. When he got back home, Galileo began to write his great defense of the Copernican model, finally completing it on December 24, 1629. After some delay, the book was approved by both the local censor in Florence and the head censor of the Vatican in Rome. It was printed in February 1632.

Called *Dialogo Dei Due Massimi Sistemi (Dialogue Concerning the Two Chief World Systems),* it confronts the ancient astronomy of Aristotle and Ptolemy with the Copernican model and with telescopic observations as evidence. Galileo wrote the book as a debate among three friends. Salviati, a swift-tongued defender of Copernicus, dominates the book; Sagredo is intelligent but largely uninformed. Simplicio is the dismal defender of Ptolemy. In fact, he does not seem very bright.

■ Figure 4-13

(a) If Venus moved in an epicycle centered on the Earth–sun line (see page 65), it would always appear as a crescent. (b) Galileo's telescope showed that Venus goes through a full set of phases, proving that it must orbit the sun.

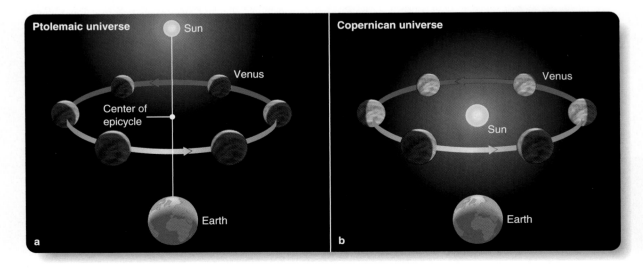

■ **Figure 4-14**

Galileo's telescope made him famous, and he demonstrated his telescope and discussed his observations with powerful people. Some thought the telescope was the work of the devil and would deceive anyone who looked. In any case, Galileo's discoveries produced intense, and in some cases, angry debate. (Yerkes Observatory)

The publication of *Dialogo* created a sensation, and it was sold out by August 1632, when the Inquisition ordered sales stopped. The book was a clear defense of Copernicus, and, either intentionally or unintentionally, Galileo exposed the pope's authority to ridicule. Urban VIII was fond of arguing that, as God was omnipotent, He could construct the universe in any form while making it appear to humans to have a different form, and thus its true nature could not be deduced by mere observation. Galileo placed the pope's argument in the mouth of Simplicio, and Galileo's enemies showed the passage to the pope as an example of Galileo's disrespect. The pope thereupon ordered Galileo to face the Inquisition.

The Trial of Galileo

The trial of Galileo was one of the turning points in the history of science and human learning, but historians still argue about what happened and why. The trial involved the highest religious principles and the lowest of behind-the-scenes political maneuvering. One thing you can be sure of: The trial changed the way humanity thought about the world and marked the beginning of modern science as a way to understand nature.

Galileo was interrogated by the Inquisition four times and was threatened with torture. He must have thought often of Giordano Bruno, tried, condemned, and burned at the stake in Rome in 1600. One of Bruno's offenses had been Copernicanism. But the trial did not center on Galileo's belief in Copernicanism. After all, *Dialogo* had been approved by two censors. Rather, the trial

centered on the instructions given Galileo in 1616. From his file in the Vatican, his accusers produced a record of the meeting between Galileo and Cardinal Bellarmine that included the statement that Galileo was "not to hold, teach, or defend in any way" the principles of Copernicus. Some historians believe that this document, which was signed neither by Galileo nor by Bellarmine nor by a legal secretary, was a forgery. Others suspect it may be a draft that was never used. In any case, it is possible that Galileo's true instructions were much less restrictive. But Bellarmine was dead and could not testify at Galileo's trial.

The Inquisition condemned him not for heresy but for disobeying the orders given him in 1616. On June 22, 1633, at the age of 70, kneeling before the Inquisition, Galileo read a recantation admitting his errors. Tradition has it that as he rose he whispered, *"E pur si muove"* ("Still it moves"), referring to Earth. Although he was sentenced to life imprisonment, he was actually confined at his villa for the next 10 years, perhaps through the intervention of the pope. He died there on January 8, 1642, 99 years after the death of Copernicus.

Galileo was not condemned for heresy, nor was the Inquisition interested when he tried to defend Copernicanism. He was tried and condemned on a charge you might call a technicality. Then why is his trial so important that historians have studied it for almost four centuries? Why have some of the world's greatest authors, including Bertolt Brecht, written about Galileo's trial? Why in 1979 did Pope John Paul II create a commission to reexamine the case against Galileo?

To understand the trial, you must recognize that it was the result of a conflict between two ways of understanding our universe. Plato had argued that observation was deceptive, and that the only way to find truth was through pure thought. Since the Middle Ages, scholars had taught that the only path to true understanding was through religious faith. St. Augustine (AD 354–430) wrote *"Credo ut intelligame,"* which can be translated as, "Believe in order to understand." But Galileo and other scientists of the Renaissance used their own observations to try to understand the universe; and, when their observations contradicted Scripture, they assumed their observations of reality were correct and that Scripture was not being correctly understood (■ Figure 4-15). Galileo paraphrased Cardinal Baronius in saying, "The Bible tells us how to go to heaven, not how the heavens go."

Galileo's discoveries produced intense, and in some cases angry, debate. Various passages of Scripture seemed to contradict observation. For example, Joshua is said to have commanded the sun to stand still, not Earth to stop rotating (Joshua 10:12–13). In response to such passages, Galileo argued that you should "read the book of nature"—that is, you should observe the universe with your own eyes.

The trial of Galileo was not about the place of the Earth. It was not about Copernicanism. It wasn't really about the instructions Galileo received in 1616. It was about the birth of modern science as a rational way to understand our universe. The

How were Galileo's observations of the moons of Jupiter evidence against the Ptolemaic model?
Modern scientific arguments depend critically on evidence, and that started with Galileo. He presented his arguments in the form of evidence and conclusions, and the moons of Jupiter were key evidence. Ptolemaic astronomers argued that Earth could not move or it would lose its moon, but even in the Ptolemaic universe Jupiter moved, and Galileo's telescope showed that it kept its moons. Evidently, Earth could move and not leave its moon behind. Furthermore, moons circling Jupiter did not fit the classical belief that all motion was centered on Earth. Obviously there could be other centers of motion. Finally, the orbital periods of the moons were related to their distance from Jupiter, just as the orbital periods of the planets were, in the Copernican system, related to their distance from the sun. This similarity suggested that the sun rules its harmonious family of planets just as Jupiter rules its harmonious family of moons.

Of all of Galileo's telescopic observations, the moons of Jupiter caused the most debate. But there was more. Use the evidence to build an argument to answer the following: **How did craters on the moon and the phases of Venus weigh against the Ptolemaic model?**

■ Figure 4-15

Although he did not invent it, Galileo will always be associated with the telescope because it was the source of the observational evidence he used to try to understand the universe. By depending on evidence instead of first principles, Galileo led the way to the invention of modern science as a way to know about the natural world.

Connections: Galileo faced the Inquisition because of a conflict between two ways of knowing and understanding the world. The Church taught faith and understanding through revelation, but the scientists of the age were inventing a new way to understand nature that relied on evidence, a way of knowing now called science. In astronomy, evidence means observations, so it is time to turn your attention from the philosophical problem of the place of the Earth to the observational problem of the motion of the planets.

commission appointed by John Paul II in 1979, reporting its conclusions in October 1992, said of Galileo's inquisitors, "This subjective error of judgment, so clear to us today, led them to a disciplinary measure from which Galileo 'had much to suffer.'" Galileo was not found innocent in 1992 so much as the Inquisition was forgiven for having charged him in the first place.

The gradual change from reliance on personal faith to reliance on scientific observation came to a climax with the trial of Galileo. Since that time, scientists have increasingly reserved Scripture and religious faith for ethical guidance and personal comfort and have depended on systematic observation to describe the physical world. The final verdict of 1992 was an attempt to bring some balance to this conflict. In his remarks on the decision, Pope John Paul II said, "A tragic mutual incomprehension has been interpreted as the reflection of a fundamental opposition between science and faith. . . . this sad misunderstanding now belongs in the past." Galileo's trial is over, but it continues to echo through history as part of humanity's struggle to understand the place of the Earth in our universe.

4-3 The Puzzle of Planetary Motion

WHILE GALILEO WAS TEACHING mathematics at Pisa and Padua, two other astronomers were working in northern Europe, beyond the sway of the Inquisition. Although they, too, struggled to understand the place of the Earth, they approached the problem differently. They used the most accurate observations of the positions of the planets to try to discover the rules that govern planetary motion. When that puzzle was finally solved, astronomers understood why the Ptolemaic model of the universe does not work well and how to make the Copernican universe a precise predictor of planetary motion.

Tycho the Observer

The great observational astronomer of this story is a Danish nobleman, Tycho Brahe, born December 14, 1546, only three years after the publication of *De Revolutionibus* (■ Figure 4-16). As a son of a powerful noble family, he was well known for his vanity and lordly manners.

Tycho's college days were eventful. He was officially studying law with the expectation that he would enter Danish politics, but he made it clear to his family that his real interest was astronomy and mathematics. It was also during his college days that he fought a duel and received a wound that disfigured his nose. For the rest of his life, he wore false noses made of gold and silver and stuck on with wax. The disfigurement probably did little to improve his disposition.

Tycho's first astronomical observations were made while he was a student. In 1563, Jupiter and Saturn passed very near each other in the sky, nearly merging into a single point on the night of August 24. Tycho found that the *Alfonsine Tables* were a full month in error and that the *Prutenic Tables* were in error by a number of days. These discrepancies dismayed Tycho and sparked his interest in the motions of the planets.

In 1572, a brilliant "new star" (now called Tycho's supernova) appeared in the sky. Such changes in the sky puzzled classically trained astronomers. Aristotle had argued that the starry sphere was perfect and unchanging, and therefore such new stars had to lie closer to Earth than the moon. Layers below the moon were thought to be less perfect and thus more changeable than the starry sphere.

Tycho carefully measured the position of the star time after time through the night and found that it displayed no parallax. To understand the significance of this observation, you must note that Tycho, like most astronomers of his time, believed Earth was fixed at the center of the universe. Consequently, he believed the heavens rotated westward around Earth once a day. The new star, according to classical astronomy, represented a change in the heavens and therefore had to lie below the sphere of the moon. In that case, Tycho reasoned, the new star would appear slightly too far east when it was in the eastern sky and slightly too far west later in the night when it was carried into in the western sky (■ Figure 4-17). That is an example of parallax. Tycho could detect no parallax in the position of the new star, so he concluded that it must lie above the sphere of the moon and was probably on the starry sphere itself. That was an astonishing discovery because it contradicted Aristotle's belief that the starry sphere was perfect and unchanging.

No one before Tycho could have made this discovery because no one had ever measured the positions of stars so accurately. Tycho had great confidence in the precision of his measurements; so, when he failed to detect parallax for the new star, he knew it was important evidence against the Ptolemaic theory. He announced his discovery in a small book, *De Stella Nova (The New Star)*, published in 1573.

The book attracted the attention of astronomers throughout Europe, and soon Tycho was famous. Although Tycho had finished at the university, he was unemployed, idling at the family home and embarrassing his powerful family. His fame allowed his family to introduce him to the court of the Danish King Frederik II, and there the king offered Tycho funds to build an observatory on the island of Hveen just off the Danish coast. Tycho also received a steady source of income as landlord of a coastal

■ **Figure 4-16**

Tycho Brahe (1546–1601) was, during his lifetime, the most famous astronomer in the world. Proud of his noble rank, he wears the elephant medal awarded him by the king of Denmark. His artificial nose is suggested in this engraving.

■ **Figure 4-17**

If an object lay lower than the celestial sphere, reasoned Tycho Brahe, then it should be seen east of its average position as it was rising and west of its average position when it was setting. Because he did not detect this daily parallax, he concluded the new star of 1572 had to lie on the celestial sphere.

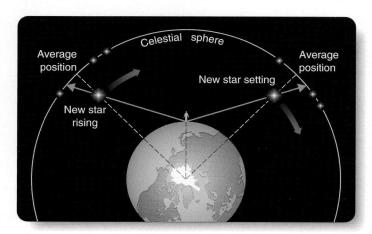

district from which he collected rents. (He was not a popular landlord.) On Hveen, Tycho constructed a luxurious home with six towers specially equipped for astronomy and populated it with servants, assistants, and a dwarf to act as jester. Soon Hveen was an international center of astronomical study.

Tycho Brahe's Legacy

Tycho Brahe made no lasting contribution to astronomical theory. Because he could measure no parallax for the stars, he concluded that Earth had to be stationary, and that led him to reject the Copernican hypothesis. However, he also rejected the Ptolemaic model because of its inaccurate predictions. Instead, he devised a complex model in which Earth was the immobile center of the universe around which the sun and moon moved. The other planets circled the sun. You might find this arrangement familiar; it is really the Copernican model with Earth held stationary and the sun allowed to move around Earth. In this way, Tycho preserved the central, immobile Earth that most astronomers believed was described in scripture (■ Figure 4-18a). Although Tycho's model, the Tychonic Universe, was very popular at first, the Copernican model replaced it within a century.

The true value of Tycho's work was observational. Because he was able to devise new and better instruments, he was able to make highly accurate observations of the positions of the stars, sun, moon, and planets. Tycho had no telescopes—they were not invented until the next century—so his observations were made by the naked eye peering along sights. All of his instruments were designed to measure angles in the sky. For example, his quadrant could measure angles up to 90° above the horizon (Figure 4-18b). By designing and building large instruments with great care, he

was able to measure angles to high precision. He measured the positions of 777 stars to better than 4 minutes of arc and regularly measured the positions of the sun, moon, and planets during the 20 years he stayed on Hveen.

Unhappily for Tycho, King Frederik II died in 1588, and his young son took the throne. Suddenly Tycho's temper, vanity, and noble presumptions threw him out of favor. In 1596, taking most of his instruments and books of observations, he went to Prague, the capital of Bohemia, and became imperial mathematician to the Holy Roman Emperor Rudolph II. His goal was to revise *The Alfonsine Tables* and publish the revision as a monument to his new patron. It would be called *The Rudolphine Tables*.

Tycho did not intend to base *The Rudolphine Tables* on the Ptolemaic system but rather on his own Tychonic system, proving once and for all the validity of his hypothesis. To assist him, he hired a few mathematicians and astronomers, including one Johannes Kepler. Then in November 1601, Tycho collapsed at a nobleman's home. Before he died, 11 days later, he asked Rudolph II to make Kepler imperial mathematician. Thus the newcomer, Kepler, became Tycho's replacement (though at one-sixth Tycho's salary).

Kepler the Analyst

No one could have been more different from Tycho Brahe than Johannes Kepler (■ Figure 4-19). He was born December 27, 1571, to a poor family in a region now included in southwestern Germany. His father was unreliable and shiftless, principally employed as a mercenary soldier fighting for whoever paid enough. He finally failed to return from a military expedition, either because he was killed or because he found other circum-

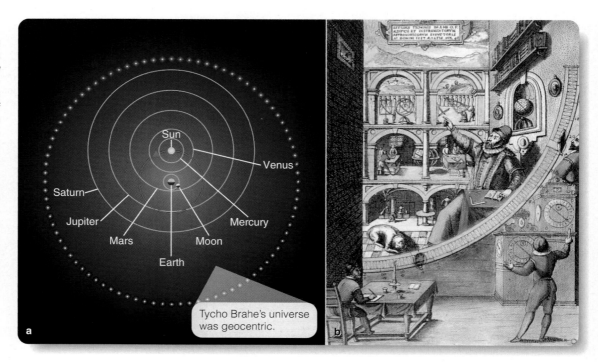

■ Figure 4-18

(a) Tycho Brahe's model of the universe held that Earth was fixed at the center of the starry sphere. The moon and sun circled Earth, while the planets circled the sun. (b) Much of Tycho's success was due to his skill in designing large, accurate instruments. In this engraving of his mural quadrant, the figure of Tycho, his dog, and the background scene are painted on the wall within the arc of the quadrant. The observer (Tycho himself) at the extreme right peers through a sight out the loophole in the wall at the upper left to measure an object's angular distance above the horizon. (Granger Collection, New York)

Sun

Venus

Saturn

Jupiter

Mercury

Mars

Moon

Earth

a

Tycho Brahe's universe was geocentric.

b

stances more to his liking. Kepler's mother was apparently an unpleasant and unpopular woman. She was accused of witchcraft in her later years, and Kepler had to defend her in a trial that dragged on for three years. She was finally acquitted but died the following year.

Kepler was the oldest of six children, and his childhood was no doubt unhappy. The family was not only poor but suffered from an absentee father whom Kepler described as "vicious, inflexible, quarrelsome, and doomed to a bad end." In addition, Kepler was never healthy, even as a child, so it is surprising that he did well in school, eventually winning promotion to a Latin school and finally a scholarship to the university at Tübingen, where he studied to become a Lutheran pastor.

During his last year of study, Kepler accepted a job in Graz teaching mathematics and astronomy. Evidently, he was not a good teacher. He had few students his first year and none at all his second. His superiors put him to work teaching a few introductory courses and preparing an annual almanac that contained astronomical, astrological, and weather predictions. Through good luck, in 1595 some of his weather predictions were fulfilled, and he gained a reputation as an astrologer and seer; even in later life he earned money by publishing almanacs.

While still a college student, Kepler had become a believer in the Copernican theory, and at Graz he used his extensive spare time to study astronomy. By 1596, the same year Tycho left Hveen, Kepler was ready to solve the mystery of the universe. That year he published a book called *The Forerunner of Dissertations on the Universe, Containing the Mystery of the Universe.* Like nearly all scientific works of the time, the book was in Latin, and it is now known as *Mysterium Cosmographicum.*

The book begins with a long appreciation of Copernicanism and then goes on to speculate on the spacing of the planetary orbits. Kepler, as a Copernican, knew of six planets circling the sun—Mercury, Venus, Earth, Mars, Jupiter, and Saturn. According to his model, the spheres containing the six planets were separated from one another, and their relative sizes were fixed by the five regular solids*—the cube, tetrahedron, dodecahedron, icosahedron, and octahedron (Figure 4-19). Kepler gave astrological, numerological, and musical arguments for his theory and so followed the tradition of Pythagoras in believing that the order of the universe is underlain by musical (meaning mathematical) principles.

In the second half of the book, Kepler tried to fit the five solids to the planetary circles of the Copernican theory, a complex problem involving three-dimensional trigonometry and geometry. He sent copies to Tycho and to Galileo, and neither seemed very impressed with his theory, but his mathematical abilities shown brightly in his calculations.

Life was unsettled for Kepler because of the persecution of Protestants in the region, so when Tycho invited him to Prague in 1600, he went readily, eager to work with the famous astronomer. Tycho's sudden death in 1601 left Kepler in a position to use the observations from Hveen to analyze the motions of the planets and complete *The Rudolphine Tables.* Tycho's family, recognizing that Kepler was a Copernican and guessing that he would not follow the Tychonic system in completing *The Rudolphine Tables,* sued to recover the instruments and books of observations. The legal wrangle went on for years. The family did recover the instruments Tycho had brought to Prague; Kepler seems to have had little interest in them, perhaps because

■ **Figure 4-19**

Johannes Kepler (1571–1630) was Tycho Brahe's sucessor. This diagram, based on one drawn by Kepler, shows how he believed the sizes of the celestial spheres carrying the outer three planets are determined by spacers (blue) consisting of the regular solids. Inside the sphere of Mars, the remaining regular solids (not shown here) separate the spheres of the Earth, Venus, and Mercury. The sun lay at the very center of this Copernican universe based on geometrical spacers.

POSTA ROMANA
55ᴮ
400 ANI DE LA NASTERE
1971
KEPLER
VLASTO

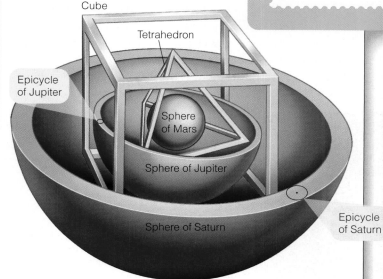

Cube

Tetrahedron

Epicycle of Jupiter

Sphere of Mars

Sphere of Jupiter

Sphere of Saturn

Epicycle of Saturn

---

*A regular solid is a three-dimensional body each of whose faces is the same. For example, a cube is a regular solid each of whose faces is a square.

of his poor eyesight. However, Kepler had the books of observations, and he kept them.

Whether Kepler had any legal right to Tycho's records is debatable, but he put them to good use. He began by studying the motion of Mars, trying to deduce from the observations how the planet moves. By 1606, he had solved the puzzle of planetary motion. The orbit of Mars is an ellipse and not a circle, he said, and with that he abandoned the 2000-year-old belief in the circular motion of the planets. But he discovered that the mystery was even more complex: The planets do not move at a uniform speed along their elliptical orbits. Kepler's analysis showed that they move faster when closer to the sun and slower when farther away. With those two brilliant discoveries, Kepler abandoned both circular motion and uniform motion.

Kepler published his results in 1609 in a book called *Astronomia Nova (The New Astronomy)*. Like Copernicus's book, *Astronomia Nova* did not become an instant best seller. It is written in Latin for other scientists and is highly mathematical. In some ways, the book is surprisingly advanced. For instance, Kepler discusses the force that holds the planets in their orbits, a question that Isaac Newton considered later in his recognition of gravity as a natural force.

Despite the abdication of Kepler's patron Rudolph II in 1611, Kepler continued his astronomical studies. He wrote about a supernova that had appeared in 1604 (now known as Kepler's supernova) and about comets, and he wrote a textbook about Copernican astronomy. In 1619, he published *Harmonice Mundi (The Harmony of the World),* in which he returned to the cosmic mysteries of *Mysterium Cosmographicum.* The main thing of note in *Harmonice Mundi* is his discovery that the radii of the planetary orbits are related to the planets' orbital periods. That and his two previous discoveries are now recognized as Kepler's three laws of planetary motion.

Kepler's Three Laws of Planetary Motion

Although Kepler dabbled in the philosophical arguments of the day, he was a mathematician, and his triumph was the solution of the problem of the motion of the planets. The key to his solution was the ellipse.

An **ellipse** is a figure drawn around two points called the foci in such a way that the distance from one focus to any point on the ellipse and back to the other focus equals a constant. You can easily draw ellipses with two thumbtacks and a loop of string. Press the thumbtacks into a board, loop the string about the tacks, and place a pencil in the loop. If you keep the string taut as you move the pencil, it traces out an ellipse (■ Figure 4-20a).

The geometry of an ellipse is described by two simple numbers. The **semimajor axis, *a*,** is half of the longest diameter. The **eccentricity** of an ellipse, *e,* is the distance from either focus to the center of the ellipse divided by the semimajor axis. If you want to draw a circle with the string and tacks as shown in Figure 4-20a, you would move the two thumbtacks together, which shows that a circle is really just an ellipse with eccentricity equal to zero. As you move the thumbtacks farther apart, the ellipse becomes flatter, and the eccentricity moves closer to 1.

Kepler used ellipses to describe the motion of the planets in three fundamental rules that have been tested and confirmed so many times that astronomers now refer to them as natural laws **Window on Science 4-2**). They are commonly called Kepler's laws of planetary motion (■ Table 4-1).

Kepler's first law states that the orbits of the planets around the sun are ellipses with the sun at one focus. Thanks to the precision of Tycho's observations and the sophistication of Kepler's mathematics, Kepler was able to recognize the elliptical shape of the orbits, even though they are nearly circular. Of the planets known to Kepler, Mercury has the most elliptical orbit, but even it deviates only slightly from a circle (■ Figure 4-21).

Kepler's second law states that a line from the planet to the sun sweeps over equal areas in equal intervals of time. This means

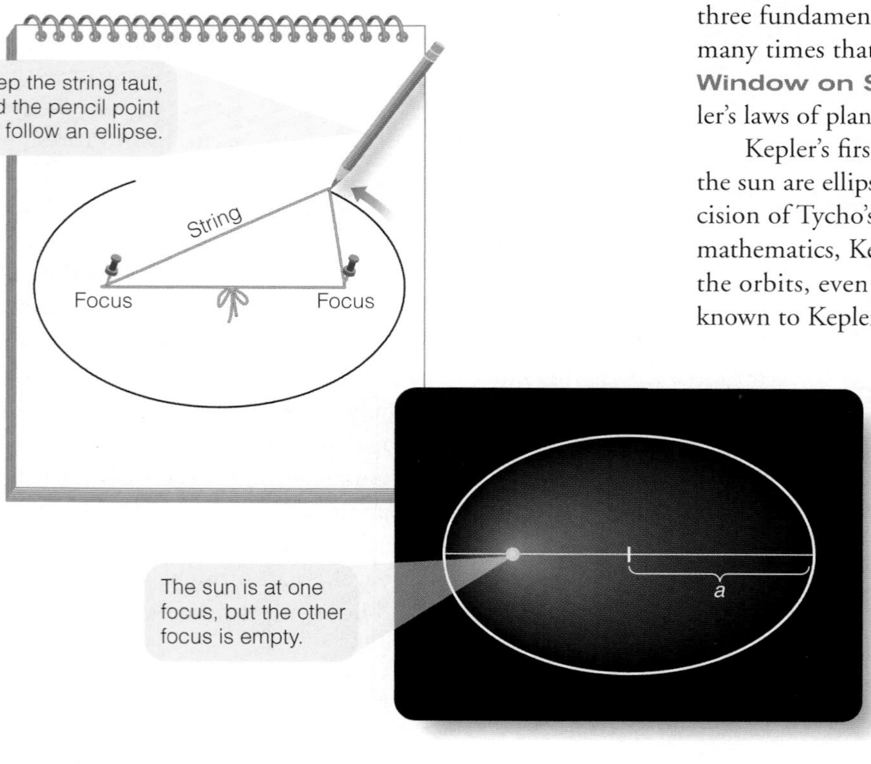

Keep the string taut, and the pencil point will follow an ellipse.

String

Focus   Focus

The sun is at one focus, but the other focus is empty.

a

■ **Figure 4-20**

The geometry of elliptical orbits: Drawing an ellipse with two tacks and a loop of string is easy. The semimajor axis, *a,* is half of the longest diameter. The sun lies at one of the foci of the elliptical orbit of a planet.

Hypothesis, Theory, and Law: Levels of Confidence

Even scientists misuse the words *hypothesis*, *theory*, and *law*. You must try to distinguish these terms from one another because they are key elements in science.

A **hypothesis** is a single assertion or conjecture that must be tested. It could be true or false. "All Texans love chili" is a hypothesis. To know whether it is true or false, you need to test it against reality by making observations or performing experiments. Copernicus asserted that the universe was heliocentric; his assertion was a hypothesis subject to testing.

In Chapter 2, you saw that a model (Window on Science 2-1) is a description of some natural phenomenon; it can't be right or wrong. A model is not a conjecture of truth but merely a convenient way to think about a natural phenomenon. Consequently, a model such as the celestial sphere cannot be a hypothesis. Copernicus used his hypothesis to build a model, but they are not the same thing.

A **theory** is a system of rules and principles that can be applied to a wide variety of circumstances. A theory may have begun as one or more hypotheses, but it has been tested, expanded, and generalized. Many textbooks refer to the "Copernican theory," but some historians argue that it was not complete and had not been tested enough to be a theory. It is probably better to call it the Copernican hypothesis.

A **natural law** is a theory that has been refined, tested, and confirmed so often scientists have great confidence in it. Natural laws are the most fundamental principles of scientific knowledge. Kepler's laws are good examples.

Confidence is the key to understanding these terms. Scientists have more confidence in a theory than in a hypothesis and great confidence in a natural law. Nevertheless, scientists are not always consistent about these words. For example, Einstein's theory of relativity is much more accurate than Newton's laws of gravity and motion, but tradition dictates that no one refers to Einstein's laws of relativity and Newton's theories of gravity and motion. Darwin's theory of evolution has been tested many times, and scientists have great confidence in it, but no one refers to Darwin's law of evolution. These distinctions are subtle and sometimes depend more on custom than on levels of confidence.

A fossil of a 500-million-year-old trilobite: Darwin's theory of evolution has been tested many times and is widely accepted, but by custom it is called a theory and not a law. (From the collection of John Coolidge III)

■ **Table 4-1** | **Kepler's Laws of Planetary Motion**

I. The orbits of the planets are ellipses with the sun at one focus.

II. A line from a planet to the sun sweeps over equal areas in equal intervals of time.

III. A planet's orbital period squared is proportional to its average distance from the sun cubed:

$$P^2_{yr} = a^3_{AU}$$

The time that a planet takes to travel around the sun once is its orbital period, P, and its average distance from the sun equals the semimajor axis of its orbit, a. Kepler's third law says these two quantities are related: The orbital period squared is proportional to the semimajor axis cubed. If you measure P in years and a in astronomical units, you can summarize the third law as:

$$P^2_{yr} = a^3_{AU}$$

The subscripts are reminders that you must measure the period in years (yr) and the semimajor axis in astronomical units (AU).

You can use Kepler's third law to make simple calculations. For example, Jupiter's average distance from the sun is 5.20 AU. What is its orbital period? If a equals 5.20, then a^3 equals 140.6. The orbital period must be the square root of 140.6, which equals about 11.8 years.

You should note that Kepler's three laws are empirical. That is, they describe a phenomenon without explaining why it occurs. Kepler derived them from Tycho's extensive observations, not from any fundamental assumption or theory. In fact, Kepler never knew

that when the planet is closer to the sun and the line connecting it to the sun is shorter, the planet moves more rapidly to sweep over the same area that is swept over when the planet is farther from the sun. So the planet in Figure 4-21 (inset) would move from point A to point B in one month, sweeping over the area shown. But when the planet is farther from the sun, one month's motion would be shorter, from A' to B'.

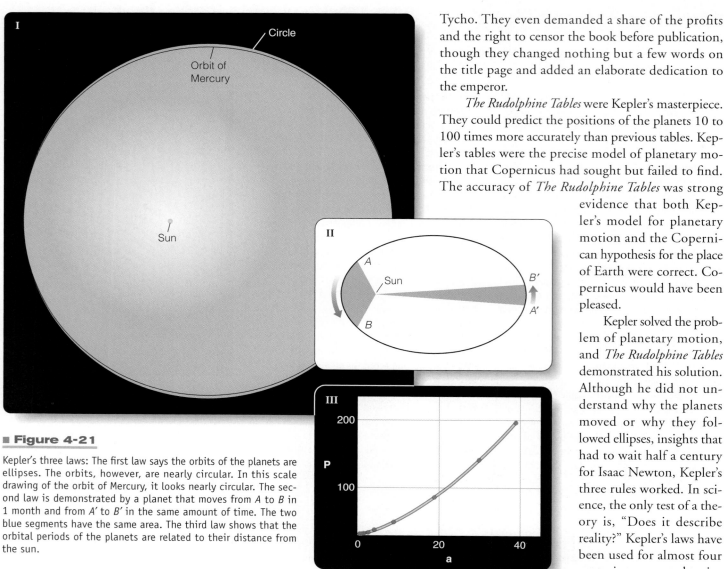

■ Figure 4-21

Kepler's three laws: The first law says the orbits of the planets are ellipses. The orbits, however, are nearly circular. In this scale drawing of the orbit of Mercury, it looks nearly circular. The second law is demonstrated by a planet that moves from *A* to *B* in 1 month and from *A'* to *B'* in the same amount of time. The two blue segments have the same area. The third law shows that the orbital periods of the planets are related to their distance from the sun.

Tycho. They even demanded a share of the profits and the right to censor the book before publication, though they changed nothing but a few words on the title page and added an elaborate dedication to the emperor.

The Rudolphine Tables were Kepler's masterpiece. They could predict the positions of the planets 10 to 100 times more accurately than previous tables. Kepler's tables were the precise model of planetary motion that Copernicus had sought but failed to find. The accuracy of *The Rudolphine Tables* was strong evidence that both Kepler's model for planetary motion and the Copernican hypothesis for the place of Earth were correct. Copernicus would have been pleased.

Kepler solved the problem of planetary motion, and *The Rudolphine Tables* demonstrated his solution. Although he did not understand why the planets moved or why they followed ellipses, insights that had to wait half a century for Isaac Newton, Kepler's three rules worked. In science, the only test of a theory is, "Does it describe reality?" Kepler's laws have been used for almost four centuries as a true description of orbital motion.

The Rudolphine Tables

In spite of Kepler's recurrent involvement with astrology and numerology, he continued to work on *The Rudolphine Tables*. At last, in 1627, they were ready, and he financed their printing himself, dedicating them to the memory of Tycho Brahe. In fact, Tycho's name appears in larger type on the title page than Kepler's own. This is especially surprising when you recall that the tables were based on the heliocentric model of Copernicus and the elliptical orbits of Kepler and not on the Tychonic system. The reason for Kepler's evident deference was Tycho's family, still powerful and still intent on protecting the memory of

what held the planets in their orbits or why they continued to move around the sun. His books are a fascinating blend of careful observation, mathematical analysis, and mystical theory.

Building Scientific Arguments

What were the main differences among *The Alfonsine Tables*, *The Prutenic Tables*, and *The Rudolphine Tables*?

Each of these tables was an expression of a different theory, and comparing theory with evidence is the heart of scientific arguments. All three of these tables predicted the motions of the sun, moon, and planets, but only *The Rudolphine Tables* proved accurate. *The Alfonsine Tables*, produced in Toledo around AD 1250, were based on the Ptolemaic model, so they were geocentric and used uniform circular motion; consequently, they were not very accurate. *The Prutenic Tables*, published in 1551, were based on the Copernican model, and so were heliocentric. But because *The Prutenic Tables* included the classical principle of uniform circular motion, they were no more accurate than *The Alfonsine Tables*. Kepler's *Rudol-*

phine Tables were Copernican in that they were heliocentric, but they included Kepler's three laws of planetary motion and consequently could predict the positions of the planets 10 to 100 times more accurately than previous tables.

One of the main reasons for the success of the Copernican hypothesis was not that it was accurate but that it was elegant. Compare the motions of the planets and the explanation of retrograde motion in the Copernican model with those in the Ptolemaic and Tychonic models. **How was the Copernican model more elegant?**

■   ■   ■

Connections: Kepler died on November 15, 1630. During his life he had been an astrologer, a mystic, a numerologist, and a seer, but he became one of the world's great astronomers. His work not only overthrew the concept of uniform circular motion that had shackled the minds of astronomers for over 2000 years but also opened the way to the empirical study of planetary motion, a study that lies at the heart of modern astronomy.

4-4 Modern Astronomy

YOU CAN DATE THE ORIGIN of modern astronomy from the 99 years between the deaths of Copernicus and Galileo (1543 to 1642); it was an age of transition. That period marked the change from the Ptolemaic model of the universe to the Copernican model with the attendant controversy over the place of the Earth. But that same period also marked a transition in the nature of astronomy in particular and science in general, a transition illustrated in the resolution of the puzzle of planetary motion. The puzzle was not solved by philosophical arguments about the perfection of the heavens or by debate over the meaning of scripture. It was solved by precise observation and careful computation, techniques that are the foundation of modern science.

The discoveries of Kepler and Galileo found acceptance in the 1600s because the world was in transition. Astronomy was not the only thing changing during this period. The Renaissance is commonly taken to be the period between 1300 and 1600, and these 99 years of astronomical history lie at the culmination of the reawakening of learning in all fields (■ Figure 4-22). Ships

■ **Figure 4-22**

The 99 years between the death of Copernicus in 1543 and the death of Galileo in 1642 marked the transition from the ancient astronomy of Ptolemy and Aristotle to the revolutionary theory of Copernicus, and, simultaneously, the invention of science as a way of understanding the world.

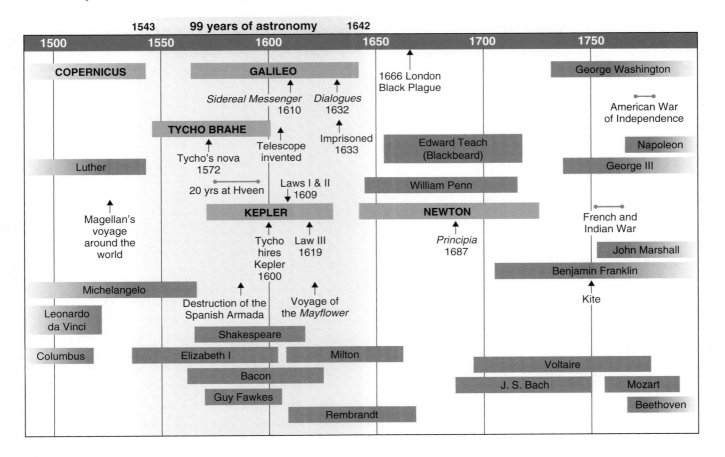

were sailing to new lands and encountering new cultures. The world was open to new ideas and new observations. Martin Luther remade religion, and other philosophers and scholars reformed their areas of human knowledge. Had Copernicus not published his hypothesis, someone else would have suggested that the universe is heliocentric. History was ready to shed the Ptolemaic system.

In addition, this period marks the beginning of the modern scientific method. Beginning with Copernicus, scientists such as Tycho, Kepler, and Galileo depended more and more on evidence, observation, and measurement. This, too, is coupled to the Renaissance and its advances in metalworking and lens making. Before the story told in this chapter began, no astronomer had looked through a telescope, because one could not be made. By 1642, not only telescopes but also other sensitive measuring instruments had transformed science into something new and precise. As you can imagine, scientists were excited by these discoveries, and they founded scientific societies that increased the exchange of observations and hypotheses and stimulated more and better work. The most important advance, however, was the application of mathematics to scientific questions. Kepler's work demonstrated the power of mathematical analysis; and, as the quality of these numerical tools improved, the progress of science accelerated. This story of the birth of modern astronomy is actually the story of the birth of modern science as well.

Building Scientific Arguments

Why was it so hard for astronomers to abandon the Ptolemaic model?

The central position of Earth and uniform circular motion were part of a paradigm—a set of ideas that people accepted as obvious. Because they all worked within that paradigm, they could not see its faults as well as you can today. It did not occur to them to test obvious ideas. Also, astronomers before the time of the Copernican Revolution were not accustomed to thinking in scientific arguments based on theory and evidence. Observations that did not fit the paradigm were not taken as seriously as they would be today. Only after the time of Galileo did scholars begin to think and reason like modern scientists. Scientific thinking is so common today, you take it for granted. When you read about new music players in a magazine like *Consumer Reports,* hear about blood tests at a trial, or try two different kinds of toothpaste to see which you like best, you are thinking scientifically and depending on evidence. Perhaps you are surprised that something as obvious as scientific thinking had to be invented.

There was yet another reason why astronomers found it hard to let go of the Ptolemaic model. The central place of the Earth had serious theological meaning. **Why did moving Earth away from the center make some astronomers hesitate to adopt the Copernican theory?**

■   ■   ■

Connections: You have read that Kepler solved the puzzle of planetary motion, but his solution seems incomplete. Why do the planets move as they do? The story will continue in the next chapter with the life of Isaac Newton. You will see how his accomplishments had their origins in the work of Galileo and led to Einstein's deeper understanding of gravity.

Summary

4-1 | The Roots of Astronomy

How did the ancients describe the place of the Earth?

■ Archaeoastronomy is the study of the astronomy of ancient peoples.

■ Many cultures around the world observed the sky and marked important alignments. Structures such as Stonehenge, Newgrange, the Sun Dagger, and Egyptian temples have astronomical alignments.

■ In most cases, ancient cultures, having no written language, left no detailed records of their astronomical beliefs.

■ Greek astronomy, derived in part from Babylon and Egypt, is better known because written documents have survived.

■ Classical philosophers accepted as a first principle that Earth was the unmoving center of the universe. Another first principle was that the heavens were perfect, so philosophers such as Plato argued that, because the sphere was the most perfect geometrical form, the heavens must be made up of spheres in uniform rotation. This led to the belief in uniform circular motion.

■ The geocentric universe became part of the teachings of the great philosopher Aristotle, who argued that the sun, moon, and stars were carried around Earth on rotating crystalline spheres.

■ Aristotle's estimate for the size of Earth was only about one-third of its true size. Eratosthenes used the well at Syene to measure the diameter of Earth and got an accurate estimate.

■ About 140 AD, Ptolemy gave mathematical form to Aristotle's model in the *Almagest*. Ptolemy preserved the principles of geocentrism and uniform circular motion, but he added epicycles, deferents, and equants to better predict the motions of the planets.

4-2 | The Copernican Revolution

How did Copernicus change the place of the Earth?

■ Copernicus devised a heliocentric model. He preserved the principle of uniform circular motion, but he argued that Earth rotates on its axis and revolves around the sun once a year. His theory was controversial because it contradicted Church teaching.

■ Copernicus published his theory in his book *De Revolutionibus* in 1543, the same year he died.

■ Because Copernicus kept uniform circular motion as part of his theory, his model did not predict the motions of the plants well, but it did offer a simple explanation of retrograde motion without using big epicycles.

■ One reason the Copernican model won converts was that it was more elegant. Venus and Mercury were treated the same as all the other planets, and the velocity of each planet was related to its distance from the sun.

Why was Galileo condemned by the Inquisition?

■ Galileo used the newly invented telescope to observe the heavens, and he recognized the significance of what he saw there. His discoveries of the phases of Venus, the satellites of Jupiter, the mountains of the moon, and other phenomena helped undermine the Ptolemaic universe.

■ Galileo based his analysis on observational evidence. In 1633, he was condemned before the Inquisition for refusing to halt his defense of Copernicanism.

4-3 | The Puzzle of Planetary Motion

How did Copernican astronomers solve the puzzle of planetary motion?

■ Tycho Brahe developed his own model in which the sun and moon circled Earth and the planets circled the sun.

■ Tycho's great contribution was to compile detailed observations over a period of 20 years, observations that were later used by Kepler.

■ Kepler inherited Tycho's books of observations in 1601 and used them to uncover three laws of planetary motion. He found that the planets follow ellipses with the sun at one focus, that they move faster when near the sun, and that a planet's orbital period squared is proportional to its orbital radius cubed.

■ Kepler's final book, *The Rudolphine Tables* (1627), combined heliocentrism with elliptical orbits and predicted the positions of the planets well.

4-4 | Modern Astronomy

■ The 99 years from the death of Copernicus to the death of Galileo marked the birth of modern science. From that time on, science depended on evidence to test theories and relied on the analytic methods first demonstrated by Kepler.

New Terms

archaeoastronomy (p. 58)

uniform circular motion (p. 60)

geocentric universe (p. 61)

eccentric (p. 62)

heliocentric universe (p. 63)

parallax (p. 64)

retrograde motion (p. 64)

epicycle (p. 65)

deferent (p. 65)

equant (p. 65)

paradigm (p. 68)

ellipse (p. 76)

semimajor axis, *a* (p. 76)

eccentricity, *e* (p. 76)

hypothesis (p. 77)

theory (p. 77)

natural law (p. 77)

Review Questions

Ace◎Astronomy™ Assess your understanding of this chapter's topics with additional quizzing and animations at **http://ace .brookscole.com/sf9**

1. What evidence is there that early human cultures observed astronomical phenomena?

2. Why did Plato propose that all heavenly motion was uniform and circular?

3. How do the epicycles of Mercury and Venus differ from those of Mars, Jupiter, and Saturn?

4. Why did Copernicus have to keep small epicycles in his model?

5. Explain how each of Galileo's telescopic discoveries contradicted the Ptolemaic theory.

6. Galileo was condemned, but Kepler, also a Copernican, was not. Why not?

7. When Tycho observed the new star of 1572, he could detect no parallax. Why did that result undermine belief in the Ptolemaic system?

8. Does Tycho's model of the universe explain the phases of Venus that Galileo observed? Why or why not?

9. How do the first two of Kepler's three laws overthrow one of the basic beliefs of classical astronomy?

10. What is the difference between a hypothesis, a theory, and a law?

11. How did *The Alfonsine Tables, The Prutenic Tables,* and *The Rudolphine Tables* differ?

12. What three astronomical objects are represented here? What are the two rings?

13. Whose observatory is shown here? Why are there no telescopes?

Discussion Questions

1. Historian of science Thomas Kuhn has said that *De Revolutionibus* was a revolution-making book but not a revolutionary book. How was it an old-fashioned, classical book?

2. Why might Tycho Brahe have hesitated to hire Kepler? Why do you suppose he appointed Kepler his scientific heir?

3. How does the modern controversy over creationism and evolution reflect two ways of knowing about the physical world?

Problems

1. Draw and label a diagram of the eastern horizon from northeast to southeast and label the rising point of the sun at the solstices and equinoxes. (See page 27 and Figure 4-1).

2. If you lived on Mars, which planets would exhibit retrograde motion? Which would never be visible as crescent phases?

3. Galileo's telescope showed him that Venus has a large angular diameter (61 seconds of arc) when it is a crescent and a small angular diameter (10 seconds of arc) when it is nearly full. Use the small-angle formula to find the ratio of its maximum distance to its minimum distance. Is this ratio compatible with the Ptolemaic universe shown on page 65?

4. Galileo's telescopes were not of high quality by modern standards. He was able to see the moons of Jupiter, but he never reported seeing features on Mars. Use the small-angle formula to find the angular diameter of Mars when it is closest to Earth. How does that compare with the maximum angular diameter of Jupiter?

5. If a planet has an average distance from the sun of 4 AU, what is its orbital period?

6. If a space probe is sent into an orbit around the sun that brings it as close as 0.5 AU and as far away as 5.5 AU, what will be its orbital period?

7. Pluto orbits the sun with a period of 247.7 years. What is its average distance from the sun?

Media Cluster

ACTIVE FIGURES

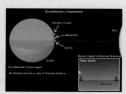

Eratosthenes' Experiment
Instead of light in a well, watch the shadow of a vertical stick from three different places on Earth. From that, you could calculate the size of Earth.

Epicycles
Watch Mars move against the background of stars as its epicycle carries it closest to Earth. You can see the classical explanation of retrograde motion.

ASTRONOMY EXERCISES

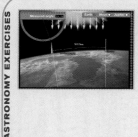

Eratosthenes' Calculations
Reproduce Eratosthenes' experiment and find the size of Earth. Then take the experiment to Earth's moon and to Jupiter.

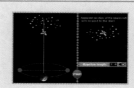

Parallax I
As you learned in this chapter, parallax is the apparent motion of an object because of the motion of the observer. In this animation, see how a spaceship appears to move against a background of stars due to Earth's motion.

Critical Inquiries for the Web

1. The trial of Galileo is an important event in the history of science. You now know, and the Roman Catholic Church now recognizes, that Galileo's view was correct, but what were the arguments on both sides of the issue as it was unfolding? Research the Internet for documents that chronicle the trial, Galileo's observations and publications, and the position of the Church. Use this information to outline the case for and against Galileo in the context of the times in which the trial occurred.

2. Take a "virtual tour" of Stonehenge by browsing the Web for information related to this megalithic site and the possibility that it was used for astronomical purposes. (Be careful to use legitimate scientific and historical websites in your survey.) Summarize the various astronomical alignments evident in the layout of the site.

Exploring *TheSky*

1. Go to Stonehenge in southern England and watch the sunrise on the morning of the summer solstice. Where does it rise on the morning of the winter solstice?

2. Observe Mars going through its retrograde motion. (*Hint:* Use **Reference Lines** under the **View** menu to turn on the ecliptic. Be sure you are in **Free Rotation** under the **Orientation** menu. Locate Mars and use the time skip arrows to watch it move.)

3. Compare the size of the retrograde loops made by Mars, Jupiter, and Saturn.

4. Can you recognize the effects of Kepler's second law in the orbital motion of any of the planets? (*Hint:* Use **3D Solar System Mode** under the **View** menu.)

5. Can you recognize the effects of Kepler's third law in the orbital motion of the planets?

Go to the Brooks/Cole Astronomy Resource Center **(http://astronomy.brookscole.com)** for critical thinking exercises, articles, and additional readings from InfoTrac College Edition, Brooks/Cole's online student library.

5 | Newton, Einstein, and Gravity

Nature and Nature's laws
lay hid in night:
God said, "Let Newton be!"
and all was light.

ALEXANDER POPE

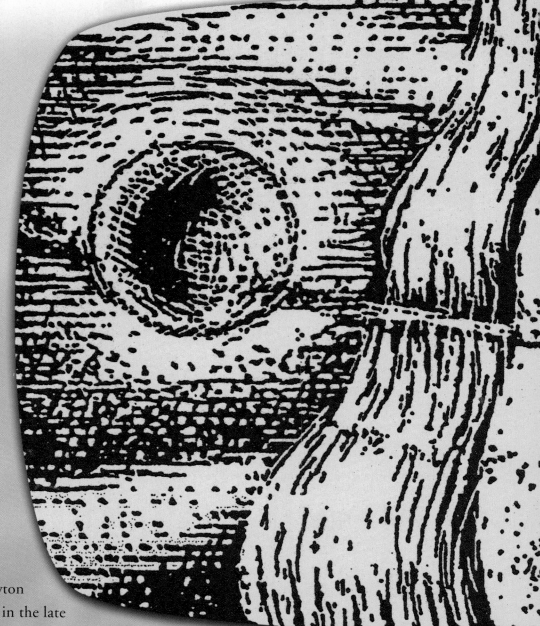

ISN'T IT WEIRD THAT Isaac Newton is said to have "discovered" gravity in the late 17th century—as if people didn't have gravity before that, as if they floated around holding onto tree branches? Newton's accomplishment is more impressive than just a discovery. Everyone experienced gravity without noticing it. Newton realized that a force had to exist that made things fall, and that changed the way people thought about nature (■ Figure 5-1). ▮ Isaac Newton was born in Woolsthorpe, England, on December 25, 1642, and on January 4, 1643. This was not a biological anomaly but a calendrical quirk. Most of Europe, following the lead of the Catholic countries, had adopted the Gregorian calendar, but Protestant England continued to use the Julian calendar. So December 25 in England was January 4 in Europe. If you take the

▮ Continued on page 86 ▮

84

The motion of the planets was not well understood until Newton discovered gravity. Much later, Einstein found an even better way to understand gravity.

Guidepost

Ace ⊛ Astronomy™ The AceAstronomy icon throughout the text indicates an opportunity for you to test yourself on key concepts and to explore animations and interactions on the AceAstronomy website at: http://ace .brookscole.com/sf9

Looking Back

In the last chapter, you saw how Renaissance astronomers struggled to understand the place of the Earth and the motion of the planets. Although Johannes Kepler seemed to solve the puzzle of planetary motion, his three laws only *described* how the planets moved. He never really understood why the planets moved in elliptical orbits.

This Chapter

Now you are ready to finish the adventure. In this chapter you will see how Galileo tried to understand motion and how Newton discovered a wonderful way to understand gravity. But then you will see how Einstein found an even better explanation for motion and gravity. Here you will find answers to five essential questions:

What happens when an object falls?

How did Newton discover gravity?

How does gravity explain orbital motion?

How does gravity explain the tides?

How did Einstein refine our understanding of motion and gravity?

Looking Ahead

Gravity rules. Every object in the universe attracts every other object. The moon orbiting Earth, matter falling into black holes, and the overall structure of the universe are dominated by gravity. The rest of this book will tell the story of matter and gravity, and you will see how modern astronomers have carried on the discoveries made by Galileo, Newton, and Einstein. The universe is a swirling waltz of matter dancing to the music of gravity, and you are going along for the ride.

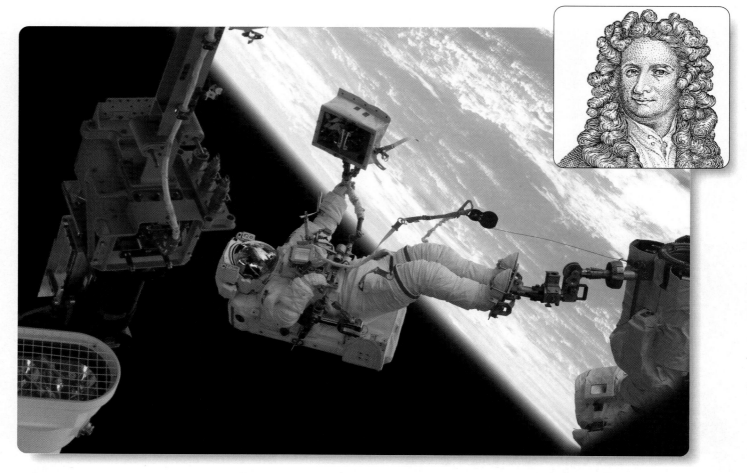

■ Figure 5-1

Space stations and astronauts, as well as planets, moons, stars, and galaxies, follow paths called orbits that are described by three simple laws of motion and a theory of gravity first understood by Isaac Newton (1642–1727). Newtonian physics is adequate to send astronauts to the moon and analyze the rotation of the largest galaxies. (NASA/JSC)

English date, then Newton was born in the same year that Galileo Galilei died.

Newton went on to become one of the greatest scientists who ever lived, but even he admitted the debt he owed to those who had studied nature before him. He said, "If I have seen farther than other men, it is because I stood on the shoulders of giants." One of those giants was Galileo. Although Galileo is remembered as the defender of Copernicanism, he was also a talented scientist who studied the motions of falling bodies.

For over two centuries following the publication of Newton's works, astronomers used Newton's laws to describe the universe. Then, early in the 20th century, Albert Einstein proposed a new way to describe gravity. The new theory did not replace Newton's laws but rather showed that they were only approximately correct and could be seriously in error under special circumstances. Einstein's theories further extend the scientific understanding of the nature of gravity. Just as Newton had stood on the shoulders of Galileo, Einstein stood on the shoulders of Newton.

5-1 Galileo and Newton

JOHANNES KEPLER DISCOVERED three laws of planetary motion, but he never understood why the planets move along their orbits. At one place in his writings, he wonders if they are pulled along by magnetic forces emanating from the rotating sun. At another place, he considers and dismisses the idea that the planets are pushed along their orbits by angels.

Newton refined Kepler's model of planetary motion but did not perfect it. In science, a model is an intellectual conception of how nature works (Window on Science 2-1). No model is perfect. Kepler's model was better than Aristotle's, but Newton improved Kepler's model by expanding it into a general theory of motion and gravity. In fact, most scientists now refer to Newton's *law* of gravity. Whether it is called a model, a theory, or a law, it is not perfect. Newton never understood what gravity was. It was as mysterious as an angel pushing the moon inward toward Earth instead of forward along the moon's orbit.

To understand science, you must understand the nature of scientific descriptions. The scientist studies nature by either creating new theories or refining old theories. Yet a theory can never be perfect, because it can never represent the universe in all its intricacies. Instead, a theory must be a limited approximation of a single phenomenon, such as orbital motion. It is fitting that Newton's discoveries all began with Kepler's fellow Copernican, Galileo.

Galileo and Motion

Even before Galileo built his first telescope, he had begun studying the motion of freely moving bodies (■ Figure 5-2). After the Inquisition condemned and imprisoned him in 1633, he continued his study of motion. He seems to have realized that he would have to understand motion before he could truly understand the Copernican system. That he was eventually able to formulate principles that later led Newton to the laws of motion and the theory of gravity is a tribute to Galileo's ability to set aside the authority of the ancients and think for himself.

The authority of the age was Aristotle, whose ideas on motion were hopelessly confused. Aristotle said that the world is made up of four classical elements: earth, water, air, and fire, each located in its proper place. The proper place for the element earth was the center of the universe, and the proper place for water was just above earth. Air and then fire formed higher layers and above them was the realm of the planets and stars. (You can see the four layers of the classical elements in the diagram at the top of page 64.) If displaced, the four elements were believed to have a natural tendency to move toward their proper place in the cosmos. Things made up mostly of air or fire—smoke, for instance—tend to move upward. Things composed mostly of earth and water—wood, rock, flesh, bone, and so on—tend to move downward toward the proper place of earth and water. According to Aristotle, objects fall downward because they are moving toward their proper place.*

Aristotle called these motions **natural motions** to distinguish them from **violent motions** produced, for instance, when you push on an object and make it move other than toward its proper place. According to Aristotle, such motions stop as soon as the force is removed. To explain how an arrow could continue to move upward even after it had left the bowstring, he said currents in the air around the arrow carried it forward even though the bowstring was no longer pushing it.

These ideas about natural and violent motion, and the necessity of a force to preserve motion, were still accepted theory in Galileo's time. In fact, in 1590, when Galileo was 26, he wrote a short work called *De Motu (On Motion)* that deals with the proper places of objects and their natural motions.

---

* This is one reason why Aristotle had to have a geocentric universe. If Earth's center had not also been the center of the cosmos, his explanation of gravity would not have worked.

■ **Figure 5-2**

Although Galileo is often associated with the telescope, as on this Italian stamp, he also made systematic studies of the motion of falling bodies and made discoveries that led to the law of inertia.

In Galileo's time and for the two preceding millennia, scholars had commonly tried to resolve problems of science by referring to authority. To analyze the flight of a cannonball, for instance, they would turn to the writings of Aristotle and other classical philosophers and try to deduce what those philosophers would have said on the subject. This generated a great deal of discussion but little real progress. Galileo broke with this tradition and conducted his own experiments.

He began by studying the motions of falling bodies, but he quickly discovered that the velocities were so great and the times so short that he could not measure them accurately. Consequently, he began using polished bronze balls rolling down gently sloping inclines. In that instance, the velocity is lower and the time longer. Using an ingenious water clock, he was able to measure the time the balls took to roll given distances down the incline, and he correctly recognized that these times are proportional to the times taken by falling bodies.

He found that falling bodies do not fall at constant rates, as Aristotle had said, but are accelerated. That is, they move faster with each passing second. Near Earth's surface, a falling object will have a velocity of 9.8 m/s (32 ft/s) at the end of 1 second, 19.6 m/s (64 ft/s) after 2 seconds, 29.4 m/s (96 ft/s) after 3 seconds, and so on. Each passing second adds 9.8 m/s (32 ft/s) to the object's velocity (■ Figure 5-3). In modern terms, this is called the **acceleration of gravity** at Earth's surface.

Galileo also discovered that the acceleration does not depend on the weight of the object. This, too, is contrary to the teachings of Aristotle, who believed that heavy objects, containing more earth and water, fall with higher velocity. Galileo found that the acceleration of a falling body is the same whether it is heavy or light. According to some accounts, he demonstrated this by dropping balls of iron and wood from the top of the Leaning Tower of Pisa to show that they would fall together and hit the ground at the same time (■ Figure 5-4a). In fact, he probably didn't perform this experiment. It would not have been conclusive anyway because of air resistance. More than 300 years later, Apollo 15 astronaut David Scott, standing on the airless moon, demonstrated

■ Figure 5-3

Galileo found that a falling object is accelerated downward. Each second, its velocity increases by 9.8 m/s (32 ft/s).

Air resistance would have slowed the wooden ball more and ruined Galileo's demonstration.

On the airless moon, there is no air resistance to slow the feather.

Hammer   Feather

■ Figure 5-4

(a) According to tradition, Galileo demonstrated that the acceleration of a falling body is independent of its weight by dropping balls of iron and wood from the Leaning Tower of Pisa. In fact, air resistance would have confused the result. (b) In a historic television broadcast from the moon on August 2, 1971, David Scott dropped a hammer and a feather at the same instant. They fell with the same acceleration and hit the surface together. (NASA)

Galileo's discovery by dropping a feather and a steel geologist's hammer. They hit the lunar surface at the same time (Figure 5-4b).

Having described natural motion, Galileo turned his attention to violent motion—that is, motion directed other than toward an object's proper place in the cosmos. He pointed out that an object rolling down an incline is accelerated and that an object rolling up the same incline is decelerated. If the incline were perfectly horizontal and frictionless, he reasoned, there could be no acceleration or deceleration to change the object's velocity, and, in the absence of friction, the object would continue to move forever. In his own words, "any velocity once imparted to a moving body will be rigidly maintained as long as the external causes of acceleration or retardation are removed."

This is contrary to Aristotle's belief that motion can continue only if a force is present to maintain it. In fact, Galileo's statement is a perfectly valid summary of the law of inertia, which became Newton's first law of motion.

Galileo published his work on motion in 1638, two years after he had become entirely blind and only four years before his death. The book was called *Mathematical Discourses and Demonstrations Concerning Two New Sciences, Relating to Mechanics and to Local Motion.* It is known today as *Two New Sciences.*

The book is a brilliant achievement for a number of reasons. To understand motion, Galileo had to abandon the authority of the ancients, devise his own experiments, and draw his own conclusions. In a sense, this was the first example of experimental science. But Galileo also had to generalize his experiments to discover how nature worked. Though his apparatus was finite and plagued by friction, he was able to imagine an infinite, frictionless plane on which a body moves at constant velocity. In his workshop, the law of inertia was obscure; but, in his imagination, it was clear and precise.

Ace Astronomy™ Log into AceAstronomy and select this chapter to see Astronomy Exercise "Falling Bodies." Try one of the most famous experiments Galileo never did.

Newton and the Laws Of Motion

Newton's three laws of motion (■ Table 5-1) are critical to understanding gravity and orbital motion. They apply to any moving object, from an automobile driving along a highway to galaxies colliding with each other.

The first law is really a restatement of Galileo's law of inertia. An object continues at rest or in uniform motion in a straight line unless acted upon by some force. Astronauts drifting in space will travel at constant rates in straight lines forever if no forces act on them (■ Figure 5-5a).

Newton's first law also explains why a projectile continues to move after all forces have been removed—for instance, why an arrow continues to move after leaving the bowstring. The object continues to move because it has momentum. You can think of an object's **momentum** as a measure of its amount of motion.

Momentum equals velocity times mass. Obviously velocity is important. A low-velocity object such as a paper clip tossed across a room has little momentum, and you could easily catch it in your hand. But the same paper clip fired at the speed of a rifle bullet would have tremendous momentum, and you would not dare try to catch it. Momentum also depends on the mass of an object (**Focus on Fundamentals 1**). To see how, imag-

■ **Table 5-1** | **Newton's Three Laws of Motion**

I. A body continues at rest or in uniform motion in a straight line unless acted upon by some force.
II. The acceleration of a body is inversely proportional to its mass, directly proportional to the force, and in the same direction as the force.
III. To every action, there is an equal and opposite reaction.

ine that, instead of tossing a paper clip, someone tosses you a bowling ball. A bowling ball contains much more mass than a paper clip and therefore has much greater momentum at the same velocity.

Newton's first law explains the consequences of the conservation of momentum. When Newton said that momentum is conserved, he meant that it remains constant until something acts to change it. A moving object has a given amount of momentum. To change that momentum, you must exert some force on the object to change either the speed or the direction. Newton's first law and the concept of momentum came from the work of Galileo.

Newton's second law of motion discusses forces, and Galileo did not talk about forces. Galileo spoke instead of accelerations. Newton saw that an acceleration is the result of a force acting on a mass (Figure 5-5b). Newton's second law is commonly written as:

$$F = ma$$

As always, you must carefully define terms when you look at an equation. An **acceleration** is a change in velocity, and a **velocity** is a directed rate of motion. Rate of motion means, of course, a speed, but the word *directed* has a special meaning. Speed itself does not have any direction associated with it, but velocity does. If you drive a car in a circle at 55 mph, your speed is constant, but your velocity is changing because your direction of motion is changing. So an object experiences an acceleration if its speed changes or if its direction of motion changes. Every

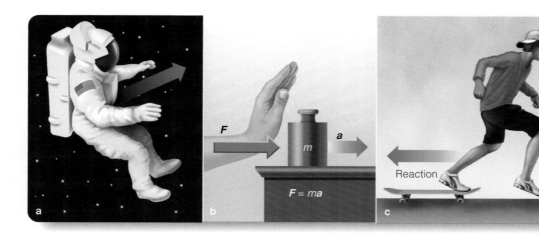

■ **Figure 5-5**

Newton's three laws of motion.

Mass

One of the most fundamental parameters in science is **mass,** a measure of the amount of matter in an object. A bowling ball, for example, contains a large amount of mass, but a child's rubber ball contains less matter than the bowling ball, and it is said to be less massive.

Mass is not the same as weight. Your weight is the force that Earth's gravity exerts on the mass of your body. Because gravity pulls you downward, you press against the bathroom scale, and you can measure your weight. Floating in space, you would have no weight at all; a bathroom scale would be useless. But your body would still contain the same amount of matter, so you would still have mass.

Sports analogies illustrate the importance of mass in dramatic ways. A bowling ball, for example, must be massive in order to have a large effect on the pins it strikes. Imagine trying to knock down all the pins with a bowling ball no more massive than a balloon. Even in space, where the bowling ball would be weightless, a low-mass bowling ball would have little effect on the pins. On the other hand, runners want track shoes that have low mass and thus are easy to move. Imagine trying to run a 100-meter dash wearing track shoes that were as massive as bowling balls. They would be very hard to move, and it would be difficult to accelerate away from the starting block. The shot put takes muscle because the shot is massive,

not because it is heavy. Imagine throwing the shot in space where it would have no weight. It would still be massive, and it would take great effort to start it moving.

Mass is a unique measure of the amount of material in an object. Using the metric system (Appendix A), mass is measured in kilograms.

Mass is not the same as weight.

100 kg

PRESSURE | **MASS** | ENERGY | TEMPERATURE AND HEAT | DENSITY

automobile has three accelerators—the gas pedal, the brake pedal, and the steering wheel. All three change the car's velocity.

In a way, the second law is just common sense; you experience it every day. The acceleration of a body is proportional to the force applied to it. This is reasonable. If you push gently against a grocery cart, you expect a small acceleration. The second law of motion also says that the acceleration depends on the mass of the body. This, too, is reasonable. If your grocery cart were filled with bricks and you pushed it gently, you would expect very little result. If it were full of inflated balloons, however, it would move easily in response to a gentle push. Finally, the second law says that the resulting acceleration is in the direction of the force. This is also what you would expect. If you push on a cart that is not moving, you expect it to begin moving in the direction you push.

The second law of motion is important because it establishes a precise relationship between cause and effect (**Window on Science 5-1**). Objects do not just move. They accelerate due to the action of a force. Moving objects do not just stop. They decelerate due to a force. Also, moving objects don't just change direction for no reason. Any change in direction is a change in velocity and requires the presence of a force. Aristotle said that objects move because they have a tendency to move. Newton said that objects move due to a specific cause, a force.

Newton's third law of motion specifies that for every action there is an equal and opposite reaction. In other words, forces must occur in pairs directed in opposite directions. For example, if you stand on a skateboard and jump forward, the skateboard will shoot away backward. As you jump, your feet exert a force

against the skateboard, which accelerates it toward the rear. But forces must occur in pairs, so the skateboard must exert an equal but opposite force on your feet that accelerates your body forward (Figure 5-5c).

Mutual Gravitation

The three laws of motion led Newton to consider the force that causes objects to fall. The first and second laws tell you that falling bodies accelerate downward because some force must be pulling downward on them. Newton wondered what that force could be.

Newton was also aware that some force has to act on the moon. The moon follows a curved path around Earth, and motion along a curved path is accelerated motion. The second law says that an acceleration requires a force, so a force must be making the moon follow that curved path.

Newton wondered if the force that holds the moon in its orbit could be the same force that causes apples to fall—gravity. He was aware that gravity extends at least as high as the tops of mountains, but he did not know if it could extend all the way to the moon. He believed that it could, but he thought it would be weaker at greater distances, and he guessed that its strength would decrease as the square of the distance increased.

This relationship, the **inverse square law,** was familiar to Newton from his work on optics, where it applied to the intensity of light. A screen set up 1 meter from a candle flame receives a certain amount of energy on each square meter. However, if that screen is moved to a distance of 2 meters, the light that originally illuminated one square meter must cover four square me-

The Unstated Assumption of Science: Cause and Effect

One of the most often used and least often stated principles of science is cause and effect, and you could argue that Newton's second law of motion was the first clear statement of the principle.

Ancient philosophers such as Aristotle argued that objects moved because of tendencies. They said that earth and water, and objects made mostly of earth and water, had a natural tendency to move toward the center of the universe. This natural motion had no cause but was inherent in the nature of the objects. But Newton's second law says $F = ma$. If an object (of mass m) changes its motion (a in the equation), then it must be acted on by a force

(*F* in the equation). Any effect *(a)* must be the result of a cause *(F)*.

The principle of cause and effect goes far beyond motion. The principle of cause and effect gives scientists confidence that every effect has a cause. Hearing loss in certain laboratory rats, color changes in certain chemical dyes, and explosions on certain stars are all effects that must have causes. All of science is focused on understanding the causes of the effects you see around you. If the universe were not rational, then you could never expect to discover causes. Newton's second law of motion was arguably the first clear statement that the behavior of the universe depends rationally on causes.

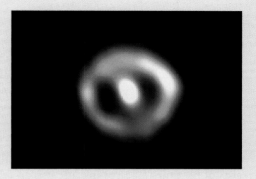

Cause and effect: Why did this star explode in 1992? There must have been a cause. (ESA/STScI and NASA)

---

ters (■ Figure 5-6). Consequently, the intensity of the light is inversely proportional to the square of the distance to the screen.

Newton made two assumptions that enabled him to predict the strength of Earth's gravity at the distance of the moon. He assumed that the strength of gravity follows the inverse square law and that the critical distance is not the distance from Earth's surface but the distance from Earth's center. Because the moon is about

■ Figure 5-6

As light radiates away from a source, it is spread thinner and becomes less intense. Here the light falling on one square meter on the inner sphere must fall on four square meters on a sphere twice as big. This shows how the intensity of light is inversely proportional to the square of the distance.

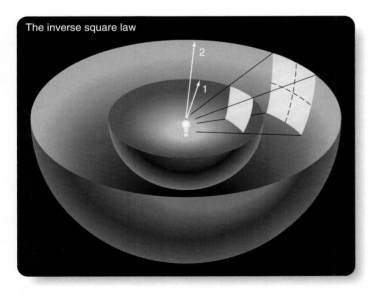

The inverse square law

60 Earth radii away, Earth's gravity at the distance of the moon should be about 60^2 times less than at Earth's surface. Instead of being 9.8 m/s² at Earth's surface, it should be about 0.0027 m/s² at the distance of the moon.

Now, Newton wondered, could this acceleration keep the moon in orbit? He knew the moon's distance and its orbital period, so he could calculate the actual acceleration needed to keep it in its curved path. The answer is 0.0027 m/s². To the accuracy of Newton's data for Earth's radius, it was exactly what his assumptions predicted. The moon is held in its orbit by gravity, and gravity obeys the inverse square law.

Newton's third law says that forces always occur in pairs, and this leads to the conclusion that gravity is mutual. If Earth pulls on the moon, then the moon must pull on Earth. Gravitation is a general property of the universe. The sun, the planets, and all their moons must also attract each other by mutual gravitation. In fact, every particle of mass in the universe must attract every other particle, and Newtonian gravity is often called universal mutual gravitation.

Clearly the force of gravity depends on mass. Your body is made of matter, and you have your own personal gravitational field. But your gravity is weak and does not attract personal satellites orbiting around you. Larger masses have stronger gravity. From an analysis of the third law of motion, Newton realized that the mass that resists acceleration in the first law must be the same as the mass associated with gravity. Then when two bodies attract one another and accelerate toward each other, the two masses must be equally involved. Newton performed precise experiments with pendulums and confirmed this equivalence between the mass that resists acceleration and the mass that causes gravity.

From this, combined with the inverse square law, he was able to write the famous formula for the gravitational force between two masses, M and m:

$$F = -\frac{GMm}{r^2}$$

The constant G is the gravitational constant, and r is the distance between the masses. The negative sign means that the force is attractive, pulling the masses together and making r decrease. In plain English, Newton's law of gravitation says: The force of gravity between two masses M and m is proportional to the product of the masses and inversely proportional to the square of the distance between them.

Newton's description of gravity was a difficult idea for physicists of his time to accept because it is an example of action at a distance. Earth and moon exert forces on each other although there is no physical connection between them. Modern scientists resolve this problem by referring to gravity as a **field.** Earth's presence produces a gravitational field directed toward Earth's center. The strength of the field decreases according to the inverse square law. Any particle of mass in that field experiences a force that depends on the mass of the particle and the strength of the field at the particle's location. The resulting force is directed toward the center of the field.

The field is an elegant way to describe gravity, but it does not say what gravity is. Later in this chapter, when you learn about Einstein's theory of curved space-time, you will get a better idea of what gravity really is.

Building Scientific Arguments

What do the words *universal* and *mutual* mean when you say "universal mutual gravitation"?
Newton argued that the force that makes an apple accelerate downward is the same as the force that accelerates the moon and holds it in its orbit. You can learn more by thinking about Newton's third law of motion, which says that forces always occur in pairs. If Earth attracts the moon, then the moon must attract Earth. That is, gravitation is *mutual* between any two objects.

Furthermore, if Earth's gravity attracts the apple and the moon, then it must attract the sun, and the third law says that the sun must attract Earth. But if the sun attracts Earth, then it must also attract the other planets and even distant stars, which, in turn, must attract the sun and each other. Step by step, Newton's third law of motion leads to the conclusion that gravitation must apply to all masses in the universe. That is, gravitation must be *universal*.

Aristotle explained gravity in a totally different way. Use Aristotle's argument that accounted for a falling apple: **Could that explanation account for a hammer falling on the surface of the moon?**

■   ■   ■

Connections: Newton first came to understand gravity by thinking about the orbital motion of the moon. Gravitation is tremendously important in astronomy. It can explain two different kinds of motions.

(5-2) Orbital Motion and Tides

ORBITAL MOTION AND TIDES are two different kinds of gravitational phenomena. As you think about the orbital motion of the moon and planets, you need to think about how gravity pulls on an object. When you think about tides, you must think about how gravity pulls on different parts of an object. Analyzing these two kinds of phenomena will give you a deeper insight into how gravity works.

Orbits

To understand how an object can orbit another object, you must see orbital motion as Newton did. Objects in orbit are falling. You can explore Newton's insight by analyzing the motion of objects orbiting Earth.

Study **Orbiting Earth** on pages 94–95 and notice three important ideas:

1 An object orbiting Earth is actually falling (being accelerated) toward Earth's center. The object continuously misses Earth because of its orbital velocity.

2 Also, notice that objects orbiting each other actually revolve around their center of mass.

3 Finally, notice the difference between closed orbits and open orbits. If you want to leave Earth never to return, you must give your spaceship a high enough velocity so it will follow an open orbit.

Ace ◗ Astronomy™ Log into AceAstronomy and select this chapter to see Astronomy Exercise "Orbital Motion." Experiment with an object in orbit.

Orbital Velocity

If you were about to ride a rocket into orbit, you would have a critical question. "How fast must I go to stay in orbit?" An object's circular velocity is the lateral velocity the object must have to remain in a circular orbit. If you assume that the mass of your spaceship is small compared with the mass of the object you expect to orbit, Earth in this case, then the circular velocity is:

$$V_c = \sqrt{\frac{GM}{r}}$$

In this formula, M is the mass of the central body in kilograms, r is the radius of the orbit in meters, and G is the gravitational constant, 6.67×10^{-11} m^3/s^{2}kg. This formula is all you need to calculate how fast an object must travel to stay in a circular orbit.

For example, how fast does the moon travel in its orbit? Earth's mass is 5.98×10^{24} kg, and the radius of the moon's orbit is 3.84×10^8 m. Then the moon's velocity is:

$$V_c = \sqrt{\frac{6.67 \times 10^{-11} \times 5.98 \times 10^{24}}{3.84 \times 10^8}} = \sqrt{\frac{39.9 \times 10^{13}}{3.84 \times 10^8}}$$

$$= \sqrt{1.04 \times 10^6} = 1020 \text{ m/s} = 1.02 \text{ km/s}$$

This calculation shows that the moon travels 1.02 km along its orbit each second. That is the circular velocity at the distance of the moon.

A satellite just above Earth's atmosphere is only about 200 km above Earth's surface, or 6578 km from Earth's center, so Earth's gravity is much stronger, and the satellite must travel much faster to stay in a circular orbit. You can use the formula above to find that the circular velocity just above Earth's atmosphere is about 7790 m/s, or 7.79 km/s. This is about 17,400 miles per hour and shows why putting satellites into Earth orbit takes such large rockets. Not only must the rocket lift the satellite above Earth's atmosphere, but the rocket must tip over and accelerate the satellite to circular velocity.

Calculating Escape Velocity

If you launch a rocket upward, it will consume its fuel in a few moments and reach its maximum speed. From that point on, it will coast upward. How fast must a rocket travel to coast away from Earth and escape? Of course, no matter how far it travels, it can never escape from Earth's gravity. The effects of Earth's gravity extend to infinity. It is possible, however, for a rocket to travel so fast initially that gravity can never slow it to a stop. That means it could leave Earth.

The escape velocity is the velocity required to escape from the surface of an astronomical body. Here you are interested in escaping from Earth or a planet; later chapters will consider the escape velocity from stars, galaxies, and even a black hole.

The escape velocity, V_e, is given by a simple formula:

$$V_e = \sqrt{\frac{2GM}{r}}$$

Here G is the gravitational constant 6.67×10^{-11} m^3/s^{2}kg, M is the mass of the astronomical body in kilograms, and r is its radius in meters. (Notice that this formula is very similar to the formula for circular velocity.)

You can find the escape velocity from Earth by looking up its mass, 5.98×10^{24} kg, and its radius, 6.38×10^6 m. Then the escape velocity is:

$$V_e = \sqrt{\frac{2 \times 6.67 \times 10^{-11} \times 5.98 \times 10^{24}}{6.38 \times 10^6}} = \sqrt{\frac{7.98 \times 10^{14}}{6.38 \times 10^6}}$$

$$= \sqrt{1.25 \times 10^8} = 11,200 \text{ m/s} = 11.2 \text{ km/s}$$

This is equal to about 25,000 miles per hour.

Notice that the formula says that the escape velocity from a body depends on both its mass and radius. A massive body might have a low escape velocity if it has a very large radius. You will meet such objects in the discussion of giant stars. On the other hand, a rather low-mass body could have a very large escape velocity if it had a very small radius, a condition you will meet in the discussion of black holes.

Circular velocity and escape velocity are two aspects of Newton's laws of gravity and motion. Once Newton understood gravity and motion, he could do what Kepler had failed to do—he could explain why the planets obey Kepler's laws of planetary motion.

Ace★Astronomy™ Log into AceAstronomy and select this chapter to see Astronomy Exercise "Escape Velocity." Try to escape from a planet.

Kepler's Laws Reexamined

Now that you understand Newton's laws, gravity, and orbital motion, you can understand Kepler's laws of planetary motion in a new way.

Kepler's first law says that the orbits of the planets are ellipses with the sun at one focus. Kepler wondered why the planets keep moving along these orbits, and now you know the answer. They move because there is nothing to slow them down. Newton's first law says that a body in motion stays in motion unless acted on by some force. The gravity of the sun accelerates the planets inward toward the sun and holds them in their orbits, but it doesn't pull backward on the planets, so they don't slow to a stop. With no friction, they must continue to move.

The orbits of the planets are ellipses because gravity follows the inverse square law. In one of his most famous problems, Newton proved that if a planet moves in a closed orbit under the influence of an attractive force that follows the inverse square law, then the planet must follow an elliptical path.

Kepler's second law says that a planet moves faster when it is near the sun and slower when it is farther away. Once again, Newton's discoveries explain why. Earlier you saw that a body moving on a frictionless surface will continue to move in a straight line until it is acted on by some force; that is, the object has momentum. In a similar way, an object set rotating on a frictionless surface will continue rotating until something acts to speed it up or slow it down. Such an object has **angular momentum,** a measure of the rotation of the body about some point. A planet circling the sun has a given amount of angular momentum; and, with no outside influences to alter its motion, it must conserve its angular momentum. That is, its angular momentum must remain constant.

Mathematically, a planet's angular momentum is the product of its mass, velocity, and distance from the sun. This explains why a planet must speed up as it comes closer to the sun along an elliptical orbit. Its angular momentum is conserved, so as its distance from the sun decreases, its velocity must increase. In the

Orbiting Earth

1 You can understand orbital motion by thinking of a cannonball falling around Earth in a circular path. Imagine a cannon on a high mountain aimed horizontally as shown at right. A little gunpowder gives the cannonball a low velocity, and it doesn't travel very far before falling to Earth. More gunpowder gives the cannonball a higher velocity, and it travels farther. With enough gunpowder, the cannonball travels so fast it never strikes the ground. Earth's gravity pulls it toward Earth's center, but Earth's surface curves away from it at the same rate it falls. It is in orbit. The velocity needed to stay in a circular orbit is called the **circular velocity**. Just above Earth's atmosphere, circular velocity is 7790 m/s or about 17,400 miles per hour, and the orbital period is about 90 minutes.

North Pole

A satellite above Earth's atmosphere feels no friction and will fall around Earth indefinitely.

Earth satellites eventually fall back to Earth if they orbit too low and experience friction with the upper atmosphere.

Ace ☯ Astronomy™

Go to AceAstronomy and click Active Figures to see "Newton's Cannon" and fire your own version of Newton's cannon.

1a A **geosynchronous satellite** orbits eastward with the rotation of Earth and remains above a fixed spot — ideal for communications and weather satellites.

A Geosynchronous Satellite

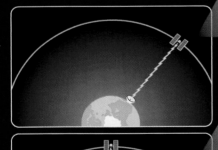

At a distance of 42,250 km (26,260 miles) from Earth's center, a satellite orbits with a period of 24 hours.

The satellite orbits eastward, and Earth rotates eastward under the moving satellite.

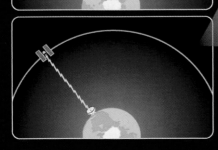

The satellite remains fixed above a spot on Earth's equator.

Ace ☯ Astronomy™

Go to AceAstronomy and click Active Figures to see "Geosynchronous Orbit" and place your own satellite into geosynchronous orbit.

1b According to Newton's first law of motion, the moon should follow a straight line and leave Earth forever. Because it follows a curve, Newton knew that some force must continuously accelerate it toward Earth — gravity. Each second the moon moves 1020 m (3350 ft) eastward and falls about 1.6 mm (about 1/16 inch) toward Earth. The combination of these motions produces the moon's curved orbit. The moon is falling.

Straight line motion of the moon

Motion toward Earth

Curved path of moon's orbit

Earth

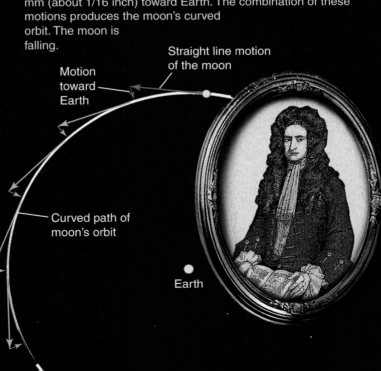

1c Astronauts in orbit around Earth feel weightless, but they are not "beyond Earth's gravity," to use a term from old science fiction movies. Like the moon, the astronauts are accelerated toward Earth by Earth's gravity, but they travel fast enough along their orbits that they continually "miss the Earth." They are literally falling around Earth. Inside or outside a spacecraft, astronauts feel weightless because they and their spacecraft are falling at the same rate. Rather than saying they are weightless, you should more accurately say they are in free fall.

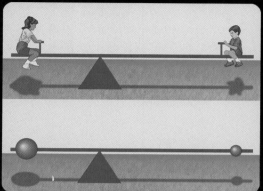

2 To be precise you should not say that an object orbits Earth. Rather the two objects orbit each other. Gravitation is mutual, and if Earth pulls on the moon, the moon pulls on Earth. The two bodies revolve around their common **center of mass,** the balance point of the system.

2a Two bodies of different mass balance at the center of mass, which is located closer to the more massive object. As the two objects orbit each other, they revolve around their common center of mass as shown at right. The center of mass of the Earth–moon system lies only 4708 km (2926 miles) from the center of Earth — inside the Earth. As the moon orbits the center of mass on one side, the Earth swings around the center of mass on the opposite side.

Center of mass

3 **Closed orbits** return the orbiting object to its starting point. The moon and artificial satellites orbit Earth in closed orbits. Below, the cannonball could follow an elliptical or a circular closed orbit. If the cannonball travels as fast as **escape velocity,** the velocity needed to leave a body, it will enter an open orbit. An **open orbit** does not return the cannonball to Earth. It will escape.

Ace ⊘ Astronomy™ Go to AceAstronomy and click Active Figures to see "Center of Mass." Change the mass ratio to move the center of mass.

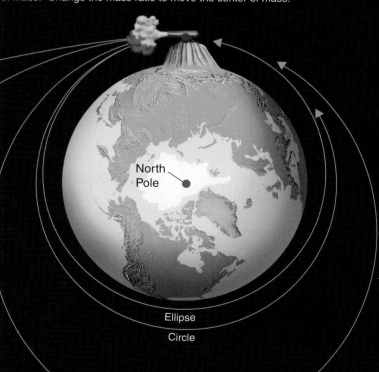

Hyberbola

A cannonball with a velocity greater than escape velocity will follow a hyperbola and escape from Earth.

Parabola

A cannonball with escape velocity will follow a parabola and escape.

North Pole

Ellipse

Circle

3a As described by Kepler's Second Law, an object in an elliptical orbit has its lowest velocity when it is farthest from Earth (apogee), and its highest velocity when it is closest to Earth (perigee). Perigee must be above Earth's atmosphere, or friction will rob the satellite of energy and it will eventually fall back to Earth.

Energy

One of the most fundamental ideas in science is **energy.** Physicists define energy as the ability to do work, but you might paraphrase that definition as the ability to produce a change. A moving body has energy called **kinetic energy.** A planet moving along its orbit, a cement truck rolling down the highway, and a golf ball sailing down the fairway all have the ability to produce a change. Imagine colliding with any of these objects!

Energy need not be represented by motion. Sunlight falling on a green plant, on photographic film, or on unprotected skin can produce chemical changes, and thus light is a form of energy. Batteries and gasoline are examples of chemical energy, and uranium fuel rods contain nuclear energy. A tank of hot water contains thermal energy.

Potential energy is the energy an object has because of its position in a gravitational field. A bowling ball on a shelf above your desk has potential energy. It is only potential, however, and doesn't produce any changes until the bowling ball descends onto your desk. Obviously the higher the shelf, the more potential energy the ball has.

Much of science is the study of how energy flows from one place to another place producing changes. A biologist might study the way a nerve cell transmits energy along its length to a muscle, while a geologist might study how energy flows as heat from Earth's interior and deforms Earth's surface. In such processes, you see energy being transformed from one state to another. Sunlight (energy) is absorbed by ocean plants and stored as sugars and starches (energy). When the plant dies, it and other ocean life are buried and become oil (energy), which gets pumped to the surface and burned in automobile engines to produce motion (energy).

Aristotle believed that all change originated in the motion of the starry sphere and flowed down to Earth. Modern science has found a more sophisticated description of the continual change you see around you. In a way, science is the study the way energy flows through the world and produces change. Energy is the pulse of the natural world.

Using the metric system (Appendix A), you express energy in **joules** (abbreviated **J**). One joule is about as much energy as that given up when an apple falls from a table to the floor.

Energy is the ability to cause change.

PRESSURE | MASS | **ENERGY** | TEMPERATURE AND HEAT | DENSITY

---

same way, the planet's velocity must decrease as its distance from the sun increases.

This conservation of angular momentum is actually a common human experience. Skaters spinning slowly can draw their arms and legs closer to their axis of rotation and, through conservation of angular momentum, spin faster (■ Figure 5-7). To slow their rotation, they again extend their arms. Similarly, divers can spin rapidly in the tuck position and slow their rotation by stretching into the extended position.

Do you like this explanation of the second law? You may not be very comfortable with it if you are not fully familiar with angular momentum. Scientists often use a well-understood idea like the conservation of angular momentum as a stepping-stone to understand a new concept, but you should hop back one stepping-stone and look at the second law again. Imagine you are in an elliptical orbit around the sun. As you round the most distant part of the ellipse, aphelion, you begin to move back closer to the sun, and the sun's gravity pulls you slightly forward in your orbit. You pick up speed as you fall closer to the sun, so, of course, you go faster as you approach the sun. As you round the closest point to the sun, perihelion, you begin to move away from the sun, and the sun's gravity pulls slightly backward on you, slowing you down as you climb away from the sun. So the second law and the conservation of angular momentum make sense when you analyze them in terms of forces and motions.

Kepler's third law is also explained by a conservation law, but in this case it is the law of conservation of energy (**Focus on Fundamentals 2**). A planet orbiting the sun has a spe-

■ Figure 5-7

Skaters demonstrate conservation of angular momentum when they spin faster by drawing their arms and legs closer to their axis of rotation.

cific amount of energy that depends only on its average distance from the sun. That energy can be divided between energy of motion and energy stored in the gravitational attraction between the planet and the sun. The energy of motion depends on how fast the planet moves, and the stored energy depends on the size of its orbit. The relation between these two kinds of energy is fixed by Newton's laws. That means there has to be a fixed relationship between the rate at which a planet moves around its orbit and the size of the orbit—between its orbital period P and the orbit's semi-major axis a. This is just Kepler's third law.

Newton's Version of Kepler's Third Law

The equation for circular velocity is actually a version of Kepler's third law, as you can prove with three lines of simple algebra. The result is one of the most useful formulas in astronomy.

The equation for circular velocity, as you have seen, is:

$$V_c = \sqrt{\frac{GM}{r}}$$

The orbital velocity of a planet is simply the circumference of its orbit divided by the orbital period:

$$V = \frac{2\pi r}{P}$$

If you substitute this for V in the equation for circular velocity and solve for P^2, you will get:

$$P^2 = \frac{4\pi^2}{GM} r^3$$

Here M is just the total mass of the system in kilograms. For a planet orbiting the sun, you can use the mass of the sun for M, because the mass of the planet is negligible compared to the mass of the sun. (In a later chapter, you will apply this formula to two stars orbiting each other, and then the mass M will be the sum of the two masses.) For a circular orbit, r equals the semi-major axis a, so this formula is a general version of Kepler's third law, $P^2 = a^3$. In Kepler's version, you used the units AU and years, but in Newton's version of the formula, you should use units of meters, seconds, and kilograms. G, of course, is the gravitational constant.

This is a powerful formula. Astronomers use it to find the masses of bodies by observing orbital motion. If, for example, you observed a moon orbiting a planet and you could measure the size of the moon's orbit, r, and the orbital period, P, you could use this formula to solve for M, the total mass of the planet plus the moon. There is no other way to find masses in astronomy, and, in later chapters, you will see this formula used to find the masses of stars, galaxies, and planets.

This discussion is a good illustration of the power of Newton's work. By carefully defining motion and gravity and by giving them mathematical expression, Newton was able to derive new truths, among them Newton's version of Kepler's third law.

His work transformed the mysterious wanderings of the planets into understandable motions that follow simple rules. In fact, his discovery of gravity explained something else that had mystified philosophers for millennia—the ebb and flow of the oceans.

Tides and Tidal Forces

Newton understood that gravity is mutual—Earth attracts the moon, and the moon attracts Earth—and that means the moon's gravity can explain the ocean tides. But Newton also realized that gravitation is universal, and that means there is much more to tides than just Earth's oceans.

Tides are caused by small differences in gravitational forces. For example, Earth's gravity attracts your body downward with a force equal to your weight. The moon is less massive than Earth and more distant, so the moon attracts your body with a force equal to roughly 0.0003 percent of your weight. You don't notice that tiny force, but Earth's oceans respond dramatically.

The side of Earth facing the moon is about 4000 miles closer to the moon than is the center of Earth. Consequently, the moon's gravity, tiny though it is at the distance of Earth, is just a bit stronger when it acts on the near side of Earth than on the center. It pulls on the oceans on the near side of Earth a bit more strongly than on Earth's center, and the oceans respond by flowing into a bulge of water on the side of Earth facing the moon. There is also a bulge on the side away from the moon, which develops because the moon pulls more strongly on Earth's center than on the far side. Thus the moon pulls Earth away from the far-side oceans, which flow into a bulge on the far side as shown at the top of ■ Figure 5-8.

You might wonder: If Earth and moon accelerate toward each other, why don't they smash together? In fact, they would smash together in just a couple weeks, except that they are moving sideways, and they keep missing. That is, they are orbiting around their common center of mass. The ocean tides are caused by the accelerations Earth and the oceans feel as they move around that center of mass.

The moon's tidal forces are not confined to water. The rocky bulk of Earth also responds to these tidal forces. Although you don't notice, Earth flexes, and the mountains and plains rise and fall by a few centimeters in response to the moon's gravitational acceleration.

You can see dramatic evidence of this effect if you watch the ocean shore for a few hours. Though Earth rotates on its axis, the tidal bulges remain fixed with respect to the moon. As the turning Earth carries you and your beach into a tidal bulge, the ocean water deepens, and the tide crawls up the sand. The tide does not so much "come in" as you are carried into the tidal bulge. Later, when Earth's rotation carries you out of the bulge, the ocean becomes shallower, and the tide falls. Because there are two bulges on opposite sides of Earth, the tides rise and fall twice a day on an ideal coast.

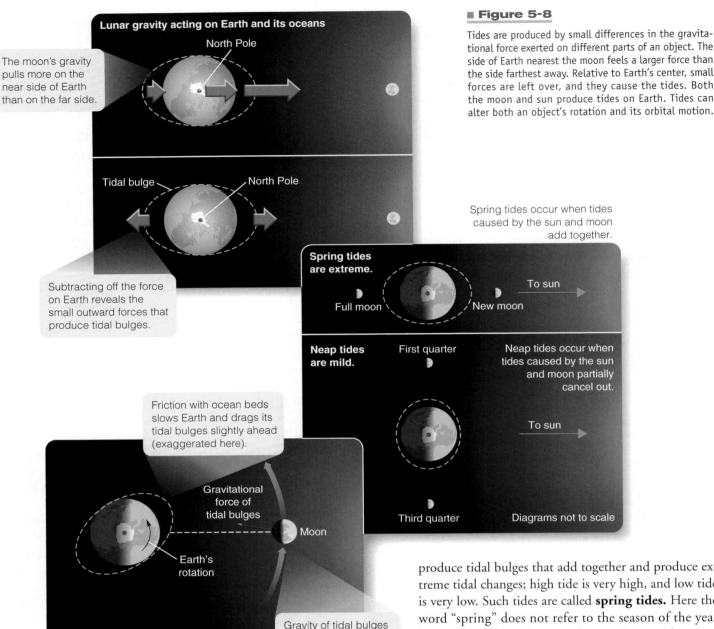

Figure 5-8

Tides are produced by small differences in the gravitational force exerted on different parts of an object. The side of Earth nearest the moon feels a larger force than the side farthest away. Relative to Earth's center, small forces are left over, and they cause the tides. Both the moon and sun produce tides on Earth. Tides can alter both an object's rotation and its orbital motion.

Lunar gravity acting on Earth and its oceans

North Pole

The moon's gravity pulls more on the near side of Earth than on the far side.

Tidal bulge

North Pole

Subtracting off the force on Earth reveals the small outward forces that produce tidal bulges.

Spring tides occur when tides caused by the sun and moon add together.

Spring tides are extreme.

Full moon

New moon

To sun

Neap tides are mild.

First quarter

Neap tides occur when tides caused by the sun and moon partially cancel out.

To sun

Third quarter

Diagrams not to scale

Friction with ocean beds slows Earth and drags its tidal bulges slightly ahead (exaggerated here).

Gravitational force of tidal bulges

Moon

Earth's rotation

Gravity of tidal bulges pulls the moon forward and alters its orbit.

In reality, the tidal cycle at any given location can be quite complex because of the latitude of the site, shape of the shore, winds, and so on. Tides in the Bay of Fundy (New Brunswick, Canada), for example, occur twice a day and can exceed 40 feet. In contrast, the northern coast of the Gulf of Mexico has only one tidal cycle a day of roughly 1 foot.

Gravity is universal, so the sun, too, produces tides on Earth. The sun is roughly 27 million times more massive than the moon, but it lies almost 400 times farther from Earth. Consequently, tides on Earth caused by the sun are less than half those caused by the moon. At new moon and at full moon, the moon and sun

produce tidal bulges that add together and produce extreme tidal changes; high tide is very high, and low tide is very low. Such tides are called **spring tides.** Here the word "spring" does not refer to the season of the year but to the rapid welling up of water. Spring tides occur twice a month, at new and full moon. At first- and third-quarter moons, the sun and moon pull at right angles to each other, and the sun's tides cancel out some of the moon's tides. These less-extreme tides are called **neap tides,** and they do not rise very high or fall very low. The word *neap* comes from an obscure Old English word, *nep,* that seems to have meant "lacking power to advance." Spring tides and neap tides are illustrated in Figure 5-8.

Galileo tried to understand tides, but not until Newton described gravity could astronomers analyze tidal forces and recognize their surprising effects. For example, the friction of the ocean waters with the ocean beds slows Earth's rotation and makes the length of a day grow by 0.0023 seconds per century. Fossils of an-

Testing a Theory by Prediction

When you read about any science, you should notice that scientific theories face in two directions. They look back into the past and explain phenomena previously observed. For example, Newton's laws of motion and gravity explained how the planets moved. But theories also face forward in that they make predictions about what you should find as you explore further. In this way, Newton's laws allowed astronomers to calculate the orbits of comets, predict their return, and eventually understand their origin.

Scientific predictions are important in two ways. First, if a theory leads to a prediction and scientists later discover the prediction was true, the theory is confirmed, and scientists gain confidence that it is a true description of nature. But predictions are important in science in a second way. Using an existing theory to make a prediction may lead to an unexplored avenue of knowledge. For example, the first theories of genetics made predictions that confirmed the genetic theory of inheritance, but those predictions also created a new understanding of how living creatures evolve.

As you read about any scientific theory, think about both what it can explain and what it can predict.

The motion of comets can be predicted using Newton's laws. (Nigel Sharp/NOAO/AURA/NSF)

---

cient corals confirm that only 900 million years ago Earth's day was 18 hours long. In addition, Earth's gravitation exerts tidal forces on the moon, and although there are no bodies of water on the moon, friction within the flexing rock has slowed the moon's rotation to the point that it now keeps the same face toward Earth.

Tidal forces can also affect orbital motion. Friction with the ocean beds drags the tidal bulges eastward out of a direct Earth–moon line. These tidal bulges contain a large amount of mass, and their gravitational field pulls the moon forward in its orbit, as shown at the bottom of Figure 5-8. As a result, the moon's orbit is growing larger by about 4 cm a year, an effect that astronomers can measure by bouncing laser beams off reflectors left on the lunar surface by the Apollo astronauts.

Newton's gravitation is much more than just the force that makes apples fall. In later chapters, you will see how tides can pull gas away from stars, rip galaxies apart, and melt the interiors of small moons orbiting near massive planets. Tidal forces produce some of the most surprising and dramatic processes in the universe.

Astronomy after Newton

Newton published his work in July 1687 in a book called *Philosophiae Naturalis Principia Mathematica* (*Mathematical Principles of Natural Philosophy*), now known simply as *Principia* (■ Figure 5-9). It is one of the most important books ever written. *Principia* (pronounced *Prin KIP ee uh*) changed astronomy, changed science, and changed the way people think about nature.

Principia changed astronomy and ushered in a new age. No longer did astronomers appeal to the whim of the gods to explain things in the heavens. No longer did they speculate on why the planets wander across the sky. *Principia* says that the motions of the heavenly bodies are governed by simple, universal rules that describe the motions of everything from planets to falling apples. Suddenly the universe was understandable in simple terms.

Newton's laws of motion and gravity made it possible for astronomers to calculate the orbits of planets and moons. Not only could they explain how the heavenly bodies move, they could predict future motions (**Window on Science 5-2**). This subject, known as gravitational astronomy, dominated

■ Figure 5-9

Newton, working from the discoveries of Galileo and Kepler, derived three laws of motion and the principle of mutual gravitation. He and some of his discoveries are honored on this English pound note. Notice the diagram of orbital motion in the background and the open copy of *Principia* in Newton's hands.

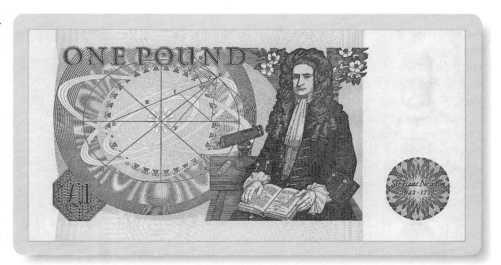

astronomy for almost 200 years and is still important. It included the calculation of the orbits of comets and asteroids and the theoretical prediction of the existence of two planets, Neptune and Pluto.

Principia also changed science in general. The works of Copernicus and Kepler had been mathematical, but no book before had so clearly demonstrated the power of mathematics as a language of precision. Newton's arguments in *Principia* were so powerful an illustration of the quantitative study of nature that scientists around the world adopted mathematics as their most powerful tool.

Also, *Principia* changed the way people thought about nature. Newton showed that the rules that govern the universe are simple. Particles move according to three rules of motion and attract each other with a force called gravity. These motions are predictable, and that makes the universe a vast machine based on a few simple rules. It is complex only in that it contains a vast number of particles. In Newton's view, if he knew the location and motion of every particle in the universe, he could, in principle, derive the past and future of the universe in every detail. This mechanical determinism has been undermined by modern quantum mechanics, but it dominated science for more than two centuries during which scientists thought of nature as a beautiful clockwork that would be perfectly predictable if they knew how all the gears meshed.

Most of all, Newton's work broke the last bonds between science and formal philosophy. Newton did not speculate on the good or evil of gravity. He did not debate its meaning. Not more than a hundred years before, scientists would have argued over the "reality" of gravity. Newton didn't care for these debates. He wrote, "It is enough that gravity exists and suffices to explain the phenomena of the heavens."

Building Scientific Arguments

How do Newton's laws of motion explain the orbital motion of the moon?

Natural laws are powerful because they explain so much. The key is often to build an argument step by step. If Earth and the moon did not attract each other, the moon would move in a straight line in accord with Newton's first law of motion and vanish into space in a few months. Instead, gravity pulls the moon toward Earth's center, and the moon accelerates toward Earth. This acceleration is just enough to pull the moon away from its straight-line motion and cause it to follow a curve around Earth. In fact, it is correct to say that the moon is falling, but because of its lateral motion it continuously misses Earth.

Every orbiting object is falling toward the center of its orbit but is moving laterally fast enough to compensate for the inward motion, and it follows a curved orbit. That is an elegant argument, but it raises a question: **How can astronauts float inside spacecraft in a "weightless" state? Why might "free fall" be a more accurate term?**

■   ■   ■

Connections: Newton's laws of motion and gravitation are critical in astronomy not only because they describe orbital motion and tides but also because they describe nearly all interactions between astronomical bodies. Newton made other discoveries, and you will meet those in later chapters. Now you are ready to jump forward a bit more than two centuries to see how Einstein described gravity in a new and powerful way.

(5-3) Einstein and Relativity

IN THE EARLY YEARS of the last century, Albert Einstein (1879–1955) (■ Figure 5-10) began thinking about how motion and gravity interact. He soon gained international fame by showing that Newton's laws of motion and gravity were only partially correct. The revised theory became known as the theory of relativity. As you will see, there are really two theories of relativity.

Special Relativity

Einstein began by thinking about how moving observers see events around them. His analysis led him to the first postulate of relativity, also known as the principle of relativity:

> **First postulate** (the principle of relativity): Observers can never detect their *uniform* motion except relative to other objects.

You may have experienced the first postulate while sitting on a train in a station. You suddenly notice that the train on the next track has begun to creep out of the station. However, after several moments you realize that it is your own train that is moving and that the other train is still motionless on its track.

■ Figure 5-10

Einstein has become a symbol of the brilliant scientist, but his fame began when he was a young man and thought deeply about the nature of motion. That led him to revolutionary insights into the meaning of space and time and a new understanding of gravity.

Consider another example. Suppose you are floating in a spaceship in interstellar space and another spaceship comes coasting by (■ Figure 5-11a). You might conclude that it is moving and you are not, but someone in the other ship might be equally sure that you are moving and it is not. Of course, you could just look out a window and compare the motion of your spaceship with a nearby star, but that just expands the problem. Which is moving, your spaceship or the star? The principle of relativity says that there is no experiment you can perform to decide which ship is moving and which is not. This means that there is no such thing as absolute rest—all motion is relative.

Because neither you nor the people in the other spaceship could perform any experiment to detect your absolute motion through space, the laws of physics must have the same form in both spaceships. Otherwise, experiments would give different results in the two ships, and you could decide who was moving. So, a more general way of stating the first postulate refers to these laws of physics:

First postulate (alternate version): The laws of physics are the same for all observers, no matter what their motion, so long as they are not *accelerated.*

The words *uniform* and *accelerated* are important. If either spaceship were to fire its rockets, then its velocity would change. The crew of that ship would know it because they would feel the acceleration pressing them into their couches. Accelerated motion, therefore, is different—the pilots of the spaceships can always tell which ship is accelerating and which is not. The postulates of relativity discussed here apply only to observers in *uniform* motion, which means *unaccelerated* motion. That is why the theory is called **special relativity.**

The first postulate led to the conclusion that the speed of light must be constant for all observers. No matter how you are moving, your

measurement of the speed of light has to give the same result (Figure 5-11b). This became the second postulate of special relativity:

Second postulate: The velocity of light is constant and will be the same for all observers independent of their motion relative to the light source.

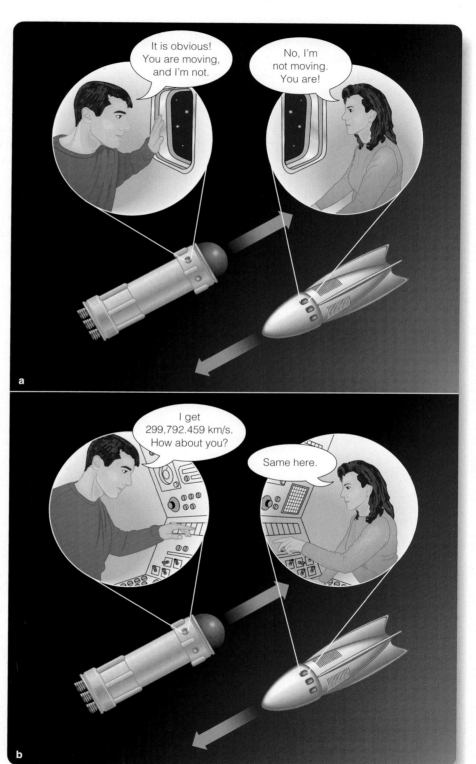

■ Figure 5-11

(a) The principle of relativity says that observers can never detect their uniform motion, except relative to other objects. Thus, neither of these travelers can decide who is moving and who is not. (b) If the velocity of light depended on the motion of the observer through space, then these travelers could perform measurements inside their spaceships to discover who was moving. If the principle of relativity is correct, then the velocity of light must be a constant for all observers.

Remember, this is required by the first postulate; if the velocity of light were not constant, then the pilots of the spaceships could decide who was moving.

Once Einstein had accepted the basic postulates of relativity, he was led to some startling discoveries. Newton's laws of motion and gravity worked well as long as distances were small and velocities were low. But when you begin to think of very large distances or very high velocities, Newton's laws are no longer adequate to describe what happens. Instead, you must use relativistic physics. For example, special relativity shows that the observed mass of a moving particle depends on its velocity. The higher the velocity, the greater the mass of the particle. This is not significant at low velocities, but it becomes very important as the velocity approaches the velocity of light. Such increases in mass are observed whenever physicists accelerate atomic particles to high velocities (■ Figure 5-12).

This discovery led to yet another insight. The relativistic equations that describe the energy of a moving particle predict that the energy of a motionless particle is not zero. Rather, its energy at rest is m_0c^2. This is of course the famous equation:

$$E = m_0c^2$$

The c is the speed of light, and the m_0 is the mass of the particle when it is at rest. This simple formula suggests that mass and energy are related, and you will see in later chapters how nature can convert one into the other inside stars.

For example, suppose that you convert 1 kg of matter into energy. You must express the velocity of light as 3×10^8 m/s, and your result is 9×10^{16} joules (J) (approximately equal to a 20-megaton nuclear bomb). (Recall that a joule is a unit of energy roughly equivalent to the energy given up when an apple falls from a table to the floor.) This simple calculation shows that the energy equivalent of even a small mass is very large.

■ **Figure 5-12**

The observed mass of moving electrons depends on their velocity. As the ratio of their velocity to the velocity of light, v/c, gets larger, the mass of the electrons in terms of their mass at rest, m/m_0, increases. Such relativistic effects are quite evident in particle accelerators, which accelerate atomic particles to very high velocities.

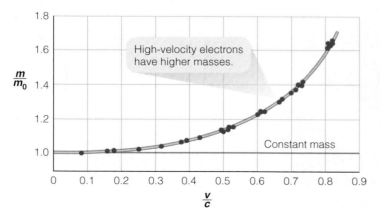

Other relativistic effects include the slowing of moving clocks and the shrinkage of lengths measured in the direction of motion. A detailed discussion of the major consequences of the special theory of relativity is beyond the scope of this book. Instead, let's consider Einstein's second advance, the general theory.

The General Theory of Relativity

In 1916, Einstein published a more general version of the theory of relativity that dealt with accelerated as well as uniform motion. This **general theory of relativity** contained a new description of gravity.

Einstein began by thinking about observers in accelerated motion. Imagine an observer sitting in a windowless spaceship. Such an observer cannot distinguish between the force of gravity and the inertial forces produced by the acceleration of the spaceship (■ Figure 5-13). This led Einstein to conclude that gravity and acceleration are related, a conclusion now known as the equivalence principle:

> **Equivalence principle:** Observers cannot distinguish locally between inertial forces due to acceleration and uniform gravitational forces due to the presence of a massive body.

This should not surprise you. Earlier in this chapter you read that Newton concluded that the mass that resists acceleration is the same as the mass that exerts gravitational forces. He even performed an elegant experiment with pendulums to test the equivalence of the mass related to motion and the mass related to gravity.

The importance of the general theory of relativity lies in its description of gravity. Einstein concluded that gravity, inertia, and acceleration are all associated with the way space is related to time. This relation is often referred to as curvature, and a one-line description of general relativity explains a gravitational field as a curved region of space-time:

> **Gravity according to general relativity:** Mass tells space-time how to curve, and the curvature of space-time (gravity) tells mass how to accelerate.

So you feel gravity because Earth's mass causes a curvature of space-time. The mass of your body responds to that curvature by accelerating toward Earth's center. According to general relativity, all masses cause curvature, and the larger the mass, the more severe the curvature. That's gravity.

Confirmation of the Curvature of Space-Time

Einstein's general theory of relativity has been confirmed by a number of experiments, but two are worth mentioning here because they were among the first tests of the theory. One involves Mercury's orbit, and the other involves eclipses of the sun.

Johannes Kepler understood that the orbit of Mercury is elliptical, but only since 1859 have astronomers known that the

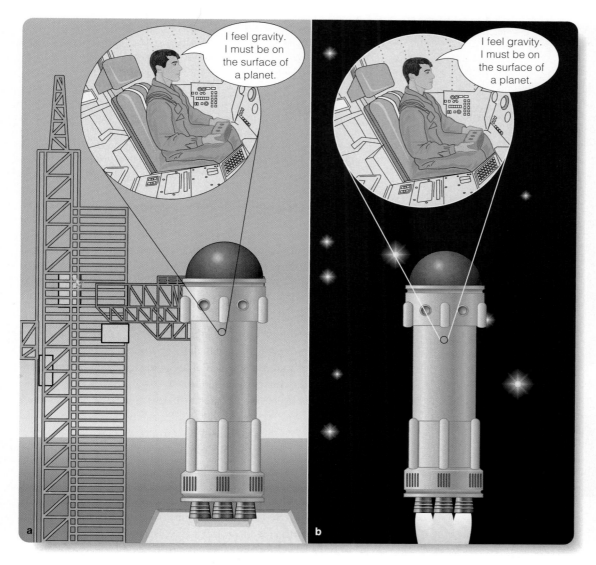

■ **Figure 5-13**

(a) An observer in a closed spaceship on the surface of a planet feels gravity. (b) In space, with the rockets smoothly firing and accelerating the spaceship, the observer feels inertial forces that are equivalent to gravitational forces.

long axis of the orbit sweeps around the sun in a motion called precession (■ Figure 5-14). The total observed precession is 5600.73 seconds of arc per century (as seen from Earth), which equals about 1.5° per century. This precession is produced by the gravitation of Venus, Earth, and the other planets. However, when astronomers used Newton's description of gravity, they calculated that the precession should amount to only 5557.62 seconds of arc per century. So Mercury's orbit is advancing 43.11 seconds of arc per century faster than Newton's law predicts.

This is a tiny effect. Each time Mercury returns to perihelion, its closest point to the sun, it is about 29 km (18 miles) past the position predicted by Newton's laws. This is such a small distance compared with the planet's diameter of 4850 km that it could never have been detected had it not been cumulative. Each orbit, Mercury gains 29 km, and in a century it gains over 12,000 km—more than twice its own diameter. This tiny effect, called the advance of perihelion of Mercury's orbit, accumulated into a serious discrepancy in the Newtonian description of the universe.

The advance of perihelion of Mercury's orbit was one of the first problems to which Einstein applied the principles of general relativity. First he calculated how much the sun's mass curves space-time in the region of Mercury's orbit, and then he calculated how Mercury moves through the space-time. The theory predicted that the curved space-time should cause Mercury's orbit to advance by 43.03 seconds of arc per century, well within the observational accuracy of the excess (Figure 5-14b).

Einstein was elated with this result, and he would be even happier with modern studies that have shown that Mercury, Venus, Earth, and even Icarus, an asteroid that comes close to the sun, have orbits observed to be slipping forward due to the curvature of space-time near the sun (■ Table 5-2). This same effect has been detected in pairs of stars that orbit each other.

A second test of the curvature of space-time was directly related to the motion of light through the curved space-time near the sun. The equations of general relativity predicted that light would be deflected by curved space-time, just as a rolling golf

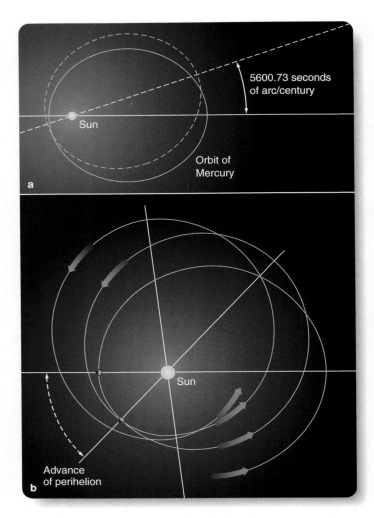

■ Figure 5-14

(a) Mercury's orbit precesses 5600.73 seconds of arc per century—43.11 seconds of arc per century faster than predicted by Newton's laws. (b) Even when you ignore the influences of the other planets, Mercury's orbit is not a perfect ellipse. Curved space-time near the sun distorts the orbit from an ellipse into a rosette. The advance of Mercury's perihelion is exaggerated about a million times in this figure.

ball is deflected by undulations in a putting green. Einstein predicted that starlight grazing the sun's surface would be deflected by 1.75 seconds of arc (■ Figure 5-15). Starlight passing near the sun is normally lost in the sun's glare, but during a total solar eclipse stars beyond the sun could be seen. As soon as Einstein published his theory, astronomers rushed to observe such stars and test the curvature of space-time.

The first solar eclipse following Einstein's announcement in 1916 was June 8, 1918. It was cloudy at some observing sites, and results from other sites were inconclusive. The next occurred on May 29, 1919, only months after the end of World War I, and was visible from Africa and South America. British teams went to both Brazil and Príncipe, an island off the coast of Africa. Months before the eclipse, they photographed that part of the sky where the sun would be located during the eclipse and mea-

■ Table 5-2 | Precession in Excess of Newtonian Physics

PLANET	Observed Excess Precession (Sec of arc per century)	Relativistic Prediction (Sec of arc per century)
Mercury	43.11 ± 0.45	43.03
Venus	8.4 ± 0.48	8.6
Earth	5.0 ± 1.2	3.8
Icarus	9.8 ± 0.8	10.3

sured the positions of the stars on the photographic plates. Then, during the eclipse, they photographed the same star field with the eclipsed sun located in the middle. After measuring the plates, they found slight changes in the positions of the stars. During the eclipse, the positions of the stars on the plates were shifted outward, away from the sun (■ Figure 5-16). If a star had been located at the edge of the solar disk, it would have been shifted outward by about 1.8 seconds of arc. This represents good agreement with the theory's prediction.

This test has been repeated at many total solar eclipses since 1919, with similar results. The most accurate results were obtained in 1973 when a Texas–Princeton team measured a deflection of 1.66 ± 0.18 seconds of arc—good agreement with Einstein's theory.

The general theory of relativity is critically important in modern astronomy. You will meet it again in the discussion of black holes, distant galaxies, and the big bang universe. The theory revolutionized modern physics by providing a theory of gravity based on the geometry of curved space-time. Thus,

■ Figure 5-15

Like a depression in a putting green, the curved space-time near the sun deflects light from distant stars and makes them appear to lie slightly farther from the sun than their true positions.

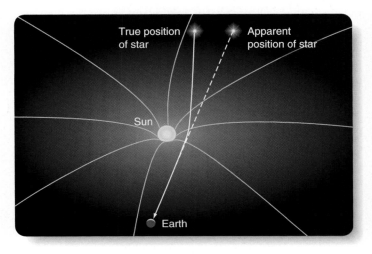

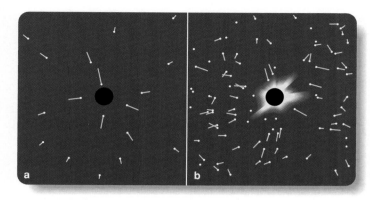

■ Figure 5-16

(a) Schematic drawing of the deflection of starlight by the sun's gravity. Dots show the true positions of the stars as photographed months before the eclipse. Lines point toward the positions of the stars during the eclipse. (b) Actual data from the eclipse of 1922. Random uncertainties of observation cause some scatter in the data, but in general the stars appear to move away from the sun by 1.77 seconds of arc at the edge of the sun's disk. The deflection of stars is magnified by a factor of 2300 in both (a) and (b).

Galileo's inertia and Newton's mutual gravitation are shown to be not just descriptive rules but fundamental properties of space and time.

Building Scientific Arguments

What does the equivalence principle tell you?
The equivalence principle says that there is no observation you can make inside a closed spaceship to distinguish between uniform acceleration and gravitation. Of course, you could open a window and look outside, but then you would no longer be in a closed spaceship. As long as you make no outside observations, you can't tell whether your spaceship is firing its rockets and accelerating through space or resting on the surface of a planet where gravity gives you weight.

Einstein took the equivalence principle to mean that gravity and acceleration through space-time are somehow related. The general theory of relativity gives that relationship mathematical form and shows that gravity is really a distortion in space-time that physicists refer to as curvature. Consequently, you can say "mass tells space-time how to curve, and space-time tells mass how to move." The equivalence principle led Einstein to an explanation for gravity.

Einstein began his work by thinking carefully about common things such as what you feel when you are moving uniformly or accelerating. This led him to deep insights now called postulates. Special relativity sprang from two postulates. **Why does the second postulate have to be true if the first postulate is true?**

■   ■   ■

Connections: Your study of the origin of astronomy began with the builders of Stonehenge and reaches the modern day with Einstein's general theory of relativity. Now that you have seen where astronomy came from, you are ready to see how it helps you understand your place the universe. Your first question could be "How do astronomers get information?" The answer involves the astronomer's most basic tool, the telescope, and that is the subject of the next chapter.

Study and Review Tools

Summary

5-1 | Galileo and Newton

What happens when an object falls?
- Galileo found that a falling object is accelerated; that is, it falls faster and faster with each passing second. The rate at which it accelerates, termed the acceleration of gravity, is 9.8 m/s² (32 ft/s²) at Earth's surface and does not depend on the weight of the object.
- According to tradition, Galileo demonstrated this by dropping balls of iron and wood from the Leaning Tower of Pisa to show that they would fall together.
- Galileo stated the law of inertia. In the absence of friction, a moving body on a horizontal plane will continue moving forever.

How did Newton discover gravity?
- Newton adopted Galileo's law of inertia as the first of three laws of motion. The second law says that a change in motion, an acceleration, must be caused by a force, and the third law says that forces occur in pairs acting in opposite directions.
- Newton realized that the curved path of the moon meant that it was being accelerated away from a straight-line path, and that required the presence of a force—gravity.
- From his mathematical analysis, Newton was able to show that the force of gravity between two masses is proportional to the product of their masses and inversely proportional to the square of the distance between them.

5-2 | Orbital Motion and Tides

How does gravity explain orbital motion?
- An object in space near Earth would move along a straight line and quickly leave Earth were it not for Earth's gravity accelerating the object toward Earth's center and forcing it to follow a curved path, an orbit. Objects in orbit around Earth are falling (being accelerated) toward Earth's center.

- If there is no friction, the object will fall around its orbit forever.
- Newton's laws explain Kepler's three laws of planetary motion. The planets follow elliptical orbits because gravity follows the inverse square law. The planets move faster when closer to the sun and slower when farther away because they conserve angular momentum. A planet's orbital period squared is proportional to its orbital radius cubed because the moving planet conserves energy.
- An object in a closed orbit follows an elliptical path. A circle is just a special ellipse of zero eccentricity.
- If a body's velocity equals or exceeds the escape velocity, V_e, it will follow a parabola or hyperbola. These orbits are termed "open" because the object never returns to its starting place.

How does gravity explain the tides?
- Tides are caused by differences in the force of gravity acting on different parts of a body.
- Tides on Earth occur because the moon's gravity pulls more strongly on the near side of Earth than on the center of Earth. A tidal bulge occurs on the far side because the moon's gravity is slightly weaker there than at the center of Earth.
- Tides produced by the moon combine with tides produced by the sun to cause extreme tides (called spring tides) at new and full moons. The moon and sun work against each other to produce less-extreme tides (neap tides) at quarter moons.
- Friction from tides can slow the rotation of a rotating world, and the gravitational pull of tidal bulges can make orbits change slowly.

5-3 | Einstein and Relativity

How did Einstein refine the understanding of motion and gravity?
- Einstein published two theories that extended Newton's laws of motion and gravity, the special theory of relativity and the general theory of relativity.
- The special theory of relativity says that uniform (unaccelerated) motion is relative. Observers cannot detect their uniform motion through space except relative to outside objects. This is known as the first postulate.
- This leads to the second postulate: the speed of light is a constant for all observers.
- A consequence of special relativity is that mass and energy are related.
- The general theory of relativity says that a gravitational field is a curvature of space-time caused by the presence of a mass. For example, Earth's mass curves space-time, and the mass of your body responds to that curvature by accelerating toward Earth's center.
- The curvature of space-time was confirmed by the slow advance in perihelion of the orbit of Mercury and by the deflection of starlight observed during a 1919 total solar eclipse.

New Terms

natural motion (p. 87)	angular momentum (p. 93)
violent motion (p. 87)	circular velocity (p. 94)
acceleration of gravity (p. 87)	geosynchronous satellite (p. 94)
momentum (p. 89)	center of mass (p. 95)
acceleration (p. 89)	closed orbit (p. 95)
velocity (p. 89)	escape velocity (p. 95)
mass (p. 90)	open orbit (p. 95)
inverse square law (p. 90)	energy (p. 96)
field (p. 92)	kinetic energy (p. 96)

potential energy (p. 96)	neap tide (p. 98)
joule (J) (p. 96)	special relativity (p. 101)
spring tide (p. 98)	general theory of relativity (p. 102)

Review Questions

Ace ⊘ Astronomy™ Assess your understanding of this chapter's topics with additional quizzing and animations at **http://ace .brookscole.com/sf9**

1. Why wouldn't Aristotle's explanation of gravity work if Earth was not the center of the universe?
2. According to the principles of Aristotle, what part of the motion of a baseball pitched across the home plate is natural motion? What part is violent motion?
3. If you drop a feather and a steel hammer at the same moment, they should hit the ground at the same instant. Why doesn't this work on Earth, and why does it work on the moon?
4. What is the difference between mass and weight? between speed and velocity?
5. Why did Newton conclude that some force had to pull the moon toward Earth?
6. Why did Newton conclude that gravity has to be mutual and universal?
7. How does the concept of a field explain action at a distance? Name another kind of field also associated with action at a distance.
8. Why can't a spacecraft go "beyond Earth's gravity"?
9. What is the center of mass of the Earth–moon system? Where is it?
10. How do planets orbiting the sun and skaters doing a spin conserve angular momentum?
11. Why is the period of an open orbit undefined?
12. How does the first postulate of special relativity imply the second?
13. When you ride a fast elevator upward, you feel slightly heavier as the trip begins and slightly lighter as the trip ends. How is this phenomenon related to the equivalence principle?
14. From your knowledge of general relativity, would you expect radio waves from distant galaxies to be deflected as they pass near the sun? Why or why not?
15. Why can the object shown at the right be bolted in place and used 24 hours a day without adjustment?
16. Why is it a little bit misleading to say that this astronaut is weightless?

Larry Mulvehill/The Image Works

(NASA/JSC)

Discussion Questions

1. How did Galileo idealize his inclines to conclude that an object in motion stays in motion until it is acted on by some force?
2. Give an example from everyday life to illustrate each of Newton's laws.
3. People who lived before Newton may not have believed in cause and effect as strongly as you do. How do you suppose they saw their daily lives?

Problems

1. Compared with the strength of Earth's gravity at its surface, how much weaker is gravity at a distance of 10 Earth radii from Earth's center? at 20 Earth radii?
2. Compare the force of lunar gravity on the surface of the moon with the force of Earth's gravity at Earth's surface.
3. If a small lead ball falls from a high tower on Earth, what will be its velocity after 2 seconds? after 4 seconds?
4. What is the circular velocity of an Earth satellite 1000 km above Earth's surface? (*Hint:* Earth's radius is 6380 km.)

5. What is the circular velocity of an Earth satellite 36,000 km above Earth's surface? What is its orbital period? (*Hint:* Earth's radius is 6380 km.)

6. What is the orbital period of an imaginary satellite orbiting just above Earth's surface? Ignore friction with the atmosphere.

7. Repeat the previous problem for Mercury, Venus, the moon, and Mars.

8. Describe the orbit followed by the slowest cannonball on page 94 on the assumption that the cannonball could pass freely through Earth. (Newton got this problem wrong the first time he tried to solve it.)

9. If you visited an asteroid 30 km in radius with a mass of 4×10^{17} kg, what would be the circular velocity at its surface? A major league fastball travels about 90 mph. Could a good pitcher throw a baseball into orbit around the asteroid?

10. What is the orbital period of a satellite orbiting just above the surface of the asteroid in Problem 9?

11. What would be the escape velocity at the surface of the asteroid in Problem 9? Could a major league pitcher throw a baseball off of the asteroid?

Media Cluster

ACTIVE FIGURES

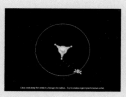

Geosynchronous Orbit
Can you put your satellite into geosynchronous orbit so that it can communicate with a ground antenna 24 hours a day? Can you move it from one antenna to the next?

Center of Mass
Adjust the masses of the stars in a binary star system and see how the orbits change. Do the sizes of the orbits correctly reflect the ratio of the masses?

ASTRONOMY EXERCISES

Falling Bodies
According to tradition, Galileo demonstrated that the acceleration of a falling body is independent of its weight by dropping various objects from the Leaning Tower of Pisa. You can repeat his experiment with this interactive exercise.

Orbital Motion
Explore how the orbit of a hypothetical planet is affected by changing the mass of the sun, distance to the sun, and eccentricity of orbit in this animation.

Escape Velocity
Escape velocity is the initial velocity an object needs to escape from the surface of a celestial body. See if you can determine the correct escape velocity for the rocket in this animation.

Critical Inquiries for the Web

1. Einstein's general theory of relativity predicts the curvature of space-time, but here on Earth you have little opportunity to observe such effects. Find an astronomical situation in which space-time curvature is evident from observations and describe the effect of the curvature on what astronomers see when they view these objects.

2. Communications satellites are obvious uses of the geosynchronous orbit, but can you think of other uses for such orbits? Find an Internet site that uses or displays information gleaned from geosynchronous orbit that provides a useful service.

 Go to the Brooks/Cole Astronomy Resource Center **(http://astronomy.brookscole.com)** for critical thinking exercises, articles, and additional readings from InfoTrac College Edition, Brooks/Cole's online student library.

6 | Light and Telescopes

DO YOU ENJOY BEING outdoors on a clear, starry night? Astronomers appreciate the beauty of the night like everyone else, but they have an additional insight. They know that starlight is going to waste. Every night, light from the stars falls on trees, oceans, roofs, and parking lots, and it is all wasted. To an astronomer, nothing is so precious as starlight. It is the only link to the sky, and the astronomer's quest is to gather as much starlight as possible and extract from it the secrets of the stars. ▌ The telescope is the emblematic tool of the astronomer because its purpose is to gather and concentrate light for analysis. Nearly all of the interesting objects in the sky are faint sources of light, so modern astronomers are driven to build the largest possible telescopes to gather the maximum amount of light (▪ Figure 6-1).

▌ Continued on page 110 ▌

Guidepost

Looking Back

In the early chapters of this book, you looked at the sky the way ancient astronomers did, with the unaided eye. In the last chapter, you got a glimpse through Galileo's telescope, and it revealed astonishing things about Jupiter, the moon, and Venus. Now it is time to examine the tools of the modern astronomer.

This Chapter

You should begin by noting that these tools gather and focus light and its related forms of radiation. For that reason, the first of the essential questions is about mysterious light:

What is light?

How do telescopes work, and how are they limited?

What kind of instruments do astronomers use to record and analyze light?

Why do astronomers use radio telescopes?

Why must some telescopes go into space?

As you answer these questions, you will discover how important it is to understand light in all its forms.

Looking Ahead

Astronomy is almost entirely an observational science. Astronomers cannot visit distant galaxies and far-off worlds, so they must observe using astronomical telescopes. Twenty chapters remain, and every one will discuss information gathered by telescopes.

Ace ☯Astronomy™ The AceAstronomy icon throughout the text indicates an opportunity for you to test yourself on key concepts and to explore animations and interactions on the AceAstronomy website at: http://ace .brookscole.com/sf9

Light-gathering optical surface

Telescope technician

■ **Figure 6-1**

Astronomical telescopes are often very large to gather large amounts of starlight. The Northern Gemini Telescope stands over 19 m (60 ft) high when pointed straight up. Its main mirror is 8.1 m (26.5 ft) in diameter—larger than some classrooms. The dome of this telescope is shown at the left of the photo on the opening page of this chapter. (NOAO/AURA/NSF)

If you wish to gather visible light, a normal telescope will do; but, as you will soon see, visible light is only one kind of radiation from the stars. Astronomers can extract information from other forms of radiation by using specialized telescopes. Radio telescopes, for example, provide an entirely different view of the sky. Some of these specialized telescopes can be used from Earth's surface, but some must go into orbit above Earth's atmosphere. Telescopes that observe X rays, for instance, must be placed in space.

As you study the sophisticated telescopes and instruments that modern astronomers use, keep in mind Robert Frost's suggestion: In every town, someone should keep a telescope. Astronomy is more than technology and scientific analysis. It tells us what we are, and every town should have a telescope to keep us looking upward.

6-1 Radiation: Information from Space

JUST AS A BOOK on bread baking might begin with a discussion of flour, this chapter on telescopes begins with a discussion of light—not just visible light, but the entire range of radiation from the sky.

Light as a Wave and a Particle

If you have admired the colors in a soap bubble, you have seen light behave as a wave. But when light enters the light meter on your camera, it behaves as a particle. How it behaves depends on how you observe it—it is both wave and particle.

You experience waves whenever you hear sound. Sound waves are a mechanical disturbance that travels through the air from source to ear. Sound requires a medium; so, on the moon, where there is no air, there can be no sound. In contrast, light is made up of electric and magnetic fields that can travel through empty space. Unlike sound, light does not require a medium, and so it can travel through a perfect vacuum. There is no sound on the moon, but there is plenty of sunlight.

Light is merely one form of radiation, called **electromagnetic radiation** because it is associated with changing electric and magnetic fields that travel through space and transfer energy from one place to another. When light enters your eye, the fluctuating electric and magnetic fields carry energy that stimulates nerve endings, and you see light.

The oscillating electric and magnetic fields that constitute electromagnetic radiation move through space at about 300,000 km/s (186,000 mi/s). This speed is commonly referred to as the speed of light, c, but it is in fact the speed of all such radiation in a vacuum.

It may seem odd to use the word *radiation* when you speak of light. The word can be used to refer to high-energy particles emitted from radioactive atoms, and you have learned to be a little bit concerned when you see the word *radiation*. But it really refers to anything that spreads outward from a source. Light radiates from a source, so you can correctly refer to light as a form of radiation.

Electromagnetic radiation is a wave phenomenon; that is, it is associated with a periodically repeating disturbance, or wave. You are familiar with waves in water. If you disturb a quiet pool of water, waves spread across the surface. Imagine that you use a meter stick to measure the distance between the successive peaks of a wave. This distance is the **wavelength,** usually represented by the Greek letter lambda (λ). If you were measuring ripples in a pond, you might find that the wavelength is a few centimeters, whereas the wavelength of ocean waves might be a hundred meters or more. There is no restriction on the wavelength of electromagnetic radiation. Wavelengths can range from smaller than the diameter of an atom to larger than that of Earth.

Because all electromagnetic radiation travels at the speed of light, wavelength is related to **frequency,** the number of cycles that pass in one second. Short-wavelength radiation has a high frequency; long-wavelength radiation has a low frequency. To understand this, imagine watching an electromagnetic wave race past while you count its peaks (■ Figure 6-2). If the wavelength is short, you will count many peaks in one second; if the wavelength is long, you will count few peaks per second. The dials on radios are marked in frequency, but they could just as easily be marked in wavelength. The relation between wavelength and frequency is a simple one:

$$\lambda = \frac{c}{f}$$

That is, the wavelength equals the speed of light c divided by the frequency f. Notice that the larger (higher) the frequency, the smaller (shorter) the wavelength. In most cases, astronomers use wavelength rather than frequency.

Radio waves can have wavelengths from a few millimeters for microwaves to kilometers. In contrast, the wavelength of light is so short that you will need more convenient units. This book uses **nanometers (nm)** because this unit is consistent with the International System of units. One nanometer is 10^{-9} meter, and visible light has wavelengths that range from about 400 nm to about 700 nm. Another unit that astronomers commonly use, and a unit that you will see in many references on astronomy, is the **Angstrom (Å).** One Angstrom is 10^{-10} meter, and visible light has wavelengths between 4000 Å and 7000 Å.

You may find radio astronomers describing wavelengths in centimeters or millimeters, and infrared astronomers often refer to wavelengths in micrometers (or microns). One micrometer (μm) is 10^{-6} meter. Whatever unit is used to describe the wavelength, you must keep in mind that all electromagnetic radiation is the same phenomenon.

■ **Figure 6-2**

All electromagnetic waves travel at the speed of light. The wavelength is the distance between successive peaks. The frequency of the wave is the number of peaks that pass you in one second.

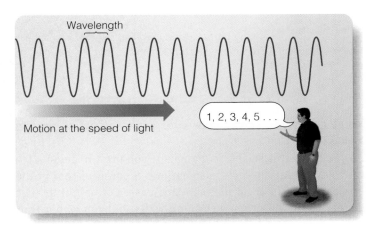

Wavelength

Motion at the speed of light

1, 2, 3, 4, 5 . . .

What exactly is electromagnetic radiation? Is it a particle or a wave? Throughout his life, Newton believed that light was made up of particles, but modern physicists now recognize that light can behave as both particle and wave. The modern model of light is more complete than Newton's, and it refers to "a particle of light" as a **photon.** You can recognize its dual nature by thinking of it as a bundle of waves.

The amount of energy a photon carries depends on its wavelength. The shorter the wavelength, the more energy the photon carries; the longer the wavelength, the less energy it contains. This is easy to remember because short wavelengths have high frequencies, and you would naturally expect rapid fluctuations to be more energetic. A simple formula expresses the relationship between energy and wavelength:

$$E = \frac{hc}{\lambda}$$

Here h is Planck's constant (6.6262×10^{-34} joule s), c is the speed of light (3×10^8 m/s), and λ is the wavelength in meters. A photon of visible light carries a very small amount of energy, but a photon with a very short wavelength can carry much more.

The Electromagnetic Spectrum

A spectrum is an array of electromagnetic radiation in order of wavelength. You are most familiar with the spectrum of visible light, which you see in rainbows. The colors of the spectrum differ in wavelength, with red having the longest wavelength and violet the shortest. The visible spectrum is shown at the top of ■ Figure 6-3.

The average wavelength of visible light is about 0.00005 cm. You could put 50 light waves end to end across the thickness of a sheet of household plastic wrap. Measured in nanometers, the wavelength of visible light ranges from about 400 to 700 nm. Just as you sense the wavelength of sound as pitch, you sense the wavelength of light as color. Light near the short-wavelength end of the visible spectrum (400 nm) looks violet to your eyes, and light near the long-wavelength end (700 nm) looks red.

Figure 6-3 shows how the visible spectrum makes up only a small part of the entire electromagnetic spectrum. Beyond the red end of the visible spectrum lies **infrared radiation,** where wavelengths range from 700 nm to about 0.1 cm. Your eyes are not sensitive to this radiation, but your skin senses it as heat. A "heat lamp" is just a bulb that gives off principally infrared radiation.

Beyond the infrared part of the electromagnetic spectrum lie radio waves. Microwaves have wavelengths of a millimeter to a few centimeters and are used for radar and long-distance telephone communication. Longer wavelengths are used for UHF and VHF television transmissions. FM, military, governmental, and ham radio signals have wavelengths up to a few meters, and AM radio waves can have wavelengths of kilometers.

The distinction between the wavelength ranges is not sharp. Long-wavelength infrared radiation and the shortest microwave

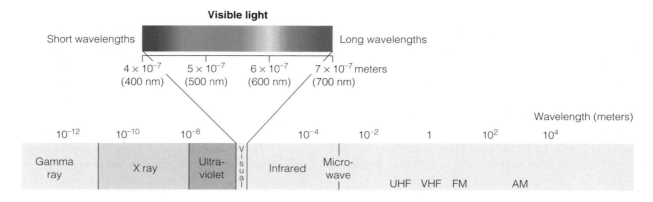

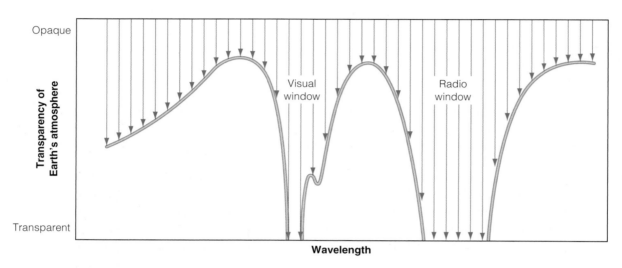

▪ Figure 6-3

The spectrum of visible light, extending from red to violet, is only part of the electromagnetic spectrum. Most radiation is absorbed in Earth's atmosphere, and only radiation in the visual window and the radio window can reach Earth's surface.

radio waves are the same. Similarly, there is no clear division between the short-wavelength infrared and the long-wavelength part of the visible spectrum. It is all electromagnetic radiation.

Look once again at the electromagnetic spectrum in Figure 6-3 and notice that electromagnetic waves shorter than violet are called **ultraviolet.** Electromagnetic waves even shorter are called X rays, and the shortest are gamma rays. Again, the boundaries between these wavelength ranges are not clearly defined.

X rays and gamma rays can be dangerous, and even ultraviolet photons have enough energy to do harm. Small doses of ultraviolet produce a suntan and larger doses sunburn and skin cancers. Contrast this to the lower-energy infrared photons. Individually they have too little energy to affect skin pigment, a fact that explains why you can't get a tan from a heat lamp. Only by concentrating many low-energy photons in a small area, as in a microwave oven, can you transfer significant amounts of energy.

Astronomers are interested in electromagnetic radiation because it carries clues to the nature of stars, planets, and other celes-

tial objects. Earth's atmosphere is opaque to most electromagnetic radiation, as shown by the graph at the bottom of Figure 6-3. Gamma rays, X rays, and some radio waves are absorbed high in Earth's atmosphere, and a layer of ozone (O_3) at an altitude of about 30 km absorbs ultraviolet radiation. Water vapor in the lower atmosphere absorbs the longer wavelength infrared radiation. Only visible light, some shorter wavelength infrared, and some radio waves reach Earth's surface through two wavelength regions called **atmospheric windows.** Obviously, if you wish to study the sky from Earth's surface, you must look out through one of these windows.

$(Ace \circledS Astronomy^{™})$ Log into AceAstronomy and select this chapter to see Astronomy Exercise "The Electromagnetic Spectrum." Explore different wavelength regions.

Building Scientific Arguments

What could you see if your eyes were sensitive only to X rays?

As you build this scientific argument, you must imagine a totally new situation. That is sometimes a powerful tool in the critical analysis of an idea. In this case, you might at first expect to be able to see through walls, but remember that your eyes detect only light that already exists. There are almost no X rays bouncing around at Earth's surface, so if you had X-ray eyes, you would be in the dark and would be unable to see anything. Even when you looked up at the sky, you would see nothing because Earth's atmosphere is not transparent to X rays. If Superman can see through walls, it is not because his eyes can detect X rays.

But now imagine a slightly different situation and modify your argument. **Would you be in the dark if your eyes were sensitive only to radio wavelengths?**

■   ■   ■

Connections: Now that you know something about electromagnetic radiation, you are ready to meet the tools astronomers use to gather and analyze that radiation.

ASTRONOMERS BUILD OPTICAL TELESCOPES to gather light and focus it into sharp images. This requires sophisticated optical and mechanical designs, and it leads astronomers to build gigantic telescopes on the tops of high mountains. As you begin, you will need to learn some technical terms that describe telescopes, but remember your real goal. You want to understand how different kinds of telescopes work and why some are better than others.

Two Kinds of Telescopes

Astronomical telescopes focus light into an image in one of two ways, as shown in ■ Figure 6-4. A lens bends (refracts) the light as it passes through the glass and brings it to a focus to form a small

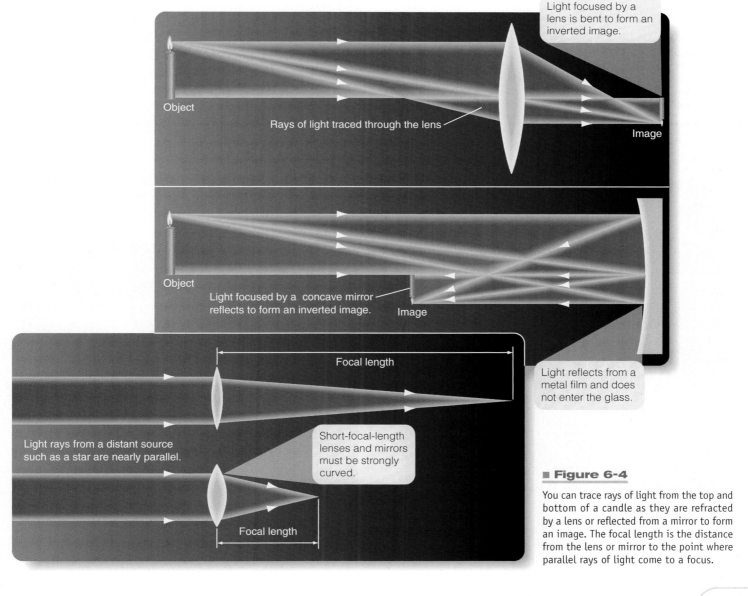

Light focused by a lens is bent to form an inverted image.

Object

Rays of light traced through the lens

Image

Light focused by a concave mirror reflects to form an inverted image.

Image

Object

Light reflects from a metal film and does not enter the glass.

Focal length

Light rays from a distant source such as a star are nearly parallel.

Short-focal-length lenses and mirrors must be strongly curved.

Focal length

■ Figure 6-4

You can trace rays of light from the top and bottom of a candle as they are refracted by a lens or reflected from a mirror to form an image. The focal length is the distance from the lens or mirror to the point where parallel rays of light come to a focus.

inverted image. A mirror—a concave piece of glass with a reflective surface—forms an image by reflecting the light. In either case, the **focal length** is the distance from the lens or mirror to the image formed of a distant light source, such as a star. Short-focal-length lenses and mirrors must be strongly curved, and long-focal-length lenses and mirrors are less strongly curved. Grinding the proper shape on a lens or mirror is a delicate, time-consuming, and expensive process.

Because there are two ways to focus light, there are two kinds of astronomical telescopes. **Refracting telescopes** use a large lens to gather and focus the light, whereas **reflecting telescopes** use a concave mirror. The advantages of the reflecting telescope have made it the preferred design for modern observatories.

The main lens in a refracting telescope is called the **primary lens,** and the main mirror in a reflecting telescope is called the **primary mirror.** These are also called the **objective lens** and **mirror.** Both kinds of telescopes form a very small, inverted image that is difficult to observe directly, so astronomers use a small lens called the **eyepiece** to magnify the image and make it convenient to view (■ Figure 6-5).

Refracting telescopes suffer from a serious optical distortion that limits their usefulness. When light is refracted through glass, shorter wavelengths bend more than longer wavelengths, and blue light, having shorter wavelengths, comes to a focus closer to the lens than does red light (■ Figure 6-6a). If you focus the eyepiece on the blue image, the red light is out of focus, and you see a red blur around the image. If you focus on the red image, the blue light blurs. The color separation is called **chromatic aberration.** Telescope designers can grind a telescope lens of two components made of different kinds of glass and so bring two different wavelengths to the same focus (Figure 6-6b). This does improve the image, but these **achromatic lenses** are not totally free of chromatic aberration, because other wavelengths still blur. Telescopes made with such lenses were popular until the end of the 19th century.

The primary lens of a refracting telescope is very expensive to make because it must be achromatic, and the glass must be pure and flawless because the light passes through the lens. The four surfaces must be ground precisely, and the lens can be supported only along its edge. The largest refracting telescope in the world was completed in 1897 at Yerkes Observatory in Wisconsin. Its lens is 1 m (40 in.) in diameter and weighs half a ton. Larger refracting telescopes are prohibitively expensive.

Reflecting telescopes are much less expensive because the light reflects from the front surface of the mirror. Consequently

■ Active Figure 6-5

(a) A refracting telescope uses a primary lens to focus starlight into an image that is magnified by a lens called an eyepiece. The primary lens has a long focal length, and the eyepiece has a short focal length. (b) A reflecting telescope uses a primary mirror to focus the light by reflection. A small secondary mirror reflects the starlight back down through a hole in the middle of the primary mirror to the eyepiece.

Ace◐Astronomy™ Log into AceAstronomy and select this chapter to see the Active Figure "Refractors and Reflectors." Watch light pass through the optics.

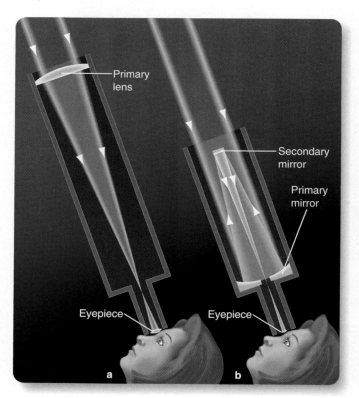

■ Figure 6-6

(a) A normal lens suffers from chromatic aberration because short wavelengths bend more than long wavelengths. (b) An achromatic lens, made in two pieces of two different kinds of glass, can bring any two colors to the same focus, but other colors remain slightly out of focus.

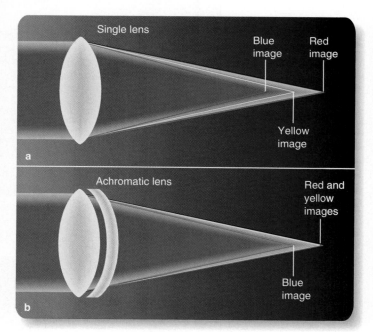

only the front surface need be ground to precise shape. Also, the glass of the mirror need not be perfectly transparent, and the mirror can be supported over its back surface to reduce sagging. Most important, reflecting telescopes do not suffer from chromatic aberration because the light is reflected from the metallic film on the front surface of the mirror and never enters the glass. For these reasons, every large astronomical telescope built since the beginning of the 20th century has been a reflecting telescope.

Ace✱Astronomy™ Log into AceAstronomy and select this chapter to see Astronomy Exercise "Lenses: Focal Length." Create different lenses and observe their focal lengths.

Ace✱Astronomy™ Log into AceAstronomy and select this chapter to see Astronomy Exercise "Telescopes: Objective Lens and Eyepiece." See how the two lenses interact to produce an image.

The Powers of a Telescope

A telescope can aid your eyes in only three ways—the three powers of a telescope. They make images brighter, more detailed, and larger.

Most interesting celestial objects are faint sources of light, so you need a telescope that can gather large amounts of light to produce a bright image. **Light-gathering power** refers to the ability of a telescope to collect light. Catching light in a telescope is like catching rain in a bucket—the bigger the bucket, the more rain it catches (■ Figure 6-7). Light-gathering power is proportional to the area of the telescope objective. A lens or mirror with a large area gathers a large amount of light. The area of a circular lens or mirror of diameter D is just πr^2 or, written in terms of the diameter, the area is $\pi(D/2)^2$. To compare the relative light-gathering powers *(LGP)* of two telescopes A and B, you can calculate the ratio of the areas of their objectives, which reduces to the ratio of their diameters *(D)* squared.

$$\frac{LGP_A}{LGP_B} = \left(\frac{D_A}{D_B}\right)^2$$

For example, suppose you compared a telescope 24 cm in diameter with a telescope 4 cm in diameter. The ratio of the diameters is 24/4, or 6, but the larger telescope does not gather 6 times as much light. Light-gathering power increases as the ratio of diameters squared, so it gathers 36 times more light than the smaller telescope. This example shows the importance of diameter in astronomical telescopes. Even a small increase in diameter produces a large increase in light-gathering power and allows astronomers to study much fainter objects.

The second power, **resolving power,** refers to the ability of the telescope to reveal fine detail. Because light acts as a wave, it produces a small **diffraction fringe** around every point of light in the image, and you cannot see any detail smaller than the fringe (■ Figure 6-8). Astronomers can't eliminate diffraction fringes,

■ **Figure 6-7**

Gathering light is like catching rain in a bucket. A large-diameter telescope gathers more light and has a brighter image than a smaller telescope of the same focal length.

but the larger a telescope is in diameter, the smaller the diffraction fringes are. That means the larger the telescope, the better its resolving power.

As you will see later, resolving power is worse for longer wavelengths, but if you consider only optical telescopes, you can estimate the resolving power by calculating the angular distance between two stars that are just barely visible through the telescope as two separate images. Astronomers say the two images are "resolved," meaning they are separated from each other. The resolving power, α, in seconds of arc, equals 11.6 divided by the diameter of the telescope in centimeters:

$$\alpha = \frac{11.6}{D}$$

For example, the resolving power of a 25-cm telescope is 11.6 divided by 25, or 0.46 second of arc. No matter how perfect the telescope optics, this is the smallest detail you can see through that telescope.

In addition to resolving power, two other factors—lens quality and atmospheric conditions—limit the detail you can see through a telescope. A telescope must contain high-quality optics to achieve its full potential resolving power. Even a large telescope reveals little detail if its optics are marred with imperfections. Also, when you look through a telescope, you are looking up through miles of turbulent air in Earth's atmosphere, which makes the image dance and blur, a condition called **seeing.** On a

Window on Science | 6-1

Resolving Power and the Resolution of a Measurement

Have you ever seen a movie in which the hero magnifies a newspaper photo and reads some tiny detail? It isn't really possible, because newspaper photos are made up of tiny dots of ink, and no detail smaller than a single dot will be visible no matter how much you magnify the photo. In fact, all images are made up of elements of some sort, and that means there is a limit to the amount of detail you can see in an image. In an astronomical image, the resolution is often set by seeing. It is foolish to attempt to see a detail in the image that is smaller than the resolution.

This limitation is true of all measurements in science. A zoologist might be trying to measure the length of a live snake, or a sociologist might be trying to measure the attitudes of people toward drunk driving, but both face limits to the resolution of their measurements. The zoologist might specify that the snake was 43.28932 cm long, and the sociologist might say that 98.2491 percent of people oppose drunk driving, but a critic might point out that it isn't possible to make these measurements that accurately. The resolution of the techniques does not justify the accuracy implied by all of those digits.

Science is based on measurement, and whenever you make a measurement you should ask yourself how accurate that measurement can be. The accuracy of the measurement is limited by the resolution of the measurement technique, just as the amount of detail in a photograph is limited by its resolution.

A high-resolution image of Mars reveals details such as mountains, craters, and the southern polar cap. (NASA)

---

night when the atmosphere is unsteady and the images are blurred, the seeing is bad (■ Figure 6-9). Even under good seeing conditions, the detail visible through a large telescope is limited, not by its diffraction fringes, but by the air through which the telescope must look. A telescope performs better on a high mountaintop where the air is thin and steady, but even there Earth's atmosphere limits the detail the best telescopes can reveal to about 0.5 second of arc.

This limitation on the amount of information in an image is related to the limitation on the accuracy of a measurement. All measurements have some built-in uncertainty (**Window on Science 6-1**), and scientists must learn to work within those limitations.

The third and least important power of a telescope is **magnifying power,** the ability to make the image bigger. Because the amount of detail you can see is limited by the seeing conditions and the resolving power, very high magnification does not necessarily show more detail. Also, you can change the magnification by changing the eyepiece, but you cannot alter the telescope's light-gathering power or resolving power without changing the diameter of the objective lens or mirror, and that would be so expensive you might as well build a whole new telescope.

■ Active Figure 6-8

(a) Stars are so far away that their images are points, but the wave nature of light surrounds each star image with diffraction fringes (much magnified in this computer model). (b) Two stars close to each other have overlapping diffraction fringes and become impossible to detect separately. (Computer model by M. A. Seeds)

Ace◐Astronomy™ Log into AceAstronomy and select this chapter to see Active Figure "Resolution and Telescopes." You can control telescope diameter and watch resolution change.

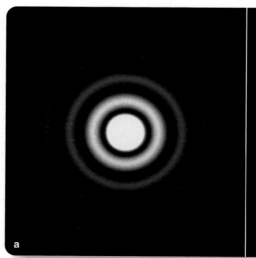
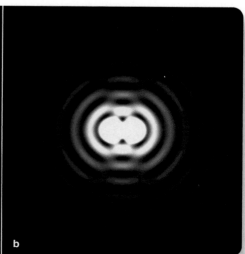

a   b

Visual-wavelength image

You can calculate the magnification of a telescope by dividing the focal length of the objective by the focal length of the eyepiece:

$$M = \frac{F_o}{F_e}$$

For example, if a telescope has an objective with a focal length of 80 cm and you use an eyepiece whose focal length is 0.5 cm, the magnification is 80/0.5, or 160 times.

Notice that the two most important powers of the telescope, light-gathering power and resolving power, depend on the diameter of the telescope. This explains why astronomers refer to telescopes by diameter and not by magnification. Astronomers will refer to a telescope as an 8-meter telescope or a 10-meter telescope, but they would never identify a telescope as a 200-power telescope.

The search for light-gathering power and high resolution explains why nearly all major observatories are located far from big cities and usually on high mountains. Astronomers avoid cities because **light pollution,** the brightening of the night sky by light scattered from artificial outdoor lighting, can make it impossible to see faint objects (■ Figure 6-10). In fact, many residents of cities are unfamiliar with the beauty of the night sky because they can

Astronomers no longer build large observatories in populous areas.

A number of major observatories are located on mountaintops in the southwest.

a Visual-wavelength image

Paranal Observatory
Altitude: 2635 m (8660 ft)
Location: Atacama desert of northern Chile
Nearest city: Antofagasta 120 km (75 mi)

b

■ **Figure 6-10**

(a) This satellite view of the continental United States at night shows the light pollution and energy waste produced by outdoor lighting. Observatories cannot be located near large cities. (NOAA) (b) The domes of four giant telescopes are visible at upper left at Paranal Observatory, built by the European Southern Observatory. The Atacama Desert is believed to be the driest place on Earth. (ESO)

see only the brightest stars. Nevertheless, nature's own light pollution, the moon, is so bright it drowns out fainter objects, and astronomers are often unable to observe on the nights near full moon when faint objects cannot be observed even with the largest telescopes on high mountains.

Astronomers prefer to place their telescopes on carefully selected high mountains. The air there is thin and more transparent. The air is very dry at high altitudes and is more transparent to infrared radiation. Most important, astronomers select mountains where the air flows smoothly and is not turbulent. This produces the best seeing. Building an observatory on top of a high mountain far from civilization is difficult and expensive, as you can imagine from the photo in Figure 6-10, but the dark sky and steady seeing make it worth the effort.

Ace✷Astronomy™ Log into AceAstronomy and select this chapter to see Astronomy Exercise "Telescopes and Resolution I." What affects resolution most in the design of a telescope?

Ace✷Astronomy™ Log into AceAstronomy and select this chapter to see Astronomy Exercise "Telescopes and Resolution II." How does wavelength affect the resolution of a telescope?

Ace✷Astronomy™ Log into AceAstronomy and select this chapter to see Astronomy Exercise "Particulate, Heat, and Light Pollution." What happens when the air near a city is polluted by dust, heat, or light?

Buying a Telescope

Thinking about how to shop for a new telescope will not only help you if you decide to buy one but will also illustrate some important points about astronomical telescopes.

Assuming you have a fixed budget, you should buy the highest-quality optics and the largest-diameter telescope you can afford. Of the two things that limit what you see, optical quality is under your control. You can't make the atmosphere less turbulent, but you should buy good optics. If you buy a telescope from a toy store and it has plastic lenses, you shouldn't expect to see very much. Also, you want to maximize the light-gathering power of your telescope, so you want to purchase the largest-diameter telescope you can afford. Given a fixed budget, that means you should buy a reflecting telescope rather than a refracting telescope. Not only will you get more diameter per dollar, but your telescope will not suffer from chromatic aberration.

You can safely ignore magnification. Department stores and camera stores may advertise telescopes by quoting their magnification, but it is not an important number. What you can see is fixed by light-gathering power, optical quality, and Earth's atmosphere. Besides, you can change the magnification by changing eyepieces.

Other things being equal, you should choose a telescope with a solid mounting that will hold the telescope steady and allow it

to point at objects easily. Computer-controlled pointing systems are available for a price on many small telescopes. A good telescope on a poor mounting is almost useless.

You might be buying a telescope to put in your backyard, but you must think about the same issues astronomers consider when they design giant telescopes to go on mountaintops. Designing new, giant telescopes has led astronomers to solve some traditional problems in new ways, as you will see in the next section.

New-Generation Telescopes

For most of the 20th century, astronomers faced a serious limitation on the size of astronomical telescopes. Traditional telescope mirrors were made thick to avoid sagging that would distort the reflecting surface, but those thick mirrors were heavy. The 5-m (200-in.) mirror on Mount Palomar weighs 14.5 tons. These traditional telescopes were big, heavy, and expensive.

Modern astronomers have solved these problems in a number of ways. Study **Modern Astronomical Telescopes** on pages 120–121 and notice four important advances in telescope design made possible by high-speed computers:

1 Traditional telescopes use large, solid, heavy mirrors to focus starlight.

2 Astronomers can now build simpler, lighter-weight telescope mountings and depend on computers to move the telescope and follow the westward motion of the stars as Earth rotates.

3 Notice that computer control of the shape of telescope mirrors allows the use of thin, lightweight mirrors—either "floppy" mirrors or segmented mirrors. Lowering the weight of the mirror lowers the weight of the rest of the telescope and makes it stronger and less expensive. Also, thin mirrors cool faster at nightfall and produce better images.

4 Also notice the way astronomers use high-speed computers to reduce seeing distortion caused by Earth's atmosphere. Only a few decades ago, many astronomers argued that it wasn't worth building more large telescopes on Earth's surface because of the limitations set by seeing. Now a number of new giant telescopes have been built, and more are in development that can partially overcome the seeing problem.

Did you notice that astronomical telescopes must be aligned with the north celestial pole? Polaris, the North Star, is one of our favorite stars in the list at the beginning of this book. It marks the location of the north celestial pole. Equatorial mountings have an axis that points toward Polaris, and alt-azimuth telescopes are run by computers, which align their motion with Polaris. Even telescopes in the southern hemisphere, where the north celestial pole lies below the horizon, must tip their hats toward Polaris. That's one reason Polaris deserves to be one of your favorite stars; whenever you notice Polaris in the night sky, think of all the

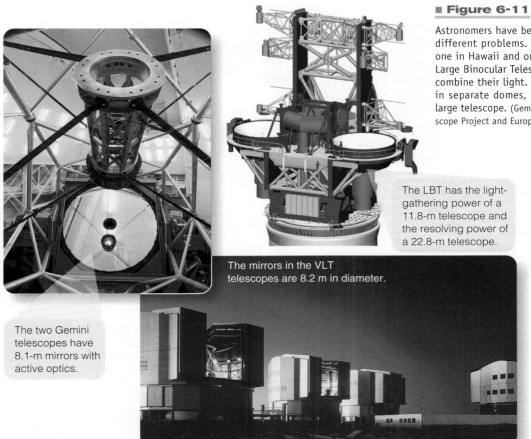

■ **Figure 6-11**

Astronomers have begun building multiple telescopes to solve different problems. Two Gemini telescopes have been built, one in Hawaii and one in Chile, to observe the entire sky. The Large Binocular Telescope (LBT) carries two 8.4-m mirrors that combine their light. The four telescopes of the VLT are housed in separate domes, but they can combine to work as a very large telescope. (Gemini: NOAO/AURA/NSF; LBT: Large Binocular Telescope Project and European Industrial Engineer; VLT: ESO)

The LBT has the light-gathering power of a 11.8-m telescope and the resolving power of a 22.8-m telescope.

The mirrors in the VLT telescopes are 8.2 m in diameter.

The two Gemini telescopes have 8.1-m mirrors with active optics.

astronomical telescopes in backyards and observatories all over the world that bow toward Polaris.

High-speed computers have allowed astronomers to build new, giant telescopes as shown in ■ Figure 6-11. An international collaboration of astronomers have built the Gemini telescopes with 8.1-m thin mirrors. One is located in the northern hemisphere and one in the southern hemisphere to cover the entire sky. The European Southern Observatory has built the Very Large Telescope (VLT) high in the remote Andes Mountains of northern Chile. The VLT consists of four telescopes with computer-controlled mirrors 8.2 m in diameter and only 17.5 cm (6.9 in.) thick. The four telescopes can work singly or can combine their light to work as one large telescope. Italian and American astronomers are building the Large Binocular Telescope, which carries a pair of 8.4-m mirrors on a single mounting.

High-speed computers have improved astronomical telescopes in another way that might surprise you. Computer control and data handling have made possible huge surveys of the sky in which millions of objects are observed. The Sloan Digital Sky Survey, for example, is mapping the sky, measuring the position and brightness of 100 million stars and galaxies at a number of wavelengths. The Two-Micron All Sky Survey (2MASS) has mapped the entire sky at three wavelengths in the infrared. Other surveys are being made at many other wavelengths. Every night large telescopes scan the sky, and billions of bytes of data are compiled automatically in immense sky atlases. Astronomers will study those data banks for decades to come.

The days when astronomers worked beside their telescopes through long, dark, cold nights are nearly gone. The complexity and sophistication of telescopes require a battery of computers, and almost all research telescopes are run from control rooms that astronomers call "warm rooms." Astronomers don't need to be kept warm, but computers demand comfortable working conditions (■ Figure 6-12).

■ **Figure 6-12**

In the control room of the 4-meter telescope atop Kitt Peak National Observatory, the telescope operator at left manages the operation and safety of the telescope. The astronomer at right operates the instruments, records data, and makes decisions on the observing program. Astronomers work through the night controlling the computers that control the telescope and its instruments. (NOAO/AURA/NSF)

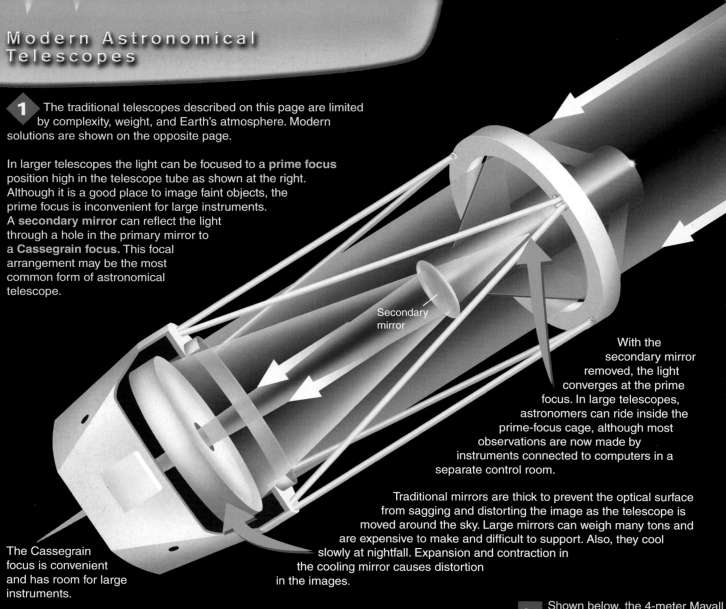

1 The traditional telescopes described on this page are limited by complexity, weight, and Earth's atmosphere. Modern solutions are shown on the opposite page.

In larger telescopes the light can be focused to a **prime focus** position high in the telescope tube as shown at the right. Although it is a good place to image faint objects, the prime focus is inconvenient for large instruments. A **secondary mirror** can reflect the light through a hole in the primary mirror to a **Cassegrain focus**. This focal arrangement may be the most common form of astronomical telescope.

Secondary mirror

With the secondary mirror removed, the light converges at the prime focus. In large telescopes, astronomers can ride inside the prime-focus cage, although most observations are now made by instruments connected to computers in a separate control room.

Traditional mirrors are thick to prevent the optical surface from sagging and distorting the image as the telescope is moved around the sky. Large mirrors can weigh many tons and are expensive to make and difficult to support. Also, they cool slowly at nightfall. Expansion and contraction in the cooling mirror causes distortion in the images.

The Cassegrain focus is convenient and has room for large instruments.

1a Smaller telescopes are often found with a **Newtonian focus**, the arrangement that Isaac Newton used in his first reflecting telescope. The Newtonian focus is inconvenient for large telescopes as shown at right.

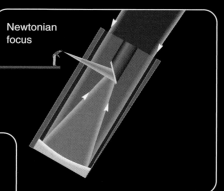

Newtonian focus

Thin correcting lens

Schmidt-Cassegrain telescope

1b Many small telescopes such as the one on your left use a **Schmidt-Cassegrain focus**. A thin correcting plate improves the image but is too slightly curved to introduce serious chromatic aberration.

1c Shown below, the 4-meter Mayall Telescope at Kitt Peak National Observatory in Arizona can be used at either the prime focus or the Cassegrain focus. Note the human figure at lower right.

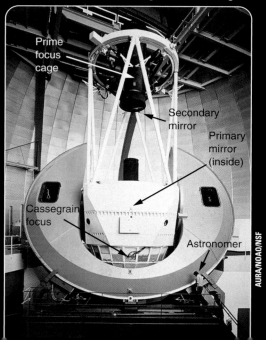

Prime focus cage

Secondary mirror

Primary mirror (inside)

Cassegrain focus

Astronomer

AURA/NOAO/NSF

Equatorial mounting

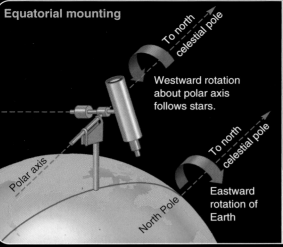

To north celestial pole

Westward rotation about polar axis follows stars.

Polar axis

North Pole

Eastward rotation of Earth

Alt-azimuth mounting

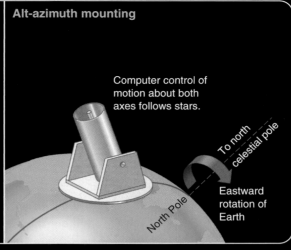

Computer control of motion about both axes follows stars.

To north celestial pole

North Pole

Eastward rotation of Earth

2 Telescope mountings must contain a **sidereal drive** to move smoothly westward and counter the eastward rotation of Earth. The traditional **equatorial mounting** (far left) has a **polar axis** parallel to Earth's axis, but the modern **alt-azimuth mounting** (near left) moves like a cannon — up and down and left to right. Such mountings are simpler to build but need computer control to follow the stars.

3 Unlike traditional thick mirrors, thin mirrors, sometimes called floppy mirrors as shown at right, weigh less and require less massive support structures. Also, they cool rapidly at nightfall and there is less distortion from uneven expansion and contraction.

Floppy mirror

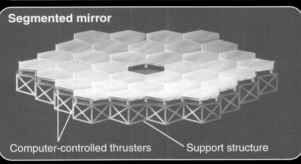

Computer-controlled thrusters   Support structure

3a Grinding a large mirror may remove tons of glass and take months, but new techniques speed the process. Some large mirrors are cast in a rotating oven that causes the molten glass to flow to form a concave upper surface. Grinding and polishing such a preformed mirror is much less time consuming.

3b Mirrors made of segments are economical because the segments can be made separately. The resulting mirror weighs less and cools rapidly. See image at right.

Segmented mirror

Computer-controlled thrusters   Support structure

3c Both floppy mirrors and segmented mirrors sag under their own weight. Their optical shape must be controlled by computer-driven thrusters under the mirror in what is called **active optics.**

3d As shown below, the two largest telescopes in the world, the Keck I and Keck II telescopes in Hawaii, contain segmented mirrors 10 m in diameter.

W. M. Keck Observatory

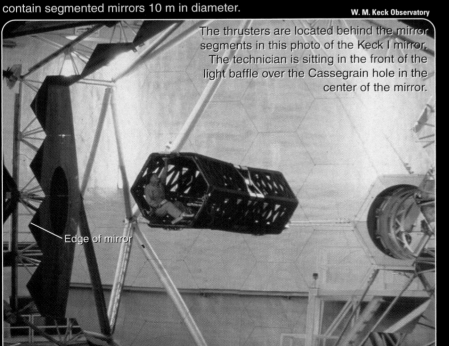

The thrusters are located behind the mirror segments in this photo of the Keck I mirror. The technician is sitting in the front of the light baffle over the Cassegrain hole in the center of the mirror.

Edge of mirror

Adaptive optics in telescopes

Adaptive optics off

Object appears to be a single star.

Adaptive optics on

Object revealed as a pair of stars.

1 second of arc

Paul Kalas

4 **Adaptive optics** uses high-speed computers to monitor the image distortion caused by Earth's atmosphere and adjust the optics many times a second to compensate. This can reduce the blurring due to seeing and dramatically improve image quality in Earth based telescopes.

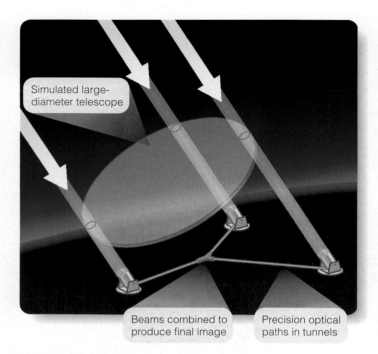

■ Figure 6-13

In an astronomical interferometer, smaller telescopes can combine their light through specially designed optical tunnels to simulate a larger telescope with a resolution set by the separation of the smaller telescopes.

Interferometry

One of the reasons astronomers build big telescopes is to increase resolving power, and astronomers have been able to achieve very high resolution by connecting multiple telescopes together to work as if they were a single telescope. This method of synthesizing a larger telescope is known as **interferometry** (■ Figure 6-13).

To work as an interferometer, the separate telescopes must combine their light through a network of mirrors, and the path that each light beam travels must be controlled so that it does not vary more than some small fraction of the wavelength. Turbulence in Earth's atmosphere constantly distorts the light, and high-speed computers must continuously adjust the light paths. Recall that the wavelength of light is very short, roughly 0.0005 mm, so building optical interferometers is one of the most difficult technical problems that astronomers face. Infrared- and radio-wavelength interferometers are slightly easier to build because the wavelengths are longer. In fact, as you will discover later in this chapter, the first astronomical interferometers worked at radio wavelengths.

The VLT shown in Figure 6-11 consists of four 8.2-m telescopes that can operate separately, but they can be linked together through underground tunnels with three 1.8-m telescopes on the same mountaintop. The resulting optical interferometer provides the resolution of a telescope 200 meters in diameter. Other telescopes can work as interferometers. The two Keck 10-m telescopes can be used as an interferometer. The CHARA array

on Mt. Wilson combines six 1-meter telescopes to create the equivalent of a telescope one-fifth of a mile in diameter. The Large Binocular Telescope shown in Figure 6-11 can be used as an interferometer.

Although turbulence in Earth's atmosphere can be partially averaged out in an interferometer, plans are being made to put interferometers in space. The Space Interferometry Mission, for example, will work at visual wavelengths and study everything from the cores of erupting galaxies to planets orbiting nearby stars.

Building Scientific Arguments

Why do astronomers build observatories at the tops of mountains?

To build this argument you need to think about the powers of a telescope. Astronomers have joked that the hardest part of building a new observatory is constructing the road to the top of the mountain. It certainly isn't easy to build a large, delicate telescope at the top of a high mountain, but it is worth the effort. A telescope on top of a high mountain is above the thickest part of Earth's atmosphere. There is less air to dim the light, and there is less water vapor to absorb infrared radiation. Even more important, the thin air on a mountaintop causes less disturbance to the image, and consequently the seeing is better. A large telescope on Earth's surface has a resolving power much better than the distortion caused by Earth's atmosphere. So it is limited by seeing, not by its own diffraction. It really is worth the trouble to build telescopes atop high mountains.

Astronomers not only build telescopes on mountaintops, they also build gigantic telescopes many meters in diameter. Revise your argument to focus on telescope design. **What are the problems and advantages in building such giant telescopes?**

■   ■   ■

Connections: Astronomers sometimes refer to a telescope that produces distorted images as a "light bucket." In a sense, all astronomical telescopes are light buckets because the light they focus into images reveals very little until it is recorded and analyzed by special instruments attached to the telescopes.

(6-3) Special Instruments

LOOKING THROUGH A TELESCOPE doesn't tell you much. A star looks like a point of light. A planet looks like a little disk. A galaxy looks like a hazy patch. To use an astronomical telescope to learn about the universe, you must be able to analyze the light the telescope gathers. Special instruments attached to the telescope make that possible.

Galaxy NGC 891 in true color. It is edge-on and contains thick dust clouds.

Visual-wavelength image

In this image, color shows brightness. White and red are brightest, and yellow and green are dimmer.

Visual image in false color

In these negative images of NGC 891, the sky is white and the stars are black.

Visual-wavelength negative images

■ Figure 6-14

Astronomical images can be manipulated in many ways to bring out details. The photo of the galaxy at upper left is dark, and the details of the dust clouds in the disk of the galaxy do not show well. The two negative images of the galaxy have been produced to show the dust clouds more clearly. (C. Hawk, B. Savage, N. A. Sharp NOAO/WIYN/NSF) The image at upper right shows two interacting galaxies known as Arp 273. The visual-wavelength image has been given false color according to brightness. (NOAO/WIYN/NSF)

Imaging Systems

The original imaging device in astronomy was the photographic plate. It could record faint objects in long time exposures and could be stored for later analysis. But photographic plates have been almost entirely replaced in astronomy by electronic imaging systems.

Most modern astronomers use **charge-coupled devices (CCDs)** to record images. A CCD is a specialized computer chip containing roughly a million microscopic light detectors arranged in an array about the size of a postage stamp. These devices can be used like small photographic plates, but they have dramatic advantages. They can detect both bright and faint objects in a single exposure, are much more sensitive than photographic plates, and can be read directly into computer memory for later analysis. Although CCDs for astronomy are extremely sensitive and therefore expensive, less sophisticated CCDs are used in video and digital cameras.

The image from a CCD is stored as numbers in computer memory, so it is easy to manipulate the image to bring out details that would not otherwise be visible. For example, astronomical images are often reproduced as negatives with the sky white and the stars dark. This makes the faint parts of the image easier to see (■ Figure 6-14). Astronomers also manipulate images to produce **false-color images** in which the colors represent different levels of intensity and are not related to the true colors of the object. You can see an example in Figure 6-14. In fact, false-color images are common in many fields (**Window on Science 6-2**).

Measurements of intensity and color were made in the past using a photometer, a highly sensitive light meter attached to a telescope. Today, however, most such measurements are made directly on CCD images. Because the CCD image is easily digitized, brightness and color can be measured more easily and more accurately than on photographic plates.

The Spectrograph

To analyze light in detail, astronomers need to spread the light out according to wavelength to form a spectrum, a task performed by a **spectrograph.** You can understand how this works if you imagine reproducing an experiment performed by Isaac Newton in 1666. Newton bored a small hole in the window shutter of his bedroom to admit a thin beam of sunlight. When he placed a prism in the beam, it spread the light into a beautiful spectrum splashed across his bedroom wall. From this Newton concluded that white light was made of a mixture of all the colors.

Newton didn't think in terms of wavelength, but you can use that modern concept to see that the light passing through the prism is bent at an angle that depends on the wavelength. Violet (short wavelength) bends most, and red (long wavelength) least. In this way, the white light entering the prism is spread into a spectrum (■ Figure 6-15). You could build a spectrograph with a prism to spread the light and a lens to guide the light into a camera.

False-Color Images and Reality

Astronomical images are usually recorded in digital form directly into computer memory, so astronomers often add false color to images to reveal things otherwise hard to see. They might exaggerate colors already present, but more often they add colors to represent levels of brightness. Of course, the colors used are entirely arbitrary. One astronomer might like shades of blue, and another might use all of the colors ranging from violet to red. To understand the image, you need to know the meaning of the colors—that white and red are bright areas, for instance, and that yellow and green are dimmer.

Astronomers also use false color to create images recorded at wavelengths not visible to the human eye. Radio, infrared, ultraviolet, and X-ray images are often displayed as false-color images. You might think of them as maps in which the colors show the relative intensity of the radiation in different parts of the image. Again the choice of colors is up to the astronomer analyzing the image.

False-color images are common in astronomy, but they are also used in other sciences. Doctors often analyze medical X-ray images, CAT scans, and so on by converting the images to false color. Biologists use false color to analyze microscopic photographs, and geologists use false color to study photographs of Earth recorded by a satellite at various wavelengths.

Even a simple photograph of a galaxy or a glowing cloud of gas in space is a kind of false-color image. The object is much too faint to see with the human eye, and even if you flew in a spaceship to within a short distance, it would still be too faint to see with your eyes. The photograph you see is a time exposure in which the light was allowed to accumulate over an extended period ranging from seconds to hours. That makes the galaxy or gas cloud visible. Even a black-and-white photograph of a glowing cloud of gas shows an aspect of the universe you can never see with unaided eyes. In that sense, all astronomical images are false-

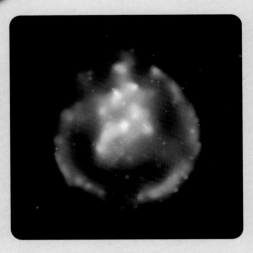

A false-color X-ray image of the expanding cloud of gas left behind by the explosion of a massive star. (NASA/CXC/Rutgers/J. Hughes)

color images. Their virtue is not that they are colorful but that they are meaningful.

Nearly all modern spectrographs use a grating in place of a prism. A **grating** is a piece of glass with thousands of microscopic parallel grooves scribed onto its surface. Different wavelengths of light reflect from the grating at slightly different angles, so white light is spread into a spectrum. You have probably noticed this effect when you look at the closely spaced lines etched onto a compact disk; as you move the disk about, different colors flash across its surface. You could build a modern spectrograph by using a high-quality grating to spread the light into a spectrum and a CCD camera to record the spectrum.

The spectrum of an astronomical object can contain hundreds of spectral lines produced by the atoms in the object. Because astronomers must identify these lines and measure the wavelengths, they use a **comparison spectrum** as a calibration of their spectrograph. Special bulbs built into the spectrograph produce bright lines given off by such atoms as thorium and argon or neon. The wavelengths of these spectral lines have been measured to high precision in the laboratory, so astronomers can use spectra of these light sources as guides to measure wavelengths and identify spectral lines in the spectrum of a star, galaxy, or planet.

Because astronomers understand how light interacts with matter, a spectrum carries a tremendous amount of information (as you will see in the next chapter), and that makes a spectrograph the astronomer's most powerful instrument. An astronomer recently remarked, "We don't know anything about an object till we get a spectrum," and that is only a slight exaggeration.

Building Scientific Arguments

What is the difference between light going through a lens and light passing through a prism?

When you think about natural processes, it is often helpful to compare similar things and scientific arguments often make such comparisons. A few simple rules explain most natural events, so the similarities are often revealing. A refracting telescope producing chromatic aberration and a prism dispersing light into a spectrum are two examples of the same thing, but one is bad and one is good. When light passes through the curved surfaces of a lens, different wavelengths are bent by slightly different amounts, and the different colors of light come to focus at different focal lengths. This produces the color fringes in an image called chromatic aberration, and that's bad. But the surfaces of a prism are made to be precisely flat, so all of the light enters the prism at the same angle, and any given wavelength is bent by the same amount. Consequently, white light is dispersed into a spectrum. You could call the dispersion of light by a prism "controlled chromatic aberration," and that's good.

Now you can build your own argument comparing similar things. CCDs have been very good for astronomy, and they have almost completely replaced photographic plates. **How are CCD chips similar to photographic plates, and how are they better?**

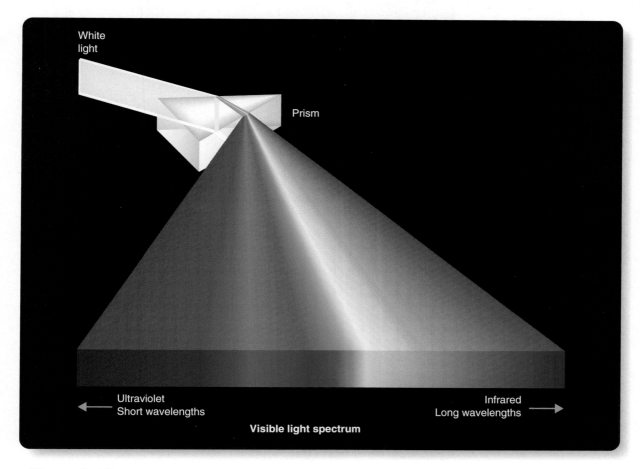

White light

Prism

Ultraviolet
Short wavelengths

Infrared
Long wavelengths

Visible light spectrum

■ **Figure 6-15**

A prism bends light by an angle that depends on the wavelength of the light. Short wavelengths bend most and long wavelengths least. Thus, white light passing through a prism is spread into a spectrum.

Connections: So far, this discussion has been limited to visual wavelengths. Now it is time for you to explore the rest of the electromagnetic spectrum.

(6-4) Radio Telescopes

ALL THE TELESCOPES and instruments you have discussed so far look out through the visible light window in Earth's atmosphere, but there is another window running from a wavelength of 1 cm to about 1 m (see Figure 6-3). By building the proper kinds of instruments, astronomers can study the universe through this radio window.

Operation of a Radio Telescope

A radio telescope usually consists of four parts: a dish reflector, an antenna, an amplifier, and a recorder (■ Figure 6-16). The components, working together, make it possible for astronomers to detect radio radiation from celestial objects.

The dish reflector of a radio telescope, like the mirror of a reflecting telescope, collects and focuses radiation. Because radio waves are much longer than light waves, the dish need not be as smooth as a mirror; wire mesh will reflect all but the shortest wavelength radio waves. But don't be surprised if you see a photo of a radio telescope that doesn't have a dish shape. There are a number of ways to collect radio energy from the sky. Nevertheless, dish-shaped reflectors are very common at radio observatories.

Though a radio telescope's dish may be many meters in diameter, the antenna may be as small as your hand. Like the antenna on a TV set, its only function is to absorb the radio energy and direct it along a cable to an amplifier. After amplification, the signal goes to some kind of recording instrument. Most radio observatories record data directly into computer memory.

However it is recorded, an observation with a radio telescope measures the amount of radio energy coming from a specific point on the sky, and that causes two problems. For one thing, the intensity at one spot doesn't tell you much, so the radio telescope must be scanned over an object, a cloud of gas, for example, to produce a map of the radio intensity at different points. The

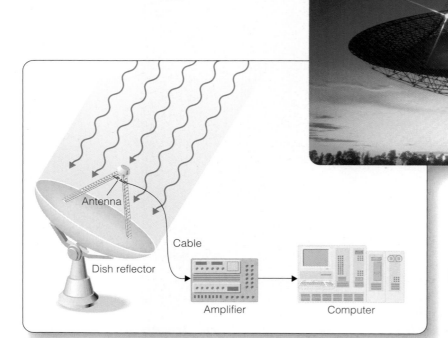

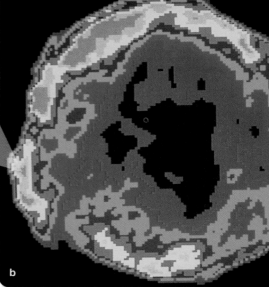

Seat prices in a
   baseball stadium
Red most expensive
Violet least expensive

■ **Figure 6-17**

(a) A contour map of a baseball stadium shows regions of similar admission prices. The most expensive seats are those behind home plate. (b) A false-color-image radio map of Tycho's supernova remnant, the expanding shell of gas produced by the explosion of a star in 1572. The radio contour map has been color-coded to show intensity. (Courtesy NRAO)

Radio energy map
Red strongest
Violet weakest

second problem is that humans can't see radio waves, so astronomers draw maps in which contours mark areas of similar radio intensity. You could compare such a map to a seating diagram for a baseball stadium in which the contours mark areas in which the seats have the same price (■ Figure 6-17a). Contour maps are very common in radio astronomy and are often reproduced using false colors (Figure 6-17b)

Limitations of a Radio Telescope

A radio astronomer works under three handicaps: poor resolution, low intensity, and interference. You remember that the resolving power of an optical telescope depends on the diameter of the objective lens or mirror. It also depends on the wavelength of the radiation. At very long wavelengths, like those of radio waves, images become fuzzy because the diffraction fringes are very large. As with an optical telescope, there is no way to improve the resolving power without building a bigger telescope. Consequently, radio telescopes must be quite large.

Even so, the resolving power of a radio telescope is not good. A dish 30 m in diameter receiving radiation with a wavelength of 21 cm has a resolving power of about 0.5°. Such a radio telescope would be unable to reveal any details in the sky smaller than the moon. Fortunately, radio astronomers can combine two or more radio telescopes to form a **radio interferometer** capable of much higher resolution. For example, the Very Large Array (VLA) consists of 27 dish antennas spread in a Y-shape across the New Mexico desert (■ Figure 6-18). In combination, they have the resolving power of a radio telescope 36 km (22 mi) in diameter. The VLA can resolve details smaller than 1 second of arc. Eight new dish antennas being added across New Mexico will give the VLA 10 times better resolving power. Another large radio interferometer, the Very Long Baseline Array (VLBA), consists of matched radio dishes spread from Hawaii to the Virgin Islands and has an effective diameter almost as large as Earth.

The second handicap radio astronomers face is the low intensity of the radio signals. You saw earlier that the energy of a photon depends on its wavelength. Photons of radio energy have such long wavelengths that their individual energies are quite low. In order to get strong signals focused on the antenna, the radio astronomer must build large collecting dishes.

■ **Figure 6-18**

The Very Large Array uses 27 radio dishes, which can be moved to different positions along a Y-shaped set of tracks across the New Mexico desert. They are shown here in the most compact arrangement. Signals from the dishes are combined to create very-high-resolution radio maps of celestial objects. (NRAO/AUI/NSF)

The largest fully steerable radio telescope in the world is at the National Radio Astronomy Observatory in Green Bank, West Virginia (■ Figure 6-19a). The telescope has a reflecting surface 100 meters in diameter, big enough to hold an entire football field, and can be pointed anywhere in the sky. Its surface consists of 2004 computer-controlled panels that adjust to maintain the shape of the reflecting surface.

The largest radio dish in the world is 300 m (1000 ft) in diameter. So large a dish can't be supported in the usual way, so it is built into a mountain valley in Arecibo, Puerto Rico.

■ **Figure 6-19**

(a) The largest steerable radio telescope in the world is the GBT located in Green Bank, West Virginia. With a diameter of 100 m, it stands higher than the Statue of Liberty. (Mike Bailey: NRAO/AUII) (b) The 300-m (1000-ft) radio telescope in Arecibo, Puerto Rico, hangs from cables over a mountain valley. The Arecibo Observatory is part of the National Astronomy and Ionosphere Foundation operated by Cornell University and the National Science Foundation. (David Parker/SPL/Photo Researchers, Inc.)

The reflecting dish is a thin metallic surface supported above the valley floor by cables attached near the rim, and the antenna hangs above the dish on cables from three towers built on three mountain peaks that surround the valley (Figure 6-19b).

Although this telescope can look only overhead, the operators can change its aim slightly by moving the antenna and by waiting for Earth's rotation to point the telescope in the proper direction. This may sound clumsy, but the telescope's ability to detect weak radio sources, together with its good resolution, makes it one of the most important radio observatories in the world.

The third handicap the radio astronomer faces is interference. A radio telescope is an extremely sensitive radio receiver listening to radio signals thousands of times weaker than artificial radio and TV transmissions. Such weak signals are easily drowned out by interference. Sources of such interference include everything from poorly designed transmitters in Earth satellites to automobiles with faulty ignition systems. To avoid this kind of interference, radio astronomers locate their telescopes as far from civilization as possible. Hidden deep in mountain valleys, they are able to listen to the sky protected from human-made radio noise.

Advantages of a Radio Telescope

Building large radio telescopes in isolated locations is expensive, but three factors make it all worthwhile. First, and most important, a radio telescope can reveal clouds of cool hydrogen in space. Because 90 percent of the atoms in the universe are hydrogen, that is important information. Large clouds of cool hydrogen are completely invisible to normal telescopes because they produce no visible light of their own and reflect too little to be detected on photographs. However, cool hydrogen emits a radio signal at the specific wavelength of 21 cm. (You will see how the hydrogen produces this radiation in the discussion of the gas clouds in space in Chapter 10.) These hydrogen clouds are important because, for one thing, they are the places where stars are born. The only way astronomers can detect these clouds of gas is with a radio telescope that receives the 21-cm radiation, so that is one reason that radio telescopes are important.

The second reason is related to dust in space. Because radio signals have relatively long wavelengths, they can penetrate the vast clouds of dust that obscure astronomers' view at visual wavelengths. Light waves are short, and they interact with tiny dust grains floating in space; as a result, the light is scattered and never gets through the dust to reach optical telescopes on Earth. However, radio signals from far across the galaxy pass unhindered through the dust, giving radio astronomers an unobscured view.

Finally, radio telescopes are important because they can detect objects that are more luminous at radio wavelengths than at visible wavelengths. This includes everything from the coldest clouds of gas to the hottest stars. Some of the most distant objects in the universe, for instance, are detectable only at radio wavelengths.

Building Scientific Arguments

Why do optical astronomers build big telescopes, while radio astronomers build groups of widely separated smaller telescopes?
Once again you can learn a lot by building a scientific argument based on comparison. Optical astronomers build large telescopes to maximize light-gathering power, but the problem for radio telescopes is resolving power. Because radio waves are so much longer than light waves, a single radio telescope can't see details in the sky much smaller than the moon. By linking radio telescopes miles apart, radio astronomers build a radio interferometer that can simulate a radio telescope miles in diameter and thus increase the resolving power.

The difference between the wavelengths of light and radio waves makes a big difference in building the best telescopes. Keep that difference in mind as you build a new argument: **Why don't radio astronomers want to build their telescopes on mountaintops as optical astronomers do?**

■   ■   ■

Connections: Earth's atmosphere causes trouble for astronomers in two ways. It distorts images, and it absorbs radiation at many wavelengths. The only way to avoid these limitations completely is to send telescopes above the atmosphere, into space.

6-5 Astronomy from Space

YOU HAVE LEARNED about the observations that ground-based telescopes can make through the two atmospheric windows in the visible and radio parts of the electromagnetic spectrum. Most of the rest of the electromagnetic radiation—infrared, ultraviolet, X ray, and gamma ray—never reaches Earth's surface. To observe at these wavelengths, telescopes must fly above the atmosphere in high-flying aircraft, rockets, balloons, and satellites. The only exceptions are observations that can be made in the near-infrared and the near-ultraviolet.

The Ends of the Visual Spectrum

Astronomers can observe in the near-infrared just beyond the red end of the visible spectrum. You can't see this light, but some of it leaks through the atmosphere in narrow, partially open atmospheric windows scattered from 1200 nm to about 40,000 nm. Infrared astronomers usually measure wavelength in micrometers (10^{-6} meters), so they refer to this wavelength range as 1.2 to 40 micrometers (or microns for short). In this range, much of the radiation is absorbed by water vapor, but carbon dioxide and oxygen molecules also absorb infrared. As you saw earlier in this chapter, it is an advantage to place telescopes on mountaintops

Infrared astronomers can often observe with the dome lights on. Their instruments are not usually sensitive to visible light.

SOFIA will fly at roughly 12 km (over 40,000 ft) to get above most of Earth's atmosphere.

Adding liquid nitrogen to the camera on a telescope is a familiar task for astronomers.

■ Figure 6-20

Comet Hale–Bopp hangs in the sky over the 3-meter NASA Infrared Telescope Facility (IRAF) atop Mauna Kea. The air at high altitudes is so dry that it is transparent to shorter infrared photons. SOFIA will fly so high it will be able to observe infrared wavelengths that cannot be observed from mountaintops. Most astronomical CCD cameras must be cooled to low temperatures, and this is especially true for infrared cameras. (Hale–Bopp over observatory: William Keel/IRAF; SOFIA: NASA; Camera: Kris Koenig/Coast Learning Systems)

where the air is thin and dry. For example, a number of important infrared telescopes observe from the 4150-m (13,600-ft) summit of Mauna Kea in Hawaii. At this altitude, the telescopes are above much of the water vapor in Earth's atmosphere (■ Figure 6-20).

The far-infrared range, which includes wavelengths longer than 40 micrometers, carries clues to the nature of comets, planets, forming stars, and other cool objects, but these wavelengths are absorbed high in Earth's atmosphere—much higher than mountaintops. Infrared telescopes have flown to high altitudes under balloons and in airplanes. NASA is now building the Stratospheric Observatory for Infrared Astronomy (SOFIA), a Boeing 747 that will carry a 2.5-m telescope, control systems, and a team of technicians and astronomers to the fringes of the atmosphere. Once at that altitude, they can open a door above the telescope and make extended infrared observations as the plane flies a precisely calculated path. You can see the door in the photo in Figure 6-20. Some infrared wavelengths are too strongly absorbed by the atmosphere to be observed from even the highest-flying aircraft. The only way to make those infrared observations is to put the telescope in space above the atmosphere.

Wherever infrared telescopes are based, they have one thing in common—cooling. If a telescope observes at far-infrared wavelengths, then it must be cooled to low temperature. Infrared radiation is emitted by heated objects, and if the telescope is warm it will emit many times more infrared radiation than that coming from a distant object. Imagine trying to look for rabbits at night through binoculars that are themselves glowing. In a tele-

scope observing in the near-infrared, only the detector, the element on which the infrared radiation is focused, must be cooled, usually with liquid nitrogen, as shown in Figure 6-20. To observe in the far-infrared, however, the entire telescope must be cooled.

At the short-wavelength end of the spectrum, astronomers can observe in the near-ultraviolet. Your eyes don't detect this radiation, but it can be recorded by photographic plates and CCDs. Wavelengths shorter than about 290 nm, the far-ultraviolet, are completely absorbed by the ozone layer extending from 20 km to about 40 km above Earth's surface. No mountaintop is that high, and no airplane can fly to such an altitude. To observe in the far-ultraviolet or beyond at X-ray or gamma-ray wavelengths, telescopes must be in space above the atmosphere.

Telescopes in Space

To observe far beyond the ends of the visible spectrum, astronomical telescopes must go above Earth's atmosphere into space. This is very expensive and difficult, but it is the only way to study some processes. Stars, for example, are born inside clouds of dust and gas, and visible wavelengths cannot escape from these dust clouds. Only observations in the infrared can reveal the secrets of star formation. Black holes are small and hard to detect, but matter falling into a black hole emits X rays. Telescopes in space can explore such processes as these that are invisible from within Earth's atmosphere.

One of the most successful space telescopes was the International Ultraviolet Explorer (IUE), launched in 1978. It carried a

1 The Hubble Space Telescope was carried into orbit by the Space Shuttle in 1990. The telescope contains a 2.4-m (96-in.) mirror and can observe from the near-infrared to the near-ultraviolet.

Orbiting above Earth's blurring atmosphere, Hubble is limited only by diffraction in its optics. It can detect details 10 times smaller than Earth-based telescopes.

The telescope, as big as a large bus, has been visited twice by astronauts, who repaired equipment and installed new instruments. Named after Edwin Hubble, the astronomer who discovered the expansion of the universe, the telescope has been tremendously productive observing everything from the weather on Mars to the most distant galaxies visible in the universe.

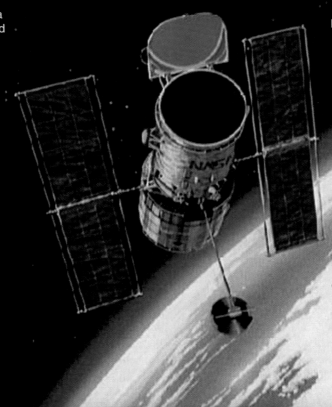

Hubble image of Mars and its polar cap.

Visual

Hubble image of a nebula around an aging star.

Visual

Hubble image of a dust-filled galaxy.

Visual

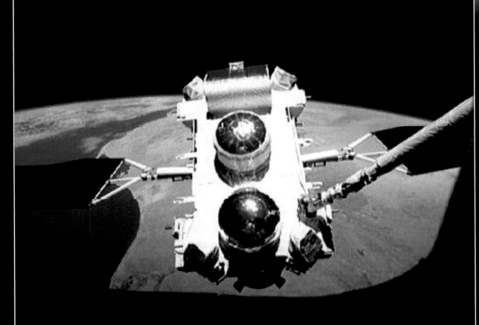

2 The Compton Gamma Ray Observatory at left was in orbit from 1991 to 2000. It made observations of very-high-energy photons, helping astronomers understand such violently active objects as neutron stars and black holes.

The Chandra X-Ray Observatory was placed in orbit 1/3 of the way to the moon in 1999. It is nearly 14 m (45 ft) long and carries highly precise mirrors 1.2 m (47 in.) in diameter. X rays would penetrate into regular mirrors, so Chandra's mirrors are designed as cylinders polished on the inside so that X rays just graze the surface and are focused onto detectors. The telescope was named after the late Indian-American Nobel laureate Subrahmanyan Chandrasekhar, who was a pioneer in many branches of theoretical astronomy.

Chandra can detect X-ray emitting objects 50 times fainter and resolve details 10 times smaller than any previous X-ray telescope.

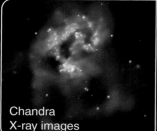
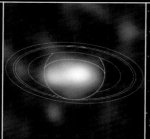

Chandra
X-ray images

Two galaxies collide and trigger the birth of stars.

Saturn emits X-rays from near its equator.

Very hot gas is trapped in a cluster of galaxies.

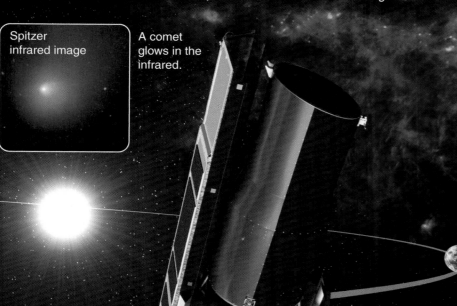

Spitzer infrared image

A comet glows in the infrared.

2b The Spitzer Space Telescope (at left) observes in the infrared. It was launched in 2003 and named in honor of the late astronomer Lyman Spitzer, a leader in space astronomy. The telescope is cooled to –273°C (–459°F) so it cannot orbit the warm Earth. Instead it is in an orbit around the sun and will drift slowly away from Earth during its lifetime. Protected from sunlight by a heat screen, it can observe a wide range of astronomical objects.

Dust warmed by hot young stars glows in the disk of this spiral galaxy.

Spitzer infrared image

Heat screen

Spitzer infrared image

An infrared image penetrates a dusty nebula to reveal newborn stars just beginning to shine.

3 The James Webb Space Telescope (at right) is planned as the next great observatory in space. Named after the early director of NASA who oversaw the planning of the Apollo moon landings, the telescope will carry a 6.5-m (256-in.) segmented mirror made of the metal beryllium. It will observe without a telescope tube from behind a multilayered sunscreen. Launch is planned for 2011.

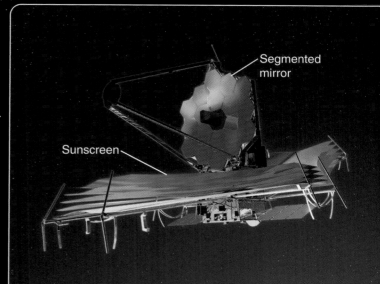

Segmented mirror

Sunscreen

telescope only 45 cm (18 in.) in diameter and was expected to last only a year or two, but it became the little telescope that could. It made many observations and exciting discoveries until it finally failed in 1996.

Many space telescopes are small satellites designed to make specific observations for a short period, but some are large general-purpose telescopes. Over two decades ago, astronomers developed a plan to place a series of great observatories in space. Those space telescopes have revolutionized human understanding of what we are and where we are in the universe. Study **The Great Observatories in Space** on pages 130–131 and notice three points:

1 Not only can a telescope in space observe at a wide range of wavelengths, but it is above the atmospheric blurring called seeing. The Hubble Space Telescope observes mostly at visual wavelengths and has the advantage of sharp images undistorted by seeing.

2 Notice how telescopes must be specialized for their wavelength range. The Compton Gamma Ray Observatory had special detectors, the Chandra X-Ray Observatory must have cylindrical mirrors, and the Spitzer infrared observatory must have cooled optics.

3 The Hubble Space Telescope has been maintained by visits from astronauts, but such visits are expensive, and the future of Hubble is in doubt. Astronauts cannot reach the Chandra and Spitzer telescopes, and the Compton Observatory was removed from orbit in 2000. Space observatories have limited lifetimes, and astronomers are already planning the next great observatory in space. The new James Webb Space Telescope will not be available for many years.

These great observatories in space are controlled from research centers on Earth and are open to proposals from any astronomer with a good idea; but competition is fierce, and only the most worthy projects win approval.

Cosmic Rays

All of the radiation you have read about in this chapter has been electromagnetic radiation. **Cosmic rays,** however, are not really rays; they are subatomic particles traveling at tremendous velocities that strike Earth's atmosphere from space. Almost no cosmic rays reach the ground, but they do smash gas atoms in the upper atmosphere, and fragments of those atoms shower down on you day and night over your entire life. These secondary cosmic rays are passing through you as you read this sentence.

Some cosmic-ray research can be done from high mountains or high-flying aircraft; but, to study cosmic rays in detail, detectors must go into space. A number of cosmic-ray detectors have been carried into orbit, but this area of astronomical research is just beginning to bear fruit.

Astronomers can't be sure what produces cosmic rays. Because they are atomic particles with electric charges, they are deflected by the magnetic fields spread through our galaxy, and that means you can't tell where they are coming from. The space between the stars is a glowing fog of cosmic rays. Some lower-energy cosmic rays come from the sun, but observations show that at least some cosmic rays are produced by the violent explosions of dying stars.

At present, cosmic rays largely remain an exciting mystery. You will meet them again in future chapters.

Building Scientific Arguments

Why can infrared astronomers observe from high mountain-tops, while X-ray astronomers must observe from space? Once again, you can analyze this question by building a scientific argument based on comparison. Infrared radiation is absorbed by water vapor in Earth's atmosphere. If you built an infrared telescope on top of a high mountain, you would be above most of the water vapor in the atmosphere, and you could collect some infrared radiation from the stars. The longer-wavelength infrared radiation is absorbed much higher in the atmosphere, so you couldn't observe it from our mountaintop. Similarly, X rays are absorbed in the uppermost layers of the atmosphere, and you would not be able to find any mountain high enough to get an X-ray telescope above those absorbing layers. To observe the stars at X-ray wavelengths, you would need to put your telescope in space, above Earth's atmosphere.

You can see why X-ray and far-infrared telescopes must observe from space. Now build another argument based on comparison: **Why must the Hubble Space Telescope be in space when it observes in the visual-wavelength range?**

■   ■   ■

Connections: The tools of the astronomer are designed to gather radiation from the sky and extract information. Perhaps no tool is as important as the spectrograph because no form of observation is as loaded with information as a spectrum. In the next chapter, you will see how astronomers can harvest the information in a star's spectrum.

Summary

6-1 | Radiation: Information from Space

What is light?
■ Light is the visible form of electromagnetic radiation, an electric and magnetic disturbance that transports energy at the speed of light. The electromagnetic spectrum includes gamma rays, X rays, ultraviolet radiation, visible light, infrared radiation, and radio waves.

■ You can think of a particle of light, a photon, as a bundle of waves that acts sometimes as a particle and sometimes as a wave.

■ The energy a photon carries depends on its wavelength. The wavelength of visible light, usually measured in nanometers (10^{-9} m), ranges from 400 nm to 700 nm. Infrared and radio photons have longer wavelengths and carry less energy. Ultraviolet, X-ray, and gamma-ray photons have shorter wavelengths and carry more energy.

6-2 | Optical Telescopes

How do telescopes work, and how are they limited?
■ Astronomers use telescopes to gather light, resolve fine detail, and magnify the image. The first two of these three powers of the telescope depend on the telescope's diameter. Consequently, astronomers strive to build telescopes with large diameters.

■ A refracting telescope uses a lens to bend the light and focus it into an image. Because of chromatic aberration, refracting telescopes cannot bring all colors to the same focus, resulting in color fringes around the images. An achromatic lens partially corrects for this, but such lenses are expensive and cannot be made much larger than about 1 m in diameter.

■ Reflecting telescopes use a mirror to focus the light and are less expensive than refracting telescopes of the same diameter. Also, reflecting telescopes do not suffer from chromatic aberration. Most recently built large telescopes are reflectors.

■ Astronomers build observatories on high mountains for two reasons. Turbulence in Earth's atmosphere blurs the image of an astronomical telescope, a phenomenon that astronomers refer to as seeing. Atop a mountain, the air is steady, and the seeing is better; the air on a mountaintop is also thin and dry and is more transparent, especially in the infrared.

■ Sometimes astronomical telescopes can be linked together to form an interferometer, which has a resolution equivalent to that of a telescope as large in diameter as the separation between the telescopes.

6-3 | Special Instruments

What kind of instruments do astronomers use to record and analyze light?
■ For many decades astronomers used photographic plates to record images at the telescope, but modern electronic systems such as CCD cameras have replaced photographic plates in most applications.

■ Spectrographs using prisms or a grating spread starlight out according to wavelength to form a spectrum revealing hundreds of spectral lines produced by atoms in the object being studied.

6-4 | Radio Telescopes

Why do astronomers use radio telescopes?
■ Astronomers use radio telescopes for three reasons: They can detect cool hydrogen in space; they can see through dust clouds that block visible light; and they can detect certain objects invisible at other wavelengths.

■ Most radio telescopes contain a dish reflector, an antenna, an amplifier, and a data recorder. Such a telescope can record the intensity of the radio energy coming from a spot on the sky. Scans of small regions are used to produce radio maps.

■ Because of the long wavelength, radio telescopes have very poor resolution, and astronomers often link separate radio telescopes together to form a radio interferometer capable of resolving much finer detail.

6-5 | Astronomy from Space

Why must some telescopes go into space?
■ Earth's atmosphere is transparent in two wavelength ranges called atmospheric windows, the visual window and the radio window. At other wavelengths, the atmosphere absorbs radiation. To observe at other wavelengths, telescopes must go into space.

■ Earth's atmosphere distorts and blurs images. Telescopes in orbit are above this seeing distortion and are limited only by diffraction in their optics.

■ Cosmic rays are not electromagnetic radiation; they are subatomic particles such as electrons and protons traveling at nearly the speed of light. They can best be studied from above Earth's atmosphere.

New Terms

electromagnetic radiation (p. 110)

wavelength (p. 110)

frequency (p. 111)

nanometer (nm) (p. 111)

Angstrom (Å) (p. 111)

photon (p. 111)

infrared radiation (p. 111)

ultraviolet radiation (p. 112)

atmospheric window (p. 112)

focal length (p. 114)

refracting telescope (p. 114)

reflecting telescope (p. 114)

primary lens, mirror (p. 114)

objective lens, mirror (p. 114)

eyepiece (p. 114)

chromatic aberration (p. 114)

achromatic lens (p. 114)

light-gathering power (p. 115)

resolving power (p. 115)

diffraction fringe (p. 115)

seeing (p. 115)

magnifying power (p. 116)

light pollution (p. 117)

prime focus (p. 120)

secondary mirror (p. 120)

Cassegrain focus (p. 120)

Newtonian focus (p. 120)

Schmidt–Cassegrain focus (p. 120)

sidereal drive (p. 121)

equatorial mounting (p. 121)

polar axis (p. 121)

alt-azimuth mounting (p. 121)

active optics (p. 121)

adaptive optics (p. 121)

interferometry (p. 122)

charge-coupled device (CCD) (p. 123)

false-color image (p. 123)

spectrograph (p. 123)

grating (p. 124)

comparison spectrum (p. 124)

radio interferometer (p. 127)

cosmic ray (p. 132)

Review Questions

Ace◐Astronomy™ Assess your understanding of this chapter's topics with additional quizzing and animations at
http://ace.brookscole.com/sf9

1. Why would you not plot sound waves in the electromagnetic spectrum?

2. If you had limited funds to build a large telescope, which type would you choose, a refractor or a reflector? Why?

3. Why do nocturnal animals usually have large pupils in their eyes? How is that related to astronomical telescopes?

4. Why do optical astronomers sometimes put their telescopes at the tops of mountains, while radio astronomers sometimes put their telescopes in deep valleys?

5. Optical and radio astronomers both try to build large telescopes but for different reasons. How do these goals differ?

6. What are the advantages of making a telescope mirror thin? What problems does this cause?

7. Small telescopes are often advertised as "200 power" or "magnifies 200 times." As someone knowledgeable about astronomical telescopes, how would you improve such advertisements?

8. Not long ago an astronomer said, "Some people think I should give up photographic plates." Why might she change to something else?

9. What purpose do the colors in a false-color image or false-color radio map serve?

10. How is chromatic aberration related to a prism spectrograph?

11. Why would radio astronomers build identical radio telescopes in many different places around the world?

12. Why do radio telescopes have poor resolving power?

13. Why must telescopes observing in the far-infrared be cooled to low temperatures?

14. What might you detect with an X-ray telescope that you could not detect with an infrared telescope?

15. The moon has no atmosphere at all. What advantages would you have if you built an observatory on the lunar surface?

16. The two images at the right show a star before and after an adaptive optics system was switched on. What causes the distortion in the first image, and how does adaptive optics correct the image?

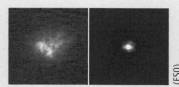
(ESO)

17. The X-ray image at right shows the remains of an exploded star. Explain why images recorded by telescopes in space are often displayed in false color rather than in the "colors" received by the telescope.

(NASA/CXC/PSU/ S. Park)

Discussion Questions

1. Why does the wavelength response of the human eye match so well the visual window of Earth's atmosphere?

2. Most people like beautiful sunsets with brightly glowing clouds, bright moonlit nights, and twinkling stars. Most astronomers don't. Why?

Problems

1. The thickness of the plastic in plastic bags is about 0.001 mm. How many wavelengths of red light is this?

2. What is the wavelength of radio waves transmitted by a radio station with a frequency of 100 million cycles per second?

3. Compare the light-gathering powers of one of the Keck telescopes and a 0.5-m telescope.

4. How does the light-gathering power of one of the Keck telescopes compare with that of the human eye? (Hint: Assume that the pupil of your eye can open to about 0.8 cm.)

5. What is the resolving power of a 25-cm telescope? What do two stars 1.5 seconds of arc apart look like through this telescope?

6. Most of Galileo's telescopes were only about 2 cm in diameter. Should he have been able to resolve the two stars mentioned in Problem 5?

7. How does the resolving power of the 5-m telescope compare with that of the Hubble Space Telescope? Why does the HST outperform the 5-m telescope?

8. If you build a telescope with a focal length of 1.3 m, what focal length should the eyepiece have to give a magnification of 100 times?

9. Astronauts observing from a space station need a telescope with a light-gathering power 15,000 times that of the human eye, capable of resolving detail as small as 0.1 second of arc, and having a magnifying power of 250. Design a telescope to meet their needs. Could you test your design by observing stars from Earth?

10. A spy satellite orbiting 400 km above Earth is supposedly capable of counting individual people in a crowd. Roughly what minimum-diameter telescope must the satellite carry? (Hint: Use the small-angle formula.)

Media Cluster

Ace◐Astronomy™ To access the resources in the Media Cluster, log into AceAstronomy at http://ace.brookscole .com/sf9 and select Chapter 6.

ACTIVE FIGURES

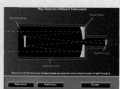

Refractors and Reflectors
Compare the two basic types of telescope designs and the paths light takes through each with this animation.

Resolution and Telescopes
This animation allows you to contrast the high resolution of larger-aperture telescopes with the lower resolution of smaller ones.

Lenses: Focal Length
Astronomers use lenses to focus starlight into an image. You can change the shape and material of the lens in this animation to study how these parameters affect the focal length.

Telescopes: Objective Lens and Eyepiece
Use this animation to study how the objective lens and eyepiece work together to form an image in a telescope. You can change parameters for the diameter of the lens and the focal lengths of the objective lens and eyepiece.

Telescopes and Resolution I
In this simulation, you can vary the focal length of the objective lens, the focal length of the eyepiece lens, and the diameter of the objective lens. Try them all and see which factor (or factors) determines how sharp the image is.

Telescopes and Resolution II
Get a feel for how wavelength affects the resolution of a telescope by comparing the lens diameter needed to observe craters on the moon with a visible-light telescope versus other types such as radio and microwave telescopes.

Particulate, Heat, and Light Pollution
These three animations let you study how environmental variables can affect viewing quality. How does this fit in with what you learned in this chapter about the locations of major astronomical telescopes?

The Electromagnetic Spectrum
In this exercise, study the properties of radiation, such as wavelength and sources of various parts of the electromagnetic spectrum.

Lab 4: Solar Wind and Cosmic Rays
This lab begins with an overview of the properties of the sun's atmosphere and how energetic particles escape and travel through the solar system. The lab ends with a discussion of cosmic rays.

Critical Inquiries for the Web

1. Research in chemistry, physics, and biology is supported in part by industry. Because astronomy has few industrial applications, it is not well supported by industry. Visit websites for major observatories and find out who pays their bills.

2. Visit websites for the major observatories in space, such as Hubble Space Telescope and Chandra, and check on their latest status. What kinds of observations are they making, and what kinds of discoveries are they making?

3. Astronomers are leaders in efforts to reduce electric power wasted on outdoor lighting because it causes light pollution. Find websites on light pollution and see if your town is making any effort to preserve the beauty of the night sky and simultaneously save electrical power.

Exploring *TheSky*

1. Astronomical telescopes using equatorial mountings must be aligned precisely with the north celestial pole. Locate Polaris and determine how far it is from the north celestial pole. (*Hint:* Use **Reference Lines** under the **View** menu and check **Grid** under Equatorial. Be sure the spacing is set to auto/fine. Then locate the Little Dipper and zoom in on Polaris.)

Go to the Brooks/Cole Astronomy Resource Center **(http:// astronomy.brookscole.com)** for critical thinking exercises, articles, and additional readings from InfoTrac College Edition, Brooks/Cole's online student library.

7 | Starlight and Atoms

Awake! for Morning in the

Bowl of Night

Has flung the Stone that puts

the Stars to Flight:

And Lo! the Hunter of the

East has caught

The Sultan's Turret in a

Noose of Light.

THE RUBÁIYÁT OF OMAR KHAYYÁM,
TRANS. EDWARD FITZGERALD

Visual-wavelength image

THE UNIVERSE IS FILLED with fabulously beautiful clouds of glowing gas illuminated by brilliant stars, but it is all hopelessly beyond reach. No laboratory jar on Earth holds a sample labeled "star stuff," and no space probe has ever visited the inside of a star. The stars are far away, and the only information you can obtain about them comes hidden in starlight (■ Figure 7-1). Whatever you want to know about stars you must catch in a noose of light. ▌ Earthbound humans knew almost nothing about stars until the early 19th century, when the Munich optician Joseph von Fraunhofer studied the solar spectrum and found it interrupted by some 600 dark lines. As scientists realized that the lines were related to the various atoms in the sun and found that stellar spectra had similar patterns of lines, the door to an understanding of stars finally opened.

▌ Continued on page 138 ▌

Clouds of glowing gas illuminated by hot, bright stars lie thousands of light years across space, but clues hidden in starlight tell a story of star birth and star death. (ESO)

Guidepost

Looking Back

In the last chapter you read how hard astronomers work to gather light from the stars using giant telescopes on mountaintops and in space. You also read how spectrographs can spread light out into spectra. Now you are ready to see what all the fuss is about.

This Chapter

This chapter explains how light interacts with matter and how astronomers must understand that interaction to understand stars. Here you will find answers to five essential questions:

How do stars produce light?

What is an atom?

How do atoms interact with light?

What kind of spectra do you see when you look at celestial objects?

What can you learn from a star's spectrum?

This chapter marks a change in the way you will look at nature. Up to this point, you have been thinking about what you can see with your eyes alone or aided by telescopes. In this chapter, you begin using modern astrophysics, the application of physics to study the sky. Now you can search out secrets of the stars that lie beyond what you can see.

Looking Ahead

The analysis of spectra is a powerful tool, and in the next chapter you will use that tool to study the sun. In the chapters that follow, you will study other suns—the stars.

Ace ⑤Astronomy™ The AceAstronomy icon throughout the text indicates an opportunity for you to test yourself on key concepts and to explore animations and interactions on the AceAstronomy website at: http://ace .brookscole.com/sf9

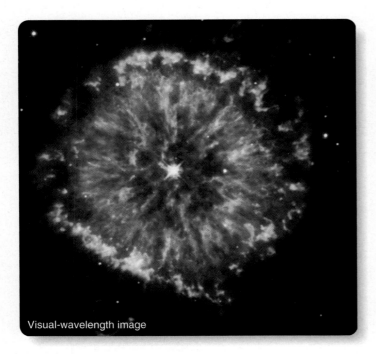

■ Figure 7-1

What's going on here? The sky is filled with beautiful and mysterious objects that lie far beyond your reach—in the case of the nebula NGC 6751, about 6500 ly beyond your reach. The only way to understand such objects is by analyzing their light. Such an analysis reveals that this object is a dying star surrounded by the expanding shell of gas it ejected a few thousand years ago. You will learn more about this phenomenon in Chapter 13. (NASA Hubble Heritage Team/STScI/AURA)

In this chapter, you will go through that door by considering how stars produce light and how atoms interact with light to produce spectral lines. The first step is to consider the hydrogen atom because it is the most common atom in the universe as well as the simplest. Other atoms are larger and more complicated, but in many ways their properties resemble those of hydrogen.

Once you understand how an atom's structure can interact with light to produce spectral lines, you will recognize certain patterns in stellar spectra. By classifying the spectra according to these patterns, you can arrange the stars in a sequence according to temperature. One of the most important pieces of information revealed in a star's spectrum is its temperature.

But, properly analyzed, a stellar spectrum can tell you much more. The spectrum contains information about the chemical composition of the star and the star's motion relative to Earth.

7-1 Starlight

IF YOU LOOK at the stars in the constellation Orion, you will notice that they are not all the same color (see Figure 2-4). One of our Favorite Stars, Betelgeuse, in the upper left corner, is quite red; another Favorite Star, Rigel, in the lower right corner, is blue.

These differences in color arise from the way the stars produce light, and they provide an important clue to the temperatures of stars.

Temperature and Heat

Temperature is one of the defining characteristics of a star. That is, if you know a star's temperature and a few other properties, such as size, you can understand the star. But if you don't know a star's temperature, you can know almost nothing about it.

A gas is made up of particles—atoms and molecules—that are in constant motion, colliding with one another millions of times a second. The **temperature** of a gas is a measure of the average kinetic energy of the particles. Recall that kinetic energy is energy of motion, so if a gas is hot, the particles must be moving very rapidly; if the gas is cool, the particles are moving more slowly. The motion of the particles represents stored energy, and you feel that energy as heat. Read **Focus on Fundamentals 3** and notice the distinction between the temperature, the stored energy, and the heat you feel.

Astronomers express the temperature of stars on the **Kelvin temperature scale.** The Kelvin scale sets its zero point at **absolute zero,** the temperature at which the particles of a gas have no remaining kinetic energy that could be extracted as heat. Absolute zero is −273.2°C, which is −459.7°F. (See Appendix A.) The temperatures of the stars range from less than 2000 K to 40,000 K or more. To see how astronomers can estimate the temperatures of the stars from starlight, you must understand how stars produce light.

The Origin of Starlight

The starlight you see comes from gases in the outer surface layers of the star—the photosphere. The gases deep inside the star also emit light, but it is absorbed before it can escape, and the low-density gas above the photosphere is too thin to emit significant amounts of light. Starlight comes from the photosphere of a star, that layer of gases dense enough to emit significant amounts of light but thin enough to allow the light to escape. (Recall that you met the photosphere of the sun in Chapter 3.) So when astronomers refer to a star as "hot" they are referring to the temperature of its visible surface—its photosphere.

To see how the photosphere of a star can produce light, you must consider two things, how photons are produced and how the temperature of a material is related to motion among its atoms.

First, a photon can be produced by a changing electric field. An **electron** is a negatively charged subatomic particle, and if you disturb the motion of an electron, the sudden change in the electric field around it can produce an electromagnetic wave. For example, if you run a comb through your hair while standing near a radio tuned to an AM station, you can produce popping noises on the radio. The moving comb disturbs the electrons in both the comb and your hair, building static electricity. The sudden sparks of static electricity produce electromagnetic waves that the radio

Focus on Fundamentals ③

Temperature, Heat, and Thermal Energy

One of the most common misconceptions in science involves temperature. People often say "temperature" when they really mean "heat," and sometimes they say "heat" when they mean something entirely different. This is a fundamental idea, so you need to understand the differences.

When something is hot, the particles in the object, be they atoms or molecules, are moving rapidly. Temperature is a measure of the average motion of the particles. (Mathematically, temperature is proportional to the square of the average velocity.) In a hot object, the particles race around at higher speeds than in a cool object. If you have your temperature taken, it will probably be 98.6°F, an indication that the atoms and molecules in your body are moving about at a normal pace. If you measure the temperature of a month-old baby, the thermometer should register the same temperature, showing that the atoms and molecules in the baby's body are moving at the same average velocity as the atoms and molecules in your body.

The energy of the moving particles in a body is called **thermal energy.** You have much more mass than the baby, so you must contain more thermal energy even though you have the same temperature. The thermal energy in your body and in the baby's body have the same intensity (temperature) but different amounts. People often confuse temperature and thermal energy, so you must be careful to distinguish between them. Temperature is an intensity, and thermal energy is an amount.

Many people say "heat" when they should say thermal energy. Heat is the thermal energy that moves from a hot object to a cool object. If two objects have the same temperature, you and the infant for example, there is no transfer of thermal energy and no heat. This is a fine distinction, but when you hear someone say "heat," check to see if he or she doesn't really mean thermal energy.

What's the difference between temperature and heat?

You may have burned yourself on cheese pizza, but you probably haven't burned yourself on green beans. At the same temperature, cheese holds more thermal energy than green beans. It isn't the temperature that burns your tongue, but the flow of thermal energy, and that's heat.

PRESSURE | MASS | ENERGY | **TEMPERATURE AND HEAT** | DENSITY

---

picks up as pops and crackles. This illustrates an important principle: Any change in the motion of an electron can generate an electromagnetic wave.

Second, recall that temperature is related to the motion among the particles in a material. In the gases of a hot star, the atoms move faster, on average, than the atoms in a cool star.

Now put these two ideas together, and you can understand why a hot object glows. The hotter an object is, the more violent the motion among its particles. The agitated particles collide with electrons, and when electrons are accelerated, part of the energy is carried away as electromagnetic radiation in the form of a photon. The radiation emitted by a heated object is called **black body radiation,** a name that refers to the way a perfect emitter of radiation would behave. A perfect emitter would also be a perfect absorber and, at room temperature, would look black. You will often find the term *black body radiation* referring to objects that glow brightly.

Black body radiation is quite common. In fact, it is responsible for the light emitted by an incandescent lightbulb. Electricity flowing through the wire filament of the lightbulb heats the wire to high temperature, and it glows. You can also recognize the light emitted by a heated horseshoe in the blacksmith's forge as black body radiation. Many objects in astronomy, including stars, emit radiation approximately as if they were black bodies.

Hot objects emit black body radiation, but so do cold objects. Ice cubes are cold, but their temperature is higher than absolute zero, so they contain some thermal energy and must emit some black body radiation. The coldest gas drifting in space has a temperature only a few degrees above absolute zero, but it too emits black body radiation.

Two important features of black body radiation will help you understand what you see when you look at a star. First, the hotter an object is, the more black body radiation it emits. Hot objects emit more radiation because their agitated particles travel faster and collide more often. So, of course, you expect a glowing coal from a fire to emit more total energy than an ice cube of the same size.

The second feature is the relationship between the temperature of the object and the wavelengths of the photons it emits. The wavelength of a photon emitted when a particle collides with an electron depends on the violence of the collision. Only a violent collision can produce a short-wavelength (high-energy) photon. Because extremely violent collisions don't occur very often, short-wavelength photons are rare. Similarly, most collisions are not extremely gentle, so long-wavelength (low-energy) photons are also rare. Consequently, black body radiation is made up of photons with a distribution of wavelengths, and very short and very long wavelengths are rare. The **wavelength of maximum** intensity (λ_{max}), the wavelength at which the object emits the most radiation, occurs at some intermediate wavelength.

■ Figure 7-2 shows the intensity of radiation versus wavelength for three objects of different temperatures. The curves are high in the middle and low at either end, which tells you that these objects emit most intensely at intermediate wavelengths. The total area under each curve is proportional to the total energy emitted, and you can see that the hotter object emits more total energy than the cooler objects. Look closely at the curves, and you will notice that the wavelength of maximum intensity depends on temperature. The hotter the object is, the shorter the wavelength

■ Figure 7-2

Black body radiation from three bodies at different temperatures demonstrates that a hot body radiates more total energy and that the wavelength of maximum intensity is shorter for hotter objects. The hotter object here will look blue to your eyes, while the cooler object will look red.

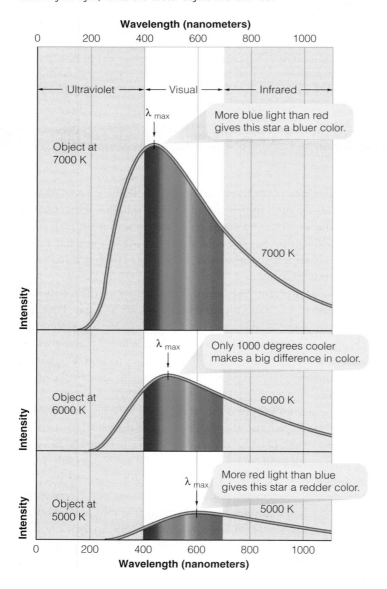

of maximum intensity. Notice in this figure how temperature determines the color of a glowing black body. The hotter object emits more blue light than red and thus looks blue, and the cooler object emits more red light than blue and consequently looks red. Now you can understand why two of our Favorite Stars, Betelgeuse and Rigel, have such different colors. Betelgeuse is cool and looks red, but Rigel is hot and looks blue

Notice that cool objects may emit little visible radiation but are still producing black body radiation. For example, the human body has a temperature of 310 K and emits black body radiation mostly in the infrared part of the spectrum. Infrared security cameras can detect burglars by the radiation they emit, and mosquitoes can track you down in total darkness by homing in on your infrared radiation. Although humans emit lots of infrared radiation, you rarely emit higher energy photons; and you almost never emit X-ray or gamma-ray photons. Your wavelength of maximum intensity lies in the infrared part of the spectrum.

Two Radiation Laws

The two features of black body radiation that you have just considered can be given precise mathematical form, and they have proven so dependable, they are known as laws. One law is related to energy and one to color.

As you saw in the previous section, a hot object emits more black body radiation than a cool object. That is, it emits more energy. Recall from Chapter 5 that energy is expressed in units called joules (J); 1 joule is about the energy of an apple falling from a table to the floor. The total radiation given off by 1 square meter of the object in joules per second equals a constant number, represented by σ, times the temperature raised to the fourth power.* This relationship is called the Stefan–Boltzmann law:

$$E = \sigma T^4 \ (\text{J/s/m}^2)$$

How does this help you understand stars? Suppose a star the same size as the sun had a surface temperature that was twice as hot as the sun's surface. Then each square meter of that star would radiate not twice as much energy but 2^4, or 16, times as much energy. From this law you can see that a small difference in temperature can produce a very large difference in the amount of energy emitted.

The second radiation law is related to the color of stars. In the previous section, you saw that hot stars look blue and cool stars look red. Wien's law tells you that the wavelength at which a star radiates the most energy, its wavelength of maximum intensity (λ_{max}), depends only on the star's temperature:

$$\lambda_{max} = \frac{3,000,000}{T}$$

---

*For the sake of completeness, you can note that the constant σ equals 5.67×10^{-8} J/m²s degree⁴.

That is, the wavelength of maximum radiation in nanometers equals 3 million divided by the temperature on the Kelvin scale.

This is a powerful tool in astronomy, because it means you can relate the temperature of a star and its wavelength of maximum intensity. For example, you might find a star that has a surface temperature of 3000 K. Then its wavelength of maximum intensity would be 3,000,000/3000, or 1000 nm—in the near-infrared. Later you will meet objects much hotter than most stars; such objects radiate most of their energy at very short wavelengths. The hottest stars, for instance, radiate most of their energy in the ultraviolet.

Now you can understand how astronomers can estimate the temperature of a star's surface from its color. The astronomers use the same technique that doctors and nurses use to measure a patient's body temperature. Medical personnel use a small device that detects the infrared radiation emerging from the patient's ear. You might suspect the device depends on the Stephan–Boltzmann law and measures the intensity of the infrared radiation. A person with a fever will emit more energy than a healthy person. However, a healthy person with a large ear canal would emit more energy than a person with a small ear canal, so measuring intensity wouldn't be accurate. The device actually depends on Wien's law in that it measures the "color" of the infrared radiation. A patient with a fever will emit at a slightly shorter wavelength of maximum intensity, and the infrared radiation emerging from that person's ear will be a tiny bit "bluer" than normal. Remember that the wavelength of maximum intensity depends on the temperature, so even a small fever produces changes in the color of the infrared radiation. Astronomers use the same technique to find the temperatures of the stars; they observe the color of the starlight.

Ace ⊙ Astronomy™ Log into AceAstronomy and select this chapter to see Astronomy Exercise "Black Body." Change the temperature and watch the black body curve change.

Ace ⊙ Astronomy™ Log into AceAstronomy and select this chapter to see Astronomy Exercise "Stefan–Boltzmann Law." Watch the brightness of a star change as you adjust its temperature.

Building Scientific Arguments

Why does the wavelength of maximum intensity depend on temperature?
Remember as you build any scientific argument that it must proceed logically, step by step. No gaps are allowed. In this case, you can begin with the temperature of an object. The hotter the object is, the more rapidly its particles move. That means that the typical collision between a particle and an electron will be more violent in a hotter body, and more violent collisions will accelerate the electrons more violently

and, on average, produce higher-energy, shorter-wavelength photons. So the wavelength of maximum intensity depends on the temperature of the body. That's Wien's law. It all makes sense when you think about it step by step.

Now change the argument to answer a different question: **Why does the Stefan–Bolzmann law work?**

■   ■   ■

Connections: You have been thinking about atoms in a general way, but now it is time to get specific. What is an atom, and how can it interact with light?

7-2 Atoms

THE ATOMS IN THE SURFACE LAYERS of stars leave their marks on the light the stars emit. By understanding what atoms are and how they interact with light, you can decode the spectra of the stars.

A Model Atom

To think about atoms and how they can interact with light, you need a working model of an atom. In Chapter 2, you used a working model of the sky, the celestial sphere. You identified and named the important parts and described how they were located and how they interacted. So you should begin your study of atoms by creating a model of an atom.

Your model atom contains a positively charged **nucleus** at the center; this nucleus consists of two kinds of particles. **Protons** carry a positive electrical charge, and **neutrons** have no charge. That means the nucleus has a net positive charge.

The nucleus in this model atom is surrounded by a whirling cloud of orbiting electrons, low-mass particles with a negative charge. In a normal atom, the number of electrons equals the number of protons, and the positive and negative charges balance to produce a neutral atom. Because protons and neutrons each have a mass 1836 times greater than that of an electron, most of the mass of the atom lies in the nucleus. The hydrogen atom is the simplest of all atoms. The nucleus is a single proton orbited by a single electron, with a total mass of only 1.67×10^{-27} kg, about a trillionth of a trillionth of a gram.

An atom is mostly empty space. To see this, imagine constructing a simple scale model. The nucleus of a hydrogen atom is a proton with a diameter of about 0.0000016 nm, or 1.6×10^{-15} m. If you multiply this by one trillion (10^{12}), you can represent the nucleus of your model atom with a grape seed, which is about 0.16 cm in diameter. The region of a hydrogen atom containing the whirling electron has a diameter of about 0.4 nm, or 4×10^{-10} m. Multiplying by a trillion magnifies the diameter to about 400 m, or about 4.5 football fields laid end to end (■ Figure 7-3). When you imagine a grape seed in the midst of a sphere

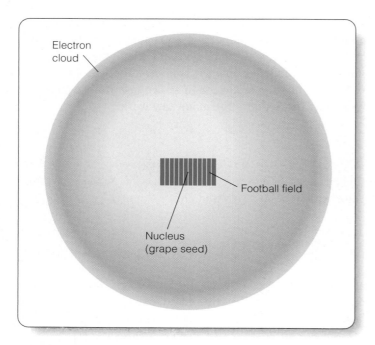

■ Figure 7-3

Magnifying a hydrogen atom by 10^{12} makes the nucleus the size of a grape seed and the diameter of the electron cloud about 4.5 times longer than a football field. The electron itself is still too small to see.

4.5 football fields in diameter, you can see that an atom is mostly empty space.

Different Kinds of Atoms

There are over a hundred kinds of atoms, called chemical elements. Which element an atom is depends only on the number of protons in the nucleus. For example, carbon has six protons in its nucleus. An atom with one more proton than this is nitrogen, and an atom with one proton fewer is boron.

Although the number of protons in an atom of an element is fixed, you could change the number of neutrons in an atom's nucleus without changing the atom significantly. For instance, if you added a neutron to a carbon nucleus, you would still have carbon, but it would be slightly heavier than normal carbon. Atoms that have the same number of protons but a different number of neutrons are **isotopes.** Carbon has two stable isotopes. One form contains six protons and six neutrons for a total of 12 particles and is thus called carbon-12. Carbon-13 has six protons and seven neutrons in its nucleus (■ Figure 7-4).

Protons and neutrons are bound tightly into the nucleus, but the electrons are held loosely in the electron cloud. Running a comb through your hair creates a static charge by removing a few electrons from their atoms. This process is called **ionization,** and the atom that has lost one or more electrons is an **ion.** A carbon atom is neutral if it has six electrons to balance the positive charge of the six protons in its nucleus. If you ionize the atom by remov-

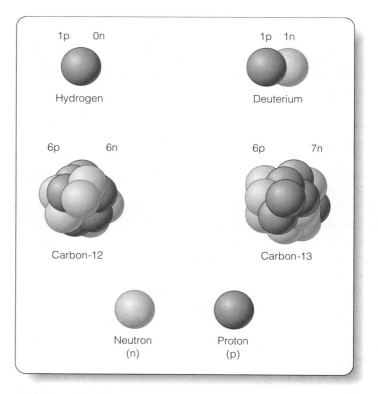

■ Figure 7-4

Some common isotopes. A rare isotope of hydrogen, deuterium, contains a proton and a neutron in its nucleus. Two isotopes of carbon are carbon-12 and carbon-13.

ing one or more electrons, the atom is left with a net positive charge. Under some circumstances, an atom may capture one or more extra electrons, giving it more negative charges than positive. Such a negatively charged atom is also considered an ion.

Atoms that collide may form bonds with each other by exchanging or sharing electrons. Two or more atoms bonded together form a **molecule.** Atoms do collide in stars, but the high temperatures cause violent collisions that are unfavorable for chemical bonding. Only in the coolest stars are the collisions gentle enough to permit the formation of chemical bonds. You will see later that the presence of molecules such as titanium oxide (TiO) in a star is a clue that the star is very cool. In later chapters, you will see that molecules can form in cool gas clouds in space and in the atmospheres of planets.

Electron Shells

So far you have been thinking of the cloud of the whirling electrons in a general way, but now it is time to be more specific as to how the electrons behave within the cloud.

The electrons are bound to the atom by the attraction between their negative charge and the positive charge on the nucleus. This attraction is known as the **Coulomb force,** after the French physicist Charles-Augustin de Coulomb (1736–1806). To ionize

Quantum Mechanics: The World of the Very Small

Quantum mechanics is the set of rules that describe how atoms and sub-atomic particles behave. When you think about large objects such as stars, planets, aircraft carriers, and hummingbirds, you don't have to think about quantum mechanics, but on the atomic scale, particles behave in ways that seem unfamiliar.

One of the principles of quantum mechanics specifies that you cannot know simultaneously the exact location and motion of a particle. This is why physicists refer to the electrons in an atom as if they were a cloud of negative charge surrounding the nucleus. Because you can't know the position and motion of the electron, you can't really describe it as a small

particle following an orbit. You can use that image as a model to help your imagination, but the reality is much more interesting, and describing the electrons as a charge cloud provides a better and more sophisticated model of an atom.

This raises some serious questions about reality. Is an electron really a particle at all? Quantum mechanics can describe particles as waves and waves as particles. If you can't know simultaneously the position and motion of a specific particle, how can you know how it will react to a collision with a photon or another particle? The answer is that you can't know, and that seems to violate the principle of cause and effect (Window on Science 5-1).

Needless to say, you can't expect to explore these discrepancies here. Scientists and philosophers of science continue to struggle with the meaning of reality on the quantum-mechanical level. Here you should note that the reality you see on the scale of stars and hummingbirds is only part of nature. You have constructed a model to help think about nature on the scale of atoms, but the truth is much more interesting and much more exciting than anything you see on larger scales. Although you will use models of atoms to study stars, there is still much to learn about the atoms themselves.

---

an atom, you need a certain amount of energy to pull an electron away from the nucleus. This energy is the electron's **binding energy,** the energy that holds it to the atom.

An electron may orbit the nucleus at various distances. If the orbit is small, the electron is close to the nucleus, and a large amount of energy is needed to pull it away. Consequently, its binding energy is large. An electron orbiting farther from the nucleus is held more loosely, and less energy will pull it away. That means it has less binding energy. The size of an electron's orbit is related to the energy that binds it to the atom.

Nature permits atoms only certain amounts (quanta) of binding energy, and the laws that describe how atoms behave are called the laws of **quantum mechanics (Window on Science 7-1)**. Much of this discussion of atoms is based on the laws of quantum mechanics.

Because atoms can have only certain amounts of binding energy, your model atom can have orbits of only certain sizes, called **permitted orbits.** These are like steps in a staircase: You can stand on the number-one step or the number-two step, but not on the number-one-and-one-quarter step. The electron can occupy any permitted orbit but not orbits in between.

The arrangement of permitted orbits depends primarily on the charge of the nucleus, which in turn depends on the number of protons. Consequently, each kind of element has its own pattern of permitted orbits (■ Figure 7-5). Isotopes of the same ele-

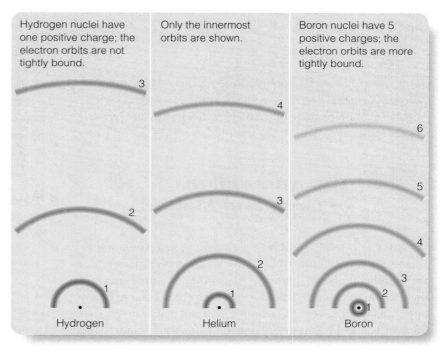

■ **Figure 7-5**

The electron in an atom may occupy only certain, permitted orbits. Because different elements have different charges on their nuclei, the elements have different, unique patterns of permitted orbits.

ments have nearly the same pattern because they have the same number of protons. However, ionized atoms have orbital patterns that differ from their un-ionized forms. Thus the arrangement of permitted orbits differs for every kind of atom and ion.

How many hydrogen atoms would it take to cross the head of a pin?

This is not a frivolous question. In answering it, you will discover how small atoms really are, and you will see how powerful physics and mathematics can be as a way to understand nature. Many scientific arguments are convincing because they have the precision of mathematics. To begin, assume that the head of a pin is about 1 mm in diameter. That is 0.001 m. The size of a hydrogen atom is represented by the diameter of the electron cloud, roughly 0.4 nm. Because 1 nm equals 10^{-9} m, you can multiply and discover that 0.4 nm equals 4×10^{-10} m. To find out how many atoms would stretch 0.001 m, you can divide the diameter of the pinhead by the diameter of an atom. That is, divide 0.001 m by 4×10^{-10} m, and you get 2.5×10^6. It would take 2.5 million hydrogen atoms lined up side by side to cross the head of a pin.

Now you can see how tiny an atom is and also how powerful a bit of physics and mathematics can be. It reveals a view of nature beyond the capability of your eyes. Now build an argument using another bit of arithmetic: **How many hydrogen atoms would you need to add up to the mass of a paper clip (1 g)?**

■   ■   ■

Connections: Astronomers are experts on electron orbits because the electrons in those orbits can interact with light. Such interactions fill starlight with clues to the secrets of the stars.

(7-3) The Interaction of Light and Matter

IF LIGHT DID NOT INTERACT with matter, you would not be able to see these words. In fact, you would not exist, because, among other problems, photosynthesis would be impossible, and there would be no grass, wheat, bread, beef, cheeseburgers, or any other kind of food. The interaction of light and matter makes your life possible, and it also makes it possible for you to understand the universe.

You should begin your study of light and matter by considering the hydrogen atom. As you read earlier, hydrogen is both simple and common. Roughly 90 percent of all atoms in the universe are hydrogen.

The Excitation of Atoms

Each orbit in an atom represents a specific amount of binding energy, so physicists commonly refer to the orbits as **energy levels.** Using this terminology, you can say that an electron in its small-est and most tightly bound orbit is in its lowest permitted energy level. You could move the electron from one energy level to another by supplying enough energy to make up the difference between the two energy levels. It would be like moving a flowerpot from a low shelf to a high shelf; the greater the distance between the shelves, the more energy you would need to raise the pot. The amount of energy needed to move the electron is the energy difference between the two energy levels.

If you move the electron from a low energy level to a higher energy level, you can call the atom an **excited atom.** That is, you have added energy to the atom in moving its electron. If the electron falls back to the lower energy level, that energy is released.

An atom can become excited by collision. If two atoms collide, one or both may have electrons knocked into a higher energy level. This happens very commonly in hot gas, where the atoms move rapidly and collide often.

Another way an atom can get the energy that moves an electron to a higher energy level is to absorb a photon. Only a photon with exactly the right amount of energy can move the electron from one level to another. If the photon has too much or too little energy, the atom cannot absorb it. Because the energy of a photon depends on its wavelength, only photons of certain wavelengths can be absorbed by a given kind of atom. ■ Figure 7-6 shows the lowest four energy levels of the hydrogen atom along with three photons the atom could absorb. The longest-wavelength photon has only enough energy to excite the electron to the second energy level, but the shorter-wavelength photons can excite the electron to higher levels. A photon with too much or too little energy cannot be absorbed. Because the hydrogen atom has many more energy levels than shown in Figure 7-6, it can absorb photons of many different wavelengths.

■ **Figure 7-6**

A hydrogen atom can absorb only those photons that move the atom's electron to one of the higher-energy orbits. Here three different photons are shown along with the change they would produce if they were absorbed.

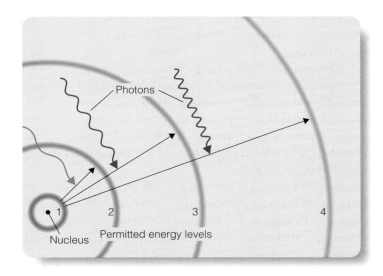

Atoms, like humans, cannot exist in an excited state forever. The excited atom is unstable and must eventually (usually within 10^{-6} to 10^{-9} seconds) give up the energy it has absorbed and return its electron to the lowest energy level. Because the electrons eventually tumble down to this bottom level, physicists call it the **ground state.**

When the electron drops from a higher to a lower energy level, it moves from a loosely bound level to one more tightly bound. The atom then has a surplus of energy—the energy difference between the levels—that it can emit as a photon. Study the sequence of events in ■ Figure 7-7 to see how an atom can absorb and emit photons. Because each type of atom or ion has its unique set of energy levels, each type absorbs and emits photons with a unique set of wavelengths. As a result, you can identify the elements in a gas by studying the characteristic wavelengths of light absorbed or emitted.

The process of excitation and emission is a common sight in urban areas at night. A neon sign glows when atoms of neon gas in the glass tube are excited by electricity flowing through the tube. As the electrons in the electric current flow through the gas, they collide with the neon atoms and excite them. As you have seen, immediately after an atom is excited, its electron drops back to a lower energy level, emitting the surplus energy as a photon of a certain wavelength. The photons emitted by excited neon produce a reddish-orange glow. Signs of other colors, erroneously called "neon," contain other gases or mixtures of gases instead of pure neon.

The Formation of a Spectrum

The spectrum of a star is formed as light passes outward through the gases near its surface. Study **Atomic Spectra** on pages 146–147. Notice three important properties of spectra:

1 There are three kinds of spectra described by three simple rules. When you see one of these types of spectra, you can recognize the kind of matter that emitted the light.

2 Notice that the wavelengths of the photons that are absorbed or emitted are determined by the atomic energy levels in the atoms. The emitted photons coming from a hot cloud of hydrogen gas have the same wavelengths as the photons absorbed by hydrogen atoms in the gases of a star. Although the hydrogen atom produces many spectral lines from the ultraviolet to the infrared, only three are visible to human eyes.

3 Most modern astronomy books display spectra as graphs of intensity versus wavelength. Be sure you see the connection between dark absorption lines and dips in the graphed spectrum.

Spectra are filled to bursting with information about the sources of the light; but, to extract that information, astronomers must be experts on the interaction of light and matter. Electrons moving among the orbits within atoms can reveal the secrets of the stars.

Ace◯Astronomy™ Log into AceAstronomy and select this chapter to see Astronomy Exercise "Emission and Absorption Spectra." This exercise will give you an inside look at atoms as they interact with photons.

Building Scientific Arguments

What spectrum would you see if you observed molten iron? To analyze this situation, you can begin your argument with the well-understood Kirchhoff's laws. Molten iron is a dense liquid, and the atoms and molecules collide so often they emit all wavelengths, so you would see a continuous spectrum. That is Kirchhoff's first law. But in order to see the molten iron, you would have to look through the hot vapors rising from it. Photons on their way to your spectrograph would pass through these gases, and atoms in the gases would absorb certain wavelengths. So what you would really see would be the continuous spectrum of the molten iron with weak absorption lines caused by the gases above the iron. This is what Kirchhoff's third law describes.

Now expand this argument. Suppose you had a very sensitive spectrograph that could look at the hot gases above the molten iron from the side so as to avoid looking directly at the molten iron. **What kind of spectrum would the hot gases emit?**

■　　　■　　　■

Connections: Whatever kind of spectrum astronomers look at, the most common spectral lines are the Balmer lines of hydrogen, the only hydrogen lines that can be studied from Earth's surface. In the next section, you will see how the Balmer lines work like a thermometer to measure a star's temperature.

■ **Figure 7-7**

An atom can absorb a photon only if the photon has the correct amount of energy. The excited atom is unstable and within a fraction of a second returns to a lower energy level, reradiating the photon in a random direction.

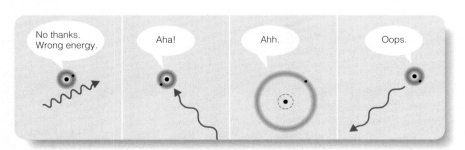

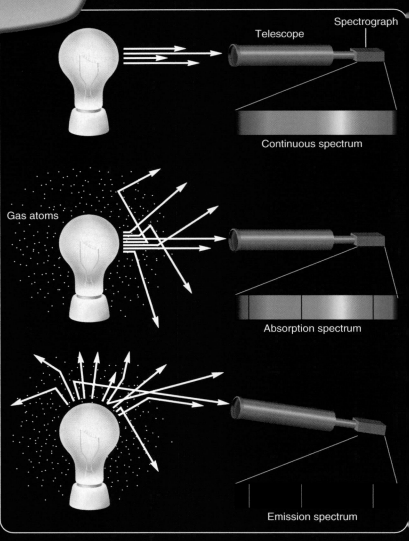

Telescope

Spectrograph

Continuous spectrum

Gas atoms

Absorption spectrum

Emission spectrum

1 To understand how to analyze a spectrum, begin with a simple incandescent lightbulb. The hot filament emits black body radiation, which forms a **continuous spectrum.**

An **absorption spectrum** results when radiation passes through a cool gas. In this case you can imagine that the lightbulb is surrounded by a cool cloud of gas. Atoms in the gas absorb photons of certain wavelengths, which are missing from the spectrum, and you see their positions as dark **absorption lines.** Such spectra are sometimes called **dark-line spectra.**

An **emission spectrum** is produced by photons emitted by an excited gas. You could see **emission lines** by turning your telescope aside so that photons from the bright bulb did not enter the telescope. The photons you would see would be those emitted by the excited atoms near the bulb. Such spectra are also called **bright-line spectra.**

1a The spectrum of a star is an absorption spectrum. The denser layers of the photosphere emit black body radiation. Gases in the atmosphere of the star absorb their specific wavelengths and form dark absorption lines in the spectrum.

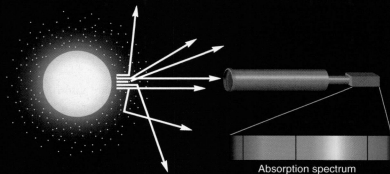

Absorption spectrum

KIRCHHOFF'S LAWS

Law I: **The Continuous Spectrum**

A solid, liquid, or dense gas excited to emit light will radiate at all wavelengths and thus produce a continuous spectrum.

Law II: **The Emission Spectrum**

A low-density gas excited to emit light will do so at specific wavelengths and thus produce an emission spectrum.

Law III: **The Absorption Spectrum**

If light comprising a continuous spectrum passes through a cool, low density gas, the result will be an absorption spectrum.

1b In 1859, long before scientists understood atoms and energy levels, the German scientist Gustav Kirchhoff formulated three rules, now known as **Kirchhoff's laws,** that describe the three types of spectra.

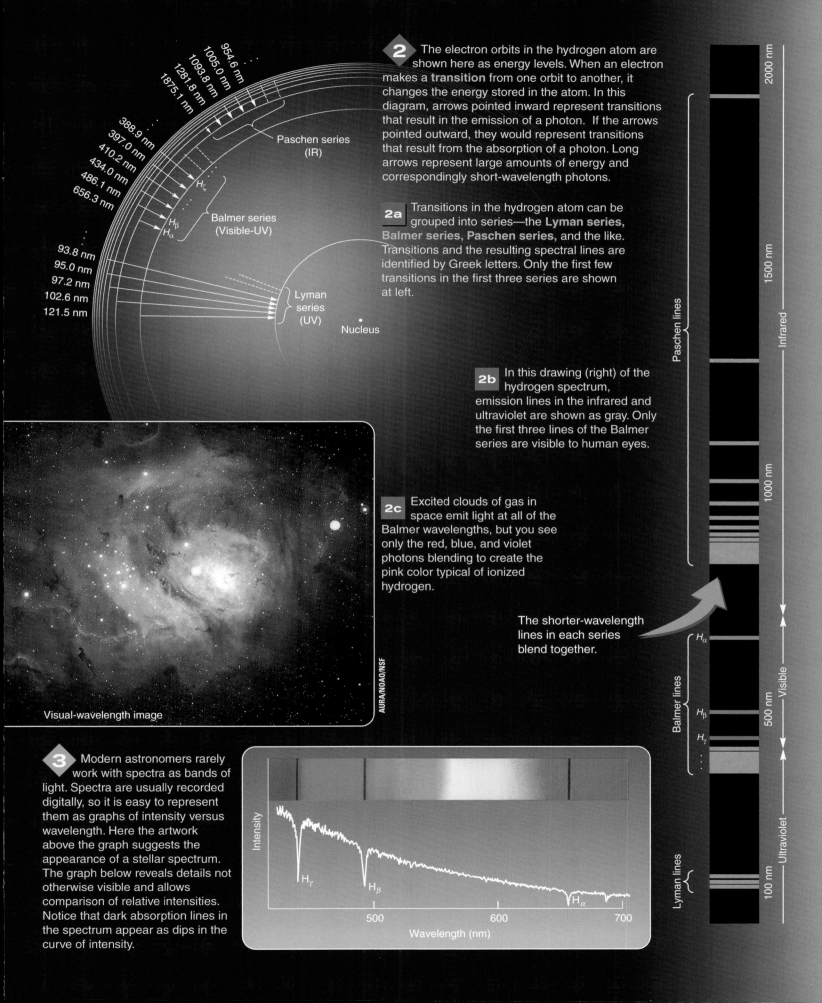

954.6 nm
1005.0 nm
1093.8 nm
1281.8 nm
1875.1 nm

388.9 nm
397.0 nm
410.2 nm
434.0 nm
486.1 nm
656.3 nm

93.8 nm
95.0 nm
97.2 nm
102.6 nm
121.5 nm

Paschen series
(IR)

H_ζ

Balmer series
(Visible-UV)

H_β
H_α

Lyman
series
(UV)

Nucleus

2 The electron orbits in the hydrogen atom are shown here as energy levels. When an electron makes a **transition** from one orbit to another, it changes the energy stored in the atom. In this diagram, arrows pointed inward represent transitions that result in the emission of a photon. If the arrows pointed outward, they would represent transitions that result from the absorption of a photon. Long arrows represent large amounts of energy and correspondingly short-wavelength photons.

2a Transitions in the hydrogen atom can be grouped into series—the **Lyman series, Balmer series, Paschen series,** and the like. Transitions and the resulting spectral lines are identified by Greek letters. Only the first few transitions in the first three series are shown at left.

2b In this drawing (right) of the hydrogen spectrum, emission lines in the infrared and ultraviolet are shown as gray. Only the first three lines of the Balmer series are visible to human eyes.

2c Excited clouds of gas in space emit light at all of the Balmer wavelengths, but you see only the red, blue, and violet photons blending to create the pink color typical of ionized hydrogen.

Visual-wavelength image

AURA/NOAO/NSF

The shorter-wavelength lines in each series blend together.

3 Modern astronomers rarely work with spectra as bands of light. Spectra are usually recorded digitally, so it is easy to represent them as graphs of intensity versus wavelength. Here the artwork above the graph suggests the appearance of a stellar spectrum. The graph below reveals details not otherwise visible and allows comparison of relative intensities. Notice that dark absorption lines in the spectrum appear as dips in the curve of intensity.

Intensity

H_γ
H_β
H_α

500   600   700
Wavelength (nm)

2000 nm

1500 nm

Infrared

1000 nm

Paschen lines

500 nm

Visible

H_α

H_β

H_γ

Balmer lines

100 nm

Ultraviolet

Lyman lines

7-4 Stellar Spectra

IN LATER CHAPTERS, you will use spectra to study galaxies and planets, but here you can begin by studying the spectra of stars. Such spectra are the easiest to understand, and the nature of stars is central to the study of all celestial objects.

The Balmer Thermometer

You can use the Balmer absorption lines as a thermometer to find the temperatures of stars. From the discussion of black body radiation, you know how to estimate temperature from color—red stars are cool, and blue stars are hot. (Remember that stellar spectra tell you about the surface of the star; that's where the light comes from.) You can estimate temperature from color, but the Balmer lines give you much greater accuracy.

The Balmer thermometer works because the Balmer absorption lines are produced only by atoms whose electrons are in the second energy level. If the star is cool, there are few violent collisions between atoms to excite the electrons, and most atoms have their electrons in the ground state. If most electrons are in the ground state, they can't absorb photons in the Balmer series. As a result, you should expect to find weak Balmer absorption lines in the spectra of cool stars.

In the surface layers of stars hotter than about 20,000 K, on the other hand, there are many violent collisions between atoms, exciting electrons to high energy levels or knocking the electron clear out of some atoms; that is, some atoms are ionized. Consequently, few hydrogen atoms have their electron in the second orbit to form Balmer absorption lines, and you should expect hot stars, like cool stars, to have weak Balmer absorption lines.

At an intermediate temperature, roughly 10,000 K, the collisions are just right to excite large numbers of electrons into the second energy level. With many atoms excited to the second level, the gas absorbs Balmer wavelength photons strongly and produces strong Balmer lines.

To summarize, the strength of the Balmer lines depends on the temperature of the star's surface layers. Both hot and cool stars have weak Balmer lines, but medium-temperature stars have strong Balmer lines.

Theoretical calculations can predict just how strong the Balmer lines should be for stars of various temperatures. Such calculations are the key to finding temperatures from stellar spectra. The curve in ■ Figure 7-8a shows the strength of the Balmer lines for

various stellar temperatures. You could use this as a temperature indicator, except that the curve gives two answers. A star with Balmer lines of a certain strength might have either of two temperatures, one high and one low. How do you know which is the right answer? You must examine other spectral lines to choose the correct temperature.

You have seen how the strength of the Balmer lines depends on temperature. Temperature has a similar effect on the spectral lines of other elements, but the temperature at which the lines reach maximum strength differs for each element (Figure 7-8b).

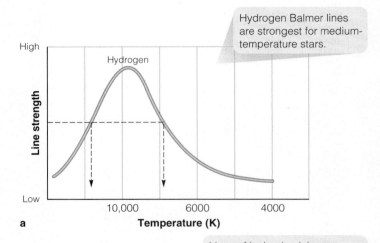

Hydrogen Balmer lines are strongest for medium-temperature stars.

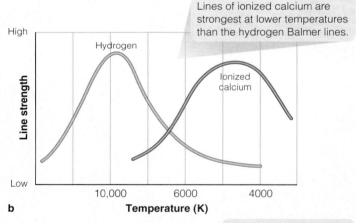

Lines of ionized calcium are strongest at lower temperatures than the hydrogen Balmer lines.

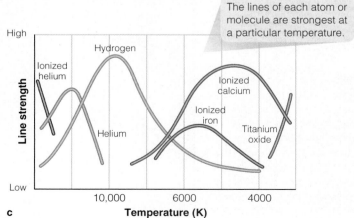

The lines of each atom or molecule are strongest at a particular temperature.

■ Figure 7-8

The strength of spectral lines can tell you the temperature of a star. (a) Balmer hydrogen lines alone are not enough because they give two answers. Balmer lines of a certain strength could be produced by a hotter star or a cooler star. (b) Adding another atom to the diagram helps, and (c) adding many atoms and molecules to the diagram creates a precise tool to find the temperatures of stars.

If you add these elements to your graph, you get a powerful tool for finding the stars' temperatures (Figure 7-8c).

How do you use this tool? You determine a star's temperature by comparing the strengths of its spectral lines with your graph. For instance, if you recorded the spectrum of a star and found medium-strength Balmer lines and strong helium lines, you could conclude it had a temperature of about 20,000 K. But if the star had weak hydrogen lines and strong lines of ionized iron, you would assign it a temperature of about 5800 K, similar to that of the sun.

The spectra of stars cooler than about 3000 K contain dark bands produced by molecules such as titanium oxide (TiO). Because of their structure, molecules can absorb photons at many wavelengths, producing numerous, closely spaced spectral lines that blend together to form bands. These molecular bands appear only in the spectra of the coolest stars because, as mentioned before, molecules in cool stars are not subject to the violent collisions that would break up molecules in hotter stars. Consequently, the presence of dark bands in a star's spectrum indicates that the star is very cool.

From stellar spectra, astronomers have found that the hottest stars have surface temperatures above 40,000 K and the coolest about 2000 K. Compare these with the surface temperature of the sun, which is about 5800 K.

Spectral Classification

You have seen that the strengths of spectral lines depend on the surface temperature of the star. From this you can conclude that all stars of a given temperature should have similar spectra. If you learn to recognize the pattern of spectral lines produced by a 6000 K star, for instance, you need not use Figure 7-8c every time you see that kind of spectrum. You can save time by classifying stellar spectra rather than analyzing each one individually.

The first widely used classification system was devised by astronomers at Harvard during the 1890s and 1900s. One of them, Annie J. Cannon, personally inspected and classified the spectra of over 250,000 stars. The spectra were first classified in groups labeled A through Q, but some groups were later dropped, merged with others, or reordered. The final classification includes the seven major **spectral classes,** or **types,** still used today: O, B, A, F, G, K, M.*

This sequence of spectral types, called the **spectral sequence,** is important because it is a temperature sequence. The O stars are the hottest, and the temperature decreases down to the M stars, the coolest. For maximum precision, astronomers divide each spectral class into 10 subclasses. For example, spectral class A consists of the subclasses A0, A1, A2, . . . A8, A9. Next comes F0, F1, F2, and so on. This finer division gives a star's temperature to an accuracy within about 5 percent. The sun, for example, is not just a G star, but a G2 star, with a temperature of 5800 K.

Astronomers classify a star by the lines and bands in its spectrum, as shown in ■ Table 7-1. For example, if it has weak Balmer lines and lines of ionized helium, it must be an O star. This table is based on the same information shown in Figure 7-8c.

■ Figure 7-9 shows 13 stellar spectra ranging from the hottest at the top to the coolest at the bottom. Notice how special features change gradually from hot to cool stars. Although these spectra are attractive, astronomers rarely work with spectra as color images. Rather, they display spectra as graphs of intensity versus wavelength with dark absorption lines as dips in the graph (■ Figure 7-10). Such graphs show more detail than photographs. Notice also that the overall curves are similar to black body curves. The wavelength of maximum intensity is in the infrared for the coolest stars and in the ultraviolet for the hottest stars.

Compare Figures 7-9 and 7-10 and notice how the strength of spectral lines depends on temperature. Note that the Balmer lines are strongest in A stars, where the temperature is moderate but still high enough to excite the electrons in hydrogen atoms

---

*Generations of astronomy students have remembered the spectral sequence using the mnemonic "Oh, Be A Fine Girl (Guy), Kiss Me." More recent suggestions from students include, "Oh Boy, An F Grade Kills Me," and "Only Bad Astronomers Forget Generally Known Mnemonics."

■ Table 7-1 I Spectral Classes

Spectral Class	Approximate Temperature (K)	Hydrogen Balmer Lines	Other Spectral Features	Naked-Eye Example
O	40,000	Weak	Ionized helium	Meissa (O8)
B	20,000	Medium	Neutral helium	Achernar (B3)
A	10,000	Strong	Ionized calcium weak	Sirius (A1)
F	7,500	Medium	Ionized calcium weak	Canopus (F0)
G	5,500	Weak	Ionized calcium medium	Sun (G2)
K	4,500	Very weak	Ionized calcium strong	Arcturus (K2)
M	3,000	Very weak	TiO strong	Betelgeuse (M2)

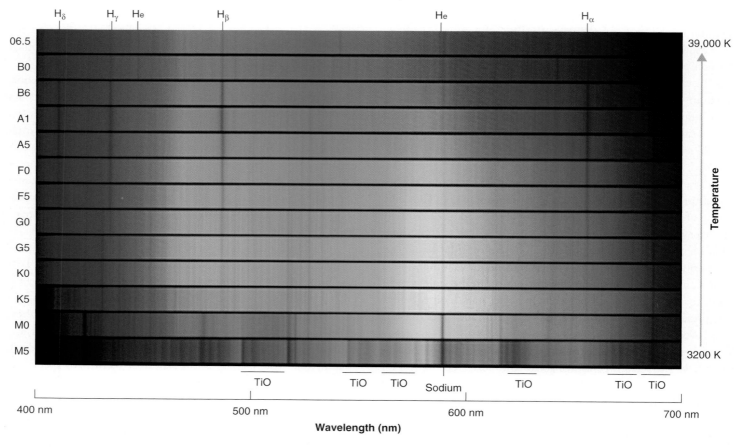

H$_\delta$ H$_\gamma$ He   H$_\beta$ He   H$_\alpha$

06.5   39,000 K

B0

B6

A1

A5

F0

F5

G0

G5

K0

K5

M0

M5   3200 K

Temperature

              TiO   TiO   TiO   Sodium   TiO   TiO   TiO

400 nm   500 nm   600 nm   700 nm

Wavelength (nm)

■ **Figure 7-9**

These spectra show stars from hot O stars at the top to cool M stars at the bottom. The Balmer lines of hydrogen are strongest about A0, but the two closely spaced lines of sodium in the yellow are strongest for very cool stars. Helium lines appear only in the spectra of the hottest stars. Notice that the helium line visible in the top spectrum has nearly but not exactly the same wavelength as the sodium lines visible in cooler stars. Bands produced by the molecule titanium oxide are strong in the spectra of the coolest stars. (AURA/NOAO/NSF)

to the second energy level, where they can absorb Balmer wavelength photons. In the hotter stars (O and B), the Balmer lines are weak because the higher temperature excites the electrons to energy levels above the second or ionizes the atoms. The Balmer lines in cooler stars (F through M) are also weak but for a different reason. The lower temperature cannot excite many electrons to the second energy level, so few hydrogen atoms are capable of absorbing Balmer wavelength photons.

The spectral lines of other atoms also change from class to class. Helium is visible only in the spectra of the hottest classes, and titanium oxide bands only in the coolest. Two lines of ionized calcium increase in strength from A to K and then decrease from K to M. Because the strength of these spectral lines depends on temperature, it requires only a few moments to study a star's spectrum and determine its temperature.

Now you can learn something new about our Favorite Stars. Sirius, brilliant in the winter sky, is an A1 star; and Vega, bright

overhead in the summer sky, is an A0 star. They have nearly the same temperature and color and strong Balmer lines in their spectra. The bright red star in Orion is Betelgeuse, a cool M2 star, but blue-white Rigel is a hot B8 star. Polaris, the North Star, is an F8 star a bit hotter than our sun, and Alpha Centauri, the closest star to the sun, seems to be a G2 star just like the sun.

The study of spectral types is a century old, but astronomers continue to discover new types of stars. The **L dwarfs,** found in 1998, are cooler and fainter than M stars. The L dwarfs are

■ **Figure 7-10** ▶

Modern digital spectra are often represented by graphs of intensity versus wavelength with dark absorption lines appearing as sharp dips in the curves. The hottest stars are at the top and the coolest at the bottom. Hydrogen Balmer lines are strongest at about A0, while lines of ionized calcium (CaII) are strong in K stars. Titanium oxide (TiO) bands are strongest in the coolest stars. Compare these spectra with Figures 7-8c and 7-9. (Courtesy NOAO, G. Jacoby, D. Hunter, and C. Christian)

clearly a different type of star. The spectra of M stars contain bands produced by metal oxides such as titanium oxide, but L dwarf spectra contain bands produced by molecules such as iron hydride (FeH) (■ Figure 7-11). The **T dwarfs,** discovered in 2000, are even cooler and fainter than L dwarfs. Their spectra show absorption by methane (CH_4) and water vapor. The development of giant telescopes and highly sensitive infrared cameras and spectrographs is allowing astronomers to find and study these coolest of stars.

The Composition of the Stars

It seems as though it should be easy to find the composition of the sun and stars just by looking at their spectra, but it turns out to be a difficult task that wasn't well understood until the 1920s.

The story of how astronomers first discovered the composition of the stars is worth telling, not only because it is the story of an important American astronomer who never got proper credit, but also because the story illustrates the temperature dependence of spectral features. The story begins in England.

As a child in England, Cecilia Payne (1900–1979) excelled in classics, languages, mathematics, and literature, but her first love was astronomy. After finishing Newnham College in Cambridge, she left England, sensing that there were no opportunities in England for a woman of science. In 1922, Payne arrived at Harvard, where she eventually earned her Ph.D., although her degree was awarded by Radcliffe because Harvard did not then admit women.

In her thesis, Payne attempted to relate the strength of the absorption lines in stellar spectra to the physical conditions in the atmospheres of the stars. This was not easy because, as you have seen in this chapter, a given spectral line can be weak because the atom is rare or because the temperature is too high or too low for that atom to be able to absorb efficiently. If you see sodium lines in a star's spectrum, you can be sure that the star contains sodium atoms, but if you see no sodium lines, you must consider the possibility that the star is too hot or too cool for sodium to produce spectral lines.

Payne's problem was to untangle these two factors and find the true temperatures of the stars and the true abundance of the atoms in their atmospheres. Recent advances in atomic physics gave her the theoretical tools she needed. About the time Payne left Newnham College, Indian physicist Meghnad Saha published his work on the ionization of atoms. Drawing from such theoretical work, Payne was able to show

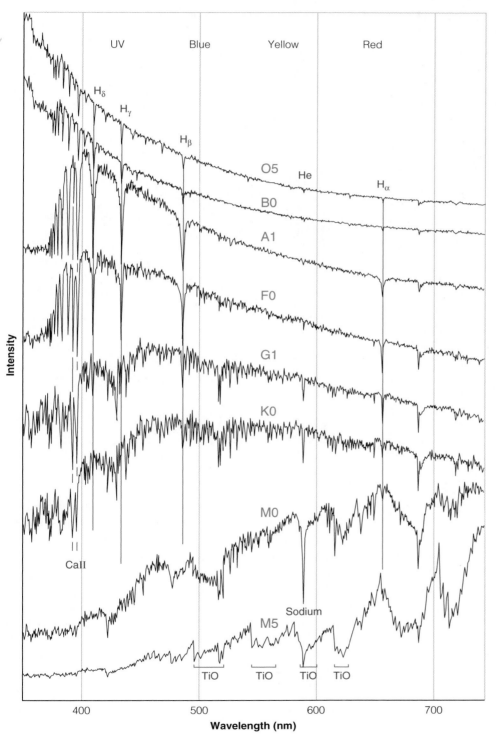

■ **Figure 7-11**

These six infrared spectra show the dramatic differences between L dwarfs and T dwarfs. Spectra of M stars show titanium oxide bands (TiO), but L and T dwarfs are so cool that TiO molecules do not form. Other molecules such as iron hydride (FeH), water (H_2O), and methane (CH_4) can form in these very cool stars. (Adapted from Thomas R. Geballe, Gemini Observatory, from a graph that originally appeared in *Sky and Telescope Magazine*, February 2005, p. 37.)

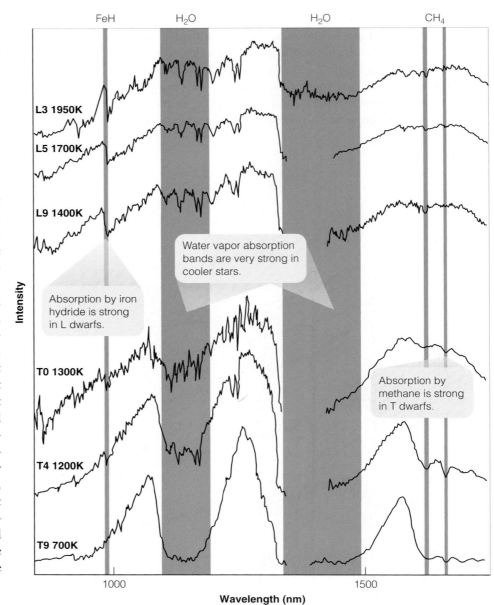

that over 90 percent of the atoms in stars (including the sun) were hydrogen and most of the rest helium (■ Table 7-2). The heavier atoms like calcium, sodium, and iron seem more abundant only because they are better at absorbing photons at the temperatures of stars.

At the time, astronomers found it hard to believe Payne's abundances of hydrogen and helium. They especially found the abundance of helium unacceptable. After all, hydrogen lines are at least visible in most stellar spectra, but helium lines are almost invisible in the spectra of all but the hottest stars. Rather, nearly all astronomers assumed that the stars had roughly the same composition as Earth's surface; that is, they believed that the stars were composed mainly of heavier atoms such as carbon, silicon, iron, aluminum, and so on. Even the most eminent astronomers dismissed Payne's result as illusory. Faced with this pressure and realizing the limited opportunities available to women in science in the 1920s, Payne could not press her discovery.

It was 1929 before astronomers generally understood the importance of temperature in measurements of composition derived from stellar spectra. At that point, astronomers recognized that stars are mostly hydrogen and helium, but Payne received no credit.

Payne worked for many years as a staff astronomer at the Harvard College Observatory with no formal position on the faculty. She married Russian astronomer Sergei Gaposchkin in 1934 and was afterward known as Cecilia Payne-Gaposchkin. In 1956, when Harvard accepted women to its faculty, she was appointed a full professor and chair of the Harvard astronomy department.

Cecilia Payne-Gaposchkin's work on the chemical composition of the stars illustrates the importance of fully understanding the interaction between light and matter. Only a detailed understanding of the physics could lead her to the correct composi-

■ **Table 7-2 I The Most Abundant Elements in the Sun**

Element	Percentage by Number of Atoms	Percentage by Mass
Hydrogen	91.0	70.9
Helium	8.9	27.4
Carbon	0.03	0.3
Nitrogen	0.008	0.1
Oxygen	0.07	0.8
Neon	0.01	0.2
Magnesium	0.003	0.06
Silicon	0.003	0.07
Sulfur	0.002	0.04
Iron	0.003	0.1

tion. As you turn your attention to other information that can be derived from stellar spectra, you will again discover the importance of understanding light.

Ace ☯ Astronomy™ Log into AceAstronomy and select this chapter to see Astronomy Exercise "Stellar Atomic Absorption Lines." You can change the chemical composition of a star and watch its spectrum change.

The Doppler Effect

Surprisingly, one of the pieces of information hidden in a spectrum is the velocity of the light source. Astronomers can measure the wavelengths of lines in a star's spectrum and find the velocity of the star. The **Doppler effect** is the apparent change in the wavelength of radiation caused by the motion of the source.

You can detect the Doppler shift in sound. Sound is not electromagnetic radiation, of course; it is a mechanical wave transmitted through air, but because it is a wave, it is subject to the Doppler effect. Sounds with long wavelengths have low pitches, and sounds with short wavelengths have higher pitches. When a car is approaching you on the highway, you hear its sound waves with a slightly shorter wavelength, and they have a higher pitch. As the car passes you, the pitch drops, and you hear slightly longer wavelengths. That's the Doppler effect.

If a star is moving toward Earth, the lines in its spectrum will be shifted slightly toward shorter wavelengths. That is, they are shifted toward the blue end of the spectrum—a **blueshift.** If a star is moving away from Earth, the lines are shifted slightly toward the red end of the spectrum—a **redshift.** These Doppler shifts are small and don't change the color of the star. But it is easy to detect these changes in wavelength in a star's spectrum.

■ Figure 7-12a shows the Doppler effect in two spectra of the star Arcturus. The lines in the top spectrum are slightly blueshifted because the spectrum was recorded when Earth, following its orbit, was moving toward Arcturus. The lines in the bottom spectrum are redshifted because it was recorded six months later, when Earth was moving away from Arcturus. The greater the shift, the greater the velocity, so astronomers can measure velocities this way.

You may be quite familiar with the Doppler effect. Meteorologists use Doppler radar to measure the velocities of clouds and weather systems. A large redshift close to a large blueshift is a warning of a possible tornado. Or perhaps you have had your own velocity measured by police radar as you drive down the highway. Radar can even be used in sports to measure the speed of a baseball or tennis ball. It's just redshifts and blueshifts.

The Doppler effect tells you how rapidly the distance between you and the source of light is increasing or decreasing. It does not matter whether you are moving or the star is moving. Only the relative velocity is important. Also, the Doppler shift is sensitive only to the part of the velocity directed away from you or toward you. This part of the velocity is called the **radial velocity (V_r),** as shown in ■ Figure 7-13.

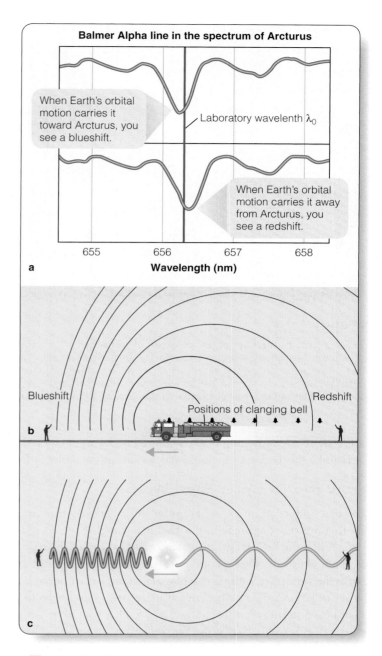

a Balmer Alpha line in the spectrum of Arcturus

When Earth's orbital motion carries it toward Arcturus, you see a blueshift.

Laboratory wavelenth λ_0

When Earth's orbital motion carries it away from Arcturus, you see a redshift.

Wavelength (nm)

b Blueshift Redshift Positions of clanging bell

c

■ **Figure 7-12**

The Doppler effect. (a) Earth's orbital motion causes small Doppler sifts in a star's spectrum. (b) The clanging bell on a moving fire truck produces sounds that move outward (black circles). An observer ahead of the truck hears the clangs closer together, while an observer behind the truck hears them farther apart. (c) A moving source of light emits waves that move outward (black circles). An observer in front of the light source observes a shorter wavelength (a blueshift), and an observer behind the light source observes a longer wavelength (a redshift).

The Doppler effect occurs for any kind of radiation, not just light. Radio astronomers, for instance, observe the Doppler effect when they measure the wavelength of radio waves coming from objects moving toward or away from Earth. Whatever wavelength range you consider—radio, X rays, gamma rays—you can

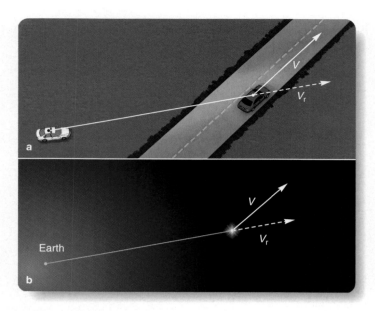

■ Figure 7-13

(a) Police radar can measure only the radial part of your velocity (V_r) as you drive down the highway, not your true velocity along the pavement (V). That is why police using radar never park far from the highway. (b) From Earth, astronomers can use the Doppler effect to measure the radial velocity (V_r) of a star, but they cannot measure its true velocity, V, through space.

still refer to a lengthening of the observed wavelength as a redshift and a shortening of the observed wavelength as a blueshift.

How the Doppler Shift Works

You can understand how the Doppler shift works by thinking about a fire truck approaching with a bell clanging once a second. When the bell clangs, the sound travels ahead of the truck to reach your ears. One second later, the bell clangs again, but not at the same place. During that one second, the fire truck moved closer to you, so the bell is closer at its second clang. Now the sound has a shorter distance to travel and reaches your ears a little sooner than it would have if the fire truck were not approaching. The third time the bell clangs, it is even closer. By timing the bell, you could observe that the clangs are slightly less than 1 second apart, all because the fire truck is approaching. If the fire truck were moving away from you, you would hear the clangs sounding more than one second apart, because each successive clang of the bell would occur farther away.

Figure 7-12b shows a fire truck moving toward one observer and away from another observer. The position of the bell at each clang is shown by the small black bells with the sound spreading outward as black circles. You can see that the clangs are squeezed together ahead of the fire truck and stretched apart behind.

Now you can substitute a source of light waves for the clanging of the bell. If the source is approaching, then each time the source emits the peak of a wave it will be slightly closer to you, and you will observe a shorter wavelength. If it is moving away, you

will observe a longer wavelength. This is shown in Figure 7-12c, where the peaks of the waves appear compressed in front of the moving light source and stretched out behind the source.

Of course, how great the change in wavelength is depends on the velocity. Just as a slowly moving car has a smaller Doppler effect in sound than does a high-speed airplane, a slowly moving star has a smaller Doppler effect in light than does a high-speed star. Consequently, you can measure the velocity by measuring the amount by which the spectral lines are shifted in wavelength. A large Doppler shift means a high radial velocity.

Calculating the Doppler Velocity

It is easy to calculate the radial velocity of an object from its Doppler shift. The formula is a simple ratio relating the radial velocity V_r divided by the speed of light c to the change in wavelength, λ, divided by the unshifted wavelength, λ_0:

$$\frac{V_r}{c} = \frac{\Delta\lambda}{\lambda_0}$$

For example, suppose you observed a line in a star's spectrum with a wavelength of 600.1 nm. Laboratory measurements show that the line should have a wavelength of 600 nm. That is, its unshifted wavelength is 600 nm. What is the star's radial velocity? First note that the change in wavelength is 0.1 nm:

$$\frac{V_r}{c} = \frac{0.1}{600} = 0.000167$$

Multiplying by the speed of light, 3×10^5 km/s, gives the radial velocity, 50 km/s. Because the wavelength is shifted to the red (lengthened), the star must be receding.

Now that you understand the Doppler shift you can understand a final illustration of the information hidden in stellar spectra. Even the shapes of the spectral lines can reveal secrets about the stars.

Ace ◉ Astronomy™ Log into AceAstronomy and select this chapter to see Astronomy Exercise "Doppler Shift." Compare the Doppler shift for a source of sound with that for a source of light.

The Shapes of Spectral Lines

When astronomers refer to the shape of a spectral line, they mean the variation of intensity across the line. An absorption line, for instance, is darkest in the center and brighter to each side. Two examples are shown in ■ Figure 7-14.

The exact shape of a line can reveal a great deal about a star, but the most important characteristic is the width of the line. Spectral lines are not perfectly narrow; if they were, they would be undetectable. They have a natural width because nature allows an atom some leeway in the energy it may absorb or emit. In the absence of all other effects, spectral lines have a natural width of about 0.001 to 0.00001 nm—very narrow indeed.

The natural widths of spectral lines are not important in most branches of astronomy because other effects smear out the

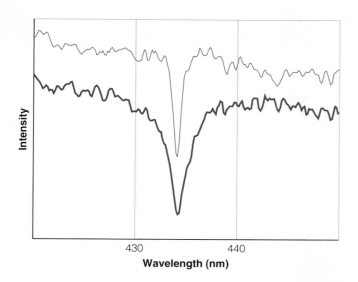

■ Figure 7-14

Here you see two dark absorption lines magnified from the spectra of two A1 stars. The upper line is quite narrow, but the bottom line is much broader. Because the two stars have the same spectral type, you can trust that they must have the same temperature. The stars differ not in temperature but in gas density. The star with the narrow spectral lines has a very low-density atmosphere. Precise observations of the shapes of spectral lines can reveal a great deal about stars. (Courtesy NOAO, G. Jacoby, D. Hunter, and C. Christian)

lines and make them much broader. For example, if a star spins rapidly, the Doppler effect will broaden the spectral lines. As the star rotates, one side will recede from Earth, and the other side will approach Earth. Light from the receding side will be redshifted, and light from the approaching side will be blueshifted, so any spectral lines will be broadened. Astronomers can measure a star's rotation rate from the width of its spectral lines.

Another important process is called **Doppler broadening.** To consider this process, imagine that you photograph the spectrum of a jar full of hydrogen atoms (■ Figure 7-15). Because the gas has some thermal energy (it is not at absolute zero), the

gas atoms are in motion. Some will be coming toward your spectrograph, and some will be receding. Most, of course, will not be traveling very fast, but some will be moving very quickly. The photons emitted by the atoms approaching you will have slightly shorter wavelengths because of the Doppler effect, and photons emitted by atoms receding from you will have slightly longer wavelengths. Thus, the Doppler shifts due to the motions of the individual atoms will smear the spectral line out and make it broader. This summary describes the Doppler broadening of an emission line, but the effect is the same for absorption lines.

The extent of Doppler broadening depends on the temperature of the gas. If the gas is cold, the atoms travel at low velocities, and the Doppler shifts are small (Figure 7-15a). If the gas is hot, however, the atoms travel faster, Doppler shifts are larger, and the lines will be wider (Figure 7-15b). Sometimes astronomers will estimate the temperature of a cloud of gas in space by looking at the widths of its spectral lines.

Another form of broadening, **collisional broadening,** is caused by collisions between atoms, and consequently it depends on the density of the gas (**Focus on Fundamentals 4**). Densities in astronomy cover an enormous range, from one atom per cubic centimeter in space to millions of tons of atoms per cubic centimeter inside dead stars. Clearly, you would expect such densities to affect the way atoms collide with one another and how they absorb and emit photons.

Collisional broadening spreads out spectral lines when the atoms absorb or emit photons while they are colliding with other

■ Figure 7-15

Doppler broadening. The atoms of a gas are in constant motion. Photons emitted by atoms moving toward the observer will have slightly shorter wavelengths, and those emitted by atoms moving away will have slightly longer wavelengths. This broadens the spectral line. If the gas is cool (a), the atoms do not move very fast, the Doppler shifts are small, and the line is narrow. If the gas is hot (b), the atoms move faster, the Doppler shifts are larger, and the line is broader.

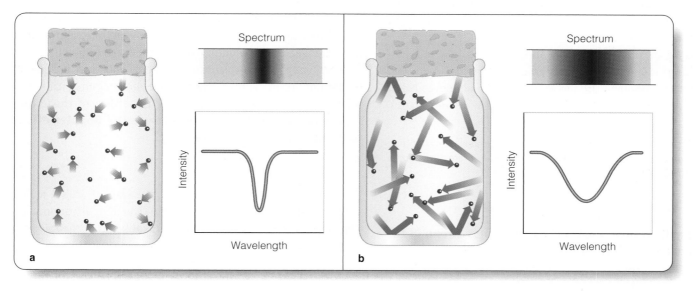

Density

You are about as dense as an average star. What does that mean? As you study astronomy, you will use the term *density* often, so you should be sure to understand this fundamental concept. **Density** is a measure of the amount of matter in a given volume. Density is expressed as mass per volume, such as grams per cubic centimeter. The density of water, for example, is about 1 g/cm³, and you are almost as dense as water.

To get a feel for density, imagine holding a brick in one hand and a similar-sized block of Styrofoam in the other hand. You can easily tell that the brick contains more matter than the Styrofoam block, even though both are the same size. The brick weighs more than the Styrofoam, but it isn't really the weight that you should consider. Rather, you should think about the mass of the two objects. In space, where they have no weight, the brick and the Styrofoam would still have mass, and you could tell just by moving them around that the brick contains more mass than the Styrofoam. For example, imagine tapping each object gently against your ear. The massive brick would be easy to distinguish from the low-mass Styrofoam block, even in weightlessness.

When you think of density, you divide mass by volume, and your mind makes that comparison at an instinctive level. You sense the density of an object just by handling it. Gift shops sometimes sell imitation rocks made of Styrofoam as humorous gifts. "Rocks" made of Styrofoam seem odd when you handle them because your brain expects rocks to be dense.

A brick would be dense even in space where it had no weight.

Density is a fundamental idea in science because it is a general property of materials. Metals tend to be dense; lead, for example, has a density of about 7 g/cm³. Rock, in contrast, has a density of 3 to 4 g/cm³. Water and ice have densities of about 1 g/cm³. If you knew that a small moon orbiting Saturn had a density of 1.5 g/cm³, you could immediately draw some conclusions about what kinds of materials the little moon might be made of—ice and a little rock, but not much metal. The density of an object is a basic clue to its composition.

Astronomical bodies can have dramatically different densities. The gas in a nebula can have a very low density, but the same kind of gas in a star can have a much higher density. The sun, for example, has an average density of about 1 g/cm³, about the same as your body. As you study astronomical objects, pay special attention to their densities. Density is fundamental.

PRESSURE | MASS | ENERGY | TEMPERATURE AND HEAT | **DENSITY**

---

atoms, ions, or electrons. The collisions disturb the energy levels in the atoms, making it possible for the atoms to absorb a slightly wider range of wavelengths. Because of this, the spectral lines are wider. Because atoms in a dense gas collide more often than atoms in a low-density gas, collisional broadening depends on the density of the gas. Temperature is also an important factor. Atoms in a hot gas travel faster and collide more often and more violently than atoms in a cool gas. The two spectral lines in Figure 7-14 illustrate the effect of density.

Once again, the physics of the interaction of light and matter provides a tool to understand starlight. In later chapters, you will see how astronomers use the widths of spectral lines to better understand clouds of gas in space, stars, and even distant galaxies.

(Ace◐Astronomy™) Log into AceAstronomy and select this chapter to see Astronomy Exercise "Stellar Rotation." See an interesting application of the Doppler shift.

Building Scientific Arguments

Why are helium lines weak and calcium lines strong in the visible spectrum of the sun?

To analyze this problem, you must recall that the ability of an atom or ion to absorb light depends on temperature. Helium is quite abundant in the sun, but the surface of the sun is too cool to excite helium atoms and enable them to easily absorb visible-wavelength photons. On the other hand, calcium is easily ionized, and calcium atoms that have lost an electron are very good absorbers of photons at temperatures like those at the sun's surface. Consequently, calcium lines are strong in the solar spectrum even though calcium ions are rare, and helium lines are weak even though helium atoms are common.

When you create a scientific argument, you must include all of the important factors. In this case, both composition and temperature are important. **But why is neither of these factors important when astronomers use the Doppler effect to measure a star's radial velocity?**

■   ■   ■

Connections: The spectra of the stars are filled with clues about the gas emitting the light. Much of astronomy is based on unraveling these clues. You will begin in the following chapter by studying the nearest star—the sun.

Summary

7-1 | Starlight

How do stars produce light?

■ Stars emit light from the hot gases at their surfaces. Light from deeper layers does not escape.

■ Temperature is a measure of the average kinetic energy of the particles in a gas. In a hot gas, the particles move fast: in a cool gas, they move slowly.

■ Thermal energy is the total kinetic energy in the particles of a material. Heat is the flow of thermal energy from hot regions to cool regions.

■ In a hot material, the rapidly moving atoms or molecules collide with electrons, and when electrons are accelerated, part of the energy is emitted as electromagnetic radiation. This is known as black body radiation.

■ The hotter a black body is, the more it radiates and the shorter is its wavelength of maximum intensity, λ_{max}. This allows astronomers to estimate the temperatures of stars from their colors.

7-2 | Atoms

What is an atom?

■ An atom consists of a nucleus surrounded by a cloud of electrons. The nucleus is made up of positively charged protons and uncharged neutrons.

■ The number of protons in an atom determines which element it is. Atoms of the same element (that is, having the same number of protons) with different numbers of neutrons are called isotopes.

■ A neutral atom is surrounded by a number of negatively charged electrons equal to the number of protons in the nucleus. An atom that has lost or gained an electron is called an ion.

■ The electrons in an atom may occupy various permitted orbits around the nucleus but not orbits in between.

7-3 | The Interaction of Light and Matter

How do atoms interact with light?

■ The size of an electron's orbit depends on the energy stored in the motion of the electron.

■ An electron may be excited to a higher orbit during a collision between atoms, or it may move from one orbit to another by absorbing or emitting a photon of the proper energy.

■ Kirchhoff's laws describe the formation of spectra. A hot solid, liquid or dense gas will emit black body radiation, producing a continuous spectrum. An excited low-density gas will produce an emission (bright-line) spectrum. Light passing through a cool, low-density gas will form an absorption (dark-line) spectrum.

7-4 | Stellar Spectra

What kind of spectra do you see when you look at celestial objects?

■ Because orbits of only certain energies are permitted in an atom, photons of only certain wavelengths can be absorbed or emitted. Each kind of atom has its own characteristic set of spectral lines. The hydrogen atom has the Lyman series of lines in the ultraviolet, the Balmer series in the visible, and the Paschen series (and others) in the infrared.

■ If light passes through a low-density gas on its way to your telescope, the gas can absorb photons of certain wavelengths, and you see dark lines in the spectrum at those positions. Such a spectrum is called an absorption spectrum.

■ If you look at a low-density gas that is excited to emit photons, you see bright lines in the spectrum. Such a spectrum is called an emission spectrum.

What can you learn from a star's spectrum?

■ Nearly all stellar spectra are absorption spectra, and the hydrogen lines seen there are Balmer lines. In cool stars, the Balmer lines are weak because atoms are not excited out of the ground state. In hot stars, the Balmer lines are weak because atoms are excited to higher orbits or are ionized. Only at medium temperatures are the Balmer lines strong.

■ The strength of spectral lines in a star's spectrum can tell you its temperature. In its simplest form, this amounts to classifying the star's spectrum in the spectral sequence: O, B, A, F, G, K, M.

■ Long after the spectral sequence was created, astronomers found the L and T stars at temperatures even cooler than the M stars.

■ A spectrum can reveal the chemical composition of the stars. The presence of spectral lines of a certain element shows that that element must be present in the star. But astronomers must proceed with care. Lines of a certain element may be weak or absent if the star is too hot or too cool.

■ The wavelengths of spectral lines provide clues to the motions of the stars. When a star is approaching, you observe slightly shorter wavelengths (a blueshift); when it is receding, you observe slightly longer wavelengths (a redshift). This Doppler effect reveals a star's radial velocity, that part of its velocity directed toward or away from Earth.

■ The shape of a spectral line can reveal the rotation of a star because one side of the star approaches Earth while the other side recedes. The Doppler shift broadens the star's spectral lines.

■ Doppler broadening can make spectral lines wider because the atoms in a hot gas move very rapidly and at any moment some are receding and some are approaching.

■ Collisional broadening occurs in a dense gas where the atoms collide often. This widens spectral lines.

New Terms

temperature (p. 138)	quantum mechanics (p. 143)
Kelvin temperature scale (p. 138)	permitted orbit (p. 143)
absolute zero (p. 138)	energy level (p. 144)
electron (p. 138)	excited atom (p. 144)
thermal energy (p. 139)	ground state (p. 145)
black body radiation (p. 139)	continuous spectrum (p. 146)
wavelength of maximum intensity (λ_{max}) (p. 139)	absorption spectrum (dark-line spectrum) (p. 146)
nucleus (p. 141)	absorption line (p. 146)
proton (p. 141)	emission spectrum (bright-line spectrum) (p. 146)
neutron (p. 141)	
isotope (p. 142)	emission line (p. 146)
ionization (p. 142)	Kirchhoff's laws (p. 146)
ion (p. 142)	transition (p. 147)
molecule (p. 142)	Lyman series (p. 147)
Coulomb force (p. 142)	Balmer series (p. 147)
binding energy (p. 143)	Paschen series (p. 147)

Review Questions

Ace◐Astronomy™ Assess your understanding of this chapter's topics with additional quizzing and animations at **http://ace .brookscole.com/sf9**

1. Why can a good blacksmith judge the temperature of a piece of heated iron by its color?

2. Why do hot stars look bluer than cool stars?

3. Use black body radiation to explain why you could hold a cigarette between your lips and light it with a match, but you couldn't hold the cigarette in your lips while you light it by sticking the tip into a bonfire. The match and bonfire have about the same temperature. (This is just one of the hazards of smoking.)

4. What is the difference between a neutral atom, an ion, and an excited atom?

5. How do the energy levels in an atom determine which wavelength photons it can absorb or emit?

6. What kind of spectrum would you expect to record if you observed molten lava? In practice, to view molten lava you must look through gases boiling out of the lava. What kind of spectrum might you see in that case?

7. Why do the strengths of the Balmer lines depend on the temperature of the star?

8. Why does a stellar spectrum tell you about the surface layers but not the deeper layers of the star?

9. Explain the similarities among Figure 7-8, Figure 7-9, Figure 7-10, and Table 7-1.

10. Why would you not expect to see TiO bands in the spectra of hot stars?

11. Imagine that you observed a star's spectrum and found that the lines of a certain element were not present. Would you be safe in concluding that the star did not contain this element? Why or why not?

12. If a star moves exactly perpendicular to a line connecting it to Earth, will the Doppler effect change its spectrum? Why or why not?

13. Why would you expect a star that rotates very rapidly to have broad spectral lines? What else could broaden the lines?

(T. Rector, University of Alaska, and WIYN/NURO/AURA/NSF)

13. The nebula shown at right is mostly low-density hydrogen excited to emit photons. What kind of spectrum would you expect this nebula to produce?

14. If the nebula in the image here crosses in front of the star and the nebula and star have different radial velocities, what might the spectrum of the star look like?

Discussion Questions

1. In what ways is the model of an atom used in this chapter a scientific model? How can you use it when it is not a completely correct description of an atom?

2. Can you think of classification systems commonly used to simplify what would otherwise be very complex measurements? Consider foods, movies, cars, grades, clothes, and so on.

Problems

1. Human body temperature is about 310 K (98.6°F). At what wavelength do humans radiate the most energy? What kind of radiation do we emit?

2. If a star has a surface temperature of 20,000 K, at what wavelength will it radiate the most energy?

3. Infrared observations of a star show that it is most intense at a wavelength of 2000 nm. What is the temperature of the star's surface?

4. If you double the temperature of a black body, by what factor will the total energy radiated per second per square meter increase?

5. If one star has a temperature of 6000 K and another star has a temperature of 7000 K, how much more energy per second will the hotter star radiate from each square meter of its surface?

6. Transition A produces light with a wavelength of 500 nm. Transition B involves twice as much energy as A. What wavelength light does it produce?

7. Determine the temperatures of the following stars based on their spectra. Use Figure 7-8.
   a. medium-strength Balmer lines, strong helium lines
   b. medium-strength Balmer lines, weak ionized calcium lines
   c. TiO bands very strong
   d. very weak Balmer lines, strong ionized calcium lines

8. To which spectral classes do the stars in Problem 7 belong?

9. In a laboratory, the Balmer beta line has a wavelength of 486.1 nm. If the line appears in a star's spectrum at 486.3 nm, what is the star's radial velocity? Is it approaching or receding?

10. The highest-velocity stars an astronomer might observe have velocities of about 400 km/s. What change in wavelength would this cause in the Balmer gamma line? (*Hint:* Wavelengths are given on page 147).

Media Cluster

Kirchhoff's Laws
Kirchhoff studied and identified three different spectra in his experiments. Use this animation to study the conditions for producing each of the types of spectra.

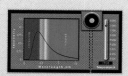

Black Body
In this animation, you can change the temperature of an object and see how its black body curve changes.

Stefan–Boltzmann Law
In this introduction to the Stefan–Boltzmann law, watch the luminosity of a star change as you adjust its temperature.

Emission and Absorption Spectra
This exercise will give you an inside look at atoms as they interact with photons. The photons are absorbed by the atom, sending the electron to a higher energy level.

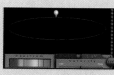

Doppler Shift
Recall that the Doppler effect is the change in wavelength of radiation due to relative radial motion of source and observer. You can compare the Doppler shift for a source of sound with that for a source of light in this animation.

Stellar Rotation
This animation lets you increase or decrease the rotation speed of a star and see the effect on the star's spectrum.

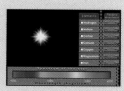

Stellar Atomic Absorption Lines
In this exercise, you can build your own star. Select from the type and ratio of element mixes in your new star.

Lab 2: Properties of Light and Its Interaction with Matter
This lab examines the wave properties of electromagnetic radiation and how different regions of the electromagnetic spectrum are related. It ends by looking at the interaction between matter and radiation on the atomic scale.

Lab 11: The Spectral Sequence and the H–R Diagram
This lab introduces the tools astronomers use to identify the stars of the main sequence and how that information is used to estimate the distance to the stars and the age of a star cluster.

Lab 3: The Doppler Effect
This lab provides a brief review of some basic properties of waves and investigates the shift in observed wavelengths of sound and light waves caused by the motion of an emitting source with respect to an observer.

Critical Inquiries for the Web

1. The name for the element helium has astronomical roots. Search the Internet for information on the discovery of helium. How and when was it discovered, and how did it get its name? Why do you suppose it took so long for helium to be recognized?

2. This chapter contains a model for an atom that may seem familiar to you. How was that model developed? What was the plum pudding model? Search the Web for information on historical models of the atom and compile a time line of important developments leading to our current understanding.

3. What can you find out about Anne J. Cannon and Cecilia Payne-Gaposchkin?

Exploring *TheSky*

1. Locate the following stars, click on them, and determine their spectral types: Antares in Scorpius, Betelgeuse in Orion, Aldebaran in Taurus, Sirius in Canis Major, Rigel in Orion.

2. How are spectral types correlated with the colors of stars? (*Hint:* Locate Orion and choose **Spectral Colors** under the **View** menu.)

Go to the Brooks/Cole Astronomy Resource Center **(http://astronomy.brookscole.com)** for critical thinking exercises, articles, and additional readings from InfoTrac College Edition, Brooks/Cole's online student library.

8 | The Sun

UV image

A WIT ONCE REMARKED that solar astronomers would know a lot more about the sun if it were farther away, and that contains a grain of truth; the sun is only a humdrum star, and although there are billions like it in the sky, the sun is the only one close enough to show surface detail. Solar astronomers can see so much detail in the swirling currents of gas and arching bridges of magnetic force that present theories seem inadequate to describe it. Yet the sun is not a complicated object. It is just a star. ▎ In their general properties, stars are very simple. They are great balls of hot gas held together by their own gravity. Their gravity would make them collapse into small, dense bodies were they not so hot. The tremendously hot gas inside stars has such a high pressure that the stars would surely explode were it not for their own confining gravity.

▎ Continued on page 162 ▎

160

Guidepost

Looking Back
The interaction of light and matter, which you studied in Chapter 7, is a powerful tool. It can reveal the secrets of the stars. The next step is to apply that tool to the sun.

This Chapter
In this chapter, you will discover how little astronomers could know about the sun were it not for the analysis of its spectrum. Just a bit of spectrographic ingenuity reveals that the brilliance of the sun hides a complex atmosphere of hot gases, powered by nuclear reactions and churned by powerful storms.

This chapter will help you answer four essential questions:

What do you see when you look at the sun?

How does the sun make its energy?

What are the dark sunspots and other forms of solar activity?

Why does the sun go through a cycle of activity?

Perhaps the most important question answered here is more general: How can astronomers learn about the sun and stars?

Looking Ahead
This is the first chapter that applies the methods of science to understand a celestial body. The tools developed in this chapter will help you explore stars, galaxies, and planets through the rest of this book.

Like soap bubbles, stars are simple structures balanced between opposing forces that individually would destroy them. You can study our sun as a close-up example of a star.

Another reason to study the sun is that life on Earth depends critically on the sun. Very small changes in the sun's luminosity can alter Earth's climate, and a slightly larger change might make Earth uninhabitable. Nearly all of our energy comes from the sun—oil and coal are merely stored sunlight. Furthermore, the sun's atmosphere of very thin gas reaches out past Earth's orbit, and any change in the sun, such as an eruption or a magnetic storm, can have a direct effect on Earth.

Finally, you should study the sun because it is beautiful. Your analysis of sunlight will reveal that the sun is both powerful and delicate. So you should study the sun not only because it is *a star,* not only because it is *our star,* but because it is the sun.

8-1 The Solar Atmosphere

WHEN YOU WATCH the sun set in the west, you see a glowing disk. What are you seeing? Here you will see the sun sketched in its general properties and describe the surface you see. Later you can plunge inside to see how it makes its energy and then return to the surface to study the storms and eruptions not generally visible to the unaided eye. What do you see when you look at the sun? You see a star.

The sun is 109 times Earth's diameter and 333,000 times Earth's mass. This seems dramatic, but look at **Celestial Profile 1** and notice the sun's density. It is only a little bit denser than water. So, although the sun is very large and very massive, it must be a gas from its surface to its center. When you look at the sun you see only the outer layers of this vast sphere of gas. In fact, these outer layers, the solar atmosphere, extend high above the visible surface of the sun.

Heat Flow in the Sun

When you look at the sun you see a hot, glowing surface, and simple logic tells you that energy in the form of heat is flowing outward from the sun's interior. The solar spectrum reveals that the sun is a G2 star with a temperature of about 5800 K. At that temperature, every square millimeter of the sun's surface must be radiating more energy than a 60-watt lightbulb. With all that energy radiating into space, the sun's surface would cool rapidly if energy did not flow up from the interior to keep the surface hot.

Not until the 1930s did astronomers understand how the sun makes its energy. Nuclear reactions occur in the core of the sun and generate energy, which flows upward as heat and keeps the surface hot. These nuclear reactions are discussed in detail later in this chapter, but here you should notice the importance of the energy flowing outward through the sun's surface.

As you study the atmosphere of the sun, you will find many phenomena that are driven by the energy flowing outward. Like a pot of boiling soup on a hot stove, the surface of the sun is in constant activity as the heat flows up from below.

For now, you can consider the sun in its quiescent, unchanging state, and explore the layers in its atmosphere.

The Photosphere

The visible surface of the sun looks like a smooth layer of gas marked by a few dark **sunspots.** Although the photosphere seems to be a distinct surface, it is not solid. In fact, the sun is gaseous from its outer atmosphere right down to its center. The photosphere is the thin layer of gas from which Earth receives most of the sun's light. It is less than 500 km deep and has an average temperature of about 5800 K. If the sun magically shrank to the size of a bowling ball, the photosphere would be no thicker than a layer of tissue paper wrapped around the ball. For comparison, the chromosphere lies above the photosphere and is only a few times thicker in extent, but the corona, beginning above the chromosphere, extends far above the visible surface (■ Figure 8-1). Recall that you first met the terms *photosphere, chromosphere,* and *corona* in connection with solar eclipses in Chapter 3.

Below the photosphere, the gas is denser and hotter and therefore radiates plenty of light, but that light cannot escape from the sun because of the outer layers of gas. So you cannot detect light from these deeper layers. Above the photosphere, the gas is less dense and so is unable to radiate much light. The photosphere is the layer in the sun's atmosphere that is dense enough to emit plenty of light but not so dense that the light can't escape.

One reason the photosphere is so shallow is related to the hydrogen atom. Because the temperature of the photosphere is sufficient to ionize some atoms, there are a large number of free electrons in the gas. Neutral hydrogen atoms can add an extra electron and become an H⁻ (H-minus) ion, but this extra electron is held so loosely that almost any photon has energy enough to free it. In the process, of course, the photon is absorbed. That makes the H⁻ ions very good absorbers of photons and makes the gas of the photosphere opaque. Light from below cannot escape easily, and you see a well-defined surface glowing like a layer of hot fog—the photosphere.

Although the photosphere appears to be substantial, it is really a very-low-density gas. Even in the deepest and densest layers visible, the photosphere is 3400 times less dense than the air you breathe. To find gases as dense as the air you breathe, you would have to descend about 70,000 km below the photosphere, about 10 percent of the way to the sun's center. With fantastically efficient insulation, you could fly a spaceship right through the photosphere.

The spectrum of the sun is an absorption spectrum, and that can tell you a great deal about the photosphere. You know from Kirchhoff's third law that an absorption spectrum is produced when a source of a continuous spectrum is viewed through a gas. In the case of the photosphere, the deeper layers are dense enough to produce a continuous spectrum, but atoms in the photosphere

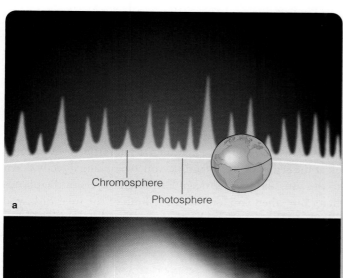

b Visual-wavelength image

■ **Figure 8-1**

(a) A cross section at the edge of the sun shows the relative thickness of the photosphere and chromosphere. Earth is shown for scale. On this scale, the disk of the sun would be more than 1.5 m (5 feet) in diameter. (b) The corona extends from the top of the chromosphere to great height above the photosphere. (b) This photograph, made during a total solar eclipse, shows only the inner part of the corona. (Daniel Good)

absorb photons of specific wavelengths, producing the absorption lines you see.

In good photographs, the photosphere has a mottled appearance because it is made up of dark-edged regions called granules, and the visual pattern is called **granulation** (■ Figure 8-2a). Each granule is about the size of Texas and lasts for only 10 to 20 minutes before fading away. Faded granules are continuously replaced by new granules. Spectra of these granules show that the centers are a few hundred degrees hotter than the edges, and Doppler shifts reveal that the centers are rising and the edges are sinking at speeds of about 0.4 km/second.

From this evidence, astronomers recognize granulation as the surface effects of convection just below the photosphere.

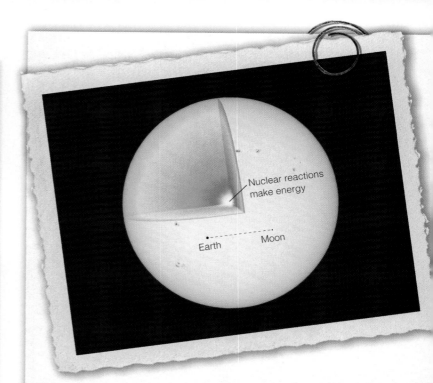

This visual-wavelength image of the sun shows a few sunspots and is cut away to show the location of energy generation at the sun's center. The Earth–moon system is shown for scale. (Dan Good)

Celestial Profile 1: The Sun

From Earth:

Average distance from Earth	1.00 AU (1.495979 × 10⁸ km)
Maximum distance from Earth	1.0167 AU (1.5210 × 10⁸ km)
Minimum distance from Earth	0.9833 AU (1.4710 × 10⁸ km)
Average angular diameter	0.53° (32 minutes of arc)
Period of rotation	25.38 days at equator
Apparent visual magnitude	−26.74

Characteristics:

Radius	6.9599×10^5 km
Mass	1.989×10^{30} kg
Average density	1.409 g/cm³
Escape velocity at surface	617.7 km/s
Luminosity	3.826×10^{26} J/s
Surface temperature	5800 K
Central temperature	15×10^6 K
Spectral type	G2 V
Absolute visual magnitude	4.83

Personality Profile:

In Greek mythology, the sun was carried across the sky in a golden chariot pulled by powerful horses and guided by the sun-god Helios. When Phaeton, son of Helios, drove the chariot one day, he lost control of the horses, and Earth was nearly set ablaze before Zeus smote Phaeton from the sky. Even in classical times, people understood that life on Earth depends critically on the sun.

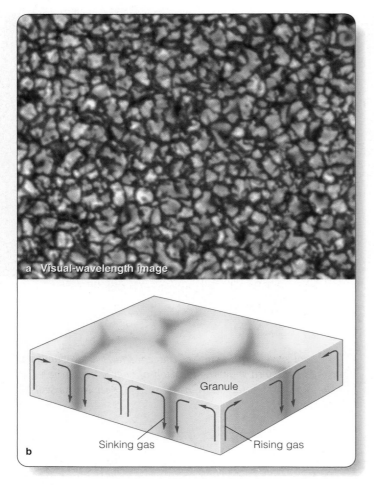

a Visual-wavelength image

b

Granule

Sinking gas

Rising gas

■ **Figure 8-2**

(a) This ultra-high-resolution image of the photosphere shows granulation. The largest granules here are about the size of Texas. (P. N. Brandt, G. Scharmer, G. W. Simon, Swedish Vacuum Solar Telescope) (b) This model explains granulation as the tops of rising convection currents just below the photosphere. Heat flows upward as rising currents of hot gas and downward as sinking currents of cool gas. The rising currents heat the solar surface in small regions seen from Earth as granules.

Convection occurs when hot fluid rises and cool fluid sinks, as when, for example, a convection current of hot gas rises above a candle flame. You can create convection in a liquid by adding a bit of cool nondairy creamer to an unstirred cup of hot coffee. The cool creamer sinks, warms, rises, cools, sinks again, and so on, creating small regions on the surface of the coffee that mark the tops of convection currents. Viewed from above, these regions look much like solar granules.

In the sun, rising currents of hot gas heat small regions of the photosphere, which, being slightly hotter, emit more black body radiation and look brighter. The cool sinking gas of the edges emits less light and thus looks darker (Figure 8-2b). The presence of granulation is clear evidence that energy is flowing upward through the photosphere.

Spectroscopic studies of the solar surface have revealed another kind of granulation. **Supergranules** are regions about 30,000 km in diameter (about 2.3 times Earth's diameter) and include about 300 granules. These supergranules are regions of very slowly rising currents that last a day or two. They may be the surface traces of larger currents of rising gas deeper under the photosphere.

The edge, or **limb,** of the solar disk is dimmer than the center (see the figure in Celestial Profile 1). This **limb darkening** is caused by the absorption of light in the photosphere. When you look at the center of the solar disk, you are looking directly down into the sun, and you see deep, hot, bright layers in the photosphere. But when you look near the limb of the solar disk, you are looking at a steep angle and cannot see as deeply. The photons you see come from shallower, cooler, dimmer layers in the photosphere. Limb darkening proves that the temperature in the photosphere increases with depth, as you would expect if energy is flowing up from below.

The Chromosphere

Above the photosphere lies the chromosphere. Solar astronomers define the lower edge of the chromosphere as lying just above the visible surface of the sun with its upper regions blending gradually with the corona. You can think of the chromosphere as being an irregular layer with a depth on average less than Earth's diameter (see Figure 8-1). Because the chromosphere is roughly 1000 times fainter than the photosphere, you can see it with your unaided eyes only during a total solar eclipse when the moon covers the brilliant photosphere. Then, the chromosphere flashes into view as a thin line of pink just above the photosphere. The word *chromosphere* comes from the Greek word *chroma,* meaning "color." The pink color is produced by the combined light of three bright emission lines—the red, blue, and violet Balmer lines of hydrogen.

Astronomers know a great deal about the chromosphere from its spectrum. The chromosphere produces an emission spectrum, and Kirchhoff's second law tells you the chromosphere must be an excited, low-density gas. The density is about 10^8 times less dense than the air you breathe.

Atoms in the lower chromosphere are ionized, and atoms in the higher layers of the chromosphere are even more highly ionized. That is, they have lost more electrons. From this, astronomers can find the temperature in different parts of the chromosphere. Just above the photosphere the temperature falls to a minimum of about 4500 K and then rises rapidly (■ Figure 8-3). The region where the temperature increases fastest is called the **transition region** because it makes the transition from the lower temperatures of the photosphere and chromosphere to the extremely high temperatures of the corona.

Solar astronomers can take advantage of some elegant physics to study the chromosphere. The gases of the chromosphere are transparent to nearly all visible light, but atoms in the gas are

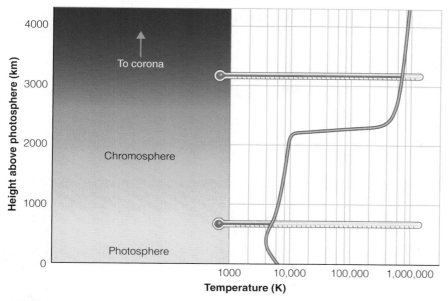

■ Figure 8-3

The chromosphere. If you could place thermometers in the sun's atmosphere, you would discover that the temperature increases from 5800 K at the photosphere to 10^6 K at the top of the chromosphere.

■ Figure 8-4 shows a filtergram made at the wavelength of the H$_\alpha$ Balmer line. This image shows complex structure in the chromosphere including long, dark **filaments** silhouetted against the brighter surface. **Spicules** are flamelike jets of gas extending upward into the chromosphere and lasting 5 to 15 minutes. Seen at the limb of the sun's disk, these spicules blend together and look like flames covering a burning prairie (Figure 8-1a), but they are not flames at all. Spectra show that spicules are cooler gas from the lower chromosphere extending upward into hotter regions. Images at the center of the solar disk show that spicules spring up around the edge of supergranules like weeds around flagstones (Figure 8-4). Although spicules are not yet well understood, they are clearly driven by the outward flow of energy in the sun.

Spectroscopic analysis of the chromosphere alerts you that it is a low-density gas in constant motion where the temperature increases rapidly with height. Just above the chromosphere lies even hotter gas.

The Solar Corona

The outermost part of the sun's atmosphere is called the corona, after the Greek word for crown. The corona is so dim that it is not visible in Earth's daytime sky because of the glare of scattered light from the brilliant photosphere. During a total solar eclipse, however, when the moon covers the photosphere, you can see the innermost parts of the corona, as shown in Figure 8-1b. Observations made with specialized telescopes called **coronagraphs** on Earth or in space can block the light of the photosphere and image the corona out beyond 20 solar radii, almost 10 percent of the

very good at absorbing photons of specific wavelengths. This produces certain dark absorption lines in the spectrum of the photosphere. A photon at one of those wavelengths is very unlikely to escape from deep layers. A **filtergram** is a photograph made using light in one of those dark absorption lines. Those photons can only have escaped from higher in the atmosphere. In this way filtergrams reveal detail in the upper layers of the chromosphere. In a similar way, an image recorded in the far-ultraviolet or in the X-ray part of the spectrum reveals other structures in the solar atmosphere.

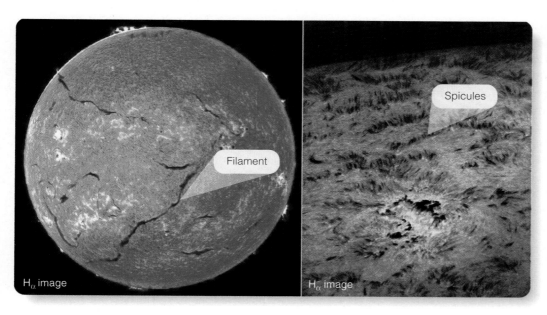

■ Figure 8-4

H$_\alpha$ filtergrams reveal complex structure in the chromosphere, including long, dark filaments and spicules springing from the edges of supergranules twice the diameter of Earth. (NOAA/SEL/USAF; NOAO/NSO)

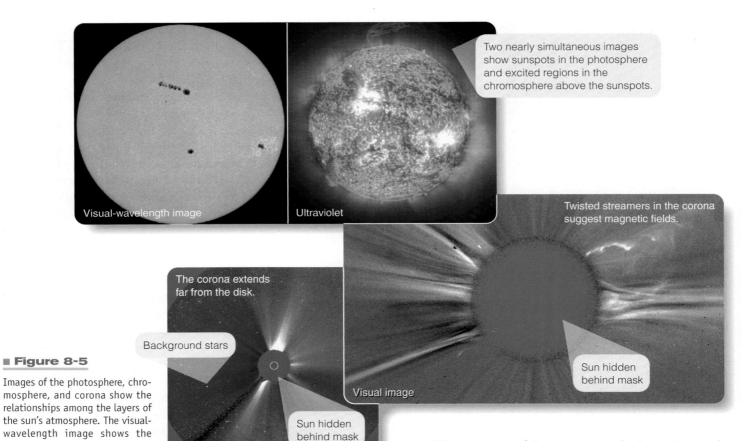

Two nearly simultaneous images show sunspots in the photosphere and excited regions in the chromosphere above the sunspots.

Visual-wavelength image

Ultraviolet

Twisted streamers in the corona suggest magnetic fields.

The corona extends far from the disk.

Background stars

Sun hidden behind mask

Visual image

Sun hidden behind mask

Visual image

■ **Figure 8-5**

Images of the photosphere, chromosphere, and corona show the relationships among the layers of the sun's atmosphere. The visual-wavelength image shows the sun in white light—that is, as you would see it with your eyes. (SOHO/ESA/NASA)

way to Earth. Such images reveal that magnetic fields link the sunspots with features in the chromosphere and corona (■ Figure 8-5).

The spectrum of the corona can tell you a great deal about the coronal gases and simultaneously illustrate how astronomers can analyze a spectrum. Some of the light from the outer corona produces a spectrum with absorption lines the same as the sun's spectrum. This light is just sunlight reflected from dust particles in the corona. In contrast, some of the light from the corona produces a continuous spectrum that lacks absorption lines, and that happens when sunlight from the photosphere is scattered off free electrons in the ionized coronal gas. Because the coronal gas has a temperature over 1 million K and the electrons travel very fast, the reflected photons suffer large, random Doppler shifts that smear out solar absorption lines to produce a continuous spectrum.

Superimposed on the corona's continuous spectrum are emission lines of highly ionized gases. In the lower corona, the atoms are not as highly ionized as they are at higher altitudes, and this tells you that the temperature of the corona rises with altitude. Just above the chromosphere, the temperature is about 500,000 K; but in the outer corona the temperature can be as high as 2 million K or more.

The spectrum of the corona reveals that it is exceedingly hot gas, but it is not very bright. Its density is very low, only 10^6 atoms/cm^3 in its lower regions. That is about a trillion times less dense than the air you breath. In its outer layers the corona contains only 1 to 10 atoms/cm^3, better than the best vacuum on Earth. Because of this low density, the hot gas does not emit much radiation.

Astronomers have wondered for years how the corona and chromosphere can be so hot. Heat flows from hot regions to cool regions, never from cool to hot. So how can the heat from the photosphere, with a temperature of only 5800 K, flow out into the much hotter chromosphere and corona? Observations made by the SOHO satellite have mapped a **magnetic carpet** of looped magnetic fields extending up through the photosphere (■ Figure 8-6). Turbulence below the surface may be whipping these fields about and heating the gases of the chromosphere and corona. Remember that the gas of the chromosphere and corona has very low density, so it can't resist the moving magnetic fields. The gas gets whipped about as the magnetic fields flick back and forth, and that heats the gas. In this instance, energy appears to flow outward as the agitation of the magnetic fields.

Gas from the solar atmosphere follows the magnetic fields pointing outward and flows away from the sun in a breeze called the **solar wind.** Like an extension of the corona, the low-density gases of the solar wind blow past Earth at 300 to 800 km/s with gusts as high as 1000 km/s. Earth is bathed in the corona's hot breath.

■ **Figure 8-6**

This extreme-ultraviolet image of a section of the sun's lower corona has been given a green color. White and black areas are regions of opposite magnetic polarity. Computer models show the location of the magnetic fields that connect these regions. The largest loops shown here could encircle Earth. The entire surface of the sun is covered by this magnetic carpet. (Stanford-Lockheed Institute for Space Research, Palo Alto, CA, and NASA GSFC)

the far-ultraviolet that could only have formed in the low-density, high-temperature gases of a chromosphere and corona. Also, many stars are sources of X rays, which appear to have been produced by coronae. This observational evidence gives astronomers good reason to believe that the sun, for all its complexity, is a typical star.

Helioseismology

Almost no light emerges from below the photosphere, so you can't see into the solar interior. However, solar astronomers can use the vibrations in the sun to explore its depths in a process called **helioseismology.** Random motions in the sun constantly produce vibrations—rumbles too low to hear with human ears. Some of these vibrations resonate in the sun like sound waves in organ pipes. A vibration with a period of 5 minutes is strongest, but the periods range from 3 to 20 minutes. These are very, very low-pitched rumbles!

Astronomers can detect these vibrations by observing Doppler shifts in the solar surface. As a vibrational wave travels down into the sun, the increasing density and temperature curve its path and it returns to the surface, where it makes the photosphere heave up and down by small amounts—roughly plus or minus 15 km (■ Figure 8-7a). Short-wavelength waves penetrate less deeply and travel shorter distances than longer-wavelength waves. This covers the surface of the sun with a pattern of rising and falling regions that can be mapped using the Doppler effect (Figure 8-7b). By observing these motions, astronomers can determine which vibrations resonate the strongest. Just as

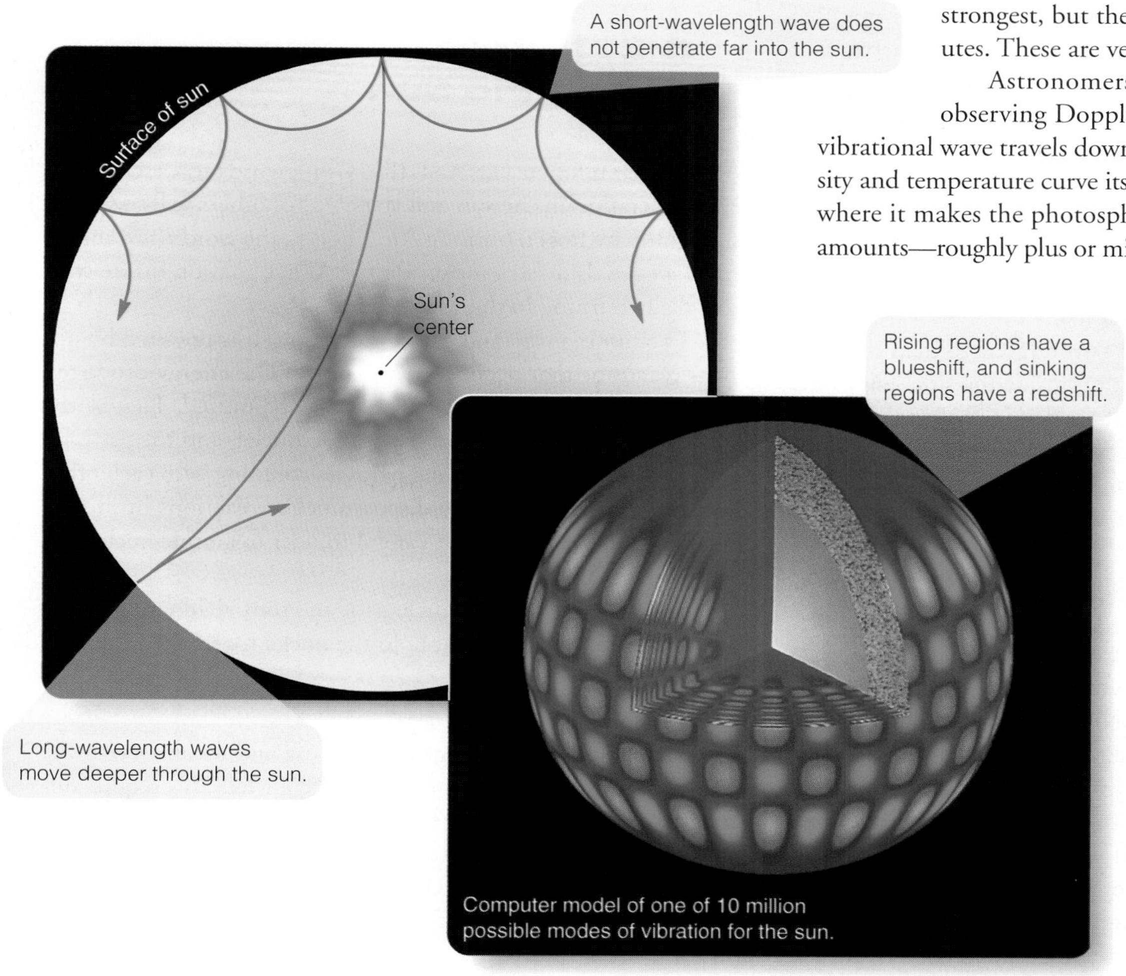

Because of the solar wind, the sun is slowly losing mass, but this is only a minor loss for an object as massive as the sun. The sun loses about 10^7 tons per year, but that is only 10^{-14} of a solar mass per year. Later in life, the sun, like many other stars, will lose mass rapidly. You will see in future chapters how this affects stars.

Do other stars have chromospheres, coronae, and stellar winds like the sun? Ultraviolet and X-ray observations suggest that the answer is yes. The spectra of many stars contain emission lines in

A short-wavelength wave does not penetrate far into the sun.

Surface of sun

Sun's center

Rising regions have a blueshift, and sinking regions have a redshift.

Long-wavelength waves move deeper through the sun.

Computer model of one of 10 million possible modes of vibration for the sun.

■ **Figure 8-7**

Helioseismology: The sun can vibrate in millions of different patterns or modes, and each mode corresponds to a different wavelength vibration penetrating to a different level. By measuring Doppler shifts as the surface moves gently up and down, astronomers can map the inside of the sun. (AURA/NOAO/NSF)

geologists can study Earth's interior by analyzing vibrations from earthquakes, so solar astronomers can use helioseismology to explore the sun's interior.

Helioseismology sounds almost magical, but you can understand it better if you think of a duck pond. If you stood at the shore of a duck pond and looked down at the water, you would see ripples arriving from all parts of the pond. Because every duck on the pond contributes to the ripples, you could, in principle, study the ripples near the shore and draw a map showing the position and velocity of every duck on the pond. Of course, it would be difficult to untangle the different ripples, so you would need lots of data and a big computer. Nevertheless, all of the information would be there, lapping at the rocks at your feet.

The sun can oscillate in about 10 million different ways, and each mode of oscillation has its own characteristic wavelength and its own unique pattern on the solar surface. The waves producing different modes penetrate to different depths where conditions can weaken or strengthen a wave. By discovering which wavelength waves are actually present, solar astronomers can determine the temperature, density, pressure, composition, and motion at different depths inside the sun.

Of course, with 10 million possible wavelengths, the observations and analysis are difficult. A single wave produces a complicated pattern of motion on the solar surface. Large amounts of data are necessary, so helioseismologists have used a network of telescopes around the world operated by the Global Oscillation Network Group (GONG). The network can observe the sun continuously for weeks at a time as Earth rotates. The sun never sets on GONG. The SOHO satellite in space can observe solar oscillations continuously and can detect motions as slow as 1 mm/s (0.002 mph). Solar astronomers can then use high-speed computers to separate the different patterns on the solar surface and measure the strength of the waves at many different wavelengths.

Helioseismology has allowed astronomers to map the temperature, density, and rate of rotation inside the sun. They have been able to detect great currents of gas flowing below the photosphere and the emergence of sunspots before they appear in the photosphere. Helioseismology can even locate sunspots on the back side of the sun, sunspots that are not yet visible from Earth.

Building Scientific Arguments

How deeply into the sun can you see?
Scientific arguments usually involve observations, and it is always important to know how observations are made. When you look into the layers of the sun, your sight does not really penetrate into the sun. Rather, your eyes record photons that have escaped from the sun and traveled outward through the layers of the sun's atmosphere. If you observe at a wavelength at the center of a dark absorption line, then the photosphere and lower chromosphere are opaque, photons can't escape to your eyes, and the only photons you can see come from the upper chromosphere. What you see are the details of the upper chromosphere. On the other hand, if you observe at a wavelength that is not easily absorbed (a wavelength between spectral lines), the atmosphere is more transparent, and photons from deep inside the photosphere can escape to your eyes. There is a limit, however, set by the H$^-$ ion, a hydrogen atom with an extra electron. At a certain depth, there is so much of this ion that the sun's atmosphere is opaque for almost all wavelengths, few photons can escape, and you can't see deeper.

By choosing the proper wavelength, solar astronomers can observe to different depths. But the corona is so thin and the gas below the photosphere so dense that this method doesn't work in these regions. Now it is time to build a new argument. **How can you observe the corona and the deeper layers of the sun?**

■   ■   ■

Connections: So far you have thought of the sun as a static, unchanging ball of gas with energy flowing outward from the interior and through the atmospheric layers. Now you are ready to plunge into the sun and find out how that energy is made.

(8-2) Nuclear Fusion in the Sun

ASTRONOMERS OFTEN USE the wrong words to describe energy generation in the sun and stars. Astronomers will say, "The star ignites hydrogen burning." In English, the word *ignite* means *catch on fire*, and *burn* means *on fire*. What goes on inside stars isn't really burning in the usual sense.

The sun is a star, and it is not burning. It is powered by nuclear reactions that occur near its center. The energy produced keeps the interior hot, and the gas is totally ionized. That is, the electrons are not attached to the atomic nuclei, and the gas is an atomic soup of rapidly moving particles colliding with each other at high velocity. When you discuss nuclear reactions inside the sun and stars, you should be careful to refer to atomic nuclei and not to atoms.

How exactly can the nucleus of an atom yield energy? The answer lies in the forces that hold the nuclei together.

Nuclear Binding Energy

The sun generates its energy by breaking and reconnecting the bonds between the particles *inside* atomic nuclei. This is quite different from the way you would generate energy by burning wood in a fireplace. The process of burning wood extracts energy by breaking and reconnecting chemical bonds between atoms in the wood. Chemical bonds are formed by the electrons in atoms, and you saw in Chapter 7 that the electrons are bound to the atoms by the electromagnetic force. So chemical energy originates in the electromagnetic force.

There are only four forces in nature: the force of gravity, the electromagnetic force, the **weak force,** and the **strong force.** The weak force is involved in the radioactive decay of certain kinds of nuclear particles, and the strong force binds together atomic nuclei. That means nuclear energy must come from the strong force.

Nuclear power plants on Earth generate energy through **nuclear fission** reactions that split uranium nuclei into less massive fragments. A uranium nucleus contains a total of 235 protons and neutrons, and it splits into a range of fragments containing roughly half as many particles. Because the fragments produced are more tightly bound than the uranium nuclei, binding energy is released during uranium fission.

Stars don't use nuclear fission. They make energy in **nuclear fusion** reactions that combine light nuclei into heavier nuclei. The most common reaction, including that in the sun, fuses hydrogen nuclei (single protons) into helium nuclei (two protons and two neutrons). Because the nuclei produced are more tightly bound than the original nuclei, energy is released. Notice in ■ Figure 8-8

■ Figure 8-8

The red line in this graph shows the binding energy per particle, the energy that holds particles inside an atomic nucleus. The horizontal axis shows the atomic mass number of each element, the number of protons and neutrons in the nucleus. Both fission and fusion nuclear reactions move downward in the diagram (arrows) toward more tightly bound nuclei. Iron has the most tightly bound nucleus, so no nuclear reactions can begin with iron and release energy.

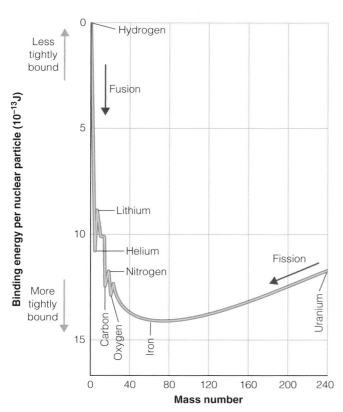

that both fusion and fission reactions move downward in the diagram toward more tightly bound nuclei. They both produce energy by releasing the binding energy of atomic nuclei.

Hydrogen Fusion

Not until the 1930s did astronomers realize how the sun generates energy. The sun fuses together four hydrogen nuclei to make one helium nucleus. Because one helium nucleus has 0.7 percent less mass than four hydrogen nuclei, it seems that some mass vanishes in the process. To see this, subtract the mass of a helium nucleus from the mass of four hydrogen nuclei.

$$\begin{aligned} 4 \text{ hydrogen nuclei} &= 6.693 \times 10^{-27} \text{ kg} \\ - 1 \text{ helium nucleus} &= 6.645 \times 10^{-27} \text{ kg} \\ \hline \text{Difference in mass} &= 0.048 \times 10^{-27} \text{ kg} \end{aligned}$$

It seems from this subtraction that a small amount of mass disappears, but it doesn't really vanish; it merely changes form. The equation $E = m_0 c^2$ reminds you that mass and energy are related, and under certain circumstances mass may become energy and vice versa. So the 0.048×10^{-27} kg does not vanish but merely becomes energy. To see how much, you use Einstein's equation:

$$\begin{aligned} E &= m_0 c^2 \\ &= (0.048 \times 10^{-27} \text{ kg})(3 \times 10^8 \text{ m/s})^2 \\ &= 0.43 \times 10^{-11} \text{ J} \end{aligned}$$

This is a very small amount of energy, hardly enough to raise a housefly one-thousandth of an inch. Because one reaction produces such a small amount of energy, it is obvious that many reactions are necessary to supply the energy needs of a star. The sun, for example, needs 10^{38} reactions per second, transforming 5 million tons of mass into energy every second, just to stay hot enough to resist its own gravity.

It seems from this that nuclear fusion is very powerful, especially if you calculate that the fusion of a milligram of hydrogen (roughly the mass of a match head) produces as much energy as burning 30 gallons of gasoline. However, the nuclear reactions in the sun are spread through a large volume in its core, and any single gram of matter produces only a little energy. A person of normal mass eating a normal diet produces about 4000 times more heat per gram than the matter in the core of the sun. The sun produces a lot of energy because it contains a lot of grams of matter in its core.

Fusion reactions can occur only when the nuclei of two atoms get very close to each other. Because atomic nuclei carry positive charges, they repel each other with an electrostatic force called the Coulomb force. Physicists commonly refer to this repulsion between nuclei as the **Coulomb barrier.** To overcome this barrier and get close together, atomic nuclei must collide violently. Violent collisions are rare unless the gas is very hot, in which case the nuclei move at high speeds and collide violently. (Remember, an object's temperature is related to the speed with which its particles move.)

So nuclear reactions in the sun take place only near the center, where the gas is hot and dense. A high temperature insures that some of the collisions between nuclei are violent enough to overcome the Coulomb barrier, and a high density ensures that there are enough collisions, and thus enough reactions, to meet the sun's energy needs.

You can symbolize the fusion reactions in the sun with a simple nuclear reaction:

$$4\ ^1H \rightarrow\ ^4He + energy$$

In this equation, ^1H represents a proton, the nucleus of the hydrogen atom, and ^4He represents the nucleus of a helium atom. The superscripts indicate the approximate weight of the nuclei (the number of protons plus the number of neutrons). The actual steps in the process are more complicated than this convenient summary suggests. Instead of waiting for four hydrogen nuclei to collide simultaneously, a highly unlikely event, the process can proceed step by step in a chain of reactions—the proton–proton chain.

The **proton–proton chain** is a series of three nuclear reactions that builds a helium nucleus by adding together protons. This process is efficient at temperatures above 10,000,000 K. The sun, for example, manufactures over 90 percent of its energy in this way.

The three steps in the proton–proton chain entail these reactions:

$$^1H +\ ^1H \rightarrow\ ^2H + e^+ + \nu$$
$$^2H +\ ^1H \rightarrow\ ^3He + \gamma$$
$$^3He +\ ^3He \rightarrow\ ^4He +\ ^1H +\ ^1H$$

In the first reaction, two hydrogen nuclei (two protons) combine to form a heavy hydrogen nucleus called **deuterium,** emitting a particle called a **positron,** e^+ (a positively charged electron), and a **neutrino,** ν (a subatomic particle having an extremely low mass and a velocity nearly equal to the velocity of light). In the second reaction, the heavy hydrogen nucleus absorbs another proton and, with the emission of a gamma ray, γ, becomes a lightweight helium nucleus. Finally, two lightweight helium nuclei combine to form a common helium nucleus and two hydrogen nuclei. Because the last reaction needs two ^3He nuclei, the first and second reactions must occur twice (■ Figure 8-9). The net result of this chain reaction is the transformation of four hydrogen nuclei into one helium nucleus plus energy.

The energy appears in the form of gamma rays, positrons, the energy of motion of the particles, and neutrinos. The gamma rays are photons that are absorbed by the surrounding gas before they can travel more than a fraction of a millimeter. This heats the gas and helps maintain the pressure. The positrons produced in the first reaction combine with free electrons, and both particles

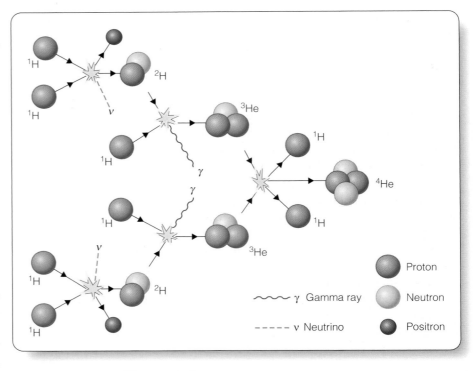

■ **Figure 8-9**

The proton–proton chain combines four protons (at far left) to produce one helium nucleus (at right). Energy is produced mostly as gamma rays and positrons, which combine with electrons and convert their mass into energy. Neutrinos escape, carrying away about 2 percent of the energy produced.

vanish, converting their mass into gamma rays, which are absorbed and also help keep the center of the star hot. In addition, when fusion produces new nuclei, they fly apart at high velocity. This energy of motion helps raise the temperature of the gas. The neutrinos resemble photons except that they almost never interact with other particles. The average neutrino could pass unhindered through a lead wall a light-year thick. Consequently, the neutrinos do not help heat the gas but race out of the star at nearly the speed of light, carrying away roughly 2 percent of the energy produced.

Energy Transport in the Sun

Now you are ready to follow the energy from the core of the sun to the surface. The surface is cool, only about 5800 K, and the center is over 10 million Kelvin, so energy must flow outward from the core.

Because the core is so hot, the photons being emitted are gamma rays. Each time a gamma ray encounters an electron, it is deflected or scattered in a random direction; and, as it bounces around, it slowly drifts outward toward the surface. That carries energy outward in the form of radiation, and astronomers refer to the inner parts of the sun as the **radiative zone.**

To examine this process, imagine picking a single gamma ray and following it to the surface. As your gamma ray is scattered over and over by the hot gas, it drifts outward into cooler layers, and the cooler gas tends to emit photons of longer wavelength. Your gamma ray will eventually be absorbed by the gas and reemitted as two X rays. Now you must follow those two X rays as they bounce around, and soon you will see them drifting outward into even cooler gas, where they will become a number of longer wavelength photons. The packet of energy that began as a single gamma ray gets broken down into a large number of lower-energy photons, and it eventually emerges from the sun's surface as about 1800 photons of visible light.

But something else happens along the way. The packet of energy that you began following from the core eventually reaches the outer layers of the sun where the gas is so cool it is not very transparent to radiation. The energy backs up like water behind a dam, and the gas begins to churn in convection. Hot blobs of gas rise, and cool blobs sink. In this region, known as the **convective zone,** the energy is carried outward as circulating gas. These two zones are shown in ■ Figure 8-10. The granulation visible on the photosphere is clear evidence that the sun has a convective zone just below the photosphere carrying energy upward to the surface.

Sunlight is nuclear energy produced in the core of the sun. The energy of a single gamma ray can take a million years to work its way outward, first as radiation and then as convection on its journey to the photosphere.

It is time to ask the critical question that lies at the heart of science. What is the evidence to support this theoretical explanation of how the sun makes its energy? The search for that evidence will introduce you to one of the great problems of modern astronomy.

Ace◐Astronomy™ Log into AceAstronomy and select this chapter to see Astronomy Exercise "Nuclear Fusion" and take control of fusion in the sun's core.

The Solar Neutrino Problem

The center of a star seems forever hidden, but the sun is transparent to neutrinos because these subatomic particles almost never interact with normal matter. Nuclear reactions in the sun's core produce floods of neutrinos that rush out of the sun and off into space. If you could detect these neutrinos, you could probe the sun's interior.

Because neutrinos almost never interact with atoms, you never feel the flood of over 10^{12} solar neutrinos that flow through your body every second. Even at night, neutrinos from the sun rush through Earth as if it weren't there, up through your bed, through you, and onward into space. Obviously you are lucky to be transparent to neutrinos, but it means that neutrinos are extremely hard to detect. Certain nuclear reactions, however, can be triggered by a neutrino of the right energy; and, in the late 1960s, chemist Raymond Davis, Jr., began using such a reaction to detect solar neutrinos.

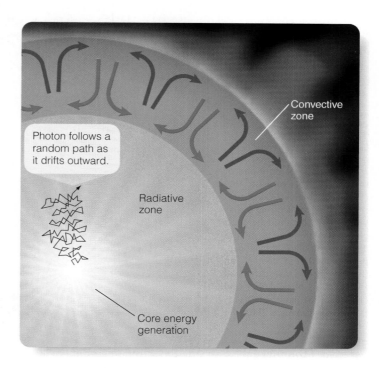

Active Figure 8-10

A cross section of the sun. Near the center, nuclear fusion reactions generate high temperatures. Energy flows outward through the radiative zone as photons are randomly deflected over and over by electrons. In the cooler, more opaque outer layers, the energy is carried by rising convection currents of hot gas (red arrows) and sinking currents of cooler gas (blue arrows).

Ace◐Astronomy™ Log into AceAstronomy and select this chapter to see Active Figure "The Sun." Watch energy flow outward from the core to the surface.

Davis filled a 100,000-gallon tank with the cleaning fluid perchloroethylene (C_2Cl_4). Theory predicts that about once a day, a solar neutrino will convert a chlorine atom in the tank into radioactive argon, which can be detected later by its radioactive decay. To protect the detector from cosmic rays from space, the tank was buried nearly a mile deep in a South Dakota gold mine (■ Figure 8-11a). Of course, the mile of rock overhead had no effect on the neutrinos.

The result of the Davis experiment startled astronomers. The cleaning fluid detected too few neutrinos—not one neutrino per day as predicted by models of the sun but about one every three days. The experiment was refined, tested, and calibrated for three decades; but it did not find the missing neutrinos. Other detectors were built, and they too counted too few neutrinos coming from the sun.

The missing solar neutrinos were one of the great mysteries of modern astronomy. Some scientists argued that astronomers didn't correctly understand how the sun and stars make their energy, but other scientists wondered if there was something about neutrinos that could explain the problem. Astronomers have great confidence in their theories of the sun's interior, and helio-

■ **Figure 8-11**

(a) The Davis solar neutrino experiment used cleaning fluid and could detect only one of the three flavors of neutrinos. (Brookhaven National Laboratory) (b) The Sudbury Neutrino Observatory is a 12-meter-diameter globe containing water rich in deuterium in place of hydrogen. Buried 6800 feet deep in an Ontario mine, it can detect all three flavors of neutrinos and confirms that neutrinos oscillate. (Photo courtesy of SNO)

seismology confirmed those theories, so astronomers did not abandon their theories immediately (**Window on Science 8-1**).

As the 21st century began, scientists were able to solve the mystery. Physicists know of three kinds of neutrinos, which they call flavors. The Davis experiment could detect (or taste) only one flavor, electron-neutrinos. Theory hinted that the electron-neutrinos produced in the core of the sun might oscillate among the three flavors as they rushed out through the sun and across space to Earth. Observations begun in 2000 confirm this theory (Figure 8-11b). Some of the electron-neutrinos produced in the sun transform into tau- and muon-neutrinos, which most detectors cannot detect.

This solution to the solar neutrino problem is exciting because neutrinos can't oscillate unless they have mass. Neutrinos were long thought to be massless, but if they have even a small mass, they are so common their gravity could affect the evolution of the universe as a whole—something you will read about in Chapter 18. The detection of neutrino oscillation excites astronomers for another reason. It confirms the theories that describe the interior of the sun and stars.

Building Scientific Arguments

Why does nuclear fusion require that the gas be very hot? This argument has to include some basic physics of atoms and thermal energy. Inside a star, the gas is so hot it is ionized, which means the electrons have been stripped off the atoms, and the nuclei are bare and have a positive charge. For hydrogen fusion, the nuclei are single protons. These atomic nuclei repel each other because of their positive charges, so they must collide with each other at high velocity to overcome that repulsion and get close enough together to fuse. If the atoms in a gas are moving rapidly, then it must have a

high temperature, and so nuclear fusion requires that the gas have a very high temperature. If the gas is cooler than about 10 million K, hydrogen can't fuse because the protons don't collide violently enough to overcome the repulsion of their positive charges.

It is easy to see why nuclear fusion in the sun requires high temperature, but now expand your argument. **Why does it require high density?**

■   ■   ■

Connections: You have been to the center of the sun and seen how it makes its energy, and you have followed that energy up to the surface. Now it is time for you to see how outrageous the sun can be as that energy bursts through the photosphere.

8-3 Solar Activity

THE SUN IS UNQUIET. It is home to slowly changing spots larger than Earth and vast eruptions that dwarf human imagination. All of these seemingly different forms of solar activity have one thing in common—magnetic fields. The weather on the sun is magnetic.

Observing the Sun

Solar activity is often visible with even a small telescope, but you should be very careful about observing the sun. Sunlight is intense, and when it enters your eye it is absorbed and converted into thermal energy. Equally dangerous is the infrared radiation in sunlight. Your eyes can't detect the infrared, but it is converted to thermal energy in your eyes and can burn and scar the retina.

Avoiding Hasty Judgments: Scientific Confidence

Scientists like to claim that every scientific understanding is based on evidence, that every theory has been tested, and that the moment a theory fails a test, it is discarded or revised. The truth is much more complicated than that, and the solar neutrino problem is a good illustration. If the detection of too few solar neutrinos contradicted the theory of stellar structure, why wasn't the theory abandoned?

While scientists do indeed have tremendous respect for evidence, they also have confidence in theories that have been tested successfully many times. If a theory has been tested and confirmed over and over, they may even begin to call it a natural law, and that means scientists have great confidence that the theory or law is a good description of how nature works.

Nevertheless, it is not unusual for an experiment or an observation to contradict well-established theories. In many cases, the experiments and observations are simple mistakes, or they have not been interpreted correctly. Scientists resist abandoning a well-tested theory even when an observation continues to contradict it. If confidence in the theory is stronger than the evidence, scientists begin by testing the evidence. Can it be right? Do we understand it correctly? Of course, if the evidence cannot be impeached and it continues to contradict the theory, scientists must eventually abandon or modify the theory no matter how many times it has previously been tested and confirmed. This ultimate reliance on evidence is the distinguishing characteristic of science.

It is only human nature to hang on to the principles you have come to trust, and that confidence in well-tested scientific principles helps scientists avoid rushing to faulty judgments. For example, claims for perpetual motion machines occasionally crop up in the news, but the world's scientists don't instantly abandon the laws of energy and motion pending an analysis of the latest claim. Of course, if such a claim did prove true, the entire structure of scientific knowledge would come crashing down, but because the known laws of energy and motion have been well tested and no perpetual motion machine has ever been successful, scientists know which way to bet. Like the keel on a ship, confidence in well-tested theories and laws keeps the scientific boat from rocking before every little breeze.

Ultimately scientific confidence must be open to change. If, eventually, a single experiment or observation conclusively contradicts a cherished law of nature, scientists must abandon that law and find a new way to understand nature. You can see this scientific confidence at work in many controversies, from the origin of the human race, to the meaning of IQ measurements; but one of the best examples is the solar neutrino problem. While astronomers struggled to understand the origin of solar neutrinos, they continued to have confidence that they do indeed understand how stars make energy.

It is not safe to look directly at the sun, and it is even more dangerous to look at the sun through any optical instrument such as a telescope, binoculars, or even the viewfinder of a camera. The light-gathering power of such an optical system concentrates the sunlight and can cause severe injury. Never look at the sun with any optical instrument unless you are certain it is safe. ■ Figure 8-12 shows a safe way to observe the sun with a small telescope.

In the early 17th century, Galileo observed the sun and saw spots on its surface; day by day he saw the spots moving across the sun's disk. He rightly concluded that the sun was a sphere and was rotating. You could repeat his observations, and you would probably see something that looks like Figure 8-12b. You would see sunspots.

Sunspots

The dark sunspots that you see at visible wavelengths only hint at the complex processes that go on in the sun's atmosphere. To explore those processes, you must turn to the analysis of images and spectra at a wide range of wavelengths.

Study **Sunspots and the Sunspot Cycle** on pages 174–175 and notice five important points:

1 Sunspots are cool spots on the sun's surface caused by strong magnetic fields.

2 Also notice that sunspots follow an 11-year cycle not only in the number of spots visible but in their location on the sun.

3 The Zeeman effect gives astronomers a way to measure the strength of magnetic fields on the sun.

4 Notice that the sunspot cycle can vary over centuries and appears to affect Earth's climate.

5 Finally, notice the clear evidence that sunspots are part of a larger magnetic process that involves all layers of the sun's atmosphere.

The sunspot groups are merely the visible traces of magnetically active regions. But what causes this magnetic activity? The answer appears to be linked to the waxing and waning of the sun's overall magnetic field.

(Ace ✆Astronomy™) Log into AceAstronomy and select this chapter to see Astronomy Exercise "Zeeman Effect" and see how astronomers measure magnetic fields.

(Ace ✆Astronomy™) Log into AceAstronomy and select this chapter to see Astronomy Exercise "Sunspot Cycle I"; you can explore the link between sunspots and the sun's magnetic field.

(Ace ✆Astronomy™) Log into AceAstronomy and select this chapter to see Astronomy Exercise "Sunspot Cycle II." You can observe sunspots over the centuries.

Sunspots and the Sunspot Cycle

1 The dark spots that appear on the sun are only the visible traces of complex regions of activity. Observations over many years and at a range of wavelengths tell you that sunspots are clearly linked to the sun's magnetic field.

Spectra show that sunspots are cooler than the photosphere with a temperature of about 4240 K. The photosphere has a temperature of about 5800 K. Because the total amount of energy radiated by a surface depends on its temperature raised to the fourth power, sunspots look dark in comparison. Actually, a sunspot emits quite a bit of radiation. If the sun were removed and only an average-size sunspot were left behind, it would be brighter than the full moon.

A typical sunspot is about twice the size of Earth, but there is a wide range of sizes. They appear, last a few weeks to as long as two months, and then shrink away. Usually, sunspots occur in pairs or complex groups.

Size of Earth

NASA

Royal Swedish Academy of Sciences

Umbra

Penumbra

Sunspots are not shadows, but astronomers refer to the dark core of a sunspot as its umbra and the outer, lighter region as the penumbra.

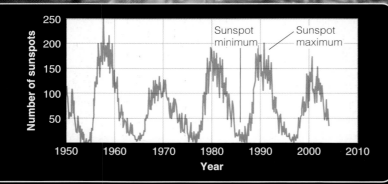

2 The number of spots visible on the sun varies in a cycle with a period of 11 years. At maximum, there are often over 100 spots visible. At minimum, there are very few.

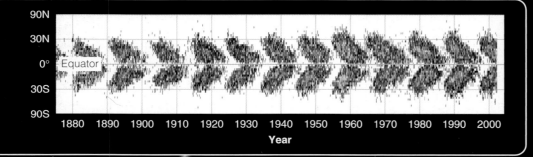

2a Early in the cycle, spots appear at high latitudes north and south of the sun's equator. Later in the cycle, the spots appear closer to the sun's equator. If you plot the latitude of sunspots versus time, the graph looks like butterfly wings, as shown in this **Maunder butterfly diagram**, named after E. Walter Maunder of Greenwich Observatory.

3 Astronomers can measure magnetic fields on the sun using the **Zeeman effect** as shown below. When an atom is in a magnetic field, the electron orbits are altered, and the atom is able to absorb a number of different wavelength photons even though it was originally limited to a single wavelength. In the spectrum, you see single lines split into multiple components, with the separation between the components proportional to the strength of the magnetic field.

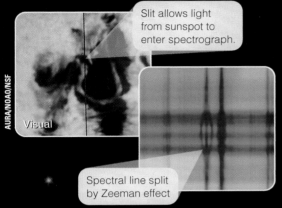

AURA/NOAO/NSF

Slit allows light from sunspot to enter spectrograph.

Visual

Spectral line split by Zeeman effect

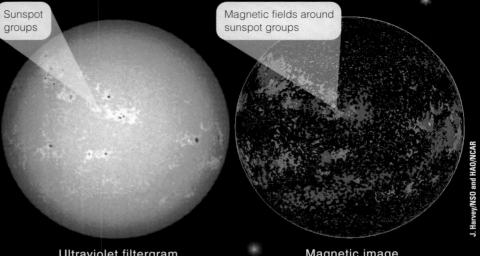

Sunspot groups

Magnetic fields around sunspot groups

J. Harvey/NSO and HAO/NCAR

Ultraviolet filtergram

Magnetic image

Simultaneous images

3a Images of the sun above show that sunspots contain magnetic fields a few thousand times stronger than Earth's. The strong fields are believed to inhibit gas motion below the photosphere; consequently, convection is reduced below the sunspot, and the surface there is cooler. Heat prevented from emerging through the sunspot is deflected and emerges around the sunspot, which can be detected in infrared images.

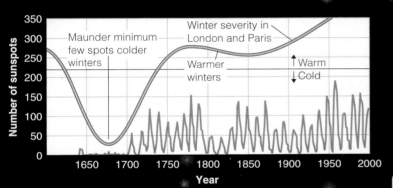

Maunder minimum few spots colder winters

Winter severity in London and Paris

Warmer winters

↑ Warm
↓ Cold

Number of sunspots

350
300
250
200
150
100
50
0

1650 1700 1750 1800 1850 1900 1950 2000

Year

4 Historical records show that there were very few sunspots from about 1645 to 1715, a phenomenon known as the **Maunder minimum.** This coincides with a period called the "little ice age," a period of unusually cool weather in Europe and North America from about 1500 to about 1850, as shown in the graph at left. Other such periods of cooler climate are known. The evidence suggests that there is a link between solar activity and the amount of solar energy Earth receives. This link has been confirmed by measurements made by spacecraft above Earth's atmosphere.

M. Seeds

Magnetic fields can reveal themselves by their shape. For example, iron filings sprinkled over a bar magnet reveal an arched shape.

The complexity of an active region becomes visible at short wavelengths.

Far-UV image

SOHO/EIT, ESA and NASA

5 Observations at nonvisible wavelengths reveal that the chromosphere and corona above sunspots are violently disturbed in what astronomers call **active regions.** Spectrographic observations show that active regions contain powerful magnetic fields.
Arched structures above an active region are evidence of gas trapped in magnetic fields.

Visual-wavelength image

Simultaneous images

Far-UV image

NASA/TRACE

(a) Looking through a telescope at the sun is dangerous, but you can always view the sun safely with a small telescope by projecting its image on a white screen. (b) If you sketch the location and structure of sunspots on successive days, you will see the rotation of the sun and gradual changes in the size and structure of sunspots just as Galileo did in 1610.

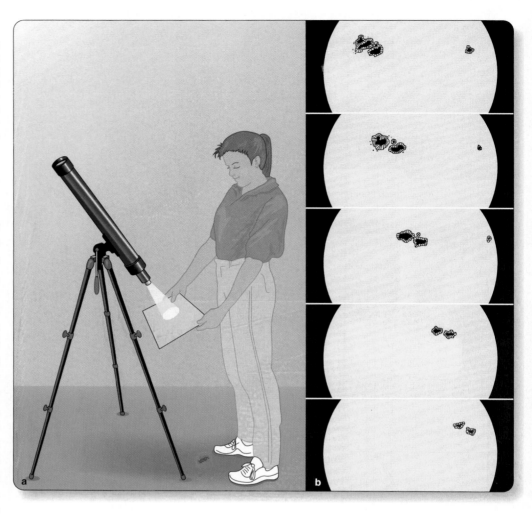

The Sun's Magnetic Cycle

Sunspots are magnetic phenomena, so the 11-year cycle of sunspots must be caused by cyclical changes in the sun's magnetic field. To explore that idea, begin with the sun's rotation.

The sun does not rotate as a rigid body. It is a gas from its outermost layers down to its center, so some parts of the sun rotate faster than other parts. The equatorial region of the photosphere rotates faster than do regions at higher latitudes (■ Figure 8-13a). At the equator, the photosphere rotates once every 25 days, but at latitude 45° one rotation takes 27.8 days. Helioseismology can map the rotation throughout the interior (Figure 8-13b). This phenomenon is called **differential rotation,** and it is clearly linked with the magnetic cycle.

■ Figure 8-13

(a) In general, the photosphere of the sun rotates faster at the equator than at higher latitudes. If you started five sunspots in a row, they would not stay lined up as the sun rotates. (b) Detailed analysis of the sun's rotation from helioseismology reveals regions of slow rotation (blue) and rapid rotation (red). Such studies show that the interior of the sun rotates differentially and that currents similar to the trade winds in Earth's atmosphere flow through the sun. (NASA/SOI)

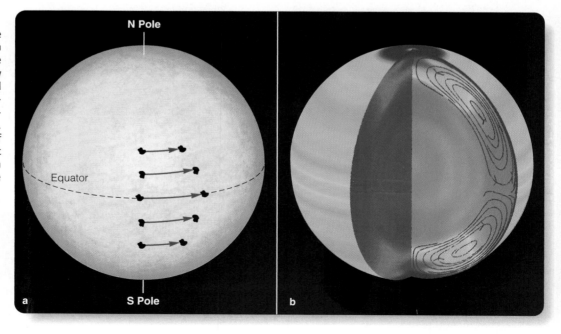

The sun's magnetic field appears to be powered by the energy flowing outward through the moving currents of gas. The gas is highly ionized, so it is a very good conductor of electricity. When an electrical conductor rotates rapidly and is stirred by convection, it can convert some of the energy flowing outward as convection into a magnetic field. This process is called the **dynamo effect,** and it is believed to produce Earth's magnetic field as well. Helioseismologists have found evidence that the sun's magnetic field is generated at the bottom of the convection zone deep under the photosphere. The details of this process are still poorly understood, but the sun's magnetic cycle is clearly related to the creation of its magnetic field.

The magnetic behavior of sunspots provides an insight into how the magnetic cycle works. Sunspots tend to occur in groups or pairs, and the magnetic field around the pair resembles that around a bar magnet with one end magnetic north and the other end magnetic south. At any one time, sunspot pairs south of the sun's equator have reversed polarity compared to those north of the sun's equator. ■ Figure 8-14 illustrates this by showing sunspot pairs south of the sun's equator with magnetic south poles leading and sunspots north of the sun's equator with magnetic north poles leading. At the end of an 11-year sunspot cycle, the new spots appear with reversed magnetic polarity.

This magnetic cycle is not fully understood, but the **Babcock model** (named for its inventor) explains the magnetic cycle as a progressive tangling of the solar magnetic field. Because the electrons in an ionized gas are free to move, the gas is a very good conductor of electricity, and any magnetic field in the gas is "frozen" into the gas. If the gas moves, the magnetic field must move with it. The sun's magnetic field is frozen into its gases, and the differential rotation wraps this field around the sun like a long string caught on a hubcap. Rising and sinking gas currents twist the field into ropelike tubes, which tend to float upward. Where these magnetic tubes burst through the sun's surface, sunspot pairs occur (■ Figure 8-15).

The Babcock model explains the reversal of the sun's magnetic field from cycle to cycle. As the magnetic field becomes tangled, adjacent regions of the sun are dominated by magnetic fields that point in different directions. After about 11 years of tangling, the field becomes so complex that adjacent regions of the sun begin changing their magnetic field to agree with neighboring regions. The entire field quickly rearranges itself into a simpler pattern, and differential rotation begins winding it up to start a new cycle. But the newly organized field is reversed, and the next sunspot cycle begins with magnetic north replaced by magnetic south. Evidently the complete magnetic cycle is 22 years long, and the sunspot cycle is 11 years long.

This magnetic cycle may explain the Maunder butterfly diagram. As a sunspot cycle begins, the twisted tubes of magnetic force first begin to float upward and produce sunspot pairs at higher latitude. Consequently the first sunspots in a cycle appear further north and south of the equator. Later in the cycle, when the field is more tightly wound, the tubes of magnetic force arch up through the surface closer to the equator. As a result, the later sunspot pairs in a cycle appear closer to the equator.

Notice the power of a scientific model. The Babcock model may in fact be incorrect in some or all details, but it provides a framework on which to organize all of the complex solar activity. Even though the models of the sky in Chapter 2 and the atom in Chapter 7 were only partially correct, they served as organizing themes to guide your thinking. Similarly, although the precise details of the solar magnetic cycle are not yet understood, the Babcock model gives you a general picture of the behavior of the sun's magnetic field (**Window on Science 8-2**).

Ace ⑤Astronomy™ Log into AceAstronomy and select this chapter to see Astronomy Exercise "Convection and Magnetic Fields." Discover the link between convection and magnetic fields.

Spots and Magnetic Cycles on Other Stars

The sun seems to be a representative star, so you should expect other stars to have similar cycles of starspots and magnetic fields. This is a difficult topic, because, except for the sun, the stars are so far away that no surface detail is visible. Some stars, however, vary in brightness in ways that suggest they are mottled by dark

■ Figure 8-14

In sunspot groups, here simplified into pairs of major spots, the leading spot and the trailing spot have opposite magnetic polarity. Spot pairs in the southern hemisphere have reversed polarity from those in the northern hemisphere.

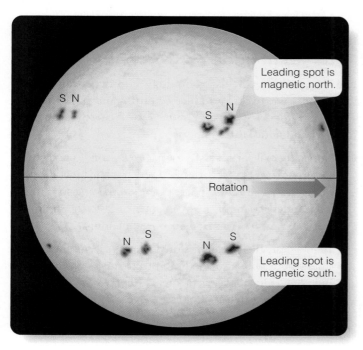

Leading spot is magnetic north.

Rotation

Leading spot is magnetic south.

spots. As the star rotates, its total brightness changes slightly, depending on the number of spots facing Earth. High-precision spectroscopic analysis has allowed astronomers to map the loca-

■ **Figure 8-15**

■ **Figure 8-15**

The Babcock model of the solar magnetic cycle explains the sunspot cycle as a consequence of the sun's differential rotation gradually winding up the magnetic field.

The Solar Magnetic Cycle

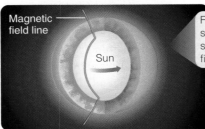

Magnetic field line

Sun

For simplicity, a single line of the solar magnetic field is shown.

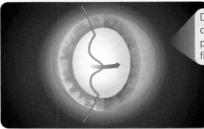

Differential rotation drags the equatorial part of the magnetic field ahead.

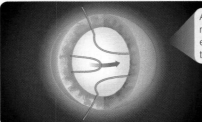

As the sun rotates, the magnetic field is eventually dragged all the way around.

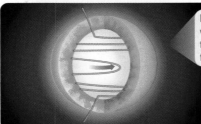

Differential rotation wraps the sun in many turns of its magnetic field.

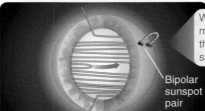

Where loops of tangled magnetic field rise through the surface, sunspots occur.

Bipolar sunspot pair

tion of spots on the surfaces of certain stars (■ Figure 8-16a). Such results tell you that the sunspots you see on our sun are not unusual.

Certain features in stellar spectra are associated with magnetic fields. Regions of strong magnetic fields on the solar surface emit strongly at the central wavelengths of the two strongest lines of ionized calcium. This calcium emission appears in the spectra of other sunlike stars and suggests that these stars, too, have strong magnetic fields on their surfaces. In some cases, the strength of this calcium emission varies over periods of days or weeks and suggests that the stars have active regions and are rotating with periods similar to that of the sun. These stars presumably have starspots as well.

In 1966, astronomers began a long-term project to monitor the strength of this calcium emission in the spectra of stars. The stars to be observed were selected to be similar to the sun. With temperatures ranging from 1000 K hotter than the sun to 3000 K cooler, these stars were considered most likely to have sunlike magnetic activity on their surfaces.

The observations show that the strength of the calcium emission varies over periods of years. The calcium emission averaged over the sun's disk varies with the sunspot cycle, and similar periodic variations can be seen in the spectra of some of the stars studied (Figure 8-16b). The star 107 Piscium, for instance, appears to have a starspot cycle lasting nine years. From this kind of evidence it appears that stars like the sun do have magnetic cycles.

These observations confirm that the sun is an average sort of star, that is, not peculiar. Most other stars like our sun have magnetic fields and starspots and go through magnetic cycles.

Ace ⊙ Astronomy™ Log into AceAstronomy and select this chapter to see Astronomy Exercise "Convection and Magnetic Fields." You can adjust the sun's central temperature and change its magnetic field.

Chromospheric and Coronal Activity

The solar magnetic fields extend high into the chromosphere and corona, where they produce beautiful and powerful phenomena. Study **Magnetic Solar Phenomena** on pages 180–181 and notice three important points:

1 Notice that all solar activity is magnetic. You do not experience such events on Earth because Earth's magnetic field is weak, and Earth's atmosphere is not ionized and so is free to move independent of the magnetic field. On the sun, however, the weather is a magnetic phenomenon.

2 Tremendous energy can be stored in arches of magnetic field. These are visible near the limb of the sun as prominences, and, seen from above, as filaments (Figure 8-4). When that stored energy is released, it can trigger powerful eruptions, and, although these eruptions occur far from Earth, they can affect us in dramatic ways.

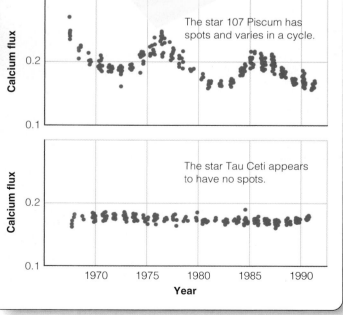

Building Confidence by Confirmation and Consolidation

While many textbooks describe science as the process of testing hypotheses by observation and experiment, you should not think that every astronomer approaches the telescope with the expectation of making an observation that will disprove long-held beliefs and trigger a revolution in science. Then what is the daily grind of science really about?

First, many observations and experiments merely confirm already tested hypotheses. The biologist knows that all worker bees in a hive are sisters, but a careful study of the DNA from different workers further confirms that hypothesis. By repeatedly confirming a hypothesis,

scientists build confidence in the hypothesis and may be able to extend it to a wider application. Of course, there is always the chance that a new observation or experiment will disprove the hypothesis, but that is usually very unlikely. Much of the daily grind of science is confirmation.

Another aspect of routine science is consolidation, the linking of a hypothesis to other well-studied phenomena. Chemists may understand certain kinds of carbon molecules shaped like rings, but by repeated study they find a carbon molecule shaped like a hollow sphere. To consolidate their findings, they must show that the chemical bonding in the two mol-

ecules follows the same rules and that the molecules have certain properties in common. No hypothesis is overthrown, but the chemists consolidate their knowledge and understand carbon molecules better.

The Babcock model of the solar magnetic cycle is an astronomical example of the scientific process. Solar astronomers know that the model explains some solar features but has shortcomings. Although most astronomers don't expect to discard the entire model, they work through confirmation and consolidation to better understand how the solar magnetic cycle works and how it is related to cycles in other stars.

Computer model of HD 12545

Dark spot
3500 K

Average
temperature
4500 K

Sun

Bright spot
4800 K

■ **Figure 8-16**

Astronomers have found clear evidence that other stars have spots. Rotation broadens the shapes of absorption lines in the spectrum of the star HD12545, and variations in the line shapes allow astronomers to map the location of large spots. In addition, long-term studies of calcium emission show that some stars have active regions like those around sunspot groups on our sun. (Model: K. Strassmeier, Vienna, AURA/NOAO/NSF; CaII emission adapted from data by Baliunas and Saar)

Emission by ionized calcium is associated with sunspots and can be detected in spectra of other stars.

The star 107 Piscum has spots and varies in a cycle.

Calcium flux

0.2

0.1

The star Tau Ceti appears to have no spots.

Calcium flux

0.2

0.1

1970   1975   1980   1985   1990

Year

3 In some regions of the solar surface, the magnetic field does not loop back. High-energy gas from these regions flows outward and produces much of the solar wind.

Solar astronomers use every part of the electromagnetic spectrum to study the sun. Some things are easily visible to human eyes, but many phenomena on the sun can only be studied at nonvisible wavelengths. To understand our sun, astronomers need every piece of evidence they can find, and that means using the entire spectrum.

Magnetic Solar Phenomena

1 Magnetic phenomena in the chromosphere and corona, like magnetic weather, result as constantly changing magnetic fields on the sun trap ionized gas to produce beautiful arches and powerful outbursts. Some of this solar activity can affect Earth's magnetic field and atmosphere.

This ultraviolet image of the solar surface was made by the NASA TRACE spacecraft. It shows hot gas trapped in magnetic arches extending above active regions. At visual wavelengths, you would see sunspot groups in these active regions.

The gas in prominences may be 60,000 to 80,000 K, quite cold compared with the low-density gas in the corona, which may be as hot as a million Kelvin.

Trace/NASA

1a A **prominence** is composed of ionized gas trapped in a magnetic arch rising up through the photosphere and chromosphere into the lower corona. Seen during total solar eclipses at the edge of the solar disk, prominences look pink because of the three Balmer emission lines. The image below shows the arch shape suggestive of magnetic fields. Seen from above against the sun's bright surface, prominences form dark filaments.

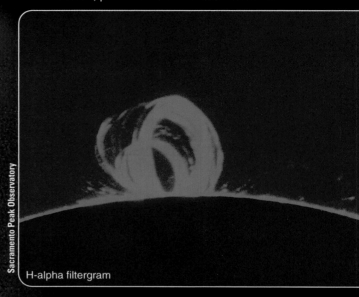

Sacramento Peak Observatory

H-alpha filtergram

1b Quiescent prominences may hang in the lower corona for many days, whereas eruptive prominences burst upward in hours. The eruptive prominences below are many Earth diameters long.

Far-UV image

SOHO, EIT, ESA and NASA

An ultraviolet image shows an active region experiencing a flare.

Far-UV image

NASA

2 Solar **flares** rise to maximum in minutes and decay in an hour. They occur in active regions where oppositely directed magnetic fields meet and cancel each other out in what astronomers call **reconnections**. Energy stored in the magnetic fields is released as short-wavelength photons and as high-energy protons and electrons. X-ray and ultraviolet photons reach Earth in 8 minutes and increase ionization in our atmosphere, which can interfere with radio communications. Particles from flares reach Earth hours or days later as gusts in the solar wind, which can distort Earth's magnetic field and disrupt navigation systems. Solar flares can also cause surges in electrical power lines and damage to Earth satellites.

Helioseismology image

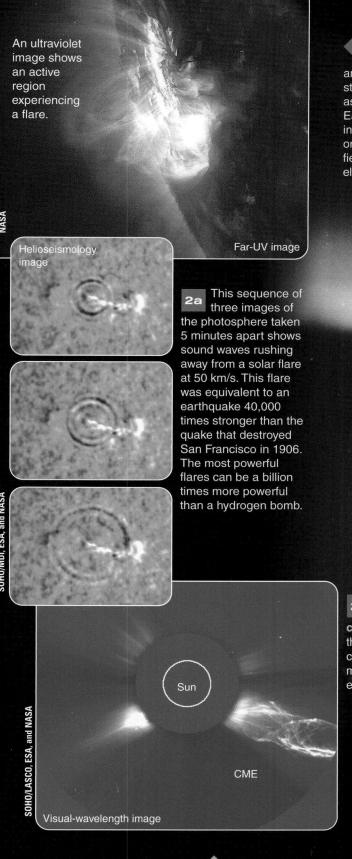

SOHO/MDI, ESA, and NASA

2a This sequence of three images of the photosphere taken 5 minutes apart shows sound waves rushing away from a solar flare at 50 km/s. This flare was equivalent to an earthquake 40,000 times stronger than the quake that destroyed San Francisco in 1906. The most powerful flares can be a billion times more powerful than a hydrogen bomb.

2b **Auroras** occur about 130 km above Earth's surface when energy in the solar wind guided by Earth's magnetic field excites gases in the upper atmosphere.

Copyright Jan Curtis

Sun

CME

SOHO/LASCO, ESA, and NASA

Visual-wavelength image

2c Magnetic reconnections can release enough energy to blow large amounts of ionized gas outward from the corona in **coronal mass ejections (CMEs).** These produce violent gusts in the solar wind, and if one blows past Earth it can create electrical currents up to a million megawatts, which flow down into Earth's magnetic poles and excite atoms in Earth's upper atmosphere to emit photons, which are seen as an aurora, as shown above.

X-ray image

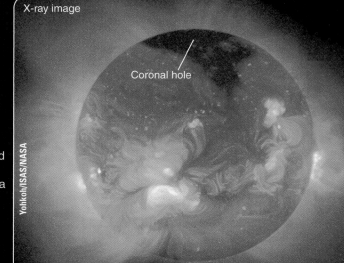

Coronal hole

Yohkoh/ISAS/NASA

3 Much of the solar wind comes from **coronal holes,** where the magnetic field does not loop back into the sun. These open magnetic fields allow ionized gas in the corona to flow away as the solar wind. The dark area in this X-ray image at right is a coronal hole.

The Solar Constant

Even a small change in the sun's energy output could produce dramatic changes in Earth's climate. The continued existence of the human species depends on the constancy of the sun, but we humans know very little about the variation of the sun's energy output.

The energy production of the sun can be measured by adding up all of the energy falling on 1 square meter of Earth's surface during 1 second. Of course, some correction for the absorption of Earth's atmosphere is necessary, and you must count all wavelengths from X rays to radio waves. The result, which is called the **solar constant,** amounts to about 1360 joules per square meter per second. A change in the solar constant of only 1 percent could change Earth's average temperature by 1 to 2°C (about 1.8 to 3.6°F). For comparison, during the last ice age Earth's average temperature was about 5°C cooler than it is now.

Some of the best measurements of the solar constant were made by instruments aboard the Solar Maximum Mission satellite. These have shown variations in the energy received from the sun of about 0.1 percent that lasted for days or weeks. Superimposed on that random variation is a long-term decrease of about 0.018 percent per year that has been confirmed by observations made by sounding rockets, balloons, and satellites. This long-term decrease may be related to a cycle of activity on the sun with a period longer than the 22-year magnetic cycle.

Small, random fluctuations will not affect Earth's climate, but a long-term decrease over a decade or more could cause worldwide cooling. History contains some evidence that the solar constant may have varied in the past. As you saw on page 175, the "Little Ice Age" was a period of unusually cool weather in Europe and America that lasted from about 1500 to 1850.* The average temperature worldwide was about 1°C cooler than it is now. This period of cool weather corresponded to the Maunder minimum, a period of reduced solar activity—few sun spots, no auroral displays, and no solar coronae visible during solar eclipses.

In contrast, an earlier period called the Grand Maximum, lasting from about 1100 to about 1250 AD, saw a warming of Earth's climate. The Vikings were able to explore and colonize Greenland, and native communities in parts of North America were forced to abandon their settlements because of long droughts.

Other minima and maxima have been found in climate data taken from studies of the growth rings of trees. In good years, trees add a thicker growth ring than in poor years, so measuring tree rings can reveal the climate in the past. Evidently, solar activity can increase or decrease the solar constant very slightly and affect Earth's climate in dramatic ways. The future of our civilization on Earth may depend on our learning to understand the solar constant.

Building Scientific Arguments

What kind of activity would the sun have if it didn't rotate differentially?
This is a really difficult question because only one star is visible close up. Nevertheless, you can construct a scientific argument by thinking about the Babcock model. If the sun didn't rotate differentially, with its equator traveling faster than do the higher latitudes, then the magnetic field might not get twisted up, and there might not be a solar cycle. Twisted tubes of magnetic field might not form and rise through the photosphere to produce prominences and flares, although convection might tangle the magnetic field and produce some activity. Is the magnetic activity that heats the chromosphere and corona driven by differential rotation or by convection? It is hard to guess, but without differential rotation, the sun might not have a strong magnetic field and high-temperature gas above its photosphere.

This is very speculative, but sometimes in the critical analysis of ideas it helps to imagine a change in a single important factor and try to understand what might happen. For example, redo the argument above. **What do you think the sun would be like if it had no convection inside?**

■ ■ ■

Connections: The sun is beautiful and complex, with great eruptions, prominences, active regions, and dark spots sweeping across its surface like magnetic weather. All of this activity is driven by the outward flow of energy generated by hydrogen fusion in the core. Presumably other stars are like the sun, but other stars are so far away their surfaces are invisible. Through the largest telescopes, they look like nothing more than points of light. How can astronomers learn about the stars? You will begin to answer that question in the next chapter.

---

*Ironically, the Maunder minimum coincides with the reign of Louis XIV of France, the "Sun King."

Summary

8-1 | The Solar Atmosphere

What do you see when you look at the sun?

■ The solar atmosphere consists of three layers of hot, low-density gas: the photosphere, chromosphere, and corona. The photosphere, or visible surface, is the level in the sun from which visible photons most easily escape.

■ The granulation of the photosphere is produced by convection currents of gas rising from below.

■ The chromosphere is most easily visible during total solar eclipses, when it flashes into view for a few seconds. It is a thin, hot layer of gas just above the photosphere, and its pink color is caused by the Balmer emission lines in its spectrum.

■ Filtergrams of the chromosphere reveal spicules and filaments.

■ The corona is the sun's outermost atmospheric layer. It is composed of a very-low-density, very hot gas extending many solar radii from the visible sun. Its high temperature—up to three million Kelvin—is believed to be maintained by the magnetic field extending up through the photosphere—the magnetic carpet.

■ Parts of the corona give rise to the solar wind, a breeze of low-density ionized gas streaming away from the sun.

■ Solar astronomers can study the motion, density, and temperature of gases inside the sun by analyzing the way the solar surface oscillates. Known as helioseismology, this process requires large amounts of data and extensive computer analysis.

8-2 | Nuclear Fusion in the Sun

How does the sun make its energy?

■ The sun generates its energy near its center, where the temperature and density are high enough for nuclear fusion reactions to combine hydrogen nuclei to make helium nuclei. This is known as the proton–proton chain.

■ Observations of too few neutrinos coming from the sun's core are now explained by the oscillation of neutrinos among three different types. The neutrinos confirm that the sun makes its energy by hydrogen fusion.

8-3 | Solar Activity

What are the dark sunspots and other forms of solar activity?

■ Sunspots seem dark because they are slightly cooler than the rest of the photosphere. The average sunspot is about twice the size of Earth and contains magnetic fields a few thousand times stronger than Earth's. Sunspots are thought to form because the magnetic field inhibits rising currents of hot gas and allows the surface to cool.

■ Astronomers can use the Zeeman effect to measure magnetic fields on the sun.

Why does the sun go through a cycle of activity?

■ The sun is very bright, and its light and infrared radiation can burn your eyes, so you must take great care in observing the sun.

■ The average number of sunspots varies over a period of about 11 years and appears to be related to a magnetic cycle. The sunspot cycle does not repeat exactly each cycle, and the decades from 1645 to 1715, known as the Maunder minimum, seem to have been a time when solar activity was very low and Earth's climate was slightly colder.

■ The sun rotates differentially, with regions far from the equator rotating slower than equatorial regions.

■ Alternate sunspot cycles have reversed magnetic polarity, which has been explained by the Babcock model, in which the differential rotation of the sun winds up the magnetic field. Tangles in the field rise to the surface and cause active regions visible to your eyes as sunspot pairs. When the field becomes strongly tangled, it reorders itself into a simpler but reversed field, and the cycle starts over.

■ Other stars are too far away for starspots to be visible, but spectroscopic observations reveal that many other stars have spots and magnetic fields that follow long-term cycles like the sun's.

■ Prominences occur in the chromosphere; their arched shapes show that they are formed of ionized gas trapped in the magnetic field. You see prominences in filtergrams as dark filaments silhouetted against the bright chromosphere.

■ Flares are sudden eruptions of X-ray, ultraviolet, and visible radiation plus high-energy atomic particles produced when magnetic fields on the sun interact and reconnect. Flares are important because they can have dramatic effects on Earth such as communications blackouts and auroras.

■ Spacecraft images show long streamers extending from the corona out into space. Coronal mass ejections occur when magnetic fields on the surface of the sun eject bursts of ionized gas that flow outward in the solar wind. Such bursts can produce auroras and other phenomena if they strike Earth.

■ Small changes in the solar constant over decades can affect Earth's climate. The Maunder minimum, a time of low solar activity, coincides with the little ice age. Other climate fluctuations have occurred and appear to be linked to small changes in solar activity.

New Terms

sunspot (p. 162)	deuterium (p. 170)
granulation (p. 163)	positron (p. 170)
convection (p. 164)	neutrino (p. 170)
supergranule (p. 164)	radiative zone (p. 170)
limb (p. 164)	convective zone (p. 171)
limb darkening (p. 164)	Maunder butterfly diagram (p. 174)
transition region (p. 164)	Zeeman effect (p. 175)
filtergram (p. 165)	Maunder minimum (p. 175)
filament (p. 165)	active region (p. 175)
spicule (p. 165)	differential rotation (p. 176)
coronagraph (p. 165)	dynamo effect (p. 177)
magnetic carpet (p. 166)	Babcock model (p. 177)
solar wind (p. 166)	prominence (p. 180)
helioseismology (p. 167)	flare (p. 181)
weak force (p. 169)	reconnection (p. 181)
strong force (p. 169)	aurora (p. 181)
nuclear fission (p. 169)	coronal hole (p. 181)
nuclear fusion (p. 169)	coronal mass ejection (CME) (p. 181)
Coulomb barrier (p. 169)	solar constant (p. 182)
proton–proton chain (p. 170)	

Review Questions

Assess your understanding of this chapter's topics with additional quizzing and animations at **http://ace .brookscole.com/sf9**

1. Why can't you see deeper than the photosphere?
2. What evidence can you give that granulation is caused by convection?
3. How are granules and supergranules related? How do they differ?
4. How can astronomers detect structure in the chromosphere?
5. What evidence can you give that the corona has a very high temperature?
6. What heats the chromosphere and corona to high temperature?
7. How are astronomers able to explore the layers of the sun below the photosphere?
8. Why does nuclear fusion require high temperatures?
9. Why does nuclear fusion in the sun occur only near the center?
10. How can astronomers detect neutrinos from the sun?
11. How can neutrino oscillation explain the solar neutrino problem?
12. What evidence can you give that sunspots are magnetic?
13. How does the Babcock model explain the sunspot cycle?
14. What does the spectrum of a prominence reveal? What does its shape reveal?
15. How can solar flares affect Earth?
16. The upper two images here show two solar phenomena. What are they, and how are they related? How do they differ?
17. The lower image was recorded in the extreme ultraviolet by the SOHO spacecraft. Explain the features you see.

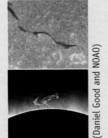

(Daniel Good and NOAO)

(NASA/SOHO)

Discussion Questions

1. What energy sources on Earth cannot be thought of as stored sunlight?
2. What would the spectrum of an auroral display look like? Why?
3. What observations would you make if you were ordered to set up a system that could warn astronauts in orbit of dangerous solar flares? Such a warning system exists.

Problems

1. The radius of the sun is 0.7 million km. What percentage of the radius is taken up by the chromosphere?
2. The smallest detail visible with ground-based solar telescopes is about 1 second of arc. How large a region does this represent on the sun? (*Hint:* Use the small-angle formula.)
3. What is the angular diameter of a star like the sun located 5 ly from Earth? Is the Hubble Space Telescope able to detect detail on the surface of such a star?
4. How much energy is produced when the sun converts 1 kg of mass into energy?
5. How much energy is produced when the sun converts 1 kg of hydrogen into helium? (*Hint:* How does this problem differ from Problem 4?)
6. A 1-megaton nuclear weapon produces about 4×10^{15} J of energy. How much mass must vanish when a 5-megaton weapon explodes?
7. Use the luminosity of the sun, the total amount of energy it emits each second, to calculate how much mass it converts to energy each second.
8. If a sunspot has a temperature of 4240 K and the solar surface has a temperature of 5800 K, how many times brighter is the surface compared to the sunspot? (*Hint:* Use the Stefan–Boltzmann law, Chapter 7.)
9. A solar flare can release 10^{25} J. How many megatons of TNT would be equivalent? (*Hint:* A 1-megaton bomb produces about 4×10^{15} J.)
10. The United States consumes about 2.5×10^{19} J of energy in all forms in a year. How many years could you run the United States on the energy released by the solar flare in Problem 9?
11. Neglecting energy absorbed or reflected by Earth's atmosphere, the solar energy hitting 1 square meter of Earth's surface is 1360 J/s (the solar constant). How long does it take a baseball diamond (90 ft on a side) to receive 1 megaton of solar energy?

Media Cluster

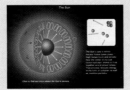

The Sun
Take a look inside the sun in this animation. You can click on various parts of the sun to see things like an animation of nuclear fusion and a movie of a solar flare.

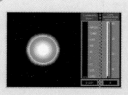

Nuclear Fusion

Nuclear fusion is a reaction that joins the nuclei of atoms to form more massive nuclei. Vary the temperature of the star in this animation to see how this affects the rate of nuclear fusion and luminosity.

Zeeman Effect

The phenomenon known as the Zeeman effect allows astronomers to measure magnetic fields. In this animation, you can adjust the strength of the magnetic field around a star to study the effects on the star's spectrum.

Sunspot Cycle I

Sunspots are relatively dark spots on the sun that contain intense magnetic fields. This animation lets you study the number and location of sunspots occurring over an 11-year cycle.

Sunspot Cycle II

In this animation, you can track sunspot activity over the past 300 years.

Convection and Magnetic Fields

You can vary the temperature of the sun's core in this animation. As you do so, notice how the amount of convection changes, as well as what happens to the magnetic field around the sun.

Auroras

This animation lets you vary the strength of Earth's magnetic field as well as the speed and number of particles in the solar wind. See if you can determine what factors cause the most- and least-intensive auroras.

Lab 4: Solar Wind and Cosmic Rays

This lab begins with an overview of the properties of the sun's atmosphere and how energetic particles escape and travel through the solar system. It ends with a discussion of cosmic rays.

Lab 10: Helioseismology

This lab examines how helioseismology, the study of the sun's vibrations, allows you to obtain detailed information about its interior. Such information allows you to test your understanding of the sun very precisely.

Critical Inquiries for the Web

1. Do disturbances in one layer of the solar atmosphere produce effects in other layers? You have seen that filtergrams are useful in identifying the layers of the solar atmosphere and the structures within them. Visit a website that provides daily solar images. Then, choose today's date (or one near it) and examine the sun in several wavelengths to explore the relation between disturbances in various layers.

2. Explore the web to find out how auroral activity is affected as solar activity rises and falls through the solar cycle. What changes in auroral visibility occur during this cycle? In what other ways can the increased activity associated with a solar maximum affect Earth?

3. Explore the Web to find photos and observations of auroras. From what places on Earth are auroras most often seen?

4. What can you find on the Web about Earth-based efforts to generate energy through nuclear fusion? How do nuclear fusion power experiments attempt to trigger and control nuclear fusion? So-called cold fusion has been abandoned as a false trail. How did it resemble nuclear fusion?

Exploring *TheSky*

1. Locate the six photos of the sun provided in *TheSky* and attempt to draw in the sun's equator in each photo. (*Hint:* In the sun's information box, choose **More Information** and then **Multimedia.** What features are visible in these images that help us recognize the orientation of the sun's equator?)

Go to the Brooks/Cole Astronomy Resource Center **(http://astronomy.brookscole.com)** for critical thinking exercises, articles, and additional readings from InfoTrac College Edition, Brooks/Cole's online student library.

Perspective: Origins

Visual-wavelength image

EVERYTHING HAS TO COME from somewhere. Look at your thumb. The atoms in your thumb are billions of years old. Only a hundred million years ago, they were inside dinosaurs. Five billion years ago, those atoms were part of a cloud of gas floating in space, and not long before that those atoms were inside stars. The atoms inside your body are old, but the matter the atoms are made of is even older. That matter had its beginning within minutes of the beginning of the universe 13.7 billion years ago. If your atoms could tell their stories, you would be stunned to know where they came from and where they have been. ▍ As you study the solar system, remember the history of the matter it is made of. The iron inside planets, the oxygen and nitrogen in Earth's atmosphere, the calcium in your bones, exist because stars have lived and

▍ Continued on page 188 ▍

Like lanterns, stars push back the darkness, but they also produce the atoms of which planets and people are made. These stars illuminate the Cone Nebula at left, and the Fox Fur nebula at right. (T. A. Rector and B. A. Wolpa, NRAO/NOAO/AUI/AURA/NSF)

Guidepost

Looking Back

The preceding chapters have introduced you to the sky, to the history of astronomy, to the tools of the astronomer, and to the second most important astronomical body in the universe, the sun. Now you are ready to study number one, Earth. To understand what Earth is, you need to study its fellow planets in our solar system; but, before you begin, you need to look around and get a perspective on where and what our solar system is.

This Chapter

This chapter will give you a quick tour of the universe of stars and galaxies, and that will help you understand our solar system. Not only will you see how our solar system is located in the universe, but you will see how our solar system fits into the origin and evolution of the universe. Your tour of the cosmos will answer four essential questions:

How are stars born, and how do they die?

What are galaxies?

How did the universe begin and form galaxies?

How were the atoms in our bodies formed?

This chapter will give you a perspective on the origins of the matter you and the solar system are made of.

Looking Ahead

The rest of this book is about the solar system—the sun and its planets. You will learn about a world where it is hot enough to melt lead, a world where it rains methane, and a world where sulfurous volcanoes erupt continuously. Keep in mind as you visit these worlds that they are made of the same stuff as Earth, the same stuff that is in your body, and that those atoms had their birthdays long ago and far away.

died. You are no more isolated from the rest of the universe than a raindrop is isolated from the sea.

P-1 The Birth of Stars

THE STARS ABOVE SEEM PERMANENT fixtures of the sky, but astronomers know that stars are born and stars die. Their lives are long compared to a human life; but, if you know where to look, you can see all of the stages of stellar birth, aging, and death represented in the sky. Astronomers have put those stages into the proper order and can tell the life story of the stars, a story that begins in the darkness of interstellar space.

Although space seems empty, it is actually filled with thinly spread gas and dust, the **interstellar medium.** The gas atoms are mostly hydrogen, roughly a centimeter apart, and the dust is made of microscopic grains of heavier atoms such as carbon and iron. The dust makes up only a few percent of the matter between the stars.

The interstellar medium is tenuous in the extreme, yet you can see clear evidence that it exists. In some places, the interstellar medium is collected into great clouds of dusty gas that obscure the stars beyond. In other cases, a nearby hot star can ionize the gas and create a glowing **nebula** (■ Figure P-1).

These nebulae not only adorn the sky, but they mark the birthplace of stars. As the cold gases of the nebula grow denser, gravity can pull parts of the nebula together to form warmer, denser bodies that eventually become **protostars**—objects destined to become stars.

Astronomers can't see these protostars easily because they are hidden deep inside the dusty gas clouds from which they form, but they are easily visible in the infrared. As the more massive protostars become hot, luminous stars, the light and gas flowing away from the newborn stars blow the nebula away to reveal the new stars (■ Figure P-2). In this way, a single gas cloud

■ **Figure P-1**

Young stars are found in clouds of gas and dust from which they have been born. The nebula N44 is 170,000 ly from Earth in a nearby galaxy, and the Horsehead Nebula is only about 1500 ly distant in our own galaxy. Gas in both nebulae is excited to glow by hot, young stars, and dust is visible as dark, twisted clouds seen against the bright background gas. (N44: ESO; Horsehead Nebula: NOAO and Nigel Sharp)

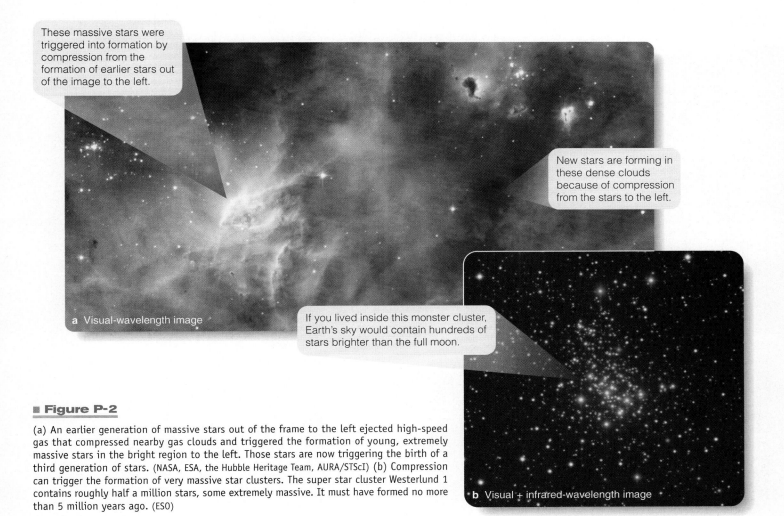

■ Figure P-2

(a) An earlier generation of massive stars out of the frame to the left ejected high-speed gas that compressed nearby gas clouds and triggered the formation of young, extremely massive stars in the bright region to the left. Those stars are now triggering the birth of a third generation of stars. (NASA, ESA, the Hubble Heritage Team, AURA/STScI) (b) Compression can trigger the formation of very massive star clusters. The super star cluster Westerlund 1 contains roughly half a million stars, some extremely massive. It must have formed no more than 5 million years ago. (ESO)

can give birth to a cluster of stars. Our sun was probably born in such a nebula about 5 billion years ago.

If you look at the constellation Orion, you can see one of these regions with your unaided eye. The patch of haze in Orion's sword is the Great Nebula in Orion. Study **Star Formation in the Orion Nebula** on pages 190–191 and notice four points:

1 The nebula you see is only a small part of a vast, dusty cloud. You see the nebula because the stars born within it have ionized the gas and driven it outward, breaking out of the cloud.

2 Also notice that a single very hot star (classified O6) is almost entirely responsible for ionizing the gas.

3 Notice how infrared observations reveal clear evidence of active star formation deeper in the cloud just to the northwest of the Trapezium.

4 Finally, notice that many stars visible in the Orion Nebula are surrounded by disks of gas and dust. Planets form in such disks, but the disks do not last long and are clear evidence that the stars are very young.

As protostars contract, something happens that is quite important to planet walkers like you—planets form. You will see in the next chapter how our solar system formed from a disk of gas and dust that orbited the protostar that became our sun.

You could identify the birth of a star as the moment when it begins fusing hydrogen into helium. That nuclear fusion reaction releases energy, supports the star, and stops its contraction.

P-2 The Deaths of Stars

STARS SPEND MOST OF THEIR LIVES fusing hydrogen into helium, and, when the hydrogen is exhausted, they fuse helium into carbon. The more massive stars can fuse carbon into even heavier atoms, but iron is the limit beyond which no star can go.

tar Formation
n the Orion Nebula

1 The visible Orion Nebula shown below is a pocket of ionized gas on the near side of a vast, sty molecular cloud that fills much of the southern rt of the constellation Orion. The molecular cloud n be mapped by radio telescopes. To scale, the ud would be many times larger than this page. As e stars of the Trapezium were born in the cloud, their liation has ionized the gas and pushed it away. here the expanding nebula pushes into the larger lecular cloud, it is compressing the gas (see gram at right) and may be triggering the formation the protostars that can be detected at infrared velengths within the molecular cloud.

ndreds of stars lie within the nebula, only the four brightest, those in the pezium, are easy to see with a all telescope. A fifth star, at the row end of the Trapezium, y be visible on nights good seeing.

e cluster of stars in nebula is less than nillion years old. s must mean e nebula is nilarly young.

Side view of Orion Nebula

Hot Trapezium stars   Protostars

To Earth

Expanding
ionized
hydrogen

Molecular
cloud

Trapezium

Visual-wavelength image
Daniel Good

Infrared

The near-infrared image above reveals about 50 low-mass, very cool stars that must have formed recently.

X-ray

Roughly 1000 young stars with hot chromospheres appear in this X-ray image of the Orion Nebula.

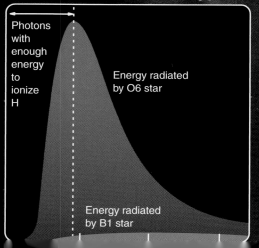

Photons
with
enough
energy
to
ionize
H

Energy radiated
by O6 star

Energy radiated
by B1 star

2 Of all the stars in the Orion Nebula, only one is hot enough to ionize the gas. Only photons with wavelengths shorter than 91.2 nm can ionize hydrogen. The second-hottest stars in the nebula are B1 stars, and they emit little of this ionizing radiation. The hottest star, however, is an O6 star 30 times the mass of the sun. At a temperature of 40,000 K, it emits plenty of photons with wavelengths short enough to ionize hydrogen. Remove that one star, and the nebula would turn off its emission

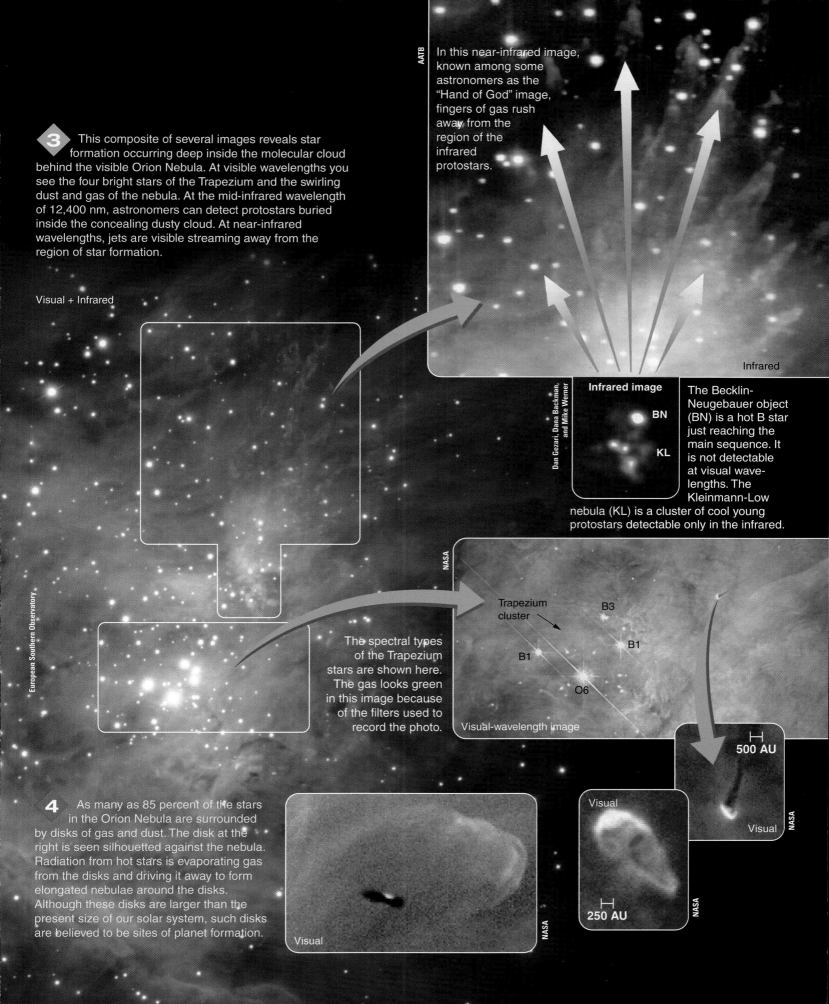

3 This composite of several images reveals star formation occurring deep inside the molecular cloud behind the visible Orion Nebula. At visible wavelengths you see the four bright stars of the Trapezium and the swirling dust and gas of the nebula. At the mid-infrared wavelength of 12,400 nm, astronomers can detect protostars buried inside the concealing dusty cloud. At near-infrared wavelengths, jets are visible streaming away from the region of star formation.

Visual + Infrared

In this near-infrared image, known among some astronomers as the "Hand of God" image, fingers of gas rush away from the region of the infrared protostars.

AATB

Infrared

Infrared image

Dan Gezari, Dana Backman, and Mike Werner

BN

KL

The Becklin-Neugebauer object (BN) is a hot B star just reaching the main sequence. It is not detectable at visual wave-lengths. The Kleinmann-Low nebula (KL) is a cluster of cool young protostars detectable only in the infrared.

European Southern Observatory

NASA

The spectral types of the Trapezium stars are shown here. The gas looks green in this image because of the filters used to record the photo.

Trapezium cluster

B3

B1

B1

O6

Visual-wavelength image

500 AU

Visual

Visual

NASA

4 As many as 85 percent of the stars in the Orion Nebula are surrounded by disks of gas and dust. The disk at the right is seen silhouetted against the nebula. Radiation from hot stars is evaporating gas from the disks and driving it away to form elongated nebulae around the disks. Although these disks are larger than the present size of our solar system, such disks are believed to be sites of planet formation.

Visual

250 AU

NASA

Understanding Science: Separating Facts from Theories

The fundamental work of science is testing theories by comparing them with facts. As you think about science, be careful to distinguish clearly between facts and theories. The facts are the evidence against which scientists test theories.

Scientific facts are those observations or experimental results of which scientists are confident. An astronomer makes observations of stars, and a botanist collects samples of related insects. A fact could be a precise measurement, such as the mass of a star expressed as a specific number, or it could be a simple observation, such as that a certain butterfly no longer visits a certain mountain valley. In each case, the scientist is gathering facts.

A theory, however, is a conjecture as to how nature works. If you are uncertain of a theory, you might call it a hypothesis. In any case, these conjectures are not facts; they are attempts to explain how nature works. In a sense, a theory or hypothesis is a story that scientists have made up to explain how nature works in some specific case. These stories can be wonderfully detailed and ingenious, but

without evidence they are nothing more than hunches.

When one of these stories is tested against the facts and confirmed, scientists have more confidence that the story is more or less right. The more a story is tested successfully, the more confidence scientists have in it. The facts represent reality, and every theory or hypothesis must be repeatedly tested against reality.

You can't test one theory against another theory. Theories are not evidence; they are conjectures. If you were allowed to test one theory against another theory, you might fall into the trap of circular reasoning. "Elves make the flowers bloom. I know that elves exist because the flowers bloom." That is using two theories to confirm each other, and it leads to nonsense. Only facts can be evidence.

Nor can you use a theory to deduce facts. You can use a theory to make predictions, which you can then test against facts; but the predictions themselves can never be certainties, so they can't be facts. The only way to arrive at facts is to consult nature and make direct measurements or observations.

In science, evidence is made up of facts, which could range from precise numerical measurements to the observation of the shape of a flower. (M. Seeds)

As you study different problems in astronomy, carefully distinguish between facts and theories. Facts are the basic building blocks of all science, and they can come only from the careful study of nature. As Galileo would say, you must read the book of nature.

How long a star can live depends on its mass. The smallest stars are very common, but they are not very luminous. These cool, **red dwarf** stars are hardly massive enough to fuse hydrogen. The sun is a common kind of star in that it fuses hydrogen into helium, and astronomers refer to it as a **main sequence** star. It is a medium-mass star that will live a total of about 10 billion years. In contrast, the most massive stars, up to almost 100 times the mass of the sun, are tremendously luminous and fuse their fuels so rapidly that they can live only a few million years.

This chapter tells the story of the birth and death of stars in a few paragraphs, but modern astronomers know a great deal about the lives of the stars. Solving the mystery of the evolution of stars is one of the greatest accomplishments of modern astronomy. Astronomers have compared observational facts with theories and gradually unraveled the story of the stars (**Window on Science P-1**). Many details remain to be understood, but the quick summary here gives you background for your study of our solar system. You will discover that the birth of our solar system is linked in more than one way to the deaths of stars.

How a star dies depends on its mass. A medium-mass star like the sun has enough hydrogen fuel to survive for billions of years,

but it must eventually exhaust its fuel, swell to become a **giant** star, and then expel its envelope in an expanding nebula (■ Figure P-3). These **planetary nebulae** were named for their planet-like appearance in small telescopes, but they are, in fact, the remains of dying stars. Once the outer layers of a star are ejected into space, the hot core contracts to form a small, dense cooling star—a **white dwarf.**

Study **The Formation of Planetary Nebulae** on pages 194–195 and notice four things:

1 First, you can understand what planetary nebulae are like by using simple observational methods such as Kirchhoff's laws and the Doppler effect.

2 Second, notice the model astronomers have developed to explain planetary nebulae. The real nebulae are more complex than the simple model of a slow wind and a fast wind, but the model provides a way to organize the observed phenomena.

3 The third thing to notice is how oppositely directed jets and multiple shells produce many of the asymmetries seen in planetary nebulae.

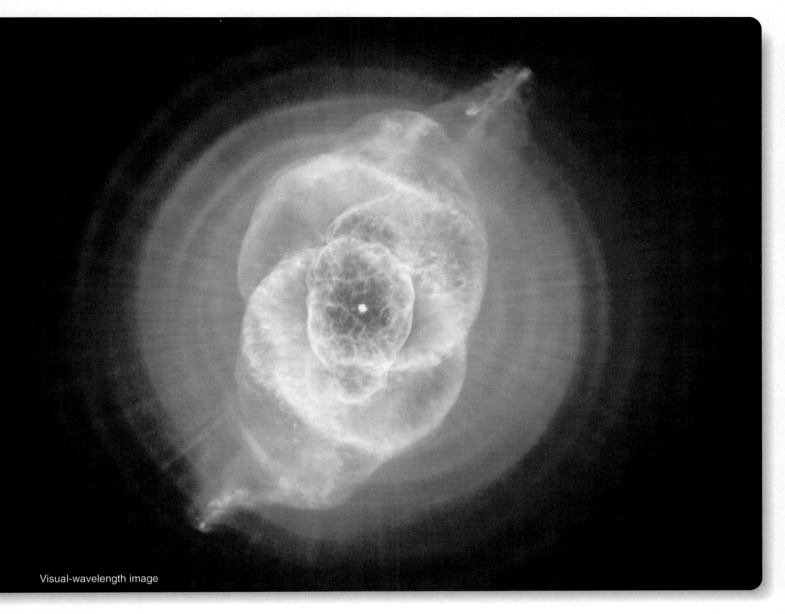

Visual-wavelength image

■ Figure P-3

Planetary nebulae are shells of gas expelled from dying stars. The Cat's Eye Nebula is complex with multiple shells and a pair of jets pointing in opposite directions. Such nebulae are produced by the deaths of sunlike stars. (NASA, ESA, HEIC, and the Hubble Heritage Team)

4 Finally, notice the fate of the star itself; it must contract into a white dwarf.

When medium-mass stars like the sun die, they expel their outer layers back into space, and any atoms that may have been cooked up inside the stars get mixed back into the interstellar medium.

In contrast with stars like the sun, massive stars become very large giants or even larger **supergiants.** They live very short lives,

perhaps only millions of years, before they develop dead-end iron cores and explode as supernovae (■ Figure P-4). The core of such a dying massive star may form a neutron star or a black hole, but the outer parts of the star, newly enriched with the atoms made inside the star, are returned to the interstellar medium.

The violence of the supernova explosion can fuse atoms together to build elements heavier than iron. Gold, silver, platinum, uranium, and other elements heavier than iron are rare and valuable because they are made only in the moments of a

The Formation of Planetary Nebulae

1 Simple observations tell astronomers what planetary nebulae are like. Their angular size and their distance indicate that their radii range from 0.2 to 3 light-years. The presence of emission lines in their spectra assures astronomers they are excited, low-density gas. Doppler shifts show they are expanding at 10 to 20 km/s. If you divide radius by velocity, you find planetary nebulae are no more than 10,000 years old. Older planetary nebulae evidently become mixed into the interstellar medium.

1a Astronomers find about 1500 planetary nebulae in the sky. Because planetary nebulae are short-lived formations, you can conclude that they must be a common part of stellar evolution. Medium-mass stars up to a mass of about 8 solar masses are destined to die by forming planetary nebulae.

Visual

NGC6369   Visual

This nearly spherical planetary nebula has a low-luminosity outer envelope and a highly excited inner region.

The Ring Nebula in the constellation Lyra is visible even in small telescopes. Note the hot blue star its center and the radial texture in the gas, suggesting outward motion.

2 The process that produces planetary nebulae involves two stellar winds. First, as an aging giant, the star gradually blows away its outer layers in a slow breeze of low-excitation gas that is not easily visible. Once the hot interior of the star is exposed, it ejects a high-speed wind that overtakes and compresses the gas of the slow wind like a snowplow, while ultraviolet radiation from the hot remains of the central star excites the gases to glow like a giant neon sign.

Slow stellar wind from a red giant

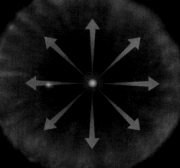

The gases of the slow wind are not easily detectable.

Fast wind from exposed interior

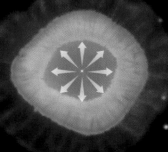

You see a planetary nebula where the fast wind

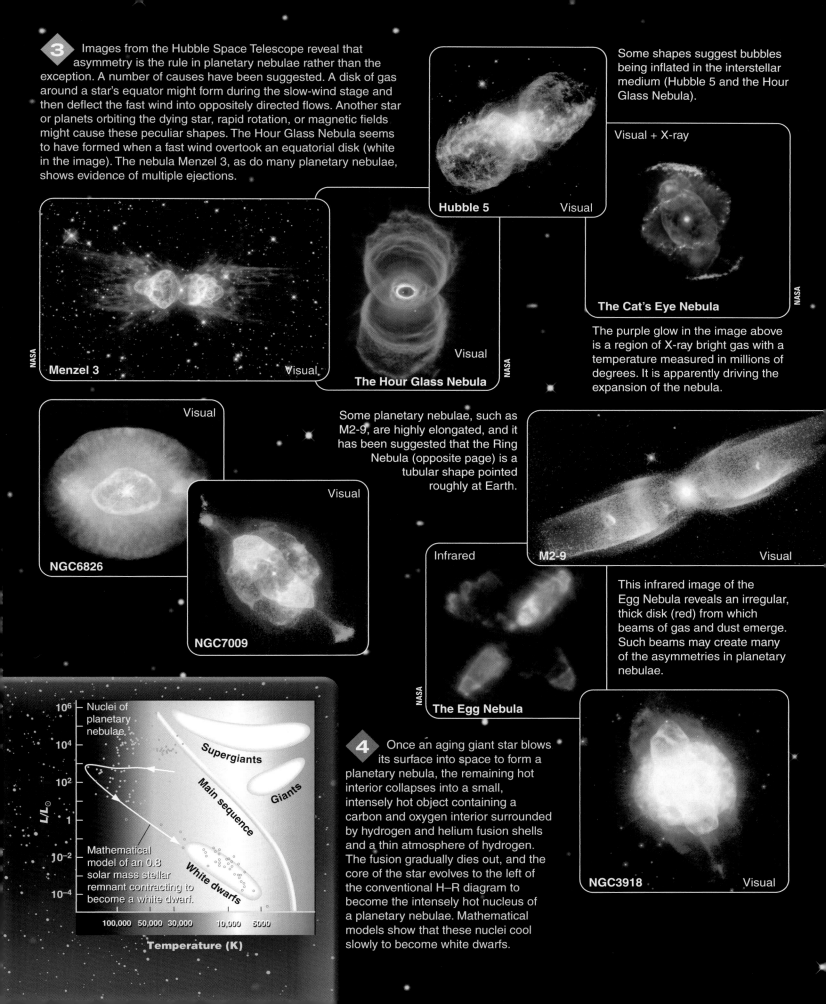

3 Images from the Hubble Space Telescope reveal that asymmetry is the rule in planetary nebulae rather than the exception. A number of causes have been suggested. A disk of gas around a star's equator might form during the slow-wind stage and then deflect the fast wind into oppositely directed flows. Another star or planets orbiting the dying star, rapid rotation, or magnetic fields might cause these peculiar shapes. The Hour Glass Nebula seems to have formed when a fast wind overtook an equatorial disk (white in the image). The nebula Menzel 3, as do many planetary nebulae, shows evidence of multiple ejections.

Some shapes suggest bubbles being inflated in the interstellar medium (Hubble 5 and the Hour Glass Nebula).

Hubble 5   Visual

Visual + X-ray

The Cat's Eye Nebula

NASA

The purple glow in the image above is a region of X-ray bright gas with a temperature measured in millions of degrees. It is apparently driving the expansion of the nebula.

Menzel 3   Visual

NASA

Visual

The Hour Glass Nebula

NASA

Visual

NGC6826

Some planetary nebulae, such as M2-9, are highly elongated, and it has been suggested that the Ring Nebula (opposite page) is a tubular shape pointed roughly at Earth.

Visual

NGC7009

Infrared

M2-9   Visual

This infrared image of the Egg Nebula reveals an irregular, thick disk (red) from which beams of gas and dust emerge. Such beams may create many of the asymmetries in planetary nebulae.

NASA

The Egg Nebula

10^6 — Nuclei of planetary nebulae.

Supergiants

10^4

10^2

Main sequence

Giants

1

Mathematical model of an 0.8 solar mass stellar remnant contracting to become a white dwarf.

10^{-2}

White dwarfs

10^{-4}

$L/L_\odot$

100,000 50,000 30,000   10,000   5000

Temperature (K)

4 Once an aging giant star blows its surface into space to form a planetary nebula, the remaining hot interior collapses into a small, intensely hot object containing a carbon and oxygen interior surrounded by hydrogen and helium fusion shells and a thin atmosphere of hydrogen. The fusion gradually dies out, and the core of the star evolves to the left of the conventional H–R diagram to become the intensely hot nucleus of a planetary nebulae. Mathematical models show that these nuclei cool slowly to become white dwarfs.

NGC3918   Visual

Visual-wavelength image

■ Figure P-4

The Crab Nebula is a supernova remnant produced by the explosion of a star, an event seen by Chinese astronomers in AD 1054. Such expanding shells of gas mix elements manufactured inside stars back into the interstellar medium. (NASA, ESA, J. Hester and A. Loll, Arizona State University)

supernova explosion. The iodine in your thyroid gland was made in supernovae.

The atoms of which you are made had their birth inside stars. That process is common in the universe because stars are common. Our galaxy contains billions.

P-3 Our Home Galaxy

FROM A DARK LOCATION away from city lights, you can see the **Milky Way** stretching across the sky. The winter Milky Way is especially dramatic (■ Figure P-5a). It is actually the disk of the galaxy that we live in—the **Milky Way Galaxy.** Of course, no one has ever journeyed out into space to look back and take a picture of our galaxy, but the nearby Andromeda Galaxy shown in Figure P-5b must look much like our own. Our galaxy contains roughly 100 billion stars, and we live about two-thirds of the way from the center to the edge.

Astronomers estimate that our galaxy is about 75,000 light-years in diameter. That is, light takes 75,000 years to travel from one edge to the other. If you had a photograph of the Milky Way Galaxy as big as North America, the entire solar system would be about the size of a small cookie, and the sun and planets would all be too small to see without a powerful microscope.

■ Active Figure P-5

(a) Nearby stars look bright, and peoples around the world group them into constellations. Nevertheless, the vast majority of the stars in our galaxy merge into a faintly luminous path that circles the sky, the Milky Way. This artwork shows the location of a portion of the Milky Way near a few bright winter constellations. (See the star charts at the end of this book to further locate the Milky Way in your sky.) (b) This photograph of the Andromeda Galaxy, a spiral galaxy about 2.3 million ly from Earth, shows what our own galaxy would look like if you could view it from a distance. (AURA/NOAO/NSF)

Ace ◐ Astronomy™   Log into AceAstronomy and select this chapter to see the Active Figure "M31 at Many Wavelengths." See what the Andromeda Galaxy looks like in X rays.

Visual-wavelength image

■ **Figure P-6**

The beautiful symmetry of the spiral pattern is clear in this image of the galaxy NGC1300. Young star clusters containing hot, bright stars and ionized hydrogen are located along the arms, while dark lanes of dust mark the inner edges of the arms. (NASA and the Hubble Heritage Team; STScI/AURA)

Large astronomical telescopes reveal other galaxies scattered across the sky. You have already met the Andromeda Galaxy in Figure P-5. It is so close you can see it with the unaided eye as a hazy patch in the constellation Andromeda. Other galaxies are all around us, and some are dramatically beautiful (■ Figure P-6).

 **The Universe of Galaxies**

OUR GALAXY IS OUR HOME, but ours is only one of billions of galaxies. Some, like our Milky Way Galaxy, are disk shaped with graceful spiral arms marked by clouds of gas and bright newborn stars. But many galaxies are great swarms of stars with little remaining gas and dust.

Study **Galaxy Classification** on pages 200–201 and notice three important points:

1 Many galaxies have no disk, no spiral arms, and almost no gas and dust. These elliptical galaxies range from huge giants to small dwarfs and contain large clusters of stars called **globular clusters.**

2 Notice that disk-shaped galaxies usually have spiral arms and contain gas and dust. Many spiral galaxies have a barred structure, but a few disk galaxies contain little gas and dust.

3 Finally, notice that some galaxies are highly irregular in shape and tend to be rich in gas and dust.

You might expect such titanic objects as galaxies to be rare, but large telescopes reveal that the sky is filled with galaxies. Like leaves on the forest floor, galaxies carpet the sky. They fill the universe in every direction as far as telescopes can see (■ Figure P-7). Grouped in clusters and superclusters, galaxies are the homes of the billions of stars that illuminate the universe and create the chemical elements.

When two galaxies collide, the stars swirl past each other without bumping, but the great clouds of gas and the magnetic fields in the galaxies do collide and compress each other. Furthermore, tidal forces twist and distort the colliding galaxies. Images of such colliding galaxies show their twisted shapes and far-flung streamers of stars and gas, and more detailed images reveal

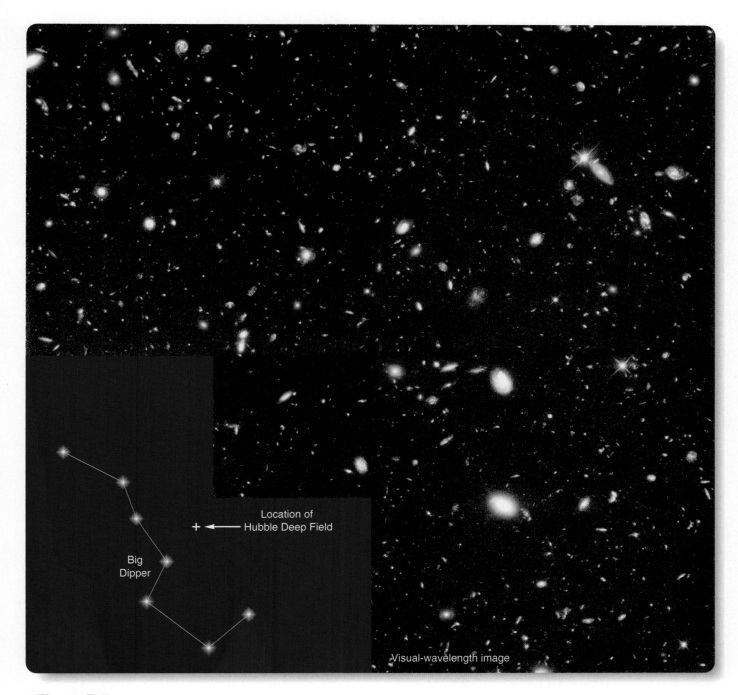

Location of
+ ◄── Hubble Deep Field

Big
Dipper

Visual-wavelength image

■ Figure P-7

An apparently empty spot on the sky only $\frac{1}{30}$ the diameter of the full moon contains over 1500 galaxies in this extremely long time exposure known as the Northern Hubble Deep Field. Presumably the entire sky is similarly filled with galaxies. (R. Williams and the Hubble Deep Field Team, STScI, NASA)

an even more impressive consequence of galaxy collisions—star formation (■ Figure P-8). The compression of the gas clouds can stimulate two colliding galaxies to form vast numbers of stars and massive star clusters. Such bursts of star formation salt the galaxies with newly formed heavy atoms. Our own Milky Way Galaxy has probably collided with more than one smaller galaxy, and

the atoms of which you are made may have been formed because of those collisions.

Ace◐Astronomy™ Log into AceAstronomy and select this chapter to see Astronomy Exercise "Milky Way Galaxy." Look at our home galaxy from outside.

Galaxy Classification

1 **Elliptical galaxies** are round or elliptical, contain no visible gas and dust, and lack hot, bright stars. They are classified with a numerical index ranging from 1 to 7; E0s are round, and E7s are highly elliptical. The index is calculated from the largest and smallest diameter of the galaxy used in the following formula and rounded to the nearest integer.

$$\frac{10(a - b)}{a}$$

Outline of an
E6 galaxy

Visual-wavelength image

The Leo 1 dwarf elliptical galaxy is not many times bigger than a globular cluster.

Anglo-Australian Telescope Board

.Visual

M87 is a giant elliptical galaxy classified E1. It is a number of times larger in diameter than our own galaxy and is surrounded by a swarm of over 500 globular clusters.

2 **Spiral galaxies** contain a disk and spiral arms. Their halo stars are not visible, but presumably all spiral galaxies have halos. Spirals contain gas and dust and hot, bright O and B stars, as shown at right. The presence of short-lived O and B stars alerts us that star formation is occurring in these galaxies. Sa galaxies have larger nuclei, less gas and dust, and fewer hot, bright stars. Sc galaxies have small nuclei, lots of gas and dust, and many hot, bright stars. Sb galaxies are intermediate.

Anglo-Australian Telescope Board

Sa

Visual   NGC 3623

BAR

Sb

NGC 3627   Visual

Ace ◉ Astronomy™

Log into AceAstronomy and select this chapter to see Active Figure "Galaxy Types" and review the classification of galaxies.

Anglo-Australian Telescope Board

2a Roughly 2/3 of all spiral galaxies are **barred spiral galaxies** classified SBa, SBb, and SBc. They have an elongated nucleus with spiral arms springing from the ends of the bar, as shown at left. Our own galaxy is a barred spiral.

NGC 1365   Visual

Sc

NGC 2997   Visual

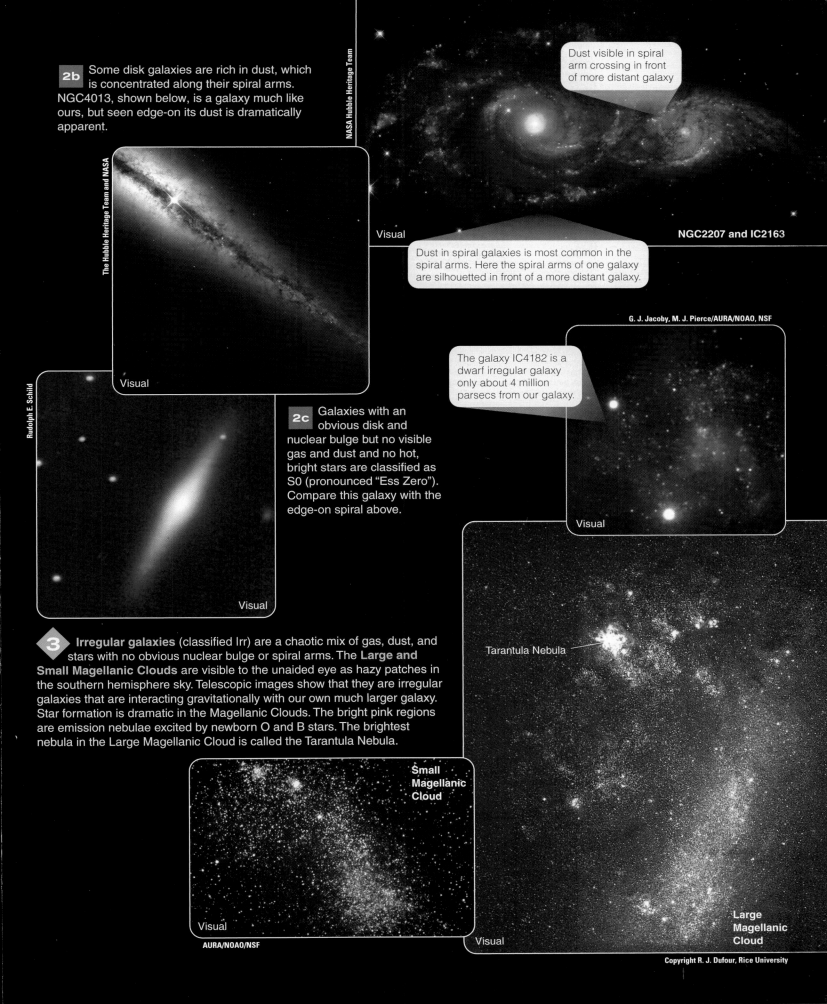

2b Some disk galaxies are rich in dust, which is concentrated along their spiral arms. NGC4013, shown below, is a galaxy much like ours, but seen edge-on its dust is dramatically apparent.

NASA Hubble Heritage Team

The Hubble Heritage Team and NASA

Visual

Dust visible in spiral arm crossing in front of more distant galaxy

Visual

NGC2207 and IC2163

Dust in spiral galaxies is most common in the spiral arms. Here the spiral arms of one galaxy are silhouetted in front of a more distant galaxy.

Rudolph E. Schild

2c Galaxies with an obvious disk and nuclear bulge but no visible gas and dust and no hot, bright stars are classified as S0 (pronounced "Ess Zero"). Compare this galaxy with the edge-on spiral above.

G. J. Jacoby, M. J. Pierce/AURA/NOAO, NSF

The galaxy IC4182 is a dwarf irregular galaxy only about 4 million parsecs from our galaxy.

Visual

Visual

3 Irregular galaxies (classified Irr) are a chaotic mix of gas, dust, and stars with no obvious nuclear bulge or spiral arms. The **Large and Small Magellanic Clouds** are visible to the unaided eye as hazy patches in the southern hemisphere sky. Telescopic images show that they are irregular galaxies that are interacting gravitationally with our own much larger galaxy. Star formation is dramatic in the Magellanic Clouds. The bright pink regions are emission nebulae excited by newborn O and B stars. The brightest nebula in the Large Magellanic Cloud is called the Tarantula Nebula.

Tarantula Nebula

Small Magellanic Cloud

Visual

AURA/NOAO/NSF

Large Magellanic Cloud

Visual

Copyright R. J. Dufour, Rice University

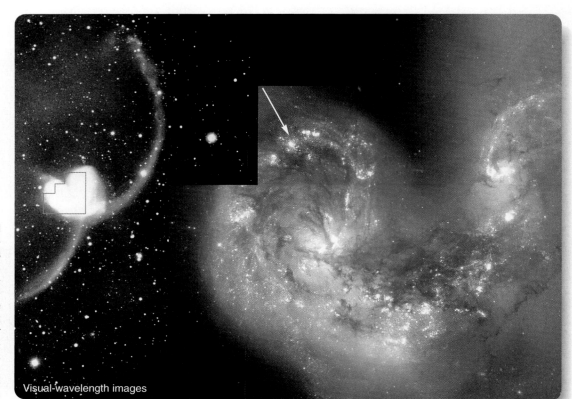

The colliding galaxies NGC4038 and NGC4039 are known as the Antennae because their long curving tails resemble the antennae of an insect. Earth-based photos (left) show little detail, but a Hubble Space Telescope image (right) reveals the collision of two galaxies producing thick clouds of dust and raging star formation, creating roughly a thousand massive star clusters such as the one at the top (arrow). Such collisions between galaxies are common. (Brad Whitmore, STScI, and NASA)

Visual-wavelength images

P-5 The Origin of the Universe

A LITTLE LESS THAN A CENTURY AGO, astronomers made an astonishing discovery. All of the galaxies in the universe are rushing away from each other. The spectrum of a galaxy is the combined spectrum of billions of stars, and the spectral lines visible in galaxy spectra are shifted slightly toward the red end of the spectrum. This redshift is proportional to distance. That is, the farther away a galaxy is, the larger is the redshift that is visible in its spectrum (■ Figure P-9). This result, first discovered by the American astronomer Edwin Hubble in 1929, is clear evidence that the universe is expanding.

■ Active Figure P-9 ▶

These galaxy spectra extend from the near-ultraviolet at left to the blue part of the visible spectrum at right. The two dark absorption lines of once-ionized calcium are prominent in the near-ultraviolet. The redshifts in galaxy spectra are expressed here as velocities of recession. Note that the apparent velocity of recession is proportional to distance. (Caltech)

Ace◐Astronomy™ Log into AceAstronomy and select this chapter to see the Active Figure "Cosmic Redshift." Plot your data to discover the expansion of the universe.

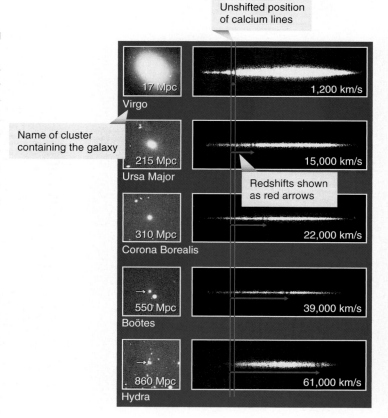

Unshifted position of calcium lines

Name of cluster containing the galaxy

Redshifts shown as red arrows

17 Mpc Virgo	1,200 km/s
215 Mpc Ursa Major	15,000 km/s
310 Mpc Corona Borealis	22,000 km/s
550 Mpc Boötes	39,000 km/s
860 Mpc Hydra	61,000 km/s

If the galaxies are all rushing away from each other, then you can imagine viewing a video running backward. You would see the galaxies rushing toward each other, and eventually you would see the galaxies pushing into each other, compressing and heating the gas until your video screen was filled with the glare of an intensely hot, fantastically dense gas filling the universe. This, the beginning of the universe, is a state called the **big bang.**

How can anyone know the big bang really happened? The evidence is conclusive—astronomers can see it. To understand how you might see an event that occurred billions of years ago, you need to look once again at the galaxies. Nearby galaxies, such as the Andromeda Galaxy, are only a few million light-years away. The hazy light you see in the sky in the constellation Andromeda has been traveling through space for only a few million years, and you see the Andromeda Galaxy as it was a few million years ago when the light began its journey. If you used a telescope and looked at more distant galaxies, you would see them as they were a billion years ago or more; light from those galaxies has been traveling that long. If you used a big telescope and looked at the most distant galaxies, you would be looking back in time and seeing them as they were over 10 billion years ago when the universe was young.

What would you see if you looked at the empty places on the sky between the most distant galaxies? You could see a glow that was emitted by the hot, dense clouds of gas of the big bang, but you couldn't see that glow with your unaided eyes because the redshift is so great that the photons of light are shifted into the long-wavelength infrared part of the spectrum. Nevertheless, astronomers can "see" the radiation from the big bang with infrared and radio detectors. This is called the **cosmic microwave background radiation,** and it pours in on Earth from all directions, telling you that you are part of a universe that was very hot and very dense only about 13.7 billion years ago (■ Figure P-10). You and your planet are part of the big bang.

Notice that the background radiation is visible in any direction in the sky. The big bang did not occur in a specific place, but it filled the entire volume of the universe. As the universe expanded, the total amount of volume increased, and the hot gases of the big bang cooled and formed galaxies. The expansion continues, and the galaxies continue to move away from each other. They are not fragments ejected from an explosion but parts of a whole that fills all of the volume of the universe and expands continuously as the total volume increases.

The big bang theory seems fantastic, but it has been tested and confirmed over and over, and modern astronomers have great confidence that there really was a big bang (**Window on Science P-2**). Of course, there are more details to understand. How did the first galaxies form? Why did the galaxies form

■ **Figure P-10**

Data from the WMAP spacecraft were used to make this far-infrared map of the sky. This infrared radiation was emitted when the universe was very young and filled with the hot gas of the big bang. The statistical distribution of the tiny irregularities in brightness allows astronomers to determine the age of the universe plus its rate of expansion. (NASA/WMAP Science Team)

The Impossibility of Proof in Science

No scientific theory or hypothesis can be proved correct. You can test a theory over and over by performing experiments or making observations, but you can never prove that the theory is absolutely true. It is always possible that you have misunderstood the theory or the evidence, and the next observation you make might disprove the theory.

For example, you might propose the theory that the sun is mostly iron. You could test the theory by looking at the iron lines in the solar spectrum, and the strength of the iron lines would suggest that your theory is right. Although your observation has confirmed your theory, it has not proven the theory is right. You might confirm the theory many times before you realized that iron absorbs photons much more efficiently than hydrogen. Although the hydrogen lines are weak in the sun's spectrum, they show that most of the atoms in the sun are hydrogen and not iron.

The nature of scientific thinking can lead to two kinds of misconceptions. Sometimes nonscientists will say, "You scientists just want to tear everything down—you don't believe in anything." Scientists test a theory over and over to test its worth. If a theory survives many tests, scientists begin to have confidence it is true, but they continue to test their understanding of nature at every opportunity.

The second misconception nonscientists have is revealed when they say, "You scientists are never sure of anything." Again, the scientist knows that no theory can be proven correct. That the sun will rise tomorrow is very likely, and scientists have great confidence in that theory. But in the end it is still a theory.

People will say of an idea they dislike, "That is only a theory," as if a theory were simply a random guess. In fact, a theory can be a well-tested truth in which all scientists have great confidence. Yet you can never prove that any theory is absolutely true.

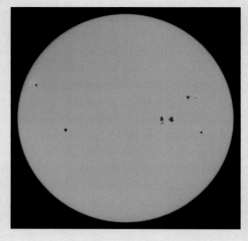

It is only a theory, but astronomers have tremendous confidence that the sun is made almost entirely of hydrogen and helium. (SOHO/MDI)

---

in giant clusters? You live in an exciting age when astronomers are able to ask these questions and expect to discover answers.

These ideas strain human imagination and challenge the most sophisticated mathematics, but the lesson is a simple one. The universe had a beginning not so long ago, and the matter you are made of had its birthday in that moment of cosmic beginnings. You are small, but you are part of something vast.

P-6 The Story of Matter

ASTRONOMERS CAN TELL the story of the matter in the universe with few gaps. Models of the big bang show that the first few minutes in the history of the universe were unimaginably hot. Energy was so intense that it could create particles of matter, and that is when the first protons and electrons formed. Protons and electrons, even when they are not attached to each other, are hydrogen, so the first gas of the big bang was hydrogen. The temperature and density were so high that nuclear fusion reactions could occur, fusing some of the hydrogen into helium.

By the time the universe was 3 minutes old, expansion had cooled its gases, and nuclear fusion stopped. In those first 3 minutes, about 25 percent of the mass had been converted into helium, but few heavier atoms could be made. Because there are

no stable atomic nuclei with masses of 5 or 8 times that of hydrogen, the fusion process could not go past helium.

As the universe expanded and cooled, it became dark. Gravity drew matter together to form great clouds that contracted to form the first galaxies. Eventually, stars began to form, and as those first stars began to shine, they lit up the universe and ionized the hydrogen gas in great shells around the galaxies (■ Figure P-11).

Since the formation of the first stars, galaxies have collided and merged; generation after generation of stars have lived and died, spreading heavy elements out into the thin gas in space. Nearly all of the atoms of which you are made except for hydrogen have been made inside stars (■ Figure P-12). When the sun began to form from a cloud of gas, the atoms now inside your body were there. They were part of Earth as it formed, and now they are part of you. Every atom has a history that stretches back through the birth and death of stars to the first moment in time. You exist because stars have died.

■ Figure P-11 ▶

In this artist's conception, the first stars produce floods of ultraviolet photons that ionize the gas in expanding bubbles. Such a storm of star formation ended an age when the universe had expanded in darkness. As star formation began in the universe, the production of the chemical elements also began. (K. Lanzetta, SUNY, A. Schaller for STScI, and NASA)

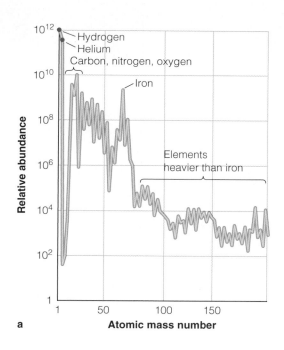

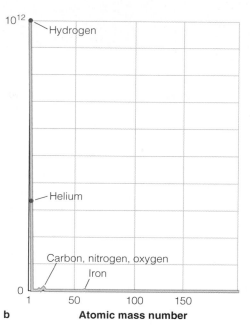

a

b

■ **Figure P-12**

The abundance of the chemical elements is usually plotted as in the graph in part (a). Hydrogen is most common, followed by helium. Carbon, nitrogen, oxygen, and iron also seem abundant, but elements heavier than iron are quite rare because they are made only in supernova explosions. Plotting these abundances on a linear scale as in the graph in part (b) provides a more realistic impression. Most of the universe is hydrogen and helium. The atoms of which you are made are hardly more than a tiny impurity in the cosmos.

Summary

P-1 | The Birth of Stars

How are stars born, and how do they die?

- The interstellar medium, the gas and dust between the stars, can be seen as nebulae. Stars are born when such nebulae contract to form protostars.
- Protostars are cool and hidden inside clouds of gas and dust, so they are most easily observed in the infrared.
- Planets form in the disks of gas and dust around protostars.

P-2 | The Deaths of Stars

- The sun is a medium-mass star and can live for about 10 billion years before it expands to become a giant, expels its outer layers to form a planetary nebula, and collapses to form a white dwarf.
- The most massive stars live only a few million years and die in supernova explosions that can form elements heavier than iron.
- Stars cook hydrogen into heavier elements and return them to the interstellar medium when the star dies.

P-3 | Our Home Galaxy

What are galaxies?

- Our Milky Way Galaxy contains about 100 billion stars and is about 75,000 ly in diameter. We live about two-thirds of the way from the center to the edge.

P-4 | The Universe of Galaxies

- Billions of galaxies fill the sky, and some are elliptical, some spiral, and some irregular.
- Collisions between galaxies are common and can trigger star formation.

P-5 | The Origin of the Universe

How did the universe begin and form galaxies?

- The redshifts of the galaxies show that the universe is expanding, and running the expansion backward in time leads to the high-temperature, high-density beginning of the universe called the big bang.
- The cosmic microwave background radiation is light from the big bang redshifted into the infrared and radio parts of the spectrum. It is direct evidence that the big bang really happened. Because Earth is part of the big bang, the radiation is detectable in every direction in the sky.

P-6 | The Story of Matter

How were the atoms in our bodies formed?

- The energy of the big bang produced protons and electrons (hydrogen gas), and some of that hydrogen was fused into helium during the first 3 minutes, but no heavier elements could be formed.
- Gravity drew the gas into great clouds that formed galaxies, and stars fused hydrogen into heavier elements up to iron. Supernova explosions can make elements heavier than iron.
- A number of generations of stars have produced the chemical elements in Earth and in you.

New Terms

interstellar medium (p. 188)	Milky Way Galaxy (p. 197)
nebula (p. 188)	globular cluster (p. 198)
protostar (p. 188)	elliptical galaxy (p. 200)
red dwarf (p. 192)	spiral galaxy (p. 200)
main sequence (p. 192)	barred spiral galaxy (p. 200)
giant (p. 192)	irregular galaxy (p. 201)
planetary nebula (p. 192)	Large and Small Magellanic Cloud (p. 201)
white dwarf (p. 192)	
supergiant (p. 193)	big bang (p. 203)
Milky Way (p. 196)	cosmic microwave background radiation (p. 203)

Review Questions

Ace ⊚ Astronomy™   Assess your understanding of this chapter's topics with additional quizzing and animations at http://ace .brookscole.com/sf9

1. What evidence could you cite that there is an interstellar medium?
2. Why are protostars difficult to observe at visible wavelengths?
3. How are protostars related to planet formation?
4. How do stars like the sun die?
5. How do massive stars die?
6. What is the difference between the Milky Way and the Milky Way Galaxy?
7. Describe how star formation is related to gas and dust in galaxies such as ellipticals and spirals.
8. What evidence can you cite to show that the universe is expanding? that there really was a big bang?
9. How did the big bang form the matter in your body, while the stars cooked the matter into the atoms heavier than helium?

Media Cluster

ACTIVE FIGURES

Ace Astronomy™ To access the resources in the Media Cluster, log into AceAstronomy at **http://ace.brookscole.com/sf9** and select this chapter.

Galaxy Types
Explore different kinds of galaxies and their structures and motions.

M31 at Many Wavelengths
In this animation, you can compare how five astronomical satellites, including Hubble and Chandra, view the Andromeda Galaxy.

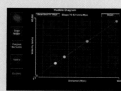

Cosmic Redshift
Graph your own observations of redshift and distance.

ASTRONOMY EXERCISES

Milky Way Galaxy
What would your home galaxy look like if you could see it from the outside? Find out with this 3-D model of the Milky Way that you can spin around to view from any angle.

Critical Inquiries for the Web

1. Search for more information about the Orion Nebula. Find images made at different wavelengths and find out what they tell you about the nebula.

2. What was the supernova of AD 1054? How do astronomers connect it with the Crab Nebula? What happened to the matter in the star that exploded?

3. Search for images of the interacting galaxies called the Mice. How do such collisions and close encounters distort the shapes of galaxies and trigger bursts of star formation?

4. Edwin Hubble and Milton Humason made the observations that revealed the expansion of the universe. Find information about these two famous astronomers and find out how they got started in astronomy.

Exploring *TheSky*

1. The Crab Nebula is M1. Locate it, zoom in, measure its angular size in seconds of arc, and compute its diameter, assuming it is about 6000 ly from Earth.

2. Locate the Andromeda Galaxy, also known as M31, and its companion galaxies. Zoom in on it and estimate its angular size compared to the full moon. (*Hint:* Use **Find** under the **Edit** menu.)

3. Take a survey of galaxies and see how many are spiral and how many are elliptical. Does your method produce a fair count, or are you biased to notice only the brighter galaxies? (*Hint:* Use **Filters** under the **View** menu to turn off everything but **Galaxies** and **Mixed Deep Sky** objects. Make sure the Messier labels are switched on using the **Labels** and **Setup** under the **View** menu.)

4. Locate the Sombrero Galaxy (M104). Study the photographs and discuss this galaxy's special properties. Zoom in on it and estimate its angular size compared to the moon.

Go to the Brooks/Cole Astronomy Resource Center **(http://astronomy.brookscole.com)** for critical thinking exercises, articles, and additional readings from InfoTrac College Edition, Brooks/Cole's online student library.

19 | The Origin of the Solar System

What place is this?

Where are we now?

CARL SANDBURG, *GRASS*

MICROSCOPIC CREATURES
LIVE in the roots of your eyelashes. Don't worry.
Everyone has them, and they are harmless.* They hatch, fight
for survival, mate, lay eggs, and die in the tiny spaces around the roots of your eyelashes without doing any harm. Some
live in renowned places—the eyelashes of a glamorous movie star—but the tiny beasts are not self-aware; they never stop
to say, "Where are we?" Humans are more intelligent; humans have the ability to wonder where we are in the universe
and how we came here. ▮ You should study the solar system for many reasons. You need to understand Earth as a
planet because six billion of us are living on Earth and changing it in ways we don't understand. You should also study

---

* Demodex folliculorum has been found in 97 percent of individuals and is a characteristic of healthy skin.

▮ Continued on page 450 ▮

The human race lives on a planet in a planetary system that appears to have formed in a nebula around the protostar that became the sun. This artist's impression shows the formation of the giant planet Jupiter. (NASA)

Guidepost

Looking Back

You have become an expert on the universe. You have studied the appearance, origin, structure, and evolution of stars, galaxies, and the universe itself. But so far, your studies have left out one important class of objects—planets. Now it is time for you to correct that omission.

This Chapter

In this chapter, you will look back on what you have learned and find your place. You are a planetwalker. What does that mean? Where do you fit into the cosmos? Most of all, you need to know how the solar system formed. That will tell you how your home planet was produced by the processes you have been studying. As you explore our solar system in space and time, you will find answers to four essential questions:

What theories account for the origin of the solar system?

What properties must these theories explain?

How do planets form?

Is our solar system unique?

Looking Ahead

One reason you should learn about the origin of the solar system is that you live here, but there is another reason. In the next three chapters you will explore in more detail each of the planets, plus the comets and asteroids. By studying the origin of the solar system first, you give yourself a framework for understanding these fascinating worlds.

Ace◐Astronomy™ The AceAstronomy icon throughout the text indicates an opportunity for you to test yourself on key concepts and to explore animations and interactions on the AceAstronomy website at: http://ace .brookscole.com/sf9

the solar system because, as you are about to discover, there are more planets in the universe than stars. Above all, you should study the solar system because it is your home in the universe. Because humans are an intelligent species, we have the right and the responsibility to wonder what we are. Our kind have inhabited this solar system for at least a million years, but only within the last hundred years have we begun to understand what a solar system is. Like sleeping passengers on a train, we waken, look out at the passing scenery, and mutter, "What place is this? Where are we now?"

19-1 Theories of Earth's Origin

YOU ARE LINKED through a great chain of origins that leads backward through time to the first instant when the universe began 13.7 billion years ago. The gradual discovery of the links in that chain is one of the most exciting adventures of the human intellect. In earlier chapters, you studied some of that story: the origin of the universe in the big bang, the formation of galaxies, the origin of stars, and the growth of the chemical elements. Here you will explore further to consider the formation of planets.

Early Hypotheses

The earliest theories for Earth's origin are myths and folktales that go back beyond the beginning of recorded history. In addition, almost all religions contain an account of the origin of the world. It was not until about the time of Galileo that philosophers began searching for rational explanations for natural phenomena. While people like Copernicus, Kepler, and Galileo tried to find logical explanations for the motions of Earth and the other planets, other philosophers began thinking about the origin of the planets.

The first rational theory for Earth's origin was proposed by the French philosopher and mathematician René Descartes (1596–1650). Because he lived and wrote before the time of Newton, Descartes did not have the concept of gravitation as the dominant force in the universe. Rather, he believed that force was communicated by contact between bodies and that the entire universe was filled with vortices of whirling invisible particles. In 1644, he proposed that the sun and planets formed when a large vortex contracted and condensed. In this way, his hypothesis explained the general properties of the solar system known at the time.

A century later, in 1745, the French naturalist Georges-Louis de Buffon (1707–1788) proposed an alternative hypothesis that the planets were formed when a passing star collided with or passed close to the sun and pulled matter out of the sun and the star. The matter condensed to form the planets, and they fell into orbit around the sun (■ Figure 19-1a). This **passing star hypothesis** was popular off and on for two centuries, but it con-

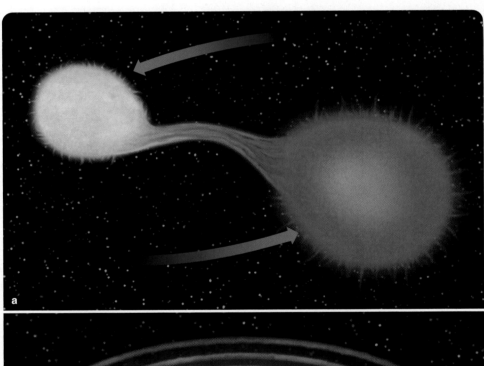

■ Figure 19-1

(a) The passing star hypothesis was catastrophic. It proposed that the sun was hit by or had a very close encounter with a passing star and that matter torn from the sun and the star formed planets orbiting the sun. The theory is no longer accepted. (b) Laplace's nebular hypothesis was evolutionary. It suggested that a contracting disk of matter conserved angular momentum, spun faster, and shed rings of matter that formed planets.

Two Kinds of Theories: Evolution and Catastrophe

Many theories in science can be classified as either evolutionary, in that they involve gradual processes, or catastrophic, in that they depend on specific, unlikely events. Scientists have generally preferred evolutionary theories. Nevertheless, catastrophic events do occur.

Something in people prefers catastrophic theories, perhaps because everyone likes to see spectacular violence from a safe distance, which may explain the success of movies that include lots of car crashes and explosions. Also, cataclysmic theories resonate with Old Testament accounts of catastrophic events and special acts of creation. Thus, people have an understandable interest in catastrophic theories.

Nevertheless, most scientific theories are evolutionary. Such theories do not depend on unlikely events or special acts of biblical creation. For example, geologists study theories of mountain building that are evolutionary, with the mountains being pushed up slowly as centuries pass. All the evidence of erosion and the folding of rock layers shows that the process is gradual. Because most such natural processes are evolutionary, scientists sometimes find it difficult to accept any theory that depends on catastrophic events.

You will see in this and later chapters that catastrophes do occur. The planets, for example, are bombarded by debris from space, and some of those impacts are very large. As you study astronomy or any other natural science, notice that most theories are evolutionary but that you must allow for the possibility of unpredictable catastrophic events.

Mountains evolve to great heights by rising slowly, not catastrophically. (Janet Seeds)

tains serious flaws. First, stars are very small compared to the distances between them, and thus they collide very infrequently. In the entire history of our galaxy, stars have probably collided only a few times. More important, the gas pulled from the sun and the star would be much too hot to condense to make planets. Furthermore, even if planets formed, they would not go into stable orbits.

The hypotheses of Descartes and Buffon fall into two broad categories. Descartes's hypothesis is **evolutionary** in that it calls upon common, gradual events to produce the sun and planets. If it is correct, stars with planets are very common. Buffon's hypothesis, on the other hand, is **catastrophic.** It calls on unlikely, sudden events to produce the solar system, and thus it implies that solar systems are very rare. While your imagination may be tempted by colliding stars, modern hypotheses for the origins of the planets are evolutionary, with, as you will see, a few astonishing catastrophes thrown in (**Window on Science 19-1**).

The modern theory of the origin of the solar system had its true beginning with Pierre-Simon de Laplace (1749–1827), the brilliant French astronomer and mathematician. In 1796, he combined Descartes's vortex with Newton's gravity to produce a model of a rotating cloud of matter contracting under its own gravitation and flattening into a disk—the **nebular hypothesis.** As the disk grew smaller, it had to conserve angular momentum and spin faster and faster. Laplace reasoned that, when it was spinning as fast as it could, the disk would shed its outer edge to leave behind a ring of matter. Then the disk could contract further, speed up again, and leave another ring. In this way, he imagined, the contracting disk would leave behind a series of rings, each to become a planet circling the newborn sun at the center of the disk (Figure 19-1b).

According to the nebular hypothesis, the sun should be spinning very rapidly, or, to put it another way, the sun should have most of the angular momentum of the solar system. (Recall from Chapter 5 that angular momentum is the tendency of a rotating object to continue rotating.) As astronomers studied the planets and the sun, however, they found that the sun rotated slowly and that the planets moving in their orbits had most of the angular momentum in the solar system. In fact, the rotation of the sun contains only about 0.3 percent of the angular momentum of the solar system. Because the nebular hypothesis could not explain this **angular momentum problem,** it was never fully successful, and astronomers toyed with various versions of the passing star hypothesis for over a century.

In contrast to the astronomy of earlier centuries, 20th-century astronomy applied modern physics to the stars and galaxies, and that illuminated the origins of the stars and the elements. You can now trace that great chain of origins that began with the big bang and led to the matter of which you are made. First you should review the story of the origin of matter, and then you can add the story of how that matter formed your home world.

A Review of the Origin of Matter

The matter in your body came into existence within minutes of the beginning of the universe. Astronomers have strong evidence that the universe began in an event called the big bang

(Chapter 18); and, by the time the universe was three minutes old, the protons, neutrons, and electrons in your body had come into existence. You are made of very old matter.

Although those particles formed quickly, they were not linked together to form the atoms that are common today. Most of the matter was hydrogen, and about 25 percent was helium. Very few heavier atoms were made in the big bang. Although your body does not contain helium, it does contain many of those ancient hydrogen atoms unchanged since the universe began.

Within a few hundred million years after the big bang, matter began to collect to form galaxies containing billions of stars. You have learned how nuclear reactions inside stars combine low-mass atoms such as hydrogen to make heavier atoms (Chapter 8). Generation after generation of stars cooked the original particles, linking them together to build atoms such as carbon, nitrogen, and oxygen (Chapter 15). Those are common atoms in your body. Even the calcium atoms in your bones were assembled inside stars.

Massive stars produce iron in their cores, but much of that iron core is destroyed when the star collapses and explodes as a supernova. Most of the iron in your body was produced by carbon fusion in type Ia supernovae and by the expanding matter ejected by type II supernovae. Atoms heavier than iron were created by rapid nuclear reactions that can occur only during supernova explosions (Chapter 13). Gold and silver are rare in your body, but iodine is critical in your thyroid gland, and those heavy atoms were produced during the violent deaths of massive stars.

Our galaxy contains at least 100 billion stars, of which the sun is one. It formed from a cloud of gas and dust about 5 billion years ago, and the atoms in your body were part of that cloud. How the sun took shape, how the cloud gave birth to the planets, how the atoms in your body found their way onto Earth and into you is the story of this chapter. As you explore the origin of the solar system, keep in mind the great chain of origins that created the atoms. As the geologist Preston Cloud remarked, "Stars have died that we might live."

The Solar Nebula Hypothesis

The stars gave birth to the heavy atoms of which our world is made, and the modern theory of the origin of the planets is based on star formation. The **solar nebula hypothesis** proposes that the planets were formed from the disk of gas and dust that surrounded the sun as it formed (■ Figure 19-2).

As stars form in contracting clouds, they remain surrounded by cocoons of dust and gas, and the rotation of the cloud causes that dust and gas to form a spinning disk around the protostar. When the center of the star grows hot enough to ignite nuclear reactions, its surface quickly heats up, becomes more luminous, and blows away the gas and dust cocoon.

The solar nebula hypothesis supposes that planets form in the rotating disks of gas and dust around young stars. You have seen clear evidence that disks of gas and dust are common around

The Solar Nebula Hypothesis

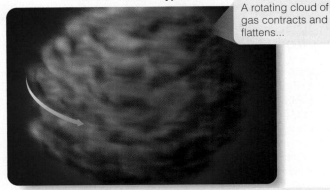
A rotating cloud of gas contracts and flattens...

to form a thin disk of gas and dust around the forming sun at the center.

Planets grow from gas and dust in the disk and are left behind when the disk clears.

■ Figure 19-2

The solar nebula hypothesis proposes that the planets formed along with the sun.

young stars. Infrared observations of T Tauri stars, for instance, show that some are surrounded by gas clouds rich in dust, and spectra show that these stars are blowing away their nebulae at speeds up to 200 km/s. Bipolar flows from protostars (Chapter 11) were the first evidence of such disks, but modern techniques can image the disks directly (see Figure 11-7 and page 253).

Our own planetary system probably formed in such a disk-shaped cloud around the sun. When the sun became luminous enough, the remaining gas and dust were blown away into space, leaving the planets orbiting the sun. Notice that the solar nebula hypothesis also suffers from the angular momentum problem. Before modern astronomers could take the theory seriously, they had to understand why the sun now rotates so slowly, a puzzle you will be able to solve later in this chapter.

If the solar nebula hypothesis is correct, then our Earth and the other planets of the solar system formed billions of years ago as the sun condensed from the interstellar medium. If that is true, then planets form as a by-product of star formation, and most stars should have planets.

Building Scientific Arguments

Why does the solar nebula theory imply planets are common?
Often, the implications of a theory are more important in building a scientific argument than the theory's own conjecture about nature. The solar nebula theory is an evolutionary theory; and, if it is correct, the planets of our solar system formed from the disk of gas and dust that surrounded the sun as it condensed from the interstellar medium. That suggests it is a common process. Most stars form with disks of gas and dust around them, and planets should form in such disks. Planets should be very common in the universe.

Now build a new scientific argument to consider the old catastrophic theory. **Why did the passing star hypothesis suggest that planets are very rare?**

■   ■   ■

Connections: The solar nebula hypothesis seems to fit with what you know about star formation, so it is time to construct a detailed theory to describe planet formation. The first step is to survey the solar system to discover the distinguishing characteristics of the planetary system. Those are the characteristics that a successful theory must explain.

19-2 A Survey of the Solar System

TO TEST THEIR THEORIES, astronomers must search the present solar system for evidence of its past. In this section, you will survey the solar system and compile a list of its most significant characteristics, potential clues to how it formed.

You should begin with the most general view of the solar system. It is, in fact, almost entirely empty space (Figure 1-7). Imagine that you reduce the solar system until Earth is the size of a grain of table salt, about 0.3 mm (0.01 in.) in diameter. The moon is a speck of pepper about 1 cm (0.4 in.) away, and the sun is the size of a small plum 4 m (13 ft) from Earth. Mercury, Venus, and Mars are grains of salt. Jupiter is an apple seed 20 m (66 ft) from the sun, and Saturn is a smaller seed over 36 m (120 ft) away. Uranus and Neptune are slightly larger than average salt grains, and Pluto, at the edge of the solar system, is a speck of pepper over 150 m (500 ft) from the central plum. Although your model solar system would be larger than a football field, you would need a powerful microscope to detect the asteroids.

The planets are tiny specks of matter scattered around the sun—the last remains of the solar nebula.

Revolution and Rotation

The planets revolve* around the sun in orbits that lie close to a common plane. The orbit of Mercury, the closest planet to the sun, is tipped 7° to Earth's orbit, and Pluto's orbit is tipped 17.2°. The rest of the planets' orbital planes are inclined by no more than 3.4°. As you can see, the solar system is basically disk shaped.

The rotation of the sun and planets on their axes also seems related to this disk shape. The sun rotates with its equator inclined only 7.25° to Earth's orbit, and most of the other planets' equators are tipped less than 30°. The rotations of Venus, Uranus, and Pluto are peculiar, however. Venus rotates backward compared with the other planets, and both Uranus and Pluto rotate on their sides (with their equators almost perpendicular to their orbits). You will explore these planets in detail in Chapters 22 and 24, but later in this chapter you will be able to understand how they could have acquired their peculiar rotations.

Apparently, the preferred direction of motion in the solar system—counterclockwise as seen from the north—is also related to its disk shape. All the planets revolve counterclockwise around the sun; and, with the exception of Venus, Uranus, and Pluto, they rotate counterclockwise on their axes. Furthermore, nearly all of the moons in the solar system, including Earth's moon, orbit around their planets counterclockwise. With only a few exceptions, most of which are understood, revolution and rotation in the solar system follow a disk theme.

Two Kinds of Planets

Perhaps the most striking clue to the origin of the solar system comes from the division of the planets into two categories: the small Earthlike worlds and the giant Jupiterlike worlds. The difference is so dramatic that it is hard to keep from shouting, "Aha, this must mean something!" Study **Terrestrial and Jovian Planets** on pages 454–455 and notice three important points:

1 Notice how the two kinds of planets are distinguished by their location. The four inner planets are quite different from the next four outward.

2 Also notice how common craters are. Almost every solid surface in the solar system is covered with craters.

3 Finally, notice how the planets are distinguished by individual properties such as rings, clouds, and moons. Any theory of the origin of the planets needs to explain these properties.

The division of the planets into two families is a clue to how our solar system formed, but you can learn little from the

---

*Recall from Chapter 2 that the words *revolve* and *rotate* refer to different motions. A planet revolves around the sun but rotates on its axis. Cowboys in the old west didn't carry revolvers. They carried rotators.

Mercury

Sun

Venus

Earth ---------- Moon

Mars

The planets and the sun to scale. Saturn's rings would just reach from Earth to the moon.

Jupiter

Saturn

Uranus

Neptune

Pluto

1 The distinction between the terrestrial planets and the Jovian planets is dramatic. The inner four planets, Mercury, Venus, Earth, and Mars, are **terrestrial planets,** meaning they are small, dense, rocky worlds with little or no atmosphere. The outer four planets, Jupiter, Saturn, Uranus, and Neptune, are **Jovian planets,** meaning they are large, low-density worlds with thick atmospheres and liquid interiors. Pluto does not fit this scheme, being small but low density; you will see in a later chapter that it is a very special world and might not be considered a planet at all.

Planetary orbits to scale. The terrestrial planets lie quite close to the sun, whereas the Jovian planets are spread far from the sun outside the asteroid belt. The elliptical shape of Pluto's orbit is visible here.

Jupiter

Mercury

Venus

Mars

Earth

Asteroids

Saturn

Uranus

Neptune

Pluto

1a Of the terrestrial planets, Earth is most massive, but the Jovian planets are much more massive. Jupiter is over 300 Earth masses, and Saturn is nearly 100 Earth masses. Uranus and Neptune are 15 and 17 Earth masses.

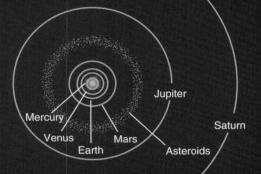

Mercury is only 40 percent larger than Earth's moon, and its weak gravity cannot retain a permanent atmosphere. Like the moon, it is covered with craters from meteorite impacts.

Mercury

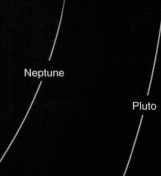

Earth's moon

2 Craters are common on all of the surfaces in the solar system that are strong enough to retain them. Earth has about 150 impact craters, but many more have been erased by erosion. Besides the planets, the asteroids and nearly all of the moons in the solar system are scarred by craters. Ranging from microscopic to hundreds of kilometers in diameter, these craters have been produced over the ages by meteorite impacts. When astronomers see a rocky or icy surface that contains few craters, they know that the surface is

Mercury is so close to the sun it is difficult to study from Earth. The Mariner 10 spacecraft flew past Mercury in 1974, but it was not able to take detailed photos of all of the planet's surface.

Mercury

Moon

The surface of Venus is not visible through its cloudy atmosphere, but radar maps reveal a dry desert world of craters and volcanoes.

These five worlds are shown in proper relative size.

Earth

3 The terrestrial planets have densities like that of rock or metal. The Jovian planets all have low densities, and Saturn's density is only 70 percent the density of water. It would float in a big-enough bathtub.

The atmospheres of the Jovian planets are turbulent, and some are marked by great storms such as the Great Red Spot on Jupiter. But the atmospheres are not deep. If Jupiter were shrunk to a few centimeters in diameter, its atmosphere would be no deeper than the fuzz on a badly worn tennis ball.

Mars

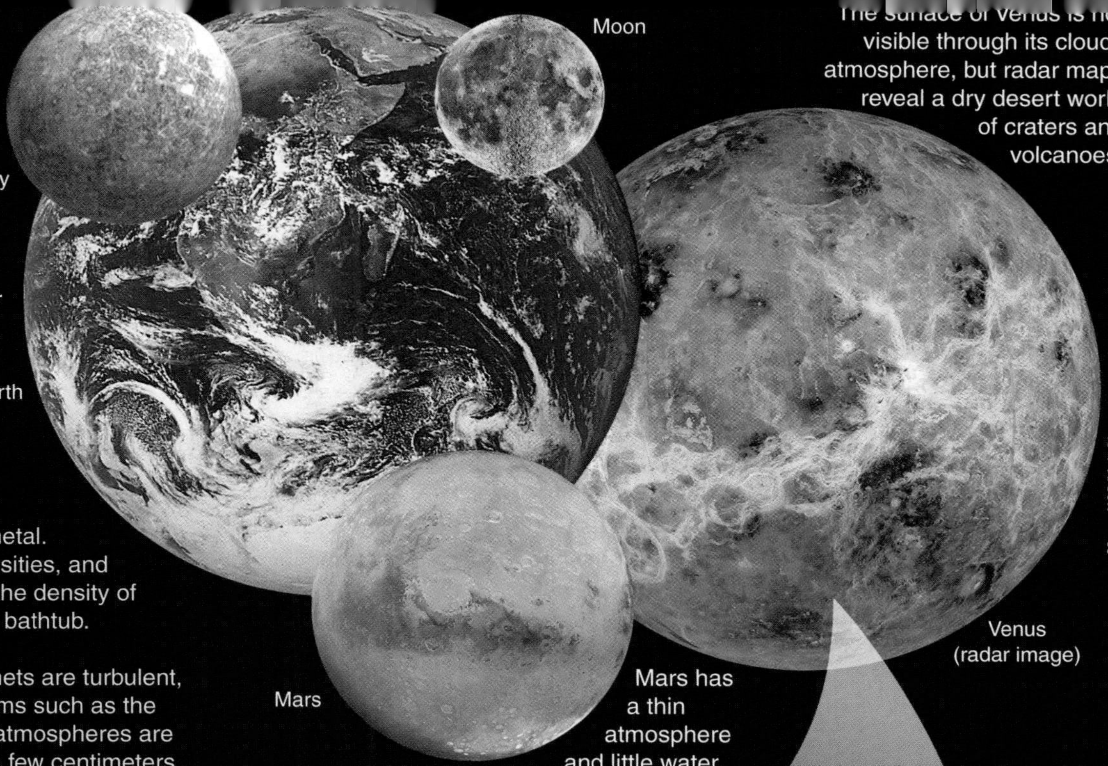

Mars has a thin atmosphere and little water. Craters and volcanoes are common on its desert surface.

Venus (radar image)

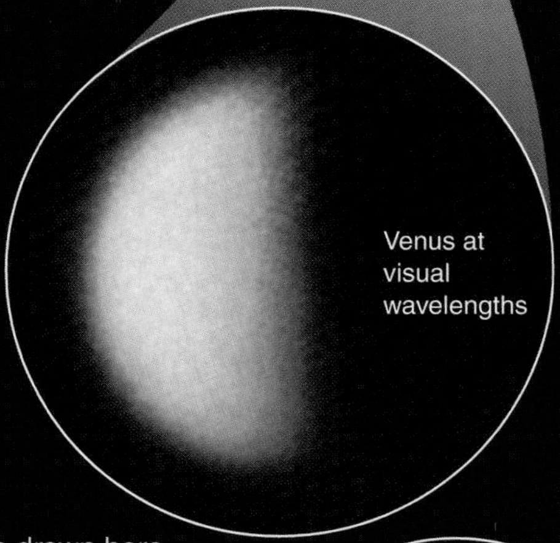

These Jovian worlds are shown in proper relative size.

3a The interiors of the Jovian planets contain small cores of heavy elements such as metals, surrounded by a liquid. Jupiter and Saturn contain hydrogen forced into a liquid state by the high pressure. Less-massive Uranus and Neptune contain heavy-element cores surrounded by partially frozen water mixed with heavy material such as rocks and minerals.

Venus at visual wavelengths

Jupiter

Great Red Spot

The terrestrial planets are drawn here to the same scale as the Jovian planets.

The Jovian planets have large systems of satellites, and Jupiter is orbited by four large moons known as the **Galilean satellites** because they were discovered by Galileo in 1610.

Saturn's rings seen through a small telescope.

Neptune

Uranus

Saturn

3b All four Jovian planets have ring systems. Saturn's rings are made of ice particles. The rings of Jupiter, Uranus and Neptune are made of dark rocky particles. Terrestrial planets have no rings.

individual planets because they have evolved since they formed. For further clues you must look at smaller objects that have remained largely unchanged since the birth of the solar system.

Space Debris

The sun and planets are not the only remains of the solar nebula. The solar system is littered with three kinds of space debris: asteroids, comets, and meteoroids. Although these objects represent a tiny fraction of the mass of the system, they are a rich source of information about the origin of the planets.

The **asteroids,** sometimes called minor planets, are small rocky worlds, most of which orbit the sun in a belt between the orbits of Mars and Jupiter. Roughly 20,000 asteroids have been charted, of which about 2000 follow orbits that bring them into the inner solar system, where they can occasionally collide with a planet. Earth has been struck many times in its history. Some asteroids are located in Jupiter's orbit, while some have been found beyond the orbit of Saturn.

About 200 asteroids are more than 100 km (60 mi) in diameter, and tens of thousands are estimated to be more than 10 km (6 mi) in diameter. There are probably a million or more that are larger than 1 km (0.6 mi) and billions that are smaller. Because even the largest are only a few hundred kilometers in diameter, Earth-based telescopes can detect no details on their surfaces, and the Hubble Space Telescope can image only the largest features.

Astronomers do, however, have clear evidence that asteroids are irregularly shaped cratered worlds. A number of spacecraft have visited asteroids and sent back photos. For instance, the NEAR spacecraft rendezvoused with the asteroid Eros, went into orbit, and studied the asteroid in detail. Like most asteroids, Eros is an irregular, rocky body pocked by craters (■ Figure 19-3). These observations will be discussed in detail in Chapter 25, but in this quick survey of the solar system you can note that all the evidence suggests the asteroids have suffered many impacts from collisions with other asteroids.

Older theories proposed that the asteroids are the remains of a planet that broke up, but modern astronomers recognize the asteroids as the debris left over by a planet that failed to form at a distance of 2.8 AU from the sun. When you study the formation of planets later in this chapter, you will discover why material in the solar nebula failed to form a planet at a distance of 2.8 AU.

Since 1992, astronomers have discovered roughly a thousand small, dark, icy bodies orbiting in the outer fringes of the solar system beyond Neptune and Pluto. This collection of objects is called the **Kuiper belt** after the Dutch-American astronomer Gerard Kuiper (KI-per), who predicted their existence in the 1950s. There are probably 70,000 of these small, asteroidlike bodies in the Kuiper belt, and any successful theory should explain how they came to orbit so far from the sun.

In contrast to the rocky asteroids and dark Kuiper belt objects, the brightest **comets** are impressively beautiful objects (■ Figure 19-4). Most comets are faint, however, and are difficult to locate even at their brightest. A comet may take months to sweep through the inner solar system, during which time it appears as a glowing head with an extended tail of gas and dust.

The beautiful tail of a comet can be longer than an astronomical unit, but it is produced by an icy nucleus only a few tens of kilometers in diameter. In its long, elliptical orbit, the nucleus remains frozen and inactive while it is far from the sun. As its orbit carries the nucleus into the inner solar system, the sun's heat begins to vaporize the ices, releasing gas and dust. The pressure of sunlight and the solar wind pushes the gas and dust away, forming a long tail. The motion of the nucleus along its orbit, the pressure of sunlight, and the outward flow of the solar wind can create comet tails that are long and straight or gently curved, but in

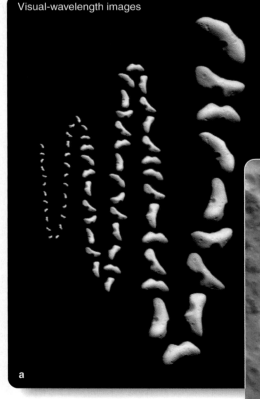

Visual-wavelength images

a

b

■ **Figure 19-3**

(a) Over a period of three weeks, the NEAR spacecraft approached the asteroid Eros and recorded a series of images arranged here in an entertaining pattern showing the irregular shape and 5-hour rotation of the asteroid. Eros is 34 km (21 mi) long. (b) This close-up of the surface of Eros shows an area about 11 km (7 mi) from top to bottom. (Johns Hopkins University, Applied Physics Laboratory, NASA)

(a) A comet may remain visible in the evening or morning sky for weeks as it moves through the inner solar system. Comet West was in the sky during March 1976. (b) A comet in a long, elliptical orbit becomes visible when the sun's heat vaporizes its ices and pushes the gas and dust away in a tail. (Celestron International)

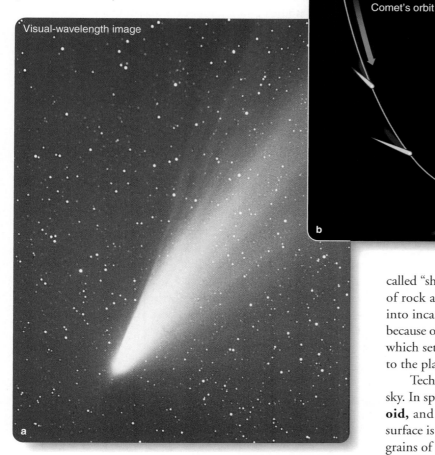

Visual-wavelength image

Comet's orbit

either case the tails of comets always point approximately away from the sun (Figure 19-4b).

For decades astronomers described comet nuclei as dirty snowballs, meaning that they were icy bodies with a little bit of embedded rock and dust. Starting with the passage of Comet Halley in 1986, astronomers have found growing evidence that comet nuclei are not made of dirty ice but rather of icy dirt. That is, the nuclei are at least 50 percent rock and dust. One astronomer has suggested replacing the dirty snowball model with the icy mudball model.

The nuclei of comets are ice-rich bodies left over from the origin of the planets. From this you can conclude that at least some parts of the solar nebula were rich in ices. You will see later in this chapter how important ices were in the formation of the Jovian planets, and you will have a chance to study comets in more detail in Chapter 25.

Unlike the stately comets, **meteors** flash across the sky in momentary streaks of light (■ Figure 19-5). They are commonly called "shooting stars." Of course, they are not stars but small bits of rock and metal falling into Earth's atmosphere and bursting into incandescent vapor about 80 km (50 mi) above the ground because of friction with the air. This vapor condenses to form dust, which settles slowly to Earth, adding about 40,000 tons per year to the planet's mass.

Technically, the word *meteor* refers to the streak of light in the sky. In space, before its fiery plunge, the object is called a **meteoroid,** and any part of it that survives its fiery passage to Earth's surface is called a **meteorite.** Most meteoroids are specks of dust, grains of sand, or tiny pebbles. Almost all the meteors you see in the sky are produced by meteoroids that weigh less than 1 g. Only rarely is one massive enough and strong enough to survive its plunge and reach Earth's surface.

Thousands of meteorites have been found, and you will learn more about their particular forms in Chapter 25. Meteorites are mentioned here for one specific clue they can give you concerning the solar nebula: Meteorites can tell you the age of the solar system.

The Age of the Solar System

If the solar nebula theory is correct, the planets should be about the same age as the sun. The most accurate way to find the age of a celestial body is to bring a sample into the laboratory and determine its age by analyzing the radioactive elements it contains.

When a rock solidifies, it incorporates known percentages of the chemical elements. A few of these elements are radioactive and can decay into another element, called the daughter element. For example, the isotope uranium-238 can decay into an isotope of lead, lead-238. The **half-life** of a radioactive element is the time

Visual-wavelength image

■ Figure 19-5

A meteor is the streak of glowing gases produced by a bit of material falling into Earth's atmosphere. Friction with the air vaporizes the material about 80 km (50 mi) above Earth's surface. (Daniel Good)

it takes for half of the atoms to decay. The half-life of uranium-238 is 4.5 billion years. The abundance of a radioactive element gradually decreases as it decays, and the abundance of the daughter element gradually increases (■ Figure 19-6). If you know the abundance of the elements in the original rock, you can measure the present abundance and find the age of the rock. For example, if you studied a rock and found that only 50 percent of the uranium-238 remained and the rest had become lead-238, you could conclude that one half-life must have passed and that the rock was 4.5 billion years old.

Uranium isn't the only radioactive element used in radioactive dating. Potassium-40 decays with a half-life of 1.3 billion years to form calcium-40 and argon-40. Rubidium-87 decays to strontium-87, with a half-life of 47 billion years. Any of these elements can be used as a radioactive clock to find the age of mineral samples.

Of course, to find a radioactive age, you need a sample in the laboratory, and the only celestial bodies from which scientists have samples are Earth, the moon, Mars, and meteorites.

The oldest Earth rocks so far discovered and dated are tiny zircon crystals from Australia 4.3 billion years old. That does not mean that Earth formed 4.3 billion years ago. The surface of Earth is active, and the crust is continually destroyed and reformed from material welling up from beneath the crust (see Chapter 20). Consequently, the age of these oldest rocks tells you only that Earth is *at least* 4.3 billion years old.

■ Active Figure 19-6

The radioactive atoms in a mineral sample (red) decay into daughter atoms (blue). Half the radioactive atoms are left after one half-life, a fourth after two half-lives, an eighth after three half-lives, and so on. Radioactive dating shows that this fragment of the Allende meteorite is 4.56 billion years old. It contains interstellar grains that formed long before our solar system. (R. Kempton, New England Meteoritical Services)

Ace ⑤ Astronomy™ Log into AceAstronomy and select this chapter to see the Active Figure "Radioactive Decay" and graph the decay of different radioactive isotopes.

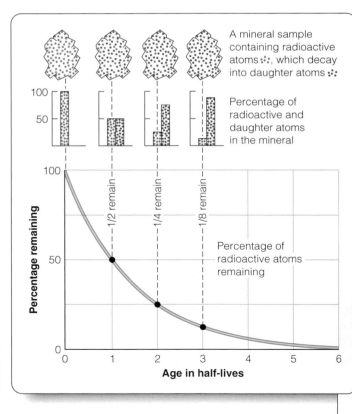

A mineral sample containing radioactive atoms ∴, which decay into daughter atoms ∵

Percentage of radioactive and daughter atoms in the mineral

Percentage of radioactive atoms remaining

One of the most exciting goals of the Apollo lunar landings was bringing lunar rocks back to Earth's laboratories, where they could be dated. Because the moon's surface is not being recycled like Earth's, some parts of it might have survived unaltered since early in the history of the solar system. Dating the rocks showed the oldest to be 4.48 billion years old. That means the solar system must be *at least* 4.48 billion years old.

Although no one has yet been to Mars, over a dozen meteorites found on Earth have been identified by their chemical composition as having come from Mars. Most of these have ages of only 1.3 billion years, but one has an age of approximately 4.6 billion years. Mars must be at least that old.

Another important source for determining the age of the solar system is meteorites. Radioactive dating of meteorites yields a range of ages, with the oldest about 4.56 billion years old. This figure is widely accepted as the age of the solar system.

Meteorites also contain tiny grains containing mostly silicon and carbon, and the composition of those grains shows that they are from the interstellar medium. They cannot be dated radioactively, but they must have formed long before our solar system and were incorporated in the nebula that contracted to form our sun and our planetary system. These tiny particles are messengers from beyond the beginning of our solar system.

One last celestial body deserves mention: the sun. Astronomers estimate the age of the sun to be about 5 billion years, but this is not a radioactive date because they cannot obtain a sample of solar material. Instead, they estimate the age of the sun from mathematical models of the sun's interior. This yields an age of about 5 billion years plus or minus 1.5 billion years, a number that is in agreement with the age of the solar system derived from the age of meteorites.

Apparently, all the bodies of the solar system formed at about the same time some 4.6 billion years ago. You can add this as the final item to your list of characteristic properties of the solar system (■ Table 19-1).

> ### ■ Table 19-1 I Characteristic Properties of the Solar System
>
> 1. Disk shape of the solar system
>   Orbits in nearly the same plane
>   Common direction of rotation and revolution
> 2. Two planetary types
>   Terrestrial—inner planets; high density
>   Jovian—outer planets; low density
> 3. Planetary ring systems and large satellite systems for Jupiter, Saturn, Uranus, and Neptune
> 4. Space debris—asteroids, comets, and meteors
>   Composition
>   Orbits
> 5. Common ages of about 4.6 billion years for Earth, the moon, Mars, meteorites, and the sun

Building Scientific Arguments

In what ways is the solar system a disk?

Notice that this argument is really a summary of pieces of evidence. First, the general shape of the solar system is that of a disk. The planets follow orbits that lie in nearly the same plane. The orbit of Mercury is inclined 7° to the plane of Earth's orbit, and the orbit of Pluto is inclined a bit more than 17°. In this way, the planets follow orbits confined to a thin disk with the sun at its center.

Second, the motions of the sun and planets also follow this disk theme. The sun and most of the planets rotate in the same direction, counterclockwise as seen from the north, with their equators near the plane of the solar system. Also, all of the planets revolve around the sun in that same direction. Our solar system seems to prefer motion in the same direction, which further reflects a disk theme.

One of the basic characteristics of our solar system is its disk shape, but another dramatic characteristic is the division of the planets into two groups. Build an argument to detail that evidence. **What are the distinguishing differences between the terrestrial and Jovian planets?**

■   ■   ■

> **Connections:** You have completed your survey of the solar system and gathered the preliminary evidence you can use to evaluate the solar nebula hypothesis. Now it is time to ask how the planets formed from the solar nebula and took on the characteristics they have today.

19-3 The Story of Planet Building

THE CHALLENGE FOR MODERN PLANETARY ASTRONOMERS is to compare the characteristics of the solar system with the solar nebula theory and tell the story of how the planets formed (**Window on Science 19-2**).

The Chemical Composition of the Solar Nebula

Everything astronomers know about the solar system and star formation suggests that the solar nebula was a fragment of an interstellar gas cloud. Such a cloud would have been mostly hydrogen with some helium and tiny traces of the heavier elements.

This is precisely what you see in the composition of the sun (see Table 7-2). Analysis of the solar spectrum shows that the sun is mostly hydrogen, with a quarter of its mass being helium and only about 2 percent being heavier elements. Of course, nuclear reactions have fused some hydrogen into helium, but this happens in the sun's core and has not affected its surface composition.

Reconstructing the Past from Evidence and Hypothesis

Scientists often face problems in which they must reconstruct the past. Some of these reconstructions are obvious, such as an archaeologist excavating the ruins of a burial tomb, but others are less obvious. In each case, success requires the interplay of hypotheses and evidence to re-create a past that no longer exists.

The reconstruction of the past is obvious when you use the chemical abundance of stars to reconstruct the story of the formation of our galaxy, but a biologist studying a centipede is also reconstructing the past. How did this creature come to have a segmented body with so many legs? How did it develop the metabolism that allows it to move quickly and hunt

prey? Although the problem might at first seem to be one of mere anatomy, the scientist must reconstruct an environment that no longer exists.

The astronomer's problem is not just to understand what the planets are like but to understand how they got that way. That means planetary astronomers must look at the evidence they can see today and reconstruct a history of the solar system, a past that is quite different from the present. If you had a time machine, it would be a fantastic adventure to go back and watch the planets form. Time machines are impossible, but scientists can use the grand interplay of evidence and theory, the distinguishing characteristic of

One way science can enrich and inform our lives is by recreating a world that no longer exists.

science, to journey back billions of years and reconstruct a past that no longer exists.

That means the composition revealed in the sun's spectrum is essentially the composition of the gases from which it formed.

This must have been the composition of the solar nebula, and you can see that composition reflected in the chemical compositions of the planets. The composition of a planet can be altered, however, depending on how it forms. The most common process of planet formation is believed to begin with the sticking together of solid bits of matter. Only when a planet has grown to a mass of about 15 Earth masses does it have enough gravitation to begin capturing gas directly from the solar nebula in a process called **gravitational collapse.** The Jovian planets began forming by the aggregation of bits of rock and ice. Once they accumulated enough mass, they began growing by gravitational collapse. That is, they began to capture large amounts of gas, mostly hydrogen and helium, directly from the solar nebula. Jupiter and Saturn grew rapidly and captured the most hydrogen and helium, whereas Uranus and Neptune apparently grew more slowly and didn't capture gas as rapidly as Jupiter and Saturn. That is how the Jovian worlds grew to be low-density, hydrogen-rich planets, with Jupiter and Saturn the largest and least dense.

The terrestrial planets, in contrast, contain very little hydrogen and helium. These small planets have such low masses they were unable to keep the hydrogen and helium from leaking into space. Indeed, because of their small masses, they were probably unable to capture very much of these lightweight gases from the solar nebula. The terrestrial planets are dense worlds because they are composed of the heavier elements from the solar nebula.

This account of planet formation is generally correct for most planets, but later in this chapter you will see how astronomers are refining the story to include other processes. Certainly, this general description of how planets form explains how the chemical composition of the solar nebula is reflected in the present composition of the sun and planets. Nevertheless, a very important question remains: How did gas and dust in the solar nebula come together to form the solid matter of the planets? You must answer that question in two stages. First, you must understand how the gas formed billions of small solid particles, and then you must explain how these particles built the planets.

The Condensation of Solids

The key to understanding the process that converted the nebular gas into solid matter is the variation in density among solar system objects. You have already noted that the four inner planets are high-density, terrestrial bodies, whereas the outermost planets are low-density, giant planets (except for Pluto, which is low in density but not a giant). This division is due to the different ways gases condensed into solids in the inner and outer regions of the solar nebula.

Even among the four terrestrial planets, you will find a pattern of subtle differences in density. Merely listing the observed densities of the terrestrial planets does not reveal the pattern because Earth and Venus, being more massive, have stronger gravity and have squeezed their interiors to higher densities. You must look at the **uncompressed densities**—the densities the planets would have if their gravity did not compress them. These densities (■ Table 19-2) show that, in general, the closer a planet is to the sun, the higher its uncompressed density.

This density variation originated when the solar system first formed solid grains. The kind of matter that condensed in a particular region would depend on the temperature of the gas there. In the inner regions, the temperature may have been 1500 K or so. The only materials that can form grains at this temperature are

■ **Table 19-2 | Observed and Uncompressed Densities**

Planet	Observed Density (g/cm³)	Uncompressed Density (g/cm³)
Mercury	5.44	5.30
Venus	5.24	3.96
Earth	5.50	4.07
Mars	3.94	3.73
(Moon)	3.36	3.40

compounds with high melting points, such as metal oxides and pure metals, which are very dense. Farther out in the nebula it was cooler, and silicates (rocky material) could condense. These are less dense than metal oxides and metals. In the cold outer regions, ices of water, methane, and ammonia could condense. These are low-density materials.

The sequence in which the different materials condense from the gas as you move away from the sun is called the **condensation sequence** (■ Table 19-3). It suggests that the planets, forming at different distances from the sun, accumulated from different kinds of materials. The inner planets formed from high-density metal oxides and metals, and the outer planets formed from low-density ices.

You must also remember that the solar nebula did not remain the same temperature throughout the formation of the planets but may have grown progressively cooler. That means a particular region of the nebula may have begun by producing solid particles of metals and metal oxides but, after cooling, began producing particles of silicates. Allowing for the cooling of the nebula makes the theory much more complex, but it also makes the processes by which the planets formed much more understandable.

The Formation of Planetesimals

In the development of a planet, three groups of processes operate. First, grains of solid matter grow larger, eventually reaching diameters ranging from a few centimeters to kilometers. The larger of these objects, called **planetesimals,** are believed to be the bodies that the second group of processes collects into planets. Finally, a third set of processes clears away the solar nebula. The study of planet building is the study of these three groups of processes.

According to the solar nebula theory, planetary development in the solar nebula began with the growth of dust grains. These specks of matter, whatever their composition, grew from microscopic size by two processes, condensation and accretion.

A particle grows by **condensation** when it adds matter one atom at a time from a surrounding gas. Snowflakes, for example, grow by condensation in Earth's atmosphere. In the solar nebula, dust grains were continuously bombarded by atoms of gas, and

some of these stuck to the grains. A microscopic grain capturing a single gas atom increases its mass by a much larger fraction than a gigantic boulder capturing a single atom. That is why condensation can increase the mass of a small grain rapidly; but, as the grain grows larger, condensation becomes less effective.

The second process is **accretion,** the sticking together of solid particles. You may have seen accretion in action if you have walked through a snowstorm with big, fluffy flakes. If you caught one of those "flakes" on your mitten and looked closely, you saw that it was actually made up of many tiny, individual flakes that had collided as they fell and accreted to form larger particles. In the solar nebula the dust grains were, on the average, no more than a few centimeters apart, so they collided frequently. Their mutual gravitation was too small to hold them to each other, but other effects may have helped. Static electricity generated by their passage through the gas could have held them together, as could compounds of carbon that might have formed a sticky surface on the grains. Ice grains might have stuck together better than some other types. Of course, some collisions might have broken up clumps of grains; on the whole, however, accretion must have increased grain size. If it had not, the planets would not have formed.

There is no clear distinction between a very large grain and a very small planetesimal, but you can consider an object a planetesimal when its diameter becomes a kilometer or so (■ Figure 19-7). Objects larger than a centimeter were subject to new processes that tended to concentrate them. One important effect was that the growing planetesimals collapsed into the plane of the solar nebula. Dust grains could not fall into the plane because the turbulent motions of the gas kept them stirred up, but the larger objects had more mass, and the gas motions could not have prevented them from settling into the plane of the spinning nebula. This would have concentrated the solid particles into a thin plane about 0.01 AU thick and would have made further planetary growth more rapid.

■ **Table 19-3 | The Condensation Sequence**

Temperature (K)	Condensate	Planet (Estimated Temperature of Formation; K)
1500	Metal oxides	Mercury (1400)
1300	Metallic iron and nickel	
1200	Silicates	
1000	Feldspars	Venus (900)
680	Troilite (FeS)	Earth (600) Mars (450)
175	H₂0 ice	Jovian (175)
150	Ammonia–water ice	
120	Methane–water ice	
65	Argon–neon ice	Pluto (65)

Visual-wavelength image

■ **Figure 19-7**

What did the planetesimals look like? You can get a clue from this photo of the 5-km-wide nucleus of Comet Wild2. Whether rocky or icy, the planetesimals must have been small, irregular bodies, pocked by craters from collisions with other planetesimals. (NASA)

This collapse of the planetesimals into the plane is analogous to the flattening of a forming galaxy. However, an entirely new process may have become important once the plane of planetesimals formed. Computer models show that the rotating disk of particles should have been gravitationally unstable and would have broken up into small clouds (■ Figure 19-8). This would fur-

■ **Figure 19-8**

Gravitational instabilities in the rotating disk of planetesimals may have forced them to collect in clumps, accelerating their growth.

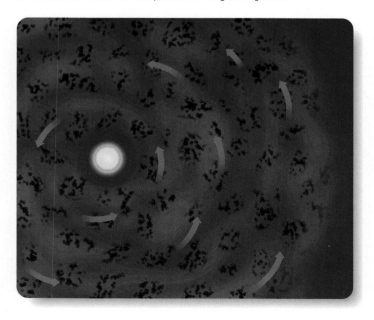

ther concentrate the planetesimals and help them coalesce into objects up to 100 km (60 mi) in diameter.

Through these processes, the theory proposes, the nebula became filled with trillions of solid particles ranging in size from pebbles to tiny planets. As the largest began to exceed 100 km in diameter, new processes began to alter them, and a new stage in planet building began, the growth of protoplanets.

The Growth of Protoplanets

The coalescing of planetesimals eventually formed **protoplanets,** massive objects destined to become planets. As these larger bodies grew, new processes began making them grow faster and altered their physical structure.

If planetesimals collided at orbital velocities, it is unlikely that they would have stuck together. The average orbital velocity in the solar system is about 10 km/s (22,000 mph). Head-on collisions at this velocity would have vaporized the material. However, the planetesimals were moving in the same direction in the nebular plane and didn't collide head on. Instead, they merely rubbed shoulders at low relative velocities. Such gentle collisions would have been more likely to fuse them than to shatter them.

In addition, some adhesive effects probably helped. Sticky coatings and electrostatic charges on the surfaces of the smaller planetesimals probably aided formation of larger bodies. Collisions would have fragmented some of the surface rock; but, if the planetesimals were large enough, their gravity would have held on to some fragments, forming a layer of soil composed entirely of crushed rock. Such a layer on the larger planetesimals may have been effective in trapping smaller bodies.

The largest planetesimals would grow the fastest because they had the strongest gravitational field. Not only could they hold on to a cushioning layer to trap fragments, but their stronger gravity could also attract additional material. These planetesimals probably grew quickly to protoplanetary dimensions, sweeping up more and more material. When massive enough, they trapped some of the original nebular gas to form primitive atmospheres. At some point, they crossed the boundary between planetesimals and protoplanets.

To trace the growth of a protoplanet, you can think of the formation of Earth. In its simplest form, the theory of protoplanet growth supposes that all the planetesimals had about the same chemical composition. The planetesimals accumulated gradually to form a planet-size ball of material that was of homogeneous composition throughout. Once the planet formed, heat began to accumulate in its interior from the decay of short-lived radioactive elements, and this heat eventually melted the planet and allowed it to differentiate. **Differentiation** is the separation of material according to density. When the planet melted, the heavy metals such as iron and nickel settled to the core, while the lighter silicates floated to the surface to form a low-density crust. The story of planet formation from planetesimals of similar composition is shown in the left half of ■ Figure 19-9.

This process depends on the presence of short-lived radioactive elements in the solar nebula. Astronomers know such elements were present because very old rock found in meteorites contains daughter isotopes such as magnesium-26. That isotope is produced by the decay of aluminum-26 in a reaction that has a half-life of only 0.74 million years. The aluminum-26 and similar short-lived radioactive isotopes are gone now, but they must have been produced in a supernova explosion that occurred shortly before the formation of the solar nebula. In fact, many astronomers suspect that the supernova explosion compressed nearby gas and triggered the formation of stars, one of which became the sun. Thus our solar system may exist because of a supernova explosion that occurred about 4.6 billion years ago.

If planets formed in this gradual way and were later melted by radioactive decay, then Earth's present atmosphere was not its first. The first atmosphere consisted of small amounts of gases trapped from the solar nebula—mostly hydrogen and helium. Those gases were later driven off by the heat, aided perhaps by outbursts from the infant sun, and new gases released from the planet's interior formed a secondary atmosphere. This creation of a planetary atmosphere from a planet's interior is called **outgassing.**

This simple theory of planet formation can be improved in two ways. First, it seems likely that the solar nebula cooled during the formation of the planets, so they did not accumulate from planetesimals of common composition. As planet building began, the first particles to condense in the inner solar system were metals and metal oxides, so the protoplanets may have begun by accreting metallic cores. Later, as the nebula cooled, more silicates could form, and the protoplanets added silicate mantles. A second improvement in the theory proposes that the planets grew so rapidly that the heat released by the violent impacts of infalling particles, the **heat of formation,** did not have time to escape. This heat rapidly accumulated and melted the protoplanets as they formed. If this is true, then the planets would have differentiated as they formed. This improved story of planet formation is shown in the right half of Figure 19-9.

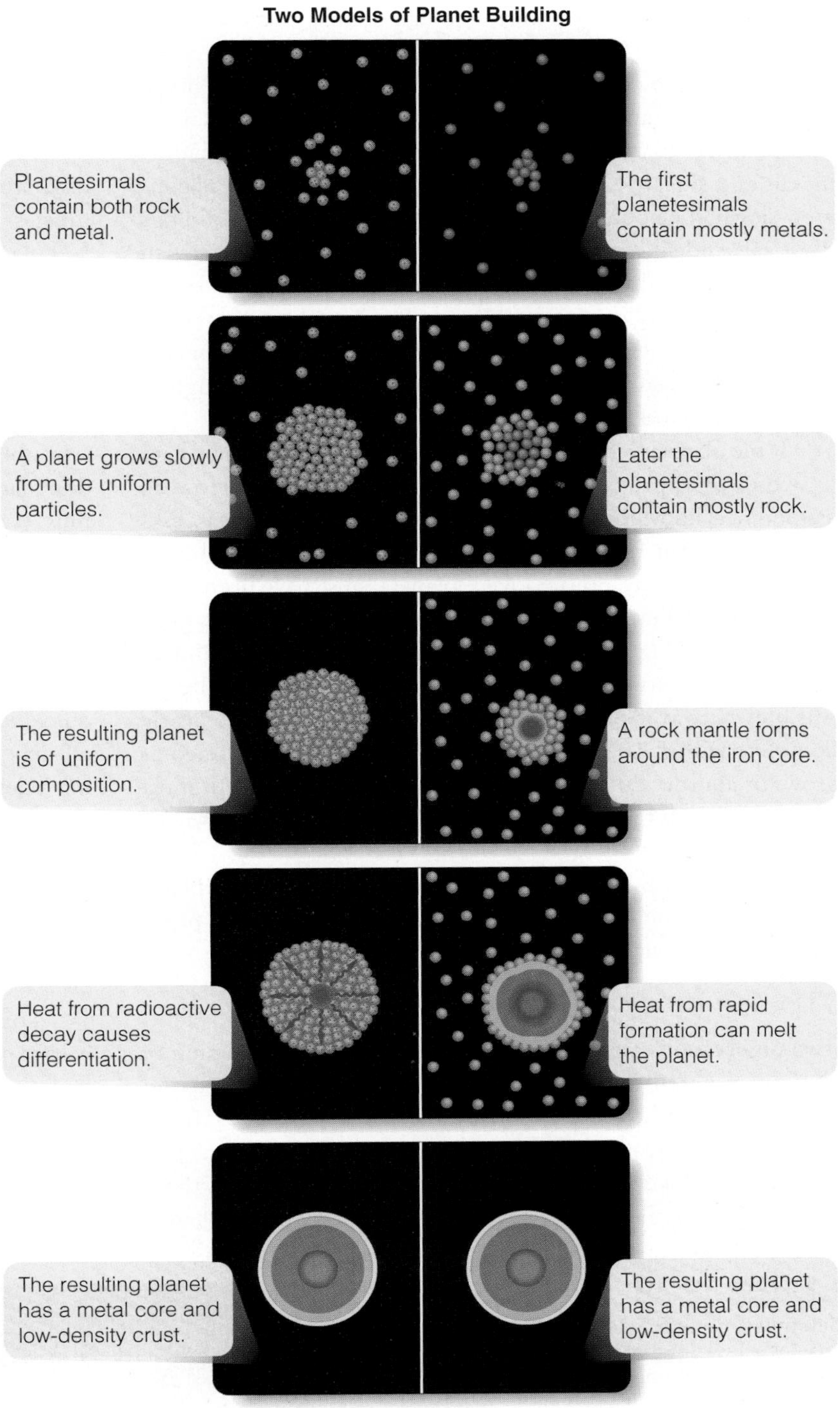

This improved theory suggests that Earth's first atmosphere was not captured from the solar nebula. If Earth formed in a molten state, then there was never a time when it had a primitive atmosphere of hydrogen and helium accumulated from the solar nebula. Rather, the gases of the atmosphere were released by the molten rock as the protoplanet grew. Those gases would not have included much water, however, so some astronomers now think that Earth's water

Two Models of Planet Building

Planetesimals contain both rock and metal.

The first planetesimals contain mostly metals.

A planet grows slowly from the uniform particles.

Later the planetesimals contain mostly rock.

The resulting planet is of uniform composition.

A rock mantle forms around the iron core.

Heat from radioactive decay causes differentiation.

Heat from rapid formation can melt the planet.

The resulting planet has a metal core and low-density crust.

The resulting planet has a metal core and low-density crust.

■ **Figure 19-9**

If the temperature of the solar nebula changed during planet building, the composition of the planetesimals may have changed. The simple model at left assumes no change occurred, but the model at the right incorporates a change from metallic to rocky planetesimals.

and much of its present atmosphere accumulated late in the formation of the planets as Earth swept up volatile-rich planetesimals forming in the cooling solar nebula. Such icy planetesimals may have formed in the outer parts of the solar nebula and been scattered by encounters with the Jovian planets in a bombardment of comets.

According to the solar nebula theory, the Jovian planets began growing by the same processes that built the terrestrial planets. But the Jovian planets grew faster. In the inner solar nebula, only metals and silicates could form solids, so the terrestrial planets grew slowly. The outer solar nebula not only contained solid bits of metals and silicates, but it was a blizzard of ices. The Jovian planets grew rapidly and quickly became massive enough to grow by gravitational collapse as they drew in large amounts of gas from the solar nebula. The terrestrial planets could not grow from ices, so they developed slowly and never became massive enough to grow by gravitational collapse.

If the planets formed in this way, the Jovian planets must have grown to their present size in about 10 million years. How do astronomers know this? Almost all T Tauri stars that have gas disks are younger than 10 million years; older T Tauri stars have blown their gas disks away. So the Jovian planets must have been complete within 10 million years or so. The terrestrial planets grew from solids and not from the gas, so they could have continued to grow by accretion from solid debris left behind when the gas was blown away. Apparently, the terrestrial planets were nearly complete within 10 million years but could have continued to grow for another 20 million years or so. All planet growth must have ended with the last of the solar nebula being blown away by 100 million years.

The solar nebula theory has been very successful in explaining the formation of the solar system. But there are some problems, and the Jovian planets are the troublemakers.

The Jovian Problem

Two observations are making it hard to explain the formation of the Jovian parents, and that is forcing astronomers to expand and revise their theories of planet formation (■ Figure 19-10).

In the final section of this chapter, you will see evidence that astronomers have found planets orbiting other stars, and those planets all have the mass of Jovian planets. There may be many terrestrial planets orbiting these stars that are too low in mass for astronomers to detect at present, but the important point is that there are lots of Jovian planets around. That is the first of the troublesome observations.

The second observation is the discovery that the gas and dust disks around newborn stars don't last long. In Chapter 11 you saw images of dusty gas disks around the young stars in the Orion Nebula (page 253). Those disks are being evaporated by the intense ultraviolet radiation from hot stars with the nebula. It seems that nearly all stars form in clusters containing massive stars, so this must happen to most disks. Even if a disk did not evaporate quickly, the gravitational influence of the crowded stars in a cluster should quickly strip away the outer parts of the disk. This is a troublesome observation because it seems to mean that disks don't last longer than about 7 million years at most, and many must evaporate within 100,000 years or so. That's not long enough to grow a Jovian planet. Yet Jovian worlds are common.

Mathematical models of the solar nebula have been computed using specially built computers running programs that take weeks to finish a calculation. The results show that the rotating gas and dust of the solar nebula may have become unstable and formed outer planets by direct gravitational collapse. That is, massive planets may have been able to form from the gas without first forming a dense core by accretion. Jupiters and Saturns form in these calculated models with a few hundred years.

If the Jovian planets formed in this way, they could have formed quickly before the solar nebula disappeared. Terrestrial planets grow from solid bits of material in the inner nebula, so they could have continued to grow by accretion even after the disk was cleared of gas and smaller particles of dust.

This new insight into the formation of the outer planets may help explain the formation of Uranus and Neptune. They are so far from the sun that accretion could not have built them rapidly. The gas and dust of the solar nebula must have been sparse out there, and Uranus and Neptune orbit so slowly they would not have swept up material very rapidly. The conventional view is that they grew by accretion so slowly that they never became quite mas-

■ **Figure 19-10**

Saturn is a beautiful planet, but the Jovian worlds are a problem for modern astronomers. The evidence suggests that accretion followed by gravitational collapse is too slow to build Jovian planets before the solar nebula was blown away. This has forced astronomers to revise their theories of how Jovian planets form. (JPL/NASA)

sive enough to begin growing by gravitational collapse. In fact, it is hard to understand how they could have reached their present mass if they started growing by accretion so far from the sun. Theoretical calculations show that they might have formed closer to the sun in the region of Jupiter and Saturn and then could have been shifted outward by gravitational interactions. In any case, the formation of Uranus and Neptune is part of the Jovian problem.

The traditional solar nebula theory proposes that the planets formed by accreting a core and then, if they became massive enough, by gravitational collapse. The new theories suggest that the outer planets may have skipped the core accretion phase.

Whatever the details of the story are, there must have been a solar nebula from which the planets formed. That is enough to allow you to explain the distinguishing characteristics of the solar system.

Explaining the Characteristics of the Solar System

Table 19-1 contains the list of distinguishing characteristics of the solar system. Any theory of the origin of the solar system should explain these characteristics.

The disk shape of the solar system is inherited from the solar nebula. The sun and planets revolve and rotate in the same direction because they formed from the same rotating gas cloud. The orbits of the planets lie in the same plane because the rotating solar nebula collapsed into a disk, and the planets formed in that disk.

The solar nebula hypothesis is evolutionary in that it calls on continuing processes to gradually build the planets. To explain the rotation of Venus, Uranus, and Pluto, however, you may need to consider catastrophic events. Uranus rotates on its side. This might have been caused by an off-center collision with a massive planetesimal when the planet was nearly formed. In Chapter 24 you will see evidence that Pluto's highly inclined rotation may be the result of a similar collision. Two theories have been proposed to explain the backward rotation of Venus. Theoretical models suggest that the sun can produce tides in the thick atmosphere of Venus and eventually reverse the planet's rotation—an evolutionary theory. It is also possible that the rotation of Venus was altered by an off-center impact late in the planet's formation, and that is a catastrophic theory.

The second item in Table 19-1, the division of the planets into terrestrial and Jovian worlds, can be understood through the condensation sequence. The terrestrial planets formed in the inner part of the solar nebula, where the temperature was high and most compounds remained gaseous. Only compounds such as the metals and silicates could condense to form solid particles. Terrestrial planets must have formed by the accumulation of solid particles because they never had enough gravitation to capture the hot gas of the inner solar nebula. That means the planets that began growing in the inner solar system had to form mainly from metals and silicates. Consequently, they became the small, dense terrestrial planets.

In contrast, the Jovian planets formed in the outer solar nebula, where the lower temperature allowed the gas to form large amounts of ices, perhaps three times more ices than silicates. That allowed the Jovian planets to grow rapidly and became massive.

The heat of formation (the energy released by infalling matter) was tremendous for these massive planets. Jupiter must have grown hot enough to glow with a luminosity of about 1 percent that of the present sun. However, because it never got hot enough to generate nuclear energy as a star would, it never generated its own energy. Jupiter is still hot inside. In fact, both Jupiter and Saturn radiate more heat than they absorb from the sun, so they are evidently still cooling.

This is an important point, so you'll want to summarize. The Jovian planets are low-density worlds because they grew in the outer solar nebula where there were large amounts of low-density ices in addition to bits of metal and silicates. Also, Jupiter and Saturn are so massive they have been able to grow by drawing in the cool gas directly from the solar nebula. The terrestrial planets could not do this because they never became massive enough and because the gas in the inner nebula was hotter and more difficult to capture.

A glance at the solar system suggests that you should expect to find a planet between Mars and Jupiter at the present location of the asteroid belt. Mathematical models show that there are asteroids there, and not a planet, because Jupiter grew into such a massive planet it was able to gravitationally disturb the motion of nearby planetesimals. The bodies that should have formed a planet just inward in the solar system from Jupiter were broken up, thrown into the sun, or ejected from the solar system. The asteroids seen today are the last remains of those rocky planetesimals.

The comets, in contrast, are evidently the last of the icy planetesimals. Some may have formed in the outer solar nebula beyond Neptune and Pluto, but many probably formed among the Jovian planets where ices could condense easily. Mathematical models show that the massive Jovian planets could have ejected some of these icy planetesimals into the far outer solar system. In a later chapter, you will see evidence that a comet appears in the sky when one of these icy bodies falls into the inner solar system.

The icy Kuiper belt objects appear to be ancient planetesimals that formed in the outer solar system but were never incorporated into a planet. They orbit slowly far from the light and warmth of the sun and, except for occasional collisions, have not changed much since the solar system was young.

The large satellite systems of the Jovian worlds may contain two kinds of moons. Some moons may have formed in orbit around the forming planet in a miniature of the solar nebula. But some of the smaller moons may be captured planetesimals and asteroids. The large masses of the Jovian planets would have made it easier for them to capture satellites.

In Table 19-1, you noted that all four Jovian worlds have ring systems, and you can understand this by considering the large mass of these worlds and their remote location in the solar system. A large mass makes it easier for a planet to hold onto or-biting ring particles; and, being farther from the sun, the ring particles are not as easily swept away by the pressure of sunlight and the solar wind. It is hardly surprising, then, that the terres-trial planets, low-mass worlds located near the sun, have no plan-etary rings.

The last entry in Table 19-1 is the common ages of solar system bodies, and the solar nebula hypothesis has no difficulty explaining that characteristic. If the hypothesis is correct, then the planets formed at the same time as the sun and should have roughly the same age. Although scientists have mineral samples for radioactive dating only from Earth, the moon, Mars, and me-teorites, so far the ages seem to agree.

The solar nebula hypothesis can account for the distinguish-ing characteristics of the solar system, but there is yet another test you should apply to the hypothesis. What about the prob-lem that troubled Laplace and his nebular hypothesis? What about the angular momentum problem? If the sun and planets formed from a contracting nebula, the sun should have been left spinning very rapidly. That is, it should have most of the angular momentum in the solar system, and instead it has very little. The solar nebula hypothesis can explain the angular momentum prob-lem if you think about the relationship between the origin of planetary systems and star formation. What you know about star formation tells you that the sun must have had a strong stellar wind when it was young. Also, a rapidly spinning star is likely to generate a strong magnetic field. Like a great paddle wheel, the sun's magnetic field must have stuck out into space, and gas flow-ing away from the sun would have been dragged around with the rotating magnetic field. This would have slowed the sun and left it rotating slowly as it is today. Astronomers now believe that the angular momentum problem is no longer an objection to the solar nebula hypothesis.

Your general understanding of the origin of the solar system gives you a new way of thinking about asteroids, meteors, and comets. They are the last of the debris left behind by the solar nebula. These objects are such important sources of information about the history of our solar system that they will be discussed in detail in Chapter 25. But for now you are probably wondering: What happened to the solar nebula?

Clearing the Nebula

The sun probably formed along with many other stars in a swirl-ing nebula. Observations of young stars in Orion suggest that radiation from nearby hot stars would have evaporated the disk of gas and dust around the sun and that the gravitational influ-ence of nearby stars would have pulled gas away. Even without the external effects, four internal processes would have gradually destroyed the solar nebula.

The most important of these internal processes was **radiation pressure.** When the sun became a luminous object, light stream-ing from its surface pushed against the particles of the solar neb-ula. Large bits of matter like planetesimals and planets were not affected, but low-mass specks of dust and individual gas atoms were pushed outward and eventually driven from the system.

The second effect that helped clear the nebula was the solar wind, the flow of ionized hydrogen and other atoms away from the sun's upper atmosphere. This flow is a steady breeze that rushes past Earth at about 400 km/s (250 mi/s). When the sun was young, it may have had an even stronger solar wind, and irregular fluc-tuations in its luminosity, like those observed in young stars such as T Tauri stars, may have produced surges in the wind that helped push dust and gas out of the nebula.

The third effect for clearing the nebula was the sweeping up of space debris by the planets. All of the old, solid surfaces in the solar system are heavily cratered by meteorite impacts (■ Fig-ure 19-11). Earth's moon, Mercury, Venus, Mars, and most of the moons in the solar system are covered with craters. A few of these craters have been formed recently by the steady rain of meteorites that falls on all the planets in the solar system, but most of the craters appear to have been formed roughly 4 billion years ago in what is called the **heavy bombardment,** as the last of the debris in the solar nebula was swept up by the planets. You will find more evidence of this bombardment in later chapters. The craters show that many of the last remaining solid objects in the solar nebula were swept up by the planets soon after they formed.

The fourth effect was the ejection of material from the solar system by close encounters with planets. If a small object such as a planetesimal passes close to a planet, it can gain energy from the planet's gravitational field and be thrown out of the solar sys-tem. Ejection is most probable in encounters with massive plan-ets, so the Jovian planets were probably very efficient at ejecting the icy planetesimals that formed in their region of the nebula.

Attacked by the radiation and gravity of nearby stars and racked by internal processes, the solar nebula could not survive very long. Once the gas and dust were gone and the last of the planetesimals were swept up, the planets could no longer gain mass, and planet building ended.

Building Scientific Arguments

Why are there two kinds of planets in our solar system?
This is an opportunity for you to build an argument that closely analyzes the solar nebula theory. Planets begin forming from solid bits of matter, not from gas. Consequently, the kind of planet that forms at a given distance from the sun depends on the kind of compounds that can condense out of the gas to form solid particles. In the inner parts of the solar nebula, the temperature was so high that most of the gas could not condense to form solids. Only metals and silicates could form solid grains, and the innermost planets grew from this dense material. Much of the mass of the solar nebula consisted of hydrogen, helium, water vapor, and other gases, and they were present in the inner solar nebula but couldn't form solid grains.

Visual-wavelength images

■ Figure 19-11

Every old, solid surface in the solar system is scarred by craters. (a) Earth's moon is scarred by craters ranging from basins hundreds of kilometers in diameter down to microscopic pits. (b) The surface of Mercury, as photographed by a passing spacecraft, shows vast numbers of overlapping craters. (NASA)

The small terrestrial planets couldn't grow from these gases, so the terrestrial planets are small and dense.

In the outer solar nebula, the composition of the gas was the same, but it was cold enough for water vapor to condense to form ice grains. Because hydrogen and oxygen are so abundant, there was lots of ice available. The outer planets grew from solid bits of metal and silicate combined with large amounts of ice. The outer planets grew so rapidly that they became massive enough to capture gas directly, and they became the hydrogen- and helium-rich Jovian worlds.

The condensation sequence combined with the solar nebula hypothesis gives you a way to understand the difference between the terrestrial and Jovian planets. Now expand your argument: **Why do some astronomers argue that the formation of the Jovian planets is a problem that needs further explanation?**

■   ■   ■

Connections: The great chain of origins leads from the first instant of the big bang through the birth of galaxies, the formation of stars, and the origin of the chemical elements to the formation of our solar system, and finally to us. You have traced the theory and compared it to evidence from our own solar system, but you must be wondering by now: Does this happen around other stars?

19-4 Planets Orbiting Other Stars

ARE THERE OTHER EARTHS? You have read earlier in this chapter that astronomers have, indeed, detected planets orbiting other stars, so now it is time to get down to the details. Are those planets like Earth? How did they form? Can astronomers see planets forming around other stars?

Your questions can be boiled down to two main unknowns. Can astronomers see planet-forming disks around other stars? And, can astronomers see actual planets orbiting other stars? The answer in both cases is an amazing yes.

Planet-Forming Disks around Other Suns

Both visible- and radio-wavelength observations detect dense disks of gas orbiting young stars. For example, at least 50 percent of the stars in the Orion nebula are encircled by dense disks of gas and dust (■ Figure 19-12 and page 253). A young star is visible at the center of each disk, and astronomers estimate that the disks contain at least a few times Earth's mass in a region a few times larger in diameter than our solar system. The Orion star-forming region is only a few million years old, so it does not seem likely that planets could have formed in these disks yet.

Many of the young stars in the Orion Nebula are surrounded by disks of gas and dust, but intense light from the brightest star in the neighborhood is evaporating the disks to form expanding clouds of gas. These disks may evaporate before they can form planets, but the large number of such disks shows that disks around young stars are common. (C. R. O'Dell, Rice, NASA; Dark disk: M. McCaughrean, Max Plank Inst. for Astronomy, C. R. O'Dell, NASA; Lower left inset: J. Bally, H. Throop, C. R. O'Dell, NASA)

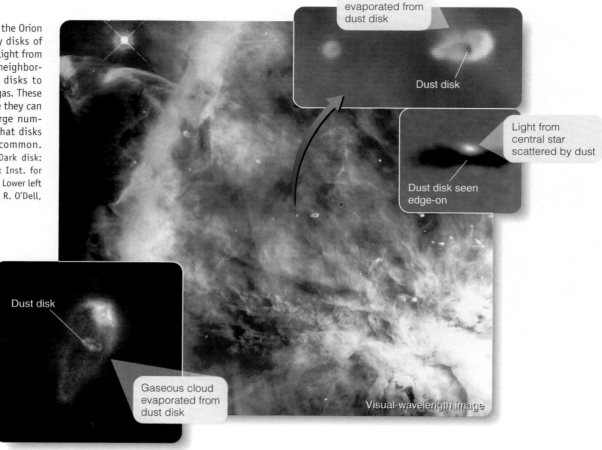

Gaseous cloud evaporated from dust disk

Dust disk

Light from central star scattered by dust

Dust disk seen edge-on

Dust disk

Gaseous cloud evaporated from dust disk

Visual-wavelength image

Furthermore, the intense radiation from the hot stars in the area is evaporating the disks so fast that planets may never have a chance to grow large. The important point for astronomers is that so many of these young stars have disks. Evidently, disks of gas and dust are a common feature of star formation. These disks appear to resemble the solar nebula from which our planetary system formed.

The Hubble Space Telescope can detect dense disks of gas and dust around young stars in a slightly different way. The disks show up by the shadows they cast in the nebulae that surround the newborn stars (■ Figure 19-13). These disks are related to the formation of bipolar flows (Figure 11-7) in that they focus the gas flowing away from a young star into two jets shooting in opposite directions. It seems likely that the sun and the solar nebula formed a bipolar flow when they were young.

Infrared astronomers have found very cold, low-density dust disks around stars such as Beta Pictoris. These stars are believed to have completed their formation, so they are clearly older than the newborn stars in Orion. The dust disk around Beta Pictoris is about 20 times the diameter of our solar system and, like the other known low-density disks, has an inner zone with even lower density. These inner regions may be places where planets have formed (■ Figure 19-14). Such tenuous dust disks are sometimes called **debris disks** because they are understood to be debris released in collisions among small bodies such as comets, asteroids,

and Kuiper belt objects. Our own solar system contains such dust, and astronomers believe it has an extensive debris disk of cold dust extending far beyond the orbits of the planets.

Any planets orbiting stars with debris disks are not detectable, but the presence of the cold dust disks assures you that small bodies like asteroids and comets are present. If those small objects are there, then you can expect that there are also planets orbiting those stars.

Remember Vega, one of our Favorite Stars? It is a very bright star high in the summer sky. Infrared observations reveal that it is surrounded by a disk of dust, and detailed studies show that some of the dust is tiny. The pressure of starlight from Vega would blow that dust way quickly, so astronomers conclude that the dust must have been produced by the collision of two large planetesimals within the last million years. Fragments from that collision are still smashing into each other now and then and producing more dust. This effect has also been found in the disk around Beta Pictoris and the faint star known as HIP8920. Such smashups happen rarely in a dust disk, but they happen often enough to keep the disk supplied with dust.

Notice the difference between the two kinds of disks that astronomers have found. The low-density dust disks such as the one around Beta Pictoris are produced by dust from collisions among comets, asteroids, and Kuiper belt objects. Such disks are evidence that planetary systems have already formed. The dense

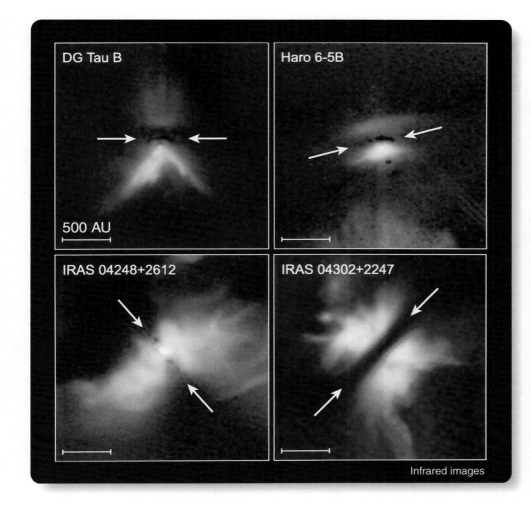

■ **Figure 19-13**

Disks around young stars are evident in these Hubble Space Telescope infrared images. The stars are so young that material is still falling inward and being illuminated by the light from the star. Dark bands across the nebulae (arrows) are caused by dense disks of gas and dust that orbit the stars. (D. Padgett, IPAC/Caltech, W. Brandner, IPAC, K. Stapelfeldt, JPL, and NASA)

The first planet detected this way was discovered in 1995. It orbits the star 51 Pegasi. As the planet circles the star, the star wobbles slightly, and this very small motion of the star is detectable as Doppler shifts in the star's spectrum (■ Figure 19-15a). From the motion of the star and estimates of the star's mass, astronomers can deduce that the planet has half the mass of Jupiter and orbits only 0.05 AU from the star. Half the mass of Jupiter amounts to 160 Earth masses, so this is a large planet. Note also that it orbits very close to its star.

Astronomers were not surprised by the announcement that a planet orbited 51 Pegasi; for years astronomers had assumed that many stars had planets. Nevertheless, they greeted the discovery with typical skepticism (**Window on Science 19-3**). That skepticism led to careful tests of the data and further observations that confirmed the discovery. In fact, over 150 planets have been discovered in this way, including at least three planets orbiting the star Upsilon Andromedae—a true planetary system (Figure 19-15b). Over a dozen such planetary systems have been found.

disks of gas and dust such as those seen round the stars in Orion are sites where planets could be forming right now.

The observational evidence gives astronomers confidence that planets orbit many stars. Of course, you are wondering if there isn't some way to see these planets directly. That's not easy, but astronomers are making progress.

Planets Orbiting Other Stars

A planet orbiting another star is called an **extrasolar planet.** Such a planet would be quite faint and difficult to detect so close to the glare of its star. But there are ways to find these planets. All you have to do is imagine walking a dog.

You will remember that Earth and its moon orbit around their common center of mass, and two stars in a binary system orbit around their center of mass. When a planet orbits a star, the star moves very slightly as it orbits the center of mass of the planet–star system. Think of someone walking a poorly trained dog on a leash; the dog runs around pulling on the leash, and even if it were an invisible dog, you could plot its path by watching how its owner is jerked back and forth. Astronomers can detect a planet orbiting another star by watching how the star moves as the planet tugs on it.

Another way to search for planets is to look for changes in the brightness of the star when the orbiting planet crosses in front of or behind the star. The decrease in light is very small, but it is detectable, and astronomers have used this technique to detect a few planets as they crossed in front of their stars. All of these planets are roughly the size of Jupiter and even more massive. The Spitzer Infrared Space Telescope has detected two planets when they passed behind their stars. The systems are young, and the planets are hot and emit significant infrared radiation. When they passed behind the stars they orbit, the total infrared brightness of the system decreased. These observations further confirm the existence of extrasolar planets.

Notice how the techniques used to detect these planets resemble techniques used to study binary stars. Most of the planets were discovered using the same observational methods used to study spectroscopic binaries, but a few were found by observing the stars as if they were eclipsing binaries (Chapter 9).

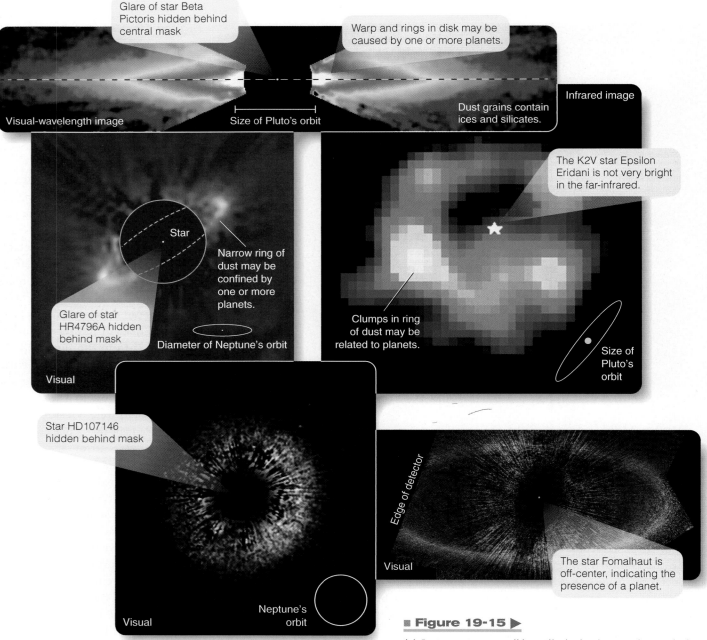

Glare of star Beta Pictoris hidden behind central mask

Warp and rings in disk may be caused by one or more planets.

Visual-wavelength image

Size of Pluto's orbit

Infrared image

Dust grains contain ices and silicates.

Star

Narrow ring of dust may be confined by one or more planets.

Glare of star HR4796A hidden behind mask

Diameter of Neptune's orbit

Visual

The K2V star Epsilon Eridani is not very bright in the far-infrared.

Clumps in ring of dust may be related to planets.

Size of Pluto's orbit

Star HD107146 hidden behind mask

Neptune's orbit

Visual

Edge of detector

Visual

The star Fomalhaut is off-center, indicating the presence of a planet.

■ Figure 19-15 ▶

(a) Just as someone walking a lively dog is tugged around, the star 51 Pegasi is tugged around by the planet that orbits it every 4.2 days. The wobble is detectable in precision observations of its Doppler shift. (b) Someone walking three dogs is pulled about in a more complicated pattern, and you can see something similar in the Doppler shifts of Upsilon Andromedae, which is orbited by three planets.

■ Active Figure 19-14

Dust disks have been detected orbiting a number of stars; but, in the visible part of the spectrum, the dust is at least 1000 times fainter than the stars, which must be hidden behind masks. In the far-infrared, the stars are not as bright as the dust. Warps, rings, clumps, and off-center rings in these disks suggest the gravitational influence of planets. (Beta Pic: NASA, Burrows and Krist; HR4769A: NASA, Weinberger, Becklin, G. Schneider; Eps Eri: Joint Astronomy Center; HD107146: NASA; Fomalhaut: NASA, ESA, Kalas, Graham, and Clampin)

Ace Astronomy™ Log into AceAstronomy and select this chapter to see the Active Figure "Circumstellar Disk" and observe a model disk at different wavelengths.

The planets discovered so far tend to be massive and have short periods because lower-mass planets or longer-period planets are harder to detect. Low-mass planets don't tug on their stars very

much, and present-day spectrographs can't detect the very small velocity changes that these gentle tugs produce. Planets with longer periods are harder to detect because Earth's astronomers have not been making high-precision observations for many years. Jupiter takes 11 years to circle the sun once, so it will take years for astronomers to see the longer-period wobbles produced by planets lying farther from their stars. So you should not be surprised that the first planets discovered are massive and have short periods.

Scientists: Courteous Skeptics

Scientists are just a bunch of skeptics who don't believe in anything." That is a common complaint about scientists, but it misinterprets the fundamental characteristic of the scientist. Yes, scientists are skeptical about new ideas and discoveries, but they also hold strong beliefs about how nature works. Ultimately, scientists try to tell stories about how nature works, and skepticism is the scientists' guide.

Some people think that "telling a story" is telling a fib. The stories that scientists tell are exactly the opposite; perhaps you could call them antifibs, because they are as true as scientists can make them. The stories are explanations that have been tested over and over to make them accurate descriptions of nature. Skepticism is the tool scientists use to test every aspect of an explanation.

When a planet was discovered orbiting 51 Pegasi, astronomers were skeptical—not because they thought the observations were wrong but because that is how science works. Every observation is tested, and every discovery is confirmed. Only an idea that survives many tests begins to be accepted as a scientific truth.

To nonscientists, this makes scientists seem like irritable skeptics, but among scientists it is not bad manners to say, "Really, how do you know that?" or "Why do you think that?" or "Show me the evidence!"

Scientists try to tell stories about how nature works, stories that are sometimes called theories. A story by John Steinbeck can be a brilliant work of art invented to help you understand some aspect of your life, and another story, such as a TV script, can be little more

than chewing gum for the mind. But scientists' stories are different in a critical way. They are true. To create such stories, scientists cannot create facts to fit their story. Rather, every link in a scientific story must be based on evidence and logic. Of course, evidence can be misunderstood, and logical errors happen. To test every aspect of their story, scientists must be continuously skeptical. That is the only way to discover those scientific truths that help you understand how nature works.

Skepticism is not a refusal to hold beliefs. Scientists often believe sincerely in their theories, once those theories have been tested over and over. Rather, skepticism is the tool scientists use to find those natural principles worthy of belief.

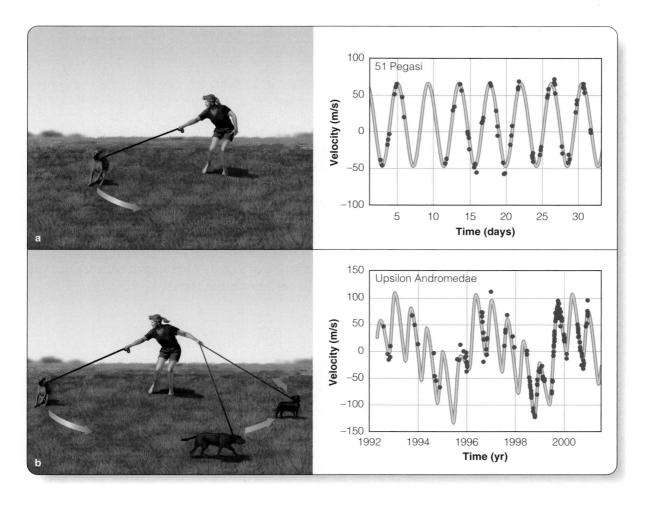

The new planets may seem odd for another reason. In our own solar system, the large planets formed farther from the sun where the solar nebula was colder and ices could condense. How could big planets form so near their stars? Theorists find that planets that form in an especially dense disk of matter could spiral inward as they sweep up planetesimals. That means it is possible for a few planets to become the massive, short-period planets that are detected most easily.

A few of the newly discovered extrasolar planets have elliptical orbits, and that seems odd compared with our solar system in which the planetary orbits are nearly circular. Theorists point out, however, that planets may interact in some young planetary systems and can be thrown into elliptical orbits. This is probably rare among planetary systems, but astronomers find these extreme systems more easily because they tend to produce big wobbles.

The preceding paragraphs should reassure you that massive planets in small orbits or in elliptical orbits are not outrageous. They do not contradict the solar nebula hypothesis.

As astronomers refine their instruments to detect smaller velocity shifts in stars, they are finding lower-mass planets. A super-Earth was found in 2005. It is about 7.5 times the mass of Earth and twice Earth's diameter. It orbits its star in only a bit over two Earth days, so it must be very hot. It does not seem that it could be a Jovian world, so astronomers suspect it is made of rock and metal, as is Earth. Evidently the extrasolar planets are not all Jovian.

Actually photographing a planet orbiting another star is about as easy as photographing a bug crawling on the bulb of a searchlight miles away. Planets are small and dim and get lost in the glare of the stars they orbit. Nevertheless, a few objects have been detected that appear to be planets (■ Figure 19-16). Searches for more are being conducted, and space telescopes are being developed that will eventually be able to image Earth-size planets orbiting nearby stars.

The discovery of extrasolar planets gives astronomers added confidence in the solar nebula theory. The theory predicts that planets are common, and astronomers are finding them orbiting many stars.

Building Scientific Arguments

Why are dust disks evidence that planets have already formed?

Sometimes a good scientific argument combines evidence, theory, and an astronomer's past experience, a kind of scien-

Planet 2M1207b orbits 77AU from its brown dwarf "sun."

Brown dwarf

Infrared image

Artist's conception

■ Figure 19-16

Infrared observations reveal a planet of about 5 Jupiter masses orbiting a brown dwarf in an orbit roughly twice as large as that of Neptune around the sun. Spectra showing water vapor and the planet's infrared colors suggest it is relatively cool. In an artist's impression, the planet orbits round a brown dwarf still surrounded by a dusty disk. (ESO)

tific common sense. Certainly the cold dust disks seen around stars like Vega are not places where planets are forming. They are not hot enough or dense enough to be young disks. Rather, the dust disks must be older, and the dust is being produced by collisions among comets, asteroids, and Kuiper belt objects. Dust would be blown away quickly, so these collisions must be a continuing process. Astronomers know from experience that where you find comets, asteroids, and Kuiper belt objects, you should also find planets, so the dust disks seem to be evidence that planets have already formed in such systems.

Now build a new argument. **What direct evidence can you cite that planets orbit other stars?**

■   ■   ■

Connections: You have watched our solar system form, and you have seen clear evidence that planets form around other stars. Now it is time to voyage out among the planets that orbit the sun and search for more clues as to how they formed and evolved. Your journey begins in the next chapter.

Summary

19-1 | Theories of Earth's Origin

What theories account for the origin of the solar system?

- René Descartes proposed that the solar system formed from a contracting vortex of matter—an evolutionary theory. Buffon later suggested that a passing star pulled matter out of the sun to form the planets—a catastrophic theory.

- Laplace's nebular hypothesis required a contracting nebula to leave behind rings that formed each planet, but it could not explain the low angular momentum of the sun.

- Modern astronomy reveals that the matter in our solar system was formed in the big bang, and the atoms heavier than helium were cooked up in a few generations of stars. The sun and planets evidently formed from a cloud of gas in the interstellar medium.

- The solar nebula hypothesis proposes that the planets formed in a disk of gas and dust around the protostar that became the sun. Observations show that these disks are common.

19-2 | A Survey of the Solar System

What properties must these theories explain?

- The solar system is disk shaped in the orbital revolution of the planets and their moons and in the rotation of the planets on their axes.

- The planets are divided into two types. The inner four planets are terrestrial—small, rocky, dense Earthlike worlds. The next four outward are Jovian planets that are large and low density.

- All four of the Jovian worlds have ring systems and large families of moons. The terrestrial planets have no rings and few moons.

- Most of the asteroids, small, irregular rocky bodies, are located between the orbits of Mars and Jupiter.

- Comets are icy bodies that fall into the inner solar system along long elliptical orbits. As the ices vaporize and release dust, the comet develops a tail that points approximately away from the sun.

- Meteoroids that fall into Earth's atmosphere are vaporized by friction and are visible as meteors. Larger and stronger meteoroids may survive to reach the ground, where they are called meteorites.

- The Kuiper belt is composed of small, icy bodies that orbit the sun beyond the orbit of Neptune.

- The age of an object can be found by radioactive dating. The oldest rocks from Earth, the moon, and Mars have ages approaching 4.6 billion years. The oldest objects in our solar system are the meteorites, which have ages of 4.56 billion years. This is taken to be the age of the solar system.

19-3 | The Story of Planet Building

How do planets form?

- Condensation in the solar nebula converted some of the gas into solid bits of matter, which accreted to form billions of planetesimals.

- Planets begin growing by accreting solid material. But once a planet approaches about 15 Earth masses, it can begin growing by gravitational collapse as it pulls in gas from the solar nebula.

- According to the condensation sequence, the inner part of the solar nebula was so hot that only metals and rocky minerals could form solid grains. The dense terrestrial planets grew from those solid particles and did not include many ices or low-density gases.

- The outer solar nebula was cold enough for metals, rocky minerals, and large amounts of ices to form solid particles. The Jovian planets grew rapidly and incorporated large amounts of low-density ices and gases.

- Evidence that the condensation sequence was important in the solar nebula can be found in the densities of the terrestrial planets compared to the Jovian planets. Even the uncompressed densities among the terrestrial planets show that the inner planets are the most dense.

- The terrestrial planets may have formed slowly from the accretion of planetesimals of similar composition and then differentiated later when radioactive decay heated the planet's interiors.

- It is also possible that the planets formed so rapidly that the heat of formation melted the planets and they differentiated as they formed. In that case, Earth's first atmosphere was not captured from the nebula but was baked out of Earth's interior.

- Disks of gas and dust may not last long enough to form Jovian planets by accretion and gravitational collapse. Some models suggest the Jovian planets could have formed directly and rapidly by gravitational collapse.

- In addition to intense light from hot nearby stars and the gravitational influence of passing stars, the solar nebula was eventually cleared away by radiation pressure, the solar wind, the sweeping up of debris by the planets, and ejection.

- All of the old surfaces in the solar system were heavily cratered by an early bombardment of the debris that filled the solar system when it was young.

19-4 | Planets Orbiting Other Stars

Is our solar system unique?

- Hot disks of gas and dust have been detected in early stages of star formation and are believed to be the kind of disk in which planets could form.

- Cold dust disks, also known as debris disks, appear to be produced by dust released by collisions among comets, asteroids, and Kuiper belt objects. Such disks may be home to planets that have already formed.

- Planets orbiting other stars have been detected by the way they tug their stars about, creating small Doppler shifts in the stars' spectra. Planets have also been detected as they cross in front of their star and dim the star's light. A few planets have been detected when they crossed behind their star and their infrared radiation was cut off.

- Nearly all extrasolar planets found so far are massive, Jovian worlds. Lower-mass, terrestrial planets are harder to detect.

New Terms

passing star hypothesis (p. 450)

evolutionary hypothesis (p. 451)

catastrophic hypothesis (p. 451)

nebular hypothesis (p. 451)

angular momentum problem (p. 451)

solar nebula hypothesis (p. 452)

terrestrial planet (p. 454)

Jovian planet (p. 454)

Galilean satellites (p. 455)

asteroid (p. 456)

Kuiper belt (p. 456)

comet (p. 456)

meteor (p. 457)

meteoroid (p. 457)

meteorite (p. 457)

half-life (p. 457)

gravitational collapse (p. 460)

uncompressed density (p. 460)

condensation sequence (p. 461)

planetesimal (p. 461)

condensation (p. 461)

accretion (p. 461)

protoplanet (p. 462)

differentiation (p. 462)

outgassing (p. 463)

heat of formation (p. 463)

radiation pressure (p. 466)

heavy bombardment (p. 466)

debris disk (p. 468)

extrasolar planet (p. 469)

Review Questions

Ace◐Astronomy™ Assess your understanding of this chapter's topics with additional quizzing and animations at **http://ace .brookscole.com/sf9**

1. What were the objections to the passing star hypothesis? to the nebular hypothesis?

2. What produced the helium now present in the sun's atmosphere? in Jupiter's atmosphere? in the sun's core?

3. What produced the iron in Earth's core and the heavier elements like gold and silver in Earth's crust?

4. What evidence can you cite that disks of gas and dust are common around young stars?

5. According to the solar nebula theory, why is the sun's equator nearly in the plane of Earth's orbit?

6. Why does the solar nebula theory predict that planetary systems are common?

7. Why do astronomers think the solar system formed about 4.6 billion years ago?

8. If you visited another planetary system, would you be surprised to find planets older than Earth? Why or why not?

9. Why is almost every solid surface in our solar system scarred by craters?

10. What is the difference between condensation and accretion?

11. Why don't terrestrial planets have rings like the Jovian planets?

12. How does the solar nebula theory help you understand the location of asteroids?

13. How does the solar nebula theory explain the dramatic density difference between the terrestrial and Jovian planets?

14. If you visited some other planetary system in the act of building planets, would you expect to see the condensation sequence at work, or was it unique to our solar system?

15. Why would you expect to find that planets are differentiated?

16. What processes cleared the nebula away and ended planet building?

17. What evidence can you cite that planets orbit other stars?

18. What is the difference between the hot disks and cold disks seen around stars?

19. What do you see in the image at the right that indi- cates that the planet formed far from the sun?

(NASA)

20. Why do astronomers conclude that the surface of Mercury, shown at right, is old? When did the majority of these craters form?

(NASA)

Discussion Questions

1. In your opinion, should all solar systems have asteroid belts? Should all solar systems show evidence of an age of heavy bombardment?

2. If the solar nebula hypothesis is correct, then there are probably more planets in the universe than stars. Do you agree? Why or why not?

3. The human race has intelligence and consequently has both the right and the responsibility to wonder about its origins. Do you agree?

Problems

1. If you observed the solar system from the nearest star (1.3 pc), what would the maximum angular separation be between Jupiter and the sun? (*Hint:* Use the small-angle formula.)

2. The brightest planet in our sky is Venus, which is sometimes as bright as apparent magnitude −4 when it is at a distance of about 1 AU. How many times fainter would it look from a distance of 1 parsec (206,265 AU)? What would its apparent magnitude be? (*Hints:* Remember the inverse square law, from Chapter 5; also see Chapter 2.)

3. What is the smallest-diameter crater you can identify in the photo of Mercury on page 454? (*Hint:* See Appendix A, Properties of the Planets, to find the diameter of Mercury in kilometers.)

4. A sample of a meteorite has been analyzed, and the result shows that out of every 1000 nuclei of potassium-40 originally in the meteorite, only 100 have not decayed. How old is the meteorite? (*Hint:* See Figure 19-6.)

5. In Table 19-2, which object's observed density differs least from its un- compressed density? Why?

6. What composition might you expect for a planet that formed in a region of the solar nebula where the temperature was about 100 K?

7. Suppose that Earth grew to its present size in 1 million years through the accretion of particles averaging 100 g each. On the average, how many particles did Earth capture per second? (*Hint:* See Appendix A to find Earth's mass.)

8. If you stood on Earth during its formation as described in Problem 7 and watched a region covering 100 m^2, how many impacts would you expect to see in an hour? (*Hints:* Assume that Earth had its present radius. The surface area of a sphere is $4\pi r^2$.)

9. The velocity of the solar wind is roughly 400 km/s. How long does it take to travel from the sun to Pluto?

ACTIVE FIGURES

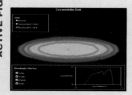

Ace ✱Astronomy™ To access the resources in the Media Cluster, log into AceAstronomy at http://ace.brookscole.com/sf9 and select Chapter 19.

Circumstellar Disk
In this animation, you can observe the disk surrounding a star in different kinds of light. Notice that what you observe depends on wavelength.

Radioactive Decay
The characteristic decay patterns of different radioactive elements provide an invaluable dating tool. In this animation, see how radioactive atoms decay into daughter atoms.

ASTRONOMY EXERCISES

Extrasolar Planets
This simulation presents a hypothetical star system with one planet orbiting the star. Study what variables help astronomers to detect extrasolar planets.

VIRTUAL ASTRONOMY LABS

Lab 8: Extrasolar Planets
This lab examines how indirect methods such as Doppler shift techniques are used to discover and study extrasolar planets. At the end of the lab, you will build your own solar system.

Critical Inquiries for the Web

1. How does our solar system compare with the others that have been found? Search the Internet for sites that give information about planetary systems around other stars. What kinds of planets have been detected by these searches so far? Discuss the selection effects (see Window on Science 25-2) that must be considered when interpreting these data.

2. The process of protoplanetary accretion is still not well understood. Search the Web for current research in this field. From the results of your search, outline the basic steps in the formation of a protoplanet through accretion. What specific factors are important in these models of planet building? Do these models produce planetary systems similar to the ones known to exist?

3. How is radioactive dating carried out on meteorites and rocks from surfaces of various bodies in the solar system? Look for websites on the details of radioactive dating and summarize the methods used to uncover the abundances of radioactive elements in a particular sample. (*Hint:* Try looking for information on how a particular meteorite—for example, the Martian meteorite ALH84001—was studied. What age range was determined, and what radioactive elements were used to arrive at the age?)

Exploring *TheSky*

1. Look at the solar system from space. Notice how thin the disk of the solar system is and how inclined the orbits of Jupiter and Mercury are. (*Hint:* Under the **View** menu, choose **3D Solar System Mode** and then zoom in or out. Tip the solar system up and down to see it edge-on.)

2. Look at the solar system from space and notice how small the orbits of the inner planets are compared to the orbits of the outer planets. They make two distinct groups. (*Hint:* Use **3D Solar System Mode.**)

3. Watch the comets orbiting around the sun. Can you locate the comet C/198 M5 (Linear)? What is its orbit like? (*Hint:* Use **3D Solar System Mode** and set the time step to 30 days [**30d**].)

Go to the Brooks/Cole Astronomy Resource Center **(http://astronomy.brookscole.com)** for critical thinking exercises, articles, and additional readings from InfoTrac College Edition, Brooks/Cole's online student library.

20 | Earth: The Standard of Comparative Planetology

Nature evolves. The world was different yesterday.

PRESTON CLOUD,
COSMOS, EARTH AND MAN

STARS DIE, BUT PLANETS DON'T. That should give you a feeling of security; the beautiful planet you live on will not explode or collapse into a black hole. But Earth is not eternal. You know that the survival of your home world is limited by the life expectancy of your star. Furthermore, you know that Earth has not always been as you see it today. Planets evolve, and our planet has changed dramatically since it formed. Mountains rise and weather away, the climate warms and cools, and the continents themselves drift and grind against each other. To fully know Earth, you need to know how planets in general evolve. Why is Venus hot and Mars cold? Why does Mercury have many craters and Earth only a few?   The best way, perhaps the only way, to answer such questions is to use **comparative planetology,** the study of planets

Continued on page 478

The beauty of planet Earth can be deceptive. It was not always as it is now, and it will not survive unchanged forever. It was born in the solar nebula and its end will come when the sun dies.

(Jean-Bernard Carillet/Getty Images)

Guidepost

Looking Back

In the previous chapter, you learned how our solar system formed as a by-product of the formation of the sun. You saw how distance from the sun determined the general character of each planet. Now you are ready to visit the planets and get to know them as individuals.

This Chapter

As you study the individual planets, you need to avoid the pitfall of learning facts instead of understanding principles. One way to focus on principles is to compare the planets with each other and search for similarities and contrasts. Like people, the terrestrial planets are more alike than they are different, but it is the differences that are most memorable.

As you explore the terrestrial planets, you can use Earth as your basis of comparison, but that means you must think of Earth as a planet. In this chapter, you will see Earth in a new way and answer four essential questions:

How can comparison help you understand the terrestrial planets?

How has Earth changed since it formed?

What is the inside of Earth like?

How has Earth's atmosphere formed and evolved?

Looking Ahead

Like a mountain climber establishing a base camp before attempting the summit, you have established your basis of comparison on Earth. In the following chapters you will visit worlds that are un-Earthly but, in many ways, familiar.

Ace★Astronomy™ The AceAstronomy icon throughout the text indicates an opportunity for you to test yourself on key concepts and to explore animations and interactions on the AceAstronomy website at: http://ace .brookscole.com/sf9

through comparison and contrast, and Earth is the basis of comparison for your study. One reason to begin with Earth is that it is the planet you know best. You can study it carefully and then draw comparisons to other worlds. But another reason is that your home is a planet of extremes. The land above the oceans is quite different from the land under the oceans (**Celestial Profile 2**). Earth's interior is molten and generates a magnetic field. Its crust is active, with moving sections that push against each other and trigger earthquakes, volcanoes, and mountain building. Even Earth's atmosphere is extreme. Processes have altered Earth's air from its original composition to a highly unusual, oxygen-rich sea of gas. The complex properties of extreme Earth will help you understand the remaining planets in our solar system.

It may seem strange to begin a journey among the planets by looking at the dirt under your feet, so you should put Earth in perspective by taking a lightning-fast tour just to see how Earth compares with the worlds on your itinerary.

20-1 A Travel Guide to the Terrestrial Planets

IF YOU VISIT the city of Granada in Spain, you will probably consult a travel guide; and, if it is a good guide, it will do more than tell you where to find museums and restrooms. It will tell you to look at the palace called the Alhambra and compare it with

■ **Figure 20-1**

Planets in comparison. Earth and Venus are similar in size, but their atmospheres and surfaces are very different. The moon and Mercury are much smaller, and Mars is intermediate in size. (Moon: UCO/Lick Observatory; All planets: NASA)

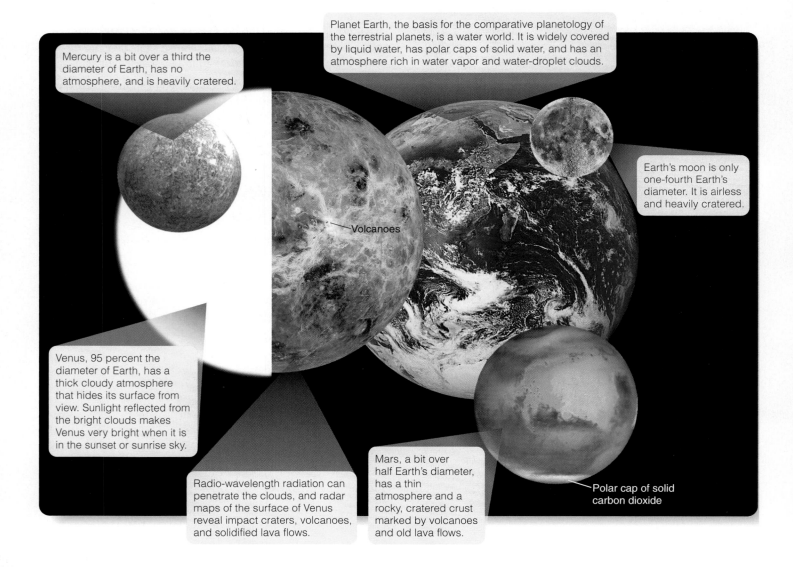

Mercury is a bit over a third the diameter of Earth, has no atmosphere, and is heavily cratered.

Planet Earth, the basis for the comparative planetology of the terrestrial planets, is a water world. It is widely covered by liquid water, has polar caps of solid water, and has an atmosphere rich in water vapor and water-droplet clouds.

Volcanoes

Earth's moon is only one-fourth Earth's diameter. It is airless and heavily cratered.

Venus, 95 percent the diameter of Earth, has a thick cloudy atmosphere that hides its surface from view. Sunlight reflected from the bright clouds makes Venus very bright when it is in the sunset or sunrise sky.

Radio-wavelength radiation can penetrate the clouds, and radar maps of the surface of Venus reveal impact craters, volcanoes, and solidified lava flows.

Mars, a bit over half Earth's diameter, has a thin atmosphere and a rocky, cratered crust marked by volcanoes and old lava flows.

Polar cap of solid carbon dioxide

buildings in Morocco. You are beginning a journey in which you will visit five Earthlike worlds, so you should consult a travel guide and see what is in store.

Five Worlds

The terrestrial planets include Mercury, Venus, Earth, Earth's moon, and Mars. It may surprise you that the moon is on your itinerary. It is, after all, just a natural satellite orbiting Earth and isn't one of the planets. But the moon is a fascinating world of its own, and it makes a striking comparison with the other worlds on your list. Just as you shouldn't tour Spain without seeing Granada, you should not tour the planets without seeing the moon.

■ Figure 20-1 compares the five worlds you will study. The first feature you should notice is diameter. The moon is small, and Mercury is hardly much bigger. Earth and Venus are large and quite similar in size, but Mars is a medium-sized world. You will discover that size is a critical factor in determining a world's personality. Small worlds tend to be cold and dead, but larger worlds can be hot and active.

Core, Mantle, and Crust

Notice that these terrestrial worlds are made up of rock and metal. They all have rocky, low-density crusts, while much of the metal in these worlds has sunk to their centers to form dense cores. You will discover that Earth has a molten iron core but that the moon lacks much iron at all. Between the crust and the core of a planet lies a deep layer of dense rock.

When the planets formed, their surfaces were cratered by the heavy bombardment of debris in the young solar system. You will see lots of craters on these worlds, especially on Mercury and the moon, but notice that cratered surfaces are old. If a lava flow covered up the craters after the end of the heavy bombardment, few new craters could be formed because most of the debris in the solar system was gone. When you see smooth plains on a planet such as those on the moon or on Mars, you can guess that the plains are younger than the cratered areas.

On your travels among the Earthlike worlds, look for signs of heat flowing up from the interior. In the preceding chapter you saw evidence that the planets probably form hot, and radioactive elements trapped in their interiors decay and generate more heat. That heat, flowing upward through the cooler crust, can make a world active with volcanoes and lava flows. Earth is a dramatic example of an active planet, but you will see evidence that Venus too is active. But watch for the small worlds that cooled fast and have little heat flowing outward now. You will find that Mercury and the moon, being small, are inactive, dead worlds.

Atmospheres

When you look at Mercury and the moon in Figure 20-1, you can see their craters and plains and mountains. But the surface of Venus is completely hidden by a cloudy atmosphere even thicker

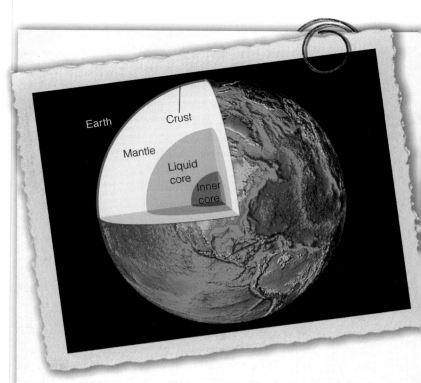

Earth's surface is marked by high continents and low sea floors, but the crust is only 10 to 60 km thick. Below that lie a deep mantle and an iron core. (NGDC)

Celestial Profile 2: Earth

Motion:

Average distance from the sun	1.00 AU (1.495979×10^8 km)
Eccentricity of orbit	0.0167
Maximum distance from the sun	1.0167 AU (1.5210×10^8 km)
Minimum distance from the sun	0.9833 AU (1.4710×10^8 km)
Inclination of orbit to ecliptic	0°
Average orbital velocity	29.79 km/s
Orbital period	1.00 y (365.25 days)
Period of rotation	$24^h00^m00^s$ (with respect to the sun)
Inclination of equator to orbit	23°27'

Characteristics:

Equatorial diameter	12,756 km
Mass	5.976×10^{24} kg
Average density	5.497 g/cm³ (4.07 g/cm³ uncompressed)
Surface gravity	1.0 Earth gravity
Escape velocity	11.2 km/s
Surface temperature	−50° to 50°C (−60° to 120°F)
Average albedo	0.39
Oblateness	0.0034

Personality Point:

Earth comes, through Old English *eorthe* and Greek *eraze,* from the Hebrew *erez,* which means ground. *Terra* comes from the Roman goddess of fertility and growth, *Terra Mater,* Mother Earth.

than Earth's. Mars, the medium planet, has a medium atmosphere of thin gases.

As you explore these worlds, you can ponder two questions. First, why do some worlds have atmospheres while others do not? You will discover that both size and temperature are important. The second question is more complex. Where did these atmospheres come from? To answer that question, you will have to study the geological history of these worlds.

The terrestrial planets are Earthlike worlds. Studying Earth will help you understand other planets, and studying the terrestrial planets will give you a new way to see Earth—as a planet.

Building Scientific Arguments

Why do you expect the inner planets to be high-density worlds?
A good scientific argument gives you a way to see nature—a way to understand why the universe behaves as it does. The best questions in science always begin with "why." In Chapter 19, you saw how the inner planets formed from hot inner parts of the solar nebula. No ice could form there, so the inner planets had to grow from the solid bits of rock and metal that condensed from the hot gas. So you expect the inner planets to be made mostly of rock and metal, which are dense materials.

As you begin studying planets one by one, keep thinking in scientific arguments. They will help you organize all of the information you will meet. Now build an argument to review an important point. **What made all of the craters that are spread through the solar system?**

■   ■   ■

Connections: Some people are satisfied just reading travel guides, but most people want to visit Granada rather than read about it. Of course, you can't really visit Venus or Mars, but one of the wonders of science is that it can take your imagination where your body cannot go. So grab your carry-on luggage. They are calling your flight, and the first world you will visit may look a little unfamiliar even though you have been there.

(20-2) The Early History of Earth

LIKE ALL THE TERRESTRIAL PLANETS, Earth formed from the inner solar nebula about 4.6 billion years ago. As it took form, it began to change, passing through four developmental stages (■ Figure 20-2). Processes that occurred during these four stages determined the present condition of Earth's interior, crust, and atmosphere.

Four Stages of Planetary Development

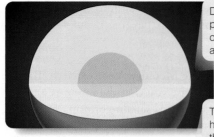

Differentiation produces a dense core, thick mantle, and low-density crust.

The young Earth was heavily bombarded in the debris-filled early solar system.

Flooding by molten rock and later by water can fill lowlands.

Slow surface evolution continues due to geological processes, including erosion.

■ **Figure 20-2**

The four stages of planetary development are illustrated for Earth.

Four Stages of Planetary Development

The first stage of planetary evolution is *differentiation,* the separation of material according to density. Earth now has a dense core and a lower-density crust, and that structure must have originated very early.

Differentiation would have occurred easily if Earth had been molten when it was young. Two sources of heat could have heated Earth. First, heat of formation would be created by infalling material. A meteorite hitting Earth at high velocity converts most of its energy of motion into heat, and the impacts of a large number of meteorites could release tremendous heat. If Earth formed rapidly, this heat would have accumulated much more rapidly than it could leak away, and Earth was probably molten when it formed. A second source of heat requires more time to develop. The decay of radioactive elements trapped in the Earth releases heat gradually; but, as soon as Earth formed, that heat would have begun to accumulate and could have helped melt Earth to facilitate differentiation. Most of Earth's radioactive elements are now concentrated in the crust, where they continue to warm and soften the rock layers.

Earth was formed in a molten state by material falling together, but meteorites could leave no trace until a crust solidified. Once Earth had a hard surface, the meteorites could form craters. This second stage in planetary evolution, *cratering,* was violent. The heavy bombardment was intense because the solar nebula was filled with rocky and icy debris, and the young Earth was battered by meteorites that pulverized the newly forming crust. The largest meteorites blasted out crater basins hundreds of kilometers in diameter. As the solar nebula cleared, the amount of debris decreased, and the level of cratering fell to its present low level. Although meteorites still occasionally strike Earth and dig craters, cratering is no longer the dominant influence on Earth's geology. As you compare other worlds with Earth, you will discover traces of this intense period of cratering, the heavy bombardment, on every old surface in the solar system.

The third stage, *flooding,* no doubt began while cratering was still intense. The fracturing of the crust and the heating caused by radioactive decay allowed molten rock just below the crust to well up through fissures and flood the deeper basins. You will find such basins filled with solidified lava flows on other worlds, such as the moon, but all traces of this early lava flooding on Earth have been destroyed by later geological activity. On Earth, flooding continued as the atmosphere cooled and water fell as rain, filling the deepest basins to produce the first oceans. Notice that on Earth flooding involves both lava and water, a circumstance that you will not find on most worlds.

The fourth stage, *slow surface evolution,* has continued for the last 3.5 billion years or more. Earth's surface is constantly changing as sections of crust slide over each other, push up mountains, and shift continents. Also, moving air and water erode the surface and wear away geological features. Almost all traces of the first billion years of Earth's geology have been destroyed by the active crust and erosion.

Earth as a Planet

You are studying Earth in this chapter in order to use it as a basis for comparison with other worlds. Of course, four of the worlds in our solar system, the Jovian planets, are dramatically different from Earth. But the four terrestrial planets are rocky worlds much like Earth, and many of the moons in the solar system have geology that you can compare with Earth and its four stages of development.

All terrestrial planets pass through these four stages, so in that respect Earth is a good basic reference planet for comparative planetology. Some planets have emphasized one stage over another, and some planets have failed to progress fully through the four stages. Nevertheless, Earth is a good standard of comparison. Every major process on any rocky world in our solar system is represented in some form on Earth. Clearly, Earth is a highly complex world; for that reason, it makes a good standard of comparison.

On the other hand, Earth is peculiar in two ways. First, it has large amounts of liquid water on its surface. Fully 75 percent of its surface is covered by this liquid, and no other planet in our solar system is known to have such extensive liquid water on its surface. Not only does water fill the oceans, but it evaporates into the atmosphere, forms clouds, and then falls as rain. Water falling on the continents flows downhill to form rivers that flow back to the sea, and, in so doing, the water produces intense erosion. Entire mountain ranges can literally dissolve and wash away in only a few tens of millions of years, less than 1 percent of Earth's total age. You will not see such intense erosion on most worlds. Liquid water is, in fact, a rare material on most planets.

Your home planet is special in a second way. Some of the matter on the surface of this world is alive, and a small part of that living matter is aware. No one is sure how the presence of living matter has affected the evolution of Earth, but this process seems to be totally missing from other worlds in our solar system. Furthermore, the thinking part of the life on Earth, humankind, is actively altering our planet. It does not seem likely that humans have made major changes in Earth yet, but your study of Earth as a standard of comparison with other worlds will also give you new insight into your own planet and how modern civilization may be altering it.

Building Scientific Arguments

Why should you think Earth went through an early stage of cratering?

When you build a scientific argument, take great care to distinguish between theory and evidence. Recall from the previous chapter that the planets formed by the accretion of planetesimals from the solar nebula. The protoearth may have been molten as it formed, but as soon as it grew cool enough to form a solid crust, the remaining planetesimal impacts would have formed craters. So you can reason from the solar nebula hypothesis that Earth should have been cratered. But you can't use a theory as evidence to support some other theory. To find real observational evidence, you need only look at the moon. The moon has craters, and so does every other old surface in our solar system. There must have been a time, when the solar system was young, when there were large numbers of objects striking all the planets and moons and blasting out

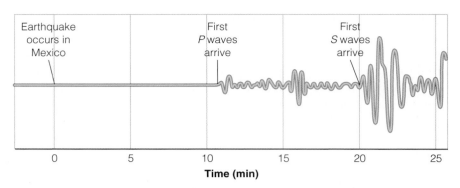

A seismograph in northern Canada made this record of seismic waves from an earthquake in Mexico. The first vibrations, *P* waves, arrived 11 minutes after the quake, but the slower *S* waves took 20 minutes to make the journey.

craters. If it happened to other worlds in our solar system, it must have happened to Earth, too.

The best evidence to support your argument would be lots of craters on Earth, but, of course, there are few craters on Earth. Extend your argument. **Why don't you see lots of craters on Earth today?**

■ ■ ■

Connections: As you explore our solar system in the chapters that follow, Earth will be your standard of comparison. But now you have to look under the hood and study a part of Earth most people comfortably ignore.

20-3 The Solid Earth

ALTHOUGH YOU MIGHT THINK OF EARTH as made of solid rock, it is in fact neither entirely solid nor entirely rock. The thin crust seems solid, but it floats and shifts on a semiliquid layer of molten rock just below the crust. Below that lies a deep, rocky mantle surrounding a core of liquid metal. Much of what you see on the surface of Earth is determined by its interior.

Earth's Interior

The theory of the origin of planets from the solar nebula predicts that Earth should have melted and differentiated into a dense metallic core, a dense mantle, and a low-density silicate crust. But did it differentiate? Where's the evidence? Clearly, Earth's surface is made of lower-density silicates, but what of the interior?

High temperature and tremendous pressure in Earth's interior make any direct exploration impossible. Even the deepest oil wells extend only a few kilometers down and don't reach through the crust. It is quite impossible to drill far enough to sample Earth's core. Yet Earth scientists have studied the interior and found clear evidence that Earth did differentiate.

This analysis of Earth's interior is possible because earthquakes produce vibrations called **seismic waves,** which travel through the interior and eventually register on sensitive detectors, known as **seismographs,** all over the world (■ Figure 20-3). Two kinds

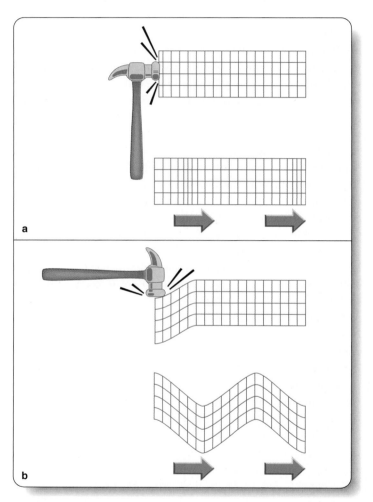

■ Figure 20-4

(a) *P,* or pressure, waves, like sound waves in air, travel as a region of compression. (b) *S,* or shear, waves, like vibrations in a bowl of jelly, travel as displacements perpendicular to the direction of travel. *S* waves tend to travel more slowly than *P* waves and cannot travel through liquids.

of seismic waves are important to us here. The *P,* or **pressure, waves** are much like sound waves in that they travel as a region of compression. As a *P* wave passes, particles of matter vibrate back and forth parallel to the direction of wave travel (■ Figure 20-4a). In contrast, the *S,* or **shear, waves** move as displacements of particles perpendicular to the waves' direction of travel (Figure 20-4b). Thus, *S* waves distort the material but do not compress it. Normal sound waves are pressure waves, whereas the vibrations in a bowl of jelly are shear waves. Because *P* waves are compression waves, they can move through a liquid, but *S* waves cannot. A glass of water can't shimmy like jelly because a liquid does not have the rigidity required to transmit *S* waves.

The *P* and *S* waves caused by an earthquake do not travel in straight lines or at constant speed within Earth. The waves may reflect off boundaries between layers of different density, or they may be refracted as they pass through such a boundary. In addition, the gradual increase in temperature and density toward Earth's center means the speed of sound is greater toward the center. Sound waves approaching such a region tend to curve away (refract), so seismic waves do not travel in straight lines through Earth's interior but are refracted to follow curved paths. Earth scientists can use the arrival times of reflected and refracted seismic waves from distant earthquakes to construct a picture of Earth's interior.

Such studies show that the interior consists of three parts: a central core, a thick mantle, and a thin crust. *S* waves provide an important clue to the nature of the core, because no such waves are observed to travel through the core. When an earthquake occurs, no direct *S* waves register on seismographs on the opposite side of Earth, as if the core were casting a shadow (■ Figure 20-5). This shows that the core is mostly liquid, and the size of the shadow fixes the size of the core at about 55 percent of Earth's radius. Mathematical models predict that the core is hot (about 6000 K), dense (about 14 g/cm³), and composed of iron and nickel.

Earth's core is as hot as the surface of the sun, but the tremendous pressure inside Earth keeps the material from vaporizing. The core is instead a liquid. Nearer the center the material is under even higher pressure, which in turn raises the melting point so high that the material cannot melt at the existing temperature (■ Figure 20-6). That is why there is an inner core of solid iron and nickel. Estimates suggest the inner core's radius is about 22 percent that of Earth.

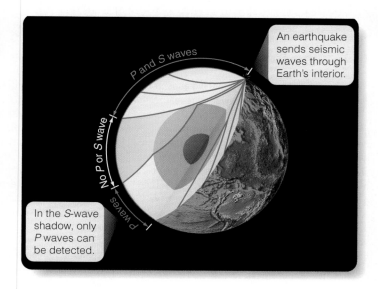

An earthquake sends seismic waves through Earth's interior.

In the *S*-wave shadow, only *P* waves can be detected.

■ Active Figure 20-5

P and *S* waves give you clues to Earth's interior. That no direct *S* waves reach the far side shows that Earth's core is liquid. The size of the *S* wave shadow tells you the size of the outer core.

Ace◗Astronomy™ Log into AceAstronomy and select this chapter to see the Active Figure "Seismic Waves" and follow *P* and *S* waves as they travel through Earth's interior.

The **mantle** is the layer of dense rock that lies between the molten core and the crust. The paths of seismic waves in the mantle show that it is not molten, but it is not precisely solid either. Mantle material behaves like a **plastic,** a material with the properties of a solid but capable of flowing under pressure. The asphalt used in paving roads is a common example of a plastic. It shatters if struck with a sledgehammer, but it bends under the steady weight of a heavy truck. Just below Earth's crust, where the pressure is less than at greater depths, the mantle is most plastic.

Earth's rocky crust is made up of low-density rocks and floats on the denser mantle. The crust is thickest under the continents,

■ Figure 20-6

Theoretical models combined with observations of the velocity of seismic waves reveal the temperature inside Earth (blue line). The melting point of the material (red line) is determined by its composition and by the pressure. In the mantle and in the inner core, the melting point is higher than the existing temperature, and the material is not molten.

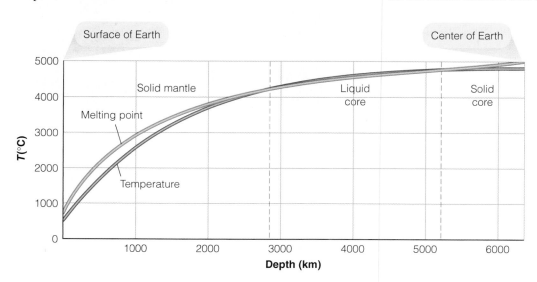

Scientific Imagination: Seeing the Unseen

One of the most fascinating aspects of science is its power to reveal the unseen. That is, it gives you an opportunity to learn about regions you can never visit. You saw this in earlier chapters when you studied the inside of the sun and stars, the surface of neutron stars, the event horizon around black holes, the cores of active galaxies, and more. In this chapter, you have seen Earth's core.

Not only does science take you to places you can never visit, but it takes you to scales you can never explore. For example, physicists can explore the inside of an atom, and biologists can study the structure of a virus. Not only can you stretch the scale of space, but you can also stretch the scale of time. Geologists can watch mountain ranges rising slowly from the plains and eroding away to grit. Astronomers can watch the 200-million-year-long whirl of two galaxies colliding.

In this respect, astronomy, geology, and psychology are particularly similar. All three sciences study realms that cannot be explored in person. The insides of stars, Earth's ancient history, and the workings of the human mind lie beyond immediate experience. Only by careful observation and careful thought can scientists learn to understand these aspects of the natural world.

Science can reveal the unseen because it can combine human imagination with the predictive power of the laws of nature and the precision of mathematics. This calls on scientists to be as creative as artists and as precise as surgeons to produce reliable theories and models, but the reward is one of the great thrills of science—exploring beyond the limits of normal human experience.

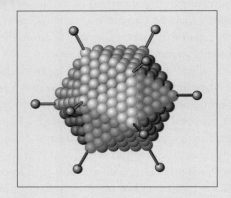

The electron microscope allows biologists to study the elegant structure of the virus that causes the common infection called pinkeye. (From *Virus Ultrastructure: Electron Micrograph Images.* Copyright 1995 by Linda M. Stannard. Reprinted with permission from Linda M. Stannard. Found at: http://www.web.uct.ac.za/depts/mmi/stannard/adeno.html)

---

up to 60 km thick, and thinnest under the oceans, where it is only about 10 km thick. Unlike the mantle, the crust is brittle and breaks much more easily than the plastic mantle.

Perhaps no region is more immediate yet more inaccessible than Earth's core. Only a few thousand kilometers from where you sit, Earth's interior is beyond your immediate reach. Yet you can send your imagination where your body can never go (**Window on Science 20-1**). Earth's seismic activity reveals some of Earth's innermost secrets. But there is another source of evidence about Earth's interior—its magnetic field.

The Magnetic Field

Apparently, Earth's magnetic field is a direct result of its rapid rotation and its molten metallic core. The internal heat forces the liquid core to churn with convection while Earth's rotation turns the core about an axis. This core is a highly conductive iron–nickel alloy, a better electrical conductor than copper, the material commonly used for electrical wiring. The rotation of this convecting, conducting liquid generates Earth's magnetic field in a process called the dynamo effect (■ Figure 20-7). This is believed to be the same process that generates the solar magnetic field in the convective layers of the sun, and you will meet it again when you explore other planets.

Earth's magnetic field protects it from the solar wind. Blowing outward from the sun at about 400 km/s, the solar wind consists of ionized gases carrying a small part of the sun's magnetic field. When the solar wind encounters Earth's magnetic field, it is deflected like water flowing around a boulder in a stream. The sur-

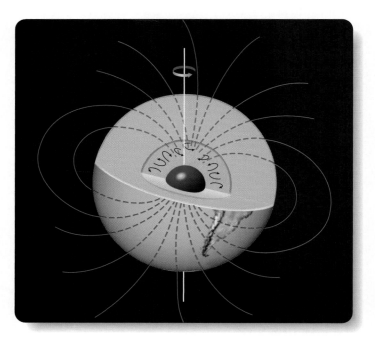

■ Figure 20-7

The dynamo effect couples convection in the liquid core with Earth's rotation to produce electric currents that are believed to be responsible for Earth's magnetic field.

face where the solar wind is first deflected is called the **bow shock,** and the cavity dominated by Earth's magnetic field is called the **magnetosphere** (■ Figure 20-8a). High-energy particles from the solar wind leak into the magnetosphere and become trapped

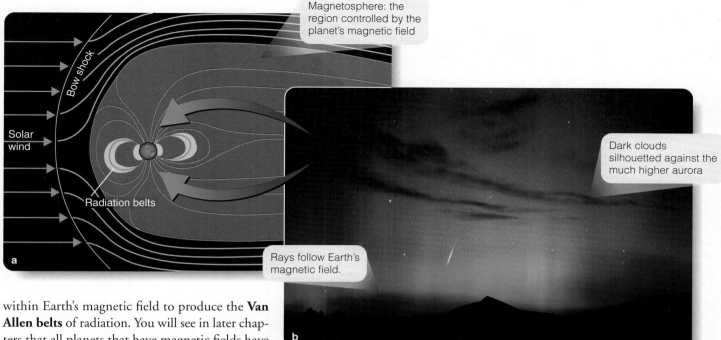

Magnetosphere: the region controlled by the planet's magnetic field

Bow shock

Solar wind

Radiation belts

a

Dark clouds silhouetted against the much higher aurora

Rays follow Earth's magnetic field.

b

within Earth's magnetic field to produce the **Van Allen belts** of radiation. You will see in later chapters that all planets that have magnetic fields have bow shocks, magnetospheres, and radiation belts.

One dramatic result of Earth's magnetic field are the auroras, glowing rays and curtains of light in the upper atmosphere (Figure 20-8b). The solar wind carries charged particles past Earth's extended magnetic field, and this generates tremendous electrical currents that flow into Earth's atmosphere near the north and south magnetic poles. The currents ionize gas atoms in Earth's atmosphere, and when the ionized atoms capture electrons and recombine, they emit light as if they were part of a vast "neon" sign. That is why the spectrum of an aurora is an emission spectrum.

Although you can be confident that Earth's magnetic field is generated within its molten core, many mysteries remain. For example, rocks retain traces of the magnetic field in which they solidify, and some contain fields that point backward. That is, they imply that Earth's magnetic field was reversed at the time they solidified. Careful analysis of such rocks indicates that Earth's field has reversed itself every million years or so, with the north magnetic pole becoming the south magnetic pole and vice versa. These reversals are poorly understood, but they may be related to changes in the core convection.

Convection in Earth's core is important because it generates the magnetic field. As you will see in the next section, convection in the mantle constantly remakes Earth's surface.

Ace ☾ Astronomy™ Log into AceAstronomy and select this chapter to see Astronomy Exercise "Convection and Magnetic Fields." Take control of a planetary interior.

Earth's Active Crust

Earth's crust is composed of lower-density rock that floats on the mantle. The image of a rock floating may seem odd, but recall that the rock of the mantle is very dense. Also, just below the crust,

■ Figure 20-8

Earth's magnetic field dominates space around Earth by deflecting the solar wind and trapping high-energy particles in radiation belts. Around the north and south magnetic poles, where the magnetic field enters Earth's atmosphere, powerful currents can flow down and excite gas atoms to emit photons, which produces auroras. Colors are produced as different atoms are excited. Note the meteor (shooting star). (Jimmy Westlake)

the mantle rock tends to be highly plastic, so great sections of low-density crust do indeed float on the semiliquid mantle like great lily pads floating on a pond.

The motion of the crust and the erosive action of water make Earth's crust highly active. Look at **The Active Earth** on pages 486–487 and notice three important points:

1 The motion of crustal plates produces much of the geological activity on Earth. Earthquakes, volcanism, and mountain building are all linked to motions in the crust and the location of plate boundaries.

2 Notice how the continents on Earth's surface have moved and changed over periods of hundreds of millions of years. A hundred million years is only 0.1 billion years, so sections of Earth's crust are in rapid motion.

3 Notice that most of the geological features you know—mountain ranges, the Grand Canyon, and even the outline of the continents—are recent products of Earth's active surface. Earth's surface is constantly renewed. The oldest rocks on Earth, small crystals called zircons from western Australia, are 4.3 billion years old. Most of the crust is much younger than that. Most of the mountains and valleys you see around you are no more than a few tens of millions of years old.

Our world is an astonishingly active planet. Not only
is it rich in water and therefore subject to rapid erosion,
crust is divided into moving sections called plates.
plates spread apart, lava wells up to form new crust;
plates push against each other, they crumple the crust
mountains. Where one plate slides over another, you
olcanism. This process is called **plate tectonics,**
g to the Greek word for "builder." (An architect is
y an arch builder.)

cal view
net Earth

Mountains
are common
on Earth, but
they erode
away rapidly
because of
the abundant
water.

William K. Hartmann
Janet Seeds

A **rift valley** forms where conti
plates begin to pull apa
Red Sea has formed
Africa has be
pull awa
the A
peni

National Geophysical Data Center

Midocean
rise

Red Sea

Midocean
rise

1a Evi
of p
tectonics
first found
ocean floors,
plates spread ap
and magma rises to fo
midocean rises made of rock called **basalt,** a rock typic
solidified lava. Radioactive dating shows that the basalt is
younger near the midocean rise. Also, the ocean floor ca
less sediment near the midocean rise. As Earth's magne
field reverses back and forth, it is recorded in the magnet
fields frozen into
basalt. This produ
a magnetic patter
the basalt that sh
that the seafloor i
spreading away fr
the midocean rise

A **subduction zone** is
a deep trench where
late slides under
er. Melting releases
ensity magma that
o form volcanoes such
se along the northwest
of North America,
ing Mt. St. Helens.

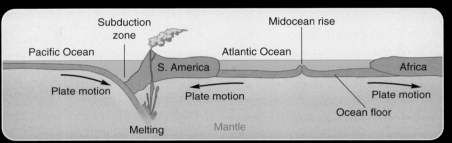

Subduction
zone

Midocean rise

Pacific Ocean

Atlantic Ocean

S. America

Africa

Plate motion

Plate motion

Plate motion

Ocean floor

Melting

Mantle

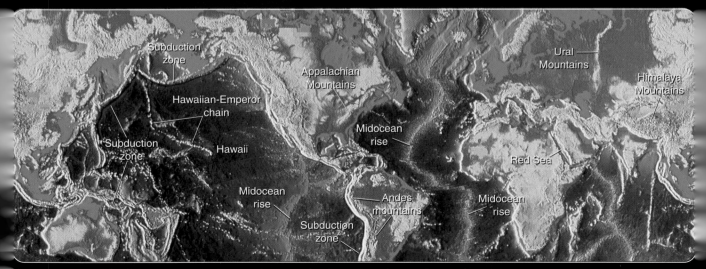

Subduction
zone

Ural
Mountains

Appalachian
Mountains

Himalaya
Mountains

Hawaiian-Emperor
chain

Midocean
rise

Subduction
zone

Hawaii

Red Sea

Midocean
rise

Andes
mountains

Midocean
rise

Subduction
zone

Hot spots caused by rising magma in the mantle can poke through a plate and cause
volcanism such as that in Hawaii. As the Pacific plate has moved northwestward, the hot
as punched through to form a chain of volcanic islands, now mostly worn below sea level.
d **mountain ranges** can form where plates push against each other. For example, the
Mountains lie between Europe and Asia, and the Himalaya Mountains are formed by India
g north into Asia. The Appalachian Mountains are the remains of a mountain range

Ace Astronomy™ Log into
AceAst
and select this chapter to see A
Figure "Hot Spot Volcanoes." N
how the moving plate can produ
chain of volcanic peaks, mostly

1d The floor of the Pacific Ocean is sliding into subduction zones in many places around its perimeter. This pushes up mountains such as the Andes and triggers earthquakes and active volcanism all around the Pacific in what is called the Ring of Fire. In places such as southern California, the plates slide past each other, causing frequent earthquakes.

Hawaii

Yellow lines on this globe mark plate boundaries. Red dots mark earthquakes since 1980. Earthquakes within the plate, such as those at Hawaii, are related to volcanism over hot spots in the mantle.

Continental Drift

Not long ago, Earth's continents came together to form one continent.

Pangaea
200 million years ago

Pangaea broke into a northern and a southern continent.

Laurasia
Gondwanaland
135 million years ago

Notice India moving north toward Asia.

65 million years ago

The continents are still drifting on the highly plastic upper mantle.

Today

2 The floor of the Atlantic Ocean is not being subducted. It is locked to the continents and is pushing North and South America away from Europe and Africa at about 3 cm per year, a motion called *continental drift*. Radio astronomers can measure this motion by timing pulsars from European and from American radio telescopes. Roughly 200 million years ago, North and South America were joined to Europe and Africa. Evidence of that lies in similar fossils and similar rocks and minerals found in the matching parts of the continents. Notice how North and South America fit against Europe and Africa like a puzzle.

3 Plate tectonics pushes up mountain ranges and causes bulges in the crust, and water erosion wears the rock away. The Colorado River began cutting the Grand Canyon only about 10 million years ago when the Colorado plateau warped upward under the pressure of moving plates. That sounds like a long time ago, but it is only 0.01 billion years. A mile down, at the bottom of the canyon, lie rocks 0.57 billion years old, the roots of an earlier mountain range that stood as high as the Himalayas. It was pushed up, worn away to nothing, and covered with sediment long ago. Many of the geological features we know on Earth have been produced by very recent events.

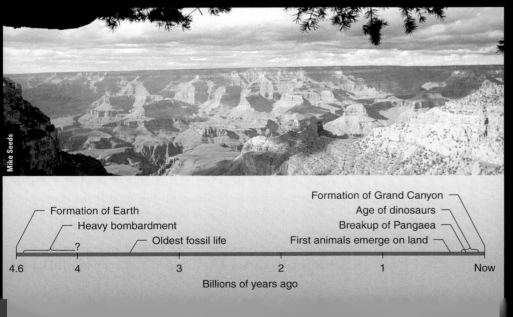

Formation of Grand Canyon
Age of dinosaurs
Formation of Earth
Breakup of Pangaea
Heavy bombardment
Oldest fossil life
First animals emerge on land
?

4.6   4   3   2   1   Now
Billions of years ago

The continents move slowly on average, but plate margins can stick and accumulate stress and then release it in sudden motions. That's what happened on December 26, 2004, along a major subduction zone in the Indian Ocean. The total motion was as much as 15 m, and the resulting earthquake caused devastating tidal waves. Minor earthquakes occur every day, and stress builds in faults all around the world that will eventually be released in major earthquakes.

Earth's active crust explains why Earth contains so few impact craters. The moon is richly cratered, but Earth contains only about 150 impact craters. Plate tectonics and erosion have destroyed all but the most recent craters on Earth.

You can see that Earth's geology is dominated by two dramatic forces. Heat rising from the interior drives plate tectonics; just below the thin crust of solid rock lies a churning molten interior that rips the crust to fragments and pushes the pieces about like bits of algae on a pond. The second force modifying the crust is water. It falls as rain and snow and tears down the mountains, erodes the river valleys, and washes any raised ground into the sea. Like angry children, tectonics and erosion build mountains and continents and then rip them to nothing.

Ace◐Astronomy™ Log into AceAstronomy and select this chapter to see Astronomy Exercise "Convection and Plate Tectonics." Compare the interiors of the four terrestrial planets.

Building Scientific Arguments

What evidence can you cite that Earth has a molten metallic core?
In a scientific argument, the critical analysis of ideas must eventually return to evidence. In this case, the evidence is indirect because you can never visit Earth's core. Seismic waves from distant earthquakes pass through Earth, but a certain kind of wave, the S waves, cannot pass through the core. Because the S waves cannot propagate through a liquid, you can conclude that Earth's core is a liquid. Earth's magnetic field gives you further evidence of a metallic core. The theory for the generation of magnetic fields, the dynamo effect, requires a rotating liquid core composed of a conducting material and stirred by convection. If the core were not a liquid, it would not be able to generate a magnetic field. That gives you two different kinds of evidence that our planet has a liquid core.

Now build a new argument again focusing on evidence. **Why do scientists conclude that Earth's crust is broken into moving plates?**

■   ■   ■

Connections: For the astronomer, accustomed to the longest view of cosmic time, Earth's geology seems not so much the history of slow-moving rock as it does the churning surface of a pot of boiling soup. And just as steam rises from boiling soup, vapors rise from the crust. Earth's atmosphere not only acts to erode the rocky surface, but some of it came from the crust as a by-product of geological activity.

20-4 Earth's Atmosphere

YOU CAN'T TELL THE STORY OF EARTH without mentioning the atmosphere. Not only is it necessary for life, but it is also intimately related to the crust. It affects the surface by eroding geological features through wind and water, and in turn the chemistry of Earth's surface affects the composition of the atmosphere.

Origin of the Atmosphere

Until a few decades ago, Earth scientists believed that Earth's atmosphere had developed slowly. According to this picture, Earth formed by the slow accumulation of planetesimals into a large, cold ball of matter whose composition was uniform from surface to center (Figure 19-9). Early Earth might have attracted small amounts of gases such as hydrogen, helium, and methane from the solar nebula to form a **primeval atmosphere.** According to the old theory, the slow decay of radioactive elements heated Earth's interior, melted it, caused it to differentiate into core, mantle, and crust, and triggered widespread volcanism.

When a volcano erupts, 50 to 80 percent of the gas released is water vapor. The rest is mostly carbon dioxide, nitrogen, and smaller amounts of sulfur gases such as hydrogen sulfide—the rotten-egg gas that you smell if you visit geothermal pools and geysers such as those at Yellowstone National Park. These gases could have diluted the primeval atmosphere and eventually produced a **secondary atmosphere** rich in carbon dioxide, nitrogen, and water vapor. That's the old theory.

A modern understanding of how the solar system formed shows that Earth formed so rapidly that it was melted by the heat released by infalling material. If Earth's surface was molten as it formed, then outgassing would have been continuous, and the atmosphere would have been rich in carbon dioxide, nitrogen, and water vapor from the beginning. That means Earth never had a hydrogen-rich primeval atmosphere.

For some years, astronomers have suspected that the abundance of water on Earth arrived late in the formation process as a bombardment of volatile-rich planetesimals. These icy bodies, the theory goes, were scattered by the growing mass of the outer planets, and the inner solar system was bombarded by a storm of comets. This theory has faced a serious objection. Spectroscopic studies of a few comets revealed that the ratio of deuterium to hydrogen in comets does not match the ratio in the water on Earth. Some astronomers think this means that the water now on Earth could not have arrived in cometlike planetesimals. However, studies of Comet Linear, which broke up in 1999 as it passed the sun, show that the water in that comet had a composition of isotopes similar to water on Earth. There may be differences among comets. Icy planetesimals that formed far from the sun may be richer in deuterium, while comets that formed closer to the orbit of Jupiter than to the orbit of Neptune may contain water with isotopes more like those in Earth's water. This is an interesting controversy, and it shows that the origin of Earth's atmosphere and oceans is yet to be fully understood.

However the atmosphere originated, the mix of gases must have changed. The young atmosphere must have been rich in water vapor, carbon dioxide, and other gases. As the atmosphere cooled, the water condensed to form the first oceans; and, as the oceans grew bigger, carbon dioxide began to dissolve in the water. Carbon dioxide is easily soluble in water—which is why carbonated beverages are so easy to manufacture—but the first oceans could not have absorbed all of the atmospheric carbon dioxide if it had not reacted with dissolved substances in the seawater to form silicon dioxide, limestone, and other mineral sediments. Thanks to those chemical reactions in the oceans the carbon dioxide was transferred from the atmosphere to the seafloor.

Earth's lower atmosphere is now protected from ultraviolet radiation by an **ozone layer** about 25 km above the surface. However, the atmosphere of the young Earth did not contain free oxygen, so an ozone layer could not form, and the sun's ultraviolet radiation penetrated deep into the atmosphere. The energetic ultraviolet photons would have broken up weaker molecules such as water (H_2O). The hydrogen escaped to space, and the oxygen formed oxides in the crust. Earth's atmosphere could not reach its present composition (■ Table 20-1) until it was protected by an ozone layer, and that required oxygen.

The origin of the atmospheric oxygen is linked to the origin of life, the subject of Chapter 26, but it is sufficient here to note that life must have originated within a billion years of Earth's formation. This life did not significantly alter the atmosphere, however, until nature invented photosynthesis about 2.7 to 2.4 billion years ago. Photosynthesis absorbs carbon dioxide from the air or water and utilizes it for plant growth, releasing oxygen. Because oxygen is a very reactive element, it combines easily with other compounds, and the oxygen abundance could not grow rapidly

at first. Apparently, the development of large, shallow seas along the continental margins half a billion years ago allowed ocean plants to manufacture oxygen faster than chemical reactions could remove it. Atmospheric oxygen then increased rapidly, and it is still increasing at the rate of about 1 percent every 36 million years.

Earth's climate is critically sensitive to many different factors. For example, the **albedo** of a world is the fraction of the sunlight that hits it that gets reflected away. An albedo of 1 would be a perfectly white world, and an albedo of 0 would be perfectly black. Earth's overall albedo is 0.39, meaning it reflects back into space 39 percent of the sunlight that hits it. Much of this reflection is caused by clouds, and the formation of clouds depends critically on the presence of water vapor in the upper atmosphere, the temperature of the upper atmosphere, and the patterns of atmospheric circulation. Even a small change in any of these factors could change Earth's albedo and thus its climate. Your comfortable life on the surface of your world depends on Earth's delicate atmosphere.

Ace ⟲Astronomy™ Log into AceAstronomy and select this chapter to see Astronomy Exercise "Primary Atmospheres." You can move your planet around to see how its atmosphere is affected.

Human Effects on Earth's Atmosphere

Earth is comfortable for human life because of its atmosphere. The temperature is moderate at sea level, but if you climb to the top of a high mountain, you find the temperature to be very low (■ Figure 20-9). Most clouds form at such altitudes. Higher still, you find the air much colder and so thin it cannot protect you

■ **Figure 20-9**

Thermometers placed in Earth's atmosphere at different levels would register the temperatures shown in the graph at the right. The lower few kilometers where you live are comfortable, but higher in the atmosphere the temperature is quite low. The ozone layer lies about 25 km above Earth's surface.

■ Table 20-1	Earth's Atmosphere
Gas	**Percent by Weight**
N_2	75.5
O_2	23.1
Ar	1.29
CO_2	0.05
Ne	0.0013
He	0.00007
CH_4	0.0001
Kr	0.0003
H_2O (vapor)	1.7–0.06

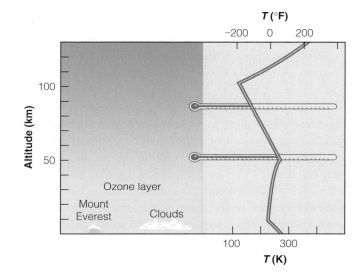

from the intense ultraviolet radiation in sunlight. If you climb mountains or even hike in high mountain meadows, you should wear strong sunscreen.

You can live on Earth's surface in safety because Earth's atmosphere protects you, but modern civilization is altering Earth's atmosphere in at least two serious ways: It is adding carbon dioxide (CO_2) and destroying ozone.

The concentration of CO_2 in Earth's atmosphere is important because CO_2 can trap heat in a process called the **greenhouse effect** (■ Figure 20-10a). When sunlight shines through the glass roof of a greenhouse, it heats the benches and plants inside. The warmed interior radiates infrared radiation, but the infrared cannot get out through the glass. Heat is trapped within the greenhouse, and the temperature climbs until the glass itself grows warm enough to radiate heat away as fast as the sunlight enters. Of course, a greenhouse retains its heat because the walls prevent the warm air from mixing with the cooler air outside. This is the same process that heats a car when it is parked in the sun with the windows rolled up.

Earth's atmosphere is transparent to sunlight, and the absorbed sunshine makes the surface grow warm. However, the CO_2 makes the atmosphere less transparent to infrared radiation, so infrared radiation from the warm surface is absorbed and cannot escape back into space. That traps heat and makes Earth warmer (Figure 20-10b). Without the greenhouse effect, Earth would be 40 K cooler than it is. That amounts to 72°F colder than at present, so Earth would be uninhabitable without the greenhouse effect. The problem is that human civilization is increasing the amount of greenhouse gases in the atmosphere, and that is making Earth warmer.

CO_2 is not the only greenhouse gas. Water vapor, methane, and other gases also help warm Earth, but CO_2 is the most important because the human race is increasing its concentration in the atmosphere. For billions of years, natural processes on Earth have removed CO_2 from the atmosphere and buried the carbon in the form of coal, oil, and natural gas. Since the beginning of the industrial revolution in the mid-19th century, civilization has been digging up this carbon, burning it to get energy, and releasing CO_2 back into the atmosphere. At the same time, many nations are cutting down parts of the forests that absorb CO_2. Estimates are that the amount of CO_2 in Earth's atmosphere will double during the 21st century.

The increased concentration of CO_2 is increasing the greenhouse effect and warming Earth in what is known as **global warming.** The actual amount of warming in the future is difficult to predict. The best mathematical models predict a warming between 1.5 and 4.5°C (3 to 8°F). Predictions are uncertain because Earth's climate depends on so many factors. A slight warming should increase water vapor in the atmosphere, and water vapor is another greenhouse gas that would enhance the warming. But increased water vapor could increase cloud cover, increase Earth's albedo, and partially reduce the warming. High icy clouds,

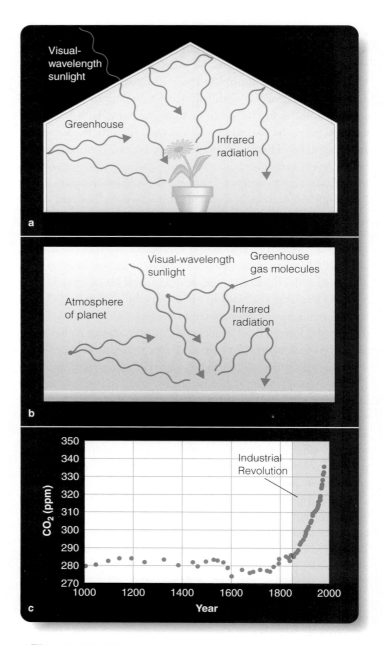

■ **Figure 20-10**

The greenhouse effect. (a) Visual-wavelength sunlight can enter a greenhouse and heat its contents, but the longer-wavelength infrared cannot get out. (b) The same process can heat a planet's surface if its atmosphere contains greenhouse gases such as CO_2. (c) The concentration of CO_2 in Earth's atmosphere as measured in Antarctic ice cores remained roughly constant until the beginning of the Industrial Revolution. (Ozone Processing Team/NASA/Goddard Space FLight Center)

however, tend to enhance the greenhouse effect. Even small changes in temperature can alter circulation patterns in the atmosphere and in the oceans, and the consequences of such changes are very difficult to predict.

The future is difficult to predict, but the evidence clearly shows that Earth is growing warmer. Studies of the growth rings in very old trees show that Earth's climate has been cooling for the last 1000 years, but the 20th century reversed that trend with a rise of 0.3 to 0.6°C (0.5 to 1.1°F). It is difficult to be sure how much of this change is due to the increased greenhouse effect and how much is natural variation in Earth's climate, but it is worrisome that general trends point toward warming. Mountain glaciers have melted back dramatically since the 19th century, and the warmest years of the last 100 years have all been recent. Measurements show that polar ice in the form of permafrost, ice shelves, and ice on the open Arctic Ocean is melting.

Although the temperature increase is small now, it is a serious issue for the future. Even a small rise in temperatures will alter climate and dramatically affect agriculture, not only through rising temperatures but also through changes in rainfall. Also, the melting of ice on polar landmasses such as Greenland can cause a rise in sea levels that will flood coastal regions and alter shore environments. A modest rise will cover huge low-lying areas such as the entire state of Florida.

There seems no doubt that civilization is warming Earth through an enhanced greenhouse effect, but a remedy is difficult to imagine. Reducing the amount of CO_2 and other greenhouse gases released to the atmosphere is difficult because modern society depends on burning fossil fuels for energy. Conserving forests is difficult because growing populations, especially in developing countries with large forest reserves, demand the wood and agricultural land produced when forests are cut. Political, business, and economic leaders argue that the issue is uncertain, but all around the world scientists of stature have reached an agreement: Global warming is real and will change Earth. What humanity can or will do about it is uncertain.

Human influences on Earth's atmosphere go beyond the greenhouse effect. Modern civilization is also reducing ozone in Earth's atmosphere. Ozone is an unstable molecule formed of three oxygen atoms (O_3). You may have heard of ozone because it is produced in auto emissions, and it is a pollutant in city air. But ozone is produced naturally in Earth's atmosphere at an altitude of about 25 km, and there it is beneficial. The ozone layer absorbs ultraviolet photons and prevents them from reaching the ground.

Certain chemicals called chlorofluorocarbons (CFCs), used in industrial processes and in refrigeration and air conditioning, can destroy ozone. As these CFCs escape into the atmosphere, they become mixed into the ozone layer and convert the ozone into normal oxygen molecules. Oxygen does not block ultraviolet radiation, so depleting the ozone layer causes an increase in ultraviolet radiation at Earth's surface. In small doses, ultraviolet radiation can produce a suntan, but in larger doses it can cause skin cancers.

The ozone layer over the Antarctic may be especially sensitive to CFCs. Since the late 1970s, the ozone concentration has been falling over the Antarctic, and a hole in the ozone layer now develops over the continent each October—spring for Antarctica (■ Figure 20-11). Satellite and ground-based measurements show that the same thing is happening at higher northern latitudes and that the amount of ultraviolet radiation reaching the ground is increasing. This is an early warning that human activity is modifying Earth's atmosphere in a potentially dangerous way.

The CO_2 problem and the ozone problem in Earth's atmosphere are paralleled on Venus and Mars. When you study Venus in Chapter 22, you will discover a runaway greenhouse effect that has made the surface of the planet hot enough to melt lead.

■ Figure 20-11

(a) Satellite observations of ozone concentrations over Antarctica are shown here as red for highest concentration and violet for lowest. Since the 1970s, a hole in the ozone layer has developed over the South Pole. (b) Although ozone depletion is most dramatic above the South Pole, ozone concentrations have declined at all latitudes. (Ozone Processing Team/NASA/Goddard Space Flight Center)

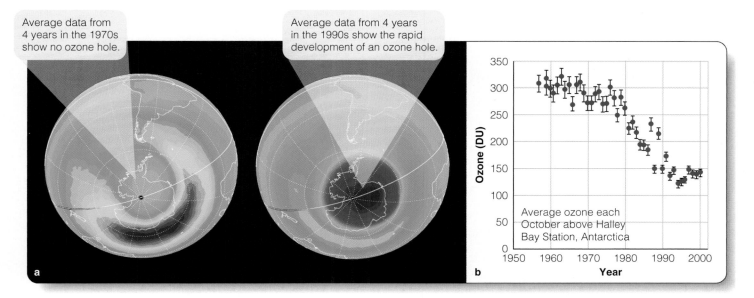

On Mars you will discover an atmosphere without an ozone layer. A few minutes of sunbathing on Mars would kill you. Once again, you will learn more about your own planet by studying other planets.

Building Scientific Arguments

Why does Earth's atmosphere contain little carbon dioxide and lots of oxygen?

Sometimes as you build a scientific argument, you must contradict what seems, at first glance, a simple truth. In this case, because outgassing releases mostly carbon dioxide, CO_2, and water vapor, you might expect Earth's atmosphere to be very rich in CO_2. Luckily for the human race, CO_2 is highly soluble in water, and Earth's surface temperature allows it to be covered with liquid water. The CO_2 dissolves in the oceans and combines with minerals in seawater to form deposits of silicon dioxide, limestone, and other mineral deposits. In this way, the CO_2 is removed from the atmosphere and buried in Earth's crust. Oxygen, in contrast, is highly reactive and forms oxides so easily you might expect it to be rare in the atmosphere. Happily it is continually replenished as green plants release oxygen into Earth's atmosphere faster than chemical reactions can remove it. Were it not for liquid-water oceans and plant life, Earth would be choking in a thick CO_2 atmosphere with no free oxygen.

Now follow up on your argument. **Why would an excess of CO_2 and a deficiency of free oxygen be harmful to all life on Earth in ways that go beyond mere respiration?**

■   ■   ■

Connections: Earth is a rich and beautiful world, and when you visit Venus and Mars, both of which have CO_2-rich atmospheres, you will appreciate Earth even more. But before you rush to compare Earth with planets that have atmospheres, you should examine two worlds that have no atmospheres at all—the moon and Mercury. As usual, the comparison of worlds will be revealing.

Summary

20-1 | A Travel Guide to the Terrestrial Planets

How can comparison help you understand the terrestrial planets?

■ Earth is the standard of comparison in the study of the terrestrial planets.

■ The terrestrial planets include Earth, the moon, Mercury, Venus, and Mars. Earth's moon is included because it is a complex world and makes a striking comparison with Earth.

■ The terrestrial worlds differ mainly in size, but they all have low-density crusts, mantles of dense rock, and metallic cores.

■ Comparative planetology warns you to expect that cratered surfaces are old, that heat flowing out of a planet drives geological activity, and that the nature of a planet's atmosphere depends of the size of the planet and its temperature.

20-2 | The Early History of Earth

How has Earth changed since it formed?

■ Earth formed rapidly from the solar nebula and was hot enough to be molten.

■ Earth has passed through four stages as it has evolved: (1) differentiation; (2) cratering; (3) flooding by lava and water; and (4) slow surface evolution.

■ Earth is peculiar in that it has large amounts of liquid water on its surface, and that water drives strong erosion that alters the surface geology.

■ Earth is also peculiar in that it is the only known home for life.

20-3 | The Solid Earth

What is the inside of Earth like?

■ Seismology shows that *S*-waves cannot pass through Earth's core, and that is evidence that Earth has a liquid core. Heat flowing outward from the interior, combined with models, reveals that the core is very hot and composed of iron and nickel.

■ Earth's magnetic field is generated by the dynamo effect in the liquid, convecting, rotating, conducting core. The magnetic field shields Earth from the solar wind by producing a magnetosphere around the planet. Radiation belts and auroras are also produced by the field.

■ Earth is dominated by plate tectonics that breaks the crust into moving plates. Plate tectonics is driven by heat flowing upward from the interior.

■ Tectonic plates are made of low-density, brittle rock that floats on the hotter plastic upper layers of the mantle. Plates spread apart along midocean rises and sink into the mantle in subduction zones. Volcanism and earthquakes are common along the edge of the plates.

■ The continents are drifting slowly on the plastic mantle, and their arrangement changes with time.

■ Most geological features on Earth, such as mountain ranges and the Grand Canyon, have been formed recently. The first billion years of Earth's geology is almost entirely erased by plate tectonics and erosion.

20-4 | Earth's Atmosphere

How has Earth's atmosphere formed and evolved?

■ Because Earth formed hot, it never had a primeval atmosphere rich in hydrogen and helium. Earth's first atmosphere was probably mostly carbon dioxide, but that gas was dissolved in seawater, and plant life has created oxygen.

■ The greenhouse effect can warm a planet if gases in the atmosphere are transparent to light but opaque to infrared. The natural greenhouse effect warms Earth and makes it comfortable for life, but greenhouse gases added by industrial civilization are responsible for global warming.

■ The ozone layer high in Earth's atmosphere protects the surface from ultraviolet radiation. Certain chemicals called chlorofluorocarbons released in industrial processes attack the ozone layer and thin it. This is allowing more harmful ultraviolet radiation to reach Earth's surface.

New Terms

comparative planetology (p. 476)	midocean rise (p. 486)
seismic wave (p. 482)	subduction zone (p. 486)
seismograph (p. 482)	basalt (p. 486)
pressure *(P)* wave (p. 482)	folded mountain range (p. 486)
shear *(S)* wave (p. 482)	rift valley (p. 486)
mantle (p. 483)	primeval atmosphere (p. 488)
plastic (p. 483)	secondary atmosphere (p. 488)
bow shock (p. 484)	ozone layer (p. 489)
magnetosphere (p. 484)	albedo (p. 489)
Van Allen belts (p. 485)	greenhouse effect (p. 490)
plate tectonics (p. 486)	global warming (p. 490)

Review Questions

Ace ⓢAstronomy™ Assess your understanding of this chapter's topics with additional quizzing and animations at **http://ace .brookscole.com/sf9**

1. Why would you include the moon in a list of the terrestrial planets?

2. In what ways is Earth peculiar among the terrestrial planets?

3. What are the four stages of planetary development?

4. How do you know that Earth differentiated?

5. What evidence can you cite that Earth's metallic core is liquid?

6. How are earthquakes in Hawaii different from those in southern California?

7. What characteristics must Earth's core have in order to generate a magnetic field?

8. How does the Hawaiian-Emperor island chain help you understand plate tectonics?

9. What evidence can you cite that the Atlantic Ocean is growing wider?

10. How is your concept of Earth's first atmosphere related to the speed with which Earth formed from the solar nebula?

11. What has produced the oxygen in Earth's atmosphere?

12. How does the increasing abundance of CO_2 in Earth's atmosphere cause a rise in Earth's temperature?

13. Why would a decrease in the density of the ozone layer cause public health problems?

14. In what ways is the photo at the right a typical view of the surface of planet Earth? How is it unusual among planets in general?

15. What do you see in this photo that suggests heat is flowing out of Earth's interior?

William K. Hartmann

USGS

Discussion Questions

1. If you visited a planet in another solar system and discovered oxygen in its atmosphere, what might you guess about its surface?

2. If liquid water is rare on the surface of planets, then most terrestrial planets must have CO_2-rich atmospheres. Right? Why?

Problems

1. In Figure 20-3, the earthquake occurred 7440 km from the seismograph. How fast did the *P* waves travel in km/s? How fast did the *S* waves travel?

2. What percentage of Earth's volume is taken up by its metallic core?

3. If the Atlantic seafloor is spreading at 3 cm/year and is now 6400 km wide, how long ago were the continents in contact?

4. The Hawaiian-Emperor chain of undersea volcanoes is about 7500 km long. The Pacific plate is moving 9.2 cm a year; how old is the oldest detectable volcano in the chain? What has happened to older volcanoes in the chain?

5. From Hawaii to the bend in the Hawaiian-Emperor chain is about 4000 km. Use the speed of plate motion given in Problem 4 to estimate how long ago the direction of plate motion changed. (The San Andreas Fault in southern California became active at about the same time!)

6. Calculate the age of the Grand Canyon as a percent of Earth's age.

Media Cluster

Ace Astronomy™ To access the resources in the Media Cluster, log into AceAstronomy at **http://ace.brookscole .com/sf9** and select Chapter 20.

ACTIVE FIGURES

Seismic Waves
Use this animation to create an earthquake and see how tracking its seismic waves provides evidence about what lies beneath Earth's surface.

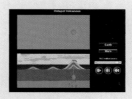

Hot Spot Volcanoes
If the crustal plate over a hot spot is moving, repeated eruptions can result in an island chain like the Hawaiian Islands. Set a series of eruptions in motion in this animation.

ASTRONOMY EXERCISES

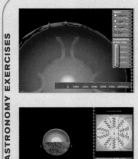

Convection and Plate Tectonics
Study the convection currents of the four Earthlike planets in this animation and see if you can determine how this relates to the amount of plate tectonic activity on each planet.

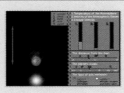

Primary Atmospheres
This exercise lets you change the distance to the sun and mass of a hypothetical planet and study how these relate to the planet's retention of its atmosphere.

Convection and Magnetic Fields
A planet's core temperature and rotation speed are factors in the strength of its magnetic field. See the effect of these variables in this animation.

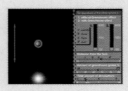

Greenhouse Effect
See how distance to the sun and amount of carbon dioxide in a planet's atmosphere affect its temperature in this animation.

VIRTUAL ASTRONOMY LABS

Lab 5: Planetary Geology
By thoroughly studying Earth's geology, you can leverage your knowledge to understand conditions on other planets. This lab includes an exercise in determining the interior characteristics of a planet by observing the travel times of its seismic waves.

Critical Inquiries for the Web

1. Search the Web to find the location of the most recent earthquakes. Earthquakes somewhere each day. Search for seismographs that will give you information about the most recent earthquakes.

2. Search for information about the arrangement of landmasses on Earth before Pangaea. Were there previous supercontinents? What was Rodinia? What is the evidence?

3. Earth's magnetic poles are not coincident with its axis of rotation, nor are their positions fixed. Search for information about the location of the magnetic poles and their motions. Why do the positions change?

4. What are the oldest known rocks on Earth? Where do they come from, and how were their ages determined?

5. What evidence can you find about global warming? Find examples of the effects of global warming. Business and political leaders do not agree with the consensus among scientists, so be prepared to analyze arguments with great care.

6. Can you locate a current image of the Antarctic ozone hole? Find data for the ozone concentration in the upper atmosphere over the place where you live. What is the latest news concerning ozone depletion?

 Go to the Brooks/Cole Astronomy Resource Center **(http:// astronomy.brookscole.com)** for critical thinking exercises, articles, and additional readings from InfoTrac College Edition, Brooks/Cole's online student library.

21 | The Moon and Mercury: Comparing Airless Worlds

That's one small step

for a man . . . one giant leap

for mankind.

NEIL ARMSTRONG,
ON THE MOON

Beautiful, beautiful.

Magnificent desolation.

EDWIN E. ALDRIN, JR.,
ON THE MOON

F YOU HAD BEEN the first person to step onto the surface of the moon, what would you have said? Neil Armstrong responded to the historic significance of the first human step onto the surface of another world. Buzz Aldrin was second, and he responded to the moon itself. It *is* desolate, and it *is* magnificent. But it is not unusual. Most planets in the universe probably look like Earth's moon, and astronauts may someday walk on such worlds and compare them with our moon. ▌ In this chapter, you begin your study of other worlds by applying comparative planetology to the study of the moon and Mercury. As you compare them with each other and with Earth, you will discover three important themes of planetary astronomy. First, you will see the importance of impact cratering. The surfaces of the moon and Mercury have been scarred by the

▌ Continued on page 498 ▌

When astronauts stepped onto the surface of the moon, they found an unearthly world with no air, no water, weak gravity, and a dusty, cratered surface. Through comparative planetology, the moon reveals a great deal about our own beautiful Earth. (JSC/NASA)

Guidepost

Looking Back

Want to fly to the moon? You will need more than your lunch. There is no air and no water, and the sunlight is strong enough to kill you. Take shelter in the shade, and you will freeze to death in moments. You have never left Earth, so our planet seems normal to you and other worlds are, well, unearthly. In the previous chapter you studied your home world and found it a comfortable place. Now it is time to explore.

This Chapter

The moon and Mercury are small and airless, and that makes them good representatives of the smaller worlds in the universe. As you compare them with Earth, you will discover a number of important principles of comparative planetology, but you should also compare the moon and Mercury with each other. The differences are revealing. As you study these two airless worlds, you will find answers to four essential questions:

Why is the moon airless and cratered?

How did the moon form and evolve?

Why is Mercury different from the moon?

How did Mercury form and evolve?

Looking Ahead

Once you feel comfortable exploring airless worlds, you will be ready for bigger planets that have kept their heat and at least some of their atmospheres. They are not necessarily more interesting places, but they are just a tiny bit less unearthly.

impact of uncountable meteorites in a process that must have affected all the planets. Second, you will see the importance of internal heat in worlds smaller than Earth. Third, you will discover evidence of giant impacts so large they could shatter planets. By concentrating on these three themes, you will be able to organize the flood of details astronomers have learned about the moon and Mercury.

Your ultimate goal in the study of any world is to tell its story. How did the planet form, how has it evolved, and why is it the way it is? Each time you study a new planet, you should conclude your analysis by trying to combine what you have learned to tell the planet's story. For the moon and Mercury, you should use the three themes—cratering, internal heat, and giant impacts—to try to understand how small worlds can become such magnificently desolate places.

21-1 The Moon

ONLY 12 PEOPLE HAVE STOOD on the moon, but planetary astronomers know it well. The photographs, measurements, and samples brought back to Earth paint a picture of an airless, ancient, battered crust.

The View from Earth

A few billion years ago, the moon must have rotated faster, but Earth is four times larger in diameter than the moon (**Celestial Profile 3**), and its tidal forces on the moon are strong. Earth's gravity raised tidal bulges on the moon, and friction in the bulges slowed the moon until it now rotates once each orbit to keep the same side facing Earth in what is called **tidal coupling.** This means that those who live on Earth always see the same side of the moon; the back of the moon is never visible from Earth. The moon's familiar face has shone down on Earth since long before there were humans (**■** Figure 21-1).

Even from Earth, you can tell that the moon is an airless world. How could you tell? First, your understanding of gravity tells you that a world as small as the moon must have a low escape velocity, and gas atoms near its surface escape easily into space. But that's theory and not evidence. What is the evidence? When you look at the moon you can see dramatic shadows near the **terminator,** the dividing line between daylight and darkness. There is no air on the moon to scatter light and soften shadows. Also, with even a small telescope you could watch stars disappear behind the **limb** of the moon—the edge of its disk—without dimming. Clearly, the moon is an airless (and soundless) world.

The most dramatic features on the moon, regions that are visible even to your unaided eyes, are **maria,** Latin for "seas." The dark maria (the singular is *mare*) that make up the face of the man in the moon were so named because early observers thought the

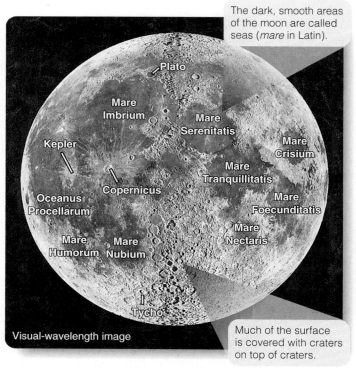

The dark, smooth areas of the moon are called seas (*mare* in Latin).

Much of the surface is covered with craters on top of craters.

Visual-wavelength image

■ Figure 21-1

The side of the moon that faces Earth is a familiar sight. Craters have been named for famous scientists and philosophers, and the so-called seas have been given romantic names. Mare Imbrium is the Sea of Rains and Mare Tranquillitatis is the Sea of Tranquillity. There is, in fact, no water on the moon. (Photo © UC Regents/Lick Observatory)

great smooth expanses were seas. Even a small telescope will show that the maria are marked by ridges, faults, smudges, and craters and therefore can't be water. The maria are actually great, smooth lowland plains covering 17 percent of the moon's surface.

Not many decades ago, astronomers argued about the nature of the maria. Some said that they were great solidified lava flows, but others argued that they were vast oceans of dust blasted from the lunar mountains by billions of years of meteorites and drawn to the lowlands by the lunar gravity. Science fiction writers wrote fascinating stories about the dust oceans. For example, *A Fall of Moon Dust* by Arthur C. Clarke tells of tourists in a lunar sightseeing boat trapped when their boat sinks in the dust to the bottom of a mare. Although it is a wonderful rescue story, it can never happen. Astronauts have walked on the maria and report that the maria are ancient, solidified lava flows with only a few centimeters of loose dust on their surfaces.

These lava flows suggest volcanism, but only small traces of past volcanic activity are visible from Earth. No major volcanic peaks are visible, and no active volcanism has ever been detected. The lava flows that created the maria happened long ago and were much too fluid to build peaks. With a good telescope and some diligent searching you can see a few small domes pushed up by lava below the surface, and you can see long, winding channels called **sinuous rilles** (■ Figure 21-2), often found near the edges of the maria. These channels were evidently cut by flowing lava. In some cases, such a channel may have had a roof of solid rock, forming a lava tube. When the lava drained away, meteorite impacts collapsed the roof to form a sinuous rille. The view from Earth provides only hints of ancient volcanic activity associated with the maria.

Even a small telescope reveals craters on the moon. The lunar surface is scarred with millions of craters formed by the impact of meteorites. Study **Impact Cratering** on pages 500–501 and notice three important points:

1 Impact craters have certain distinguishing characteristics, such as their shape and the way they eject material across the lunar surface.

2 Notice the range of sizes of impact craters from giant basins to microscopic pits.

3 Finally, notice that most of the craters on the moon are old; they were formed long ago when the solar system was young.

Meteorites strike the moon all the time, but large impacts are rare. Meteorites a few tens of meters in diameter probably strike the moon every 50 years or so, but no one has ever seen such an impact with certainty. Small flashes of light have been seen on the dark side of the moon during showers of meteors, but those impacts must have been made by very small objects. A flash of light photographed on the moon in 1953 has been tentatively linked to a small crater just visible in photographs made by spacecraft in orbit around the moon. No significant change has been seen on the moon since the invention of the telescope. Large impacts on the moon and Earth are quite rare, and nearly all of the lunar craters seen through telescopes date from the solar system's youth.

Craters are a key element in planetary study. They are one of the main themes of this chapter, and they will help you understand the moon, Mercury, and many other planets and moons throughout our solar system. You can begin here by using craters to unravel the history of the lunar maria.

Highlands and Lowlands

The surface of the moon is divided into two dramatically different kinds of terrain. The highlands are heavily cratered and are about 3 km higher than the lowlands, which are filled by the mostly uncratered, solidified lava flows known as maria. The number of craters on these surfaces will give you an important clue to understanding these two types of terrain.

There are no folded mountain ranges on the moon. No Rockies, Andes, or Urals stretch across the moon. What appears to be a mountainous region is really the highly cratered lunar highlands. The surface there is saturated with craters, meaning that it would be impossible to form a new crater without destroying the equivalent of one old crater (Figure 21-2).

The moon   Earth

Visual-wavelength images

Earth's moon is about one-fourth the diameter of Earth. Its low density indicates that it contains little iron, but the size of its iron core and the amount of remaining heat are unknown. (NASA)

Celestial Profile 3: The Moon

Motion:

Average distance from Earth	384,400 km (center to center)
Eccentricity of orbit	0.055
Maximum distance from Earth	405,500 km
Minimum distance from Earth	363,300 km
Inclination of orbit to ecliptic	5°9'
Average orbital velocity	1.022 km/s
Orbital period (sidereal)	27.321661 days
Orbital period (synodic)	29.5305882 days
Inclination of equator to orbit	6°41'

Characteristics:

Equatorial diameter	3476 km
Mass	7.35×10^{22} kg (0.0123 $M_{\oplus}$)
Average density	3.36 g/cm³ (3.35 g/cm³ uncompressed)
Surface gravity	0.167 Earth gravity
Escape velocity	2.38 km/s (0.21 $V_{\oplus}$)
Surface temperature	−170° to 130°C (−274° to 266°F)
Average albedo	0.07

Personality Point:

Lunar superstitions are very common. The words *lunatic* and *lunacy* come from *luna*, the moon. Someone who is *moonstruck* is supposed to be at least a bit nutty. Because the moon affects the ocean tides, many superstitions link the moon to water, to weather, and to women's cycle of fertility. According to legend, moonlight is supposed to be harmful to unborn children, but on the plus side, moonlight rituals are said to remove warts.

Impact Cratering

1 The craters that cover the moon and many other bodies in the solar system were produced by the high-speed impact of meteorites of all sizes. Meteorites striking the moon travel 10 to 60 km/s and can hit with the energy of many nuclear bombs.

A meteorite striking the moon's surface can deliver tremendous energy and can produce an impact crater 10 or more times larger in diameter than the meteorite. The vertical scale is exaggerated at right for clarity.

A meteorite approaches the lunar surface at high velocity.

On impact, the meteorite is deformed, heated, and vaporized.

The resulting explosion blasts out a round crater.

Slumping produces terraces in crater walls, and rebound can raise a central peak.

Ace Astronomy™

Log into AceAstronomy and select this chapter to see Active Figure "The Moon's Craters." Notice that the structure of the craters depends on their size.

1a Lunar craters such as Euler, 27 km (17 mi) in diameter, look deep when you see them near the terminator where shadows are long, but a typical crater is only a fifth to a tenth as deep as its diameter, and large craters are even shallower.

Because craters are formed by shock waves rushing outward, by the rebound of the rock, and by the expansion of hot vapors, craters are almost always round, even when the meteorite strikes at a steep angle.

Debris blasted out of a crater is called ejecta, and it falls back to blanket the surface around the crater. Ejecta shot out along specific directions can form bright rays.

Euler

NASA

Visual-wavelength image

1b Rock ejected from distant impacts can fall back to the surface and form smaller craters called **secondary craters**. The chain of craters here is a 45-km-long chain of secondary craters produced by ejecta from the large crater Copernicus 200 km out of the frame to the lower right.

Bright ejecta blankets and rays gradually darken as sunlight darkens minerals and small meteorites stir the dusty surface. Bright rays are signs of youth. Rays from the crater Tycho, perhaps only 100 million years old extend halfway around the moon.

Rays

Tycho

Visual

Visual

NASA

NASA

2 Plum Crater, 40 m (130 ft) in diameter, was visited by Apollo 16 astronauts. Note the many smaller craters visible. Lunar craters range from giant impact basins to tiny pits in rocks struck by **micrometeorites**, meteorites of microscopic size.

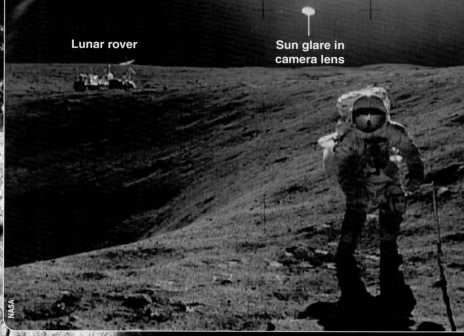

Lunar rover

Sun glare in camera lens

NASA

Visual-wavelength images

Mare Orientale

Solidified lava

2a In larger craters, the deformation of the rock can form one or more inner rings concentric with the outer rim. The largest of these craters are called **multiringed basins**. In Mare Orientale on the west edge of the visible moon, the outermost ring is almost 900 km (550 mi) in diameter.

NASA

2b The energy of an impact can melt rock, some of which falls back into the crater and solidifies. When the moon was young, craters could also be flooded by lava welling up from below the crust.

A few meteorites found on Earth have been identified chemically as fragments of the moon's surface blasted into space by cratering impacts. The fragmented nature of these meteorites indicates that the moon's surface has been battered by impact craters.

3 Most of the craters on the moon were produced long ago when the solar system was filled with debris from planet building. As that debris was swept up, the cratering rate fell rapidly, as shown below.

Rate of Crater Formation

Notice the exponential scale in this graph of the cratering rate. Only half a billion years after the origin of the solar system, the cratering rate had fallen by a factor of 10,000. Today, the rate is very low, and no crater visible from Earth is known to have appeared on the moon in historic times.

Cratering rate

10^6
10^5
10^4
10^3
10^2
10
1

4   3   2   1   Now

Time before present (billion years)

Meteorite from moon

NASA

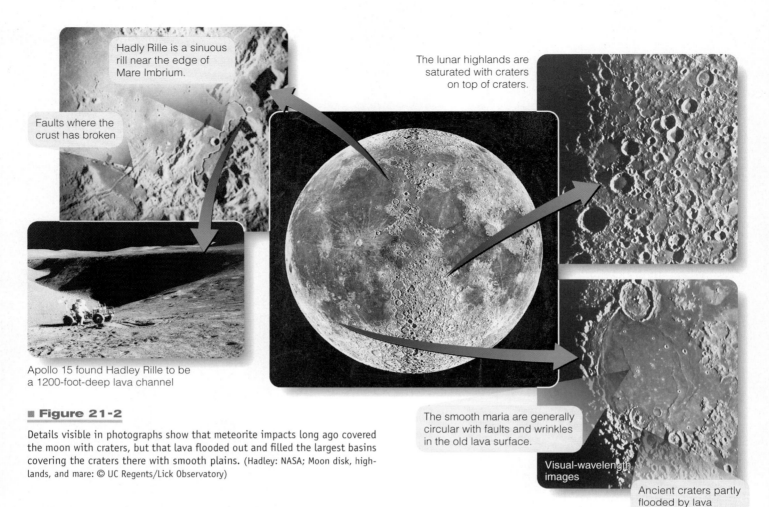

Hadly Rille is a sinuous rill near the edge of Mare Imbrium.

Faults where the crust has broken

The lunar highlands are saturated with craters on top of craters.

Apollo 15 found Hadley Rille to be a 1200-foot-deep lava channel

The smooth maria are generally circular with faults and wrinkles in the old lava surface.

Visual-wavelength images

Ancient craters partly flooded by lava

■ **Figure 21-2**

Details visible in photographs show that meteorite impacts long ago covered the moon with craters, but that lava flooded out and filled the largest basins covering the craters there with smooth plains. (Hadley: NASA; Moon disk, highlands, and mare: © UC Regents/Lick Observatory)

The maria fill the lunar lowlands, and they contain few craters, as you can see in Figure 21-2. Among the few craters that do scar the maria, many have bright rays, suggesting that they are younger craters. For examples, locate the craters Kepler and Copernicus in Figure 21-1.

You can conclude from this that the highlands are old regions where you see craters from early in the history of the solar system. Most of those craters were formed during the heavy bombardment over 4 billion years ago. The maria that fill the lowlands must have formed later, after the decline of the heavy bombardment. That means the maria are younger than the highlands. These are called **relative ages** because you can tell which features formed first, but you can't give ages in years.

The cratering curve shown on page 501 can reveal **absolute ages,** meaning ages in years. Studies of cratered surfaces in the solar system have allowed planetary scientists to calibrate the cratering curve. The lava flows that filled the lowlands and covered up the craters there could not have formed earlier than 3 to 4 billion years ago. Had they occurred earlier, the surfaces of the maria would show more craters.

Absolute ages derived from the cratering curve cannot be precise because planetary scientists suspect that the cratering rate may not have fallen as smoothly as the cratering curve suggests. Al-

though the debris in the early solar system was gradually removed, bursts of cratering may have occurred. A major impact between large planetesimals, for example, might shower the solar system with fragments and produce a storm of crater impacts. Mathematical models suggest that the final formation of Uranus and Neptune might have flung icy planetesimals through the solar system like giant shrapnel. The true cratering curve might have rises and dips superimposed on a gradual decline.

Whatever the exact timing of the heavy bombardment, it is clear that the craters in the highlands are old and that the later lava flows that formed the maria covered over any craters that had been formed in the lowlands. Using such clues, astronomers were able to begin telling the story of the moon's surface (**Window on Science 21-1**). But the view from Earth does not provide enough evidence. To really understand the lunar surface, humans had to go there.

The Apollo Missions

On May 25, 1961, President John Kennedy addressed Congress and committed the United States to landing a human being on the moon by 1970. The reasons for that decision are complicated

How Hypotheses and Theories Unify the Details

When scientists create a hypothesis, it draws together a great many observations and measurements. (Phyllis Leber)

Like any technical subject, science includes a mass of details, facts, figures, measurements, and observations. It is easy to be overwhelmed by the flood of details, but one of the most important characteristics of science comes to your rescue. The goal of science is not to discover more details but to explain the details with a unifying hypothesis or theory. A good theory is like a basket that makes it easier for you to carry a large assortment of details.

This is true of all the sciences. When a psychologist begins studying the way the human eye and brain respond to moving points of light, the data are a sea of detailed measurements and observations. Once the psychologist forms a hypothesis about the way the eye and brain interact, the details fall into place as parts of a logical story. If you understand the hypothesis, the details all fit together and make sense, and you can remember the details without blindly memorizing tables of facts and figures. Remember that the goal of science is understanding nature, not memorizing details.

Scientists are in the storytelling business. The stories are often called hypotheses or theories, but they are really just stories to explain how nature works. The difference between scientific stories and works of fiction lies in the use of facts. Scientific stories are constructed to fit all known facts and are then tested over and over against new facts obtained by observation and experiment.

When you try to tell the story of each planet in our solar system, you pull together all the hypotheses, theories, and observations and try to make them into a logical history of how the planet got to be the way it is. Of course, your stories will be incomplete because astronomers don't understand all the factors in planetary astronomy. Nevertheless, your story of each planet will draw together the known facts and details and attempt to make them a logical whole.

Memorizing a list of facts can give you a false feeling of security, just as when someone memorizes the names of things without understanding them (Window on Science 2-2). Rather than memorizing facts, you should search for the unifying hypothesis that pulls the details together into a single story. That is the goal of science.

---

and related more to economics, international politics, and the stimulation of technology than to science, but the Apollo program became a fantastic adventure in science, a flight to the moon that changed how we all think about Earth.

Flying to the moon is not particularly difficult; with powerful enough rockets and enough food, water, and air, it is a straightforward trip. Landing on the moon is more difficult but not impossible. The moon's gravity is only one-sixth that of Earth, and there is no atmosphere to disturb the trajectory of the spaceship. Getting to the moon is simple and landing is possible; the difficulty is doing both on one trip. The spaceship must carry food, water, and air for a number of days in space plus fuel and rockets for midcourse corrections and for a return to Earth. All of this adds up to a ship that is too massive to make a safe landing on the lunar surface. The solution was to take two spaceships to the moon, one to ride in and one to land in (■ Figure 21-3).

The command module was the long-term home and command center for the trip. Three astronauts had to live in it for a week, and it had to carry all the life-support equipment, navigation instruments, computers, power packs, and so on for a week's jaunt in space. The lunar landing module (LM for short) was tacked to the front of the command module like a bicycle strapped to the front of the family camper. It carried only enough fuel and supplies for the short trip to the lunar surface, and it was built to minimize weight and maximize maneuverability.

The weaker gravity of the moon made the design of the LM simpler. Landing on Earth requires reclining couches for the astronauts, but the trip to the lunar surface involved smaller accelerations. In an early version of the LM, the astronauts sat on what looked like bicycle seats, but these were later scrapped to save weight. The astronauts had no seats at all in the LM, and once they began their descent and acquired weight, they stood at the controls held by straps, riding the LM like two daredevils riding a rocket surfboard.

Lifting off from the lunar surface, the LM saved weight by leaving the larger descent rocket and support stage behind. Only the astronauts, their instruments, and their cargo of rocks returned to the command module orbiting above. Again, the astronauts in the LM blasted up from the lunar surface standing at the controls. The rocket engine that lifted them back into lunar orbit was not much bigger than a dishwasher.

The most complicated parts of the trip were the rendezvous and docking between the tiny remains of the LM and the command module, aided by radar systems and computers. The astronauts rejoined their colleague in the command module, transferred their moon rocks, and jettisoned the LM. Only the command module returned to Earth.

The first manned lunar landing was made on July 20, 1969. While Michael Collins waited in orbit around the moon, Neil Armstrong and Edwin Aldrin, Jr., took the LM down to the

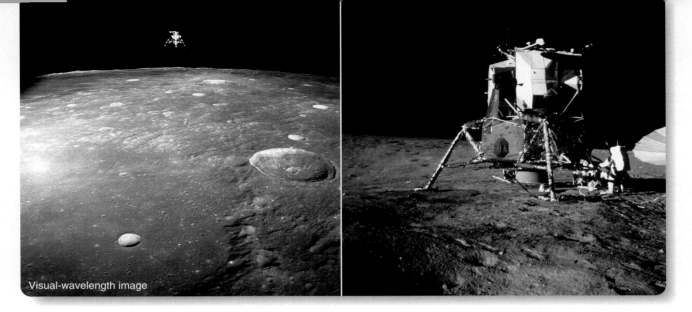

■ Figure 21-3

The Apollo 12 lunar module. The two astronauts stood in a space hardly bigger than two telephone booths. The metal skin was so thin it was easily flexible, like metal foil, and the legs of the module, designed specifically for the moon's weak gravity, could not support the lander's weight on Earth. Only the upper half of the lander blasted off from the surface to return the astronauts to the command module in orbit around the moon. (NASA)

surface. Although much of the descent was controlled by computers, the astronauts had to override a number of computer alarms and take control of the LM to avoid a boulder-strewn crater bigger than a football field. Climbing down the ladder, Armstrong and then Aldrin stepped onto the lunar surface.

Between July 1969 and December 1972, 12 people reached the lunar surface and collected 380 kg (840 lb) of rocks and soil (■ Table 21-1). The flights were carefully planned to visit different regions and develop a comprehensive history of the lunar surface.

■ Table 21-1 I Apollo Lunar Landings

Apollo Mission*	Astronauts: Commander LM Pilot CM Pilot	Date	Mission Goals	Sample Weight (kg)	Typical Samples	Ages (10^9 y)
11	Armstrong Aldrin Collins	July 1969	First manned landing; Mare Tranquillitatis	21.7	Mare basalts	3.48–3.72
12	Conrad Bean Gordon	Nov. 1969	Visit Surveyor 3; sample Oceanus Procellarum (mare)	34.4	Mare basalts	3.15–3.37
14	Shepard Mitchell Roosa	Feb. 1971	Fra Mauro, Imbrium ejecta sheet	42.9	Breccia	3.85–3.96
15	Scott Irwin Worden	July 1971	Edge of Mare Imbrium and Apennine Mountains, Hadley Rille	76.8	Mare basalts; highland anorthosite	3.28–3.44 4.09
16	Young Duke Mattingley	April 1972	Sample highland crust; Cayley formation (ejecta); Descartes	94.7	Highland basalt; breccia	3.84 3.92
17	Cernan Schmitt Evans	Dec. 1972	Sample highland crust; dark halo craters; Taurus–Littrow	110.5	Mare basalt; highland breccia; fractured dunite	3.77 3.86 4.48

*The Apollo 13 mission suffered an explosion on the way to the moon and did not land.

Apollo 17, the last Apollo mission to the moon, landed in the highlands in December 1972.

Apollo 11 landed in the lunar lowlands in July 1969.

■ **Figure 21-4**

Apollo 11, the first mission to the moon, landed on the smooth surface of Mare Tranquillitatis in the lunar lowlands and the horizon was straight and level. When Apollo 17 landed at Taurus–Littrow in the lunar highlands, the astronauts found the horizon mountainous and the terrain rugged. Landing sites for the other Apollo missions are shown. (Moon: © UC Regents/Lick Observatory; Astronauts: NASA)

The first flights went into relatively safe landing sites (■ Figure 21-4)—Mare Tranquillitatis for Apollo 11 and Oceanus Procellarum for Apollo 12. Apollo 13 was aimed at a more complicated site, but an explosion in an oxygen tank on the way to the moon ended all chances of a landing and nearly cost the astronauts their lives. They succeeded in using the life support in the LM to survive the trip around the moon, and they eventually landed safely in the crippled command module.

The last four Apollo missions, 14 through 17, sampled geologically important places on the moon. Apollo 14 visited the Fra Mauro region, which is covered by ejecta from the impact that dug the basin now filled by Mare Imbrium. Apollo 15 visited the edge of Mare Imbrium at the foot of the Apennine Mountains and examined Hadley Rille (Figure 21-2). Apollo missions 16 and 17 visited highland regions to sample the older parts of the lunar crust (Figure 21-4). Almost all of the lunar samples from these six landings are now held at the Planetary Materials Laboratory at the Johnson Space Center in Houston. They are a national treasure containing clues to the beginnings of our solar system.

Moon Rocks

If you visited the moon, where the gravity is one-sixth that on Earth, you would discover that you weighed only one-sixth what you weigh on Earth. The Apollo astronauts found it easy to dance across the lunar surface conducting experiments and collecting geological specimens.

Of all of the rock samples that the Apollo astronauts carried back to Earth, every one is igneous. That is, they formed by the cooling and solidification of molten rock. No sedimentary rocks were found, which is consistent with the moon never having had liquid water on its surface. In addition, the rocks were extremely dry. Almost all Earth rocks contain 1 to 2 percent water, either as free water trapped in the rock or as water molecules chemically bonded with certain minerals. But moon rocks contain no water at all.

Rocks from the lunar maria are dark-colored, dense basalts much like the solidified lava produced by the Hawaiian volcanoes (■ Figure 21-5). These rocks are rich in heavy elements such as iron, manganese, and titanium, which give them their dark color. Some of the basalts are **vesicular,** meaning that they contain holes caused by bubbles of gas in the molten rock. Like bubbles in a carbonated beverage, these bubbles do not form while the magma is under pressure. Only when the molten rock flows out onto the surface, where the pressure is low, do bubbles appear. The presence of vesicular basalts assures that these rocks

The Apollo astronauts found that all moon rocks are igneous, meaning they solidified from molten rock.

Rocks exposed on the lunar surface become pitted by micrometeorites.

Vesicular basalt contains bubbles frozen into the rock when it was molten.

A breccia is formed by rock fragments bonded together by heat and pressure.

■ Figure 21-5

Rocks returned from the moon show that the moon formed in a molten state, that it was heavily fractured by cratering when it was young, and that it is now affected mainly by micrometeorites grinding away at surface rock. (NASA)

formed in lava flows that reached the surface and did not solidify underground.

Absolute ages of the mare basalts can be found from the radioactive atoms they contain (Chapter 19), and these ages range from about 3.1 to 3.8 billion years. That means the lava flows that formed these rocks must have happened some time after the end of the heavy bombardment.

The highlands are composed of low-density rock containing calcium-, aluminum- and oxygen-rich minerals that would be among the first to solidify and float to the top of molten rock. The crustal rocks range in age from 4.0 to 4.5 billion years old, significantly older than the mare basalts. The highlands contain **anorthosite,** a light-colored rock that contributes to the highlands' bright contrast with the dark, iron-rich basalts of the lowlands. The rocks of the highlands, although badly shattered by impacts, represent the moon's original low-density crust, whereas the mare basalts rose as molten rock from the deep crust and upper mantle.

A large fraction of the lunar rocks are **breccias,** rocks that are made up of fragments of earlier rocks cemented together by heat and pressure. Evidently, meteorite impacts have broken up the rocks and fused them together time after time.

If you went to the moon, you would get your spacesuit dirty. Both the highlands and the lowlands of the moon are covered by a layer of powdered rock and crushed fragments called the **regolith.** It is about 10 m deep on the maria but over 100 m deep in certain places in the highlands. About 1 percent of the regolith is meteoric fragments; the rest is the smashed remains of moon rocks that have been pulverized by the constant rain of meteorites. The smallest meteorites, the micrometeorites, do the most damage by their constant sandblasting of the lunar surface, which grinds the rock down to fine dust. The Apollo astronauts found that the dust coated their spacesuits and equipment and made a mess where it got tracked into the Lunar Landing Module. From all this you can see that impact cratering, a theme of this chapter, dominates the lunar surface and is even responsible for the lunar regolith.

Understanding Natural Processes: Follow the Energy

One of the best ways to think about a scientific problem is to follow the energy.

According to the principle of cause and effect, every effect must have a cause, and every cause must involve energy (Focus on Fundamentals 2). Energy moves from regions of high concentration to regions of low concentration and in doing so produces changes. For example, coal burns to make steam in a power plant, and the steam passes through a turbine and then escapes into the air. In flowing from the burning coal to the atmosphere, the heat spins the turbine and makes electricity.

It is surprising how commonly scientists use energy as a key to understanding nature. A biologist might ask where certain birds get the energy to fly thousands of miles, and an economist might ask where the economic energy to support the creation of new investments comes from. Energy is everywhere; and when it moves, whether it is birds, money, or molten magma, it causes change. Energy is the cause in cause and effect.

In earlier chapters, the flow of energy from the inside of a star to its surface helped you understand how stars, including the sun, work. You saw that the outward flow of energy supports the star against its own weight, drives convection currents that produce magnetic fields, and causes surface activity such as spots, prominences, and flares. You were able to understand stars because you could follow the flow of energy outward from their interiors.

You can think of a planet by following the energy. The heat in the interior of a planet may be left over from the formation of the planet, or it may be heat generated by radioactive decay, but it must flow outward toward the cooler surface, where it is radiated into space. In flowing outward, the heat can cause convection currents in the mantle, magnetic fields, plate motions, quakes, faults, volcanism, mountain building, and more.

When you think about any world, be it a small asteroid or a giant planet, think of it as a source of heat that flows outward through the planet's surface into space. If you can fol-

Heat flows out of Earth's interior and generates geological activity such as that at Yellowstone National Park. (M. Seeds)

low that energy flow, you can understand a great deal about the world. A planetary astronomer once said, "The most interesting thing about any planet is how its heat gets out."

The moon rocks are old, dry, igneous, and badly shattered by impacts. You can use these facts, combined with what you know about lunar features, to tell the story of the moon.

The History of the Moon

The four-stage history of the moon is dominated by a single fact—it is small, only one-fourth the diameter of Earth. The escape velocity is low, so it has been unable to hold any atmosphere, and it cooled rapidly as its internal heat flowed outward into space. As you will see when you visit other worlds, a planet's geology is largely driven by energy in the form of heat flowing outward from its interior (**Window on Science 21-2**). Small worlds have less heat and lose it more rapidly, so the moon's small size has been critical in its history.

The Apollo moon rocks show that the moon must have formed in a molten state. Planetary geologists now refer to the newborn moon as a sea of magma. Denser materials sank to form a small core; and, as the magma ocean cooled, low-density minerals floated to the top to form a low-density crust. In this way the moon must have differentiated into core, mantle, and crust. The radioactive ages of the moon rocks show that the surface solidified from 4.6 to 4.1 billion years ago. The moon has a low density and no magnetic field, so its dense core must be small. The core may still retain enough heat to be partially molten, but it

can't contain much iron or the dynamo effect would produce a magnetic field.

The second stage, cratering, began as soon as the crust solidified, and the older highlands show that the cratering was intense during the first 0.5 billion years—during the heavy bombardment at the end of planet building.

The moon's crust was shattered to a depth of 10 kilometers or so, and the largest impacts formed giant multiringed crater basins hundreds of kilometers in diameter, such as Mare Orientale. This led to the third stage—flooding. Although Earth's moon cooled rapidly after its formation, some process such as radioactive decay heated material deep in the crust, and part of it melted. Molten rock followed the cracks up to the surface and flooded the giant basins with successive lava flows of dark basalts from 3.8 to 3.2 billion years ago. This formed the maria (■ Figure 21-6).

Make a special note that the lava that floods out on the surfaces of planets does not come from the molten core. That is a common misconception. The lava comes from the lower crust and possibly from the top of the mantle. The pressure is lower there, and that lowers the melting point of the rock. Radioactive decay can generate enough heat to melt portions of the rock, and if there is enough heat and if there are faults and cracks, the magma can reach the surface and form volcanoes and lava flows. Whenever you see lava flows on a planet, you can be sure heat is flowing

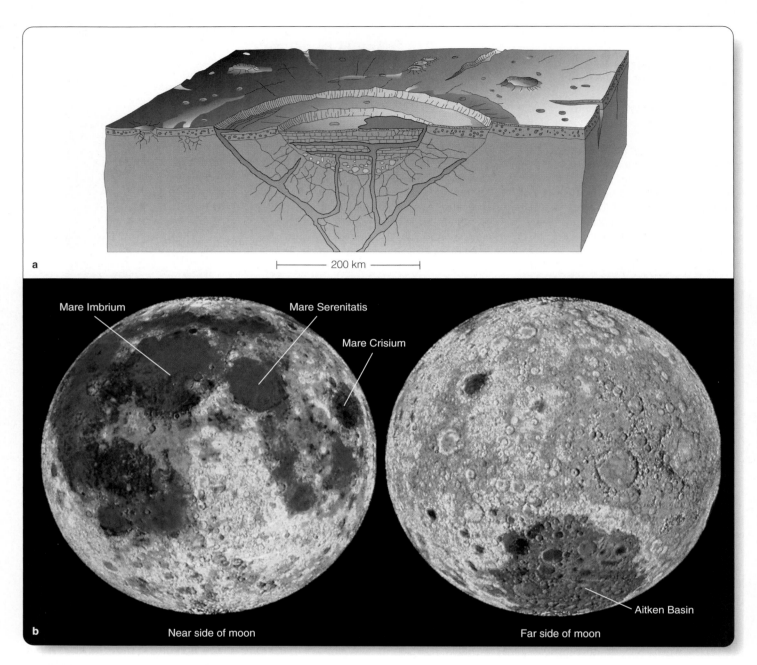

■ Figure 21-6

(a) Major impacts broke the lunar crust and produced large basins. Lava flooded up through the fractures and flooded the basins to produce maria. (Adapted from a diagram by William Hartmann) (b) In these maps, color indicates elevation, with red the highest regions and purple the lowest. Maria on the near side are generally circular, low, lava-filled basins. Note the circular outline of the labeled maria. The crust on the far side is thicker, and there is less flooding. Even the Aitken Basin contains little lava flooding. (NASA/ Clementine)

out of the interior (one of the themes of this chapter), but the lava itself did not come all the way from the core.

Some maria on the moon, such as Mare Imbrium, Mare Serenitatis, Mare Humorum, and Mare Crisium, retain their round shapes, but others are irregular because the lava overflowed the edges of the basin or because the shape of the basin was modified by further cratering. The moving lava formed channels that are seen from Earth as sinuous rills. The weight of the maria pressed the crater basins downward, and the solidified lava was compressed and formed wrinkle ridges visible even in small telescopes. The tension at the edges of the maria broke the hard lava to produce straight fractures and faults. All of these features are visible in

Apollo 14 landed on rugged terrain suspected of being ejecta from the Imbrium impact. The large boulder here is ejecta that, at some time in the past, fell here from an impact beyond the horizon. (NASA)

■ **Figure 21-8**

Lava flooding after the end of the heavy bombardment filled a giant multiringed basin and formed Mare Imbrium. (Don Davis)

Origin of Mare Imbrium

Four billion years ago, an impact forms a multiringed basin over 1000 km (700 mi) in diameter.

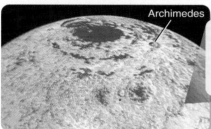

Continuing impacts crater the surface but do not erase the high walls of the multiringed basin.

Archimedes

Impacts form a few large craters, and, starting about 3.8 billion years ago, lava floods low regions.

Repeated lava flows cover most of the inner rings and overflow the basin to merge with other flows.

Kepler   Copernicus

Impacts continue, including those that formed the relatively young craters Copernicus and Kepler.

Figure 21-2. Later cratering and overlapping lava floods further modified the maria. Consequently you should think of the maria as accumulations of features reflecting the moon's complex history.

Mare Imbrium is a dramatic example of how the great basins became the maria. Near the end of the heavy bombardment, roughly 4 billion years ago, a planetesimal the size of Rhode Island struck the moon and blasted out a giant multiringed basin. The impact was so violent the ejecta blanketed 16 percent of the moon's surface. Apollo 14 landed at Fra Mauro to sample the ejecta (■ Figure 21-7). After the cratering rate fell at the end of the heavy bombardment, lava flows welled up time after time and flooded the Imbrium Basin, burying all but the highest parts of the giant multiringed basin. The Imbrium Basin is now a large, generally round mare marked by only a few craters that have formed since the last of the lava flows (■ Figure 21-8).

This story of the moon might suggest that it was a violent place during the cratering phase, but large impacts were in fact rare; and, except for a continuous rain of smaller meteorites, the moon was, for the most part, a peaceful place even during the heavy bombardment. Had you stood on the moon at that time you would have experienced a continuous rain of micrometeorites and much less common pebble-size impacts.

Centuries might pass between major impacts. Of course, a large impact far beyond the horizon might have threatened to bury you under ejecta or jolted you by seismic shocks. You could have felt the Imbrium impact anywhere on the moon, but had you been standing on the side of the moon opposite that impact, you would have been at the focus of seismic waves traveling around the moon from different directions. When the waves met under your feet, the surface would have jerked up and down by as much as 10 m. You would not have liked that. The place on the moon opposite the Imbrium Basin is a strangely disturbed landscape called **jumbled terrain.** You will see similar effects of large impacts on other worlds.

Studies of our moon show that its crust is thinner on the side toward Earth, perhaps due to tidal effects. Consequently, while lava flooded the basins on the Earthward side, it was unable to rise through the thicker crust to flood the lowlands on the far side. The largest impact basin in the solar system is the south pole Aitkin Basin on the moon (Figure 21-6b). It is about 2500 kilometers (1500 miles) in diameter and as deep as 13 kilometers (8 miles) in places, but flooding has never filled it with smooth lava flows.

The fourth stage, slow surface evolution, is limited because our moon lacks water and has cooled rapidly. Flooding on Earth included water, but the moon has never had an atmosphere and thus has never had liquid water. With no air and no water, erosion is limited to the constant bombardment of micrometeorites and rare larger impacts.

The moon is small, and small worlds cool rapidly because they have a large ratio of surface area to volume. You may have noticed that a small cupcake fresh from the oven cools more rapidly than a large cake. The moon has lost much of its internal heat, and it is the outward flow of heat that produces geological activity. That means the moon is mostly inactive. The crust of the moon rapidly grew thick and never divided into moving plates. There are no rift valleys or folded mountain chains on the moon. The last lava flows on the moon ended one or two billion years ago when the moon's temperature fell too low to maintain subsurface lava.

The overall terrain on the moon is almost fixed. On Earth a billion years from now, plate tectonics will have totally changed the shapes of the continents, and erosion will have long ago worn away the Rocky Mountains. On the moon, impacts will have formed only a few more large craters, and nearly all of the lunar scenery will be unchanged. Micrometeorites, however, will have blasted the soil, erasing the footprints left by the Apollo astronauts and reducing the equipment they left behind to peculiar chemical contamination in the regolith at the six Apollo landing sites.

You have studied the story of the moon's evolution in detail for later comparison with other planets and moons in our solar system, but the story has skipped one important question: Where did Earth get such a large satellite?

The Origin of Earth's Moon

Over the last two centuries, astronomers developed three different hypotheses for the origin of Earth's moon, but these traditional ideas have failed to survive comparison with the evidence. A the-ory proposed in the 1980s may hold the answer. You can begin by testing the three unsuccessful theories against the evidence to see why they failed.

The first of the three traditional theories, the **fission hypothesis,** supposes that the moon formed by the fission of Earth. If the young Earth spun fast enough, tides raised by the sun might make it break into two parts (■ Figure 21-9a). If this separation occurred after Earth differentiated, the moon would have formed from crust material. In that way, the theory explains the moon's low density.

But the fission theory has problems. No one knows why the young Earth should have spun so fast, nearly ten times faster than today, nor where all that angular momentum went after the fission. In addition, the moon's orbit is not in the plane of Earth's equator, as it would be if it had formed by fission.

The second traditional theory is the **condensation** (or double-planet) **hypothesis.** It supposes that Earth and the moon condensed as a double planet from the same cloud of material (Figure 21-9b). However, if they formed from the same material, they should have the same chemical composition and density, which they don't. The moon is very poor in iron and related elements such as titanium and in volatiles such as water vapor and sodium. Yet the moon contains almost exactly the same ratios of oxygen isotopes as does Earth's mantle. The condensation theory cannot explain these compositional differences.

The third theory is the **capture hypothesis.** It supposes that the moon formed somewhere else and was later captured by Earth (Figure 21-9c). You might imagine the moon forming inside the orbit of Mercury, where the heat would prevent the condensation of solid metallic grains and only high-melting-point metal oxides could have solidified. A later encounter with Mercury could "kick" the moon out to Earth.

The capture theory was not popular because it requires highly unlikely events involving interactions with Mercury and Earth to move the moon from place to place. Scientists are always suspicious of explanations that require a chain of unlikely coincidences. Also, upon encountering Earth, the moon would have been moving so rapidly that Earth's gravity would have been unable to capture it without ripping the moon to fragments through tidal forces.

Until recently, astronomers were left with no acceptable theory to explain the origin of the moon, and they occasionally joked that the moon could not exist. But during the 1980s, plan-

■ **Figure 21-9**

Three traditional theories for the moon's origin. (a) Fission theories suppose that Earth and the moon were once one body and broke apart. (b) Condensation theories suppose that the moon formed at the same time and from the same material as Earth. (c) Capture theories suggest that the moon formed elsewhere and was captured by Earth. None of these theories explains all the facts.

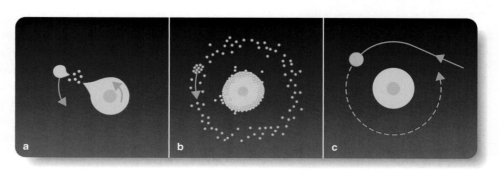

etary astronomers developed a new theory that combines the best aspects of the fission hypothesis and the capture hypothesis.

The **large-impact hypothesis** supposes that the moon formed from debris ejected into a disk around Earth by the impact of a large body. The impacting body may have been twice as large as Mars. Instead of saying that Earth was hit by a large body, it may be more nearly correct to say that Earth and the moon resulted from the collision and merger of two very large planetesimals. The resulting large body became Earth, and the ejected debris formed the moon (■ Figure 21-10). Such an impact would have melted the proto-Earth, and the material falling together to form the moon would have been heated hot enough to melt. This fits the evidence from moon rocks that the moon formed as a sea of magma.

This hypothesis would explain a number of things. The collision had to occur at a steep angle to eject enough matter to make the moon. That is, the objects did not collide head-on. Such a glancing collision would have spun the resulting material rapidly enough to explain the observed angular momentum in the Earth–moon system. If the two colliding planetesimals had already differentiated, the ejected material would be mostly iron-poor mantle and crust. Calculations show that the iron core of the impacting body could have fallen into the larger body that became Earth. This would explain why the moon is so poor in iron and why the abundances of other elements are so similar to rocks from Earth's mantle. Also, the material that eventually became the moon would have remained in a disk long enough for volatile elements, which the moon lacks, to be lost to space.

The moon may be the remains of a giant impact. Until recently, astronomers have been reluctant to consider such catastrophic events, but a number of lines of evidence suggest that planets have been affected by giant impacts. Consequently, the third theme identified in the introduction to this chapter, giant impacts, has the potential to help you understand other worlds. Catastrophic events are rare, but they can occur.

Building Scientific Arguments

If the moon was intensely cratered by the heavy bombardment and then formed great lava plains, why didn't the same thing happen on Earth?

Is this argument obvious? It is still worth reviewing as a way to test your understanding. In fact, the same thing did happen on Earth. Although the moon has more craters than Earth, the moon and Earth are the same age, and both were battered by meteorites during the heavy bombardment. Some of those impacts on Earth must have been large and dug giant multiringed basins. Lava flows must have welled up through Earth's crust and flooded the lowlands to form great lava plains much like the lunar maria.

Earth, however, is a larger world and has more internal heat, which escapes more slowly than the moon's heat did. The moon

The Large-Impact Hypothesis

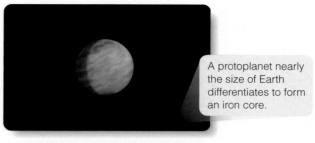

A protoplanet nearly the size of Earth differentiates to form an iron core.

Another body that has also formed an iron core strikes the larger body and merges, trapping most of the iron inside.

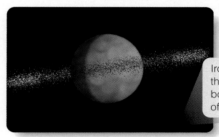

Iron-poor rock from the mantles of the two bodies forms a ring of debris.

Volatiles are lost to space as the particles in the ring begin to accrete larger bodies.

Eventually the moon forms from the iron-poor and volatile-poor matter in the disk.

■ **Figure 21-10**

Sometime before the solar system was 50 million years old, a collision produced Earth and the moon in its inclined orbit.

is now geologically dead, but Earth is very active, with heat flowing outward from the interior to drive plate tectonics. The moving plates long ago erased all evidence of the cratering and lava flows dating from Earth's youth.

Comparative planetology is a powerful tool in that it allows you to see similar processes occurring under different circumstances. For example, expand your argument to explain a different phenomenon. **Why doesn't the moon have a magnetic field?**

■   ■   ■

Connections: The moon's evolution has been restricted by its size. It is too small to have an atmosphere, and it lost its internal heat so rapidly that it never developed plate motion. Another world in our solar system has also suffered from its diminutive size. Mercury, like the moon, is a small, rocky world, cratered by impacts and flooded by ancient lava flows.

21-2 Mercury

EARTH'S MOON AND MERCURY are good subjects for comparative planetology. They are similar in a number of ways. Their rotation has been altered by tides, their surfaces are heavily cratered, their lowlands are flooded in places by ancient lava flows, and both are small, airless, and have ancient, inactive surfaces. Yet the differences between them are impressive and will help you understand the nature of these airless worlds.

Mercury is the innermost planet in the solar system (**Celestial Profile 4**), and thus its orbit keeps it near the sun in the sky. It is sometimes visible near the horizon in the evening sky after sunset or in the dawn sky just before sunrise. An Earth-based telescope shows Mercury passing through phases like the moon, but little detail is visible. Most of what is known about the surface of Mercury is based on photographs and measurements made by the Mariner 10 spacecraft when it looped through the inner solar system in 1974 and 1975 (■ Figure 21-11). Until then, Mercury was a mystery world.

Rotation and Revolution

During the 1880s, the Italian astronomer Giovanni Schiaparelli sketched the faint features he thought he saw on the disk of Mercury and concluded that the planet was tidally coupled to the sun just as the moon is tidally coupled to Earth. That is, he concluded that Mercury rotated on its axis to keep the same side facing the sun throughout its orbit. This was actually a very good guess because, as you will see, tidal coupling between rotation and revolution is common in the solar system. But the rotation of Mercury is more complex than Schiaparelli thought.

In 1962, radio astronomers detected radio emissions from the planet and concluded that the dark side was not as cold as it should have been if the planet kept one side perpetually in darkness. In 1965, radio astronomers made radar contact with Mercury. They used the 305-m Arecibo dish (Figure 6-19) to transmit a pulse of radio energy at Mercury and then waited for the re-

■ Figure 21-11

Photomosaic of Mercury made by the Mariner 10 spacecraft. Caloris Basin is in shadow at left. Almost no surface detail is visible from Earth (inset). (NASA; inset courtesy Lowell Observatory)

flected signal to return. Doppler shifts in the reflected radio pulse showed that the planet was rotating with a period of only about 59 days, much shorter than the orbital period of 87.969 days.

Apparently, Mercury is tidally coupled to the sun such that it rotates not once per orbit but 1.5 times per orbit. That is, its period of rotation is two-thirds its orbital period, or 58.65 days. This means that a mountain on Mercury directly below the sun at one place in its orbit will point away from the sun one orbit later and toward the sun after the next orbit (■ Figure 21-12).

If you flew to Mercury and landed your spaceship in the middle of the day side, the sun would be high overhead, and it would be noon. Your watch would show that almost 44 Earth days passed before the sun set in the west, and a total of 88 Earth days would have passed before the sun reached the midnight position. In those 88 Earth days, Mercury would have completed one orbit around the sun (Figure 21-12). It would require another entire orbit of Mercury for the sun to return to the noon position overhead. So a full day on Mercury is two Mercury years long!

The complex tidal coupling between the rotation and revolution of Mercury is an important illustration of the power of tides. Just as the tides in the Earth–moon system have slowed the moon's rotation and locked it to Earth, so have the sun–Mercury tides slowed the rotation of Mercury and coupled its rotation to its revolution. Astronomers refer to such a relationship as a **resonance.** You will see many such resonances as you explore the solar system.

Like its rotation, Mercury's orbital motion is also peculiar. Recall from Chapter 5 that Mercury's orbit precesses faster than

The Rotation of Mercury

Imagine a mountain on Mercury that points at the sun. It is noon at the mountain.

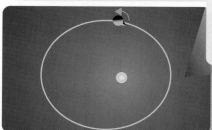

The planet orbits and rotates in the same direction, counter-clockwise as seen from the north.

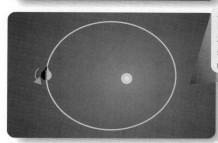

After half an orbit, Mercury has rotated 3/4 of a turn, and it is sunset at the mountain.

As the planet continues along its orbit, rotation carries the mountain into darkness.

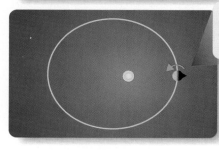

After one orbit, Mercury has rotated 1.5 times, and it is midnight at the mountain.

■ Figure 21-12

Mercury's orbital motion and rotation are related. It orbits the sun in 88 days and rotates on its axis in two-thirds of that time. One full day on Mercury from noon to noon takes two full orbits.

can be explained by Newton's laws but at just the rate predicted by Einstein's theory of general relativity. The orbital motion of Mercury is taken as strong confirmation of the curvature of space-time as predicted by general relativity.

Mercury is a bit over one-third the diameter of Earth, but its high density must mean it has a large iron core. The amount of heat it retains is unknown. (NASA)

Celestial Profile 4: Mercury

Motion:

Average distance from the sun	0.387 AU (5.79×10^7 km)
Eccentricity of orbit	0.2056
Maximum distance from the sun	0.467 AU (6.97×10^7 km)
Minimum distance from the sun	0.306 AU (4.59×10^7 km)
Inclination of orbit to ecliptic	7°00′16″
Average orbital velocity	47.9 km/s
Orbital period	87.969 days (0.24085 y)
Period of rotation	58.646 days (direct)
Inclination of equator to orbit	0°

Characteristics:

Equatorial diameter	4878 km (0.382 $D_\oplus$)
Mass	3.31×10^{23} kg (0.0558 $M_\oplus$)
Average density	5.44 g/cm³ (5.4 g/cm³ uncompressed)
Surface gravity	0.38 Earth gravity
Escape velocity	4.3 km/s (0.38 $V_\oplus$)
Surface temperature	−173° to 430°C (−280° to 800°F)
Average albedo	0.1
Oblateness	0

Personality Point:

Mercury lies very close to the sun and completes an orbit in only 88 days. For this reason, the ancients named the planet after Mercury, the fleet-footed messenger of the gods. The name is also applied to the element mercury, which is also known as quicksilver because it is a heavy, quickly flowing silvery liquid at room temperatures.

The Surface of Mercury

Because the days are so long on Mercury and because it is close to the sun and too small to keep an atmosphere, the temperatures on Mercury are extreme. If you stood in direct sunlight on Mercury, you would hear your spacesuit's cooling system cranking up to top speed as it tried to keep you cool. Daytime temperatures can exceed 700 K (800°F), although about 500 K is a more usual high temperature. If you stepped into shadow on Mercury or took a walk at night, with no atmosphere to distribute heat, your spacesuit heaters would struggle to keep you warm. The surface can cool to 100 K (−280°F). Nights on Mercury are bitter cold. Don't go to Mercury in a cheap spacesuit.

In appearance, Mercury looks much like Earth's moon. It is heavily battered, with craters of all sizes, including some large basins. Some craters are obviously old and degraded; others seem quite young and have bright rays of ejecta. However, a quick glance at photos of Mercury shows no large, dark maria like the moon's flooded basins.

The largest basin on Mercury is called Caloris Basin after the Latin word for "heat," recognition of its location at one of the two "hot poles," which face the sun at alternate perihelions. At the times of the Mariner encounters, the Caloris Basin was half in shadow (■ Figure 21-13a). Although half cannot be seen, the low angle of illumination is ideal for the study of the lighted half because it produces dramatic shadows.

The Caloris Basin is a gigantic multiringed impact basin 1300 km in diameter with concentric mountain rings up to 3 km high. The impact threw ejecta 600 to 800 km across the planet, and the focusing of seismic waves on the far side produced peculiar terrain (called lineated terrain) that looks much like the jumbled surface of the moon that lies opposite the Imbrium basin (Figure 21-13b and c). The Caloris Basin is partially filled with lava flows. Some of this lava may be material melted by the energy of the impact, but some may be lava from below the crust that leaked up through cracks. The weight of this lava and the response of the crust have produced deep cracks in the central lava plains. The geophysics of such large, multiringed crater basins is not well understood at present, but Caloris Basin seems to be the same kind of structure as the Imbrium Basin on the moon, but it has not been as deeply flooded with lava.

When planetary scientists began looking at the Mariner photographs in detail, they discovered something not seen on the moon. Mercury is marked by great curved cliffs called **lobate scarps** (■ Figure 21-14). These seem to have formed when the planet cooled and shrank by a few kilometers, wrinkling its crust as a drying apple wrinkles its skin. Some of these scarps are as high

as 3 km and reach hundreds of kilometers across the surface. Other faults in Mercury's crust are straight and may have been produced by tidal stresses generated when the sun slowed Mercury's rotation.

The Plains of Mercury

The most striking difference between Mercury and the moon is that Mercury lacks the great dark lava plains so obvious on the

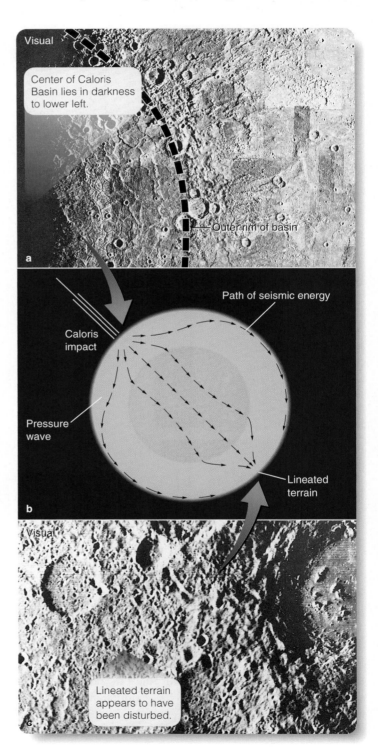

■ Figure 21-13

The huge impact that formed the Caloris Basin on Mercury sent seismic waves through the planet. Where the waves came together on the far side, they produced the lineated terrain, which resembles the jumbled terrain on Earth's moon opposite the Imbrium impact. (NASA)

■ **Figure 21-14**

A lobate scarp (arrow) crosses craters, indicating that Mercury cooled and shrank, wrinkling its crust, after many of its craters had formed. (NASA)

moon. Under careful examination, the Mariner 10 photographs show that Mercury has plains—two different kinds, in fact—but they are different from the moon's. Understanding the differences is the key to understanding the history of Mercury.

Much of Mercury's surface is old, cratered terrain (■ Figure 21-15), but other areas called **intercrater plains** are less heavily cratered. These plains are marked by meteorite craters less than 15 km in diameter and secondary craters produced by chunks of ejecta from larger impacts. Unlike the heavily cratered regions, the intercrater plains are not totally saturated with craters. As an expert of comparative planetology, you recognize that this must mean that the intercrater plains were produced by later lava flows that buried older terrain.

In contrast, smaller regions called **smooth plains** appear to be even younger than the intercrater plains. They have even fewer craters and appear to be ancient lava flows that occurred after most cratering had ended. Much of the region around the Caloris Basin is composed of these smooth plains (Figure 21-13), and they appear to have formed soon after the Caloris impact.

Given the available evidence, most planetary astronomers believe that the plains of Mercury are solidified lava flows, in which case they are much like the maria on the moon. Unlike the maria, Mercury's lava plains are not significantly darker than the rest of the crust. Except for a few bright crater rays, Mercury's surface is a uniform gray with an albedo of only about 0.1. That means Mercury's lava plains are not as dramatically obvious on photographs as the much darker maria on our own moon.

The Interior of Mercury

One of the most striking differences between Mercury and the moon is the difference between their interiors. You have seen that

the moon is a low-density world that contains no more than a small core of metals. Yet evidence suggests that Mercury has a large metallic core.

Mercury is over 60 percent denser than the moon. Yet Mercury's surface appears to be normal rock, so you must conclude that its interior contains a large core of dense metals, probably iron. In proportion to its size, Mercury must have a larger metal-

■ **Figure 21-15**

Study this region of Mercury carefully and you will notice the plains between the craters. This surface is not saturated as are the lunar highlands, but it contains more craters than the surface the maria on Earth's moon. This shows that these plains formed after the heaviest of the cratering in the early solar system. (NASA)

lic core than Earth (see the diagram in Celestial Profile 4). If Mercury had a large metallic core that remained molten, then the dynamo effect would generate a magnetic field (Chapter 20). The Mariner 10 spacecraft found, however, a magnetic field only about 0.5 percent as strong as Earth's, and this weak field makes it difficult to understand the nature of the interior. Because Mercury is a relatively small world, about halfway between the size of Earth and the size of the moon, it should have lost most of its internal heat long ago. But if its metallic core is not molten, you would not expect to detect a magnetic field as strong as 0.5 percent of Earth's field. The verdict is still out on the nature of Mercury's large metallic core and weak magnetic field.

Similarly, it is difficult to explain the proportionally large size of the metallic core inside Mercury. In Chapter 19, you saw how planets that formed near the sun should incorporate more metals and be denser, but some models suggest that this cannot account for all of Mercury's large metallic core. One hypothesis that has been proposed involves a giant impact when Mercury was young, an impact much like the planet-shattering impact proposed to explain the origin of Earth's moon. If the forming planet had differentiated and was then struck by a large planetesimal, the impact could have shattered the crust and mantle and blasted much of the lower-density material into space. The denser core could have survived, re-formed, and attracted some of the lower-density debris to form a thin mantle and crust and so leave Mercury with a deficiency of low-density crust rock. Both Mercury and the moon may be products of giant impacts.

A History of Mercury

Can you combine evidence and theory to tell the story of Mercury? Presumably, it formed in the innermost part of the solar nebula, and, as you have seen, a giant impact may have robbed it of some of its lower-density rock and left it a small, dense world with a large metallic core.

Like the moon, Mercury suffered heavy cratering by debris in the young solar system. Planetary scientists don't know accurate ages for features on Mercury, but you can assume that cratering, the second stage of planetary evolution, occurred over the same period as did the cratering on the moon—during the heavy bombardment that lasted from formation until about 4 billion years ago. This intense cratering declined rapidly as the planets swept up the last of the debris left over from planet building.

The cratered surface of Mercury is not exactly like that of the moon. Because of Mercury's stronger gravity, the ejecta from an impact on Mercury would be thrown only about 65 percent as far as on the moon, and that means the ejecta from an impact does not blanket as much of the surface. Also, the intercrater plains appear to have formed when lava flows occurred during the heavy bombardment, burying the older surface, and then accumulated more craters. Sometime near the end of cratering, a planetesimal over 100 km in diameter smashed into the planet and blasted out

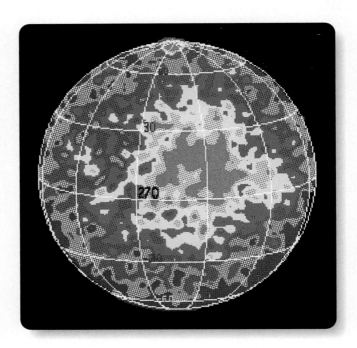

■ **Figure 21-16**

A radar map of Mercury reveals a bright region at its north pole (top). Observers suspect that ice deposits may survive just under the surface or in regions that never receive direct sunlight. (Martin Slade, JPL)

the great multiringed Caloris Basin. Only parts of that basin have been flooded by lava flows.

The smooth plains contain fewer craters and may date from the time of the Caloris impact. It may have been so big it fractured the crust and allowed lava flows to resurface some areas. Because this happened near the end of cratering, the smooth plains have few craters.

Finally, the cooling interior contracted, and the crust broke to form the lobate scarps. Lava flooding ended quickly, perhaps because the shrinkage of the planet squeezed off the lava channels to the surface. Mercury lacks a true atmosphere, so you would not expect flooding by water, but radar images show a bright spot at the planet's north pole (■ Figure 21-16) that may be caused by ice trapped in perpetually shaded crater floors where the temperature never exceeds 60 K (−350°F). This may be water from comets that occasionally collide with Mercury in a burst of water vapor. Similar water deposits have been found at the poles of Earth's moon.

The fourth stage in the story of Mercury, slow surface evolution, is now limited to micrometeorites, which grind the surface to dust; rare larger meteorites, which leave bright-rayed craters; and the slow but intense cycle of heat and cold, which weakens the rock at the surface. The planet's crust is now thick, and although its core may be partially molten, the heat flowing outward is unable to drive plate tectonics that would actively erase craters and build folded mountain ranges.

Building Scientific Arguments

Why don't Earth and the moon have lobate scarps?
Of course, this calls for a scientific argument based on comparative planetology. You might expect that any world with a large metallic interior should have lobate scarps. When the metallic core cools and contracts, the world should shrink, and the contraction should wrinkle and fracture the brittle crust to form lobate scarps. But there are other factors to consider. Earth has a fairly large metallic core; but it has not cooled very much, and the crust is thin, flexible, and active. If any lobate scarps ever did form on Earth, they would have been quickly destroyed by plate tectonics.

The moon does not have a large metal core. It may be that the rocky interior did contract slightly as the moon lost its internal heat, but that smaller contraction may not have produced major lobate scarps. Also, any surface features that formed early in the moon's history would have been destroyed by the heavy bombardment.

You can, in a general way, understand lobate scarps, but now add timing to your argument. **How do you know the lobate scarps on Mercury formed after most of the heavy bombardment was over?**

■   ■   ■

Connections: Mercury, like the moon, is now an inactive world. Impact cratering, internal heat, and possible giant impacts have left their marks on these worlds, but they are now nearly dead. To find planets where the heat flowing outward from the interior still drives geological activity, you must look at larger worlds such as Venus and Mars, the subjects of the next chapter.

Study and Review Tools

Summary

21-1 | The Moon

Why is the moon airless and cratered?
■ The moon is small and has only one-sixth the gravity of Earth. It has such a low escape velocity that it is unable to retain an atmosphere.

■ When a meteorite strikes the moon, it digs a large round impact crater and can eject debris that falls back and can form rays and secondary craters.

■ The largest impacts on the moon dug huge multiringed basins.

■ The highlands on the moon are saturated with craters. Rare meteorite impacts continue to form new craters, and micrometeorites constantly grind the surface down to dust, but most of the craters on the moon are very old.

■ Lowlands contain ancient lava flows called maria, which contain features such as sinuous rilles, faults, and wrinkles that suggest they are solidified lava flows.

■ Most of the craters on the moon were formed during the heavy bombardment at the end of planet building about 4 billion years ago. No crater is known with certainty to have been formed on the moon in historic times.

■ Between 1969 and 1972, 12 Apollo astronauts landed on the moon and returned specimens to Earth.

■ The moon rocks are all igneous, showing that they solidified from molten rock. Features such as vesicles show that some formed in lava flows, and the breccias show that much of the lunar crust was fractured by meteorite impacts.

■ The surface of the moon is covered by rock crushed and powdered by meteorite impacts to form a soil called regolith.

How did the moon form and evolve?
■ The fission, condensation, and capture hypotheses have all been abandoned. The commonly accepted theory is called the large-impact theory. The moon appears to have formed from a ring of debris ejected into space when a large planetesimal struck the protoearth after it had differentiated. This would explain, for example, the moon's low density and lack of volatiles.

■ The moon rocks show that the moon formed in a molten state.

■ The moon differentiated but contains little iron. Its low-density crust was heavily cratered and shattered to great depth.

■ Lava, flowing up from below the crust, filled the lowlands to form the smooth plains known as maria. The maria formed after the end of the heavy bombardment and contain few craters.

■ The near side of the moon, possibly due to tidal forces, has a thin crust, but the back side has a thicker crust and little lava flooding.

■ Because the moon is small, it has lost its internal heat and is geologically dead. The only slow surface evolution taking place is the blasting of micrometeorites.

21-2 | Mercury

Why is Mercury different from the moon?
■ Mercury is tidally locked to the sun but rotates 1.5 times per orbit.

■ Mercury is a small world that has been unable to retain an atmosphere and has lost most of its internal heat. It is cratered and inactive.

■ The Caloris Basin is a large multiringed basin on Mercury that has been partially flooded by lava flows.

■ The intercrater plains appear to be later lava flows that covered older craters and then accumulated more craters. The smooth plains appear to be more recent lava flows that contain few craters. All of these plains are a similar shade of gray, so dark lava flows are not visible on Mercury as they are on the moon.

How did Mercury form and evolve?

■ Mercury formed in the inner solar system and contains a larger proportion of dense metals. It is possible that a large impact shattered and drove off some of the planet's crust. This could explain why it has such a large metallic core.

■ It is not clear how much heat remains in Mercury. It is not geologically active, and it does not have a strong magnetic field. Yet it has a weak magnetic field that suggests some internal heat.

■ Mercury was heavily cratered during the heavy bombardment, but lava flows covered some of those craters and new craters formed the intercrater plains. Fractures produced by the Caloris impact may have triggered later lava flows that formed the smooth plains.

■ As the metallic core cooled and contracted, the brittle crust broke to form lobate scarps like wrinkles in the skin of a drying apple.

New Terms

tidal coupling (p. 498)

terminator (p. 498)

limb (p. 498)

mare (p. 498)

sinuous rille (p. 499)

ejecta (p. 500)

ray (p. 500)

secondary crater (p. 500)

micrometeorite (p. 501)

multiringed basin (p. 501)

relative age (p. 502)

absolute age (p. 502)

vesicular basalt (p. 505)

anorthosite (p. 506)

breccia (p. 506)

regolith (p. 506)

jumbled terrain (p. 509)

fission hypothesis (p. 510)

condensation hypothesis (p. 510)

capture hypothesis (p. 510)

large-impact hypothesis (p. 511)

resonance (p. 512)

lobate scarp (p. 514)

intercrater plain (p. 515)

smooth plain (p. 515)

Review Questions

Ace ◑Astronomy™ Assess your understanding of this chapter's topics with additional quizzing and animations at **http://ace .brookscole.com/sf9**

1. How could you find the relative ages of the moon's maria and highlands?

2. How can you tell that Copernicus is a young crater?

3. Why did the first Apollo missions land on the maria? Why were the other areas of more scientific interest?

4. Why do astronomers suppose that the moon formed with a molten surface?

5. Why are so many lunar samples breccias?

6. What do the vesicular basalts tell you about the evolution of the lunar surface?

7. What evidence would you expect to find on the moon if it had been subjected to plate tectonics? Do you find such evidence?

8. Cite objections to the fission, condensation, and capture hypotheses.

9. How does the large-impact hypothesis explain the moon's lack of iron? of volatiles?

10. How does the tidal coupling between Mercury and the sun differ from that between the moon and Earth?

11. What is the difference between the intercrater plains and the smooth plains in terms of time of formation?

12. What evidence can you cite that Mercury has a partially molten, metallic core?

13. In this photo, astronaut Alan Bean works at the Apollo 12 lander. Describe the surface you see. What kind of terrain did they land on for the second mission to the moon?

(NASA)

Discussion Questions

1. Old science-fiction paintings and drawings of colonies on the moon often showed very steep, jagged mountains. Why did the artists assume that the mountains would be more jagged than mountains on Earth? Why are lunar mountains actually less jagged than mountains on Earth?

2. From your knowledge of comparative planetology, propose a description of the view that astronauts would have if they landed on the surface of Mercury.

Problems

1. Calculate the escape velocity of the moon from its mass and diameter. (*Hint:* See Chapter 5.)

2. Why do small planets cool faster than large planets? Compare surface area to volume.

3. The smallest detail visible through Earth-based telescopes is about 1 second of arc in diameter. What size is this on the moon? (*Hint:* See Chapter 3.)

4. The trenches where Earth's seafloor slips downward are 1 km or less wide. Could Earth-based telescopes resolve such features on the moon? Why can you be sure that such features are not present on the moon?

5. The Apollo command module orbited the moon about 200 km above the surface. What was its orbital period? (*Hint:* See Chapter 5.)

6. From a distance of 200 km above the surface of the moon, what is the angular diameter of an astronaut in a space suit? Could someone have seen the astronauts from the command module? (*Hint:* See Chapter 3.)

7. If you transmitted radio signals to Mercury when it was closest to Earth and waited to hear the radar echo, how long would you wait?

8. Suppose you sent a spacecraft to land on Mercury, and it transmitted radio signals to Earth at a wavelength of 10 cm. If you saw Mercury at its greatest angular distance west of the sun, to what wavelength would you have to tune your radio telescope to detect the signals? (*Hints:* See Celestial Profile 4 and the section on the Doppler shift in Chapter 7.)

9. What would the wavelength of maximum be for infrared radiation from the surface of Mercury? How would that differ for the moon?

10. Calculate the escape velocity from Mercury. How does that compare with the escape velocity from the moon and from Earth?

Media Cluster

 To access the resources in the Media Cluster, log into AceAstronomy at **http://ace.brookscole.com/sf9** and select Chapter 21.

The Moon's Craters
The craters that cover the moon were produced by the high-speed impact of meteorites of all sizes. Explore three types of lunar craters in this animation.

ASTRONOMY EXERCISES

Cratering
Turn up your speakers for this one, where you can design collisions between a meteorite and a hypothetical planet.

VIRTUAL ASTRONOMY LABS

Lab 5: Planetary Geology
By thoroughly studying Earth's geology, you can leverage your knowledge to understand conditions on other planets. This lab includes an exercise in determining the interior characteristics of a planet by observing the travel times of its seismic waves.

Lab 7: Planetary Atmospheres and Their Retention
This lab investigates the retention of atmospheres. You will explore the factors that govern the loss of atmospheric gases, and you will see why certain bodies can retain some gases but not others.

Critical Inquiries for the Web

1. What did the Apollo astronauts do while they were on the moon? Search for websites that describe the experiments performed during the astronauts' extravehicular activities (EVAs). What types of experiments or activities were performed? What did planetary scientists learn from them?

2. Search for maps of the moon that show geological information such as chemical composition, elevation, age, and so on. Do the maps confirm your understanding of the lunar surface?

3. What would it be like to walk on the lunar surface? Apollo astronauts visited six different locations on the moon, exploring the variations in lunar terrain. Describe the horizons and general relief of the landing locations of the different missions by exploring websites that provide lunar surface photography from the missions. What differences do you see between images from landings in highlands and in maria?

4. What plans are being made to send a spacecraft to Mercury? What kind of observations will be made, and what theories will be tested?

Exploring *TheSky*

1. Locate the moon in the sky. What constellation is it in? Find the moon's phase as a percentage of full, its angular diameter, and its distance from Earth. Use **Find** to locate the moon.

2. Follow the moon through one complete orbit measuring its distance and angular diameter. How do these change? Under **Tools**, use the **Moon Phase Calendar.** Change dates by clicking on **Data** and then **Site Information.**

3. Locate Mercury and zoom in until you can see its phase. Follow it as it orbits the sun and describe how its phase changes with its orbital position. Use **Find** to locate Mercury.

 Go to the Brooks/Cole Astronomy Resource Center **(http://astronomy.brookscole.com)** for critical thinking exercises, articles, and additional readings from InfoTrac College Edition, Brooks/Cole's online student library.

22 | Comparative Planetology of Venus and Mars

There wasn't a breath

in that land of death . . .

ROBERT SERVICE,
THE CREMATION OF SAM MCGEE

WITHOUT A SPACESUIT, you could live only about 30 seconds on the surface of Mars. The temperature might not be too bad. On a hot summer day at noon, the temperature might reach 20°C (68°F). But the air is 96 percent carbon dioxide, is deadly dry, and contains almost no oxygen. Also, the air pressure at the surface of Mars is only 0.01 that at the surface of Earth. Stepping out onto the surface of Mars without a spacesuit would be about the same as opening the door of an airplane flying at 30,000 m (100,000 ft) above Earth. (Commercial jets do not generally fly above 12,000 m, or 38,000 ft.) No spacesuit would save you on Venus. The air pressure is 90 times Earth's, and the air is almost entirely carbon dioxide, with traces of various acids. Worse yet, the surface is hot enough to melt lead.

Continued on page 522

The cold, dry desert surface of Mars was photographed by Rover Spirit in 2004. This is a place, and it has not changed much since the photo was taken. You could go there and walk over and pick up that rock by the dune. The rock is there right now. (NASA)

Guidepost

Looking Back

The spacesuit you wore when you visited the moon and Mercury protected you and allowed you to explore those airless worlds. On your visits you discovered the fate that awaits small worlds. They can't keep atmospheres, become heavily cratered, lose their internal heat, and quickly become dead, inactive worlds of rock and dust. You will find that Venus and Mars are different.

This Chapter

Your spacesuit will serve you well on Mars, and you will find it an interesting world, but no spacesuit will protect you on the surface of Venus. It too is an interesting world, but it is also a very unpleasant place. You will need to observe it from a distance. As you explore you will discover answers to six essential questions:

Why is the atmosphere of Venus so thick?

What is the hidden surface of Venus really like?

How did Venus form and evolve?

Why is the atmosphere of Mars so thin?

Has Mars ever had water on its surface?

How did Mars form and evolve?

Of course, the key question of comparative planetology, the question always on your mind when you explore other worlds, is this: Why is this world so different from Earth? You will see that small initial differences can have big effects.

Looking Ahead

You are a planet walker, and you are becoming an expert on the kind of planets you can imagine walking on. But there are other worlds in our solar system so peculiar they have no surfaces to walk on, even in your imagination. Nevertheless, you will be able to use your experience with comparative planetology to explore those worlds in the next two chapters.

Ace◐Astronomy™ The AceAstronomy icon throughout the text indicates an opportunity for you to test yourself on key concepts and to explore animations and interactions on the AceAstronomy website at: http://ace .brookscole.com/sf9

Why are Venus and Mars such rotten places for picnics? In many ways, they resemble Earth. They are made of silicates and metals, have atmospheres, and lie in the inner solar system. You might expect them to be similar to Earth. How did they get to be so different from each other and from Earth? Comparative planetology will give you some clues.

In this chapter, you will discover the important principle that the size of a world determines how fast it loses its internal heat. Venus, roughly the size of Earth, has lost internal heat slowly, but Mars, intermediate in size between Earth and the moon, is nearly as cold and dead as the moon. Furthermore, these larger worlds have enough gravity to hold onto an atmosphere, and that alters the evolution of surface features.

Venus and Mars are more than just two more planets to add to your gallery of weird worlds. They will give you an entirely new perspective on comparative planetology and a new perspective on Earth.

22-1 Venus

AN ASTRONOMER ONCE BECAME ANNOYED when someone referred to Venus as a planet gone wrong. "No," she argued, "Venus is probably a fairly normal planet. It is Earth that is peculiar. The universe probably contains more planets like Venus than like Earth." To understand how unusual, how special Earth is, you have only to compare it with Venus.

In many ways, Venus is a twin of Earth, and you might expect the two planets to be quite similar. Venus is nearly 95 percent

the diameter of Earth (**Celestial Profile 5**) and has a similar average density. It formed in the same general part of the solar nebula, so you might expect it to have a composition similar to Earth's. Planets of Earth's size cool slowly, so you could expect Venus to have a molten metallic interior and an active crust.

Unfortunately, the surface of Venus is perpetually hidden by thick clouds that completely envelop the planet. From the time of Galileo to the early 1960s, astronomers could only speculate about Earth's twin, and science fiction writers supposed that Venus was a steamy swamp planet inhabited by strange creatures.

Since the 1960s, astronomers have used radar to penetrate the clouds and image the surface. At least 23 spacecraft have flown past or orbited Venus, and over a dozen have landed on its surface. The resulting picture of Venus is dramatically different from the murky swamps of fiction. In fact, the surface of Venus is drier than any desert on Earth and twice as hot as a kitchen oven set to its highest temperature.

If Venus is not a planet gone wrong, it is certainly a planet gone down a different evolutionary path. How did Earth's twin become so different?

The Rotation of Venus

Nearly all of the planets in our solar system rotate counterclockwise as seen from the north. Uranus and Pluto are exceptions, and so is Venus.

In 1962, radio astronomers were able to transmit a radio pulse of precise wavelength toward Venus and detect the echo returning some minutes later. That is, they detected Venus by radar. The echo was not a precise wavelength, however. Part of the reflected signal had a longer wavelength, and part had a shorter wavelength. Evidently the planet was rotating; radio energy re-

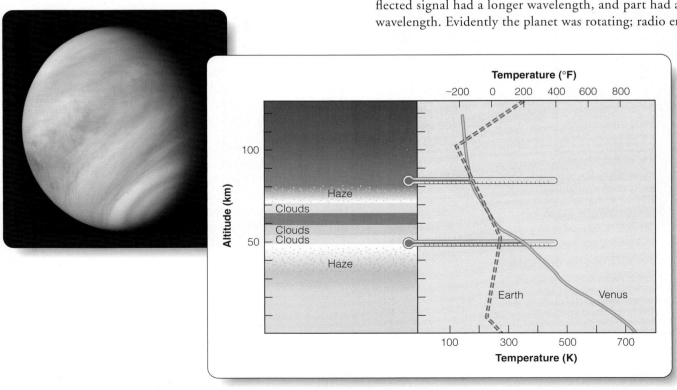

flected from the receding edge was redshifted, and radio energy reflected from the approaching edge was blueshifted. From this Doppler effect, the radio astronomers could tell that Venus was rotating once every 243.01 days. Because the west edge of Venus produced the blueshift, the planet had to be rotating in the backward direction.

This rotation period turns nearly the same face toward Earth each time the planets come closest together, and it is tempting to imagine that there is a tidal resonance locking the rotation of Venus to the orbital motion of Earth. However, a true resonance would require that Venus rotate 0.06 percent slower.

Why does Venus rotate backward? For decades, textbooks have suggested that proto-Venus was set spinning backward when it was struck off-center by a large planetesimal. That is a reasonable possibility; you have seen that a similar collision probably gave birth to Earth's moon. But there is an alternative. Sophisticated mathematical models suggest that a terrestrial planet with a molten core and a dense atmosphere can have its rotation gradually reversed by solar tides in its atmosphere. (Notice the contrast between the catastrophic theory of a giant impact and the evolutionary theory of atmospheric tides.)

Whichever theory explains the backward rotation of Venus, the near resonance with the motion of the Earth seems to be a coincidence. Earth is too small and too distant to raise tides on Venus and affect its rotation.

The Atmosphere of Venus

Although Venus is Earth's twin in size, the composition, temperature, and density of its atmosphere make it the most inhospitable of planets. About 96 percent of its atmosphere is carbon dioxide, and 3.5 percent is nitrogen. The remaining 0.5 percent is water vapor, sulfuric acid (H_2SO_4), hydrochloric acid (HCl), and hydrofluoric acid (HF). In fact, the thick clouds that hide the surface are composed of sulfuric acid droplets and microscopic sulfur crystals.

Soviet and American spacecraft have dropped probes into the atmosphere of Venus, and those probes have radioed data back to Earth as they fell toward the surface. These studies show that the cloud layers are much higher and much more stable than those on Earth. The highest layer of clouds, the layer visible from Earth, extends from 68 to 58 km above the surface (■ Figure 22-1). For comparison, the highest clouds on Earth do not

■ Active Figure 22-1 ◀

The four main cloud layers in the atmosphere of Venus are over 10 times higher above the surface than are Earth clouds. They completely hide the surface. If you could insert thermometers into the atmosphere at different levels, you would find that the lower atmosphere is much hotter than that of Earth, as indicated by the red line in the graph. (© Calvin J. Hamilton)

Ace ◈ Astronomy ™ Log into AceAstronomy and select this chapter to see the Active Figuree "Planetary Atmospheres." You can explore this figure and compare planets with each other.

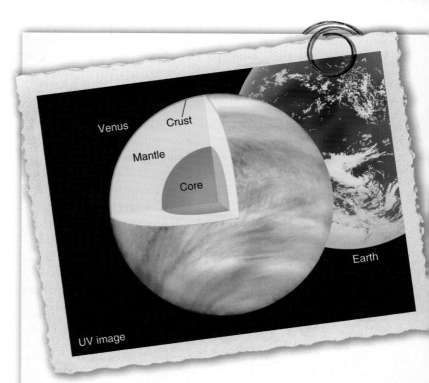

Venus is only 5 percent smaller than Earth, but its atmosphere is perpetually cloudy and its surface is hot enough to melt lead. It is believed to have a hot core about the size of Earth's.

Celestial Profile 5: Venus

Motion:

Average distance from the sun	0.7233 AU (1.082×10^8 km)
Eccentricity of orbit	0.0068
Maximum distance from the sun	0.7282 AU (1.089×10^8 km)
Minimum distance from the sun	0.7184 AU (1.075×10^8 km)
Inclination of orbit to ecliptic	3°23'40"
Average orbital velocity	35.03 km/s
Orbital period	0.61515 y (224.68 days)
Period of rotation	243.01 days (retrograde)
Inclination of equator to orbit	177°

Characteristics:

Equatorial diameter	12,104 km (0.95 $D_\oplus$)
Mass	4.870×10^{24} kg (0.815 $M_\oplus$)
Average density	5.24 g/cm³ (4.2 g/cm³ uncompressed)
Surface gravity	0.903 Earth gravity
Escape velocity	10.3 km/s (0.92 $V_\oplus$)
Surface temperature	745 K (472°C, or 882°F)
Albedo (cloud tops)	0.76
Oblateness	0

Personality Point:

Venus is named for the Roman goddess of love, perhaps because the planet often shines so beautifully in the evening or dawn sky. In contrast, the ancient Maya identified Venus as their war god Kukulkan and sacrificed human victims to the planet when it rose in the dawn sky.

extend higher than about 16 km. Another cloud layer lies from 56 to 52 km, and a third, the densest, from 52 to 48 km. Thin haze extends down to about 33 km, below which the air is surprisingly clear.

These cloud layers are highly stable because the atmospheric circulation on Venus is much more regular than that on Earth. The heated atmosphere at the **subsolar point,** the point on the planet where the sun is directly overhead, rises and spreads out in the upper atmosphere. Convection circulates this gas toward the dark side of the planet and the poles, where it cools and sinks. This circulation produces 300-km/h jet streams in the upper atmosphere, which carry the atmosphere from east to west (the same direction the planet rotates) so rapidly that the entire atmosphere seems to rotate with a period of only four days.

The details of this atmospheric circulation are not well understood, but it seems that the slow rotation of the planet is an important factor. On Earth, large-scale circulation patterns are broken into smaller cyclonic disturbances by Earth's rapid rotation. Because Venus rotates more slowly, its atmospheric circulation is not broken up into small cyclonic storms but instead is organized as a planetwide wind pattern.

Although the upper atmosphere is cool, the lower atmosphere is quite hot (Figure 22-1b). Instrumented probes that have reached the surface report that the temperature is 745K (880°F), and the atmospheric pressure is 90 times that of Earth. Earth's atmosphere is 1000 times less dense than water, but on Venus the air is only 10 times less dense than water. If you could survive the unpleasant composition, intense heat, and high pressure, you could strap wings to your arms and fly.

The present atmosphere of Venus is extremely dry, but there is evidence that it once had significant amounts of water. As a Pioneer Venus spacecraft descended through the atmosphere of Venus in 1978, it discovered that deuterium is about 150 times more abundant compared to hydrogen atoms than it is on Earth. This abundance of deuterium, the heavy isotope of hydrogen, could develop if ultraviolet radiation in sunlight broke water molecules into hydrogen and oxygen. The hydrogen would escape into space, but the heavier deuterium atoms would escape more slowly than normal hydrogen. The excess of deuterium must mean that Venus once had more water than it has now. Calculations suggest that it may have once had enough water to make a planetwide ocean up to 25 meters deep. (For comparison, the water on Earth would make a uniform planetwide ocean 3000 meters deep.) Without an ozone layer to protect the atmosphere from ultraviolet radiation, the water was broken up and much of the hydrogen was lost to space. The oxygen presumably formed oxides in the soil. Consequently, Venus is now a deadly dry world with only enough water vapor in its atmosphere to make an ocean 0.3 meter deep.

Even if Venus once had lots of water, it was probably too warm for that water to fall from the atmosphere as rain and form oceans, and that lack of oceans is the biggest difference between Earth and Venus. Earth's oceans have made it a living paradise. The lack of oceans on Venus, through the greenhouse effect, has made it a hell.

The Venusian Greenhouse

You saw in Chapter 20 how the greenhouse effect warms Earth. Carbon dioxide (CO_2) is transparent to light but opaque to infrared (heat) radiation. Consequently, energy can enter the atmosphere as light and warm the surface, but the surface cannot radiate an equal amount of energy back to space because the carbon dioxide is opaque to infrared radiation. Like Earth, Venus has a greenhouse effect, but on Venus the effect is fearsome. Whereas Earth's atmosphere contains only about 0.03 percent carbon dioxide, the atmosphere of Venus contains 96 percent carbon dioxide, and the surface of Venus is nearly twice as hot as a hot kitchen oven.

Planetary astronomers think they know how Venus got into such a jam. When Venus was young, it may have been cooler than it is now; but, because it formed 30 percent closer to the sun than did Earth, it was warmer than Earth, and that unleashed processes that made it even hotter. As you saw in the previous section, even if Venus once had water, it probably never had large liquid-water oceans. Carbon dioxide is highly soluble in water, but without large oceans to dissolve carbon dioxide and remove it from the atmosphere, Venus was trapped with a carbon dioxide–rich atmosphere, and the greenhouse effect made the planet warmer. As the planet warmed, any small oceans that did exist evaporated, and Venus lost the ability to cleanse its atmosphere of carbon dioxide. Volcanoes on Earth vent gases that are mostly carbon dioxide and water vapor, and presumably the volcanoes on Venus did the same. The water vapor was broken up by ultraviolet radiation, the hydrogen was lost to space, and the oxygen formed oxides in the soil; the carbon dioxide accumulated in the atmosphere. The high temperature baked even more carbon dioxide out of the surface, and the atmosphere became even less transparent to infrared radiation. Consequently, the temperature rose further. This runaway greenhouse effect has made the surface so hot that even sulfur, chlorine, and fluorine have baked out of the rock and formed sulfuric, hydrochloric, and hydrofluoric acid vapors.

Earth avoided this runaway greenhouse effect because it was farther from the sun and cooler. Consequently, it could form and preserve liquid-water oceans to absorb the carbon dioxide, which left a nitrogen atmosphere that was relatively transparent in some parts of the infrared. If all of the carbon dioxide in Earth's sediments was put back into the atmosphere, our air would be as dense as that of Venus, and Earth would suffer from a terrifying greenhouse effect. Recall from Chapter 20 how the use of fossil fuels and the destruction of forests is increasing the carbon dioxide concentration in our atmosphere and warming the planet. Venus warns us of what a greenhouse effect can do.

Ace Astronomy™   Log into AceAstronomy and select this chapter to see Astronomy Exercise "The Greenhouse Effect." Take control of the carbon dioxide in your experimental planet's atmosphere.

The Venera lander touched down on Venus in 1982 and carried a camera that swiveled from side to side to photograph the surface. The orange glow is produced by the thick atmosphere; when that is corrected, you can see that the rocks are dark gray. Isotopic analysis suggests they are basalts. (NASA)

The Surface of Venus

Given that the surface of Venus is hidden perpetually by clouds, is hot enough to melt lead, and suffers under crushing atmospheric pressure, it is surprising how much planetary scientists know about the geology of Venus. Early radar maps made from Earth penetrated the clouds and showed that it had mountains, plains, and some craters. The Soviet Union launched a number of spacecraft that landed on Venus, and although the spacecraft failed within an hour or so of landing, they did analyze the rock and transmit a few images back to Earth. The rock seems to be basalt, typical of volcanism, and the images revealed dark-gray rocky plains bathed in a deep-orange glow caused by sunlight filtering down through the thick atmosphere (■ Figure 22-2).

Both the United States and the Soviet Union sent spacecraft to Venus. Some probed the atmosphere, and others landed on the surface and sent back data before succumbing, within an hour or so, to the intense heat and pressure. Some of the most fruitful missions orbited the planet and used radar to penetrate the clouds and map the surface. The NASA Pioneer Venus probe orbited Venus in 1978 and made radar maps showing features as small as 25 km in diameter. Later two Soviet Venera spacecraft mapped the north polar regions with a resolution of 2 km. From 1992 to 1994, the NASA Magellan spacecraft orbited Venus and created radar maps showing details as small as 100 m. These radar maps provide a unique look below the clouds.

The color of Venus radar maps is mostly arbitrary. Human eyes can't see radio waves, so the radar maps must be given some false colors to distinguish height or roughness or composition. Some maps use blues and greens for lowlands and yellows and reds for highlands. Coloring the lowlands blue like oceans on Earth maps might be deceptive, however, because there is no liquid water on Venus. Rather, Magellan scientists chose to use yellows and oranges for their radar maps in an effort to mimic the orange color of daylight caused by the thick atmosphere (■ Figure 22-3). When you look at these orange images, you need to remind yourself that the true color of the rock is dark gray.

■ Figure 22-3

Venus without its clouds: This mosaic of Magellan radar maps has been given an orange color to mimic the sunset coloration of daylight at the surface of the planet. The image shows scattered impact craters and volcanic regions such as Beta Regio and Atla Regio. (NASA)

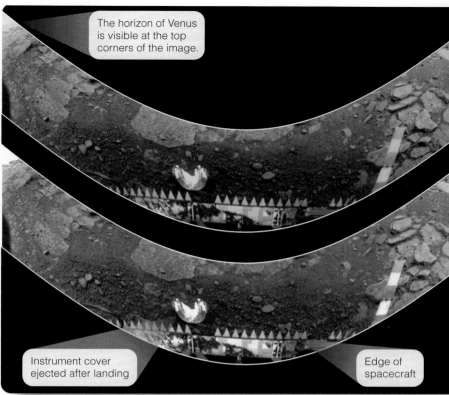

The horizon of Venus is visible at the top corners of the image.

Instrument cover ejected after landing

Edge of spacecraft

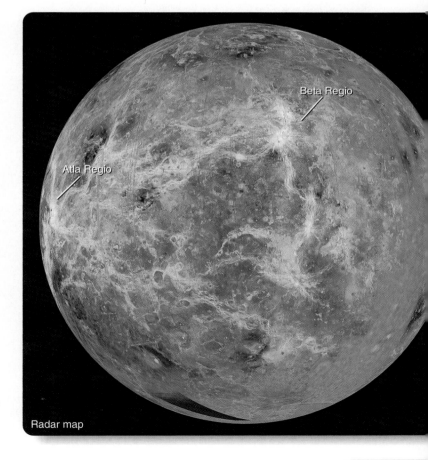

Beta Regio

Atla Regio

Radar map

■ Figure 22-4 ▲

Although it is nearly 1000 years old, this lava flow near Flagstaff, Arizona, is still such a rough jumble of sharp rock that it is dangerous to venture onto its surface. Rough surfaces are very good reflectors of radio waves and look bright in radar maps. Solidified lava flows on Venus show up as bright regions in the radar maps because they are rough. (M. Seeds)

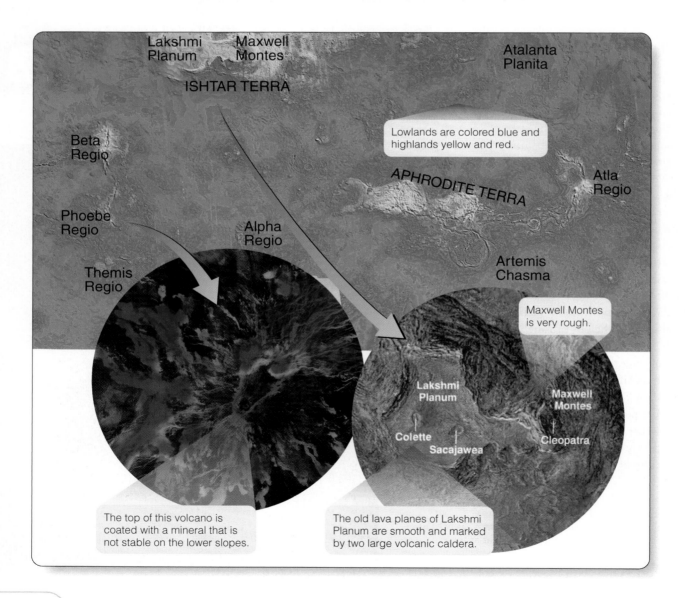

Radar maps of Venus can reveal a number of things about the surface. If you transmit a radio signal down through the clouds and measure the time until you hear the echo coming back up, you can measure the altitude of the surface. Part of the Magellan data is a detailed altitude map of the surface. You can also measure the amount of the signal that is reflected from each spot on the surface. Much of the surface of Venus is made up of old, smooth lava flows that do not look bright in radar maps, but faults and uneven terrain look brighter. Some young, rough lava flows are very rough and contain billions of tiny crevices (■ Figure 22-4) that bounce the radar signal around and shoot it back the way it came. These rough lava flows look bright in radar maps. Certain mineral deposits are also bright; consequently, radar maps do not show how the surface would look to human eyes but rather provide information about altitude, roughness, and, in some cases, chemical composition.

The big map in ■ Figure 22-5 is a map of all of Venus except the polar regions. (Note that this map has been color coded by altitude.) By international agreement, the names of celestial bodies and features on celestial bodies are assigned by the International Astronomical Union, which has decided that all names on Venus should be female. There are only a few exceptions, such as the mountain Maxwell Montes, 50 percent higher than Mt. Everest. It was discovered in early Earth-based radar maps and named for James Clark Maxwell, the 19th-century physicist who first described electromagnetic radiation. Alpha Regio and Beta Regio were also discovered by radar before the naming convention was adopted. Other names on Venus are female, such as the highland regions Ishtar Terra and Aphrodite Terra, named for the Babylonian and Greek goddesses of love. The insets in Figure 22-5 show roughness and composition of the surface, made by different satellites and color coded in different ways. All of these radar maps paint a picture of a violent, hot, desert world.

Radar maps show that the surface of Venus consists of low, rolling plains and highland regions. When comparing them with Earth, you might be tempted to think of these regions as ocean floors and continents, but there is no liquid water on Venus. Rather, the rolling plains appear to be large-scale smooth lava flows, and the highlands are regions of deformed crust.

Craters on Venus tell you that the surface is not old. With nearly 1000 impact craters on its surface, Venus has more craters than Earth but not nearly as many as the moon. The craters are uniformly scattered over the surface and look sharp and fresh

■ **Figure 22-6**

Impact crater Howe in the foreground of this Magellan radar image is 37 km in diameter. Craters in the background are 47 km and 63 km in diameter. This radar map has been digitally modified to represent the view from a spacecraft flying over the craters. (NASA)

(■ Figure 22-6). With no water and a thick, sluggish atmosphere, there is little erosion on Venus, and the thick atmosphere protects the surface from small meteorites. Consequently, there are no small craters. Putting all of this observational evidence together, planetary scientists conclude that the surface of Venus is older than Earth's surface but not nearly as old as the moon's. Unlike the moon, there are no old, cratered highlands. Lava flows have covered the entire surface within the last half-billion years.

Volcanism on Venus

Volcanism seems to dominate the surface of Venus. Much of Venus is covered by lava flows such as those photographed by the Venera landers (Figure 22-2). Also, volcanic peaks and other volcanic features are evident in the radar maps.

As usual, you can learn more about other worlds by comparing them with each other and with Earth. Study **Volcanoes** on pages 528–529 and notice three important ideas:

1 There are two main types of volcanoes found on Earth. One is associated with plate motion, and one is associated with hot spots.

2 Notice that you can recognize the volcanoes on Venus and Mars by their shapes, even when the images are manipulated in computer mapping (**Window on Science 22-1**). The shield volcanoes found on Venus and Mars are the kind produced by hot-spot volcanism and not by plate tectonics.

3 Also notice the large size of volcanoes on Venus and Mars. They have grown very large because of repeated eruptions at the same place in the crust. This is evidence that neither Venus nor Mars has been dominated by plate tectonics as on Earth.

■ **Figure 22-5** ◀

Notice how these three radar maps show different things. The main map here shows elevation over most of the surface of Venus. Only the polar areas are not shown. The inset map at left shows an electrical property of surface minerals related to chemical composition. The detailed map of Maxwell Montes and Lakshmi Planum at right is color coded to show roughness with purple smooth and orange rough. The rock on Venus is, in fact, dark gray. (Maxwell and Lakshmi Planum map: USGS; Other maps: NASA)

Volcanoes

1   Molten rock (magma) is less dense than its surroundings and tends to rise. Where it bursts through Earth's crust, you see volcanism. The two main types of volcanoes on Earth provide good examples for comparison with those on Venus and Mars.

On Earth, **composite volcanoes** form above subduction zones where the descending crust melts and the magma rises to the surface. This forms chains of volcanoes along the subduction zone, such as the Andes along the west coast of South America.

Magma rising above subduction zones is not very fluid, and it produces explosive volcanoes with sides as steep as 30°.

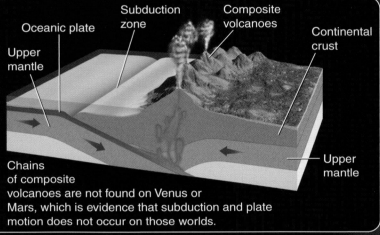

Oceanic plate   Subduction zone   Composite volcanoes   Continental crust

Upper mantle

Upper mantle

Chains of composite volcanoes are not found on Venus or Mars, which is evidence that subduction and plate motion does not occur on those worlds.

Based on *Physical Geology*, 4th edition, James S. Monroe and Reed Wicander, Wadsworth Publishing Company. Used with permission.

Mount St. Helens exploded northward on May 18, 1980, killing 63 people and destroying 600 km² (230 mi²) of forest with a blast of winds and suspended rock fragments that moved as fast as 480 km/hr (300 mph) and had temperatures as hot as 350°C (660°F). Note the steep slope of this composite volcano.

Shield volcano   Lava flow

Oceanic crust   Magma chamber

1a   A shield volcano is formed by highly fluid lava (basalt) that flows easily and creates low-profile volcanic peaks with slopes of 3° to 10°. The volcanoes of Hawaii are shield volcanoes that occur over a hot spot in the middle of the Pacific plate.

Magma collects in a chamber in the crust and finds its way to the surface through cracks.

Magma forces its way upward through cracks in the upper mantle and causes small, deep earthquakes.

A hot spot is formed by a rising convection current of magma moving upward through the hot, deformable (plastic) rock of the mantle.

The Cascade Range composite volcanoes are produced by an oceanic plate being subducted below North America and partially melting.

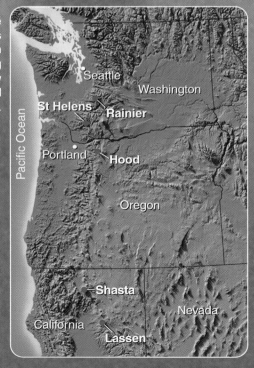

Seattle
Washington
St Helens   Rainier
Pacific Ocean
Portland   Hood
Oregon
Shasta
Nevada
California
Lassen

USGS

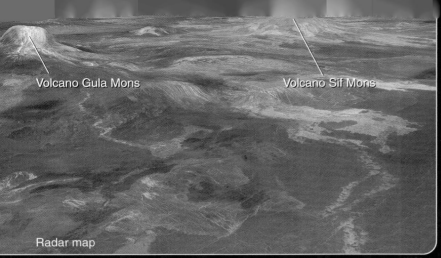

Volcano Gula Mons   Volcano Sif Mons

Radar map

NASA

2a This computer model of a mountain with the vertical scale magnified 10 times appears to have steep slopes such as those of a composite volcano.

A true profile of the computer model shows the mountain has very shallow slopes typical of shield volcanoes.

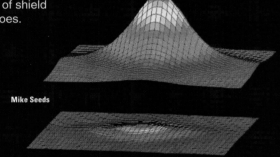

Mike Seeds

2 Volcanoes on Venus are shield volcanoes. They appear to be steep sided in some images created from Magellan radar maps, but that is because the vertical scale has been exaggerated to enhance detail. The volcanoes of Venus are actually shallow-sloped shield volcanoes.

3 Volcanism over a hot spot results in repeated eruptions that build up a shield volcano of many layers. Such volcanoes can grow very large.

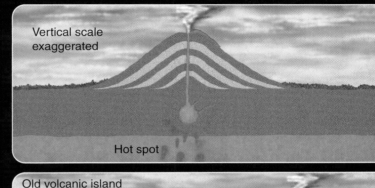

Vertical scale exaggerated

Hot spot

Old volcanic island eroded below sea level

Plate motion

Hot spot

If the crustal plate is moving, magma generated by the hot spot can repeatedly penetrate the crust to build a chain of volcanoes. Only the volcanoes over the hot spot are active. Older volcanoes slowly erode away. Such volcanoes cannot grow large because the moving plate carries them away from the hot spot.

Ace Astronomy™

Log into AceAstronomy and select this chapter to see the Active Figure "Hot Spot Volcanoes" and compare volcanism on Earth with that on Venus.

3a The volcanoes that make up the Hawaiian Islands as shown at left have been produced by a hot spot poking upward through the middle of the moving Pacific plate.

NASA

Time since last eruption (million years)

5   3   1.5   1   0

Kauai   Oahu Molokai Maui Hawaii Active volcanoes

Plate motion

Lo'ihi— Newborn underwater volcano

3b The plate moves about 9 cm/yr and carries older volcanic islands northwest, away from the hot spot. The volcanoes cannot grow extremely large because they are carried away from the hot spot. New islands form to the southeast over the hot spot.

Olympus Mons at right is the largest volcano on Mars. It is a shield volcano 25 km (16 mi) high and 700 km (440 mi) in diameter at its base. Its vast size is evidence that the crustal plate must have remained stationary over the hot spot. This is evidence that Mars has not had plate tectonics.

Olympus Mons contains 95 times more volume than the largest volcano on Earth, Mauna Loa in Hawaii.

Caldera from repeated eruptions

Digital elevation map

Window on Science | 22-1

Graphical Manipulation of Scientific Data

Because so much modern scientific data is stored inside computers, it is common to see data manipulated to exaggerate certain characteristics and suppress others. In some fields, this would be dishonest, but the goals of science make such manipulation valuable. By purposely distorting data, you may reveal things that would not have been seen otherwise.

Computers make it easy to manipulate the colors in an image and produce false-color images. Astronomers commonly change and exaggerate the colors in images and maps to reveal details invisible to our eyes. Radio maps of galaxies and X-ray maps of supernova remnants are only two examples. As long as you remember what real galaxies and real nebulae look like, you can learn a great deal from false-color images.

Geologists, too, manipulate data. Geological maps have been color coded for over a century, and now computers allow faster and more precise use of false color in maps and images. But geological data can also be manipulated to reveal other features. A relief map of the United States would look as smooth as glass if the map maker did not exaggerate the height of mountain ranges, and having such data in a computer makes it easy to "stretch" the heights to make them more visible. Astronomers often do the same with radar maps of Venus and surface maps of Mars.

Such data distortion would be inadmissible in a court of law. In fact, photographic evidence is generally inadmissible in court because it is so easy to fake a photograph and show the defendant in almost any incriminating situation. Faking a photograph in journalism is at least disreputable, and in some cases it is libelous. In fact, modern computer graphics make it possible to put Humphrey Bogart into a modern TV commercial with no trace of fakery, so photographic manipulation is unlikely ever to become acceptable in court or in journalism.

In science, however, the quest is not for a first impression but for detailed analysis. Data can be manipulated and distorted in any way that is useful, as long as the form of the manipulation is specified. Then viewers are free to form their own opinions as to the meaning of the data. Of course, this assumes that the viewer is alert and takes into consideration the nature of any data manipulation. Thus, when

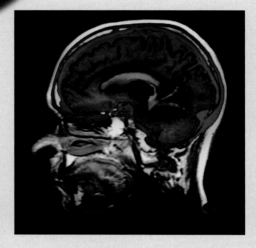

You are accustomed to seeing data manipulated and presented in convenient ways. (PhotoDisc/Getty Images)

you look at a relief map of the United States or a radar image of a mountain on Venus, it is your responsibility to take into account any data manipulation. If you don't, you might think that the Rocky Mountains are 100 miles high and that mountains on Venus are 20 times steeper than mountains on Earth.

The radar image of Sapas Mons in ■ Figure 22-7 shows a dramatic overhead view of this volcano, which is 400 km (250 mi) in diameter at its base and 1.5 km (0.9 mi) high. Many bright lava flows radiate outward, covering older, less bright flows. Remember that the colors in this image are artificial; if you could walk across these lava flows, you would find them solid stone of a dark-gray color. Sapas Mons and other nearby volcanoes are located along a system of faults where rising magma broke through the crust.

In addition to actual volcanoes, radar images reveal other volcanic features on the surface of Venus. Lava channels are common on Venus, and they appear similar to the sinuous rills visible on Earth's moon. The longest channel on Venus is also the longest known lava channel in the solar system. It stretches 6800 km (4200 miles), roughly twice the distance from Chicago to Los Angeles. These channels are 1 to 2 km wide and can sometimes be traced back to collapsed areas where lava appears to have drained from beneath the crust.

For further evidence of volcanism on Venus, you can look at circular features called **coronae**, circular bulges up to 2100 km in diameter containing volcanic peaks and lava flows and bordered by fractures. The coronae appear to be caused by rising currents

of molten magma below the crust that create an uplifted dome and then withdraw to allow the surface to subside. Coronae are sometimes accompanied by features called pancake domes, circular outpourings of viscous lava, and by domes and hills pushed up by molten rock below the surface. All of these volcanic features are shown in ■ Figure 22-8.

Volcanoes may be erupting on Venus right now, but the radar maps caught no evidence of actual eruptions in progress. Certainly there is no reason to suppose that all of these volcanoes are extinct.

If you want to learn the secrets of Venus, you might want to visit the huge landmass called Ishtar Terra. Western Ishtar consists of a high volcanic plateau called Lakshmi Planum. It rises 4 km above the rolling plains and appears to have formed from lava flows from volcanic vents such as Collette and Sacajawea (see Figure 22-5). Although folded mountain ranges do not occur on Venus as they do on Earth, the areas north and west of Lakshmi Planum appear to be ranges of wrinkled mountains where horizontal motion of the crust has pushed up against Ishtar Terra. Furthermore, faults and deep chasms are widespread over the surface of Venus, and that suggests stretching of the crust in those areas. So, although no direct evidence of plate motion is visible,

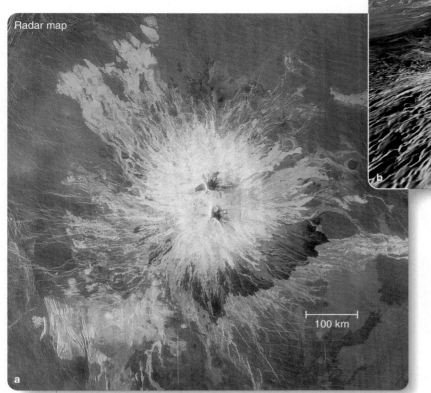

100 km

a

b

Figure 22-7

(a) Volcano Sapas Mons, lying along a major fracture zone, is topped by two lava-filled calderas and flanked by rough lava flows. The orange color of this radar map mimics the orange light that filters through the thick atmosphere. (NASA) (b) Seen by light typical of Earth's surface, Sapas Mons might look more like this computer-generated landscape. Volcano Maat Mons rises in the background. The vertical scale has been exaggerated by a factor of 15 to reveal the shape of the volcanoes and lava flows. (Copyright © 1992, David P. Anderson, Southern Methodist University)

there has been compression and wrinkling of the crust in some areas and stretching and faulting in other areas, and that suggests some limited crustal motion.

The history of Venus has been dominated by volcanism. While Earth's crust has broken into rigid moving plates, the surface of Venus seems more pliable. In any case, to tell the story of Venus you will have to tell a fiery tale.

Ace ✷ Astronomy™ Log into AceAstronomy and select this chapter to see Astronomy Exercise "Convection and Plate Tectonics." Compare the interiors of the four terrestrial planets.

A History of Venus

Earth has passed through four major stages in its history (Chapter 20), and you have seen how the moon and Mercury were affected by the same stages. Venus, however, has had a peculiar passage through the four stages of planet development, and its history is difficult to understand. Planetary scientists are not sure how the planet formed and differentiated, how it was cratered and flooded, or how its surface has evolved.

Venus formed only slightly closer to the sun than did Earth, so you might expect it to be a similar planet differentiated into a silicate mantle and a molten iron core. The density and size of Venus require that it have a dense interior much like Earth's; but, if the core is molten, then you would expect the dynamo effect to generate a magnetic field. No spacecraft has ever detected such a field around Venus. The magnetic field must be at least 25,000

times weaker than Earth's. Some theorists wonder if the core of the planet is solid. If it is solid, then how did Venus get rid of its internal heat so much faster than Earth did?

Because the planet lacks a magnetic field, it is not protected from the solar wind. The solar wind slams into the uppermost layers of the planet's atmosphere, forming a bow shock where the wind is slowed and deflected (■ Figure 22-9). Planetary scientists know little about the differentiation of the planet into core and mantle, so the size of the core shown in Figure 22-9 is estimated by analogy with Earth's. The magnetic field carried by the solar wind drapes over Venus like seaweed over a fishhook, forming a long tail within which ions flow away from the planet. You will see in Chapter 25 that comets, which also lack magnetic fields, interact with the solar wind in the same way.

Presumably, Venus and Earth formed in molten states and never had primeval atmospheres rich in hydrogen. Rather, they formed carbon dioxide atmospheres as they formed. Calculations show that Venus and Earth have outgassed about the same amount of carbon dioxide, but Earth's oceans have dissolved it and converted it to sediments such as limestone. If all of Earth's carbon were dug up and converted back to carbon dioxide, Earth's atmosphere would be about as dense as the air on Venus. This suggests that the main difference between Earth and Venus is the lack of water on Venus. Venus may have had small oceans when it was young; but, being closer to the sun, it was warmer, and the carbon dioxide in the atmosphere created a greenhouse effect that made the planet even warmer. That process could have dried

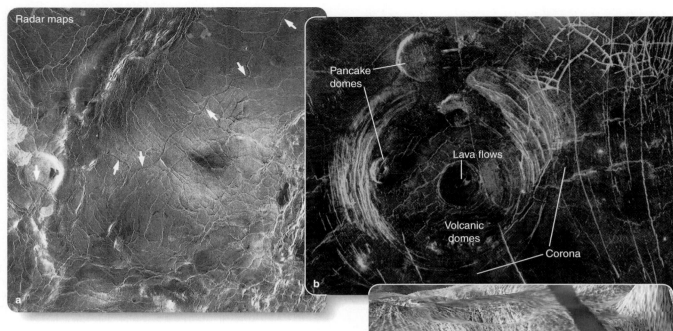

■ Figure 22-8

Volcanic features on Venus: (a) Arrows point to a 600-km segment of Baltis Vallis, the longest lava-flow channel in the solar system. It is at least 6800 km long. (b) Aine Corona, about 200 km in diameter, is marked by faults, lava flows, small volcanic domes, and pancake domes of solidified lava. (c) This perspective view shows a hill typical of those associated with a corona. Molten lava below the surface appears to have pushed up the hill, causing the network of radial faults. (d) Other regions of the crust appear to have collapsed as subsurface magma drained away. (NASA)

■ Figure 22-9

By analogy with Earth, the interior of Venus should contain a molten core (estimated here), but no spacecraft has detected a planetary magnetic field. Thus, Venus is unprotected from the solar wind, which strikes the planet's upper atmosphere and is deflected into an ion tail.

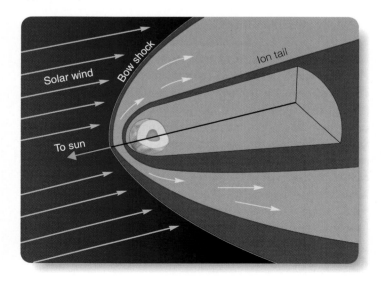

up any oceans that did exist and reduced the ability of the planet to clear its atmosphere of carbon dioxide. As more carbon dioxide was outgassed, the greenhouse effect grew even more severe. Venus seems to have been trapped in a runaway greenhouse effect.

Fully 70 percent of the heat from Earth's interior flows outward through midocean volcanism. But Venus lacks crustal rifts, and its numerous volcanoes cannot carry much of the heat out of the interior. Rather, Venus seems to get rid of its interior heat through large currents of hot magma that rise beneath the crust. Coronae, lava flows, and volcanism form above such currents. In fact, the surface rock is dark-gray basalt typical of basalts found in ocean crust on Earth.

True plate tectonics is not important on Venus. For one thing, the crust is very dry and is consequently about 12 percent less dense than Earth's crust. This low-density crust is more buoyant than Earth's crust and resists being pushed into the interior. Also, the crust is so hot it is halfway to the melting point of rock. Such hot rock is not very stiff, so it cannot form the rigid plates typical of plate tectonics on Earth.

There is no sign of plate tectonics on Venus, but rather evidence that convection currents below the crust are deforming the crust to create coronae and push up mountains such as Maxwell. Detailed measurements of the strength of gravity over the mountains on Venus show that some must be held up not by deep roots like mountains on Earth but by rising currents of magma. Other mountains, like those around Ishtar Terra, appear to be folded mountains caused by limited horizontal motions in the crust, driven perhaps by convection currents in the mantle.

The small number of craters on the surface of Venus hints that the entire crust has been replaced within the last half-billion years or so. This may have occurred in a planetwide overturning as the old crust broke up and sank and lava flows created a new crust. This could happen periodically on Venus, or the planet may have had a geology more like Earth's until a single resurfacing not too long ago. In either case, unearthly Venus may eventually reveal more about how our own world works.

Ace⊙Astronomy™ Log into AceAstronomy and select this chapter to see Astronomy Exercise "Convection and Magnetic Fields." Take control of a planetary interior.

Building Scientific Arguments

What evidence can you point to that Venus does not have plate tectonics?

Sometimes a scientific argument can be helpful by eliminating a possibility. On Earth, plate tectonics is identifiable by the worldwide network of faults, subduction zones, volcanism, and folded mountain chains that outline the plates. Although some of these features are visible on Venus, they do not occur in a planetwide network that outlines plates. Volcanism is widespread, but folded mountain ranges occur in only a few places, such as near Lakshmi Planum and Maxwell Montes, and they do not make up long mountain chains as on Earth. Also, the large size of the shield volcanoes on Venus show that the crust is not moving over the hot spots as the Pacific seafloor is moving over the Hawaiian hot spot.

At first glance, you might think that Earth and Venus should be sister worlds, but comparative planetology reveals that they can be no more than distant cousins. You can blame the thick atmosphere of Venus for altering its geology, but that calls for a new scientific argument: **Why isn't Earth's atmosphere similar to that of Venus?**

Connections: Venus is certainly an unearthly world, but the familiar principles of comparative planetology can help you understand it. As you turn your attention to Mars, you will see the same principles in action, but you will discover that Mars is a world that lacks what Venus has in abundance— internal heat and air.

22-2 Mars

MERCURY AND THE MOON are small. Venus and Earth are, for terrestrial planets, large. But Mars occupies an intermediate position (**Celestial Profile 6** on page 535). It is twice the diameter of the moon but only 53 percent Earth's diameter. Its small size has allowed it to cool faster than Earth, and much of its atmosphere has leaked away. Its present carbon-dioxide atmosphere is only 1 percent as dense as Earth's.

In some ways, Mars is much like Earth. A day on Mars is nearly the same length as an Earth day—24 hours and 40 minutes—and its year lasts 1.88 Earth years. Also, just as Earth's axis is tipped 23.5°, that of Mars is tipped 25°. As the northern and southern hemispheres turn alternately toward the sun, seasonal changes are visible even through a small telescope. As spring comes to the southern hemisphere, the white polar cap shrinks, and the grayish surface markings grow darker and, according to some observers, greener. At one time, these seasonal changes led some to believe that plant life flourished on Mars when spring thawed the polar cap. You will see later that this is not so.

In other ways, Mars is very different from Earth. Its surface has never broken into moving plates like those that create the topography on Earth. Instead, Mars is a one-plate planet, with its crust frozen into a solid layer but bearing signs of geological activity that include some of the largest volcanoes in the solar system.

The Canals on Mars

Long before the space age, the planet Mars was a mysterious landscape in the public mind. In the century following Galileo's first astronomical use of the telescope, astronomers discovered dark markings on Mars as well as bright polar caps. Timing the motions of the markings, they concluded that a Martian day was about 24 hours long, and the similarity with Earth's day further supported the belief that Mars was another Earth.

The so-called canals of Mars began life innocently enough in 1858 when the Jesuit astronomer Angelo Secchi referred to a region on Mars known later as Syrtis Major as *Atlantic Canale*. This is the first use of the Italian word *canale* (channel) to refer to a feature on Mars. Then, in the late summer of 1877 the Italian astronomer Giovanni Virginio Schiaparelli, using a telescope only 8.75 in. in diameter, thought he glimpsed fine, straight lines on

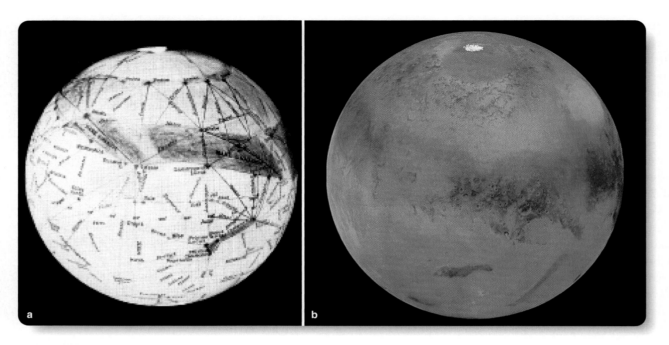

■ Figure 22-10

(a) Early in the 20th century, Percival Lowell mapped canals over the face of Mars and concluded that intelligent life resided there. (b) Modern images recorded by spacecraft reveal a globe of Mars with no canals. Instead, the planet is marked by craters and, in some places, volcanoes. Both of these images are reproduced with south at the top, as they appear in telescopes. Lowell's globe is inclined more nearly vertically and is rotated slightly to the right compared with the modern globe. (a, Lowell Observatory; b, U.S. Geological Survey)

Mars. He too used the Italian word *canali* (plural) for these lines, and the word was translated into English not as "channel," a narrow body of water, but as "canal," an artificially dug channel. The "canals of Mars" were born. Many astronomers could not see the canals at all, but others drew maps showing hundreds (■ Figure 22-10).

In the decades that followed Schiaparelli's discovery, many people assumed that the canals were watercourses built by an intelligent race on Mars to carry water from the polar caps to the lower latitudes. Much of this excitement was generated by Percival Lowell, a wealthy Bostonian who founded Lowell Observatory in 1894, principally for the study of Mars. He not only mapped hundreds of canals but also popularized his results in books and lectures. Although some astronomers claimed the canals were merely illusions, by 1907 the general public was so sure that life existed on Mars that the *Wall Street Journal* could suggest that the most extraordinary event of the previous year had been "the proof by astronomical observations . . . that conscious, intelligent human life exists upon the planet Mars." Further sightings of bright clouds and flashes of light on Mars strengthened this belief, and some urged that gigantic geometrical diagrams be traced in the Sahara Desert to signal to the Martians that Earth, too, is inhabited. All seemed to agree that the Martians were older and wiser than humans.

This fascination with men from Mars was not a passing fancy. Beginning in 1912, Edgar Rice Burroughs wrote a series of 11 novels about the adventures of the Earthman John Carter, lost on Mars. Burroughs made the geography of Mars, named by Schiaparelli after Mediterranean lands both real and mythical, into household words. He also made his Martians small and gave them green skin.

By Halloween night of 1938, people were so familiar with life on Mars that they were ready to believe that Earth had been invaded. When a radio announcer repeatedly interrupted dance music to report the landing of a spaceship in New Jersey, the emergence of monstrous beings, and their destruction of whole cities, thousands of otherwise sensible people rushed to their cars and fled in panic, not knowing that Orson Welles and other actors were dramatizing H. G. Wells's book *The War of the Worlds.*

The fascination with Mars, its canals, and its little green men lasted right up to July 15, 1965, when Mariner 4, the first spacecraft to fly past Mars, radioed back photos of the surface and proved that there are no canals and no Martians. Apparently, the canals are optical illusions produced by the human brain's astounding ability to assemble a field of disconnected marks into a coherent image. If your brain could not do this, the photos on these pages would be nothing but swarms of dots, and the images on the screen of a television set would never make sense. The brain of an astronomer looking for something at the edge of visibility is capable of connecting faint, random markings on Mars into the straight lines of canals.

Even today, Mars holds a fascination for the general public. The grocery store tabloids regularly run stories about a giant face carved on Mars by an ancient race. Although planetary scientists recognize it as nothing more than chance shadows in a photograph and dismiss the issue as a silly hoax, the stories persist. A hundred years of speculation have raised high expectations for Mars. If there really was intelligent life on Mars and its representatives came to Earth, they would probably be a big disappointment to the readers of the tabloids.

The Atmosphere of Mars

If you visited Mars, your first concern, even before you opened the door of your spaceship, would be the atmosphere. Is it breathable? Even for the astronomer observing safely from Earth, the atmosphere of Mars is a major concern. The gases that cloak Mars are critical to understanding the history of the red planet.

The air on Mars is 95 percent carbon dioxide, with a few percent each of nitrogen and argon. It contains almost no water vapor or oxygen, and its density at the surface of the planet—only 1 percent that of Earth's atmosphere—does not provide enough pressure to prevent liquid water from boiling into vapor. If you stepped outdoors on Mars without a spacesuit, your own body heat would make your blood boil.

Although the air is thin, it is dense enough to be visible in photographs (■ Figure 22-11). Haze and clouds come and go, and occasional weather patterns are visible. Winds on Mars can be strong enough to produce dust storms that envelop the entire

Mars is only half the diameter of Earth and probably retains some internal heat, but the size and composition of its core are not well known.

■ Figure 22-11

The atmosphere of Mars is evident in this image made by the Hubble Space Telescope. The haze is made up of high, water-ice crystals in the thin CO_2 atmosphere. The spot at extreme left is the volcano Ascraeus Mons, 25 km (16 mi) high, poking up through the morning clouds. Note the north polar cap at the top. (Philip James, University of Toledo; Steven Lee, University of Colorado, Boulder; and NASA)

Visual-wavelength image

Celestial Profile 6: Mars

Motion:

Average distance from the sun	1.5237 AU (2.279×10^8 km)
Eccentricity of orbit	0.0934
Maximum distance from the sun	1.6660 AU (2.492×10^8 km)
Minimum distance from the sun	1.3814 AU (2.066×10^8 km)
Inclination of orbit to ecliptic	1°51'09"
Average orbital velocity	24.13 km/s
Orbital period	1.8808 y (686.95 days)
Period of rotation	$24^h37^m22.6^s$
Inclination of equator to orbit	25°11'

Characteristics:

Equatorial diameter	6792 km (0.53 $D_\oplus$)
Mass	0.6424 X 10^{24} kg (0.1075 $M_\oplus$)
Average density	3.94 g/cm³ (3.3 g/cm³ uncompressed)
Surface gravity	0.379 Earth gravity
Escape velocity	5.0 km/s (0.45 $V_\oplus$)
Surface temperature	−140° to 20°C (−220° to 68°F)
Average albedo	0.16
Oblateness	0.009

Personality Point:

Mars is named for the god of war. Minerva was the goddess of defensive war, but Bullfinch's *Mythology* refers to Mars's "savage love of violence and bloodshed." You can see how the planet glows blood red in the evening sky because of iron oxides in its soil.

Visual-wavelength image

■ **Figure 22-12**

Mars is a red desert planet, as shown in this true-color photo made by the Rover Spirit. Rock layers are fractured by meteorite impacts, and dust is blown around the planet by winds. Dust suspended in the atmosphere colors the sky red. (NASA)

planet. The reddish color of the soil is evidently caused by iron oxides (rusts), and that warns that the oxygen humans would prefer to find in the atmosphere is locked in chemical compounds in the soil. The polar caps visible in photos are also related to the Martian atmosphere. The ices in the polar caps are frozen carbon dioxide (dry ice) with unknown amounts of frozen water underneath.

If you could visit Mars you would find it a reddish, airless, bone-dry desert (■ Figure 22-12). To understand Mars, you should ask why its atmosphere is so thin and dry and why the surface is rich in oxides. The answers to these questions lie in the origin and evolution of the Martian atmosphere.

Presumably, the gases in the Martian atmosphere were mostly outgassed from its interior. Volcanism on terrestrial planets typically releases carbon dioxide and water vapor, plus other gases. Because Mars formed farther from the sun, you might expect that it incorporated more volatiles when it formed. But Mars is smaller than Earth, so it has had less internal heat to drive geological activity, and you might suspect that it has not outgassed as much as Earth. In any case, the outgassing occurred early in the planet's history, and Mars, being small, cooled rapidly and now releases little gas.

How much atmosphere a planet has depends on how rapidly it releases internal gas and how rapidly it loses gas from its atmosphere, and the rate at which a planet loses gas depends on its mass and temperature. The more massive the planet, the higher its escape velocity (Chapter 5), and the more difficult it is for gas atoms to leak into space. Mars has a mass less than 11 percent that of Earth, and its escape velocity is only 5 km/s, less than half Earth's. Consequently, gas atoms can escape from it much more easily than they can escape from Earth.

The temperature of a planet's atmosphere is also important. If the gas is hot, its molecules have a higher average velocity and are more likely to exceed escape velocity. That means a planet near the sun and very hot is less likely to retain an atmosphere

than a more distant, cooler planet. The velocity of a gas molecule, however, also depends on the mass of the molecule. On average, a low-mass molecule travels faster than a massive molecule. For that reason, a planet loses its lowest-mass gases more easily because those molecules travel fastest.

You can see this principle of comparative planetology if you plot a diagram such as that in ■ Figure 22-13. The points show the escape velocity versus temperature of the larger objects in our solar system. The temperature used in the diagram is the temperature of the gas that is in a position to escape. For the moon, which has essentially no atmosphere, this is the temperature of the sunlit surface. For Mars, the temperature that is important is that at the top of the atmosphere. The lines in Figure 22-13 show the typical velocity of the fastest-traveling examples of various molecules. At any given temperature, some water molecules, for example, travel faster than others, and it is the highest-velocity molecules that escape from a planet. The diagram shows that the Jovian planets have escape velocities so high that very few molecules can escape. Earth and Venus can't hold hydrogen, and Mars can hold only the more massive molecules. Earth's moon is too small to keep any of the common gases. Refer to this diagram again whenever you study the atmospheres of other worlds.

Over the 4.6 billion years since Mars formed, it has lost some of the lower-mass gases. Water molecules are massive enough for Mars to keep, but ultraviolet radiation can break up the water molecules. On Earth, the ozone layer protects water vapor from ultraviolet radiation, but Mars never had an oxygen-rich atmosphere, so it never had an ozone layer. Ultraviolet photons from the sun can penetrate deep into the atmosphere and break up molecules such as water. The hydrogen escapes, and the oxygen, a very reactive element, forms more oxides in the soil—the oxides that make Mars the red planet. In this way, molecules too massive to leak into space can be lost if they break into lower-mass fragments.

The argon in the Martian atmosphere is evidence of a denser blanket of air in the past. Argon atoms are massive, almost as massive as a carbon monoxide molecule, and would not be lost easily. In addition, argon is inert and would not form compounds in the soil. The 1.6 percent argon in the atmosphere of Mars is

evidently left over from an ancient atmosphere that was 10 to 100 times denser than the present Martian air.

Finally, you should consider the interaction of the solar wind with the atmosphere of Mars. This is not an important process for Earth because Earth has a magnetic field that deflects the solar wind. Because Mars has no magnetic field, the solar wind interacts directly with the Martian atmosphere, and detailed calculations show that significant amounts of carbon dioxide could

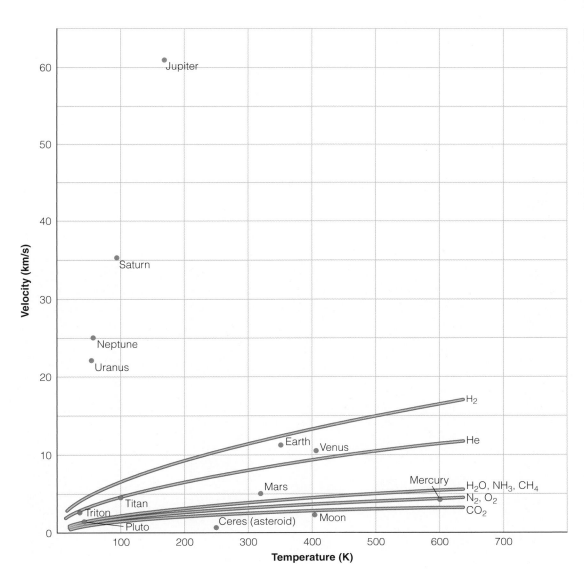

■ **Figure 22-13**

The loss of planetary gases. Dots represent the escape velocity and temperature of various solar-system bodies. The lines represent the typical highest velocities of molecules of various masses. The Jovian planets have high escape velocities and can hold on to even the lowest-mass molecules. Mars can hold only the more massive molecules, and the moon has such a low escape velocity that even the most massive molecules can escape.

have been carried away by the solar wind over the history of the planet. This process would have been most efficient long ago when the sun was more active and the solar wind was stronger. However, you should also consider that Mars probably had a magnetic field when it was younger and still retained significant internal heat. A magnetic field would have protected its atmosphere from the solar wind.

Although planetary scientists remain uncertain as to how much of an atmosphere Mars has had in its past and how much it has lost, it is a good example for your study of comparative planetology. When you look at Mars, you see what can happen to the atmosphere of a medium-size world. Like its atmosphere, the geology of Mars is typical of medium-size worlds.

Ace★Astronomy™ Log into AceAstronomy and select this chapter to see Astronomy Exercise "Primary Atmospheres." You can move your planet around to see how its atmosphere is affected.

The Geology of Mars

If you ever decide to visit another world, Mars may be your best choice. You will need a heated, pressurized spacesuit with air, water, and food, but Mars is not as inhospitable as the moon. The nights on Mars are deadly cold, but a hot summer day would be comfortable (Celestial Profile 6). It is also more interesting than the moon, with weather, complex geology, and signs that water once flowed over its surface. You might even hope to find traces of ancient life hidden in the rocks.

Spacecraft have been visiting Mars for almost 40 years, but the pace has picked up recently. A small armada of spacecraft has gone into orbit around Mars to photograph and analyze its surface, and five spacecraft have landed. Two Viking landers touched down in 1976, and three rovers have landed in recent years. Rovers have an advantage because they are wheeled robots that can be directed from Earth to travel from feature to feature and make detailed measurements. Pathfinder and its rover, Sojourner, landed in 1997. Rovers Spirit and Opportunity landed in January 2004 and carried sophisticated instruments to explore the rocky surface (■ Figure 22-14).

Photographs made by rovers on the surface of Mars, such as Figure 22-12, show reddish deserts of broken rock. These appear to be lava plains fractured by meteorite impacts, but they don't look much like the surface of Earth's moon. The atmosphere of Mars, thin though it is, protects the surface from the blast of micrometeorites that grinds moon rocks to dust. Also, the Martian dust storms may sweep fine dust away from some areas and leave larger rocks exposed.

Spacecraft orbiting Mars have imaged the surface and measured elevations to reveal that all of Mars is divided into two parts. The southern highlands are heavily cratered, and the number of craters there shows that the highlands must be old. In contrast, northern lowlands are smooth (■ Figure 22-15) and so remarkably free of craters that the lowlands must have been resurfaced roughly a billion years ago. Some astronomers have suggested that volcanic floods filled the northern lowlands and buried the

■ Figure 22-14

If you had been standing beside the small crater now known as Endurance, you might have seen the Rover Opportunity roll up to the edge and pause to await further instructions from Earth. This true-color image of the layered rock in the crater wall was taken by Opportunity as it circled the crater looking for a safe way down to the bottom. NASA engineers added a digital image of the rover to recreate the scene. (NASA/JPL/Cornell)

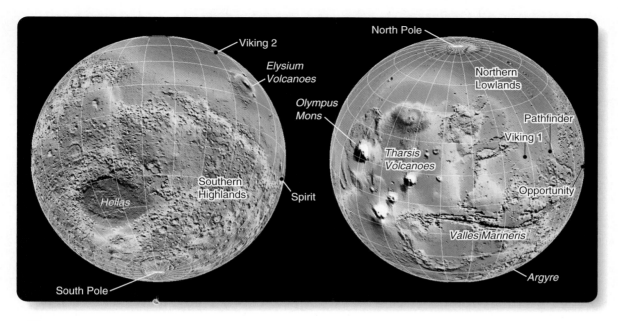

■ Figure 22-15

These globes of Mars are color coded to show elevation. The northern lowlands lie about 4 km below the southern highlands. Volcanoes are very high (white), and the giant impact basins, Hellas and Argyre, are low. Note the depth of the canyon Valles Marineris. The two Viking spacecraft landed on Mars in 1976. Pathfinder landed in 1997. Rovers Spirit and Opportunity landed in 2004. (NASA)

craters there. Growing evidence, however, suggests that the northern lowlands were once filled by an ocean of liquid water. This is a controversial theory, which will be discussed later, but it serves to illustrate how planetary scientists must work from the features seen today to reconstruct the past history of a world (**Window on Science 22-2**).

The cratering and volcanism on Mars will fit with what you know of comparative planetology. It is larger than Earth's moon, so it cooled more slowly, and its volcanism has continued longer than on the moon. But Mars is smaller than Earth and less geologically active, so some of its ancient cratered terrain has survived undamaged by volcanism and plate tectonics.

The Martian volcanoes are shield volcanoes with shallow slopes, showing that the lava flowed easily. Such volcanoes occur over hot spots of rising magma below the crust and are not related to plate tectonics. The largest volcano in the solar system is Olympus Mons on Mars (see page 529). The volcano Mauna Loa in Hawaii is so heavy it has sunk into Earth's crust to form an undersea moat. Olympus Mons is much larger than Mauna Loa but has not sunk into the crust of Mars, which is evidence that the crust of Mars is much thicker than the crust of Earth. You can see that in ■ Figure 22-16.

Other evidence shows that the Martian crust has been thinner and more active than the moon's. Valles Marineris is a network of canyons that stretches 4000 km (2500 mi) long and up to 600 km (400 mi) wide (Figure 22-16). At its deepest, it is four times deeper than the Grand Canyon on Earth. It is long enough to stretch from New York to Los Angeles. The canyon has been

produced by faults in the crust that allowed great blocks to sink. Landslides and some erosion have further modified the canyon. Although Valles Marineris is an old feature, it does show that the crust of Mars has been more active than the crusts of the moon or Mercury, worlds that lack such dramatic canyons.

The faults that created Valles Marineris seem linked at the west end to a great volcanic bulge in the crust of Mars called the Tharsis rise. Nearly as large as the United States, the Tharis rise extends 10 km (6 mi) above the mean radius of Mars. Tharsis is home to many smaller volcanoes, but on its summit lie three giants, and just off of its northwest edge lies Olympus Mons (Figure 22-16). The origin of the Tharsis rise is not well understood, but it appears that magma rising from below the crust has broken through repeatedly to build a giant bulge of volcanic deposits. This bulge is large enough to have modified the climate and seasons on Mars and may be critical in understanding the history of the planet.

A similar uplifted volcanic bulge, the Elysium region visible in Figure 22-15, lies halfway around the planet. It appears to be similar to the Tharsis rise, but it is more heavily cratered and so must be older.

The vast sizes of the Tharsis rise and of Olympus Mons show that the crust of Mars has not been broken into moving plates. If a plate were moving over a hot spot, the rising magma would produce a long chain of shield volcanoes and not a single large peak (page 529). On Earth, the hot spot that creates the volcanic Hawaiian Islands has punched through the moving Pacific plate repeatedly to produce the Hawaiian-Emperor island chain ex-

The Present Is the Key to the Past

Geologists are fond of saying, "The present is the key to the past." By that they mean that you can learn about Earth's history by looking at the present condition of Earth's surface. The position and composition of various rock layers in the Grand Canyon, for example, tell you that the western United States was once at the floor of an ocean. This principle of geology was astonishing when it was first formulated in the late 1700s, and it continues to be relevant today as you try to understand other worlds, such as Venus and Mars.

In the late 18th century, naturalists recognized that the present gave them clues to Earth's history. This was astonishing because most people assumed that Earth had no history. That is, they assumed either that Earth had been created in its present state as described in the Old Testament or that Earth was eternal. In either case, people commonly assumed that the hills and mountains they saw around them had always existed more or less as they were. The 18th-century naturalists began to see evidence that the hills and mountains were not eternal but were the result of past processes and were slowly changing. That gave birth to the idea that Earth had a history.

As the naturalists of the 18th century made the first attempts to thoughtfully and logically explain Earth's nature by looking at the evidence, they were inventing modern geology as a way of understanding Earth. What Copernicus, Kepler, and Newton did for the heavens in the 1600s, the first geologists did for Earth beginning in the late 1700s. Of course, the invention of geology as the study of Earth led directly to modern attempts to understand the geology of other worlds.

Geologists and astronomers share a common goal: They are attempting to reconstruct the past (Window on Science 19-2). Whether you study Earth, Venus, or Mars, you are looking at the present evidence and trying to reconstruct the past history of the planet by drawing on observations and logic to test each step in the story. How did Venus get to be covered with lava, and how did Mars lose its atmosphere? The final goal of planetary astronomy is to draw together all of the available evidence (the present) to tell the story (the past) of how the planet got to be the way it is. Those first geologists of the late 1700s would be fascinated by the stories planetary astronomers tell today.

Layers of rock in the Martian crater Terby hint at a time when the crater was filled with a lake. (NASA/JPL/Malin Space Science Systems)

---

tending 7500 km (4700 mi) northwest across the Pacific seafloor (page 486). No such chains of volcanoes are evident on Mars, so you can conclude that the crust is not divided into moving plates.

No spacecraft has ever photographed an erupting volcano on Mars, but it is possible that some of the volcanoes are still active. Craters in the youngest lava flows in the Elysium region show that the volcanoes may have been active as recently as a few million years ago. Mars may still retain enough heat to trigger an eruption, but the interval between eruptions may be very long.

Hidden Water on Mars

You would not expect to find liquid water on Mars. The atmospheric pressure is too low to keep it from boiling away to vapor. Nevertheless, some geological features seem to have been formed by moving water, and sophisticated measurements from orbit and on-site observations by rovers suggest that significant amounts of water are hidden in the crust.

When the two Viking spacecraft reached orbit around Mars in 1976, they found exciting hints that water once flowed over the surface. More recent spacecraft such as Mars Global Surveyor,

which reached Mars in 1997, and Mars Odyssey, which orbited in 2001, have identified additional features that seem to be related to water. You must examine those features with care, however, because the water on Mars is not obvious at all.

■ Figure 22-17 summarizes water-related features. **Outflow channels** appear to have been cut by massive floods carrying as much as 10,000 times the water flowing down the Mississippi River. In a matter of hours or days, such floods swept away geological figures and left outflow channels. One way to account for these features is to suppose that the water was frozen in the soil as permafrost. Melted by volcanic heat and pressurized under the weight of the overlying rock, water could have burst out to scour the surface in sudden floods. The number of craters formed on top of the outflow channels shows they are billions of years old.

In contrast, the **valley networks** look like meandering riverbeds that may have formed over long periods. Some streambeds appear to have flowed repeatedly over millions of years. The valley networks are located in the old, cratered, southern hemisphere, and they have about the same age. Evidently the atmosphere was denser a few billion years ago.

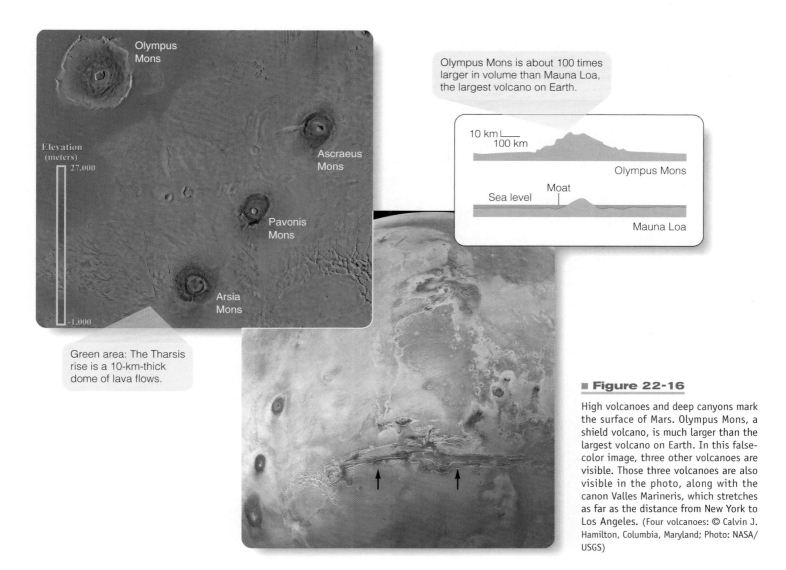

Olympus Mons

Ascraeus Mons

Pavonis Mons

Arsia Mons

Elevation (meters)

27,000

-1,000

Green area: The Tharsis rise is a 10-km-thick dome of lava flows.

Olympus Mons is about 100 times larger in volume than Mauna Loa, the largest volcano on Earth.

10 km
100 km

Olympus Mons

Moat

Sea level

Mauna Loa

■ **Figure 22-16**

High volcanoes and deep canyons mark the surface of Mars. Olympus Mons, a shield volcano, is much larger than the largest volcano on Earth. In this false-color image, three other volcanoes are visible. Those three volcanoes are also visible in the photo, along with the canon Valles Marineris, which stretches as far as the distance from New York to Los Angeles. (Four volcanoes: © Calvin J. Hamilton, Columbia, Maryland; Photo: NASA/USGS)

As shown in Figure 22-17, some features suggest water in the crust of Mars, perhaps as ice. Jumbled valleys appear to be regions where water below the surface has drained away and allowed the terrain to collapse. Some impact craters appear to be surrounded by a "splash" of material, as if the crust was rich in water at the time of the impact. Slow seepage from underground water or ice could gradually form a valley network over long periods. The low atmospheric pressure at the surface of Mars today could not prevent liquid water from boiling away to vapor, but water trapped in the crust might be under enough pressure to remain a liquid. Also, water frozen as ice could be trapped in the crust. This gives planetary scientists confidence that water is trapped in the crust of Mars.

More recent spacecraft have been able to photograph the surface of Mars at a much higher resolution than could the Viking orbiters. Those new images show evidence of water. In many places, the spacecraft found gullies leading down steep slopes, and the lack of craters formed on the gullies shows that they are young. Early interpretations suggested the gullies were cut by subsurface water released in a gush when an ice plug melted. Newer studies suggest the gullies could be debris flows caused by rock and soil sliding downhill after being released by evaporating ice. Another study suggests that the gullies formed beneath deep layers of melting snow. Choosing among these hypotheses is not easy, and you must keep in mind that more than one process may have formed gullies. In any case, the observation of young gullies is further evidence that there is still water on Mars.

Other images show the details of valley networks that contain common features such as sandbars and islands much like rivers on Earth (■ Figure 22-18a). Other features appear to show deltas of material deposited by flowing water (Figure 22-18b). The similarity with features found on Earth is dramatic.

The atmosphere of Mars contains a tantalizing clue. Recall that deuterium is an isotope of hydrogen that contains a proton and a neutron as a nucleus. Because deuterium is twice as heavy as normal hydrogen, it escapes from a planet's atmosphere more

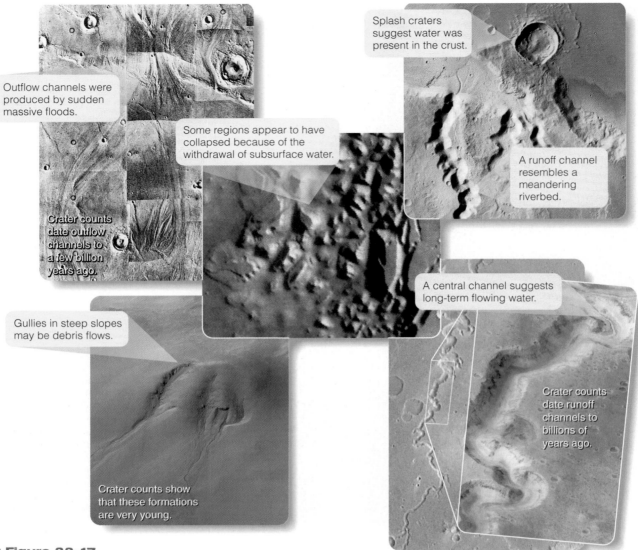

■ Figure 22-17

These visual-wavelength images made by the Viking orbiters and Mars Global Surveyor show some of the features that suggest liquid water on Mars. Outflow channels and valley networks are old, but some gullies may be quite recent. (Malin Space Science Systems and NASA)

slowly. Earlier in this chapter you saw that the deuterium in the atmosphere of Venus was evidence that Venus once had more water. Observations of the Martian atmosphere show that deuterium is 5.5 times more abundant relative to hydrogen on Mars than it is on Earth. That suggests that Mars once had about 20 times more water than it does now.

The terrain at the edges of the northern lowlands has been compared to shorelines, and many planetary scientists suspect that the northern lowlands were at least once filled with an ocean. Study Figure 22-15, where the lowlands have been color-coded blue, and notice the major outflow channels leading from the highlands into the lowlands northwest of the Viking 1 landing site and southeast of the Pathfinder landing site. The evidence suggests that the northern lowlands were filled by a major ocean a

few billion years ago. Large, generally circular depressions such as Hellas and Argyre are believed to be ancient impact basins and were probably once filled with water.

The Mars Odyssey Spacecraft may have found the remains of that ancient water. Two instruments on Mars Odyssey are capable of detecting water within the top meter of the crust, and both agree that water ice makes up a significant fraction of the crust from latitude 60° north and from latitude −60° south. It isn't possible to determine how deep the ice goes, but the amount detected so far is enough to fill Lake Michigan twice.

Rovers Spirit and Opportunity were targeted to land in areas suspected of having had water on their surfaces. Images made from orbit showed flow features at the Spirit landing site, and hematite, a mineral that forms in water, was detected from orbit at the Opportunity landing site. Both rovers reported exciting discoveries, including evidence of past water. Using its analytic instruments, Opportunity found small spherical concretions (dubbed blue-

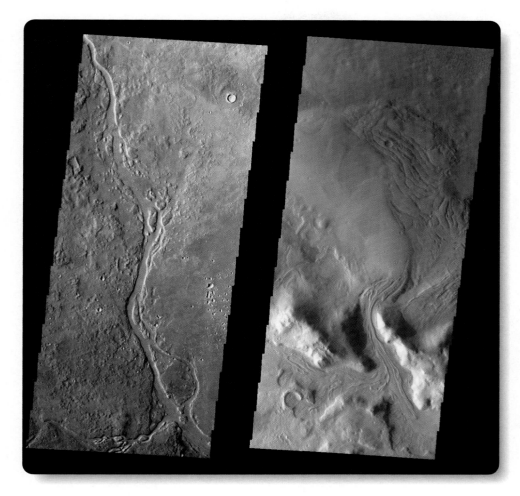

■ **Figure 22-18**

Details of past water flow are visible in these false-color images. At the left, an ancient flow channel meanders across the Apsus Vallis regions, complete with islands and bars. At the right, water flowing through a gap in the wall of the crater Moreux has deposited large amounts of sediment on the crater floor. (NASA/JPL/Arizona State University)

berries although they are, in fact, hematite) that must have formed in water. Later, Spirit found similar concretions in its area. In other rocks, Opportunity found layers of sediments with ripple marks and crossed layers showing they were deposited in moving water (■ Figure 22-19). Chemical analysis of the rocks at the Opportunity site showed the presence of sulfates much like Epsom salts, plus bromides and chlorides. On Earth, these compounds are left behind when bodies of water dry up. Halfway around Mars, Spirit found the mineral goethite, clear evidence that water was once present.

If liquid water once flowed on Mars, then its climate must have been different in the past. The atmosphere of Mars must have been thicker long ago before much of the gas had time to leak away. Also, volcanism may have been more active in the past, and outpourings of volcanic gases could have added to the atmosphere, increased the greenhouse effect, and vaporized more of the polar caps to further thicken the atmosphere.

Hints of climate change were found in the 1970s when the Viking orbiters photographed layered terrain near the polar caps. More modern Mars probes have imaged these layers in greater detail (■ Figure 22-20). These layers are laid down year after year as dust accumulates on the carbon dioxide polar caps and is then left behind when the polar caps vaporize in the spring. Over pe-

riods of thousands of years, deep layers can develop. What is significant is that newer layers are superimposed at angles over older layers, showing that the climate on Mars has changed repeatedly. Mars Global Surveyor has found similar layered features widespread on Mars. If these regions show traces of climate cycles, it supports studies that show that the climate on Mars may vary because of cyclic changes in the rotation and orbital revolution of the planet. Recall from Chapter 2 that Earth is affected by the Milankovitch cycles.

One astonishing bit of evidence of water on Mars is the analysis of rock samples from the planet. Of course, no astronaut has ever visited Mars and brought back a rock, but over the history of the solar system occasional impacts by asteroids have blasted giant craters on Mars and ejected bits of rock into space. A few of those bits of rock have fallen to Earth as meteorites, and over 30 have been found and identified (■ Figure 22-21). These meteorites include basalts, as you might expect from a planet so heavily covered by lava flows. They also contain small traces of water and minerals that are deposited by water. Chemical analysis shows that the magma from which the rocks solidified must have contained up to 1.8 percent water.

If all of the lava flows on Mars had contained that much water and it was all outgassed, it would have created a planetwide

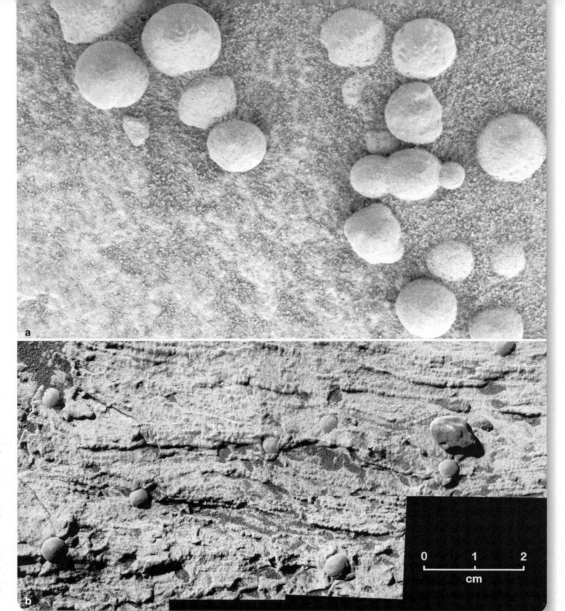

■ Figure 22-19

(a) Rover Opportunity photographed these hematite concretions in a rock near its landing site. The spheres appear to have grown as minerals collected around small crystals in the presence of water. Similar concretions are found on Earth. (b) The layers in this rock were deposited as sand and silt in rapidly flowing water. From the way the layers curve and cross each other, geologists can estimate that the water was at least ten centimeters deep. A few "blueberries" and one small pebble are also visible in this image. (NASA/JPL/Cornell/USGS)

ocean 20 m deep. That wouldn't have been quite enough to explain all the flood features on Mars, but it would have provided much more water on the surface than the planet has today.

Mars has water, but it is hidden. When humans reach Mars, they will not need to dig far to find water, probably as ice. They can use solar power to break the water into hydrogen and oxygen. Hydrogen is fuel, and oxygen is the breath of life, so the water on Mars may prove to be buried treasure.

Even more exciting is the realization that Mars once had bodies of liquid water on its surface. It is a desert world now, but someday an astronaut may scramble down an ancient Martian streambed, turn over a rock, and find a fossil.

A History of Mars

The history of Mars is like a play where all the exciting stuff happens in the first act. After the first couple billion years on Mars, most of the activity was over, and it has been downhill ever since.

You can check off differentiation, the first stage of planetary development. Planetary scientists have good evidence that Mars differentiated when it formed and had a hot, molten core. Some of the evidence comes from exquisitely sensitive Doppler-shift measurements of radio signals coming from spacecraft orbiting Mars. The shifts allowed scientists to map the gravitational field and study the shape of Mars in such detail they could detect tides on Mars caused by the sun's gravity. Those tides are less than a centimeter high; but, by comparing them with models of the interior of Mars, the scientists can show that Mars has a very dense, liquid core, a dense mantle, and a low-density crust.

Observations made from orbit show that Mars has no magnetic field, but it does have traces of magnetism frozen into some sections of old crust. That shows that soon after Mars formed, it

■ Figure 22-20 ▶

(a) The north polar cap of Mars is made of many regions of ice separated by narrow valleys free of ice. (NASA) (b) Near the edge of the polar cap, a set of light-colored layers of dust overlies a set of darker layers. Sand eroding from the lower layers is forming dunes. (NASA/JPL/Malin Space Science Systems) (c) In places, different sets of layers are superimposed, suggesting periodic changes in the Martian climate. (Adapted from a diagram by J. A. Cutts, K. R. Balasius, G. A. Briggs, M. H. Carr, R. Greeley, and H. Masursky)

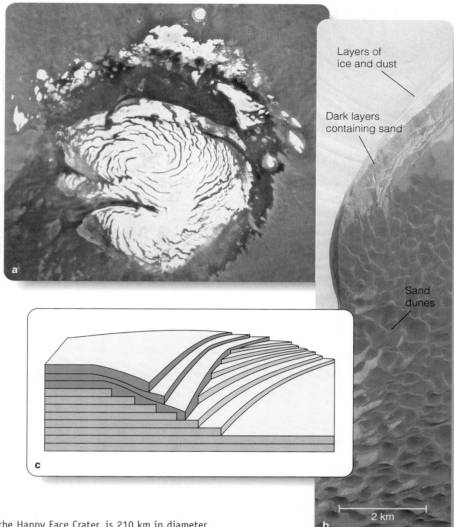

had a hot metallic core in which the dynamo effect generated a magnetic field. Because Mars is small, it lost its heat rapidly, and the central part of the core gradually froze solid; that may be why the dynamo finally shut down. Today Mars probably has a large solid core surrounded by a thin shell of liquid core material in which no dynamo can generate a magnetic field.

No one is sure what produced the southern highlands and northern lowlands. Powerful convection in the planet's mantle may have pushed crust together to form the southern highlands. The suggestion that a catastrophic impact produced the northern lowlands does not seem to fit with the structure of the highlands, but it is not impossible.

During the second stage of planetary evolution, cratering, the crust of Mars was battered

■ Figure 22-21 ▼

(a) The impact crater Galle, known for obvious reasons as the Happy Face Crater, is 210 km in diameter. Such large impacts may occasionally eject fragments of the crust into space. A few fragments from Mars have fallen to Earth as meteorites. (Malin Space Science Systems/NASA) (b) Meteorite ALH84001, found in Antarctica, has been identified as an ancient rock from Mars. Minerals found in the meteorite were deposited in water and thus suggest that the Martian crust was once richer in water than it is now. (NASA)

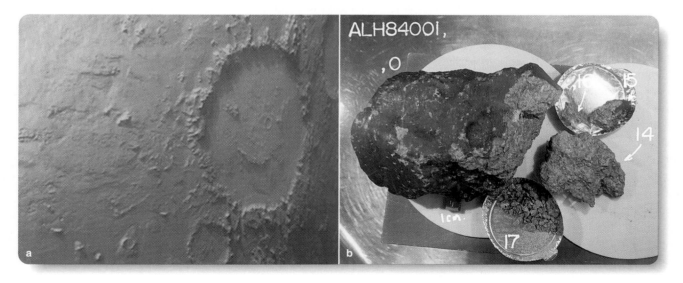

■ Figure 22-22

The Tharsis rise is a bulge on the side of Mars consisting of stacked lava flows extending up to 10 km high and over a third the diameter of Mars. It dominates this image of Mars with a few clouds clinging to the four largest volcanoes. This image was recorded in late winter for the southern hemisphere. Notice the size of the southern polar cap. (NASA/JPL/Malin Space Science Systems)

by the heavy bombardment as the last of the debris in the young solar system was swept up. The old southern hemisphere survives from this age, about 4 billion years ago. The largest impacts blasted out the great basins like Hellas and Argyre very late in the cratering, and there is no trace of magnetic field in those basins. Evidently the dynamo had shut down by then.

The third stage of planet formation, flooding, included flooding by great lava flows that smoothed some regions. Volcanism in the Tharsis and Elysium regions was very active, and the Tharsis rise grew into a huge bulge on the side of Mars (■ Figure 22-22). Some of the oldest lava flows on Mars are in the Tharsis rise, but it also contains some of the most recent lava flows. It has evidently been a major volcanic area for most of the planet's history.

Flooding included liquid water on Mars. You can check that off your list, but add a question mark because it is uncertain how much water was present and how long it lasted. Until nearly the end of cratering, Mars had a magnetic field, and that should have protected the atmosphere from the solar wind. Violent volcanism and the heat of formation should have released gases to keep the air pressure high enough to allow liquid water to exist on the surface. In fact, volcanism releases water vapor as well as gases, and impacts could have vaporized water trapped in the crust.

Most of the water features on Mars were formed in the first billion years or so. Valley networks drained water released by seeping groundwater or by rain and snow falling out of the atmosphere. Many of these valley networks flow away from the Tharsis rise, showing that they formed after the rise developed. Some very

old flood features are known, but most are a little younger than the valley networks and appear to have been formed by the sudden release huge amounts of groundwater. Water seems to have filled the northern lowlands and the deep basins to form oceans, but it's not known how long those bodies of water survived.

Because Mars is small, it lost its internal heat quickly, and atmospheric gases have escaped into space. It isn't clear how long liquid water remained stable on the surface, but it is clear that Mars dried up, chilled, and became the deadly cold desert world of today. There may have been episodes when the atmosphere grew denser and warmer and more water was available. Major volcanic eruptions could add gas and water vapor to the atmosphere. This may have produced short-lived episodes in which water flowed over the surface and collected in the lowlands to form new oceans. As the planet aged, the water became trapped in the crust as permafrost or at the poles where it froze out along with solid carbon dioxide.

The history of Mars may hinge on climate variations. Recently calculated models suggest that Mars may have once rotated at a much steeper angle to its orbit, as much as 45°. This could have supported a warmer climate and kept more of the carbon dioxide from freezing out at the poles. The rise of the Tharsis bulge could have tipped the axis to its present 25° and cooled the climate. Some calculations suggest that Mars goes through cycles as its rotational inclination fluctuates, and this may cause short-term variations in climate, much like the ice ages on Earth. That might explain the layered terrain near the polar caps.

The fourth stage of planet formation is unremarkable on Mars. The crust of Mars is now too thick to be active. The planet has lost much of its internal heat, and the core no longer generates a magnetic field. Although the crust was stressed by rising magma that produced the Tharsis and Elysium volcanic rises and the great canyon Valles Marineris, it is too thick for plate tectonics. There are no folded mountain ranges on Mars as there are on Earth. The huge size of the Martian volcanoes clearly shows that crustal plates have not moved on Mars, and repeated eruptions have built the volcanoes to vast size. Volcanism may still occur on Mars, but the crust has grown too thick for much geological activity beyond slow erosion by wind-borne dust and the occasional meteorite impact.

This history of Mars is far from complete, but it does illustrate how the size of Mars has influenced both its atmosphere and its geology. Neither small nor large, Mars is a medium world.

Building Scientific Arguments

Why doesn't Mars have coronae like those on Venus?
This argument is a good opportunity to apply the principles of comparative planetology. The coronae on Venus are caused by rising currents of molten magma in the mantle pushing upward under the crust and then withdrawing to leave the circular scars called coronae. Earth, Venus, and Mars have had signifi-

cant amounts of internal heat, and there is plenty of evidence that they have had rising convection currents of magma under their crusts. Of course, you wouldn't expect to see coronae on Earth; its surface is rapidly modified by erosion and plate tectonics. Furthermore, the mantle convection on Earth seems to produce plate tectonics rather than coronae. Mars, however, is a smaller world and must have cooled faster. There is no evidence of plate tectonics on Mars, and giant volcanoes suggest rising plumes of magma erupting up through the crust at the same point over and over. Perhaps there are no coronae on Mars because the crust of Mars rapidly grew too thick to deform easily over a rising plume. On the other hand, perhaps you could think of the entire Tharsis bulge as a single giant corona.

Planetary scientists haven't explored enough planets yet to see all the fascinating combinations nature has in store. But it does seem likely that the geology of Mars is typical of medium-size worlds. Of course, Mars is not medium in terms of its location. Of the terrestrial planets, Mars is the farthest from the sun. Build a scientific argument to analyze that factor. **How has the location of Mars affected the evolution of its atmosphere?**

Connections: If you want your history of Mars to be truly complete, it must include a phenomenon that was not part of your histories of Mercury and Venus. Unlike those worlds, Mars has moons. These moons, and the moons you will study in the outer solar system, contain important clues to the origin of the planets.

22-3 The Moons of Mars

IF YOU COULD CAMP OVERNIGHT on Mars, you might notice its two small moons, Phobos and Deimos. Phobos, shaped like a flattened loaf of bread 20 km × 23 km × 28 km, would appear less than half as large as Earth's full moon. Deimos, only 12 km in diameter and three times farther from Mars, would look only $\frac{1}{15}$ the diameter of Earth's moon.

Both moons are tidally locked to Mars, keeping the same side facing the planet as they orbit. Also, both moons revolve around Mars in the same direction that Mars rotates, but Phobos follows such a small orbit that it revolves faster than Mars rotates. That makes Phobos rise in the west and set in the east.

Origin and Evolution

Deimos and Phobos are typical of the small, rocky moons in our solar system (■ Figure 22-23). They are dark gray, with albedos of only about 0.06, making them as dark as coal. They have low densities, about 2 g/cm³.

Many of the properties of these moons hint that they are captured asteroids. In the outer parts of the asteroid belt, almost all asteroids are dark, low-density objects. Massive Jupiter, orbiting

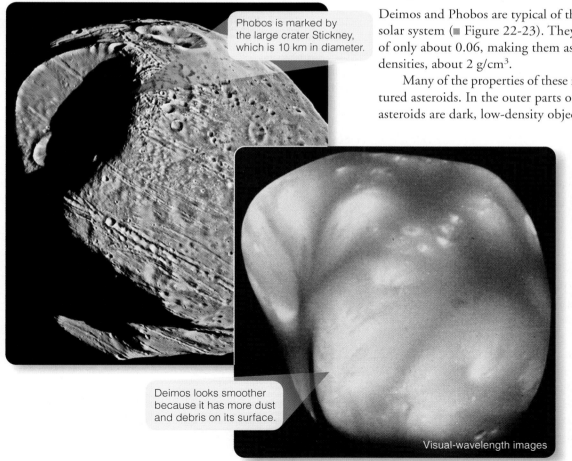

Phobos is marked by the large crater Stickney, which is 10 km in diameter.

Deimos looks smoother because it has more dust and debris on its surface.

Visual-wavelength images

■ **Figure 22-23**

The moons of Mars are too small to pull themselves into spherical shape. Deimos is about half the size of Phobos. The two moons were named for the mythical attendants of Mars, the god of war. Phobos was the god of panic, and Deimos was the god of fear. (Phobos: Damon Simonelli and Joseph Ververka, Cornell University/NASA; Deimos: NASA)

just outside the asteroid belt, can scatter such bodies throughout the solar system, so you should not be surprised if Mars, the next planet inside the asteroid belt, has captured a few of these as satellites.

However, capturing a passing asteroid into a closed orbit is not so easy that it happens often. The asteroid approaches the planet along a hyperbolic (open) orbit and, if it is unimpeded, swings around the planet and disappears back into space. To convert the hyperbolic orbit into a closed orbit, the planet must slow the asteroid as it passes. Tidal forces might do this, but they would be rather weak. Interactions with other moons or grazing collisions with a thick atmosphere might also slow the asteroid.

Both satellites have been photographed by spacecraft, and those photos show that the satellites are heavily cratered. Such cratering could have occurred while the moons were either still in the asteroid belt or in orbit around Mars. In any case, the heavy battering has broken the satellites into irregular chunks of rock, and they cannot pull themselves into smooth spheres because their gravity is too weak to overcome the structural strength of the rock. You will discover that low-mass moons are typically irregular in shape, whereas more massive moons are more spherical.

Images of Phobos reveal a unique set of narrow, parallel grooves (see Figure 22-23). Averaging 150 m wide and 25 m deep, the grooves run from Stickney, the largest crater, to an oddly featureless region on the opposite side of the satellite. One theory suggests that the grooves are deep fractures produced by the impact that formed the crater. The featureless region opposite Stickney may be similar to the jumbled terrain on Earth's moon and on Mercury. All were produced by the focusing of seismic waves from a major impact on the far side of the body. High-resolution photographs show that the grooves are lines of pits, suggesting that the pulverized rock material on the surface has drained into the fractures or that gas, liberated by the heat of impact, escaped through the fractures and blew away the dusty soil.

The Mars Global Surveyor spacecraft revealed more about the dust. Observations made with the spacecraft's infrared spectrometer show that the moon's surface cools quickly from $-4°C$ to $-112°C$ (from 25°F to $-170°F$) as it passes from sunlight into the shadow of Mars. Solid rock would retain heat and cool more slowly, so the dust must be at least a meter deep and very fine. In photos made by the spacecraft camera, the dust blankets the terrain, but the photos also show boulders a few meters in diameter, thought to be ejecta from impacts.

Deimos not only has no grooves, it also looks smoother because of a thicker layer of dust on its surface (Figure 22-23). This material partially fills craters and covers minor surface irregularities. It seems certain that Deimos experienced collisions in its past, so fractures may be hidden below the debris.

The debris on the surfaces of the moons raises an interesting question. How can the weak gravity of small bodies hold any fragments from meteorite impacts? Escape velocity on Phobos is only 12 m/s. An athletic astronaut could almost jump into space. Certainly, most fragments from impacts should escape, but some could fall back and accumulate on the surface.

Deimos, smaller than Phobos, has a smaller escape velocity. But it has more debris on its surface because it is farther from Mars. Phobos is close enough to Mars that most ejecta from impacts on Phobos will be drawn into Mars. Deimos, being farther from Mars, is able to keep a larger fraction of its ejecta. Phobos is so close to Mars that tides are making its orbit shrink, and it will fall into Mars or be ripped apart by tidal forces within about 100 million years.

Deimos and Phobos illustrate three principles of comparative planetology that you will find helpful as you explore farther from the sun. First, some satellites are probably captured asteroids. Second, small satellites tend to be irregular in shape and heavily cratered. And third, tidal forces can affect small moons and gradually change their orbits. You will find even stronger tidal effects in Jupiter's satellite system (Chapter 23).

Building Scientific Arguments

Why would you be surprised if you found volcanism on Phobos or Deimos?

This is another obvious argument, isn't it? But remember, the purpose of a scientific argument is to test your own understanding, so it is a good way to review. In discussing Earth's moon, Mercury, Venus, Earth, and Mars, you have seen illustrations of the principle that the larger a world is, the more slowly it loses its internal heat. It is the flow of that heat from the interior through the surface into space that drives geological activity such as volcanism and plate motion. A small world, like Earth's moon, cools quickly and remains geologically active for a shorter time than a larger world like Earth. Phobos and Deimos are not just small, they are tiny. However they formed, any interior heat would have leaked away very quickly; with no energy flowing outward, there can be no volcanism.

Some futurists suggest that the first human missions to Mars will not land on the surface of the planet but will build a colony on Phobos or Deimos. These plans speculate that there may be water deep inside the moons that colonists could use. Build an argument based on what you know about water on Mars. **What would happen to water released in the sunlight on the surface of such small worlds?**

■   ■   ■

Connections: Venus is nearly the same size as Earth, but it is unlikely that humans will ever colonize its surface. The smaller and colder Mars, however, may someday be home to large colonies. In the next chapter, you will explore Jupiter and Saturn, giant worlds so totally unlike Earth that you will need entirely new principles of comparative planetology to describe them.

Summary

22-1 | Venus

Why is the atmosphere of Venus so thick?

■ Venus is Earth's twin in size but is slightly closer to the sun.

■ The atmosphere of Venus is 95 times thicker than Earth's and composed almost entirely of carbon dioxide. The temperature at the surface is about 745K (880°F).

■ Venus is heated by a runaway greenhouse effect.

■ The surface is so hot that compounds have cooked out of the crust to form traces of sulfuric, hydrochloric, and hydrofluoric acids in the atmosphere. The very high clouds on Venus are composed of small droplets of sulfuric acid and sulfur crystals.

What is the hidden surface of Venus really like?

■ Although it is perpetually hidden below thick clouds, the surface of Venus can be studied by radar mapping, which reveals higher uplands and low rolling plains. Volcanoes, lava flows, and impact craters are visible.

■ Volcanic features such as coronae appear to be produced by rising currents of molten rock pushing up under the crust. Where the magma breaks through it produces volcanoes, lava flows, and lava channels.

■ Landers have analyzed the surface rock and found it to be similar to basalts on Earth.

How did Venus form and evolve?

■ Because Venus formed closer to the sun, it was warmer, and large liquid-water oceans were not able to form and remove carbon dioxide from the atmosphere. The accumulating carbon dioxide produced an intense greenhouse effect, made the planet very hot, and evaporated any small bodies of water that did form.

■ Venus rotates slowly retrograde (backward). This may have been caused by an off-center impact by a very large planetesimal as the planet was forming or by solar tides raised in its very thick atmosphere.

■ Volcanism is common on Venus, and much of the surface is solidified lava flows. Volcanoes are probably still active on Venus.

■ The surface contains more impact craters than the maria on the moon, but fewer than does the Earth. The crust appears to be about half a billion years old. Planetary scientists suspect that the entire planet was resurfaced by an outpouring of lava about half a billion years ago.

■ Venus has no detectable magnetic field, so the state of its core is unknown.

22-2 | Mars

Why is the atmosphere of Mars so thin?

■ Mars is smaller than Earth but larger than the moon. It has lost much of its atmosphere because of its low escape velocity.

■ The atmosphere on Mars is very-low-density carbon dioxide, and the pressure at the surface is too low to prevent water from boiling away.

■ Mars has no magnetic field to protect it from the solar wind, and some of its atmosphere has been blasted away over its history by the pressure of the solar wind.

Has Mars ever had water on its surface?

■ Mars is now a dry, desert world on which liquid water cannot exist.

■ The southern hemisphere is old and heavily cratered, but the northern lowlands are smooth and mostly free of craters.

■ Images from orbit reveal outflow channels that appear to have been cut by massive floods and valley networks that resemble dry river beds. Crater counts show that these features are very old.

■ The smooth lowlands of the northern hemisphere appear to have once contained a liquid water ocean. Some outflow channels lead into the lowlands and features resembling shore lines have been found.

■ Evidence suggests that whatever water Mars retains is now frozen in the crust as permafrost. If it seeps out, it may cut gullies and form similar flow features, but rivers, lakes, and oceans cannot now exist on Mars because of the low atmospheric pressure.

How did Mars form and evolve?

■ Mars formed farther from the sun than Earth did and is much smaller.

■ Mars once had a molten core that generated a magnetic field, but it has no detectable magnetic field now. Evidently the inner core has solidified, and the molten-out layer of the core cannot generate a magnetic field.

■ Volcanism has been important in the history of Mars, and the Tharsis rise is a huge volcanic uplift. Volcanoes may still erupt on Mars, but, because of its small size, it has lost much of its internal heat.

■ Olympus Mons is a very large volcano, but it has not sunk into the crust, and that shows that the crust of Mars is now quite thick.

■ Mars must have had a thicker atmosphere in the past as evidenced by the liquid water flow features visible on it surface. But it has lost much of its water, and what remains is frozen in the crust or under the carbon dioxide polar caps.

■ The development of the Tharsis rise may have altered the rotation of Mars and changed its climate, but climate cycles driven by changes in rotation and revolution may explain the layered terrain near the polar caps.

22-3 | The Moons of Mars

■ Mars has captured two asteroids into orbit as moons. Phobos and Deimos are small, irregularly shaped, and cratered.

New Terms

subsolar point (p. 524)

composite volcano (p. 528)

shield volcano (p. 528)

corona (p. 530)

outflow channel (p. 540)

valley network (p. 540)

Review Questions

Ace ⑤ Astronomy™ Assess your understanding of this chapter's topics with additional quizzing and animations at http://ace .brookscole.com/sf9

1. Why might you expect Venus and Earth to be similar?

2. What evidence can you cite that Venus and Mars once had more water than at present? Where did that water come from? Where did it go?

3. What features would you look for in high-resolution radar maps of Venus to search for plate tectonics?

4. Why doesn't Mars have mountain ranges like those on Earth? Why doesn't Earth have large volcanoes like those on Mars?

5. What were the canals on Mars? How do they differ from the dry streambeds on Mars?

6. Propose an explanation for the nearly pure carbon dioxide atmospheres of Venus and Mars. How did Earth avoid such a fate?

7. What evidence can you cite that the climate on Mars has changed?

8. How can planetary scientists estimate the ages of the streambeds on Mars?

9. Why are Phobos and Deimos nonspherical? Why is Earth's moon much more spherical?

10. Volcano Sif Mons on Venus is shown in this radar image. What kind of volcano is it, and why is it orange in this image? What color would the rock be if you could see it with your own eyes?

(NASA)

11. Olympus Mons on Mars is a very large volcano. In this image you can see multiple calderas at the top. What do those calderas and the immense size indicate about the geology of Mars?

(NASA)

Discussion Questions

1. From what you know of Earth, Venus, and Mars, do you expect the volcanoes on Venus and Mars to be active or extinct? Why?

2. If humans someday colonize Mars, the biggest problem may be finding water and oxygen. With plenty of solar energy beating down through the thin atmosphere, how might colonizers extract water and oxygen from the Martian environment?

Problems

1. How long would it take radio signals to travel from Earth to Venus and back if Venus were at its nearest point to Earth? at its farthest point from Earth?

2. The Pioneer Venus Orbiter circled Venus with a period of 24 hours. What was its average distance above the surface of Venus? (*Hint:* See Chapter 5.)

3. Calculate the velocity of Venus in its orbit around the sun.

4. What is the maximum angular diameter of Venus as seen from Earth? (*Hint:* Use the small-angle formula.)

5. If the Magellan spacecraft transmitted radio signals down through the clouds on Venus and heard an echo from a certain spot 0.000133 second before the main echo, how high is the spot above the average surface of Venus?

6. The smallest feature visible through an Earth-based telescope has an angular diameter of about 1 second of arc. If a canal on Mars was just visible when Mars was at its closest to Earth, how wide was the canal? (*Hint:* Use the small-angle formula.)

7. What is the maximum angular diameter of Phobos as seen from Earth? What surface features should you expect to see from Earth? from the surface of Mars? (*Hint:* Use the small-angle formula.)

8. What is the maximum angular diameter of Deimos as seen from the surface of Mars?

9. Deimos is about 12 km in diameter and has a density of 2 g/cm^3. What is its mass? (*Hint:* The volume of a sphere is $\frac{4}{3}\pi R^3$.)

Media Cluster

ACTIVE FIGURES

Planetary Atmospheres
Experiment with the temperature throughout the atmospheres of several planets with this animation.

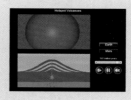

Hot Spot Volcanoes
If the crustal plate over a hot spot is moving, repeated eruptions can result in an island chain like the Hawaiian-Emperor island chain. Set a series of eruptions in motion in this animation.

ASTRONOMY EXERCISES

The Greenhouse Effect
See how distance to the sun and amount of carbon dioxide in a planet's atmosphere affect its temperature in this animation.

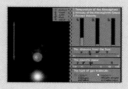

Convection and Magnetic Fields
A planet's core temperature and rotation speed are factors in the strength of its magnetic field. See the effect of these variables in this animation.

Convection and Plate Tectonics
Study the convection currents of the four Earthlike planets in this animation and see if you can determine how this relates to the amount of plate tectonic activity on each planet.

Primary Atmospheres
This animation lets you change the distance to the sun and mass of a hypothetical planet and study how these relate to the planet's retention of its atmosphere.

VIRTUAL ASTRONOMY LABS

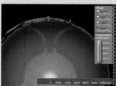

Lab 5: Planetary Geology
By thoroughly studying Earth's geology, you can extend your knowledge to understand conditions on other planets. This lab includes an exercise in determining the interior characteristics of a planet by observing the travel times of its seismic waves.

Lab 7: Planetary Atmospheres and Their Retention
This lab investigates the retention of atmospheres. You will explore the factors that govern the loss of atmospheric gases, and you will see why certain bodies can retain some gases but not others.

Critical Inquiries for the Web

1. Have you ever wondered how the various features on the planets get their names? Surface features on Venus are (mostly) named for female figures from history and mythology. Who decides how planetary features are named? Look for information on planetary nomenclature and summarize the way different types of features on Venus are assigned names.

2. "Martians" have fascinated humans for the last century or more. Many online sources chronicle the representation of life on Mars by Earthlings throughout history and literature. Read about the Martians as represented by a particular literary work or nonfiction account and discuss to what extent they are (or are not) based on realistic views of the nature of Mars—in terms of your current understanding as well as the views of the period in which the work was written.

3. Spacecraft are studying Mars right now. Search for the latest discoveries and photographs.

Exploring *TheSky*

1. Venus and Mars exhibit phases similar to the moon's phases. (a) Use *TheSky* to sketch the appearance of Venus and Mars at the present time. Also give the phase as a percentage of full. How to proceed: Use **Find** to give you a highly magnified view of a particular planet. Click on the planet to obtain **Object Information.** (b) Explain the observed phase on the basis of geometrical relationship between the sun, Earth, and the planet as shown by the **3D Solar System Mode.**

23 | Comparative Planetology of Jupiter and Saturn

There is something

fascinating about science.

One gets such wholesale returns

of conjecture out of such a

trifling investment of fact.

MARK TWAIN,
LIFE ON THE MISSISSIPPI

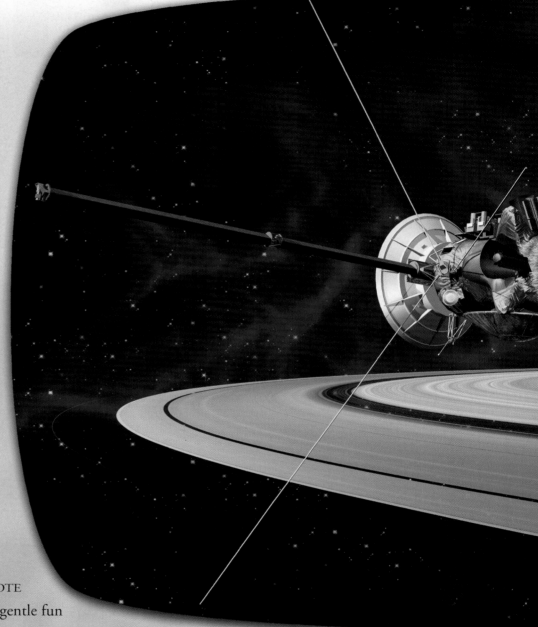

Wʜᴇɴ Mᴀʀᴋ Tᴡᴀɪɴ ᴡʀᴏᴛᴇ the sentences above, he was poking gentle fun at science, but he was right. The exciting thing about science isn't the so-called facts, the observations in which scientists have greatest confidence. Rather, the excitement lies in the understanding that scientists get by rubbing a few facts together. Science can take you to strange new worlds. Jupiter and Saturn are certainly strange, but you can combine the available observations with known principles of comparative planetology and understand these two giant planets. ▍ Jupiter and Saturn have been studied from Earth for centuries, but much of what is known has been radioed back by space probes. Pioneer 10 and Pioneer 11 explored Jupiter and Saturn in the mid-1970s, but their instruments were not very sensitive. The Voyager 1 and Voyager 2 spacecraft were

▍ Continued on page 554 ▍

Science can take you to other worlds. When the Cassini probe rounded Saturn, it fired its rockets to slow down and settle into orbit. Controlled from Earth, it studies the planet, the moons and the rings of the Saturn system. (NASA)

Guidepost

Looking Back

You have become an experienced planetwalker. You have been to Mercury, Venus, Earth, Mars, and Earth's moon. You know all about craters, volcanoes, lava flows, and plate tectonics. But your experience is limited to terrestrial planets. Now it is time to try something entirely different.

This Chapter

As you begin this chapter, you leave behind the psychological security of planetary surfaces. You can imagine standing on the moon, on Mars, or even on Venus, but Jupiter and Saturn have no surfaces. Here you face a new challenge—to use comparative planetology to study worlds so unearthly you cannot imagine being there. As you study these worlds you will find answers to five essential questions:

How do the outer planets compare with the inner planets?

How is Jupiter different from Earth?

How did Jupiter and its system of moons and rings form and evolve?

How is Saturn different from Jupiter?

How did Saturn and its system of moons and rings form and evolve?

There's no place to stand on Jupiter or Saturn, but be sure to bring your spacesuit. Both planets have big systems of moons, and you will be able to watch erupting volcanoes, stroll through a methane rain storm, and swim in Saturn's rings. It will be interesting, but it is no place like home.

Looking Ahead

Jupiter and its cousin Saturn make a good basis for comparison in your study of the rest of the outer solar system in the next chapter. Uranus and Neptune are small, peculiar relatives, and you will find that tiny Pluto may not even be a member of the family of planets.

launched in 1977 on a mission to visit all four Jovian worlds, concluding when Voyager 2 flew past Neptune in 1989. In December 1995, the Galileo spacecraft reached Jupiter. While an instrumented probe dropped into Jupiter's atmosphere, the main Galileo spacecraft went into orbit to begin a long-term study of the planet and its moons. The Cassini spacecraft went into orbit around Saturn in 2004 and dropped a probe into the atmosphere of Saturn's biggest moon, Titan. Throughout this discussion of the Jovian worlds, you will find images and data returned to Earth by these robotic explorers.

Jupiter and Saturn are unearthly; but, because they obey the same laws of nature, you can understand their peculiar personalities. They make a perfect introduction to the study of the outer solar system.

(23-1) A Travel Guide to the Outer Planets

IF YOU TRAVEL MUCH, you know that some cities make you feel at home, and some do not. In this and the next chapter, you will visit five worlds that are truly unearthly. You will not feel welcome, so this travel guide will warn you what to expect.

The Outer Planets

The outermost planets in our solar system are Jupiter, Saturn, Uranus, Neptune, and Pluto. The first four are often called the Jovian planets, meaning they are like Jupiter. In fact, they are all individuals with their own separate personalities.

■ Figure 23-1 compares the five outer worlds, and one striking feature is diameter. Jupiter is the largest of the Jovian worlds, over 11 times the diameter of Earth. Saturn is a bit smaller, but Uranus and Neptune are quite a bit smaller than Jupiter. Pluto, in contrast, is hardly visible in this diagram. It is a tiny world and doesn't seem to fit with its companions in the outer solar system. You will discover it is truly an oddball, and some astronomers argue that it is not a planet at all.

The other feature you will notice immediately when you look at Figure 23-1 is Saturn's rings. They are bright and beautiful and composed of billions of ice particles, each orbiting around the planet. Jupiter, Uranus, and Neptune also have rings, but they are not easily detected from Earth and are not visible in this figure. Nevertheless, as you visit these worlds you will be able to compare four different sets of planetary rings.

Atmospheres and Interiors

The four Jovian worlds have hydrogen-rich atmospheres filled with clouds. On Jupiter and Saturn, you can see that the clouds form belts and zones that circle the planets like the stripes on a child's ball. This form of atmospheric circulation is called **belt–zone circulation.** You will find traces of belts and zones on Uranus and Neptune, but they are not very distinct.

The atmospheres of the Jovian planets are not very deep. Jupiter's atmosphere makes up only about one percent of its radius. Below that, Jupiter and Saturn are composed of liquid hydrogen, so the older term for these planets, the *gas giants,* should probably be changed to the *liquid giants.* Only near their centers do these worlds have solid cores of dense material with the composition of rock and metal.

Uranus and Neptune are sometimes called the ice giants because they are rich in water, both as a liquid and as a solid. They too have denser material in their cores.

On your visits to these worlds, notice that they are low-density worlds that are rich in hydrogen. Jupiter and Saturn are mostly liquid hydrogen, and even Uranus and Neptune contain a much larger proportion of hydrogen than does Earth. Recall from Chapter 19 that these worlds are hydrogen-rich, low-density worlds because they formed in the outer solar nebula where water vapor could freeze to form tremendous amounts of tiny ice particles. Once these planets grew massive enough, they could draw in the hydrogen gas directly by gravitational collapse.

Satellite Systems

All of the Jovian worlds have large satellite systems, and even little Pluto has a moon. As you visit the moons of the Jovian worlds, look for two processes. In many cases, the moons interact with each other gravitationally and have adjusted their orbits. Their gravity can also influence planetary ring systems.

The second thing to notice is that a few moons are geologically active, and others show signs of past activity. You have learned that cratered surfaces are old, so when you see a section of a moon's surface that has few craters, you know that the moon must have been geologically active since the end of the heavy bombardment. Of course, geological activity depends on heat, so be alert for sources of internal heat as you visit these moons.

Building Scientific Arguments

Why do you expect the outer planets to be low-density worlds?
To build this scientific argument, you need to think about how the planets formed from the solar nebula. In Chapter 19, you discovered that the inner planets could not incorporate ice when they formed because it was too hot near the sun; but, in the outer solar nebula, the growing planets could accumulate lots of ice. Eventually they grew massive enough to grow by gravitational collapse, and that pulled in hydrogen and helium gas. That makes the outer planets low-density worlds.

The outer planets may be unearthly, but they are understandable. For example, extend your argument. **Why do you expect the outer planets to have rings and moons?**

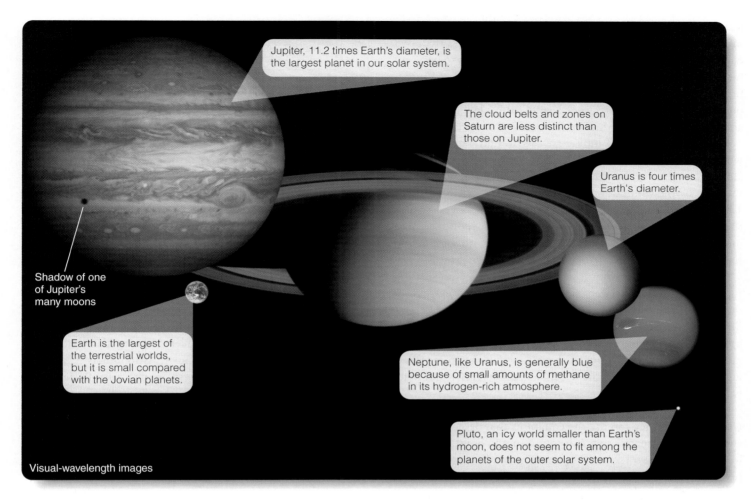

Jupiter, 11.2 times Earth's diameter, is the largest planet in our solar system.

The cloud belts and zones on Saturn are less distinct than those on Jupiter.

Uranus is four times Earth's diameter.

Shadow of one of Jupiter's many moons

Earth is the largest of the terrestrial worlds, but it is small compared with the Jovian planets.

Neptune, like Uranus, is generally blue because of small amounts of methane in its hydrogen-rich atmosphere.

Pluto, an icy world smaller than Earth's moon, does not seem to fit among the planets of the outer solar system.

Visual-wavelength images

■ **Figure 23-1**

The worlds of the outer solar system consist of the four Jovian planets, which are large, low-density worlds rich in hydrogen, and tiny, icy Pluto. (NASA/JPL/Space Science Institute/University of Arizona)

Connections: The outer solar system is so far from the sun that it is cold and dimly lit. The planets are far apart and orbit slowly. It is not a place that makes humans feel comfortable and welcome, but it is both mysterious and revealing. Exploring the outer solar system will give you a new appreciation for your home world.

23-2 Jupiter

JUPITER IS THE MOST MASSIVE of the Jovian planets. Over eleven times the diameter of Earth, it contains nearly three-fourths of all the planetary matter in our solar system. This high mass accentuates some processes that are less obvious or nearly absent on the other Jovian worlds. Just as you used Earth as the basis of comparison for your study of the terrestrial planets, you should examine Jupiter in detail so you can use it as a standard in your comparative study of the other Jovian planets.

Surveying Jupiter

Jupiter is interesting because it is big, massive, mostly liquid hydrogen, and very hot inside. The preceding facts are common knowledge among astronomers, but you should demand an explanation of how they know these facts. Often the most interesting thing about a fact isn't the fact itself but how it is known.

You can be sure that Jupiter is a big planet because it looks big. At its closest point to Earth, Jupiter is about eight times farther away than Mars, but even a small telescope will reveal that the disk of Jupiter is more than twice as big as the disk of Mars. If you use the small-angle formula, you can compute the diameter of Jupiter—a bit over 11 times Earth's diameter (**Celestial Profile 7**). That's a big planet!

You can be confident that Jupiter is massive because its moons race around it at high speed. Io is the innermost of the four Galilean moons, and its orbit is just a bit larger than the orbit of our moon around Earth. Io streaks around its orbit in less than two days, whereas Earth's moon takes a month. Jupiter has to be a very massive world to hold on to such a rapidly moving moon (■ Figure 23-2). In fact, you can put the radius of Io's orbit and its orbital period into Newton's version of Kepler's third law (Chapter 5) and calculate the mass of Jupiter—almost 318 times Earth's mass.

The size and mass of Jupiter are easy to find, but you might wonder how astronomers know that it is made of hydrogen. That fact is just a little bit harder to find, but the first step is to divide the mass by the volume and find Jupiter's average density, about 1.34 g/cm³. Of course, it is denser at the center and less dense near the surface, but this average density reveals that Jupiter can't be made totally of rock. Rock has a density of 2.5 to 4 g/cm³, so Jupiter must contain less dense material, such as hydrogen.

Spectra recorded from Earth and from spacecraft visiting Jupiter show that the composition of Jupiter is much like that of the sun—it is mostly hydrogen and helium. This fact was confirmed when a probe from the Galileo spacecraft fell into the atmosphere in 1995 and radioed its results back to Earth. Jupiter is mostly hydrogen and helium, with traces of heavier atoms forming molecules such as methane (CH₄), ammonia (NH₃), and water (■ Table 23-1).

Just as astronomers can build mathematical models of the interiors of stars, they can use the equations that describe gravity, energy, and how matter responds to pressure to build mathematical models of the interior of Jupiter. The models suggest that a planet as big and massive as Jupiter, having a low density and composed mostly of hydrogen and helium, must contain only a small core of elements heavier than helium. Astronomers often refer to this as a "rocky" core, but the pressure is so high there is nothing rocklike about the material. It would be better to call it a heavy-element core. Although the core is about the size of Earth, it must

■ **Table 23-1** | **Composition of Jupiter and Saturn (by mass)**

Molecule	Jupiter (%)	Saturn (%)
H_2	78	88
He	19	11
H_2O	0.0001	—
CH_4	0.2?	0.6
NH_3	0.5?	0.2

contain roughly 15 Earth masses. The models also reveal that the pressure below the atmosphere is so great that hydrogen cannot exist as a gas; most of Jupiter is liquid hydrogen.

Even the shape of the planet is revealing. By watching the cloud belts, astronomers can tell that Jupiter rotates rapidly, in about 10 hours. Simple measurements on a photograph show that it is slightly flattened; it is 6.37 percent larger in diameter through its equator than it is through its poles. Astronomers refer to this by saying that Jupiter's **oblateness,** its degree of flattening, is 0.0637. The amount of flattening depends on the speed of rotation and on the rigidity of the material. A rotating planet that is mostly liquid will be more oblate than a planet that is mostly rocky. In this way, astronomers can use the observed oblateness of Jupiter to check their models of the size of its heavy-element core.

It is easy to think of the heavy-element core as a ball of rock, but astronomers are quick to point out that it is very hot, about 30,000 K. That is five times hotter than the surface of the sun, but the pressure is so great that it keeps the material confined to the core of the planet. How do they know it is so hot? Because infrared observations show that Jupiter emits about 1.7 times as

■ **Figure 23-2**

It is obvious that Jupiter is a very massive planet when you compare Jupiter's moon Io with Earth's moon. Although Io is 10 percent farther from Jupiter, it travels 17 times faster in its orbit than does Earth's moon around Earth. Clearly, Jupiter's gravitational field is much stronger than Earth's, and that means Jupiter must be very massive.

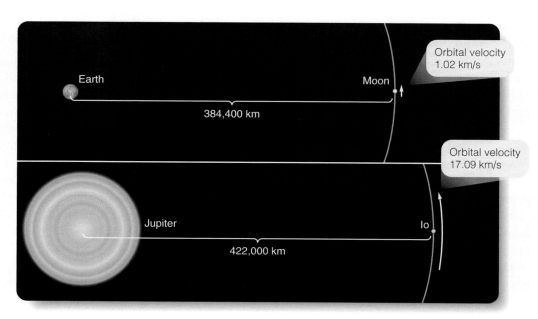

much energy as it receives from the sun. It must be very hot inside to have so much heat flowing outward into space.

The models of Jupiter's interior reveal that ordinary liquid hydrogen cannot support the pressure deep inside the planet. It is compressed into a state in which the electrons are free to move easily from atom to atom, and, although the hydrogen can flow like a liquid, it can conduct electricity like a metal. This **liquid metallic hydrogen** does not occur naturally on Earth, but it has been created under tremendous pressure in laboratories.

Basic observations and the known laws of physics can tell you a great deal about Jupiter. Its vast magnetic field can tell you even more.

Jupiter's Magnetic Field

Astronomers detected a magnetic field around Jupiter as early as the 1950s, which is surprising considering that magnetic fields are invisible to the eye. In fact, Jupiter's giant magnetic field was first detected by radio.

In 1955, radio astronomers detected bursts of synchrotron radio waves coming from Jupiter. They called it **decameter radiation** because its wavelength was tens of meters. Synchrotron radiation is electromagnetic radiation from fast electrons spiraling in a magnetic field, so it was obvious that Jupiter had a magnetic field.

Because the radio signals fluctuated with a period of 9 hours 55.5 minutes, astronomers concluded that this period was the true period of rotation of Jupiter. This result is what you would expect from the dynamo theory for the generation of magnetic fields, which requires an electrically conducting fluid, convection driven by heat flowing outward, and rapid rotation. The liquid metallic hydrogen in Jupiter is a very good electrical conductor, and its infrared brightness is clear evidence that heat flows outward from inside Jupiter. That, coupled with Jupiter's rapid rotation, should generate a powerful dynamo effect. The fluctuation of the radio signals reveals the true rotation period of the planet's liquid interior.

By the early 1960s, radio astronomers had discovered short-wavelength **decimeter radiation,** with wavelengths a few tenths of a meter long. Although this type of radiation is also synchrotron radiation, radio telescopes revealed that it comes from a region at least three times larger than the disk of Jupiter. Detailed studies of the decimeter radiation showed that Jupiter's magnetic field was at least ten times stronger than Earth's and, like Earth's field, was inclined about 10° to the axis of rotation.

Radio studies were helpful, but nothing is as good as a visit. In 1973 and 1974, two Pioneer spacecraft flew past Jupiter, followed in 1979 by two Voyager spacecraft. The Galileo spacecraft entered orbit around Jupiter in late 1995, and the Cassini spacecraft flew past in December 2000. These spacecraft found Jupiter surrounded by a magnetosphere over 100 times larger than Earth's.

The magnetic field deflects the solar wind and traps high-energy particles in radiation belts much more intense than Earth's. The radiation is more intense because Jupiter's magnetic field is stronger and can trap and hold higher-energy particles, mostly

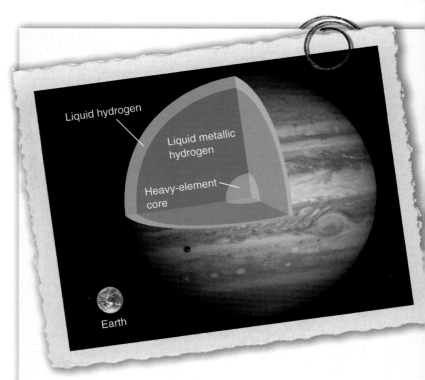

Jupiter is mostly a liquid planet with a small core of heavy elements that is not much bigger than Earth. (NASA/JPL/University of Arizona)

Celestial Profile 7: Jupiter

Motion:

Average distance from the sun	5.2028 AU (7.783 × 10⁸ km)
Eccentricity of orbit	0.0484
Maximum distance from the sun	5.455 AU (8.160 × 10⁸ km)
Minimum distance from the sun	4.951 AU (7.406 × 10⁸ km)
Inclination of orbit to ecliptic	1°18′29″
Average orbital velocity	13.06 km/s
Orbital period	11.867 y (4334.3 days)
Period of rotation	9ʰ55ᵐ30ˢ
Inclination of equator to orbit	3°5′

Characteristics:

Equatorial diameter	142,900 km (11.20 $D_\oplus$)
Mass	1.899 × 10²⁷ kg (317.83 $M_\oplus$)
Average density	1.34 g/cm³
Gravity at base of clouds	2.54 Earth gravities
Escape velocity	61 km/s (5.4 $V_\oplus$)
Temperature at cloud tops	−130°C (−200°F)
Albedo	0.51
Oblateness	0.0637

Personality Point:

Jupiter is named for the Roman king of the gods, and it is the largest planet in our solar system. It can be very bright in the night sky, and its cloud belts and four largest moons can be seen through even a small telescope. Its moons are even visible with a good pair of binoculars mounted on a tripod or braced against a wall.

electrons and protons. The spacecraft passing through the radiation belts received radiation doses equivalent to a billion chest X rays—at least 100 times a lethal dose for a human. Some of the electronics on the spacecraft were damaged by the radiation.

Jupiter's magnetic field is very strong, and it dominates a huge magnetosphere around the planet. Compare the size of Jupiter's field with that of Earth in ■ Figure 23-3.

You will recall that Earth's magnetosphere interacts with the solar wind to produce auroras, and the same is true on Jupiter. Charged particles in the magnetosphere leak downward along the magnetic field, and, where they enter the atmosphere, they produce auroras 1000 times more powerful than those on Earth. The auroras on Jupiter, like auroras on Earth, occur in rings around the magnetic poles (■ Figure 23-4).

Some of the heavier ions in the radiation belts appear to come from the inner Galilean moon, Io. As you will see later in this chapter, Io has active volcanoes that spew gas and ash. Because Io orbits with a period of 1.8 days, and Jupiter's magnetic field rotates in only 10 hours, the wobbling magnetic field rushes past Io at high speed, sweeping up stray particles, accelerating them to high energy, and spreading them around Io's orbit in a doughnut of ionized gas called the **Io plasma torus.** Jupiter's magnetic field interacts with Io to produce a powerful electric current (about a million amperes) that flows through a curving path called the **Io flux tube** from Jupiter out to Io and back to Jupiter. Small spots of bright auroras lie at the two points where the Io flux tube enters Jupiter's atmosphere (Figure 23-4).

This Io flux tube explains a mystery. Back in 1955, when radio astronomers discovered the decametric radio signals, they noticed that the signals were stronger when Io was in certain parts of its orbit. Now you can understand that. The electrical current flowing through the Io flux tube is at least partially responsible for generating the decametric radio signals and beams them out into space in certain directions.

Fluctuations in the auroras reveal that the solar wind buffets Jupiter's magnetosphere, but some of the fluctuations seem to be caused by changes in the magnetic dynamo deep inside the planet. In this way, studies of the auroras on Jupiter may help astronomers learn more about its liquid depths.

Much of the power and fascination of Jupiter is invisible to your eyes, but the swirling cloud belts are beautiful in their complexity.

Ace◐Astronomy™ Log into AceAstronomy and select this chapter to see Astronomy Exercise "Auroras." See how auroras are produced.

Ace◐Astronomy™ Log into AceAstronomy and select this chapter to see Astronomy Exercise "Convection and Magnetic Fields." How does convection in a planet generate a magnetic field?

Jupiter's Atmosphere

If you parachuted into Jupiter with a small rubber boat, expecting to go sailing on the liquid hydrogen ocean, you would be disappointed. Mathematical models of Jupiter show that there is no surface. Deep below the clouds, the temperature and pressure exceed the **critical point** for hydrogen, the temperature and pressure above which liquid hydrogen and gaseous hydrogen have the same density and are indistinguishable. As you parachuted down through Jupiter's atmosphere, the temperature, pressure, and density would rise, and gradually the gas would become a fluid. You would never splash down, and your rubber boat would be useless. Below the clouds of Jupiter lies the largest ocean in the solar system—and it has no surface and no waves.

When you look at Jupiter, you don't see a surface. You see those parts of the atmosphere that contain clouds. These cloud layers lie deep inside a nearly transparent atmosphere of hydrogen and helium. You can detect this nearly transparent hydrogen atmosphere by noticing that Jupiter has limb darkening just as the sun does (Chapter 8). When you look near the limb of Jupiter (the edge of its disk), the clouds are much dimmer (Figure 23-1) because it is nearly sunset or sunrise along the limb. If you were on Jupiter at that location, you would see the sun just above the horizon, and it would be dimmed by the atmosphere. In addition, light reflected from clouds must travel out at a steep angle through the atmosphere to reach Earth, dimming the light further. Jupiter

■ Figure 23-3

Jupiter's magnetic field is large and powerful. It traps particles from the solar wind to form powerful radiation belts. The rapid rotation of the planet forces the slightly inclined magnetic field to wobble up and down as the planet rotates. Earth's magnetosphere and radiation belts are shown to scale.

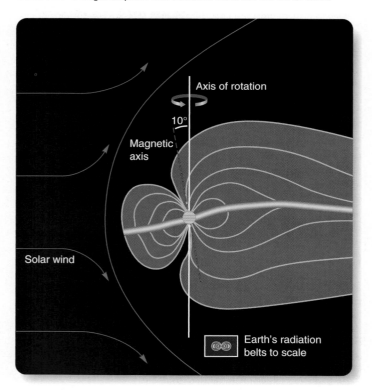

Axis of rotation

10°

Magnetic axis

Solar wind

Earth's radiation belts to scale

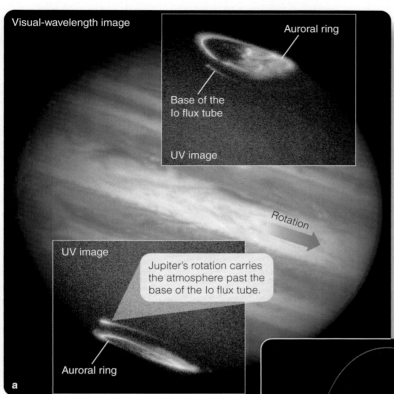

Visual-wavelength image

Auroral ring

Base of the
Io flux tube

UV image

Rotation

UV image

Jupiter's rotation carries
the atmosphere past the
base of the Io flux tube.

Auroral ring

a

The Io flux tube is an
electric circuit connecting
Io to Jupiter's magnetic field.

Io

Visual

UV

b

■ **Active Figure 23-4**

Jupiter's huge magnetic field funnels energy from the solar wind
down to form rings of auroras around its magnetic poles, which are
tipped some distance from the rotational poles. The same thing
happens on Earth. The Io flux tube connects the small moon Io to
the planet and carries a powerful electric current that creates spots
of auroras where it touches the planet's atmosphere. (Jupiter: John
Clarke, University of Michigan, NASA; Flux tube: NASA)

Ace **Astronomy**™   Log into AceAstronomy and select this
chapter to see the Active Figure "Explorable Jupiter" and observe
Jupiter at different wavelengths.

is brighter near the center of the disk because
the sunlight shines nearly straight down and
the cloud layers look brighter.

Study **Jupiter's Atmosphere** on
pages 560–561, and notice four important
ideas:

1 The atmosphere is hydrogen rich, and the
clouds are confined to a shallow layer.

2 Notice how the cloud layers lie at certain temperatures within
the atmosphere where ammonia (NH_3), ammonium hydrosulfide (NH_4SH), and water (H_2O) can condense.

3 Also notice how the belt–zone circulation is related to the
high- and low-pressure areas you find on Earth.

4 Finally, the major spots on Jupiter, although they are only
circulating storms, can remain stable for decades or even
centuries.

The circulation in Jupiter's atmosphere is not totally understood. Observations made by the Cassini spacecraft as it raced past
Jupiter on its way to Saturn revealed that the dark belts contain
small rising storm systems too small to see in previous images.
Evidently, the general circulation usually attributed to the belts
and zones is much more complex when it is observed in more
detail. Further understanding of the small-scale motions in Jupiter's atmosphere may have to await future planetary probes to
Jupiter.

The highly complex spacecraft that have visited Jupiter have
produced a new understanding of the giant planet, but don't confuse the technology of spacecraft with the science of comparative
planetology (**Window on Science 23-1**). Science is about
understanding nature, and Jupiter is an entirely new kind of planet
in your study. In fact, Jupiter has a feature that you did not find
among the terrestrial planets. Jupiter has a ring.

Ace **Astronomy**™   Log into AceAstronomy and select this chapter to see Astronomy Exercise "Convection and Turbulence." Study
the influence that rotation and temperature have on this hypothetical
planet.

Jupiter's Ring

Astronomers have known for centuries that Saturn has rings, but
it was not until 1979, when the Voyager 1 spacecraft sent back
photos, that Jupiter's ring was discovered. Less than 100 times as
bright as Saturn's rings, the ghostly ring around Jupiter is a puzzle.

Jupiter's Atmosphere

1 You probably won't ever visit Jupiter's atmosphere. Its cloud layers are deathly cold, and the deeper layers that are warmer have a crushingly high pressure. There is no free oxygen to breathe; the gases are roughly three-quarters hydrogen and a quarter helium, plus traces of water vapor, methane, ammonia, and similar molecules. Traces of sulfur and molecules containing sulfur probably make it smell bad. Of course, Jupiter has no surface, so there isn't even a place to stand. Jupiter is a nice planet to look at, but it's not a place to visit.

Belts are dark bands of clouds.

Zones are bright bands of clouds.

Shadow of Europa

Jupiter's moon Europa

NASA/JPL/Univ. of AZ

NASA

1a The only spacecraft to enter Jupiter's atmosphere was the Galileo probe. Released from the Galileo spacecraft, the probe entered Jupiter's atmosphere in December 1995. It parachuted through the upper atmosphere of clear hydrogen, released its heat shield, and then fell through Jupiter's stormy atmosphere until it was crushed by the increasing pressure.

Jupiter's atmosphere is a very thin layer of turbulent gas above the liquid interior. It makes up only about 1 percent of the radius of the planet.

Lightning bolts are common in Jupiter's turbulent clouds.

Hughes Aircraft Co

The Great Red Spot at right is a giant circulating storm in one of the southern zones. It has lasted at least 300 years since astronomers first noticed it after the invention of the telescope. Smaller spots are also circulating storms.

NASA/JPL

2 The visible clouds on Jupiter are composed of ammonia crystals, but models predict that deeper layers of clouds contain ammonia hydrosulfide crystals, and deeper still lies a cloud layer of water droplets. These compounds are normally white, so planetary scientists think the colors arise from small amounts of other molecules formed by lightning or by sunlight.

If you could put thermometers in Jupiter's atmosphere at different levels, you would discover that the temperature rises below the uppermost clouds.

Far below the clouds, the temperature and pressure climb so high the gaseous atmosphere merges gradually with the liquid hydrogen interior and there is no surface.

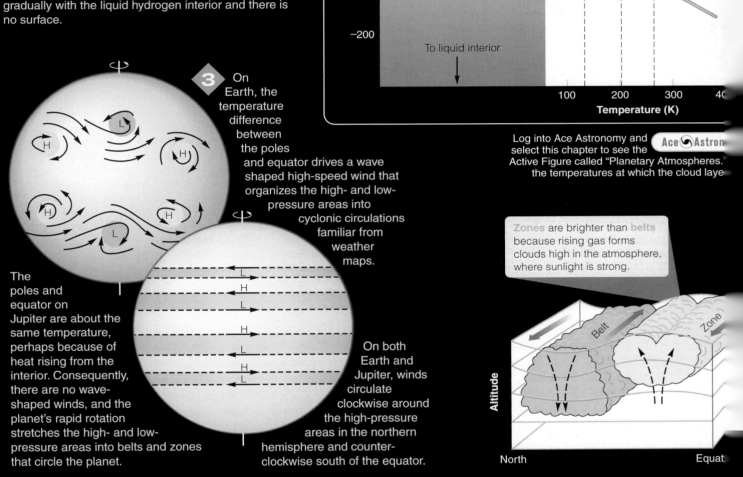

Temperature (°F)

Clear hydrogen atmosphere

Ammonia
Ammonia hydrosulfide
Water

Altitude (km)

To liquid interior

Temperature (K)

Log into Ace Astronomy and select this chapter to see the Active Figure called "Planetary Atmospheres." the temperatures at which the cloud laye

3 On Earth, the temperature difference between the poles and equator drives a wave shaped high-speed wind that organizes the high- and low-pressure areas into cyclonic circulations familiar from weather maps.

The poles and equator on Jupiter are about the same temperature, perhaps because of heat rising from the interior. Consequently, there are no wave-shaped winds, and the planet's rapid rotation stretches the high- and low-pressure areas into belts and zones that circle the planet.

On both Earth and Jupiter, winds circulate clockwise around the high-pressure areas in the northern hemisphere and counter-clockwise south of the equator.

Zones are brighter than belts because rising gas forms clouds high in the atmosphere, where sunlight is strong.

Altitude

Belt

Zone

North

Equat

4 The two white ovals here are counterclockwise, high-pressure weather systems that have been visible in Jupiter's southern hemisphere since they formed in the 1930s. These two huge storms merged into one spot in February 1998. The pear-shaped circulation visible between the two storms vanished during the merger.

Features in Jupiter's atmosphere may be stable for decades or centuries, but even the Giant Red Spot may someday vanish.

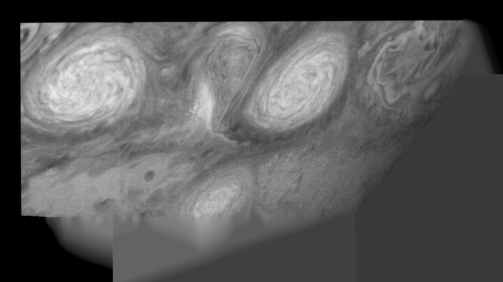

Science, Technology, and Engineering

Sending spacecraft to another world is expensive, and it may seem pointless when that world seems totally hostile to human life. What practical value is there in sending a space probe to Jupiter? To resolve that question, you need to consider the distinction between science, technology, and engineering.

Science is nothing more than the logical study of nature, and the goal of science is understanding. Although much scientific knowledge proves to have tremendous practical value, the only goal of science is a better understanding of how nature works. Technology, in contrast, is the practical application of scientific knowledge to solve a specific problem. People trying to find a faster way to paint automobiles might use all the tools and techniques of science, but if their goal has some practical outcome, you should more properly call it technology rather than science. Engineering is the most practical form of technology. An engineer is likely to use well-understood technology to find a practical solution to a problem.

Of course, there are situations in which science and technology blur together. For example, humanity has a practical and tragic need to solve the AIDS problem; we need a cure and

a preventative. Unfortunately, scientists don't understand the AIDS virus itself or viruses in general well enough to design a simple solution, and consequently much of the work involves going back to basic science and trying to better understand how viruses interact with the human body. Is this technology or science? It is hard to decide.

People often describe science that has no known practical value as basic science or basic research. The exploration of worlds such as Jupiter would be called basic science, and it is easy to argue that basic science is not worth the effort and expense because it has no known practical use. Of course, the problem is that no one has any way of knowing what knowledge will be of use until that knowledge is acquired. In the middle of the 19th century, Queen Victoria is supposed to have asked physicist Michael Faraday what good his experiments with electricity and magnetism were. He answered, "Madam, what good is a baby?" Of course, Faraday's experiments were the beginning of the electronic age. Many of the practical uses of scientific knowledge that fill your world—digital electronics, vaccines, plastics—began as basic research. Basic scientific research provides the raw materials that technology and engineering use to solve problems.

Exploring other worlds is of cultural value. It helps us as humans understand ourselves. (NASA/JPL/Space Science Institute)

Basic scientific research has yet one more important use that is so valuable it seems an insult to refer to it as merely practical. Science is the study of nature, and as you learn more about how nature works, you learn more about what your existence in this universe means. The seemingly impractical knowledge gained from space probes visiting other worlds tells you about your own planet and your own role in the scheme of nature. Science tells us where we are and what we are, and that knowledge is beyond value.

What is it made of? Why is it there? A few simple observations will help you solve some of these puzzles.

Saturn's rings are made of bright ice chunks, but the particles in Jupiter's ring are very dark and reddish. You can conclude that the ring is rocky rather than icy.

You can also conclude that the ring particles are mostly microscopic. Photos of the ring show that it is very bright when illuminated from behind (■ Figure 23-5)—it is scattering light forward. Efficient **forward scattering** occurs when particles have diameters roughly the same as the wavelength of light, a few millionths of a meter. Large particles do not scatter light forward, so a ring filled with basketball-size particles would look dark when illuminated from behind. Forward scattering tells you that the ring is made of particles about the size of those in cigarette smoke.

The location of these particles near the planet is understandable. They orbit inside the **Roche limit,** the distance from a planet within which a moon cannot hold itself together by its own gravity. If a moon stays far from its planet, then the moon's gravity will be much greater than the tidal forces caused by the planet, and the moon will be able to hold itself together. If, however, the planet's moon comes inside the Roche limit, the tidal

forces overcome its gravity and pull the moon apart. The International Space Station can orbit inside Earth's Roche limit because it is held together by bolts and welds, but a moon held together by its own gravity cannot survive inside a planet's Roche limit. If a planet and its moon have similar densities, the Roche limit is 2.44 planetary radii. Jupiter's main ring has an outer radius of 130,000 km (1.8 planetary radii) and lies inside the Roche limit, as do the rings of Saturn, Uranus, and Neptune.

Now you can understand the dust near Jupiter. If a dust speck gets knocked loose from a small moon inside the Roche limit, the moon's gravity cannot hold the dust speck. And the billions of dust specks in the ring can't pull themselves together to make a new moon because of tidal forces inside the Roche limit.

You can also be sure that the ring particles are not old. The pressure of sunlight and the powerful magnetic field alter the orbits of the particles, and they gradually spiral into the planet. Images show faint ring material extending down toward the cloud tops, and this is evidently dust specs spiraling into the planet. Dust is also lost from the ring as electromagnetic effects force it out of the plane of the ring to form a low-density halo above and below the ring (Figure 23-5b). Another reason the

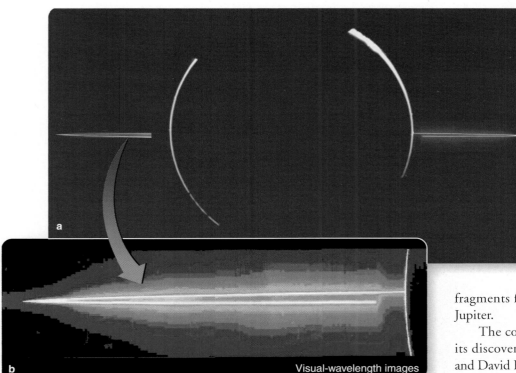

■ **Active Figure 23-5**

(a) The main ring of Jupiter, illuminated from behind, glows brightly in this visual-wavelength image made by the Galileo spacecraft from within Jupiter's shadow. (b) Digital enhancement and false color reveal the halo of ring particles that extends above and below the main ring. The halo is just visible in part a. (NASA)

Ace Astronomy™ Log into AceAstronomy and select this chapter to see the Active Figure "Roche Limit." Observe the powerful influence of tides on objects passing near planets.

Visual-wavelength images

ring particles can't be old is that the intense radiation around Jupiter will grind the dust specks down to nothing in a century or so. That means the rings seen today can't be material left over from the formation of Jupiter.

The rings of Jupiter must be continuously resupplied with new dust. Small moons that orbit near the outer edge of the rings may lose dust particles as they are hit by micrometeorite impacts. Observations made by the Galileo spacecraft show that the main ring is densest at its outer edge where the small moon Adrastea orbits and that another small moon, Metis, orbits inside the ring. Galileo images reveal much fainter rings, called the **gossamer rings,** extending twice as far from the planet as the main ring. These gossamer rings are most dense at the orbits of two small moons, Amalthea and Thebe, again providing evidence that the dust is being blasted into space by impacts on the moons.

Besides supplying the rings with particles, it is possible that the moons help confine the ring particles and keep them from spreading outward. You will find this an important process in planetary rings when you study the rings of Saturn later in this chapter.

Your exploration of Jupiter reveals that it is much more than just a big planet. It is the gravitational and magnetic center of an entire community of objects. Occasionally the community suffers an intruder.

Comet Impact on Jupiter

Comets are very common in the solar system, and Jupiter, because of its strong gravity, probably gets hit by comets more often than most planets. But no one had ever seen it happen until 1994, when fragments from a disrupted comet smashed into Jupiter.

The comet, named Shoemaker–Levy 9 after its discoverers, Eugene and Carolyn Shoemaker and David H. Levy, was captured into orbit around Jupiter sometime in the past. In 1992, it passed inside Jupiter's Roche limit and was pulled into at least 21 pieces that looped out away from Jupiter in long elliptical orbits. At this point, the objects were discovered on a photograph. Drawn out into a long strand of small comets, the pieces fell back and slammed into Jupiter over a period of six days in July 1994 (■ Figure 23-6).

The fragments were no bigger than a kilometer or so in diameter and contained a fluffy mixture of rock and ices. They hit Jupiter at speeds of about 60 km/s and released energy equivalent to a few million megatons of TNT. (At the height of the Cold War, all the nuclear weapons on Earth totaled about 80,000 megatons. The Hiroshima bomb was only 0.15 megaton.)

The impacts occurred just over Jupiter's horizon as seen from Earth, but they produced fireballs almost 3000 km high, and the rapid rotation of Jupiter brought the impact points within sight of Earth only 15 minutes later. Infrared telescopes easily spotted the glowing scars where the comets hit; and, as they cooled, the impact sites became dusty, dark spots at visible wavelengths as shown in Figure 23-6. Visible through even small telescopes, some of these dark smudges were larger than Earth.

The comet impact was an astonishing spectacle, but what does it reveal about Jupiter? In fact, the impact was revealing in two ways. First, astronomers used the impacts as probes of Jupiter's atmosphere. By making assumptions about the nature of Jupiter's atmosphere and by using the most powerful computers, astronomers created models of a high-velocity projectile penetrating into Jupiter's upper atmosphere. By comparing the observed impacts with the models, astronomers were able to fine-tune the models to better represent Jupiter's atmosphere. This method of comparing

■ Figure 23-6

In 1994, fragments of Comet Shoemaker–Levy 9 struck Jupiter. Although these were the first comet impacts ever seen by Earth's inhabitants, such events are probably common occurrences in planetary systems. (Composite and visual images: NASA; IR sequence: Mike Skrutskie; IR image: University of Hawaii)

models with reality is a critical part of science; therefore, the impacts on Jupiter resulted in a better understanding of the nature of Jupiter's atmosphere.

Second, the spectacle is revealing in a more general way. It reminds you that planets are sometimes hit by large objects such as asteroids and the heads of comets. Jupiter probably is hit every century or so. In 1690, the Italian astronomer Cassini observed a dark spot that appeared on Jupiter and changed in a pattern over a period of days. Astronomers now recognize that pattern as the scar of a comet impact. You have seen evidence for large impacts on Earth, the moon, Venus, Mercury, and Mars, and you will see further evidence as you continue your exploration of the solar system.

Comets hit planets all the time. The impact in 1994 was just the first such event to occur since the rise of modern astronomy. You will see in Chapter 25 that a comet or asteroid impact on Earth may have changed the climate and killed the dinosaurs. A solar system is a dangerous place to put an inhabited planet.

The History of Jupiter

Your goal in studying any planet is to be able to tell its story—to describe how it got to be the way it is. While you can understand part of the story of Jupiter, there is still much to learn.

If the solar nebula theory for the origin of the solar system is correct, then Jupiter formed from the colder gases of the outer solar nebula, where ices were able to condense. Thus, Jupiter grew rapidly and eventually became massive enough to trap hydrogen and helium gas directly from the solar nebula. The denser material, such as rock and metal, sank to the center to form a core, and Jupiter became a low-density world of liquid hydrogen surrounding a hot core of heavier elements.

In the interior of Jupiter, the hydrogen takes the form of liquid metallic hydrogen, which is a very good electrical conductor. The planet's rapid rotation, coupled with the outward flow of heat from its hot interior, drives a dynamo effect that produces a powerful magnetic field. That vast magnetic field traps high-energy particles from the solar wind to form intense radiation belts and auroras.

The rapid rotation and large size of Jupiter cause the weather patterns to take the form of belt–zone circulation. Heat flowing upward from the interior causes rising currents in the bright zones, and cooler gas sinks in the dark belts. As on Earth, winds blow at the margins of these regions, and large spots appear to be cyclonic disturbances.

Although the age of planet building is long past, Jupiter continues to be hit by meteorites and occasional comets, as do all the planets. Any debris left over from the formation of Jupiter would have been blown away long ago by the solar wind, so the dust trapped in Jupiter's thin ring must be young. It probably comes from meteorites hitting the innermost moons.

Your study of Jupiter has been challenging because Jupiter lacks any surface—it is difficult to imagine being there. Most of the surface features and processes you found on the terrestrial planets are missing on Jupiter, but, as the prototype of the Jovian worlds, it earns its place as the ruler of the solar system.

Building Scientific Arguments

How do astronomers know Jupiter is hot inside?
A scientific argument is a way to test ideas, and sometimes it is helpful to test even the most basic ideas. You know that something is hot if you touch it and it burns your fingers, but you can't touch Jupiter. You also know something is hot if it is glowing bright red—it is red hot. But Jupiter is not glowing red hot. You can tell that something is hot if you can feel heat when you hold your hand near it. That is, you can detect infrared radiation with your skin. In the case of Jupiter, you would need greater sensitivity than the back of your hand, but infrared telescopes reveal that Jupiter is a source of infrared radiation. It really is glowing in the infrared. Sunlight would warm it a little bit, but it is emitting 70 percent more infrared than it should. That means it must be hot inside. From models of the interior, astronomers conclude that the center must be about 30,000 K in order to heat the surface of the planet and make it glow in the infrared.

Astronomical understanding is usually based on simple observations, so build an argument to answer the following simple question. **How do astronomers know that Jupiter has a low density?**

■   ■   ■

Connections: The study of Jupiter is more meteorology than geology, but when you examine its moons, you find all the excitement of impact cratering and active volcanism. The moons have surfaces and are places you could go. Bring your spacesuit.

(23-3) Jupiter's Family of Moons

HOW MANY MOONS does Jupiter have? You will have to check the World Wide Web to get the latest figure. Astronomers are finding more and more small moons, and the count is now well over two dozen. Most of these moons are small and rocky, and many are probably captured asteroids. Four of the moons, those discovered by Galileo and now called the Galilean moons, are large and have interesting geologies (■ Figure 23-7).

■ **Figure 23-7**

The Gallilean moons of Jupiter from left to right are Io, Europa, Ganymede, and Callisto. The circle shows the size of Earth's moon. (NASA)

Size of Earth's moon

Visual-wavelength images

Your study of the moons of Jupiter will illustrate three important principles in comparative planetology. First, a body's composition depends on the temperature of the material from which it formed. This is illustrated by the prevalence of ice as a building material in the outer solar system, where sunlight is weak. The second principle is that cratering can reveal the age of a surface. Finally, you will see that internal heat has a powerful influence over the geology of these larger moons.

Callisto: The Ancient Face

The outermost of Jupiter's four large moons, Callisto is half again as large in diameter as Earth's moon. Like all of Jupiter's larger satellites, Callisto is tidally locked to its planet, keeping the same side forever facing Jupiter. From its gravitational influence on passing spacecraft, astronomers can calculate Callisto's mass, and dividing by its volume shows that its density is 1.79 g/cm³. Ice has a density of about 1 and rock 2.5 to 4 g/cm³, so Callisto must be a mixture of rock and ice.

Images from the Voyager and Galileo spacecraft show that the surface of Callisto is dark, dirty ice heavily pocked with craters (■ Figure 23-8). Old, icy surfaces in the solar system become dark because of dust added by meteorites and because meteorite impacts vaporize water, leaving the rock in the ice behind to form a dirty crust. You may have seen the same thing happen to a city snowbank. As the snow evaporates over a few days, the crud in the snow is left behind to form a dirty rind. Break through that dirty surface, and the snow is much cleaner underneath.

Spectra of the surface show that it is mostly a 50/50 mix of ice and rock, but some areas are ice free. Nevertheless, the slumped

■ **Figure 23-8**

The dark surface of Callisto is dirty ice marked by craters in these visual-wavelength images. The youngest craters look bright, because they have dug down to cleaner ice. Valhalla is the 4000-km-diameter scar of a giant impact feature, one of the largest in the solar system. Valhalla is so large and old that the icy crust has flowed back to partially heal itself, and the outer rings of Valhalla are shallow troughs marking fractures in the crust. (NASA)

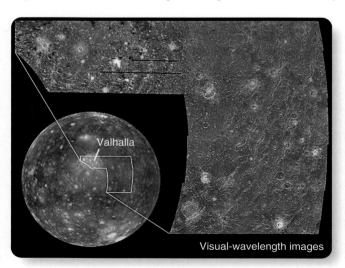

Valhalla

Visual-wavelength images

shape of craters suggests that the outer 10 km is mostly frozen water; ice isn't very strong, so big piles of it tend to slump under their own weight. The disagreement between the spectra and the shapes of craters can be understood when you recall that the spectra contain information about the outer 1 mm of the surface, which can be quite dirty, while the shapes of craters tell you about the outermost 10 km, which could be rich in ice.

Delicate measurements of the shape of Callisto's gravitational field were made by the Galileo spacecraft, and they show that Callisto has never fully differentiated to form a dense core and a lower-density mantle. Its interior is a mixture of rock and ice. This is consistent with the observation that it has only a weak magnetic field of its own. A strong magnetic field could be generated by the dynamo effect in a molten convecting core, and Callisto has no core. It does, however, interact with Jupiter's magnetic field in a way that suggests it has a layer of liquid water roughly 10 km thick about 100 km below its icy surface. Slow radioactive decay in its interior may provide enough heat to keep this layer of water from freezing.

Ganymede: A Hidden Past

The next Galilean moon inward is Ganymede, larger than Earth's moon, larger than Mercury, and over three-quarters the diameter of Mars. Its density is 1.9 g/cm³, and its influence on the Galileo spacecraft reveals that it has a rocky core, an ice-rich mantle, and a crust of ice 500 km thick. It may even have a small iron core at its center. It is large enough for radioactive decay to have melted its interior when it formed, allowing iron to drain to its center.

The surface hints at an active past. Although a third of the surface is old, dark, and cratered, the rest is marked by bright parallel grooves. Because this bright **grooved terrain** (■ Figure 23-9a) contains fewer craters, it must be younger.

Observations show that the bright terrain was produced when the icy crust broke and water flooded up from below. Some low-lying regions are smooth and appear to have been flooded by water. Spectra reveal concentrations of salts such as those that would be left behind by the evaporation of mineral-rich water. Also, some features in or near the bright terrain appear to be calderas formed when subsurface water drained away and the surface collapsed (Figure 23-9b).

The Galileo spacecraft found that Ganymede has a magnetic field about 10 percent as strong as Earth's. It even has its own magnetosphere inside the larger magnetosphere of Jupiter. Theoretical calculations find it difficult to produce a magnetic field in a water-rich mantle, and there does not appear to be enough heat in Ganymede for it to have a molten metallic core. Astronomers wonder if its magnetic field is left over from a time when it was hotter and more active.

Ganymede's magnetic field fluctuates with the 10-hour period of Jupiter's rotation. The rotation of the planet sweeps its tilted magnetic field past the moon, and the two fields interact. That interaction reveals that the moon has a layer of liquid water

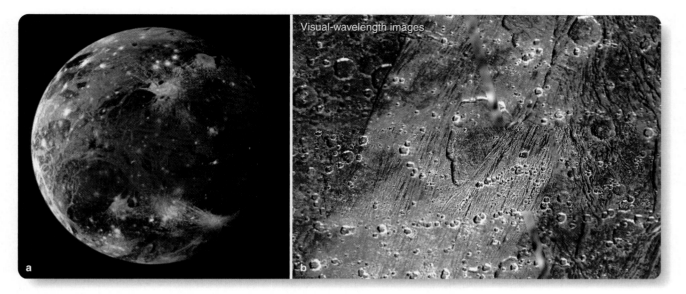

Visual-wavelength images

■ Figure 23-9

(a) This color-enhanced image of Ganymede shows the frosty poles at top and bottom, the old dark terrain, and the brighter grooved terrain. (b) A band of bright terrain runs from lower left to upper right, and a collapsed area, a possible caldera, lies at the center in this visual-wavelength image. Calderas form where subsurface liquid has drained away, and the bright areas do contain other features associated with flooding by water. (NASA)

about 170 km (100 mi) below its surface. The water layer may be about 5 km (3 mi) thick. It is possible that the water layer was thicker and closer to the surface long ago when the interior of the moon was warmer. That might explain the flooding that appears to have helped form the bright grooved terrain.

■ Figure 23-10

Two effects on planetary satellites. (a) Tidal heating occurs when changing tides cause friction within a moon. (b) The focusing of meteoroids exposes satellites in small orbits to more impacts than satellites in larger orbits receive.

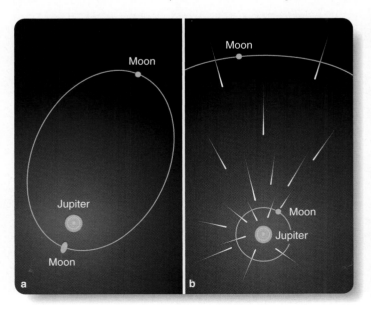

Ganymede orbits rather close to a massive planet, and that exposes it to two processes that many worlds never experience. **Tidal heating,** the frictional heating of a body by changing tides (■ Figure 23-10a) could have heated Ganymede's interior and added to the heat generated by radioactive decay. In a circular orbit, a moon experiences no tidal heating; but, at some point in the past, Ganymede could have interacted with the other moons and entered a slightly elliptical orbit. Jupiter's gravity would have deformed the moon. As Ganymede followed its elliptical orbit, tides would have flexed it, and friction would have heated it. Such an episode of tidal heating might have been enough to drive a dynamo to produce a magnetic field and break the crust to make the grooved terrain.

The second process that affects Ganymede exposes it to more than its share of meteorite impacts. Because massive planets like Jupiter draw debris inward, a moon orbiting near a massive planet should be struck by many meteorites (Figure 23-10b). Nevertheless, the bright terrain on Ganymede has few craters and appears to be only about 1 billion years old.

Europa: A Hidden Ocean

The next Galilean moon inward is Europa, a bit smaller than Earth's moon. Europa has a density of about 3 g/cm^3, so it must be mostly rock and metal. Yet its surface is ice.

Europa lies closer to Jupiter than Ganymede, so it should be exposed to more meteorite impacts than Callisto or Ganymede, but the icy crust of Europa is almost free of craters. A recent crater such as Pwyll is bright, but most are hardly more than blemishes

■ **Figure 23-11**

(a) The icy surface of Europa is shown here in natural color. Many faults are visible on its surface, but very few craters. The bright crater is Pwyll, a young impact feature. (b) This circular bull's-eye is 140 km in diameter. It is the remains of an impact by an object about the size of a mountain. Notice the younger cracks and faults that cross the older impact feature. (c) Like icebergs on the Arctic Ocean, blocks of crust on Europa appear to have floated apart. Spectra show that the blue ice is stained by salts such as those that would be left behind by mineral-rich water welling up from below and evaporating. White areas are ejecta from the impact that formed Pwyll. (NASA)

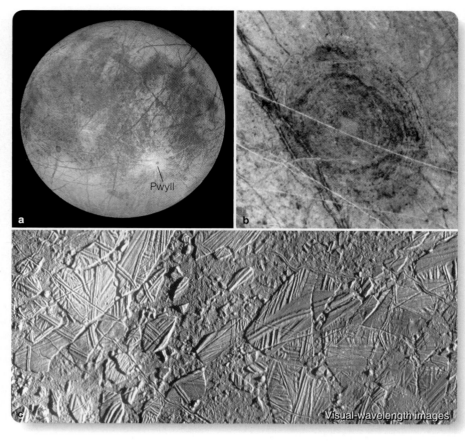

in the ice (■ Figures 23-11a and b). Evidently the surface of Europa is active and erases craters almost as fast as they form. The number of impact scars on Europa suggests that the average age of its surface is only 10 million years, and that isn't very old at all. Other signs of activity include long cracks in the icy crust and regions where the crust has broken into sections that have moved apart as if they were icebergs floating on water (Figure 23-11c).

Even Europa's clean face tells you its surface is young. The albedo of the surface is 0.69, meaning that it reflects 69 percent of the light that hits it. This high albedo is produced by clean ice. You have discovered that old, icy surfaces tend to be very dark. Europa's high albedo must mean the surface is active with fresh ice covering older surfaces.

Europa is too small to have retained much heat from its formation or from radioactive decay, and the Galileo spacecraft found

that Europa has no magnetic field of its own. It cannot have a molten conducting core. Tidal heating, however, is important for Europa and apparently provides enough heat to keep the little moon active. In fact, the curving cracks in its crust reveal the shape of the tidal forces that flex it as it orbits Jupiter.

■ **Figure 23-12**

The gravitational influence of Europa on the passing Galileo spacecraft shows that the little moon has differentiated into a dense core and rocky mantle. Magnetic interactions with Jupiter show that it has a liquid-water ocean below its icy crust. Heat produced by tidal heating could flow outward as convection in such an ocean and drive geological activity in the icy crust. (NASA)

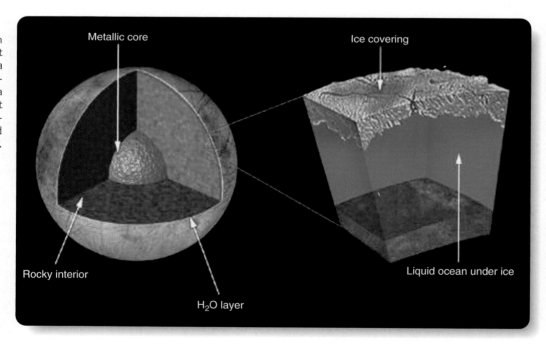

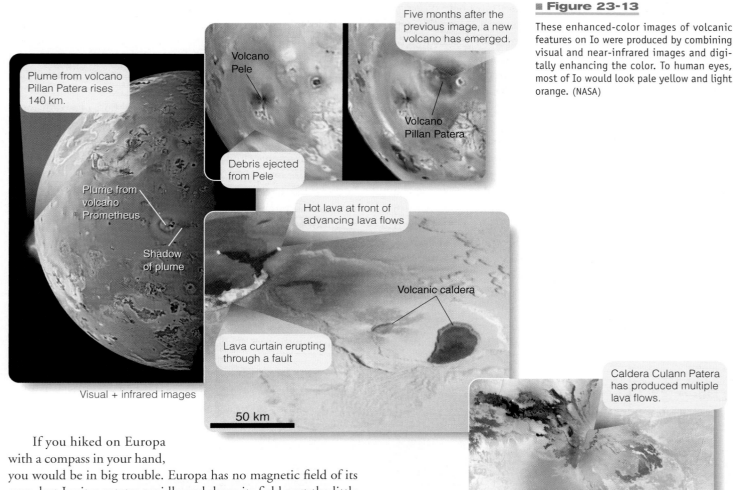

Plume from volcano Pillan Patera rises 140 km.

Plume from volcano Prometheus

Shadow of plume

Visual + infrared images

Volcano Pele

Debris ejected from Pele

Five months after the previous image, a new volcano has emerged.

Volcano Pillan Patera

Hot lava at front of advancing lava flows

Volcanic caldera

Lava curtain erupting through a fault

50 km

Caldera Culann Patera has produced multiple lava flows.

■ **Figure 23-13**

These enhanced-color images of volcanic features on Io were produced by combining visual and near-infrared images and digitally enhancing the color. To human eyes, most of Io would look pale yellow and light orange. (NASA)

If you hiked on Europa with a compass in your hand, you would be in big trouble. Europa has no magnetic field of its own, but Jupiter rotates rapidly and drags its field past the little moon. That induces a fluctuating magnetic field in Europa that would make your compass wander uselessly. Europa's interaction with Jupiter's magnetic field reveals the presence of a liquid-water ocean lying about 15 km (10 mi) below the icy surface. The ocean may be as deep as 100 miles (■ Figure 23-12) and could contain twice as much water as all the oceans on Earth. It would be rich in dissolved minerals, which would probably make it taste really bad. But those minerals would make the water a good electrical conductor and allow it to interact with Jupiter's magnetic field. No one knows what might be swimming through such an ocean, and many scientists hope for a future mission to Europa to drill through the ice crust and sample the ocean below for signs of life.

Tidal heating makes Europa geologically active. Apparently, rising currents of water can break through the icy crust or melt patches. Many of the cracks show evidence that they have spread apart and that fresh water has welled up and frozen between. In other regions, compression in the crust is revealed by networks of faults and low ridges. Compression on Earth pushes up mountain ranges, but no such ranges appear on Europa. The icy crust isn't strong enough to support ridges higher than a kilometer or so.

Orbiting deep inside Jupiter's radiation belts, Europa is bombarded by high-energy particles that damage the icy surface.

Water molecules are freed and broken up to be dispersed into a doughnut-shaped cloud spread round Jupiter and enclosing Europa's orbit. Flying past Jupiter in 2002, the Cassini spacecraft was able to image this cloud of excited gas. Europa's gas cloud serves as a warning that moons orbiting deep inside a massive planet's radiation belts are exposed to a form of erosion that is entirely lacking on Earth's moon.

Io: Bursting Energy

Geological activity is driven by heat flowing out of a planet's interior, and nothing could illustrate this principle better than Io, the innermost of Jupiter's Galilean moons. Photographs from the Voyager and Galileo spacecraft show no impact craters at all—surprising considering Jupiter's power to focus meteoroids inward (Figure 23-10b). No subtlety is needed to explain the missing craters. Over 150 active volcanoes are visible on Io's surface, blasting sulfurous gas and ash out over the surface to bury any newly formed craters (■ Figure 23-13). Unlike the dead or dormant outer Galilean moons, Io is bursting with energy.

Spectra reveal that Io has a tenuous atmosphere of gaseous sulfur and oxygen, but those gases can't be permanent. The erupting volcanoes pour out about one ton of gases per second; but, because of Io's low escape velocity, the gases leak into space easily. Also, any gas atoms that become ionized are swept away by Jupiter's rapidly rotating magnetic field. The ions produce a cloud of sulfur and sodium ions in a torus (a doughnut shape) enclosing Io's orbit (■ Figure 23-14).

You would need a very good spacesuit to visit the surface of Io. The temperature and atmospheric pressure would not be a big problem. Because of the continuous volcanism and the sulfurous gases, Io's thin atmosphere is dirty and probably smelly with sulfur. In fact, the reddish color of the small inner moon Amalthea may be caused by sulfur pollution escaping from Io. Your real problem walking across the surface of Io would be radiation. Io is deep inside Jupiter's magnetosphere and radiation belts; and, unless your spacesuit had astonishingly impressive shielding, the radiation would be lethal. Io, like Venus, may be a place that humans will never visit with any ease.

You can use basic observations to deduce the nature of Io's interior. From the density, 3.55 g/cm³, you can conclude that it is mostly rock. Spectra reveal no trace of water at all, so there is no ice on Io. It is the driest world in our solar system. The oblateness of Io caused by its rotation and by the slight distortion produced by Jupiter's gravity gives astronomers clues to its interior. Model calculations suggest it contains a modest core of metals and a deep rocky mantle.

The colors of Io have been compared to those of a badly made pizza. Certainly the reds, oranges, and browns of Io are caused by sulfur and sulfur compounds, but the crust is true rock. Infrared measurements show that volcanoes on Io erupt lava with a temperature over 1500°C (2700°F), about 300°C hotter than lavas on Earth. Sulfur on Io would boil at only 550°C. Also, a few isolated mountains exist that are as high as 18 km, twice the height of Mount Everest. Sulfur is not strong enough to support such high mountains. The lava flows and most of the crust must be silicate rock and not sulfur.

Volcanism is continuous on Io. Plumes come and go over periods of months, but some volcanic vents, such as Pele, have been active since the Voyager spacecraft first visited Io in 1979 (Figure 23-13). Earth's volcanoes eject lava and ash because of water dissolved in the lava. As rising lava reaches Earth's surface, the sudden decrease in pressure allows the water to come out of solution in the lava. It is like popping the cork on a bottle of champagne. The water flashes into vapor and blasts material out of the volcano, a process

■ **Figure 23-14**

Sulfur venting from Io is trapped in Jupiter's magnetic field to form a torus around the moon's orbit. Ions trapped in Jupiter's magnetosphere form radiation belts that enclose all four of the Galilean satellites. A fainter torus of water atoms is located around the orbit of Europa. (Sulfur cloud: NASA/JPL/Bruce A. Goldberg; Ion and neutral image: NASA/JPL/Johns Hopkins Univ. Applied Physics Lab; Radio: NASA/JPL)

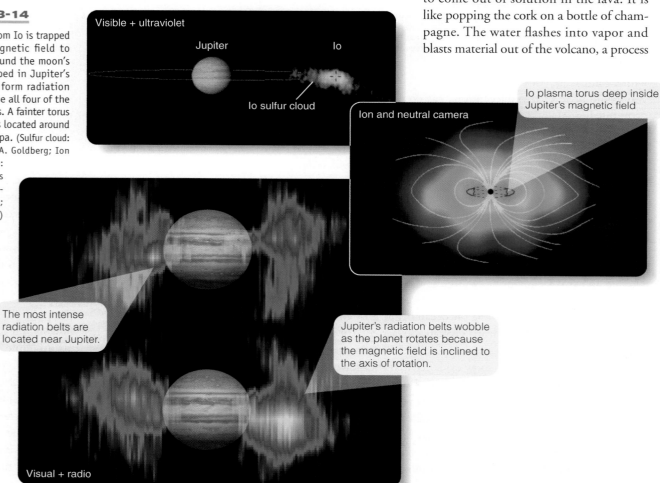

Visible + ultraviolet

Jupiter

Io

Io sulfur cloud

Ion and neutral camera

Io plasma torus deep inside Jupiter's magnetic field

The most intense radiation belts are located near Jupiter.

Jupiter's radiation belts wobble as the planet rotates because the magnetic field is inclined to the axis of rotation.

Visual + radio

that was responsible for the Mount St. Helens explosion in 1980. But Io is dry. Its volcanoes appear to be powered by sulfur dioxide dissolved in the magma. When the pressure is released, the sulfur dioxide boils out of solution and blasts gas and ash high above the surface in plumes up to 500 km high. Ash falling back to the surface produces debris layers around the volcanoes, such as that around Pele in Figure 23-13. Whitish areas on the surface are thought to be frosts of sulfur dioxide.

Great lava flows are visible carrying molten material downhill, where it buries the surface under layer after layer. Sometimes lava bursts upward through faults to form long lava curtains, a form of eruption seen in Hawaii. Both of these processes are shown in Figure 23-13.

What powers Io? It is bursting with energy, but it is only five percent bigger than Earth's moon, which is cold and dead. Io is too small to have retained heat from its formation or to remain hot from radioactive decay. In fact, the energy blasting out of its volcanoes is about three times more energy than it could make by radioactive decay in its interior.

Io is heated by the same kind of tidal heating that is suspected in Ganymede's past and in Europa. Because Io is so close to Jupiter, tides are powerful and should have forced Io's orbit to become circular long ago. Io, however, has fallen in with a bad crowd. Io, Europa, and Ganymede are locked in an orbital resonance; in the time it takes Ganymede to orbit once, Europa orbits twice and Io four times. This gravitational interaction keeps the orbits slightly elliptical; and Io, being closest to Jupiter, suffers dramatic tides, with its surface rising and falling by about 100 m. Tides on Earth move the solid ground by only a few centimeters. The resulting friction in Io is enough to melt the interior and drive volcanism. In fact, the energy flowing outward is continually recycling Io's crust. Deep layers melt, are spewed out through the volcanoes, fall back to cover the surface, and are later covered themselves until they are buried so deeply that they are again melted.

Io is the most active world in our solar system because it lives so close to Jupiter. The other Galilean moons are more distant and have had less dramatic histories of tidal heating and geological activity.

The History of the Galilean Moons

Each time you have finished studying a world, you have tried to summarize its history. Now you have a system of four small worlds. Can you tell their story? To do that you need to draw on what you have learned about the moons, but also on what you have learned about Jupiter and the origin of the solar system (Chapter 19).

The minor moons of Jupiter are probably captured asteroids, but the Galilean moons seem to be primordial. That is, they formed with Jupiter. Also, they seem to be a set in that they share common features and their densities are related to their distance from Jupiter (■ Table 23-2). This suggests that the four moons

■ **Table 23-2 | The Galilean Satellites***

Name	Radius (km)	Density (g/cm³)	Orbital Period (days)
Io	1816	3.55	1.769
Europa	1563	3.04	3.551
Ganymede	2638	1.93	7.155
Callisto	2410	1.81	16.689

*For comparison, the radius of Earth's moon is 1738 km, and its density is 3.36 g/cm³.

formed in a disk-shaped nebula around Jupiter—a mini-solar nebula—in much the same way the planets formed from the solar nebula around the sun.

As Jupiter grew massive, it should have formed a hot, dense disk of matter around its equator. The moons could have condensed inside that disk with the innermost moons, Io and Europa, forming from rocky material and the outer moons, Ganymede and Callisto, incorporating more ices. This theory follows the condensation sequence that led to rocky planets forming near the sun and ice-rich worlds forming farther away.

There are objections to this theory. The disk around Jupiter would have been dense and hot, and moons would have formed rapidly, perhaps in only 1000 years. The heat of formation released as material fell into the moons would have heated them so hot they would have lost their water. Ganymede and Callisto are rich in water, and Callisto has never been hot enough to differentiate and form a core. Also, mathematical models show that moons orbiting in the dense disk would have swept up debris and lost orbital momentum; they would have spiraled into Jupiter within a century.

A newer theory proposes that Jupiter's early disk was indeed dense and hot and may have created moons, but those moons spiraled into the planet and were lost. Only later, as the disk grew thinner and cooler, did the present Galilean moons begin to form. Additional material may have dribbled slowly into the disk, but the moons must have formed slowly enough to retain their water and avoid spiraling into Jupiter.

You can understand the interiors of the moons by thinking about tidal heating. The moons formed slowly, over perhaps 100,000 years, and were not heated severely by infalling material. The inner moons, however, were cooked by tidal heating—possibly enhanced when an orbital resonance developed between Ganymede, Europa, and Io. The innermost moon, Io, was heated so much it lost all of its water, and Europa retained only a small amount. Ganymede was heated enough to differentiate but retained much of its water. Callisto, orbiting far from Jupiter and avoiding orbital resonances, was never heated enough to differentiate. The Galilean moons as they appear today seem to be the result of slow formation and tidal heating.

Io is bursting with energy, and the Galilean satellite system is bursting with history. As you learn to understand that history, you will learn more about how worlds form and evolve.

Building Scientific Arguments

What produces Io's internal heat?

Scientific arguments commonly draw on basic principles that are well understood. In this case, you understand that small worlds lose their internal heat quickly and become geologically inactive. Io is only slightly larger than Earth's moon, which is cold and dead, but Io is bursting with energy flowing outward.

Clearly, Io must have a powerful source of heat inside, and that heat source is tides. Io's orbit is slightly elliptical, so it is sometimes closer to Jupiter and sometimes farther away. That means that Jupiter's powerful gravity sometimes squeezes Io more than at other times, and the flexing of the little moon's interior produces heat through friction. Such tides would rapidly force Io's orbit to become circular, and then tidal heating would end and the planet would become inactive—except that the gravitational tugs of the other moons keep Io's orbit elliptical. Thus, it is the influence of its companions that keeps Io in such an active state.

Io has almost no impact craters, but Callisto has many. Build a new scientific argument drawing on a different principle. **What does the distribution of craters on the Galilean satellites tell you about their history?**

∎   ∎   ∎

Connections: You have now applied the principles of comparative planetology not only to planets but also to moons. With the insight you have gained from Jupiter and its satellite system, you are ready to visit the most beautiful of the Jovian worlds.

(23-4) Saturn

SATURN HAS PLAYED SECOND FIDDLE to its own rings since Galileo first saw it in 1610. He didn't recognize the rings for what they are, but today astronomers know them as one of the great sights in the solar system. Nevertheless, Saturn itself, only slightly smaller than Jupiter (**Celestial Profile 8**), is a fascinating planet with a few mysteries of its own. Your exploration of Saturn can make use of the principles you have learned from Jupiter.

Planet Saturn

The basic characteristics of Saturn reveal its composition and interior. Only about a third of the mass of Jupiter and 16 percent smaller in radius, Saturn has an average density of 0.69 g/cm³. It

is less dense than water—it would float! Although Saturn rotates about as fast as Jupiter, it is almost twice as oblate, so you can conclude that it must not contain a large core of heavy elements. Spectra show that its atmosphere is rich in hydrogen and helium (see Table 23-1), and models predict that it is mostly liquid hydrogen.

Infrared observations show that Saturn is radiating 1.8 times as much energy as it receives from the sun, showing that heat is flowing out of its interior. It must be hot inside. In fact, it is too hot. It should have lost more heat since it formed. Astronomers suspect that helium in the liquid hydrogen interior is condensing into droplets and falling inward. The falling droplets, releasing energy as they pick up speed, heat the planet. This heating is similar to the heating produced when a star contracts.

You can learn about Saturn's interior from its magnetic field. The Voyager spacecraft found that Saturn's magnetic field is about 20 times weaker than Jupiter's. It also has correspondingly weaker radiation belts. Models predict that the pressure inside Saturn, being weaker than that inside Jupiter, produces a smaller mass of liquid metallic hydrogen. Heat flowing outward must cause convection in this conducting core, and the rapid rotation drives a dynamo effect to produce the magnetic field. Unlike most magnetic fields, Saturn's field is not inclined to its axis of rotation, something you can see in ultraviolet images that show rings of auroras around Saturn's poles (∎ Figure 23-15). This alignment

∎ Figure 23-15

Auroras on Saturn occur in rings around the planet's magnetic poles. Because the magnetic field is not inclined very much to the axis of rotation, the rings occur nearly at the planet's geometrical poles. (Compare with Figure 23-4.) (NASA, ESA, J. Clarke, Boston Univ., and X. Levay, STScI)

Visual + ultraviolet

Auroras fluctuate day to day because of changes in the solar wind.

between the magnetic axis and the axis of rotation is peculiar and isn't understood.

The Cassini spacecraft arrived at Saturn 24 years after the Voyager spacecraft had flown past, and Cassini found that the rotation period of Saturn's magnetic field had slowed by 6 minutes. Surely the planet did not slow down, so planetary scientists suspect that the circulation in the core of the planet may have changed in some way.

Saturn's atmosphere is rich in hydrogen and contains belt–zone circulation, which appears to arise in the same way as that on Jupiter. Zones are higher clouds formed by rising gas, and belts are lower clouds formed by sinking gas.

But the clouds are not very distinct on Saturn (■ Figure 23-16a). Measurements from the Voyager and Cassini spacecraft indicate that Saturn's atmosphere is much colder than Jupiter's—something you would expect because Saturn is twice as far from the sun and receives only one-fourth as much solar energy per square meter. The clouds on Saturn form at about the same temperature as the clouds on Jupiter, but those temperatures are deeper in Saturn's cold atmosphere. Compare the cloud layers in Figure 23-16b with those on page 561. The atmospheres of Jupiter and Saturn are very similar once you remember that Saturn is colder.

One dramatic difference between Jupiter and Saturn concerns the winds. On Jupiter, winds bound each of the belts and zones, but on Saturn the pattern is not the same. Saturn has fewer such winds, but they are much stronger. The eastward wind at the equator of Saturn, for example, blows at 500 m/s (1100 mph), roughly five times faster than the eastward wind at Jupiter's equator. The reason for this difference is not clear.

Saturn's Rings

Looking at the beauty and complexity of Saturn's rings, an astronomer once said, "The rings are made of beautiful physics." You could add that the physics is actually rather simple, but the result is one of the most astonishing sights in our solar system.

In 1609, Galileo was the first to see the rings of Saturn; but, perhaps because of the poor optics in his telescopes, he did not recognize the rings as a disk. He drew Saturn as three objects—a central body and two smaller ones on either side. In 1659, Christian Huygens realized that the rings were a disk surrounding but not touching the planet.

Saturn's rings are governed by simple physics, but understanding what the rings are has demanded over a century of human ingenuity. In 1859, James Clerk Maxwell (for whom the large mountain on Venus is named) proved mathematically that solid rings would be unstable. Saturn's rings, he concluded, had to be made of particles. In 1867, Daniel Kirkwood demonstrated that gaps in the rings were caused by resonances with some of Saturn's moons. Spectra of the rings eventually showed that the particles were mostly water ice.

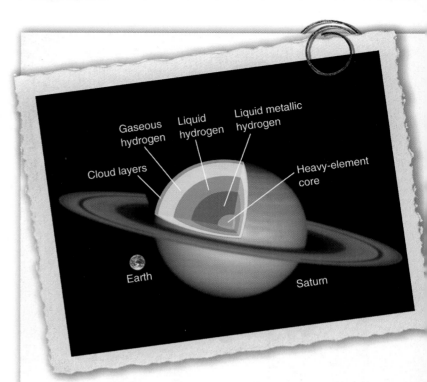

Density, oblateness, and gravity measurements made by planetary probes allow planetary astronomers to model Saturn's interior. (NASA/STScI)

Celestial Profile 8: Saturn

Motion:

Average distance from the sun	9.5388 AU (14.27 × 10⁸ km)
Eccentricity of orbit	0.0560
Maximum distance from the sun	10.07 AU (15.07 × 10⁸ km)
Minimum distance from the sun	9.005 AU (13.47 × 10⁸ km)
Inclination of orbit to ecliptic	2°29′17″
Average orbital velocity	9.64 km/s
Orbital period	29.461 y (10,760 days)
Period of rotation	$10^h39^m25^s$
Inclination of equator to orbit	26°24′

Characteristics:

Equatorial diameter	120,660 km (9.42 $D_\oplus$)
Mass	5.69 × 10²⁶ kg (95.147 $M_\oplus$)
Average density	0.69 g/cm³
Gravity at base of clouds	1.16 Earth gravities
Escape velocity	35.6 km/s (3.2 $V_\oplus$)
Temperature at cloud tops	−180°C (−292°F)
Albedo	0.61
Oblateness	0.102

Personality Point:

The Greek god Cronus was forced to flee when his son Zeus took power. Cronus fled to Italy, where the Romans called him Saturn, protector of the sowing of seed. He was celebrated in a weeklong wild party called the Saturnalia at the time of the winter solstice in late December. Early Christians took over the holiday to celebrate Christmas.

(a) Saturn's belt–zone circulation is not very distinct at visible wavelengths. These images were recorded when Saturn's southern hemisphere was tipped toward Earth. (NASA and E. Karkoschka) (b) Because Saturn is colder than Jupiter, the clouds form deeper in the hazy atmosphere. Notice that the three cloud layers on Saturn form at about the same temperature as do the three cloud layers on Jupiter.

Ace◗Astronomy™ Log into AceAstronomy and select this chapter to see the Active Figure "Planetary Atmospheres" and compare the atmospheres of the planets.

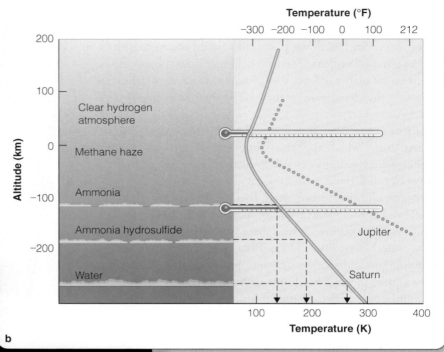

Study **The Ice Rings of Saturn** on pages 576–577 and notice three things:

1 The rings are made up of billions of ice particles, each in its own orbit around the planet. But the rings can't be as old as Saturn. The rings must be replenished now and then by impacts on Saturn's icy moons or by the disruption of a small moon that wanders too close to the planet.

2 The gravitational effects of small moons can confine some rings in narrow strands or keep the edges of rings sharp. Moons can also produce waves in the rings that are visible as tightly wound ringlets.

3 The ring particles are confined in a thin plane spread among small moons and confined by gravitational interactions with larger moons. The rings of Saturn, and the rings of the other Jovian worlds, are created by and controlled by the planet's moons. Without the moons, there would be no rings.

Modern astronomers find simple gravitational interactions producing even more complex processes in the rings. Where particles orbit in resonance with a moon, the moon's gravity triggers spiral density waves—the same bit of physics that generates spiral arms in galaxies. The spiral density waves spread outward through the rings. If the moon follows an orbit that is inclined to the ring plane, the moon's gravity causes spiral bending waves—ripples extending above and below the ring plane—which spread inward. Both of these kinds of rings are shown in the inset ring image on page 577.

Many other processes occur in the rings. Specks of dust become electrically charged by sunlight, and Saturn's magnetic field lifts them out of the ring plane. Small moonlets imbedded in the rings produce gaps, waves, and scallops in the rings.

Particle size ranges from snowlike powder to large particles and aggregates of particles. The subtle colors of the rings must arise from contamination in the ice, and some areas have unusual compositions. Cassini's Division, for instance, is made up of particles richer in rock than is most of the ring. No one knows how these differences in composition arise, but they must be related to the way the rings formed and are replenished.

Like a beautiful flower, the rings of Saturn are controlled by many different natural processes. Observations from spacecraft

Who Pays for Science?

Science is an expensive enterprise, and that raises the question of payment. Some science has direct technological applications, but some basic science is of no immediate practical value. Who pays the bill?

In Window on Science 23-1, you saw that technology is of immediate practical use; consequently, business funds much of this kind of scientific research. Auto manufacturers need inexpensive, durable, quick-drying paint for their cars, and they find it worth the cost to hire chemists to study the way paint dries. Many industries have large research budgets; and some industries, such as pharmaceutical manufacturers, depend exclusively on scientific research to discover and develop new products.

If a field of research has immediate potential to help society, it is likely that the government will supply funds. Much of the research on public health is funded by government institutions such as the National Institutes of Health and the National Science Foundation. The practical benefit of finding new ways to prevent disease, for example, is well worth the tax dollars.

Basic science, however, has no immediate practical use. That doesn't mean it is useless, but it does mean that the practical-minded stockholders of a company will not approve major investments in such research. Digging up dinosaur bones, for instance, is very poorly funded because no industry can make a profit from the discovery of a new dinosaur. Astronomy is another field of science that has few direct applications, and consequently very little astronomical research is funded by industry.

The value of basic research is twofold. Discoveries that have no known practical use today may be critically important years from now. For that reason, society needs to continue basic research to protect its own future. But basic research, such as studying Saturn's rings or digging up dinosaur bones, is also of cultural value because it tells us what we are. Each person benefits in intangible ways from such research, and that is why society needs to fund basic research for the same reason it funds art galleries and national parks—to make everyone's life richer and more fulfilling.

Because there is no immediate financial return from this kind of research, it falls to government institutions and private foundations to pay the bill. The Keck Foundation has built two giant telescopes with no expectation of financial return, and the National Science Foundation has funded thousands of astronomy research projects for the benefit of society. Debates rage as to how much money is

Sending the Cassini spacecraft to Saturn cost each American 56¢ per year over the life of the project. (NASA)

enough and how much is too much, but, ultimately, funding basic scientific research is a public responsibility that society must balance against other needs. There isn't anyone else to pick up the tab.

---

such as Voyager and Cassini will reveal even more about the rings (**Window on Science 23-2**).

The History of Saturn

The farther you journey from the sun, the more difficult it is to understand the history of the planets. Any fully successful history of Saturn should explain its low density, its peculiar magnetic field, and its beautiful rings. Planetary scientists can't tell a fully successful story yet, but you can understand a few of the principles that led to the formation of Saturn and its rings.

Saturn formed in the outer solar nebula, where ice particles were stable. It grew rapidly and became massive enough to capture hydrogen and helium gas directly from the nebula. The denser elements sank to the middle to form a denser core, and the hydrogen formed a liquid mantle containing liquid metallic hydrogen. The outward flow of heat from the core is believed to drive convection currents in this mantle that, coupled with the rapid rotation of the planet, produce its magnetic field. Because Saturn is smaller than Jupiter, it has less liquid metallic hydrogen, and its magnetic field is weaker.

You might be tempted at first to suppose that the rings of Saturn are primordial, made of material left over from the formation of the planet. That seems unlikely, however. Saturn, like Jupiter, should have been very hot when it formed, and that heat would have vaporized and driven off any leftover material. Also, such a hot Saturn would have had a very distended atmosphere, which would have slowed ring particles by friction and caused the particles to fall into the planet.

Planetary rings do not seem to be stable over 4.6 billion years, so the ring material must be more recent. Saturn's beautiful rings may have been produced within the last 100,000 years. One suggestion is that a small moon or an icy planetesimal came within Saturn's Roche limit, and tides pulled it apart. At least some of the debris would settle into the ring plane. You will see in the next two chapters that icy planetesimals are common in the Kuiper Belt in the outer solar system. Another possibility is that a comet struck one of Saturn's moons. Because both comets and moons in the outer solar system are icy, such a collision would produce icy debris. Bright planetary rings such as Saturn's may be temporary phenomena, produced when fresh ice

The Ice Rings of Saturn

1 The brilliant rings of Saturn are made up of billions of ice particles ranging from microscopic specks to chunks bigger than a house. Each particle orbits Saturn in its own circular orbit. Much of what astronomers know about the rings was learned when the Voyager 1 spacecraft flew past Saturn in 1980, followed by the Voyager 2 spacecraft in 1981. The Cassini Spacecraft reached orbit around Saturn in 2004. From Earth, astronomers see three rings labeled A, B, and C. Voyager and Cassini images reveal over a thousand ringlets within the rings.

Saturn's rings can't be leftover material from the formation of Saturn. The rings are made of ice particles, and the planet would have been so hot when it formed that it would have vaporized and driven away any icy material. Rather, the rings must be debris from collisions between passing comets and Saturn's icy moons. Such impacts should occur every 10 million years or so, and they would scatter ice throughout Saturn's system of moons. The ice would quickly settle into the equatorial plane, and some would become trapped in rings. Although the ice may waste away due to meteorite impacts and damage from radiation in Saturn's magnetosphere, new impacts could replenish the rings with fresh ice. The bright, beautiful rings you see today may be only a temporary enhancement caused by an impact that occurred since the extinction of the dinosaurs.

Encke's division

Cassini's division

A ring

B ring

C ring
The Crepe ring

As in the case of Jupiter's ring, Saturn's rings lie inside the planet's Roché limit where the ring particles cannot pull themselves together to form a moon.

Earth to scale

Visual-wavelength image

Because it is so dark, the C ring has been called the Crepe ring, referring to the black, semitransparent cloth associated with funerals.

1a An astronaut could swim through the rings. Although the particles orbit Saturn at high velocity, all particles at the same distance from the planet orbit at about the same speed, so they collide gently at low velocities. If you could visit the rings, you could push your way from one icy particle to the next. This artwork is based on a model of particle sizes in the A ring.

The C ring contains boulder-size chunks of ice, whereas most particles in the A and B rings are more like golf balls, down to dust-size ice crystals. Further, C ring particles are less than half as bright as particles in the A and B rings. Cassini observations show that the C ring particles contain less ice and more minerals.

NASA

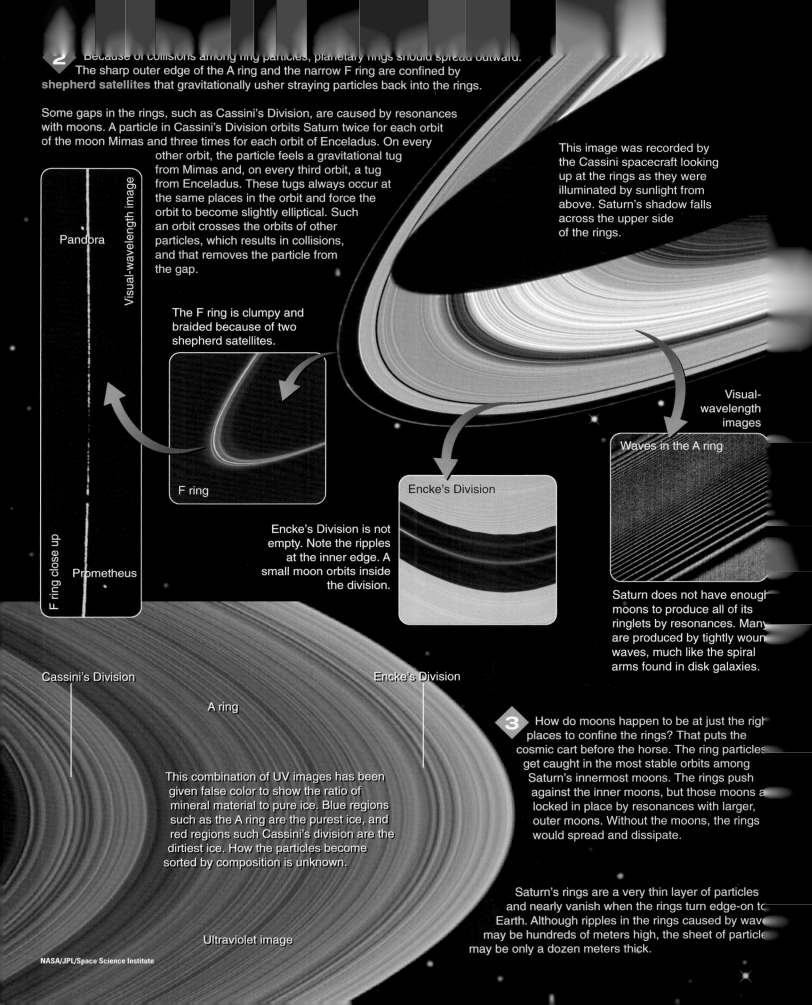

2 Because of collisions among ring particles, planetary rings should spread outward. The sharp outer edge of the A ring and the narrow F ring are confined by **shepherd satellites** that gravitationally usher straying particles back into the rings.

Some gaps in the rings, such as Cassini's Division, are caused by resonances with moons. A particle in Cassini's Division orbits Saturn twice for each orbit of the moon Mimas and three times for each orbit of Enceladus. On every other orbit, the particle feels a gravitational tug from Mimas and, on every third orbit, a tug from Enceladus. These tugs always occur at the same places in the orbit and force the orbit to become slightly elliptical. Such an orbit crosses the orbits of other particles, which results in collisions, and that removes the particle from the gap.

Pandora

Visual-wavelength image

F ring close up

Prometheus

The F ring is clumpy and braided because of two shepherd satellites.

F ring

This image was recorded by the Cassini spacecraft looking up at the rings as they were illuminated by sunlight from above. Saturn's shadow falls across the upper side of the rings.

Visual-wavelength images

Waves in the A ring

Encke's Division is not empty. Note the ripples at the inner edge. A small moon orbits inside the division.

Encke's Division

Saturn does not have enough moons to produce all of its ringlets by resonances. Many are produced by tightly wound waves, much like the spiral arms found in disk galaxies.

Cassini's Division

Encke's Division

A ring

This combination of UV images has been given false color to show the ratio of mineral material to pure ice. Blue regions such as the A ring are the purest ice, and red regions such Cassini's division are the dirtiest ice. How the particles become sorted by composition is unknown.

3 How do moons happen to be at just the right places to confine the rings? That puts the cosmic cart before the horse. The ring particles get caught in the most stable orbits among Saturn's innermost moons. The rings push against the inner moons, but those moons are locked in place by resonances with larger, outer moons. Without the moons, the rings would spread and dissipate.

Saturn's rings are a very thin layer of particles and nearly vanish when the rings turn edge-on to Earth. Although ripples in the rings caused by waves may be hundreds of meters high, the sheet of particles may be only a dozen meters thick.

Ultraviolet image

is scattered into the rings and then wasting away as the ice is gradually lost.

Building Scientific Arguments

Why do the belts and zones on Saturn look so bland?
One of the most powerful tools of critical thought is simple comparison and contrast. You can make that the theme of this scientific argument by comparing and contrasting Saturn with Jupiter. In the atmosphere of Jupiter, the dark belts form in regions where gas sinks, and zones form where gas rises. The rising gas cools and condenses to form icy crystals of ammonia, which are visible as bright clouds. Clouds of ammonia hydrosulfide and water form deeper, below the ammonia clouds, and are not as visible. Saturn is twice as far from the sun as Jupiter, so sunlight is four times dimmer. (Remember the inverse square law from Chapter 5.) The atmosphere is colder, and gas currents do not have to rise as far to reach cold levels and form clouds. That means the clouds are deeper in Saturn's atmosphere than in Jupiter's atmosphere. Because the clouds are deeper, they are not as brightly illuminated by sunlight and look dimmer. Also, a layer of methane-ice-crystal haze high above the ammonia clouds makes the clouds even less distinct.

Now build a new argument comparing the ring systems. **How is Saturn's ring system similar to and different from Jupiter's ring system?**

■   ■   ■

Connections: In any contest, Saturn should win the ribbon for most beautiful planet. Its subtly marked disk and its delicate rings are both massive and graceful. Its extensive system of moons only adds to the mystery of this beautiful and curious world.

(23-5) Saturn's Moons

SATURN HAS OVER TWO DOZEN known satellites—far too many to examine individually—but these moons share characteristics common to icy worlds. Most of them are small and dead, but one is big enough to have an atmosphere and perhaps even oceans or lakes.

Titan

Saturn's largest satellite is a giant ice moon with a thick atmosphere and a mysterious surface. From Earth it is only a dot of light, and no detail is visible. Nevertheless, a few basic observations made from Earth can tell you a great deal about this strange world.

Titan's mass could be estimated from its influence on other moons, and its mass divided by its volume revealed that its density is 1.9 g/cm^3. Its uncompressed density (Chapter 19) is only

■ **Table 23-3 | Some Organic Compounds Detected on Titan**

C_2H_6	Ethane
C_2H_2	Acetylene
C_2H_4	Ethylene
C_3H_4	Methylacetylene
C_3H_8	Propane
C_4H_2	Diacetylene
HCN	Hydrogen cyanide
HC_3N	Cyanocetylene
C_2N_2	Cyanogen

1.2 g/cm^3. Although it must have a rocky core, it must also contain a large amount of ices. Titan is a bit larger than the planet Mercury and almost as large as Jupiter's moon Ganymede. Unlike these worlds, Titan has a thick atmosphere. Its escape velocity is low, but it is so far from the sun that it is very cold, and most gas atoms don't move fast enough to escape. (See Figure 22-13.) Methane was detected spectroscopically in 1944, as were various hydrocarbons beginning in about 1970 (■ Table 23-3), but most of Titan's air is nitrogen with only 2 to 3 percent methane.

When the Voyager 1 and Voyager 2 spacecraft flew past Saturn in the early 1980s, their cameras could not penetrate Titan's hazy atmosphere (■ Figure 23-17). Measurements showed that the surface temperature was about 94 K ($-290°F$) and the surface atmospheric pressure was 50 percent greater than on Earth. Model calculations show that methane should condense from the atmosphere and fall as rain. Furthermore, sunlight converts methane (CH_4) into the gas ethane (C_2H_6) plus a collection of other organic molecules.* Some of these molecules produce the smoglike haze; and, as the smog particles gradually settle, they could deposit a layer of smelly, organic goo over the icy surface. This goo is exciting because such organic molecules may have been the precursors of life on Earth. You will examine this further in Chapter 26.

Before you imagine floundering through this gooey landscape, you should review the results of the Cassini spacecraft, which began exploring Saturn and its moons in 2005.

Infrared cameras and radar instruments on Cassini have been able to see through the hazy atmosphere. The surface is not a featureless ocean of methane, as some models had predicted. Rather, the surface consists of icy, irregular highlands and smoother dark areas (Figure 23-17). There are only a few craters, which suggests geological activity is erasing craters almost as fast as they are formed.

---

*Organic molecules are common in living things on Earth but do not have to be derived from living things. One chemist defined an organic molecule as "any molecule with a carbon backbone."

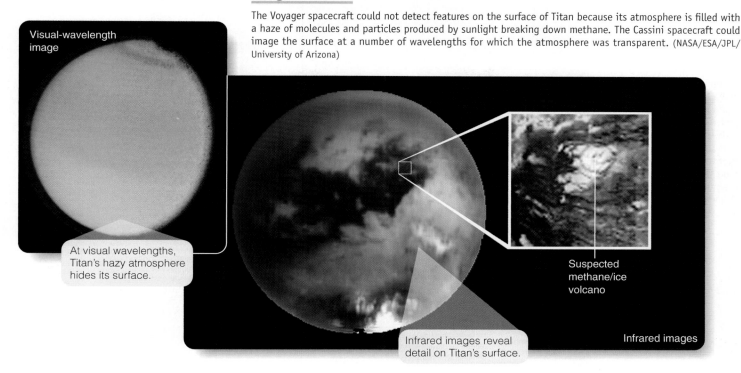

■ Figure 23-17

The Voyager spacecraft could not detect features on the surface of Titan because its atmosphere is filled with a haze of molecules and particles produced by sunlight breaking down methane. The Cassini spacecraft could image the surface at a number of wavelengths for which the atmosphere was transparent. (NASA/ESA/JPL/University of Arizona)

The Cassini spacecraft released a probe named Huygens, which parachuted down through the atmosphere of Titan and eventually landed on the surface. Huygens radioed back images of the surface as it descended under its parachute, and those images show dark drainage networks that lead into dark smooth areas (■ Figure 23-18). One astronomer said it looks like surf on a dark ocean rolling onto an icy beach, but there is no clear evidence of methane rain or liquid methane on the surface. Precipitation could wash the black goo off the highlands into the stream channels and lowlands, so that they would look smooth and dark even if all of the liquid had temporarily evaporated. If you visit Titan, you may get caught in a shower of methane rain, but it may not rain at your landing site for decades or even a century.

By the way, before you go to Titan, check your spacesuit for leaks. Nitrogen is not a very reactive gas, but methane is earthly cooking gas and is highly flammable. Of course, there is no free oxygen on Titan, so you are safe so long as your spacesuit does not leak oxygen. That might be exciting.

When the Huygens probe landed on the frigid surface, it radioed back measurements and images. The surface is mostly frozen water ice with some methane mixed in. The sunlight is orange because it has filtered down through the orange haze. Rocks littering the ground are actually steel-hard chunks of supercold ice rounded by erosion. Some rest in small depressions, suggesting that a liquid has flowed around them. You can see these depressions around the rocks in Figure 23-18.

The methane in the atmosphere is a puzzle. Sunlight can destroy methane by breaking the molecule up. The hydrogen escapes easily into space. In addition, sunlight can trigger chemical reactions that convert methane into organic compounds that settle out of the atmosphere. The 2 to 3 percent methane in Titan's atmosphere should be gone in only a million years. Obviously there is some source of fresh methane. There is no global methane ocean, so the methane must be produced by outgassing from Titan's icy crust. Detailed studies have identified at least one area where water ice and methane may be rising to the crust and forming a methane volcano (Figure 23-17). Such geological activity requires an internal heat source, and astronomers suspect that tidal heating may be helping radioactive decay keep the interior of Titan warm.

Titan is an exciting world to think about because it might harbor life in its organic deposits, but it is also exciting because it is so totally unearthly.

The Smaller Moons

In addition to Titan, Saturn has a large family of smaller moons. They are mixtures of rock and ice and are heavily cratered. Some of the smallest are probably captured objects. Some of the larger moons show traces of geological activity. You can compare the sizes of a few of these in ■ Figure 23-19.

The moon Phoebe orbits on the outer fringes of Saturn's satellite family, and it follows a retrograde orbit. That is, it orbits backward. It is quite small, only 214 km (133 mi) in diameter, and its surface is dark and cratered (■ Figure 23-20). Traces of ice are detected where impacts have excavated deeper layers or where landslides have exposed fresh ice. Nevertheless, the density

The Huygens probe descended by parachute through the thick atmosphere of Titan and photographed the surface. From an altitude of 8 km (26,000 ft), the surface showed clear signs that a liquid, almost certainly liquid methane, had drained over the surface and into the lowlands. Once Huygens landed on the surface, it radioed back photos showing a level plain and chunks of ice rounded by a moving liquid. (ESA/NASA/JPL/University of Arizona)

Drainage channels were cut by flowing liquid methane.

Possible methane fog

Icy grapefruit-size "rocks" on Titan are bathed in orange light from its hazy atmosphere.

of Phoebe is 1.6 g/cm³, which is high enough to show that it contains a significant amount of rock. It seems unlikely that it came from the asteroid belt, where ices are rare, but rather that it is a captured Kuiper Belt object that wandered in from an orbit beyond Neptune.

Larger moons such as Tethys, with a diameter over 1000 km (620 mi), are icy and cratered, but they show some signs of geological activity. Some smooth areas on Tethys appear to have been resurfaced by flowing water "lava," and long cracks and grooves may have formed when geological activity strained the icy crust (Figure 23-20).

With a diameter of 520 km (320 mi), the small moon Enceladus isn't much larger than Phoebe, but Enceladus shows dramatic signs of geological activity (■ Figure 23-21). For one thing, Enceladus has an albedo of 0.9. That is, it reflects 90 percent of

the sunlight that hits it, and that makes it the most reflective object in the solar system. You know that old icy surfaces become dark, so the surface of Enceladus must be quite young. Look closely at the surface, and you will see that some regions have few craters and that grooves and cracks are common. Observations made by the Cassini spacecraft show that Enceladus has a tenuous atmosphere. It is too small to keep such an atmosphere, so it must be releasing gas continuously, and Cassini detected a large cloud of water vapor over the moon's south pole where some cracks are actually warmer than its equator. Many planetary

■ Figure 23-19

A few of Saturn's moons compared with Earth's moon at the right. In general, larger moons are round and more likely to show signs of geological activity. Small moons such as Phoebe and Hyperion are cratered and do not have enough gravity to overcome the strength of their own ice and squeeze themselves into a spherical shape.

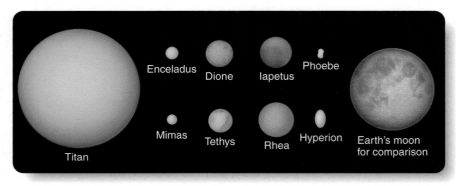

Phoebe is a small, dark, cratered moon with no sign of geological activity.

Long cracks in the cratered surface of Tethys show the larger moon has been geologically active.

Visual-wavelength images

■ Figure 23-20

Phoebe and Tethys have ancient cratered surfaces, but they differ in interesting ways. Phoebe is only 21 percent the diameter of Tethys and shows no sign of internal heat causing surface activity. Tethys has smooth areas and long cracks on its surface, showing that it has been active. In addition, Phoebe is probably a captured Kuiper Belt object, but Tethys might have formed with Saturn. (NASA/JPL/Space Science Institute)

Visual-wavelength images

The terrain at the right appears to be older than the terrain at the left.

■ Figure 23-21

The bright, clean surface of Enceladus does not look old. Some areas have few craters, and the numerous cracks and lanes of grooved terrain resemble the surface of Jupiter's moon Ganymede. (NASA/JPL/Space Science Institute)

scientists suspect that ice volcanism must be active on its surface. Ammonia dissolved in water can lower its melting point and might allow heat from inside Enceladus to produce eruptions. Cassini observations have detected tiny dust particles around Enceladus, and the E ring of Saturn is densest at the position of Enceladus. That suggests active eruptions.

Of course, you are wondering how a little moon like Enceladus can have heat flowing up from its interior. It is much too small to have remained active since it formed. A clue lies in the moon's orbit. Enceladus orbits Saturn in a resonance with the larger moon Dione. Each time Dione orbits Saturn once, Enceladus orbits twice. That means Dione's gravitational tugs on Enceladus always occur in the same places and make the orbit of the little moon into a slight ellipse. As Enceladus follows that elliptical orbit around Saturn, it feels tides flexing it, and tidal heating warms the interior. You saw how resonances can keep some of Jupiter's moons active, so you can add Enceladus to the list.

Saturn has too many moons to discuss in detail here, but you have to meet Iapetus. It is literally an odd ball. Iapetus is an asymmetric moon. Its trailing side, the side that always faces backward as it orbits Saturn, is old, cratered, icy, and about as bright as dingy snow. Its leading side, the side that always faces forward in its orbit, is also old and cratered, but it has an albedo of only 4 percent—about as dark as fresh asphalt on a highway

■ **Figure 23-22**

Like the windshield of a speeding car, the leading side of Saturn's moon Iapetus seems to have accumulated a coating of dark material. The poles and trailing side of the moon are much cleaner ice. The equatorial ridge is 20 km (12 mi) wide and up to 13 km (8 mi) high. It stretches roughly 1300 km (800 mi) along the moon's equator. (NASA/JPL/Space Science Institute)

Visual-wavelength image

(■ Figure 23-22). The origin of this dark material is unknown, but theorists suspect that the little moon has swept up dark, silicon- and carbon-rich material on its leading side. This dust could be produced by meteorites striking the outermost moon, Phoebe. Cassini images of Iapetus reveal an equatorial ridge that stands as high as 13 km (8 mi) in some places. You can see the ridge clearly in Figure 23-22. The origin of this ridge is unknown, but it is not a minor feature. At 13 km high, it is over 50 percent higher than Mount Everest, and it extends for a long distance across the surface. That is a big pile of rock and ice.

Saturn's moons illustrate a number of principles of comparative planetology. Small moons are irregular in shape, and old surfaces are dark and cratered. Resonances can trigger tidal heating, and that can resurface moons and outgas atmospheres. Small moons can't keep atmospheres, but big, cold moons can. You are an expert in all of this, so you are ready to wonder where the moons came from.

The Origin of Saturn's Moons

Jupiter's four Galilean satellites seem clearly related to one another, and you can safely conclude that they formed with Jupiter. No such systematic relationship links Saturn's satellites. Planetary scientists suspect that they are captured icy planetesimals left over from the solar nebula or that the impacts of comets have so badly fractured them that they no longer show evidence of their common origin.

Understanding the origin of Saturn's moons is also difficult because the moons interact gravitationally in complicated ways, and the orbits they now occupy may differ from their earlier orbits. These interactions are dramatically illustrated by the two small moons that shepherd the F ring. An even more peculiar pair of moons is known as the coorbital satellites. These two irregularly shaped moonlets have orbits separated by only 100 km. Because one moon is about 200 km in diameter and the other about 100 km, they cannot pass in their orbits. Instead, the innermost moon gradually catches up with the outer moon. As the moons draw closer together, the gravity of the trailing moon slows the leading moon and makes it fall into a lower orbit. Simultaneously, the gravity of the leading moon pulls the trailing moon ahead, and it rises into a higher orbit (■ Figure 23-23). The higher orbit has a longer period, so the trailing moon begins to fall behind the leading moon, which is now in a smaller, faster orbit. Through this elegant dance, the moons exchange orbits and draw apart only to meet again and again. It seems very likely that these two moons are fragments of a larger moon destroyed by a major impact.

In addition to waltzing coorbital moons, the Voyager spacecraft discovered small moonlets trapped at the L4 and L5 Lagrangian points (see Figure 13-5) in the orbits of Dione and Tethys. These points of stability lie 60° ahead of and 60° behind the two moons, and small moonlets can become trapped in these regions. (You will see in Chapter 25 that asteroids are trapped in the Lagrangian points of Jupiter's orbit around the sun.) This

Coorbital Moons

The inner moon orbits faster and overtakes the outer moon.

(Not to scale)

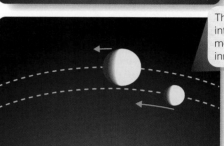

The gravitational interaction pulls the outer moon backward and the inner moon forward.

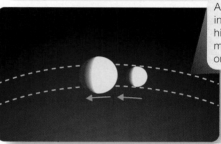

As they approach, the inner moon moves to a higher orbit, and the outer moon sinks to a lower orbit.

The moons have changed orbits, and the inner moon begins gaining on the outer moon.

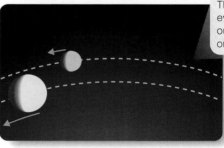

The inner moon will eventually overtake the outer moon from behind once again.

■ **Active Figure 23-23**

Saturn's coorbital moons follow nearly identical orbits. The moon in the lower orbit travels faster and always overtakes the other moon from behind. The moons interact and change orbits over and over again.

Ace◯Astronomy™ Log into AceAstronomy and select this chapter to see the Active Figure "Coorbital Moons" and animate this diagram.

gravitational curiosity is surely not unique to the Saturn system, and it warns that interactions between moons can dramatically alter their orbits.

Some of Saturn's larger moons, especially Titan, may have formed with the planet, but the orbital relationships and intense cratering suggest that the moons have interacted and may have collided with large planetesimals or even the heads of comets in the past. Many of the moons may be icy bodies that Saturn has captured since the formation of the solar system. As you continue your exploration of the outer solar system, you must be alert for the presence of such small, icy worlds.

Building Scientific Arguments

What features on Enceladus suggest that it has been active? This argument is based on comparative planetology applied, in this case, to moons. The smaller moons of Saturn are icy worlds battered by impact craters, and you can suspect that they are cold and old. Small worlds lose their heat quickly; and, with no internal heat, there is no geological activity to erase impact craters. A small, icy world covered with craters is exactly what you would expect in the outer solar system. Enceladus, however, is peculiar. Although it is small and icy, its surface is highly reflective, and some areas contain fewer craters than you would expect. In fact, some regions seem almost free of craters. Grooves and faults mark some regions of the little moon and suggest motion in the crust. These features should have been destroyed long ago by impact cratering, so you must suppose that the moon has been geologically active at some time since the end of heavy cratering at the conclusion of planet building.

If you look at Titan, Saturn's largest moon, you see quite a different world. Build an argument based on a different principle of comparative planetology. **How can such a small world as Titan keep a thick atmosphere?**

■ ■ ■

Connections: Jupiter and Saturn reign over the solar system like their namesake ancient gods. Beyond them, out where the sunlight is dim and temperatures deadly cold, orbit three more planets that were unknown to ancient astronomers. In the next chapter, you will add these worlds to your study of comparative planetology.

Summary

23-1 | A Travel Guide to the Outer Planets

How do the outer planets compare with the inner planets?

- Of the outer planets, Jupiter, Saturn, Uranus, Neptune, and Pluto, the first four are much larger than Earth and lower in density. These Jovian planets are rich in hydrogen. Pluto is small and icy and does not seem to fit its role as a planet.

- All four of the Jovian planets have rings, large satellite systems, and shallow atmospheres above liquid mantles. Jupiter and Saturn show strong belt–zone circulation, but this is harder to see on Uranus and Neptune.

- Moons in the Jovian satellite systems interact gravitationally, and some have been heated by tides to produce geological activity. Most are old and cratered.

23-2 | Jupiter

How is Jupiter different from Earth?

- Simple observations made from Earth show that Jupiter is 11 times Earth's diameter and 318 times Earth's mass. From that you can calculate its density, which is much lower than Earth's.

- The low density combined with the oblateness and mathematical models reveal that Jupiter must have only a small core of heavy elements and a deep mantle of liquid hydrogen. The atmospheric composition is much like that of the sun—mostly hydrogen and helium.

- Infrared observations show that Jupiter radiates more heat than it receives from the sun, so its interior must be five times hotter than the sun's surface and is prevented from flashing into vapor by the high pressure.

- The pressure inside Jupiter converts much of its hydrogen into liquid metallic hydrogen, and the dynamo effect generates a powerful magnetic field that produces rings of auroras around the planet's magnetic poles, interacts with the small moon Io, and traps high-energy particles to form intense radiation belts.

- Jupiter's magnetic field was first discovered by radio astronomers who detected radio signals coming from Jupiter.

- Jupiter's shallow atmosphere is rich in hydrogen, with three layers of clouds at the temperatures where ammonia, ammonium hydrosulfide, and water condense out. High- and low-pressure areas form belt–zone circulation. Spots on Jupiter are long-lasting, circulating storm systems.

How did Jupiter and its system of moons and rings form and evolve?

- Forward scattering shows that Jupiter's ring is composed of tiny dust specks orbiting inside Jupiter's Roche limit. The dust particles cannot have survived since the formation of the planet; rather, they are being produced by meteorite impacts on some of Jupiter's inner moons.

- Jupiter was hit by the fragmented head of a comet in 1992. Such impacts had not been seen before but must be common in the history of the solar system.

- Jupiter formed in the outer solar system where ice could condense from the solar nebula. It grew rapidly and captured large amounts of hydrogen and helium gas.

23-3 | Jupiter's Family of Moons

- At least some of Jupiter's small moons are captured asteroids.

- The four Galilean moons appear to have formed with Jupiter. Callisto, the outermost, is composed of ice and rock and has an old and cratered surface. Unlike the three inner moons, Calisto is not caught in an orbital resonance and has never been active.

- Ganymede, Europa, and the innermost moon, Io, are locked in an orbital resonance, and that causes tidal heating. Ganymede's surface is old and cratered in some areas, but smooth, grooved areas must have been produced by a past episode of geological activity.

- Europa is mostly rock with a thin icy crust that contains almost no craters. Cracks and lines show that the crust has broken repeatedly, and a sub-surface ocean probably vents through the crust and deposits ice to cover craters as fast as they form. The subsurface ocean might be a place to look for life.

- Io is strongly heated by tides and has no water at all. Over a hundred volcanoes erupt on its surface, and no impact craters are visible. Sulfur compounds color the surface yellow and orange and vent into space to be caught in Jupiter's magnetic field.

23-4 | Saturn

How is Saturn different from Jupiter?

- Saturn is slightly smaller than Jupiter and less dense than water. It has a hot interior but contains less liquid metallic hydrogen, so its magnetic field is weaker than Jupiter's.

- Saturn is twice as far from the sun as Jupiter and is much colder. The three cloud belts seen on Jupiter form deeper in Saturn's hazy atmosphere and are not as clearly visible.

How did Saturn and its system of moons and rings form and evolve?

- Saturn must have formed much as Jupiter did, but it is smaller and less dense.

- Saturn's rings are composed of ice particles and cannot have lasted since the formation of the planet. The rings must receive occasional additions of ice particles when a small moon wanders too close and is pulled apart by tides or when comets hit the planet's icy moons.

- Icy particles can become trapped in stable places among the orbits of the innermost small moons, those within the Roche limit. Resonances with outer moons can produce gaps in the rings and generate waves that move like ripples through the rings. Small moons can shepherd sections of the ring to produce sharp edges, ripples, or narrow ringlets.

- Without moons to confine them, the rings would have spread outward and dissipated long ago.

23-5 | Saturn's Moons

- Many of Saturn's smaller moons, such as Phoebe, are probably captured asteroids or Kuiper Belt objects. All of the moons are mixtures of rock and ice.

- Titan, the largest moon, is so cold it can retain a dense atmosphere of nitrogen with a few percent methane. The methane condenses from the atmosphere, falls as rain, and drains over the surface, washing dark, organic material into drainage channels and on into the lowlands.

- The methane in Titan's atmosphere can be destroyed by sunlight, so it must be continuously replenished. It probably vents from the icy crust and may form methane volcanoes.

- Some of the larger moons, such as Tethys, have old, dark, cratered surfaces with cracks and smoothed areas that suggest past activity. Enceladus, a rather small moon, is the most reflective object in the solar system and has large smooth areas, so it must have been very active. It orbits in a resonance with the moon Dione, and that may cause tidal heating inside Enceladus.

- The moon Iapetus has a bright icy surface on its trailing side, but its leading side is coated with very dark material, possibly debris from

Phoebe. In addition, Iapetus has a long equatorial ridge higher than Mount Everest on Earth. The origin of the ridge is not known.

■ The origin and evolution of Saturn's moons is not as clear as for Jupiter's Galilean moons. Orbital interactions and impacts have been important to these moons.

New Terms

belt–zone circulation (p. 554)

oblateness (p. 556)

liquid metallic hydrogen (p. 557)

decameter radiation (p. 557)

decimeter radiation (p. 557)

Io plasma torus (p. 558)

Io flux tube (p. 558)

critical point (p. 558)

belt (p. 561)

zone (p. 561)

forward scattering (p. 562)

Roche limit (p. 562)

gossamer ring (p. 563)

grooved terrain (p. 566)

tidal heating (p. 567)

shepherd satellite (p. 577)

Review Questions

Ace ✺ Astronomy™ Assess your understanding of this chapter's topics with additional quizzing and animations at **http://ace .brookscole.com/sf9**

1. Why is Jupiter more oblate than Earth? Do you expect all Jovian planets to be oblate? Why or why not?

2. How do the interiors of Jupiter and Saturn differ? How does this affect their magnetic fields?

3. What is the difference between a belt and a zone?

4. How can you be certain that Jupiter's ring does not date from the formation of the planet?

5. If Jupiter had a satellite the size of our own moon orbiting outside the orbit of Callisto, what would you predict for its density and surface features?

6. Why are there no craters on Io and few on Europa? Why should you expect Io to suffer more impacts per square kilometer than Callisto?

7. Why are the belts and zones on Saturn less distinct than those on Jupiter?

8. If Saturn had no moons, what do you suppose its rings would look like?

9. Where did the particles in Saturn's rings come from?

10. How can Titan keep an atmosphere when it is smaller than airless Ganymede?

11. If you piloted a spacecraft to visit Saturn's moons and wanted to land on a geologically old surface, what features would you look for? What features would you avoid?

12. Why does the leading side of some satellites differ from the trailing side?

13. Why would you expect small satellites to be irregular in shape? Give examples and exceptions.

14. This image shows a segment of the surface of Jupiter's moon Callisto. Why is the surface dark? Why are some craters dark and some white? What does this image tell you about the history of Callisto?

(NASA/JPL)

15. The Cassini spacecraft recorded this image of Saturn's A ring and Encke's division. What do you see in this photo that tells you about processes that confine and shape planetary rings?

(NASA/JPL/Space Science Institute)

Discussion Questions

1. Some astronomers argue that Jupiter and Saturn are unusual, while other astronomers argue that all solar systems should contain one or two such giant planets. What do you think? Support your argument with evidence.

2. Why don't the terrestrial planets have rings?

Problems

1. What is the maximum angular diameter of Jupiter as seen from Earth? What is the minimum? Repeat this calculation for Saturn and Titan. (*Hint:* Use the small-angle formula.)

2. The highest-speed winds on Jupiter are in the equatorial jet stream, which has a velocity of 150 m/s. How long does it take for these winds to circle Jupiter?

3. What are the orbital velocity and period of a ring particle at the outer edge of Jupiter's ring? at the outer edge of Saturn's A ring? (*Hints:* The radius of the edge of the A ring is 136,500 km. See Chapter 5.)

4. What is the angular diameter of Jupiter as seen from the surface of Callisto? (*Hint:* Use the small-angle formula.)

5. What is the escape velocity from the surface of Ganymede if its mass is 1.5×10^{26} g and its radius is 2628 km? (*Hint:* See Chapter 5.)

6. If you were to record the spectrum of Saturn and its rings, you would find light from one edge of the rings redshifted and light from the other edge blueshifted. If you observed at a wavelength of 500 nm, what difference in wavelength should you expect between the two edges of the rings? (*Hints:* See Problem 3 and Chapter 7.)

7. What is the difference in orbital velocity between particles at the outer edge of Saturn's B ring and particles at the inner edge of the B ring? (*Hint:* The outer edge of the B ring has a radius of 117,500 km, and the inner edge has a radius of 92,000 km.)

8. What is the difference in orbital velocity between the two coorbital satellites if the semimajor axes of their orbits are 151,400 km and 151,500 km? The mass of Saturn is 5.7×10^{26} kg. (*Hint:* See Chapter 5.)

Media Cluster

Ace Astronomy™ To access the resources in the Media Cluster, log into AceAstronomy at **http://ace.brookscole.com/sf9** and select Chapter 23.

ACTIVE FIGURES

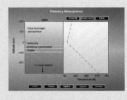

Planetary Atmospheres
Experiment with the temperature throughout the atmospheres of several planets with this animation.

Explorable Jupiter
Study Jupiter at visible, ultraviolet, radio, and X-ray wavelengths in this animation. What features can you see at the other wavelengths that are not visible to the human eye?

Coorbital Moons
Study the phenomenon of two of Saturn's moons repeatedly swapping orbits in this animation.

Roche Limit
Some moons remain intact in orbit, and some break apart to become rings like Saturn's rings. See if you can figure out why with this animation.

ASTRONOMY EXERCISES

Auroras
Use this simulation to study the formation of auroras. You can vary the strength of Earth's magnetic field and the speed and number of particles in the solar wind.

Convection and Turbulence
Core temperature and rotation speed both contribute to the amount of turbulence in a planet's atmosphere. Explore their relationship with this animation.

Convection and Magnetic Fields
A planet's core temperature and rotation speed are factors in the strength of its magnetic field. See the effect of these variables in this animation.

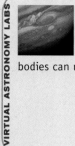

Lab 7: Planetary Atmospheres and Their Retention
This lab investigates the retention of atmospheres. You will explore the factors that govern the loss of atmospheric gases, and you will see why certain bodies can retain some gases but not others.

Critical Inquiries for the Web

1. Saturn has a great many satellites. Locate information on these moons, including discovery dates (make sure you use an updated source that includes discoveries from the past few years).

2. The collision of Comet Shoemaker–Levy 9 with Jupiter was one of the first events to be publicized directly on the Web. Search for websites dedicated to this comet and its spectacular demise. How long were the effects of the impacts visible in the atmosphere of Jupiter? How might they have appeared differently if this event had occurred at Saturn?

3. What is the status of scientists' understanding of the Jovian moon Europa? This moon was studied extensively by the Galileo mission. Visit Galileo-related websites to find the results of this exploration and discuss current ideas regarding the possibility of liquid water beneath Europa's crust. How might future missions to Europa probe such an underground ocean?

4. Cassini reached orbit around Saturn on July 1, 2004, and the Huygens probe parachuted through the atmosphere of Titan on January 14, 2005. Check the Cassini Web page to get the latest news from Saturn and Titan.

5. When will Saturn's rings next be edge-on to the Earth? When were they last edge-on? Find photographs of Saturn and its rings at different inclinations.

Exploring *TheSky*

1. Galileo recognized the significance of Jupiter's moons when he realized they were moving around Jupiter. Sketch the position of the four Galilean moons at intervals of 1 hour. Could you recognize their orbital motion in a long evening, or would you need observations over a number of evenings? (*Hint:* Use **Find** to **Center & Frame** Jupiter. Zoom in until you can see the moons. Use **Tracking Setup** to lock on Jupiter and then use the **Time Skip** buttons to make the moons move.)

2. Calculate the mass of Jupiter from your own observations of the orbital period and orbital radius of Jupiter's moons. (*Hint:* Use the equation for circular velocity in Chapter 5.)

3. Repeat Activity 2 for Saturn.

 Go to the Brooks/Cole Astronomy Resource Center **(http://astronomy.brookscole.com)** for critical thinking exercises, articles, and additional readings from InfoTrac College Edition, Brooks/Cole's online student library.

24 | Comparative Planetology of Uranus, Neptune, and Pluto

A good many things go around in the dark besides Santa Claus.

HERBERT HOOVER

O UT IN THE DARKNESS be-
yond Saturn, out where sunlight is about 1000
times fainter than on Earth, there are things going around
that Aristotle, Galileo, and Newton never imagined. They knew about Mercury, Venus, Mars, Jupiter, and Saturn, but
our solar system includes three more planets, planets that were not discovered until after the invention of the telescope. ▌ Uranus and Neptune are Jovian planets and are similar in size, about four times Earth's diameter. That
makes them about half the size of Saturn. Through most Earth-based telescopes, you can see nothing but tiny disks, and
until recently what little scientists knew of these planets was deduced from a few observations compared with Jupiter
and Saturn. But in January 1986 the Voyager 2 spacecraft flew past Uranus and returned a wealth of photographs and

▌ Continued on page 590 ▌

Pluto and its moon Charon lie so far from the sun that no spacecraft has ever visited them. No one knows what they are like, but someday a spacecraft will radio back images from the edge of the solar system. (NASA/Southwest Research Institute/ JHUAPL/SwRI)

Guidepost

Looking Back

Some people say they are seasoned travelers because they have been to Europe. You have visited six planets and a bunch of moons, braved the radiation belts of Jupiter, and swum through the rings of Saturn. Now you are ready to explore farther—out into the dark far from the sun.

This Chapter

Three worlds circle the sun in the twilight beyond Saturn. Uranus and Neptune are Jovian worlds, but you will find them strangely different from Jupiter and Saturn. Pluto is an oddity, and you will wonder if it is really a planet. As you explore you will answer five essential questions:

How is Uranus different from Jupiter?

How did Uranus, its rings, and its moons form and evolve?

How is Neptune different from Uranus?

How did Neptune, its rings, and its moons form and evolve?

Is Pluto a planet or something else?

Looking Ahead

As you finish this chapter, you will have visited all of the major worlds in our solar system. But there is more to see. Vast numbers of small rocky and icy bodies orbit among the planets, and you should visit them to complete your tour but also to search for clues to how the solar system formed.

measurements. Then, in August 1989, Voyager 2 flew past Neptune, revealing a blue world with complex weather, rings, and moons. In just a few years, Uranus and Neptune went from obscure points of light to known worlds. More recently, the Hubble Space Telescope and new-generation telescopes on Earth have been able to image Uranus and Neptune and monitor changing weather patterns.

Pluto, however, is a more difficult subject. Through an Earth-based telescope, it looks like nothing more than a faint point of light, and even through the Hubble Space Telescope it is a tiny disk. Voyagers 1 and 2 left the solar system without passing near Pluto, and, although a spacecraft is planned for a journey to Pluto, no spacecraft has been there yet. Consequently there are no photos of the little world's surface. Nevertheless, as an experienced traveler among planets, you may be able to guess Pluto's nature by combining the limited observations with the principles of comparative planetology.

Uranus, Neptune, and Pluto differ from the other planets in one special way. They were not known from antiquity. They had to be discovered. The stories of these discoveries highlight the nature of scientific discovery.

24-1 Uranus

IN MARCH 1781, BENJAMIN FRANKLIN was in France raising money, troops, and arms for the American Revolution. George Washington and his colonial army were only six months away from the defeat of Cornwallis at Yorktown and the end of the war. In England, King George III was beginning to show signs of madness. And a German-speaking music teacher in the English resort city of Bath was about to discover the planet Uranus.

The Discovery of Uranus

William Herschel (■ Figure 24-1) came from a musical family in Hanover, Germany, but immigrated to England as a young man and eventually obtained a prestigious job as the organist at the Octagon Chapel in Bath.

To compose exercises for his students and choral and organ works for the chapel, Herschel studied the mathematical prin-

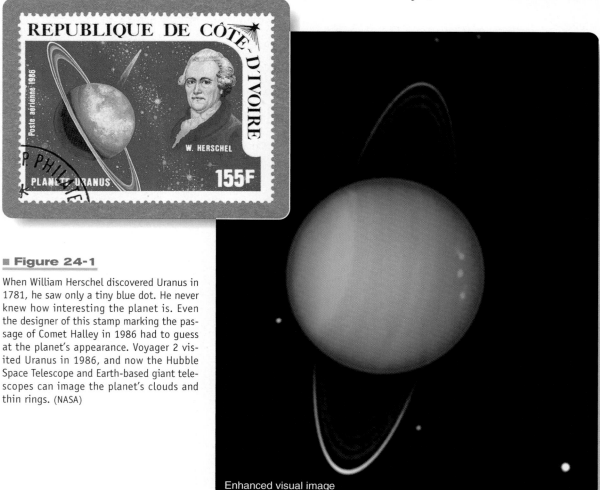

■ Figure 24-1

When William Herschel discovered Uranus in 1781, he saw only a tiny blue dot. He never knew how interesting the planet is. Even the designer of this stamp marking the passage of Comet Halley in 1986 had to guess at the planet's appearance. Voyager 2 visited Uranus in 1986, and now the Hubble Space Telescope and Earth-based giant telescopes can image the planet's clouds and thin rings. (NASA)

ciples of musical harmony from a book by Professor Robert Smith of Cambridge. The mathematics was so interesting that Herschel searched out other books by Smith, including a book on optics. Of course, an 18th-century book on optics written by a professor in England relied heavily on Isaac Newton's discoveries and described the principles of the reflecting telescope (see Chapter 6). Herschel and his brother Alexander began building telescopes, and William went on to study astronomy in his spare time.

Herschel's telescopes were similar to Newton's in that they had metal mirrors, but they were much larger. Newton's telescope had a mirror about 1 in. in diameter, but Herschel developed ways of making much larger mirrors. He soon had telescopes as long as 20 ft; although one of his favorite telescopes was 7 ft long with a mirror 6.2 in. in diameter. Using it, he began the research project that led to the discovery of Uranus.

Herschel did not set out to search for a planet; he was trying to detect stellar parallax produced by Earth's motion around the sun. No one had yet detected this effect, although by the 1700s all astronomers accepted Earth's motion. Galileo had pointed out that parallax might be detected if a nearby star and a very distant star lay so nearly in the same direction that they looked like a very close double star through a telescope. In such a case, Earth's orbital motion would produce a parallactic shift in the position of the nearby star with respect to the more distant star (■ Figure 24-2). Herschel began to examine all stars brighter than eighth magnitude to search for double stars that might show parallax. That project alone took over two years.

On the night of March 13, 1781, Herschel set up his 7-ft telescope in his back garden and continued his work. He later wrote, "In examining the small stars in the neighborhood of H Geminorum, I perceived one that appeared visibly larger than the rest." As seen from Earth, Uranus is never larger in angular diameter than 3.7 seconds of arc, so Herschel's detection of the disk illustrates the quality of his telescope and his eye. At first he suspected that the object was a comet, but other astronomers quickly showed that it was a planet orbiting the sun beyond Saturn.

The discovery of Uranus made Herschel world famous. Since antiquity, astronomers had known the five planets—Mercury, Venus, Mars, Jupiter, and Saturn—and they supposed that the universe was complete. Herschel's discovery extended the classical universe by adding a new planet. The English public accepted Herschel as their astronomer-hero, and, having named the new planet *Georgium Sidus* (George's Star) after King George III, Herschel received a royal pension. The former music teacher was welcomed into court society, where he later met and married a wealthy widow and took his place as one of the great English astronomers in history. With his new financial position, he built large telescopes on his estate and, with his sister Caroline, a talented astronomer herself, attempted to map the extent of the universe. You read about their research in Chapter 15.

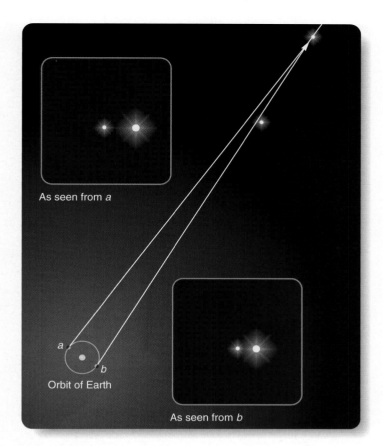

■ Figure 24-2

A double star consisting of a nearby star and a more distant star could be used to detect stellar parallax. The separation between the stars would be different as seen from different places in Earth's orbit. The effect is actually much too small to detect by eye.

Continental astronomers were less than thrilled that an Englishman had made the great discovery, and even some professional English astronomers thought Herschel a mere amateur. They called his discovery a lucky accident. Herschel responded in his own defense by making three points. First, he had built some of the finest-quality telescopes then in existence. Second, he had been conducting a systematic research project and would have found Uranus eventually because he was inspecting all of the brighter stars visible with his telescope. And third, he had great experience seeing fine detail with his telescopes. As a musician, he knew the value of practice and applied it to the business of astronomical observing. In fact, records show that astronomers had seen Uranus more than 17 times before Herschel, but each time they failed to notice that it was not a star. They plotted Uranus on their charts as if it was just another faint star.

This illustrates one of the ways in which scientific discoveries are made. Often, discoveries seem accidental, but on closer examination you find that the scientist has earned the right to the discovery through many years of study and preparation. To quote a common saying, "Luck is what happens to people who work hard."

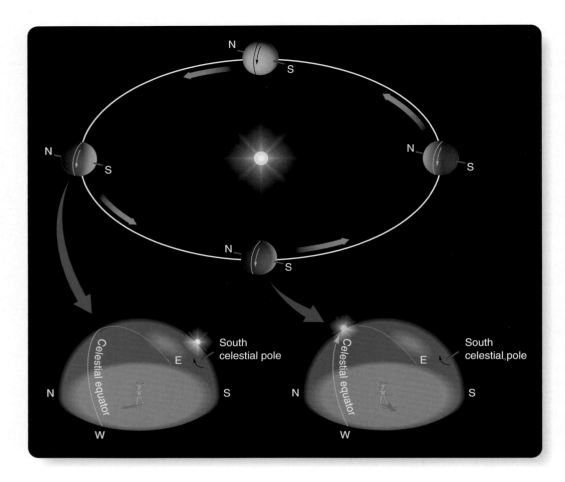

Uranus rotates on an axis that is tipped 97.9° from the perpendicular to its orbit, so its seasons are extreme. When one of its poles is pointed nearly at the sun (a solstice), a citizen of Uranus would see the sun near a celestial pole, and it would never rise or set. As it orbits the sun, the planet maintains the direction of its axis in space, and thus the sun moves from pole to pole. At the time of an equinox on Uranus, the sun would be on the celestial equator and would rise and set with each rotation of the planet. Compare with similar diagrams for Earth on page 27.

observed position of the planet. Tiny variations in the orbital motion of Uranus eventually led to the discovery of Neptune, a controversial story you will hear later in this chapter.

The Motion of Uranus

Uranus follows a slightly elliptical orbit with an average distance from the sun of 19.18 AU and an orbital period of 84.013 years (**Celestial Profile 9**). The ancients thought of Saturn as the slowest of the planets, but Saturn orbits in a bit over 29 years. Uranus, being further from the sun, has a longer orbital period.

Continental astronomers, especially the French, insisted that the new planet not be named after an English king. The French stubbornly called the planet Herschel, as did many other non-English astronomers. Some years later, German astronomer J. E. Bode suggested the name Uranus, one of the oldest of the Greek gods.

Over the half-century following the discovery of Uranus, astronomers noted that Newton's laws did not quite predict the

The rotation of Uranus is peculiar. Earth rotates approximately upright in its orbit. That is, its axis of rotation is inclined only 23.5° from the perpendicular to its orbit. Uranus, in contrast, rotates on an axis that is inclined 97.9° from the perpendicular to its orbit. It rotates on its side (■ Figure 24-3).

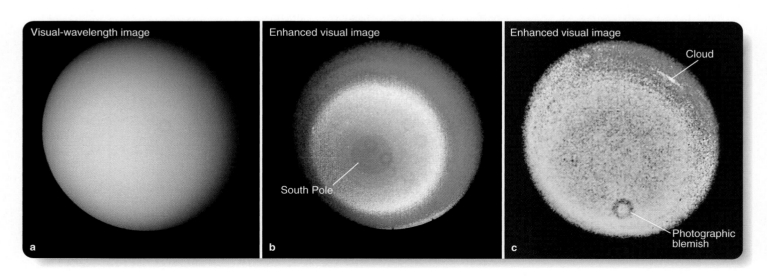

The seasons on Uranus are extreme. In 1986, when Voyager 2 flew past Uranus, the planet was in the segment of its orbit in which its south pole faced the sun. Consequently, its southern hemisphere was bathed in continuous sunlight, and a creature living on Uranus (an unlikely possibility, as you will discover later) would have seen the sun circling near the planet's south celestial pole. You can see this at lower left in Figure 24-3. Once Uranus moves about a quarter of the way around its orbit, the sun will be above its equator, and the citizen of Uranus would see the sun rise and set with the rotation of the planet. Look at the lower right in Figure 24-3 to see that geometry. As Uranus continues along its orbit, the sun will approach the planet's north celestial pole, and the southern hemisphere of the planet will experience a lightless winter lasting 21 Earth years. In other words, the ecliptic on Uranus passes very near the planet's celestial poles, and the resulting seasons are extreme.

The Atmosphere of Uranus

Like Jupiter and Saturn, Uranus has no surface. The gases of its atmosphere—mostly hydrogen, 15 percent helium, and a few percent methane, ammonia, and water vapor—blend gradually with a fluid interior.

Seen through Earth-based telescopes, Uranus is a small, featureless blue disk. The blue color arises because the atmosphere contains methane, which is a good absorber of longer-wavelength photons. As sunlight penetrates into the atmosphere and is scattered back out, the longer-wavelength (red) photons are more likely to be absorbed. Consequently, the light entering your eye is richer in blue photons and poorer in red photons, giving the planet a blue color.

As Voyager 2 drew closer to the planet in late 1985, astronomers studied the images radioed back to Earth. Uranus was a pale blue ball with no obvious clouds, and only when the images were carefully computer enhanced was any banded structure detected (■ Figure 24-4). A few very high clouds of methane ice particles were detected, and that allowed astronomers to measure the rotation period of the planet.

You can understand the appearance of the atmosphere by studying the temperature profile of Uranus shown in ■ Figure 24-5. The atmosphere of Uranus is much colder than that of Saturn or Jupiter. Consequently, the three cloud layers of ammonia, ammonia hydrosulfide, and water that occur in the atmosphere of Jupiter and Saturn would form very deep in the atmosphere of Uranus. These cloud layers are not visible because of the thick

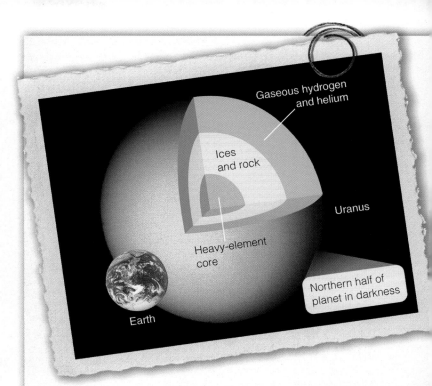

Uranus rotates on its side; when Voyager 2 flew past in 1986, the planet's south pole was pointed almost directly at the sun. (NASA)

Celestial Profile 9: Uranus

Motion:

Average distance from the sun	19.18 AU (28.69 × 10⁸ km)
Eccentricity of orbit	0.0461
Maximum distance from the sun	20.1 AU (30.0 × 10⁸ km)
Minimum distance from the sun	18.3 AU (27.4 × 10⁸ km)
Inclination of orbit to ecliptic	0°46'23"
Average orbital velocity	6.81 km/s
Orbital period	84.013 y (30,685 days)
Period of rotation	17ʰ14ᵐ
Inclination of equator to orbit	97°55'

Characteristics:

Equatorial diameter	51,118 km (4.01 $D_\oplus$)
Mass	8.69 × 10²⁵ kg (14.54 $M_\oplus$)
Average density	1.29 g/cm³
Gravity	0.919 Earth gravity
Escape velocity	22 km/s (1.96 $V_\oplus$)
Temperature above cloud tops	−220°C (−364°F)
Albedo	0.35

Personality Point:

Most creation stories begin with a separation of opposites, and Greek mythology is no different. Uranus (the sky) separated from Gaia (Earth), who was born from the void, Chaos. They gave birth to the giant Cyclops, Cronos (Saturn, father of Zeus), and his fellow Titans. Uranus is sometimes called the starry sky, but the sun (Helius), moon (Selene), and the stars were born later, so Uranus, one of the most ancient gods, began as the empty, dark sky.

■ **Figure 24-4** ◀

(a) This Voyager 2 image of Uranus was made in 1986 and shows no clouds. Only when the image is computer enhanced, as in (b), is a banded structure visible. At the time, the axis of rotation was pointed nearly at the sun. (c) Under extreme computer enhancement, small methane clouds were visible. The geometry of the banding and the clouds suggests belt–zone circulation. (NASA)

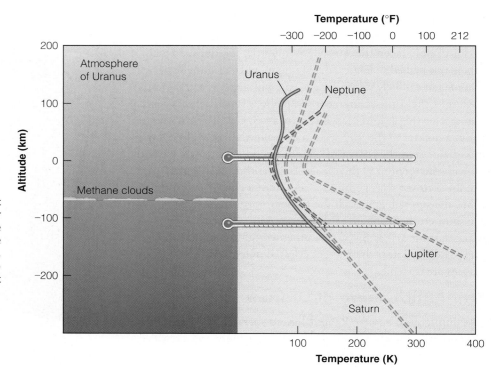

■ **Active Figure 24-5**

The atmosphere of Uranus is much colder than that
of Jupiter or Saturn, and the only visible cloud layer
is one formed of methane ice crystals deep in the
hydrogen atmosphere. Other cloud layers would be
much deeper in the atmosphere and are not visible.
The temperature profile of Neptune is similar to
that of Uranus, and it has methane clouds at about
the same place in its atmosphere.

Ace◐Astronomy™ Log into AceAstronomy
and select this chapter to see the Active Figure
"Planetary Atmospheres" and compare the atmospheres of the planets.

atmosphere of hydrogen through which an observer has to look.
The clouds that are visible on Uranus are clouds of methane ice
crystals, which form at such a low temperature that they occur
higher in the atmosphere of Uranus. Look again at Figure 24-5
and notice that there can be no methane clouds on Jupiter because
that planet is too warm. The coldest part of the atmosphere of
Saturn is just cold enough to form a thin methane haze high
above its more visible cloud layers. (See Figure 23-16b.)

The clouds and atmospheric banding faintly visible on Uranus appear to be a form of belt–zone circulation, which is a bit
surprising. Uranus rotates on its side, so solar energy strikes its
surface in a geometry quite different than it does on Jupiter and
Saturn. Evidently the belt–zone circulation is dominated by the
rotation of the planet and not by the direction of sunlight.

Is Uranus always as featureless as it was then Voyager 2 flew
past in early 1986? The Hubble Space Telescope and giant Earth-
based telescopes have detected changing clouds on Uranus (■ Figure 24-6), and those clouds appear to follow belt–zone patterns.
Furthermore, since 1986, Uranus has moved on along its orbit,
and spring has been coming to its northern hemisphere. As sunlight has reached these latitudes, clouds have appeared there,
probably generated by convection in the atmosphere. This may be
part of a seasonal cycle on Uranus, but its year lasts 85 Earth
years, so you will have to be patient.

The Interior of Uranus

Astronomers cannot describe the interiors of Uranus and Neptune as accurately as they can the interiors of Jupiter and Saturn.

Observational data are sparse, and the materials inside these planets are not as easy to model as simple liquid hydrogen.

The average density of Uranus, 1.29 g/cm³, tells you that the
planet must contain a larger share of dense materials such as rock
than does Jupiter or Saturn. Nearly all models of the interior of
Uranus contain three layers. The atmosphere is rich in hydrogen
and helium. A deep mantle must contain large amounts of ices
of water, methane, and ammonia, mixed with hydrogen and with
silicate matter. This mantle is sometimes described as "liquid" or
as "ice," but it is probably quite unlike the materials these words
suggest. The third layer in these three-layer models is a small core
of rock. Because of the high pressure and high temperature, the
material is probably not very rocklike. The term *rock* refers to its
chemical composition and not its appearance. It would be better
to call it a heavy-element core.

Planetary scientists make a subtle distinction among the Jovian planets. Jupiter and Saturn are sometimes called "gas giant
planets," because they are rich in hydrogen and helium. A growing number of scientists are referring to Uranus and Neptune as
"ice giant planets," in recognition of the larger proportion of ices
in their interiors.

Because Uranus has a much lower mass than Jupiter, its internal pressure is not high enough to produce liquid metallic hydrogen. Consequently, you might expect it to lack a strong magnetic
field, but the Voyager 2 spacecraft found that Uranus has a magnetic field about 75 percent as strong as Earth's. Furthermore,
the field is tipped 60° to the axis of rotation and is offset from
the center of the planet about 30 percent of the planet's radius

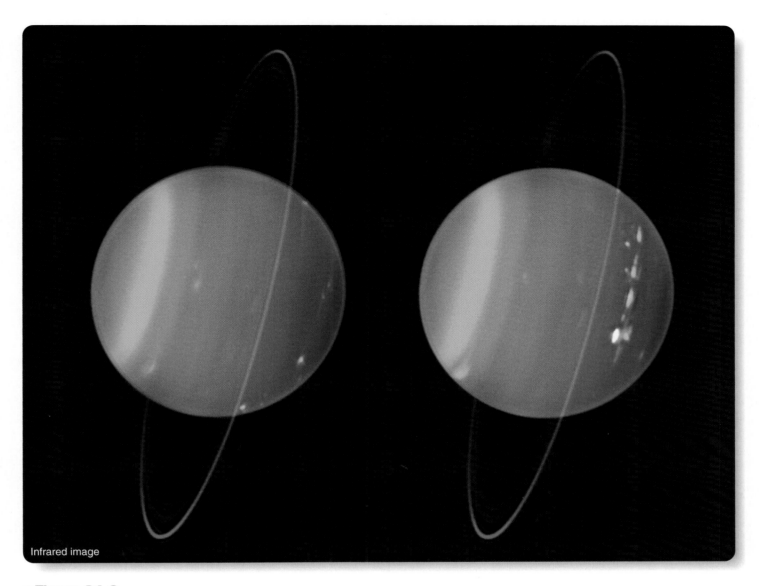

Infrared image

■ Figure 24-6

Images of Uranus recorded by the Keck near-infrared camera show both sides of Uranus. Banding and cloud features are visible, and some clouds are highly changeable. This suggests that seasonal changes are taking place on Uranus as spring comes to the northern hemisphere. The rings look red in this image because of image processing. (Lawrence Sromovsky, UW-Madison Space Science and Engineering Center)

(■ Figure 24-7). Theorists suggest that the magnetic field is produced by a dynamo effect, operating not at the center but nearer the surface in a layer of liquid water with dissolved ammonia and methane. Such a material would be a good conductor of electricity, and the rotation of the planet coupled with circulation in the fluid could generate the magnetic field.

The high inclination of the magnetic field and its offset from the center of Uranus give planetary scientists a good estimate of the true rotation period of the planet, otherwise difficult to estimate because of the lack of cloud features and high atmospheric winds. The magnetic field deflects the solar wind and traps some charged particles to create weak radiation belts trapped in a magnetosphere. High-speed electrons spiraling along the magnetic field produce synchrotron radio emission just as on Jupiter, and the Voyager 2 spacecraft detected this radiation fluctuating with a period of 17.24 hours—the period of rotation of the magnetic field and, presumably, the planetary interior.

The magnetic field and the high inclination of the planet produce some peculiar effects. As is the case for all planets, the solar wind deforms the magnetosphere and draws it out into a long tail extending away from the planet in the direction opposite the sun. The rapid rotation of Uranus and its high inclination make

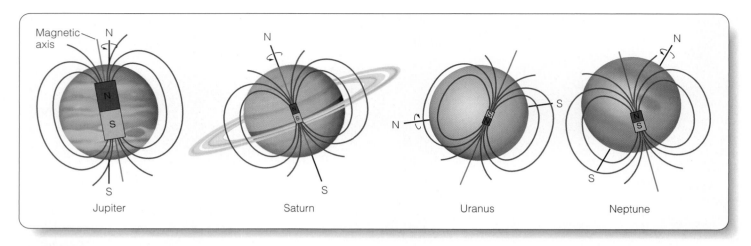

■ Figure 24-7

The magnetic fields of Uranus and Neptune are peculiar. While the magnetic axis of Jupiter is tipped only 10° from its axis of rotation and the magnetic axis of Saturn is not tipped at all, the magnetic axes of Uranus and Neptune are tipped at large angles. Furthermore, the magnetic fields of Uranus and Neptune are offset from the center of the planet. This suggests that the dynamo effect operates differently in Uranus and Neptune than in Jupiter and Saturn.

the magnetosphere and its long extension take on a corkscrew shape, which was partially mapped by Voyager 2. At the time Voyager 2 flew past in 1986, the south pole of Uranus was pointed nearly at the sun, and once each rotation the solar wind could pour directly down into the south magnetic pole. This interaction produced strong auroras at both magnetic poles, which were detected by the spacecraft in the ultraviolet (■ Figure 24-8). In the years since, Uranus has moved farther around its orbit, so the geometry

■ Figure 24-8

Auroras on Uranus were detected in the ultraviolet by the Voyager 2 spacecraft when it flew by in 1986. These maps of opposite sides of Uranus show the location of auroras near the magnetic poles. Recall that the magnetic field is highly inclined and offset from the planet's center, so the magnetic poles do not lie near the poles defined by rotation. In fact, the north magnetic pole (right) falls near the planet's equator. Also recall that Uranus is highly inclined and that in 1986 its south pole, located at the bottom of these diagrams, was pointed nearly at the sun. The white dashed line marks zero longitude. (Courtesy Floyd Herbert, LPL)

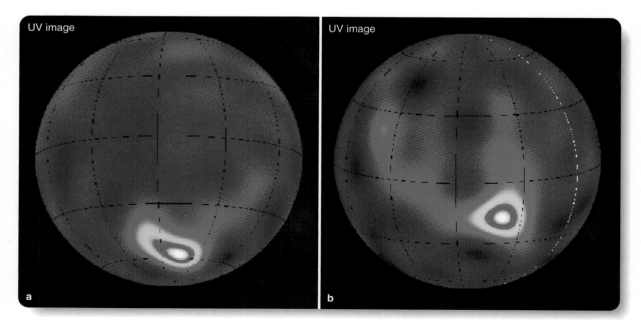

of its interaction with the solar wind is now different. Unfortunately, no spacecraft orbit Uranus at present, and there is no way to observe this interaction in detail.

The temperature of Uranus can reveal something about its interior. Jupiter and Saturn are warmer than you would expect, given the amount of energy they receive from the sun, and that must mean that heat is leaking out from their hot interiors. Uranus, in contrast, is about the temperature you would expect for a world at its distance from the sun. Apparently, it has lost much of its interior heat. Yet it must have some internal heat to cause convection in the fluid mantle and drive the dynamo effect. The decay of natural radioactive elements would generate heat, but some astronomers have suggested that the slow settling of heavier elements through the fluid mantle could release sufficient energy to warm the interior.

Laboratory studies of methane show that it can break down under the temperature and pressure found inside Uranus and form various compounds plus pure carbon in the form of diamonds. If this happens in Uranus, the diamond crystals would fall inward, warming the interior through friction. Whatever diamonds might exist in Uranus are forever beyond reach, but some of Uranus's internal heat may come from its diamond-making layers.

As you explored the atmosphere and interior of Uranus, you may have been a little disappointed. Uranus seems to be a small, modest, Jovian world. But now you are ready to visit one of its best features—its rings.

Ace◔Astronomy™ Log into AceAstronomy and select this chapter to see Astronomy Exercise "Auroras." Review the factors that affect auroras on Earth.

Ace◔Astronomy™ Log into AceAstronomy and select this chapter to see Astronomy Exercise "Convection and Magnetic Fields." How does convection in a planet produce a magnetic field?

The Rings of Uranus

Both Uranus and Neptune have rings that are more like those of Jupiter than those of Saturn. They are dark, faint, and confined by shepherd satellites. They are not easily visible from Earth, and the first hint that these planets had rings came from **occultations;** an occultation is the passage of a planet in front of a star. Nevertheless, most of what is known about these ring systems comes from observations by the Voyager 2 spacecraft.

Study **The Rings of Uranus and Neptune** on pages 598–599 and notice three important points:

1 The rings of Uranus were discovered during an occultation when Uranus crossed in front of a star.

2 The rings of Uranus are very dark, contain little dust, and are confined by small moons.

3 The rings of Uranus cannot survive for long periods, so the rings need to be resupplied with material from impacts on

moons. When you read about Neptune's rings later in this chapter, you will return to this artwork and see how closely the two ring systems compare.

As you read about planetary rings, notice their close relationship with moons. Because of collisions among ring particles, planetary rings tend to spread outward almost like an expanding gas. If a planet had no moons, its rings would spread out into a more and more tenuous sheet until they were gone. The spreading rings can be anchored by small moons, which get pushed outward very slowly. Through orbital resonances, those small moons can be anchored by larger, more distant moons that are so massive they do not get pushed outward significantly. In this way, a system of moons can confine and preserve a system of planetary rings.

Planetary rings are beautiful blossoms created when debris falls into a system of small moons and becomes trapped in the most stable orbits among the moons. Without moons, there would be no rings.

The Moons of Uranus

Uranus has five large moons that were known from Earth-based observations. Those five moons (■ Figure 24-9), from the outermost inward, are Oberon, Titania, Umbriel, Ariel, and Miranda.

■ Figure 24-9

This infrared image shows Uranus, its disk, and its five major satellites. Because of photographic effects, the images of the satellites are much larger than the moons themselves. The largest, Titania, is nearly 32 times smaller in diameter than Uranus. Faint points of light to the left and right of the rings are the smaller moons Portia and Puck. Other faint points of light in this image are background stars. (ESO)

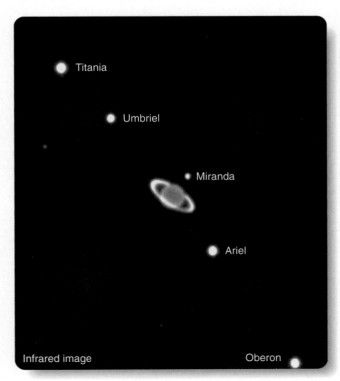

The Rings of Uranus and Neptune

1   The rings of Uranus were discovered in 1977, when Uranus crossed in front of a star. During this occultation, astronomers saw the star dim a number of times before and again after the planet crossed over the star. The dips in brightness were caused by rings circling Uranus.

More rings were discovered by Voyager 2. The rings are identified in different ways depending on when and how they were discovered.

Notice the eccentricity of the ε ring. It lies at different distances on opposite sides of the planet.

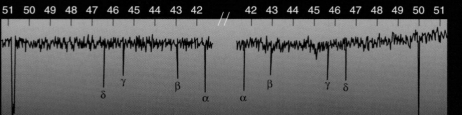

Minutes from midoccultation

51 50 49 48 47 46 45 44 43 42   42 43 44 45 46 47 48 49 50 51

Intensity

δ   γ   β   α   α   β   γ   δ

ε   ε

51,000 49,000 47,000 45,000   45,000 47,000 49,000 51,000

Distance from center of Uranus (km)

2   The albedo of the ring particles is only about 0.015, darker than lumps of coal. If the ring particles are made of methane-rich ices, radiation from the planet's radiation belts could break the methane down to release carbon and darken the ices. The same process may darken the icy surface of Uranian moons.

The narrowness of the rings suggests they are shepherded by small moons. Voyager 2 found Ophelia and Cordelia shepherding the ε ring. Other small moons must be shepherding the other narrow rings. Such moons must be structurally strong to hold themselves together inside the planet's Roche limit.

ε
λ
η γ δ
β
α
6 5 4
1986
U2R

Ophelia

Cordelia

The eccentricity of the ε ring is apparently caused by the eccentric orbits of Ophelia and Cordelia.

2a   When the Voyager 2 spacecraft looked back at the rings illuminated from behind by the sun, the rings were not bright. That is, the rings are not bright in forward-scattered light. That means they must not contain much dust. The nine main rings contain particles no smaller than meter-sized boulders.

Uranus

3   Ring particles don't last forever as they collide with each other and are exposed to radiation. The rings of Uranus may need to be resupplied with fresh particles occasionally as impacts on icy moons scatter icy debris.

Collisions among ring particles produce dust, which is thinly scattered inward from the λ ring, but the high, tenuous atmosphere of Uranus is slowing the dust particles and making them fall into the planet. The rings of Uranus actually contain very little dust.

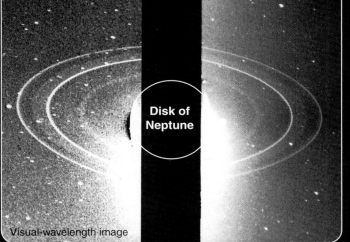

Disk of Neptune

Visual-wavelength image

The rings of Neptune are bright in forward-scattered light, as in the image above, and that indicates that the rings contain significant amounts of dust. The ring particles are as dark as those that circle Uranus, however, so they probably contain methane-rich ice darkened by radiation.

4 The brightness of Neptune is hidden behind the black bar in this Voyager 2 image. Two narrow rings are visible, and a wider, fainter ring lies closer to the planet. More ring material is visible between the two narrow rings.

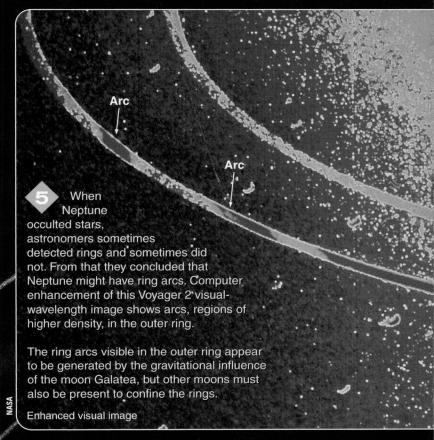

Arc

Arc

5 When Neptune occulted stars, astronomers sometimes detected rings and sometimes did not. From that they concluded that Neptune might have ring arcs. Computer enhancement of this Voyager 2 visual-wavelength image shows arcs, regions of higher density, in the outer ring.

The ring arcs visible in the outer ring appear to be generated by the gravitational influence of the moon Galatea, but other moons must also be present to confine the rings.

Enhanced visual image

NASA

4a Neptune's rings lie in the plane of the planet's equator and inside the Roche limit. The narrowness of the rings suggests that shepherd moons must confine them, and a few such moons have been found among the rings. There must be more small moons to confine the rings.

Naiad

Galetea

Thalassa

Adams

LeVerrier

Despina

Galle

4c Like the rings of Uranus, the ring particles that orbit Neptune cannot have survived since the formation of the planet. Occasional impacts on Neptune's moons must scatter debris and resupply the rings with fresh particles.

4b Neptune's rings have been given names associated with the planet's history. English astronomer Adams and French astronomer LeVerrier predicted the existence of Neptune from the motion of Uranus. The German astronomer Galle discovered the planet in 1846 based on LeVerrier's prediction.

Neptune

Umbriel and Ariel are names from Alexander Pope's *The Rape of the Lock,* and the rest are from Shakespeare's *A Midsummer Night's Dream* and *The Tempest* (an Ariel also appears in *The Tempest*). Spectra show that they contain frozen water, and their surfaces are very dark, so planetary scientists assumed they were made of dirty ices. Little more was known of the moons before Voyager 2 flew through the system.

The Voyager 2 cameras discovered ten moons too small to have been seen from Earth. Also, with the construction of new-generation telescopes and new imaging techniques, astronomers have found even more small moons orbiting Uranus, and there are almost certainly more still to be found.

The smaller moons are all as dark as coal. They are probably icy worlds with surfaces that have been darkened by impacts. Also they orbit inside the radiation belts, and the radiation may convert methane trapped in their ices into dark carbon deposits and further darken their surfaces.

The five large moons are all tidally locked to Uranus, and that means their south poles were pointed toward the sun in 1986; it was these hemispheres that were photographed by Voyager 2. Their densities suggest that they contain relatively large rock cores surrounded by icy mantles (■ Figure 24-10).

Oberon, the outermost of the large moons, has a dark, cratered surface, but it was once an active moon (■ Figure 24-11). A large fault crosses the sunlit hemisphere, and dark material, perhaps dirty water "lava," appears to have flooded the floors of some craters.

Titania is the largest of the five moons and has a heavily cratered surface, but it lacks the largest craters (Figure 24-11). This suggests that soon after it formed, after most of the large debris had been swept up, Titania underwent an active phase in which internal melting flooded the surface with water and erased the early craters. Since then, more craters have formed, but few are as large as those that were erased. Another clue to Titania's past activity is the network of faults that crosses its surface.

Umbriel, the third moon inward, is a very dark, cratered world with no sign of faults or surface activity (Figure 24-11). It is the darkest of the moons, with an albedo of only 0.16 compared with 0.25 to 0.45 for the other moons. Planetary astronomers suspect that Umbriel's crust is a dark mixture of rock and ice. A bright crater floor in one region suggests that some clean ice may lie at shallow depths in some regions.

Ariel has the brightest surface of the five major satellites and shows clear signs of geological activity. It is crossed by faults over 10 km deep, and some regions appear to have been smoothed by resurfacing (Figure 24-11). Crater counts show that the smoothed regions are younger than the other regions. Apparently, past melting in the interior has caused geological activity on the moon's surface. Ariel may have been subject to tidal heating caused by an orbital resonance with Miranda and Umbriel. If you could visit the surface of Ariel, the boots of your spacesuit would probably kick up nothing more than dirty ices; the interesting features would be the broad valleys and faults.

Miranda is a mysterious moon. As you can see in Figure 24-11, it is the smallest of the five large moons, but it appears to have been the most active. In fact, its active past appears to have been quite unusual. One astronomer referred to it as a "moon designed by a committee." Miranda is marked by oval patterns of grooves known as **ovoids** (■ Figure 24-12). These features were originally thought to have been produced when Miranda was shattered by a major impact and the mutual gravity of the fragments pulled them back together to reassemble the little moon. More recent studies of the ovoids show that they are associated with faults, ice-lava flows, and rotated blocks of crust. These features suggest that a major impact was not involved. Rather, large convection currents in Miranda's icy mantle may have forced material upward to create the ovoids.

Certainly, Miranda has had a violent past. Near the equator, a huge cliff rises 20 km. If you stood in your spacesuit at the top of the cliff and dropped a rock over the edge, it would fall for 10 minutes before hitting the bottom. The cliff and the ovoids are old, however. Miranda is no longer geologically active, and you can read the history of its past on its disturbed surface.

The peculiar geology of Miranda illustrates the problem astronomers face in understanding these icy moons. Has Miranda

■ Figure 24-10

The densities of the five major satellites of Uranus suggest that they contain relatively large rocky cores with mantles of dirty ice. The size of Earth's moon is shown for comparison.

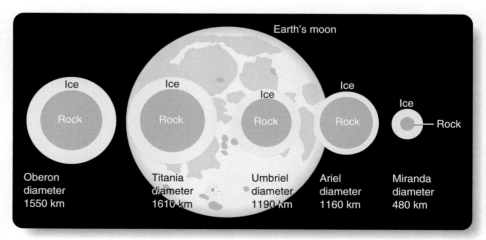

■ Figure 24-11

The five largest moons of Uranus range from the largest, Titania, 46 percent the diameter of Earth's moon, down to Miranda, only 14 percent the diameter of Earth's moon. For a better view of Miranda, see the next figure. The moons are shown to scale here. (Ariel and Miranda: U.S. Geological Survey, Flagstaff, Arizona; Other images: NASA)

been dominated by interior or exterior factors? If the ovoids were caused by convection in Miranda's icy mantle, then heat rising from its interior must be the dominant factor. Miranda is too small to have kept its heat of formation or to have been so severely heated by radioactive decay. Tidal heating could have heated Miranda if its orbit was made slightly elliptical by a resonance with other moons. Your experience with tidal heating in other moons can help you understand how internal factors could have driven slow convection currents rising through Miranda's icy mantle to create its ovoids.

But external factors are also a possibility. Comets sweep through the solar system, and they must occasionally strike the planets and moons. The Jovian planets have no surfaces and would retain no permanent scars of such impacts, as Earth's astronomers saw dramatically in the summer of 1994 when a comet struck Jupiter. The moons in the outer solar system, however, could be battered or even disrupted by such impacts, and a major impact could have broken Miranda up and allowed it to re-form.

Although planetary scientists now prefer internal heat to explain Miranda's geology, you should make a note that external factors such as major impacts do occur. Miranda is a reminder that modern astronomy recognizes the importance of catastrophic events in the history of the solar system. (See Window

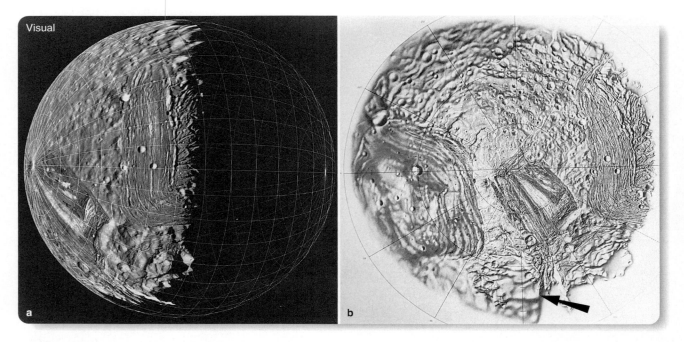

■ Figure 24-12

Miranda, the smallest of the five major moons of Uranus, is only 480 km in diameter, but its surface shows signs of activity. This photomosaic of Voyager 2 images (a) reveals that it is marked by great oval systems of grooves. In the map (b), the smallest features detected are about 1.5 km in diameter. The cliff marked by an arrow near the bottom of the map is 20 km high. (U.S. Geological Survey, Flagstaff, Arizona; map by P. Bridges)

on Science 19-1.) You will see more hints of catastrophic events when you visit the moons of Neptune.

A History of Uranus

The challenge of comparative astronomy is to tell the story of a world, and Uranus may present the biggest challenge of all the objects in our solar system. Not only is it so distant that it is difficult to study, but it is also peculiar in many ways.

Presumably, Uranus formed from the solar nebula, as did the other Jovian planets, but it must have grown more slowly, perhaps because the solar nebula was less dense beyond Saturn. Also, objects orbit slowly in the outer solar system, so protoplanets don't sweep up material rapidly out there. By the time the proto-Uranus had grown massive enough to begin trapping hydrogen, the sun had turned on, and the solar nebula was being blown away. As one astronomer said, "By the time Uranus got all dressed up, the party was over." In fact, some theorists believe Uranus and Neptune could not have grown to their present size in the slow-moving orbits they now occupy. Sophisticated mathematical models of planet formation in the solar nebula suggest that Uranus and Neptune might have formed closer to the sun, in the neighborhood of Jupiter and Saturn. Gravitational interactions among the growing planets could have moved Uranus and Neptune outward to their present locations. This is an exciting hypothesis, and it illustrates how uncertain the histories of the outer planets really are.

The highly inclined axis of Uranus may have originated late in its formation when it was struck by a planetesimal as large as Earth. That impact may have disturbed the interior of the world and caused it to lose much of its heat. Uranus now radiates less than 10 percent more heat than it receives from the sun. There is now just enough heat flowing outward to drive circulation in its slushy mantle and generate its magnetic field.

Impacts have been important in the history of the moons and rings of Uranus. All of the moons are cratered, and some show signs of large impacts. Planetesimals and the nuclei of comets striking the moons may create debris that becomes trapped among the orbits of the smaller moons to produce the narrow rings. The rings could not have lasted since the formation of the planet, so they must be replenished with fresh material now and then. When the final history of Uranus is written, it will surely include some dramatic impact events.

Building Scientific Arguments

How do astronomers know what the interior of Uranus is like?

This argument must combine observation with theory. Obviously, you can't see inside the planet, so planetary scientists are limited to a few basic observations that they must connect with models in a chain of inference to describe the interior. First, the size of Uranus can be found from its angular diameter and distance, and you can find its mass by observing the orbital radii and periods of the orbits of its moons. Its mass divided by its volume equals its density, which implies that, for a Jovian planet, it must contain a higher proportion of dense material such as ice and rock. Add to that the oblateness, rotation period, and chemical composition obtained from spectra, and you have the data needed to build a mathematical model of the interior. Such models predict a core of heavy elements and a mantle of icy slush. The models cannot make the planet as flattened as is observed if the heavy elements are highly concentrated toward the core, so the mantle must contain denser material such as rock as well as ices.

Going from observable properties to unobservable properties is the heart of astronomical research. Now use this approach to build a new argument: **What observational properties of the rings of Uranus show that small moons must orbit among the rings?**

■ ■ ■

Connections: When Voyager 2 departed Uranus, Earth's astronomers were left grappling with new planetary puzzles. Over the following years, some of the puzzles were explained, but in August 1989 Voyager 2 flew past Neptune, and the game began all over again.

(24-2) Neptune

URANUS AND NEPTUNE are often discussed together. They are about the same size and density and are probably very similar planets in some ways. They do differ, however, in certain respects. Neptune has a peculiar ring and a peculiar satellite system. Even the discovery of Neptune was different from the discovery of Uranus.

The Discovery of Neptune

The discovery of Neptune triggered one of the greatest controversies in science. For over a century people told the story and took sides, but only in the last few years has the real story become known. You can follow the story and decide which side you favor.

In 1843, the young English astronomer John Couch Adams (1819–1892) completed his degree in astronomy and immediately began the analysis of one of the great problems of 19th-century astronomy. William Herschel had discovered Uranus in 1781, but earlier astronomers had seen the planet as early as 1690 and had plotted it on charts as a star. When 19th-century astronomers tried to combine this data, they found that no planet obeying Newton's laws of motion and gravity could follow such an orbit.

Some astronomers suggested that the gravitational attraction of an undiscovered planet was causing the discrepancies. Adams began with the variation in the positions of Uranus, never more than 2 minutes of arc, and by October 1845 he had, through a laborious and difficult calculation, computed the orbit of the undiscovered planet. He sent his prediction to the Astronomer Royal, Sir George Airy, who passed it on to an observer who began a painstaking search of the area star by star.

Meanwhile, the French astronomer Urbain Jean Leverrier (1811–1877) made the same calculations and sent his predicted position of the planet to Johann Galle at the Berlin Observatory. Galle received Leverrier's prediction on the afternoon of September 23, 1846, and, after searching for 30 minutes that evening, found Neptune. It was only 1° from the position predicted by Leverrier.

The discovery of a new planet caused a sensation, but English astronomers didn't like a Frenchman getting all the credit. After all, they said, the planet was found only 2° from the position predicted by the orbit computed by their own astronomer, Adams. When the English announced Adams's work, the French suspected that he had plagiarized the calculations, and the controversy was bitter, as controversies between England and France often are. Historians of science have repeated the story for over a century of how the young English astronomer missed his chance because the astronomers in charge of the search were careless and slow.

Original papers related to Adam's calculations were lost for 30 years, but when they were found in 1998, they painted a different picture. Adams did the calculations correctly, but he only computed the orbit for the new planet. He didn't actually calculate its position in the sky; that was left to the English astronomers conducting the search. Once the new planet was discovered, the English astronomers, out of national pride, pressed Adam's case further than they should have.

For 150 years, astronomers told the story and took sides or gave both astronomers equal credit. In fact, it seems clear that Leverrier deserves credit for making an accurate, useful prediction of a position on the sky and pressing it aggressively on an astronomer who had the experience to make the search. But you should give both astronomers credit for solving a difficult problem that was one of the great challenges of the age.

Nevertheless, Leverrier and Adams could have been beaten to the discovery had Galileo paid less attention to Jupiter and more attention to what he saw in the background. Modern studies of Galileo's notebooks show that he saw Neptune on December 24, 1612, and again on January 28, 1613, but he plotted it as a star in the background of drawings of Jupiter. It is interesting to speculate about the response of the Inquisition had Galileo proposed that a planet existed beyond Saturn. Unfortunately for history (perhaps fortunately for Galileo), he did not recognize Neptune as a planet, and its discovery had to wait another 234 years.

The discovery of Neptune has been repeated by historians as a triumph for Newtonian physics: The three laws of motion and the law of gravity had proved sufficient to predict the position and orbit of an unseen planet. Thus, the discovery of Neptune was fundamentally different from the discovery of Uranus—the exis-

Neptune in 1989

Great Dark Spot

Visual-wavelength image
from Voyager 2

1996

2002

Visual-wavelength
images from
Hubble Space Telescope

1998

■ **Figure 24-13**

Because Neptune is inclined almost 29 degrees in its orbit, it experiences seasons that each last about 40 years. Since 1989, spring has come to the southern hemisphere, and the weather has clearly changed, which is surprising because sunlight on Neptune is 900 times dimmer than on Earth. (NASA, L. Sromovsky, and P. Fry, University of Wisconsin–Madison.)

tence of Neptune was predicted from basic laws. But a modern analysis shows that both Leverrier and Adams made unwarranted assumptions about Neptune's distance. By good fortune their assumptions made no difference in the 19th century, and the planet was close to the positions they predicted. Do you think they deserve less credit? They tried, and the other astronomers of the world didn't.

The Atmosphere and Interior of Neptune

Little was known about Neptune before the Voyager 2 spacecraft swept past it in August 1989. Seen from Earth, Neptune is a tiny, blue-green dot never more than 2.3 seconds of arc in diameter. Astronomers knew it was almost four times the diameter of Earth and about 4 percent smaller in diameter than Uranus (**Celestial Profile 10**). Spectra revealed that its blue-green color was caused by methane in its hydrogen-rich atmosphere. Methane absorbs red light, making Neptune look blue-green. Neptune's density showed that it was a Jovian planet rich in hydrogen, but almost no detail was visible from Earth, so even its period of rotation was poorly known.

Voyager 2 passed only 4905 km above Neptune's cloud tops, closer than any spacecraft had ever come to one of the Jovian planets. The Voyager images revealed that Neptune is marked by dramatic belt–zone circulation parallel to the planet's equator. Voyager 2 saw at least four cyclonic disturbances. The largest, dubbed the Great Dark Spot, looked similar to the Great Red Spot on Jupiter (■ Figure 24-13). The Great Dark Spot was located in the southern hemisphere and rotated counterclockwise, with a period of about 16 days. It appeared to be caused by gas rising from the planet's interior. When the Hubble Space Telescope began imaging Neptune in 1994, the Great Dark Spot was gone, and

new cloud features were seen appearing and disappearing in Neptune's atmosphere (Figure 24-13). Evidently, the cyclonic disturbances on Neptune are not nearly as long-lived as Jupiter's Great Red Spot.

The Voyager 2 images reveal other cloud features standing out against the deep blue of the methane-rich atmosphere. The white clouds are made up of crystals of frozen methane and range up to 50 km above the deeper layers, just where the temperature in Neptune's atmosphere is low enough for rising methane to freeze into crystals. (See Figure 24-5.) Presumably these features are related to rising convection currents that form clouds high in Neptune's atmosphere, where they catch sunlight and look bright. Special filters can reveal these bands in visual-wavelength images and at infrared wavelengths (■ Figure 24-14). Observations made by the Hubble Space Telescope suggest that atmospheric activity on Neptune may be related in some way to flares and other eruptions on the sun, but further observations are needed to explore this connection.

As on the other Jovian worlds, high winds circle Neptune parallel to its equator, but the winds blow at very high velocities and tend to blow backward—against the rotation of the planet. Why Neptune should have such high winds is not understood, and it is part of the larger problem of the belt–zone circulation. Now that you have seen at least traces of belts and zones on all four of the Jovian planets (assuming the faint clouds on Uranus are traces of belt–zone circulation), you can ask what drives this circulation. Because the belts and zones remain parallel to the planet's equator even when the planet rotates at a high inclination, as in the case of Uranus, it seems reasonable to believe that

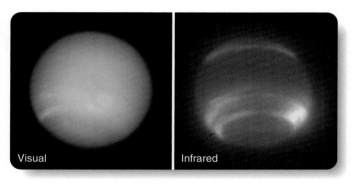

■ Figure 24-14

(a) This visual-wavelength image was made through filters that make belts of methane stand out. (NASA and Erich Karkoschka) (b) The color-corrected infrared image at right was recorded using one of the Keck 10-meter telescopes and adaptive objects. It shows the location of methane cloud belts around the planet. (U. C. Berkeley/W. M. Keck Observatory)

the atmospheric circulation is dominated by the rotation of the planet and not by solar heating.

The same observations that helped define the interior of Uranus can be applied to Neptune. The interior is believed to contain a heavy-element core, smaller than that in Jupiter, surrounded by a deep mantle of slushy ice mixed with heavier material such as rock. Neptune's magnetic field is a bit less than half as strong as Earth's and is tipped 47° from the axis of rotation. It is offset 55 percent of the way to the surface (Figure 24-7). As in the case of Uranus, the field is probably generated by the dynamo effect acting in the conducting fluid mantle.

Neptune has more internal heat than Uranus, and part of that may be generated by radioactive decay in the minerals in its interior. Some of the energy may be released by denser material falling inward, including, as in the case of Uranus, diamond crystals formed by the disruption of methane.

Neptune is a tantalizing world just big enough to be imaged by the Hubble Space Telescope but far enough away to make it difficult to study. The data from Voyager 2 revealed one detail invisible from Earth—rings.

The Rings of Neptune

Astronomers on Earth saw hints of rings when Neptune occulted stars, but the rings were not firmly recognized until Voyager 2 flew past the planet in 1989.

Look again at The Rings of Uranus and Neptune on pages 598–599 and compare the rings of Neptune with those of Uranus. Notice two additional points:

4 Neptune's rings, named after the astronomers involved in the discovery of the planet, are similar to those of Uranus but contain more dust.

5 Also notice a new way that rings can interact with moons; one of Neptune's moons is producing short arcs in the outermost ring.

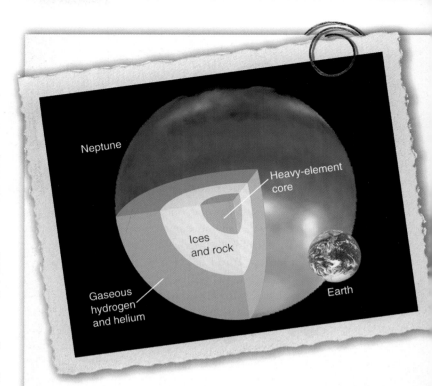

Neptune was tipped slightly away from the sun when the Hubble Space Telescope recorded this image. The interior is much like Uranus's, but there is more heat. (NASA)

Celestial Profile 10: Neptune

Motion:

Average distance from the sun	30.0611 AU (44.971×10^8 km)
Eccentricity of orbit	0.0100
Maximum distance from the sun	30.4 AU (45.4×10^8 km)
Minimum distance from the sun	29.8 AU (44.52×10^8 km)
Inclination of orbit to ecliptic	1°46′27″
Average orbital velocity	5.43 km/s
Orbital period	164.793 y (60,189 days)
Period of rotation	16^h3^m
Inclination of equator to orbit	28°48′

Characteristics:

Equatorial diameter	49,500 km (3.93 $D_\oplus$)
Mass	1.030×10^{26} kg (17.23 $M_\oplus$)
Average density	1.66 g/cm³
Gravity	1.19 Earth gravities
Escape velocity	25 km/s (2.2 $V_\oplus$)
Temperature at cloud tops	−216°C (−357°F)
Albedo	0.35
Oblateness	0.027

Personality Point:

Because the planet Neptune looked so blue, astronomers named it after the Roman god of the sea, Neptune (Poseidon to the Greeks). His wife was Amphitrite, granddaughter of Ocean, one of the Titans. Neptune controlled the storms and waves and was a powerful god not to be trifled with. His three-pronged trident became the symbol for the planet Neptune.

The rings of Neptune resemble the rings of Uranus in one very important way. They can't be primordial. That is, they can't have lasted from the formation of Neptune. Evidently, impacts on moons occasionally scatter debris through the satellite system, and some of it falls into the rings in the most stable places among the orbits of the moons. Planetary rings are constantly being remade.

The Moons of Neptune

Before Voyager 2 visited Neptune, only two moons were known, but Voyager 2 discovered six more small moons, and a few more small moons have been found since using Earth-based telescopes. The two largest moons, Triton and Nereid, have been a puzzle for years because of their peculiar orbits.

Triton has a nearly circular orbit, but it travels backward—clockwise as seen from the north. This makes Triton the only large satellite in the solar system with a retrograde (backward) orbit. The other satellite visible from Earth, Nereid, goes in the right direction, but its orbit is highly elliptical and very large (■ Figure 24-15). Nereid takes 359.4 days to orbit Neptune once. Many astronomers have speculated that the orbits of the two moons are evidence of a violent event. An encounter with a massive planetesimal may have disturbed the moons, or Triton itself may have been captured into orbit during a close encounter with Neptune.

Voyager 2 found six more moons orbiting Neptune, but they all lie close to the planet among the rings. No new moons were found beyond the orbit of Triton, and some astronomers have suggested that Triton, in its retrograde orbit, would have consumed any moons near it.

The Voyager 2 photographs show that Triton is a highly complex moon. Although it is only 1360 km in radius (78 percent the radius of Earth's moon), it is so cold (34.5 K, or −398°F) that it can hold an atmosphere of nitrogen with some methane. Its air

is 10^5 times less dense than Earth's atmosphere. Although a few wisps of haze are visible in the photographs, the atmosphere is transparent, and the surface is easily visible (■ Figure 24-16).

The surface of Triton is evidently composed of ices. The surface ice is dominated by frozen nitrogen with some methane,

■ Figure 24-16

(a) Triton's south pole (bottom) had been in sunlight for 30 years when Voyager 2 flew past in 1989. The frozen nitrogen in the polar cap appears to be vaporizing, perhaps to refreeze in the darkness at the north pole. The dark smudges are produced when liquid nitrogen in the crust vaporizes and drives nitrogen geysers. Methane carried along in the nitrogen is turned to black compounds by sunlight and makes the black smudges. (b) Roughly round basins on Triton may be old impact basins flooded repeatedly by liquids from the interior. Notice the small number of craters on Triton, a clue that it is a partially active world. (NASA)

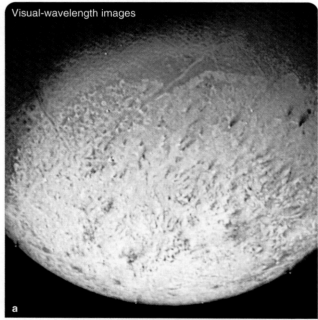

Visual-wavelength images

a

b

■ Figure 24-15

Nereid has a large, elliptical orbit around Neptune, but Triton follows a small, circular, retrograde orbit. At a distance of 30 AU from Earth, Triton is never more than 16 seconds of arc from the center of Neptune.

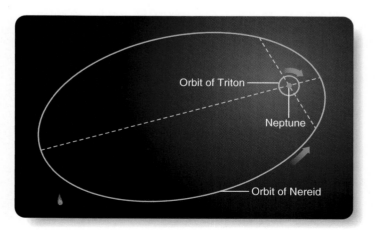

Orbit of Triton

Neptune

Orbit of Nereid

carbon monoxide, and carbon dioxide. This is consistent with the nitrogen-rich atmosphere. Some regions of the surface look dark and may be slightly older terrain that has grown dark as sunlight has converted methane into dark organic compounds. The south pole of Triton had been turned toward the sun for 30 years when Voyager 2 flew past, and deposits of nitrogen frost in a large polar cap appeared to be vaporizing there and refreezing in the darkness of the north pole (Figure 24-16a). The cycle of nitrogen on Triton resembles in some ways the cycle of carbon dioxide on Mars.

The icy surface of Triton suggests that it has been active recently and may still be active. Few craters mark the surface of the icy world, and comparison with the cratering rate you might expect for a moon located so close to the Kuiper Belt suggests the average age of the surface is no more than 100 million years and may be even less. Some process has erased older craters. Long linear features appear to be fractures in an icy crust, and some roughly round basins appear to have been flooded time after time by liquids from the interior (Figure 24-16b). Triton is much too cold for the liquid to be molten rock or even liquid water. Rather, the flooding may have been caused by water containing agents such as ammonia, which would lower the freezing point of the liquid. It isn't possible to say at present whether Triton is still active, but 100 million years isn't very long in the history of a world. Triton may still suffer periodic eruptions and flooding.

Another form of activity on Triton is the dark smudges visible in the bright nitrogen ices near the south pole. These appear to be caused by nitrogen ice beneath the crust. Warmed slightly by the sun, the nitrogen ice can change from one form of nitrogen ice to another and release heat that could vaporize some of the nitrogen. This nitrogen vapor vents through the crust to form nitrogen gas geysers. Sunlight converts some of the methane carried along with the nitrogen into dark organic material, which falls to the surface and accounts for the dark smudges on the ice.

If you want to visit Triton, you should schedule visits to sample the dirty crust contaminated by one of these nitrogen geysers and the material in one of the flooded basins. Mineral samples from those areas would help explain Triton's geological activity, but remember to refrigerate your sample boxes. The minerals there are frozen gases that are not stable at room temperature.

Active worlds must have a source of energy, so you are probably wondering where Triton gets its energy. Triton is big enough to retain some heat from low-level radioactivity in its interior. That may be enough to melt exotic ices and cause flooding on the surface. The nitrogen geysers appear to be powered not from the interior but from sunlight. In fact, Triton is very efficient at absorbing sunlight. Its thin atmosphere allows sunlight to fall on the crust, which is composed of ices that are partially transparent to light. As the light penetrates into the ice and is absorbed, it is converted to heat. The ices, however, are not very transparent to infrared radiation, and the heat is trapped in the crust. In this way, the crust of Triton appears to be heated, in part, by an icy form of the greenhouse effect. The heat could transform nitrogen ice from one form into another form and release more heat to drive the geysers.

This low-level heating would not be enough to erase nearly all of Triton's craters, so some planetary astronomers wonder if Triton might have been captured by Neptune within the last billion years. Tidal forces from such a capture would cause enough tidal heating to melt Triton and totally resurface it. Triton may still be active with residual heat.

The History of Neptune

Can you tell the story of Neptune? In some ways it seems to be a simple world, but its magnetic field is peculiar, and its moons and rings demand careful attention.

Neptune is much like Uranus, and you can assume that it formed from the solar nebula much as did Uranus. It is a mostly liquid world with a dense core. Its hazy atmosphere, marked by changing cloud patterns, hides a mantle of partially frozen ices where astronomers suspect the dynamo effect generates its off-center magnetic field.

Neptune's satellite system suggests a peculiar history. Triton, the largest moon, revolves around Neptune in a retrograde orbit, while Nereid's long-period orbit is highly elliptical. These orbital oddities suggest that the satellite system may have been disturbed by an encounter with some other object, such as a large planetesimal or the capture of Triton. You have seen evidence in other satellite systems of impacts with large objects, so such an event is not unreasonable.

At least six smaller moons orbit Neptune, but they lie near its ring system. Because the rings are bright in forward scattering, you can conclude that they contain some dust, and the shepherding of small satellites must confine their width and produce the observed arcs. Such delicate rings could not have survived since the origin of the planets, so you must suppose that the rings are occasionally supplied with fresh particles generated by collisions between comets and Neptune's moons. Once again, the evidence of major impacts in the solar system's history assures you that such impacts do occur.

Building Scientific Arguments

Why is Neptune blue but its clouds white?

To solve this problem you must build a scientific argument that follows a process step by step. When you look at something, you really turn your eyes toward it and receive light from the object. When you look at Neptune, the light you receive is sunlight that is reflected from various layers of Neptune and journeys to your eyes. Because sunlight contains a distribution of photons of all visible wavelengths, it looks white to human eyes, but sunlight entering Neptune's atmosphere must pass through hydrogen gas that contains a small amount of methane. While hydrogen is transparent, methane is a good absorber of longer wavelengths, so red photons are more likely to be absorbed than blue photons. Once the light is scattered from deeper layers, it must run this methane gauntlet again to emerge from the atmosphere, and again red photons are more likely to be absorbed. The light that finally emerges from Neptune and eventually reaches your eyes is poor in longer wavelengths and thus looks blue.

The methane-ice-crystal clouds lie at high altitudes, so sunlight does not have to penetrate very far into Neptune's atmosphere to reflect off the clouds, and consequently it loses many fewer of its red photons. The clouds look white.

This discussion shows how a careful, step-by-step analysis of a natural process can help you better understand how nature works. For example, build a step-by-step argument to answer the following: **Where does the energy come from to power Triton's surface geysers?**

■   ■   ■

Connections: Neptune was the last planet visited by Voyager 2, which is now journeying out of our solar system. It will not pass near any other worlds unless, millions of years from now, it enters some distant planetary system. Nevertheless, you have one more planet to visit. Turn up the heater in your spacesuit. It is a cold place.

24-3 Pluto

PLUTO, THE FARTHEST OUTPOST of our solar system, is a small, icy world. Discovered through a combination of ingenuity, perseverance, and luck, it is now known to have a satellite. But some astronomers are wondering if it is really a planet.

The Discovery of Pluto

Percival Lowell (1855–1916) was fascinated with the idea that an intelligent race built the canals he thought he could see on Mars (Chapter 22). Lowell founded Lowell Observatory primarily for the study of Mars. Later, perhaps to improve the reputation of his observatory, he began to search for a planet beyond Neptune.

Lowell used the same method that Adams and Leverrier had used to predict the position of Neptune. Working from the observed irregularities in the motion of Neptune, Lowell predicted the location of an undiscovered planet of about 7 Earth masses, which, he concluded, would look like a 13th-magnitude stellar image in eastern Taurus. Lowell searched for the planet photographically until his death in 1916.

In the late 1920s, a 22-year-old amateur astronomer, Clyde Tombaugh, began using a homemade 9-in. telescope to sketch Jupiter and Mars from his family's wheat farm in western Kansas. He sent his drawings to Lowell Observatory, and the observatory director hired him without an interview. The young Tombaugh bought a one-way train ticket for Flagstaff not knowing what his new job would be like.

The observatory director set Tombaugh to work photographing the sky along the ecliptic around the predicted position of the planet. The search technique was a classic method in astronomy. Tombaugh obtained pairs of 14 × 17-inch glass plates exposed two or three days apart. To search a pair of plates, he mounted them in a blink comparator, a machine that allowed him to look through a microscope at a small spot on one plate and then at the flip of a lever see the same spot on the other plate. As he blinked back and forth, the star images did not move, but a planet would have moved along its orbit during the two or three

Visual-wavelength images.

a

b

Pluto was here on the first plate.

days that elapsed before the second plate was exposed. So Tombaugh searched the giant plates, star image by star image, looking for an image that moved. A single pair of plates could contain 400,000 star images. He searched pair after pair and found nothing.

The observatory director turned to other projects, and Tombaugh, working alone, expanded his search to cover the entire ecliptic. For almost a year, Tombaugh exposed plates by night and blinked plates by day. Then on February 18, 1930, nearly a year after he had left Kansas, a quarter of the way through a pair of plates, he found a 15th magnitude image that moved (■ Figure 24-17). He later remembered, "'Oh,' I thought, 'I had better look at my watch; this could be a historic moment.' It was within about 2 minutes of 4 PM [MST]." The discovery was announced on March 13, the 149th anniversary of the discovery of Uranus and the 75th anniversary of the birth of Percival Lowell. The planet was named Pluto after the god of the underworld and, in a way, after Lowell, because the first two letters in Pluto are the initials of Percival Lowell.

The discovery of Pluto seemed a triumph of discovery by prediction, but Tombaugh sensed something was wrong from the first moment he saw the image. It was moving in the right direction by the right amount, but it was 2.5 magnitudes too faint. Clearly, Pluto was not the 7-Earth-mass planet that Lowell had predicted. The faint image implied that Pluto was a small planet with a mass too low to seriously alter the motion of Neptune.

Later analysis has shown that the variations in the motion of Neptune, which Lowell used to predict the location of Pluto, were random uncertainties of observation and could not have led to a trustworthy prediction. The discovery of the new planet only 6° from Lowell's predicted position was apparently an accident, which proves that if you search long enough and know what to expect, you are likely to find something (**Window on Science 24-1**).

Pluto as a World

Pluto is very difficult to observe from Earth. Only a bit larger than 0.1 second of arc in diameter, it is only 65 percent the diameter of Earth's moon and shows little surface detail. The best photos by the Hubble Space Telescope reveal large areas of light and dark terrain (**Celestial Profile 11**), promising that Pluto will reveal itself to be an interesting world when spacecraft finally visit it.

Most planetary orbits in our solar system are nearly circular, but Pluto's is quite elliptical (see Figure 1-7). On the average, it is the most distant planet, but it can come closer to the sun than

■ **Figure 24-17** ◀

Pluto is small and far away, so its image is indistinguishable from that of a star on most photographs. Clyde Tombaugh discovered the planet in 1930 by looking for an object that moved relative to the stars on a pair of photographs taken a few days apart. (Lowell Observatory photographs)

Pluto is a small, low-density world composed of ices and rock. No spacecraft has every visited it, so the best image available shows only bright and dark regions. (NASA)

Celestial Profile 11: Pluto

Motion:

Average distance from the sun	39.44 AU (59.00 × 10⁸ km)
Eccentricity of orbit	0.2484
Maximum distance from the sun	49.24 AU (73.66 × 10⁸ km)
Minimum distance from the sun	29.64 AU (44.34 × 10⁸ km)
Inclination of orbit to ecliptic	17°9'3"
Average orbital velocity	4.73 km/s
Orbital period	247.7 y (90,465 days)
Period of rotation	6ᵈ9ʰ21ᵐ
Inclination of equator to orbit	119.6°

Characteristics:

Equatorial diameter	2370 km (0.19 $D_\oplus$)
Mass	1.2 × 10²² kg (0.002 $M_\oplus$)
Average density	2.0 g/cm³
Surface gravity	0.06 Earth gravity
Escape velocity	1.2 km/s (0.11 $V_\oplus$)
Surface temperature	−230°C (−382°F)
Albedo	0.5

Personality Point:

Pluto (also known as Hades) was the ruler of the underworld. He kidnapped beautiful Persephone, maiden of spring, and made her his wife. But Zeus ruled that she need only spend a third of each year in the underworld; during that time, Earth grieves for her and experiences winter. Pluto possessed a helmet that made its wearer invisible—appropriate for a world so difficult to see.

Scientific Discoveries

In 1928, Alexander Fleming noticed that bacteria in a culture dish were avoiding a spot of mold. He went on to discover penicillin. In 1895, Conrad Roentgen noticed a fluorescent screen glowing in his laboratory when he experimented with other equipment. He discovered X rays. In 1896, Henri Becquerel happened to place a mineral containing uranium on a photographic plate safely wrapped in black paper. The plate was fogged, and Becquerel discovered natural radioactivity. Like many discoveries in science, these seem to be accidental; but, as you have seen in the case of the discovery of Uranus, Neptune, and Pluto, "accidental" doesn't quite describe what happened.

The most important discoveries in science are those that totally change the way people think about nature, and it is very unlikely that anyone would predict such discoveries. For the most part, scientists work within a paradigm (Window on Science 4-1), a set of models, theories, and expectations about nature, and it is very difficult to imagine natural events that lie beyond that paradigm. Ptolemy, for example, could not have imagined a universe filled with galaxies because they lie totally beyond his geocentric paradigm. That means that the most important discoveries in science are almost bound to arise unexpectedly.

An unexpected discovery, however, is not the same as an accidental discovery. Fleming discovered penicillin in his culture dish not because he was the first to see it, but because he had studied bacterial growth for many years; so, when he saw what many others must have seen, he recognized it as important. Roentgen recognized the importance of the glowing screen in his lab, and Becquerel didn't simply discard the fogged photographic plate. Long years of experience prepared them to recognize the significance of what they saw.

A study of discoveries made with new telescopes from Galileo to the present revealed that each time astronomers built a new telescope that significantly surpassed existing telescopes, they discovered new and wonderful things, such as the moons of Jupiter, binary stars, spiral galaxies, and quasars. But nearly all the discoveries made with these telescopes were unpredicted.

While it seemed possible to predict the presence of Neptune from the orbital motion of Uranus, this is not typical of scientific discoveries. Most important scientific discoveries lie

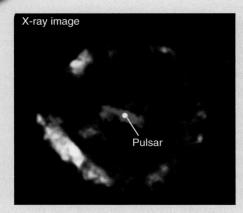

X-ray image

Pulsar

The discovery of pulsars, spinning neutron stars, was totally unexpected, as important scientific discoveries often are. (NASA/McGill/V. Kaspi et al.)

outside the current paradigm and so must be made by scientists who are searching for something else or studying some other phenomenon. That means that scientists pursuing true basic research are rarely able to explain the potential value of their work, just as astronomers can't predict what they will discover with a new telescope. That's why science is so exciting.

■ **Figure 24-18**

(a) A high-quality ground-based photo shows Pluto and its moon, Charon, badly blurred by seeing. (NASA) (b) The Hubble Space Telescope image clearly separates the planet and its moon and allows more accurate measurements of the position of the moon. (R. Albrecht, ESA/ESO Space Telescope European Coordinating Facility, NASA)

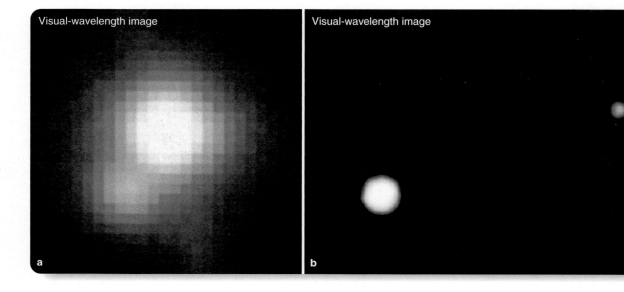

Visual-wavelength image

Visual-wavelength image

a

b

Neptune. In fact, from January 21, 1979, to March 14, 1999, Pluto was closer to the sun than Neptune. The planets will never collide, however, because Pluto's orbit is inclined 17°.

If you land on the surface of Pluto, your spacesuit will have to work hard to keep you warm. Orbiting so far from the sun, Pluto is cold enough to freeze most compounds that you think of as gases, and spectroscopic observations have found evidence of solid nitrogen ice on its surface with traces of frozen methane and carbon monoxide. The daytime temperature of about 50 K (−370° F) is enough to vaporize some of the nitrogen and carbon monoxide and a little of the methane to form a thin atmosphere around Pluto. This atmosphere was detected in 1988 when Pluto passed in front of (occulted) a distant star, and the starlight faded gradually rather than winking out suddenly.

Pluto's largest moon was discovered on photographs in 1978. It is very faint and about half the diameter of Pluto (■ Figure 24-18). The moon was named Charon after the mythological ferryman who transports souls across the river Styx into the underworld. Two small moons were found in 2005.

The discovery of Charon was important for a number of reasons. Charon orbits Pluto in a nearly circular orbit in the plane of Pluto's equator. Observations show that the moon and Pluto are tidally locked to each other, and that Pluto's axis of rotation is highly inclined (■ Figure 24-19). Furthermore, the orbital motion of Charon allows the calculation of the mass of Pluto. Charon orbits 19,640 km from Pluto with an orbital period of 6.387 days. Kepler's third law reveals that the mass of the system is 6.5×10^{-9} solar masses or about 0.002 Earth mass. Most of this mass is Pluto, which seems to be about 12 times more massive than Charon.

The mass of a body is important in astronomy because mass divided by volume is density. The density of Pluto is about 2 g/cm^3, and the density of Charon is just a bit less. Pluto and Charon must contain about 35 percent ice and 65 percent rock. Spectra of Charon show that the small moon has a surface that is mostly water ice. Perhaps it has lost its more volatile compounds because of its lower escape velocity. Water ice at Charon's surface temperature is no more volatile than is a piece of rock on Earth, so a water ice surface could last a long time.

Both Pluto and Charon go through dramatic seasons much like those on Uranus as they circle the sun in their highly inclined orbit. This may cause changes in Pluto's atmosphere, as it grows warmer when it is closest to the sun, as it was in the late 1980s, and then freezes as it draws away.

Although NASA is currently developing a mission to Pluto, no spacecraft has visited the little world. You must strain your imagination and try to visualize the surface of Pluto. Old surfaces should be cratered, so you might imagine an icy, cratered world. The tidal lock between Pluto and Charon suggests an age of tidal heating that may have driven geological activity and left traces on the surfaces of these worlds. However, the seasonal vaporization and refreezing of ices may have erased or covered much of the

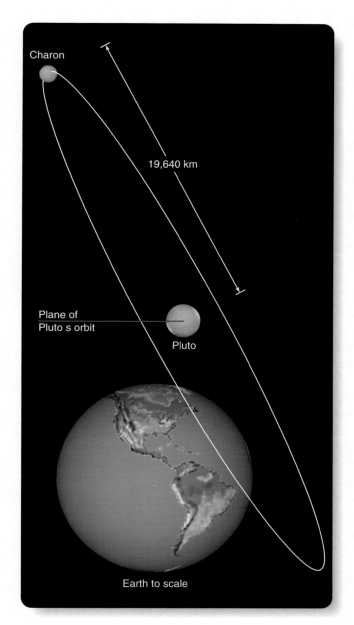

■ **Figure 24-19**

The nearly circular orbit of Charon is only a few times bigger than Earth. It is seen here nearly edge-on, and consequently it looks elliptical in this diagram. The orbit and the equator of Pluto are inclined 119.6° to the plane of Pluto's orbit around the sun.

older surface with ices. Thus Pluto and Charon may be worlds unlike any you have visited before. Yet the biggest puzzle may be not their geology, but their origin.

Pluto and the Plutinos

No, this section is not about a 1950s rock-and-roll band. It is about the history of Pluto, and it will take you back billions of years to watch the outer planets form. Pluto doesn't fit with the Jovian planets found in the outer solar system, and some astronomers have argued that Pluto is not a planet. As an expert on planets, you can vote; but, before you do, you should eliminate an old idea and review some new discoveries.

You should begin by eliminating that old idea. As soon as Pluto was discovered, astronomers realized that it was much smaller than its Jovian neighbors. Some suggested that it was a moon of Neptune that had escaped. After all, their orbits actually cross, although they will never collide because Pluto's orbit is so highly inclined. That escaped moon theory has been totally abandoned. It is hard to understand how Pluto could have escaped from Neptune and reached its present orbit. And how did it get a moon, if it was once a moon itself? Modern astronomers have a much better theory, and it links Pluto to the Kuiper Belt objects, the small icy bodies that orbit in the outer solar system beyond Neptune and Pluto.

Some of the Kuiper Belt objects are quite large. Two large objects named Sedna and 2004DW are about 63 percent the size of Pluto. Another object called Quaoarh (kwah-o-wahr) is 50 percent the diameter of Pluto. In 2005, an object roughly 50 percent bigger than Pluto was fund in a highly inclined, elliptical orbit with a period of 650 years. So at least one Kuiper Belt object is bigger than Pluto, and more large ones may be found. A few of these objects have moons of their own, or perhaps you should say that they are binaries with two icy bodies orbiting each other. In that way, they resemble Pluto and its big moon Charon.

Some Kuiper Belt objects have orbits similar to Pluto's. Over a dozen are known that are caught in a 3:2 resonance with Neptune. That is, they orbit the sun twice while Neptune orbits three times. You learned about orbital resonances when you studied Jupiter's Galilean moons. Because Pluto is also caught in the same 3:2 resonance, these Kuiper Belt objects have been named **plutinos.** Many other Kuiper Belt objects are caught in other resonances with Neptune.

If Pluto is related to the plutinos, then Pluto is related to the Kuiper Belt objects. They formed in the outer solar nebula, but how did they get caught in resonances with Neptune? Sophisticated models of the formation of the planets suggest that Uranus and Neptune may have formed closer to the sun, where the solar nebula was denser. Sometime later, gravitational interactions with Jupiter and Saturn could have gradually shifted the two ice giants outward, and, as Neptune migrated outward, its orbital resonances could have swept up icy bodies like strange nets pushed in front of a fishing boat.

Some icy Kuiper Belt objects may have been captured as moons, and astronomers have wondered if moons such as Triton and Charon could have been Kuiper Belt objects. Could Pluto have started life as a Kuiper Belt object and gotten swept up by a Neptune resonance?

Now you can vote: Is Pluto a planet? Is it the first object discovered in the Kuiper Belt? Whatever the final outcome, Pluto and the plutinos have rocked modern astronomers and forced them to reconsider their ideas on the origin of the outer worlds. As more Kuiper Belt objects are found beyond the orbit of Neptune, astronomers will see more clearly how the outer solar nebula formed planets.

Building Scientific Arguments

What evidence can you point to that cataclysmic impacts have occurred in our solar system?
To build this argument you must clearly distinguish between evidence and theory. You would expect objects that formed together from the solar nebula to rotate and revolve in the plane of the solar system, so the high inclination of Pluto and the orbit of Charon suggest a past collision or, at least, a close gravitational interaction with a passing body. And the Pluto–Charon system is not the only evidence for major impacts in the past. The peculiar orbits of Neptune's moons Triton and Nereid, the fractured and cratered state of the Jovian satellites, and the peculiar rotation of Uranus all hint that impacts and encounters with large planetesimals or the nuclei of comets have been important in the history of these worlds. Furthermore, the existence of planetary rings suggests that impacts have scattered particles into stable orbits among the small inner moons and replenished the ring systems. Even in the inner solar system, the backward rotation of Venus, the density of Mercury, and the formation of Earth's moon are seen as possible consequences of major impacts when the solar system was young.

Certainly, astronomers do not imagine that planets bounce around like billiard balls, but they do recognize that the planets are not totally isolated and may have been victims (or beneficiaries) of large impacts as they formed long ago. For example, build a new argument: **How do the size, orbits, and composition of Pluto and Charon suggest they originated in the Kuiper Belt?**

■　　■　　■

Connections: Like clues at the scene of a crime, the evidence in the outer solar system warns that all of the planets in our solar system have been and still are sitting ducks for impacts by the smaller objects that hurtle along orbits around the sun. What are these objects? What can they say about the origin and evolution of our solar system? You will explore these questions in the next chapter.

Summary

24-1 | Uranus

How is Uranus different from Jupiter?

■ Uranus was discovered by William Herschel in 1781.

■ Although Uranus is a Jovian planet, it is significantly smaller and less massive than Jupiter. Uranus is an ice giant planet about four times the diameter of Earth. It rotates on its side and has extreme seasons.

■ The atmosphere of Uranus is mostly hydrogen and helium with some methane, which absorbs longer-wavelength photons and gives the planet a blue color.

■ The atmosphere is so cold that only methane clouds are visible, and some traces of belt–zone circulation can be seen. As spring has come to the northern hemisphere, more cloud features have developed, suggesting that Uranus follows a seasonal cycle and is not always as bland as it was when Voyager 2 flew past in 1986.

■ Uranus has a small core of dense matter and a deep slushy mantle of ice, water, and rock. Convection in the mantle may produce its highly inclined offset magnetic field.

■ Uranus emits about the right amount of heat for a planet at its distance from the sun, which suggests that it is not extremely hot inside. The settling of heavy material, including diamonds, formed from the carbon in methane could generate some heat.

How did Uranus, its rings, and its moons form and evolve?

■ The rings of Uranus are boulder-size objects darkened by radiation and trapped among the orbits of small moons. They make up a layer one particle thick. The ring material was probably produced by impacts on icy moons.

■ The five large moons of Uranus appear to be icy and old, although some show signs of geological activity. Miranda appears to have suffered from dramatic activity. Tidal heating is a likely source of energy.

■ Uranus appears to have formed slowly and never became massive enough to trap large amounts of hydrogen and helium from the solar nebula. It and Neptune may have formed closer to the sun and were moved outward by gravitational interactions with Jupiter and Saturn.

■ An impact by a large planetesimal as Uranus was forming may have caused the planet to rotate around a highly inclined axis.

24-2 | Neptune

How is Neptune different from Uranus?

■ Neptune was discovered in 1846 based on a position computed from irregularities in the orbital motion of Uranus.

■ Neptune is an ice giant only slightly smaller than Uranus. It has a dense core; a mantle of ice, water, and rock; and an atmosphere of hydrogen and helium with traces of methane, which gives the planet a blue color.

■ Methane clouds come and go in the cold atmosphere and appear to follow a belt–zone circulation.

■ Circulation in the liquid mantle gives rise to the planet's magnetic field.

■ Neptune has more internal heat than Uranus, and the sinking of dense material, including diamonds, may be adding to the internal heat.

How did Neptune, its rings, and its moons form and evolve?

■ The rings of Neptune probably formed when impacts on moons scattered icy debris into stable places among the orbits of the small moons. The rings contain more dust than those of Uranus. Arcs in the rings appear to be caused by the gravitational influence of a small moon.

■ Small moons are probably captured objects. Triton, the largest moon, follows a retrograde orbit, and Nereid follows a long elliptical orbit. These two moons may have been disturbed by the gravitational influence of a massive planetesimal or by the capture of Triton.

■ Triton has an icy surface and a thin atmosphere of nitrogen. The lack of many craters and the presence of cracks and faults suggest that the moon may still be active. Sunlight appears to trigger nitrogen geysers in the crust. Radioactive decay and tidal heating during a close approach to Neptune when it was captured into orbit are other possible sources of heat.

■ Neptune formed slowly, as did Uranus, and never accumulated a deep atmosphere of hydrogen and helium before the solar nebula was blown away.

■ Neptune may have formed closer to the sun and was moved outward by interactions with Jupiter and Saturn.

24-3 | Pluto

Is Pluto a planet or something else?

■ Pluto was discovered in 1930 during a search for a large planet orbiting beyond Neptune and disturbing Neptune's orbital motion. Pluto is much too small to affect Neptune.

■ No spacecraft has visited Pluto, and it is too small and too far away to study in detail from Earth.

■ Pluto has a frigid crust of solid nitrogen ice with traces of frozen methane and carbon monoxide. Its thin atmosphere is mostly nitrogen.

■ Pluto's moon Charon orbits in a highly inclined orbit, and it and Pluto are tidally locked to face each other. That means Pluto rotates on its side much like Uranus.

■ The density of Pluto shows that it and Charon must contain about two-thirds rock and one-third ices.

■ Pluto and Charon resemble the plutinos, Kuiper Belt objects caught in a resonance with the orbital motion of Neptune. These plutinos could have been swept up as Neptune moved outward in the solar system, and some astronomers suggest that Pluto is not a planet but is a large Kuiper Belt object.

New Terms

occultation (p. 597)

ovoid (p. 600)

plutino (p. 612)

Review Questions

Ace ⑤Astronomy™ Assess your understanding of this chapter's topics with additional quizzing and animations at http://ace .brookscole.com/sf9

1. Describe the location of the equinoxes and solstices in the Uranian sky. What are seasons like on Uranus?

2. Why is belt–zone circulation difficult to detect on Uranus?

3. Discuss the origin of the rings of Uranus and Neptune. Cite evidence to support hypotheses.

4. How do the magnetic fields of Uranus and Neptune suggest that the mantles inside those planets are fluid?

5. If Neptune had no satellites at all, would you expect it to have rings? Why or why not?

6. Why might the surface brightness of ring particles and small moons orbiting Uranus and Neptune depend on whether those planets have magnetic fields?

7. Both Uranus and Neptune have a blue-green tint when observed through a telescope. What does that tell you about their composition?

8. How can small worlds like Triton and Pluto have atmospheres when a larger world such as Ganymede has none?

9. Why do you suspect that Triton has had an active past? What source of energy could power such activity?

10. If you visited the surface of Pluto and found Charon a full moon directly overhead, where would Charon be in the sky when it was new? when it was first quarter?

11. What evidence can you cite that Pluto and Charon are made of mixtures of rock and ice?

12. Why might you suspect that Pluto is a Kuiper Belt object?

13. Two images of Uranus show it as it would look to the eye and through a red filter that enhances methane clouds in the northern hemisphere. Why didn't Voyager 2 photograph the northern hemisphere? What do the visible atmospheric features tell you about circulation on Uranus?

(NASA and Heidi Hammel)

Discussion Questions

1. Why might it be unfair to describe William Herschel's discovery of Uranus as accidental? Why might it be unfair to describe the discovery of the rings of Uranus as accidental?

2. Suggest a single phenomenon that could explain the inclination of the rotation axis of Uranus, the orbits of Neptune's satellites, and the existence of Pluto's moon.

Problems

1. What is the maximum angular diameter of Uranus as seen from Earth? of Neptune? of Pluto? (*Hint:* Use the small-angle formula.)

2. One way to recognize a distant planet is by its motion along its orbit. If Uranus circles the sun in 84 years, how many seconds of arc will it move in 24 hours? (This does not include the motion of Earth. Assume a circular orbit for Uranus.)

3. What is the orbital velocity of Miranda around Uranus?

4. What is the escape velocity from the surface of Miranda? (*Hints:* Radius = 242 km. Assume that the density is 2 g/cm^3. See Chapter 5.)

5. The magnetosphere of Uranus rotates with the planetary interior in 17.24 hours. What is the velocity of the outer portion of the magnetic field just beyond the orbit of Oberon? (*Hint:* The circumference of a circle is $2\pi r$.)

6. If the ε ring is 60 km wide and the orbital velocity of Uranus is 6.81 km/s, how long a blink should you expect to see when the ring crosses in front of a star? Is this consistent with the data on page 598?

7. What is the escape velocity from the surface of an icy moon with a diameter of 20 km? (*Hints:* The density of ice is 1 g/cm^3. The volume of a sphere is $\frac{4}{3}\pi R^3$. See Chapter 5.)

8. What is the difference in the orbital velocities of the two shepherd satellites Cordelia and Ophelia? (*Hints:* Orbital radii = 49,800 km and 53,800 km. See Chapter 5.)

9. Repeat Problem 2 for Pluto. That is, ignoring the motion of Earth, how far across the sky would Pluto move in 24 hours? (Assume a circular orbit for Pluto.)

10. Given the size of Triton's orbit ($R = 355,000$ km) and its orbital period ($P = 5.877$ days), calculate the mass of Neptune. (*Hint:* See Chapter 5.)

Media Cluster

ACTIVE FIGURES

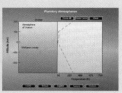

Planetary Atmospheres
Experiment with the temperature throughout the atmospheres of several planets with this animation.

Uranus's Ring Detection
See how the rings of Uranus were first discovered during an occultation.

ASTRONOMY EXERCISES

Auroras
Use this simulation to study the formation of auroras. You can vary the strength of Earth's magnetic field and the speed and number of particles in the solar wind.

Convection and Magnetic Fields
A planet's core temperature and rotation speed are factors in the strength of its magnetic field. See the effect of these variables in this animation.

VIRTUAL ASTRONOMY LABS

Lab 7: Planetary Atmospheres and Their Retention
This lab investigates the retention of atmospheres. You will explore the factors that govern the loss of atmospheric gases, and you will see why certain bodies can retain some gases but not others.

Critical Inquiries for the Web

1. Who was William Herschel? What can you find out about his life after he discovered Uranus?

2. What is weather on Uranus like now? Can you locate recent photos of cloud patterns there?

3. How many moons does Uranus have? More are discovered every year, it seems, so search for the latest discoveries.

4. What happened to John Couch Adams and Urbain Jean Leverrier after the discovery of Neptune? Did they make other important discoveries? What is the Adams Society? Can you find out what Leverrier did in the revolution of 1848?

5. Repeat Inquiries 2 and 3 for Neptune. Remember that finding new moons orbiting Neptune is difficult because it is farther away.

6. Who was Clyde Tombaugh? What can you find out about his life after he discovered Pluto?

Go to the Brooks/Cole Astronomy Resource Center **(http:// astronomy.brookscole.com)** for critical thinking exercises, articles, and additional readings from InfoTrac College Edition, Brooks/Cole's online student library.

25 | Meteorites, Asteroids, and Comets

Y OU ARE NOT AFRAID of comets, of course; but not long ago, people viewed them with terror. In 1910, Comet Halley was spectacular. On the night of May 19, Earth actually passed through the tail of the comet—and millions of people panicked. The spectrographic discovery of cyanide gas in the tails of comets led many to believe that life on Earth would end. Householders in Chicago stuffed rags around doors and windows to keep out the gas, and bottled oxygen was sold out. Con artists in Texas sold comet pills and inhalers to ward off the noxious fumes. An Oklahoma newspaper reported (in what was apparently a hoax) that a religious sect tried to sacrifice a virgin to the comet. ▮ Throughout history, bright comets have been seen as portents of doom. Even in our own time, the appearance of bright comets has generated predictions

▮ Continued on page 618 ▮

Comets can be terrifying to the superstitious, but they are dramatically beautiful and carry clues to the origin of the solar system. The Stardust spacecraft flew through the dust and gas spewing from comet Wild 2 in 2004. (NASA/JPL)

Guidepost

Looking Back

Everything has to have come from somewhere, and you began wondering where the solar system came from back in Chapter 19. You have now visited all the planets and many of the moons in the solar system, gathering evidence and making observations. But you found that the planets and moons did not offer many clues to the formation of the solar system because they have been heavily altered by cratering and by heat flowing out of their interiors. You will find the smaller objects in the solar system less altered.

This Chapter

Compared with planets, the comets and asteroids are unevolved objects. You will find them much as they were when they formed 4.6 billion years ago. The fragments of these objects that fall to Earth, the meteors and meteorites, will give you a close look at these ancient planetesimals. As you explore, you will find answers to four essential questions:

Where do meteors and meteorites come from?

What are the asteroids?

Where do comets come from?

What happens when an asteroid or comet hits Earth?

Looking Ahead

As you finish this chapter, you will have an astronomer's insight into your place in nature. You live on the surface of a planet. There are other planets. Are they inhabited too? That is the subject of the next chapter.

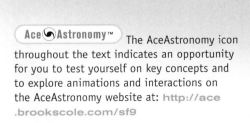
Ace Astronomy™ The AceAstronomy icon throughout the text indicates an opportunity for you to test yourself on key concepts and to explore animations and interactions on the AceAstronomy website at: http://ace.brookscole.com/sf9

of the end of the world. Comet Kohoutek in 1973, Comet Halley in 1986, and Comet Hale–Bopp in 1997 caused deep concern among the superstitious. A bright comet moving slowly through the night sky is so out of the ordinary that you should not be surprised if it generates some instinctive alarm (■ Figure 25-1).

In fact, comets are graceful and beautiful visitors to our skies. Astronomers think of comets as messengers from the age of planet building. By studying comets, you can learn about the conditions in the solar nebula from which planets formed. But comets are only the icy remains of the solar nebula; the asteroids are the rocky debris left over from planet building. Because you cannot easily visit comets and asteroids, you need to begin your analysis with the fragments of those bodies that fall to Earth— the meteorites.

25-1 Meteorites

IN THE AFTERNOON of November 30, 1954, Mrs. E. Hulitt Hodges of Sylacauga, Alabama, lay napping on her living room couch. An explosion and a sharp pain jolted her awake, and she found that a meteorite had smashed through the ceiling and bruised her left leg. Mrs. Hodges is the only person known to have been injured by a meteorite. By the way, Mrs. Hodges lived right across the street from the Comet Drive-In Theater.

Meteorite impacts on homes are not common, but they do happen. Statistical calculations show that a meteorite should damage a building somewhere in the world about once every 16 months. About two meteorites large enough to produce visible impacts strike somewhere on Earth each day, but most meteors

■ **Figure 25-1**

Comet Hyakutake swept through the inner solar system in 1996 and was dramatic in the northern sky. Seen here from Kitt Peak National Observatory, the comet passed close to the north celestial pole (behind the observatory dome in this photo). Notice the Big Dipper below and to the left of the head of the comet. (Courtesy Tod Lauer)

Visual-wavelength image

are small particles ranging from a few centimeters down to microscopic dust. Earth gains about 40,000 tons of mass per year from meteorites of all sizes.

Recall from Chapter 19 that astronomers distinguish between the words *meteoroid, meteor,* and *meteorite.* A small body in space is a meteoroid, but once it begins to vaporize in Earth's atmosphere it is called a meteor. If it survives to strike the ground, it is called a meteorite.

You probably have two questions concerning meteorites: Where in the solar system do these objects come from? What can meteorites tell you about the origin of the solar system? To answer these questions, you must consider the orbits of meteoroids and the minerals found in meteorites.

Meteoroid Orbits

Meteoroids are much too small to be visible through even the largest telescope. They are visible only when they fall into Earth's atmosphere at 10 to 30 km/s, roughly 30 times faster than a rifle bullet, and are vaporized by friction with the air. The average meteoroid is about the mass of a paper clip and vaporizes at an altitude of about 80 km above Earth's surface, but it produces a bright streak of fire that you see as a meteor. The trail of a meteor points back along the path of the meteoroid, so if you could study the direction and speed of meteors, you could get clues to the orbits of the meteoroids.

One way to backtrack meteor trails is to observe meteor showers. On any clear night, you can see 3 to 15 meteors an hour, but on some nights you can see a shower of dozens of meteors an hour that are obviously related to each other. To confirm this, try observing a meteor shower. Pick a shower from ■ Table 25-1 and on the appropriate night stretch out in a lawn chair and watch a large area of the sky. When you see a meteor, sketch its path on

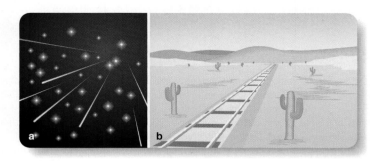

■ **Figure 25-2**

(a) Meteors in a meteor shower enter Earth's atmosphere along parallel paths, but perspective makes them appear to diverge from a radiant point. (b) Similarly, parallel railroad tracks appear to diverge from a point on the horizon.

the appropriate sky chart from the back of this book. In just an hour or so you will discover that most of the meteors you see seem to come from a single area of the sky, the **radiant** of the shower (■ Figure 25-2a). In fact, meteor showers are named after the constellation from which they seem to radiate. The Perseid shower radiates from the constellation Perseus in mid-August. The Leonids were spectacular in November of 2001, when observers saw hundreds per hour. The Leonids are not often that dramatic.

Observing a meteor shower is relaxing, and the show is natural fireworks, but it is even more exciting when you understand what a meteor shower tells you (**Window on Science 25-1**). The significant fact is that the meteors in a shower appear to come from a single point in the sky, the radiant. That must mean the meteoroids were traveling through space along parallel paths. When they encounter Earth and are vaporized in the upper atmosphere, you see their fiery tracks in perspective; they appear to

Shower	Dates	Hourly Rate	Radiant*		Associated Comet
			R. A.	Dec.	
Quadrantids	Jan. 2–4	30	15^h24^m	50°	
Lyrids	April 20–22	8	18^h4^m	33°	1861 I
η Aquarids	May 2–7	10	22^h24^m	0°	Halley?
δ Aquarids	July 26–31	15	22^h36^m	−10°	
Perseids	Aug. 10–14	40	3^h4^m	58°	1982 III
Orionids	Oct. 18–23	15	6^h20^m	15°	Halley?
Taurids	Nov. 1–7	8	3^h40^m	17°	Encke
Leonids	Nov. 14–19	6	10^h12^m	22°	1866 I Temp
Geminids	Dec. 10–13	50	7^h28^m	32°	

■ Table 25-1 I Meteor Showers

*R. A. and Dec. give the celestial coordinates (right ascension and declination) of the radiant of each shower.

Enjoying the Natural World

The beauty of flowers becomes more interesting when you know the reasons for their shapes and colors. (M. Seeds)

You can enjoy a meteor shower as Mother Nature's fireworks, but your enjoyment is much greater once you begin to understand what causes meteors and why meteors in a shower follow a pattern. When you know that meteor showers help you understand the origin of our world, the evening display of shooting stars is even more exciting. Science typically increases your enjoyment of the natural world by revealing the significance of things you might otherwise enjoy only in a casual way.

Everyone likes flowers, for example. You can enjoy them casually, admiring their colors and fragrances, their graceful petals and massed blooms. But botanists know that evolution has carefully designed flowers to welcome insects and spread pollen. The bright colors attract insects, and the shapes of the flowers provide little runways so the insect will find it easy to land. Some color patterns even guide the insect in like landing lights at an airport, and many flowers, such as orchids and snapdragons, force the insect to crawl inside in just the right way to exchange pollen and fertilize the flower. Nectar is bug bait. Once you begin to understand what flowers are for, a visit to a garden becomes not only an adventure of color and fragrance, but also an adventure in understanding the beauty of design and function, an adventure in meaning as well as an adventure of the senses.

And your understanding of a natural phenomenon can help you understand and enjoy related phenomena. For example, some flowers attract flies for pollination, and such blossoms smell like rotting meat. A few flowers depend on bats, and they open their blossoms at night. Many plants, such as pine trees, depend on the wind to spread their pollen, and those plants do not have colorful flowers at all; but flowers that depend on hummingbirds have long trumpet-shaped blossoms that just fit the hummingbirds' beaks.

The more science tells you about nature, the more enjoyable the natural world is. You can enjoy a meteor shower as nothing more than a visual treat, but the more you know about it, the more interesting and exciting it becomes. The natural world is filled with meaning, and science, as a way of discovering and understanding that meaning, gives you new ways to enjoy the world around you.

---

come from a single radiant point, just as railroad tracks seem to come from a single point on the horizon (Figure 25-2b).

Studies of meteor-shower radiants reveal that these meteors are produced by bits of matter orbiting the sun along the paths of comets. The vaporizing head of the comet releases bits of rock that become spread along the entire orbit (■ Figure 25-3). When Earth passes through this stream of material, you see a meteor shower. In some cases the comet has wasted away, has lost all of its ices, and is no longer visible, but in other cases the comet is still prominent though somewhere else along its orbit. For example, in May Earth comes near the orbit of Comet Halley, and Earthlings see the Eta Aquarids shower. Around October 20, Earth passes near the other side of the orbit of Comet Halley, and you can see the Orionids shower. These pleasant observations of meteor showers reveal that at least some meteoroids come from comets.

Even when there is no shower, you will still see meteors, which are called **sporadic meteors** because they are not part of specific showers. Many of these are produced by stray bits of matter that were released long ago by comets. Such comet debris gets spread throughout the inner solar system, and a few bits fall into Earth's atmosphere continuously even when there is no shower.

Another way to backtrack meteor trails is to photograph a meteor from two locations on Earth a few miles apart. Then astronomers can use triangulation to find the altitude, speed, and direction of the meteor as it moves through the atmosphere and work backward to find out what its orbit looked like before it entered Earth's atmosphere. These studies reveal that meteors belonging to showers and some sporadic meteors have orbits that are similar to the orbits of comets. In many cases, the comets can be identified, so the origin of these meteors is well known. But a few sporadic meteors have orbits that lead back to the asteroid belt between Mars and Jupiter. From this you can conclude that meteors have a dual source: Many come from comets, but some come from the asteroid belt. To learn more about meteors, you need to examine those that make it to Earth's surface, the meteorites. You can begin with their dramatic arrivals.

Meteorite Impacts on Earth

When a meteor is massive enough and strong enough to survive its plunge through Earth's atmosphere, it finally reaches Earth's surface, and it is called a meteorite. A large meteorite hitting Earth might dig an impact crater much like those on the moon (see page 500). Most meteorites are small and burn up in the atmosphere, but a few survive and hit Earth's surface with a bang. Big ones can dig craters.

Fresh material ejected from the comet forms a fan.

Orbit of comet

Infrared image

Orbit of comet

Earth

Sun

■ **Figure 25-3**

As a comet's ices evaporate, it releases rocky bits of material that spread along its orbit. If Earth passes through such material, it experiences a meteor shower. In this image of Comet Encke, the millimeter-size bits of rock along its orbit glow in the infrared as they are warmed by sunlight. The Taurid meteor shower occurs every October when Earth crosses the orbit. (NASA/JPL-Caltech/M. Kelley, Univ. of Minnesota)

Over 150 impact craters have been found on Earth. The Barringer Meteorite Crater near Flagstaff, Arizona, is a good example (■ Figure 25-4). It was created about 50,000 years ago by a meteorite about as large as a big building (40 m). Debris at the site shows that the meteorite was composed of iron, and astronomers can calculate from the size of the crater that the meteorite hit at a speed of about 11 km/s and released as much energy as a large thermonuclear bomb.

The Barringer crater is 1.2 km in diameter and 200 m deep. It seems large when you stand on the edge, and the hike around it, though beautiful, is long and dry. Nevertheless, the crater is actually small compared with some impact features on Earth. About 65 million years ago, an impact in the northern Yucatán of Central America produced a crater, now covered by sediment, that was between 180 and 300 km in diameter, almost as big as the state of Ohio. This impact has been blamed for changing Earth's climate and causing the extinction of the dinosaurs. Later in this chapter, you can come back to this Earth-shaking event. The 1994 comet impacts on Jupiter (shown in Figure 23-6) were even larger, but they formed no craters because Jupiter has no surface. During the Leonid meteor shower in recent years, astronomers watching the dark portion of the moon's disk have seen a number of bright flashes apparently caused by meteorite impacts. No new craters on the moon have been identified, but it is clear that meteorite impacts and the formation of impact craters are a continuing process in our solar system.

Large meteorites can produce craters, but most meteorites are small and don't create significant craters. Also, craters erode rapidly on Earth, so most meteorites are found without associated craters. The best place to look for meteorites turns out to be certain areas of Antarctica—not because more meteorites fall there but because the meteorites are easy to recognize on the icy terrain. Most meteorites look like Earth rocks, but on the ice of Antarctica there are no natural rocks. Also, the slow creep of the ice toward the sea concentrates the meteorites in certain areas where the ice runs into mountain ranges, slows down, and evaporates. Consequently, teams of scientists can recover meteorites from the ice (■ Figure 25-5). Such unbiased collections of meteorites are important in helping define the frequency of different meteorite types.

Meteorites that are seen to fall are called **falls.** A fall is known to have occurred at a given time and place, and thus the meteorite is well documented. A meteorite that is discovered but was not seen to fall is called a **find.** Such a meteorite could have fallen thousands of years ago. This distinction between falls and finds will be important as you analyze the different kinds of meteorites.

An Analysis of Meteorites

Meteorites can be divided into three broad categories: *Iron* meteorites are solid chunks of iron and nickel, *stony* meteorites are silicate masses that resemble Earth rocks, and *stony-iron* meteorites

■ **Active Figure 25-4**

(a) The Barringer Meteorite Crater (near Flagstaff, Arizona) is nearly a mile in diameter and was formed about 50,000 years ago by the impact of an iron meteorite roughly 40 m in diameter. It hit with energy equivalent to that of a 3-megaton hydrogen bomb. Notice the raised and deformed rock strata all around the crater. For scale, locate the brick building on the far rim at right. (M. Seeds) (b) Like all larger-impact features, the Barringer Meteorite Crater has a raised rim and scattered ejecta. (USGS)

Ace Astronomy™ Log into AceAstronomy and select this chapter to see the Active Figuree "Impacts on Earth." Notice that large impacts are less common than smaller ones.

are mixtures of iron and stone. These types of meteorites are illustrated in ■ Figure 25-6.

Iron meteorites are easy to recognize because they are heavy, dense lumps of iron-nickel steel—a magnet will stick to them. That explains an important bit of statistics. Iron meteorites make up 66 percent of finds (■ Table 25-2) but only 6 percent of falls. Why? Because an iron meteorite does not look like a rock. If you trip over one on a hike, you are more likely to carry it home and show it to your local museum. Also, some stone meteorites deteriorate rapidly when exposed to weather; irons survive longer. That means there is a **selection effect** that makes it more likely that iron meteorites will be found (**Window on Science 25-2**). That only 6 percent of falls are irons shows that iron meteorites are fairly rare.

When iron meteorites are sliced open, polished, and etched with nitric acid, they reveal regular bands called **Widmanstätten patterns** (Figure 25-6). These patterns are caused by alloys of iron and nickel called kamacite and taenite that formed billions of years ago as the molten iron cooled and solidified. The size and shape of the bands indicate that the molten metal cooled very slowly, no faster than 20 K per million years.

A lump of molten metal floating in space would cool very quickly, so the presence of the Widmanstätten pattern tells you that the molten metal must have been well insulated to have cooled so slowly. Such slow cooling is typical of the interiors of bodies at least 30 to 50 km in diameter. On the other hand, the

iron meteorites show no evidence of the very high pressures that would develop in larger bodies. Evidently, the iron meteorites formed from the interiors of molten objects of planetesimal size.

In contrast to irons, **stony meteorites** are common. Among falls, 92 percent are stones. Although there are many different types of stony meteorites, you can classify them into two main categories depending on their physical and chemical content—chondrites and achondrites.

Roughly 80 percent of all meteorite falls are stony meteorites called **chondrites,** which look like dark gray, granular rocks. The chemical composition of chondrites is the same as a sample of matter from the sun with the volatile gases removed. The classification of meteorites has become quite sophisticated, and there are many types of chondrites, but in general they appear to be original samples of the material that condensed from the solar nebula. Some have slight mineral differences, which show that the solar nebula was not totally uniform when this material condensed.

Most types of chondrites contain **chondrules,** round bits of glassy rock no larger than 5 mm across (Figure 25-6). To be glassy rather than crystalline, the chondrules must have cooled from a molten state quickly, in a few hours, but their origin is not clear. One theory is that they are bits of matter that were suddenly melted by shock waves spreading through the solar nebula. Whatever their origin, they are very old.

All of the different types of chondrites contain some volatiles such as water, and this shows that the meteoroids were never

Each meteorite is assigned a number and photographed as it was found.

■ **Figure 25-5**

Braving bitter cold and high winds, teams of scientists riding snowmobiles search for meteorites that fell long ago in Antarctica and are exposed as the ice evaporates. Thousands of these meteorites have been collected, including rare meteorites from the moon and Mars. (Courtesy Dr. Monika Kress)

Large or small, meteorites are sealed airtight and refrigerated until they can be studied.

heated to high temperatures. Differences among the many types of chondrites arise from their condensation in different parts of the solar nebula and from processes that altered their composition after they formed. Some, for example, appear to have been altered by the presence of water released by the melting of ice. Among the chondrites, the **carbonaceous chondrites** are rare, only about 5.7 percent of falls. These very dark gray,

rocky meteorites contain volatiles such as water and carbon compounds that would have been driven off if the meteoroid had been heated much above room temperature. It has been common for astronomers to think of the carbonaceous chondrites as

Iron meteorites are very heavy for their size and have a dark, irregular surface.

Stony meteorites tend to have a fusion crust caused by melting in Earth's atmosphere.

A stony-iron meteorite cut and polished reveals a mixture of iron and rock.

Chondrules are small, glassy spheres found in chondrites.

Cut, polished, and etched with acid, iron meteorites show a Widmanstätten pattern.

This carbonaceous chondrite contains chondrules and volatiles, including carbon, that make the rock very dark.

■ Figure 25-6

The three main types of meteorites, irons, stones, and stony-iron, have distinctive characteristics. (Lab photos courtesy of Russell Kempton, New England Meteoritical Services)

the least altered of the chondrites, but many types of chondrites are essentially unaltered samples of the solar nebula. A carbonaceous chondrite is shown in Figure 25-6.

One of the most important meteorites ever recovered was a carbonaceous chondrite that was seen falling on the night of February 8, 1969, near the Mexican village of Pueblito de Allende. The brilliant fireball was accompanied by tremendous sonic booms and showered an area about 50 km by 10 km with over 4 tons of fragments. About 2 tons were recovered.

Studies of the Allende meteorite disclosed that it contained—besides volatiles and chondrules—small, irregular inclusions rich in calcium, aluminum, and titanium. Now called **CAIs,** for calcium–

aluminum-rich inclusions, these bits of matter are highly refractory (■ Figure 25-7); that is, they can survive very high temperatures. If you could scoop out a ton of the sun's surface matter and cool it, the CAIs would be the first particles to form. As the temperature fell, other materials would condense in accord with the condensation sequence described in Chapter 19. When the material finally reached room temperature, you would find that all of the hydrogen, helium, and a few other gases like argon and neon had escaped and that the remaining lump, weighing about 18 kg (40 lb), had almost the same composition as the Allende meteorite, including CAIs. The Allende meteorite seems to be a very old sample of the solar nebula.

Unlike the chondrites, stony meteorites called **achondrites** (7.1 percent of falls) are highly modified. They contain no chondrules and no volatiles. This suggests that they have been hot enough to melt chondrules and drive off volatiles, leaving behind rock with compositions similar to Earth's basalts.

Iron meteorites and stony meteorites make up most falls, but 2 percent of falls are meteorites that are made up of mixed iron and stone. These **stony-iron meteorites** appear to have solidified from a region of molten iron and rock—the kind of environment you might expect deep inside a planetesimal with a molten iron core and a rock mantle.

■ Table 25-2 | Proportions of Meteorites

Type	Falls (%)	Finds (%)
Stony	92	26
Iron	6	66
Stony-iron	2	8

Selection Effects in Science

Many different kinds of science depend on selecting objects to study. Scientists studying insects in the rain forest, for example, must choose which ones to catch. They can't catch every insect they see, so they might decide to catch and study any insect that is red. If they are not careful, a selection effect could bias their data and lead them to incorrect conclusions without their ever knowing it. The reason selection effects are dangerous in science is that they can have powerful influences without being evident. Only by carefully designing a research project can scientists avoid selection effects.

For example, suppose you decided to measure the speed of cars on a highway. There are too many cars to measure every one, so you reduce the workload and measure only red cars. It is quite possible that this selection criterion will mislead you because people who buy red cars may be more likely to be younger and drive faster. Should you measure only brown cars? No, because you might suspect that only older, more sedate people would buy a brown car. Should you measure the speed of any car in front of a truck? Perhaps you should pick any car following a truck? Again, you may be selecting cars that are traveling a bit faster or

a bit slower than normal. Only by very carefully designing your experiment can you be certain that the cars you measure are traveling at representative speeds.

Public-opinion companies that survey voters in shopping malls must think carefully about selection effects. If they are seeking opinions about property taxes and decide to interview only men with gray hair, they will get a biased result. Interviewing only men wearing T-shirts would give them quite a different result. They must be very careful to avoid unnoticed selection effects.

Astronomers face the danger of selection effects quite often. Iron meteorites are easier to find than stone meteorites. Very luminous stars are easier to see at great distances than faint stars. Spiral galaxies are brighter, bluer, and more noticeable than elliptical galaxies. What you see through a telescope depends on what you notice, and that is powerfully influenced by selection effects.

Scientists engaged in observation must spend considerable time designing their experiments. They must be careful to observe an unbiased sample if they expect to make logical deductions from their results. The scientists in the rain forest, for example, should not catch

Visual-wavelength image

Things that are bright and beautiful, such as spiral galaxies, may attract a disproportionate amount of attention. Scientists must be aware of such selection effects. (Hubble Heritage Team/STScI/AURA/NASA)

and study only the red insects. Often, the most brightly colored insects are poisonous (or at least taste bad) to predators. Catching only brightly colored insects could produce a highly biased sample of the insect population. Careful scientists must plan their work with great care and avoid any possible selection effects.

The Origin of Meteorites

Where do meteorites come from? The properties of meteorites suggest that they formed in the solar nebula. Their radioactive ages are about 4.6 billion years (Chapter 19), and some appear not to have been modified since they formed. Others have been heated or melted sometime after formation.

Meteorites almost certainly do not come from comets. Most cometary particles are very small specks of low-density, almost fluffy, material. When these enter Earth's atmosphere, you see them incinerated as meteors, and most meteors are produced by this cometary debris. But such meteors are small and weak and do not survive to reach the ground. Although most meteors come from comets, most meteorites are stronger chunks of matter—more like fragments of asteroids.

Meteorites must come from asteroidlike planetesimals rich in metals and silicates, not the cometlike planetesimals that were rich in ices. Indeed, many of the properties of the meteorites suggest that they originated in rocky planetesimals, which evolved in complicated ways and were eventually broken up by collisions.

Chondrules
CAIs

■ Figure 25-7

A sliced portion of the Allende meteorite showing round chondrules and irregularly shaped white inclusions called CAIs. (NASA)

Of course, the solar nebula was full of rocky planetesimals, but you must be wondering about two things: How did these planetesimals evolve to produce both iron and stony meteorites? And when did these planetesimals break up?

The iron and achondritic meteorites show that at least some of the planetesimals melted and differentiated, but what produced this heat? Planets the size of Earth can accumulate a great deal of heat from the decay of radioactive atoms such as uranium, thorium, and the radioactive isotope of potassium. The heat is trapped deep underground by thousands of kilometers of insulating rock. But in a small planetesimal, the insulating layers are not as thick, and the heat leaks out into space as fast as the slowly decaying atoms can produce it. If a small planetesimal is to melt, it must have a more rapidly decaying heat source.

Modern studies have shown that some meteorites may have contained the radioactive element aluminum-26. These aluminum atoms are now gone, decayed to magnesium-26, which remains in the meteorites. If aluminum-26 was once present, its rapid decay could have melted the center of a planetesimal as small as 20 km in diameter. This is a sufficient heat source to melt the parent bodies of the meteorites.

Where did this aluminum-26 come from? Supernova explosions can manufacture aluminum-26, but it decays very rapidly, with a half-life of only 715,000 years. If the solar nebula was enriched by aluminum-26 from a nearby supernova explosion, the explosion must have occurred just before the formation of our solar system. In fact, some astronomers wonder if the shock waves from the supernova explosion compressed gas clouds and triggered the formation of the sun and planets.

If a planetesimal's interior melted, it would have differentiated as heavy metals sank to the center to form a molten metal core and the less-dense silicates floated upward to form a stony mantle. Once the aluminum-26 had decayed away, the planetesimals cooled and slowly solidified. In some cases, the outer surface may not have been melted. If such a planetesimal were broken up (■ Figure 25-8), fragments from the center would look much like iron meteorites with their Widmanstätten patterns.

The achondrites, which have been strongly heated or melted, appear to come from the mantles and surfaces of the larger planetesimals that melted and differentiated. The stony-iron meteorites probably come from the core–mantle boundaries in these bodies where iron and stone were mixed. But the chondrites are probably fragments of smaller bodies that never melted. Volatile-rich meteorites such as the carbonaceous chondrites may have formed in smaller bodies farther from the sun where temperatures were lower.

■ **Figure 25-8** ▶

Planetesimals formed when the solar system was forming may have melted and separated into layers of different density and composition. The fragmentation of such a body could produce many types of meteorites. (Adapted from a diagram by C. R. Chapman)

The evidence seems strong that the meteorites are the result of the breaking up of larger bodies. For example, some meteorites are breccias—collections of fragments cemented together. Breccias are found on Earth and are very common on the moon, but studies of the meteoric breccias show that they were produced by impacts. A collision between planetesimals produces fragments, and the slower-moving particles fall back to the surface of the planetesimal to form a regolith, a soil of broken rock fragments.

The Origin of Meteorites

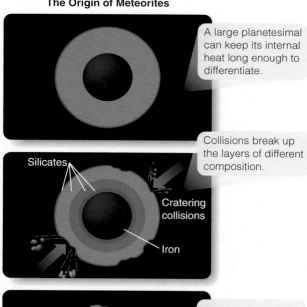

A large planetesimal can keep its internal heat long enough to differentiate.

Collisions break up the layers of different composition.

Meteorites from deeper in the planetesimal were heated to higher temperatures.

Fragments from near the core might have been melted entirely.

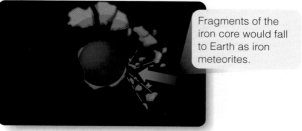

Fragments of the iron core would fall to Earth as iron meteorites.

Later impacts may add to this regolith and stir it. Still later, an impact may be so violent that the fragments are pressed together and momentarily melt where they touch. Almost instantly, the material cools, and the fragments weld themselves together to form a breccia. Most of the breccias were formed as the planetesimals formed. Much later, as the planetesimals were broken up by collisions, the layers of breccia were exposed, shattered, and became brecciated meteoroids.

When did these bodies break up? Recall from Chapter 19 that radioactive dating shows that meteorites formed about 4.6 billion years ago. That is when the parent bodies formed, melted, differentiated, and cooled. However, the collisions that broke up the planetesimals could not have happened that long ago. Small meteoroids would have been swept up by the planets in a billion years or less. Also, cosmic rays striking meteoroids in space produce isotopes such as helium-3, neon-20, and argon-38. Studies of these atoms in meteorites show that most meteorites have not been exposed to cosmic rays for more than a few tens of millions of years. The meteorites must have been buried under protective layers until recently. That means that the thousands of meteorites now in museums around the world must have broken off planetesimals somewhere in our solar system within the last billion years or less. Where are those planetesimals?

To answer that question, you need only recall that many meteoroid orbits lead back to the asteroid belt. The asteroids are evidently the planetesimals from which meteorites are born.

Building Scientific Arguments

How can meteors come from comets, but meteorites come from asteroids?
This is a revealing argument because it contains a warning that seeing is not enough in science; thinking about seeing is critical. A selection effect can determine what you notice when you observe nature, and a very strong selection effect prevents people from finding meteorites that originated in comets. Cometary particles are physically weak, and they vaporize in Earth's atmosphere easily. Very few ever reach the ground, and people are unlikely to find them. Furthermore, even if a cometary particle reached the ground, it would be so fragile that it would weather away rapidly, and, again, people would be unlikely to find it. Asteroidal particles, however, are made from rock and metal and so are stronger. They are more likely to survive their plunge through the atmosphere and more likely to survive erosion on the ground. Meteors from the asteroid belt are rare. Almost all of the meteors you see come from comets, but not a single meteorite is known to be cometary.

The meteorites are valuable because they provide hints about the process of planet building in the solar nebula. Build a new argument but, as always, think carefully about what you see. **Why are most falls stony, but most finds are irons?**

■   ■   ■

Connections: The vast majority of meteorites must be coming from the asteroids. Your next goal is to try to understand the nature and origin of these tiny rocky worlds.

25-2 Asteroids

UNTIL RECENTLY, FEW ASTRONOMERS KNEW or cared much about asteroids. They were small chunks of rock drifting between the orbits of Mars and Jupiter that occasionally marred long-exposure photographs by drifting past more interesting objects. Asteroids were more irritation than fascination.

Now you know differently. The evidence from meteorites shows that the asteroids are the last remains of the rocky planetesimals that built the planets 4.6 billion years ago. The study of the asteroids gives you a way to explore the ancient past of our planetary system.

Properties of Asteroids

In Chapter 19, you learned that most of the asteroids orbit between Mars and Jupiter and that images recorded by passing spacecraft reveal them to be small, complex worlds with irregular shapes and heavily cratered surfaces (Figure 19-3).

Study **Observations of Asteroids** on pages 628–629 and notice four important points:

1 Most asteroids are irregular in shape and battered by impact cratering. In fact, some appear to be rubble piles of broken fragments.

2 Some asteroids are double objects or have small moons in orbit around them. This is further evidence of collisions among the asteroids.

3 A few larger asteroids show signs of geological activity on their surfaces that may have been caused by volcanic activity when the asteroid was young.

4 Asteroids can be classified by their albedo and color to reveal clues to their compositions.

Collisions among asteroids must have been occurring since the formation of the solar system, and astronomers have found evidence of catastrophic impacts powerful enough to break up an asteroid. Early in the 20th century, Japanese astronomer Kiyotsugu Hirayama discovered that some groups of asteroids share similar orbits. They have the same average distance from the sun, the same eccentricity, and the same inclination. Up to 20 of these **Hirayama families** are known, and modern observations show that the asteroids in a family typically share similar spectroscopic characteristics. Evidently, a family is produced by a catastrophic collision that breaks a single asteroid into a family of fragments that continue traveling along similar orbits around the sun. Evidence shows that one family was produced only 5.8 million years ago in a collision between asteroids 3 km and 16 km in diameter at

Observations of Asteroids

1 Seen from Earth, asteroids look like faint points of light moving in front of distant stars. Not many years ago they were known mostly for drifting slowly through the field of view and spoiling long time exposures. Some astronomers referred to them as "the vermin of the sky." Spacecraft have now visited asteroids, and the images radioed back to Earth show that the asteroids are mostly small, gray, irregular worlds heavily cratered by impacts.

The Near Earth Asteroid Rendezvous (NEAR) spacecraft visited the asteroid Eros in 2000 and found it to be heavily cratered by collisions and covered by a layer of crushed rock ranging from dust to large boulders. The NEAR spacecraft eventually landed on Eros.

Eros appears to be a solid fragment of rock.

Visual-wavelength image

10 km

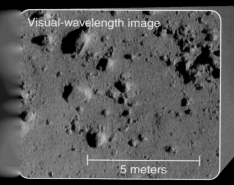

Visual-wavelength image

5 meters

Most asteroids are too small for their gravity to pull them into a spherical shape. Impacts break them into irregularly shaped fragments.

Visual

The surface of Mathilde is very dark rock.

50 km

Enhanced visual image

1a The mass of an asteroid can be found from its gravitational influence on passing spacecraft. Mathilde, at left, has a low mass, and that makes its density so low it cannot be solid rock. Like many asteroids, Mathilde may be a rubble pile of broken fragments with large empty spaces between fragments.

If you walked across the surface of an irregularly shaped asteroid such as Eros, you would find gravity very weak; and in many places, it would not be perpendicular to the surface.

Like most asteroids, Gaspra would look gray to your eyes; but, in this enhanced image at left, color differences probably indicate difference in mineralogy.

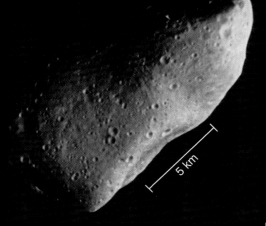

5 km

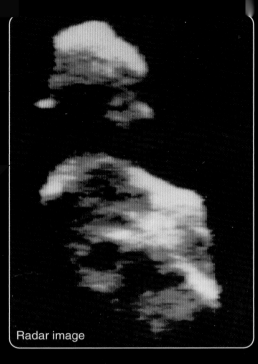

2 Asteroids that pass near Earth can be imaged by radar. The asteroid Toutatis is revealed to be a double object—two objects orbiting close to each other or actually in contact.

Radar image

Double asteroids are more common than was once thought, reflecting a history of collisions and fragmentation. The asteroid Ida is orbited by a moon Dactyl only about 1.5 km in diameter.

Ida

Dactyl

30 km

Enhanced visual + infrared

Occasional collisions among the asteroids release fragments, and Jupiter's gravity scatters them into the inner solar system as a continuous supply of meteorites.

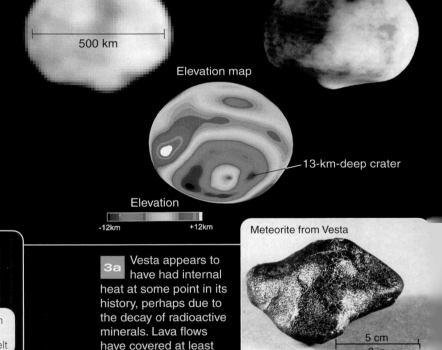

Visual-wavelength image   **Vesta**   Model

500 km

Elevation map

13-km-deep crater

Elevation

-12km   +12km

Meteorite from Vesta

5 cm
2 in.

3 The large asteroid Vesta, as shown at right, provides evidence that some have suffered geological activity. No spacecraft has visited it, but its spectrum resembles that of solidified lava. Images made by the Hubble Space Telescope allow the creation of a model of its shape. It has a huge crater at its south pole. A family of small asteroids is evidently composed of fragments from Vesta, and a certain class of meteorites, spectroscopically identical to Vesta, are believed to be fragments from the asteroid. The meteorites appear to be solidified basalt.

3a Vesta appears to have had internal heat at some point in its history, perhaps due to the decay of radioactive minerals. Lava flows have covered at least some of its surface.

4 Although asteroids would look gray to your eyes, they can be classified according to their albedos (reflected brightness) and spectroscopic colors. As shown at left, S-types are brighter and tend to be reddish. They are the most common kind of asteroid and appear to be the source of the most common chondrites.

M-type asteroids are not too dark but are also not very red. They may be mostly iron-nickel alloys.

C-type asteroids are as dark as lumps of sooty coal and appear to be carb...

Bright

0.4

0.3

0.2

Albedo (reflected brightness)

0.1

Common in the inner asteroid belt

M   S

0.06

Common in the outer asteroid belt

0.04

Grayer   Redder

C

Dark

0.8   1.0   1.2   1.4   1.6

Ultraviolet minus visual color index

about 5 km/s (11,000 mph), a typical speed for asteroid collisions. Evidently the fragmentation of asteroids is a continuing process.

In 1983, the Infrared Astronomy Satellite detected the infrared glow of sun-warmed dust scattered in bands throughout the asteroid belt. These dust bands appear to be the products of past collisions. The dust will eventually be blown away, but collisions occur constantly in the asteroid belt, so new dust bands will presumably be produced as the present bands dissipate.

What are the asteroids? You can be sure they have been modified and fragmented since they formed, but where did they come from? There are clues hidden among their orbits. You should begin that story at the beginning.

The Asteroid Belt

The first asteroid was discovered on January 1, 1801 (the first night of the 19th century), by the Sicilian monk Giuseppe Piazzi. It was later named Ceres after the Roman goddess of the harvest (thus our word *cereal*).

Astronomers were excited by Piazzi's discovery. There seemed to be a gap where a planet might exist between Mars and Jupiter at an average distance from the sun of 2.8 AU. Ceres fit right in. Its average distance from the sun is 2.766 AU. It was a bit small to be a planet, and because three more objects—Pallas, Juno, and Vesta—were discovered in the following years, astronomers realized that Ceres and the other asteroids were not true planets.

Today thousands of asteroids have accurately known orbits, and there are many more as yet undiscovered; but they are all small bodies. All of the larger asteroids in the asteroid belt have been found.

If you discovered an asteroid, you would be allowed to choose a name for it, and asteroids have been named for spouses, lovers, dogs, Greek gods, politicians, and others.* Once an orbit has been calculated, the asteroid is assigned a number listing its position in the catalog known as the *Ephemerides of Minor Planets*. Thus, Ceres is known as 1 Ceres, Pallas as 2 Pallas, and so on. Only three asteroids are larger than 400 km in diameter, and most are much smaller (■ Figure 25-9).

Most of these objects orbit the sun in the asteroid belt between Mars and Jupiter, and you might suspect that massive Jupiter was responsible for their origin. Certainly the distribution of asteroids in the belt shows the importance of Jupiter's gravitation. Certain regions of the belt, called **Kirkwood's gaps** after their discoverer, are almost free of asteroids (■ Figure 25-10). These gaps lie at certain distances from the sun where an aster-

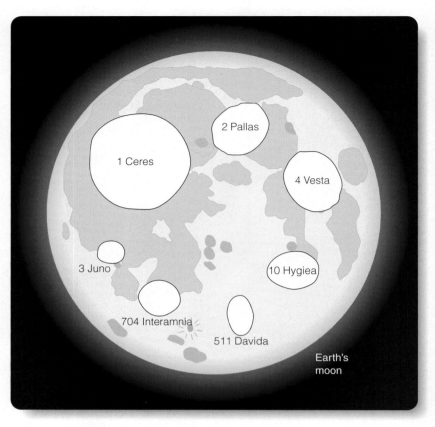

■ **Figure 25-9**

The relative size and approximate shape of the larger asteroids are shown here compared with the size of Earth's moon. Smaller asteroids can be highly irregular in shape.

oid would find itself in resonance with Jupiter. For example, if an asteroid lay 3.28 AU from the sun, it would revolve twice around its orbit in the time it took Jupiter to revolve once around the sun. On alternate orbits, the asteroid would find Jupiter at the same place in space tugging outward. The cumulative perturbations would rapidly change the asteroid's orbit until it was no longer in resonance with Jupiter. Our example is in a 2:1 resonance, but gaps occur in the asteroid belt at many resonances, including 3:1, 5:2, and 7:3. You will recognize that Kirkwood's gaps in the asteroid belt are produced in the same way as some gaps in Saturn's rings. Both were discovered by Daniel Kirkwood (Chapter 23).

Modern research shows that the motion of asteroids in Kirkwood's gaps is described by a theory in mathematics that deals with chaotic behavior. The smooth motion of water sliding over the edge of a waterfall decays rapidly into a chaotic jumble. The same mathematical theory of chaos that describes the motion of the water shows how the slowly changing orbit of an asteroid within one of Kirkwood's gaps can suddenly change into a long, elliptical orbit that carries the asteroid into the inner solar system, where it is likely to be removed by a collision with Mars, Earth, or Venus. This ejection process also explains how Jupiter's gravity can throw so many meteoroids from the asteroid belt into the inner solar system.

---

*Some sample asteroid names: Olga, Chicago, Vaticana, Noel, Ohio, Tea, Gaby, Fidelio, Hagar, Geisha, Dudu, Tata, Mimi, Dulu, Tito, Zulu, Beer, and Zappafrank (after the late musician Frank Zappa).

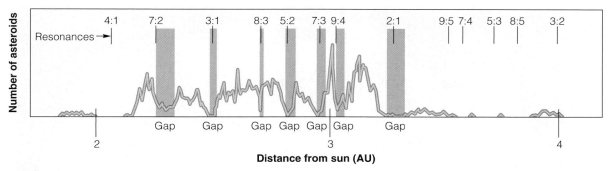

Figure 25-10

Here the red curve shows the number of asteroids at different distances from the sun. Purple bars mark Kirkwood's gaps, where there are few asteroids. Note that these gaps match resonances with the orbital motion of Jupiter.

Nonbelt Asteroids

You don't have to go all the way out to the asteroid belt if you want to visit an asteroid. Some of the most interesting follow orbits that cross the orbits of the terrestrial planets or wander among the Jovian worlds. In fact, some asteroids even share the orbits of the larger planets.

The **Apollo–Amor objects** are asteroids whose orbits carry them into the inner solar system. The Amor objects follow orbits that cross the orbit of Mars but don't reach the orbit of Earth. The Apollo objects have Earth-crossing orbits. These Apollo–Amor objects are dangerous. Jupiter's influence makes their orbits precess, so they are bound to interact with a planet eventually. About one-third will be thrown into the sun, and a few will be ejected from the solar system, but many of these objects are doomed to collide with a planet—perhaps ours. Earth is hit by an Apollo object once every 250,000 years, on average. With a diameter of up to 2 km, they hit with the power of a 100,000-megaton bomb and can dig craters 20 km in diameter.

Many Apollo objects have known orbits, and none of those will hit Earth in the foreseeable future. The bad news is that there are about 1000 of these near-Earth objects (NEOs) larger than 1 km in diameter, the minimum size of an impactor that could cause global effects on Earth. More than half a dozen teams are now searching for these NEOs. For example, LONEOS (Lowell Observatory Near Earth Object Search) is searching the entire sky visible from Lowell Observatory once a month. The LINEAR (Lincoln Near-Earth Asteroid Research) telescope in New Mexico has been very successful in finding NEOs and in finding new main-belt asteroids (■ Figure 25-11). The combined searches should be able to locate all of the largest NEOs by 2010.

This is a serious issue because even a small asteroid could do serious damage. For example, in late December 2004, an asteroid was discovered that had a 2.6 percent chance of striking Earth on April 13, 2029. The object is large enough to do significant damage over a wide area but would not be large enough to alter Earth's climate. Astronomers sprang into action and obtained further observations to refine the asteroid's orbit. The new orbit shows that the object will not hit Earth in 2029. Objects a few tens of meters or less fragment and explode in the atmosphere, but the shock waves from their explosions could cause serious damage on Earth's surface. Declassified data from military satellites show that Earth is hit about once a week by meter-size asteroids. Larger impacts produce more damage but are much less common.

It is easy to assume that the Apollo–Amor objects are rocky asteroids that have been thrown into their extreme orbits by events in the main asteroid belt. At least some of these objects, however, may be comets that have exhausted their ices and become trapped in short orbits that keep them in the inner solar system. You will see later in this chapter that the distinction between comets and asteroids is not totally clear.

While the Apollo–Amor objects rush through the inner solar system, there are also nonbelt asteroids in the outer solar system. These objects, being farther from the sun, move more slowly. The object Chiron, found in 1977, appears to be a body about 170 km in diameter. Its orbit carries it from just inside the orbit of Uranus to just inside the orbit of Saturn. Although it was first classified as an asteroid, its status is now less certain. Ten years after its discovery, Chiron surprised astronomers by suddenly brightening with a release of vapor and dust. Like a comet, Chiron was releasing a cloud of gas and dust in jets. Studies of older photographs showed that Chiron had been brighter even when it was farther from the sun, and astronomers now suspect that it may have a rocky crust covering deposits of ices such as solid nitrogen, methane, and carbon monoxide. Thus, Chiron may be more comet than asteroid, and it warns that the distinction is not clear-cut. Chiron's orbit is not stable. It cannot have remained in its present orbit for the entire history of the solar system but must have been perturbed into its present orbit within the last million years or so.

Jupiter ushers two groups of asteroids within its own orbit. These nonbelt asteroids have become trapped in the Lagrangian points along Jupiter's orbit. (See Figure 13-5.) These points lie 60° ahead of and 60° behind the planet and are regions where the gravitation of the sun and Jupiter combine to trap small bodies

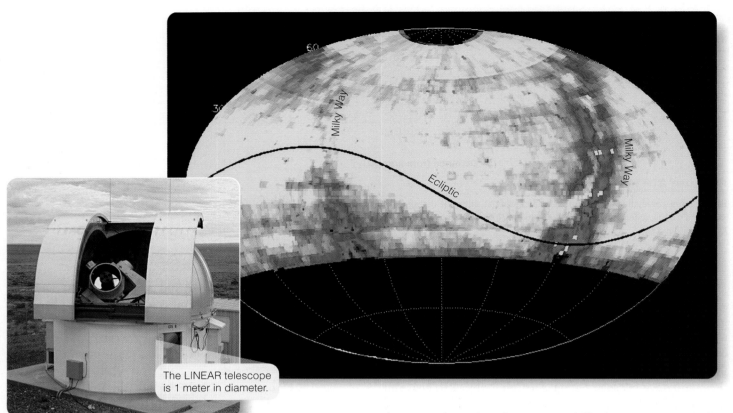

The LINEAR telescope is 1 meter in diameter.

■ Figure 25-11

The LINEAR telescope searches for asteroids every clear night when the moon is not bright. The diagram shows the thoroughness of its search over the entire sky for one year. Asteroids are hard to discover in front of the starry Milky Way. (Reprinted with permission of MIT Lincoln Laboratory, Lexington, Massachusetts)

(■ Figure 25-12). Like cosmic sinkholes, the Lagrangian points have trapped chunks of debris now called **Trojan asteroids** and named after the heroes of the Trojan War (588 Achilles, 624 Hektor, 659 Nestor, and 1143 Odysseus, for example). Slightly over 1000 Trojan asteroids are known, but only the brightest have been given names.

Astronomers have also found a few objects in the Lagrangian points of the orbits of Mars and Neptune. Other planets may have Trojan asteroids trapped in their orbits. As astronomers search for smaller objects, they are finding that our solar system still contains large numbers of these small bodies. The challenge is to explain their origin.

The Origin of the Asteroids

You have concluded your study of each planet by trying to summarize its history. Can you tell the story of the asteroids? Begin with an idea that didn't work out.

An old theory held that the asteroids are the remains of a planet that broke up, but modern astronomers discount that idea. Once formed, a planet is very difficult to tear apart. Rather, astronomers believe the asteroids are the remains of planetesimals left over from the formation of the planets. Jupiter's gravity could have stirred the planetesimals just inside its orbit and caused collisions at unusually high velocities. These impacts would have tended to break up the planetesimals rather than assemble them into a planet. Resonances and chaos would have helped eject material, and most of the planetesimal objects have now been lost—swept up by planets, captured as satellites, or ejected from the solar system. The asteroid belt today contains hardly 4 percent the mass of Earth's moon.

The C-type asteroids have albedos smaller than 0.06 and would look very dark to your eyes. They are probably made of carbon-rich material similar to that in carbonaceous chondrites. The S-type asteroids have albedos of 0.1 to 0.2, so they would look brighter and spectroscopically redder. The M-type asteroids are bright but not as red. S-type asteroids are believed to be rocky, but M-type asteroids appear to be metal rich and may be the iron cores of fragmented asteroids.

Although S-type asteroids are very common in the inner asteroid belt, their spectroscopic colors are different from the chondrites—the most common kind of meteorite. New evidence from the analysis of moon rocks and from observations of Eros, an S-type asteroid, shows that bombardment by micrometeorites can redden and darken S-type asteroids. From this result, it seems that the most common kind of meteorites come from the most common kind of asteroid.

■ Figure 25-12

This diagram plots the position of known asteroids inside or near the orbit of Jupiter on a specific day. Squares, filled or empty, show the location of known comets. Although asteroids and comets are small bodies and lie far apart, there are a great many of them in the inner solar system. (Gareth Williams/Minor Planet Center)

lava flows? One source of heat in newly formed asteroids would be the decay of short-lived radioactive elements such as aluminum-26. Such radioactive elements could have been produced in a supernova explosion occurring shortly before the formation of our solar system. The smallest asteroids lose their heat too fast to melt, but the decay of aluminum-26 occurs fast enough to melt the interiors of bodies larger than a few dozen kilometers in diameter. That makes it reasonable to suppose that Vesta and some other asteroids larger than about 100 km in diameter might have been strongly heated and modified by lava flows.

In contrast, the largest main-belt asteroid, Ceres, about 900 km in diameter, does not seem to have been modified by internal heating. The light reflected from its surface suggests a claylike surface related to carbonaceous chondrites. This is surprising, because clays form when minerals are exposed to water. In fact, spectroscopic observations reveal what seem to be traces of water ice on Ceres. Evidently some asteroids may have had significant amounts of water bound into their crusts when they were young.

All of the evidence suggests that the asteroids are the broken remains of planetesimals that formed in the solar nebula as planet building began. The largest remaining asteroids, such as Ceres and Vesta, may be largely unbroken planetesimals, and some of these may have evolved due to internal heat or the presence of water. Nevertheless, the vast majority of the asteroids are fragments, and many may consist of bodies that have been shattered and then reformed as gravity pulled the fragments back together. Some compositional differences seem to have originated when the asteroids formed in that the objects in the outer asteroid belt contain more carbon-rich compounds. The presence of massive Jupiter orbiting just beyond this region prevented the original planetesimals from accreting to build a planet. Instead, collisions fragmented the planetesimals, and nearly all of the material is now lost.

Some experts discuss the mining of the asteroid belt for its mineral riches, but the asteroid belt is nearly empty. The small, fragmented asteroids drift through space far from one another and only rarely collide with each other or with planets. A voyage to explore or exploit the asteroid belt would find the distance between bodies forbiddingly large.

A few other types of asteroids are known, and a number of individual asteroids have been found that are unique, but these three classes contain a majority of the known asteroids.

How did these three types originate? A clue lies in their distribution in the asteroid belt. The S-type asteroids are much more common in the inner belt. In fact, there are almost none beyond a distance of about 3.45 AU. The C types are rather rare in the inner belt but are very common in the outer belt. Many astronomers believe that this distribution reflects the differences in temperature during the formation of the planetesimals. It was cooler in the outer belt, and the planetesimals there tended toward the volatile-rich composition of carbonaceous chondrites. In the warmer regions of the inner belt, the composition of the growing planetesimals was more like that of the chondrites.

Earlier you saw evidence that Vesta has at some time in the past been at least partly resurfaced by lava flowing up from its interior. How could a small asteroid be heated enough to produce

What evidence makes you think that the asteroids have been fragmented?

Perhaps the best scientific arguments test the interpretation of evidence. If you understand the evidence, you hold the key to the science. To begin, you might note that the solar nebula theory of the formation of the solar system predicts that planetesimals collided and either stuck together or fragmented. This is suggestive, but it is not evidence. A theory can never be used as evidence to support some other theory or hypothesis. Evidence means observations or the results of experiments, so you need to turn to observations of asteroids. Spacecraft photographs of asteroids such as Ida, Gaspra, and Eros show irregularly shaped little worlds heavily scarred by impact craters. In fact, observations of some asteroids show what may be pairs of bodies in contact, and the Galileo image of Ida reveals its small satellite, Dactyl. Furthermore, some meteorites appear to come from the asteroid belt, and a few have been linked to specific asteroids such as Vesta. There are even families of asteroids that seem to be fragments from a single collision. All of this evidence suggests that the asteroids have been broken up by violent impacts.

The impact fragmentation of asteroids has been important, but it has not erased all traces of the original planetesimals from which the asteroids formed. Build another argument based on evidence. **What evidence can you cite that reveals what those planetesimals were like?**

■ ■ ■

Connections: The asteroids are only the last crumbs left behind by failed planet building near Jupiter. Farther out in the solar system are other remains—ice shards from the age of planet formation.

25-3 Comets

FEW THINGS IN ASTRONOMY are more beautiful than a bright comet hanging in the night sky (■ Figure 25-13). Whereas meteors shoot across the sky like demented fireflies, a comet moves with the stately grace of a great ship at sea, its motion hardly apparent. Night by night it shifts against the stars and may remain visible for months. Faint comets are common; a number are discovered every year. But a truly bright comet appears about once a decade. Comet Hyakutake in 1996 (Figure 25-1) and Comet Hale–Bopp in 1997 were both dramatic, but the later comet was so bright that you might class it with the great comets such as Comet Halley in 1910. A patient person might see half a dozen or more bright comets in a lifetime.

Visual-wavelength image

■ Figure 25-13

Comet C/2001 Q4 (NEAT) was discovered in 2001 by the Near Earth Asteroid Tracking (NEAT) system operated by NASA. The comet was brightest in 2004. A comet can remain bright in the sky for weeks or months as it sweeps along its orbit through the inner solar system. (T. Rector, Univ. of Alaska Anchorage, X. Levay and L. Frattare, STScI and WIYN/NOAO/AURA/NSF)

While everyone enjoys the beauty of comets, astronomers study them for their cargo of clues to the origin of our solar system.

Properties of Comets

As always, you should begin your study of a new kind of object by summarizing its observational properties. What do comets look like, and how do they behave? The observations are the evidence that reveals the secrets of the comets.

Study **Comet Observations** on pages 636–637 and notice three important properties of comets:

1 Comets have two kinds of tails shaped by the solar wind and solar radiation. The two kinds of tails show that the nucleus contains ices of water and other compounds plus rocky material most evident as dust.

2 Notice the importance of dust in comets. It not only produces dust tails but spreads throughout the solar system.

3 Notice the evidence that comet nuclei are fragile and can break into pieces.

Astronomers can put these and other observations together to discuss the structure of comet nuclei.

Ace◆Astronomy™ Log into AceAstronomy and select this chapter to see Astronomy Exercise "Comets." Design your own cometary nucleus and see what kind of tails it produces.

The Geology of Comet Nuclei

The nuclei of comets are quite small and cannot be studied in detail from Earth-based telescopes. Nevertheless, astronomers are beginning to understand the geology of these peculiar worlds.

Comet nuclei contain ices of water and other volatile compounds such as carbon dioxide, carbon monoxide, methane, ammonia, and so on. These ices are the kinds of compounds that should have condensed from the outer solar nebula, and that makes astronomers think that comets are ancient samples of the gases and dust from which the planets formed.

When the nuclei of comets approach the sun, the ices absorb energy from sunlight and sublime—change from a solid directly into a gas—to produce the observed tails. As the gases break down and combine chemically, they produce many compounds found in comet tails. Vast clouds of hydrogen gas observed around the heads of comets are derived from the breakup of molecules from the ices.

Five spacecraft flew past the nucleus of Comet Halley when it visited the inner solar system in 1985 and 1986. Spacecraft flew past the nuclei of Comet Borrelly in 2001 and Comet Wild 2 in 2004. The Deep Impact probe hit Comet Tempel 1 in 2005. Photos show that these comet nuclei are irregular in shape and very dark, with jets of gas and dust spewing from active regions on the nuclei (■ Figure 25-14). In general, these nuclei are darker than a lump of coal, which suggests the composition of the carbon-rich meteorites called carbonaceous chondrites.

From the gravitational influence of a nucleus on a passing spacecraft, astronomers can find the mass and density of the nucleus. Comet nuclei appear to have densities of 0.1 to 0.25 g/cm^3, much less than the density of ice. From these observations you can conclude that comet nuclei are not solid balls of ice but must be fluffy mixtures of ices and rocky dust with significant amounts of empty space.

Photographs of the comae (plural of coma) of comets often show jets springing from the nucleus and being swept back by the pressure of sunlight and by the solar wind to form the tail (Figure 25-14). Studies of the motions of these jets as the nucleus rotates reveal that the jets originate from active regions that may be faults or vents. As the rotation of a cometary nucleus carries an

■ Figure 25-14

Visual-wavelength images made by spacecraft and by the Hubble Space Telescope show how the nucleus of a comet produces jets of gases from regions where sunlight vaporizes ices. (Halley nucleus: © 1986 Max-Planck Institute; Halley coma: Steven Larson; Comets Borrelly, Hale–Bopp and Wild 2: NASA)

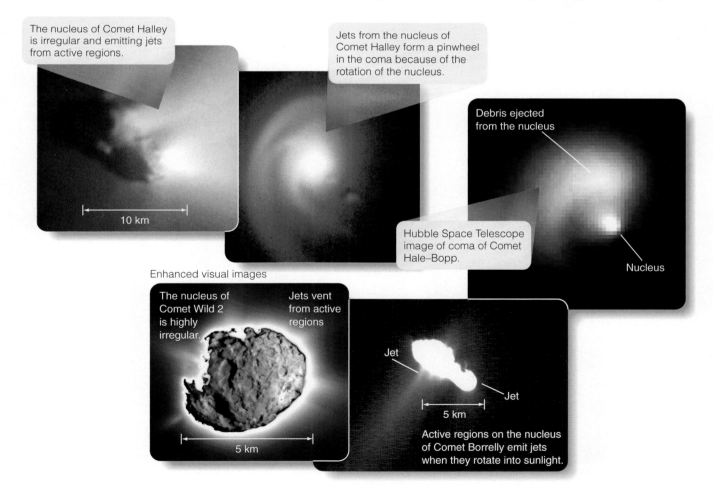

The nucleus of Comet Halley is irregular and emitting jets from active regions.

10 km

Jets from the nucleus of Comet Halley form a pinwheel in the coma because of the rotation of the nucleus.

Debris ejected from the nucleus

Hubble Space Telescope image of coma of Comet Hale–Bopp.

Nucleus

Enhanced visual images

The nucleus of Comet Wild 2 is highly irregular.

Jets vent from active regions

5 km

Jet

Jet

5 km

Active regions on the nucleus of Comet Borrelly emit jets when they rotate into sunlight.

Comet Observations

1 A **type I** or **gas tail** is produced by ionized gas carried away from the nucleus by the solar wind. The spectrum of a gas tail is an emission spectrum. The atoms are ionized by the ultraviolet light in sunlight. The wisps and kinks in gas tails are produced by the magnetic field embedded in the solar wind.

Spectra of gas tails reveal atoms and ions such as H_2O, CO_2, CO, H, OH, O, S, C, and so on. These are released by the vaporizing ices or produced by the breakdown of those molecules. Some gases, such as hydrogen cyanide (HCN), must be formed by chemical reactions.

Gas tail (Type I)

Dust tail (Type II)

1a A **type II** or **dust tail** is produced by dust from the vaporizing ices of the nucleus. The dust is pushed gently outward by the pressure of sunlight, and it reflects an absorption spectrum, the spectrum of sunlight. The dust is not affected by the magnetic field of the solar wind, so dust tails are more uniform than gas tails. Dust tails are often curved because the dust particles follow their individual orbits around the sun once they leave the nucleus.

When a spacecraft named ICE passed through the gas tail of a comet, it found a magnetic field from the solar wind draped over the nucleus like seaweed draped over a fishhook.

Nucleus ————

1b The nucleus of a comet (not visible here) is a small, fragile lump of porous rock containing ices of water, carbon dioxide, ammonia, and so on. Comet nuclei can be 10 to 100 km in diameter.

Coma

The **coma** of a comet is the cloud of gas and dust that surrounds the nucleus. It can be over 1,000,000 km in diameter, bigger than the sun.

1c Comet Mrkos in 1957 shows how the gas tail can change from night to night due to changes in the magnetic field in the solar wind.

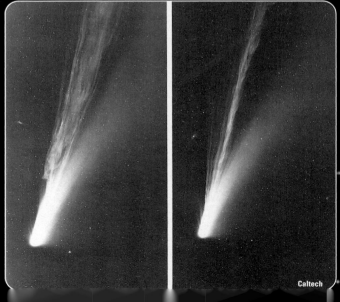

Caltech

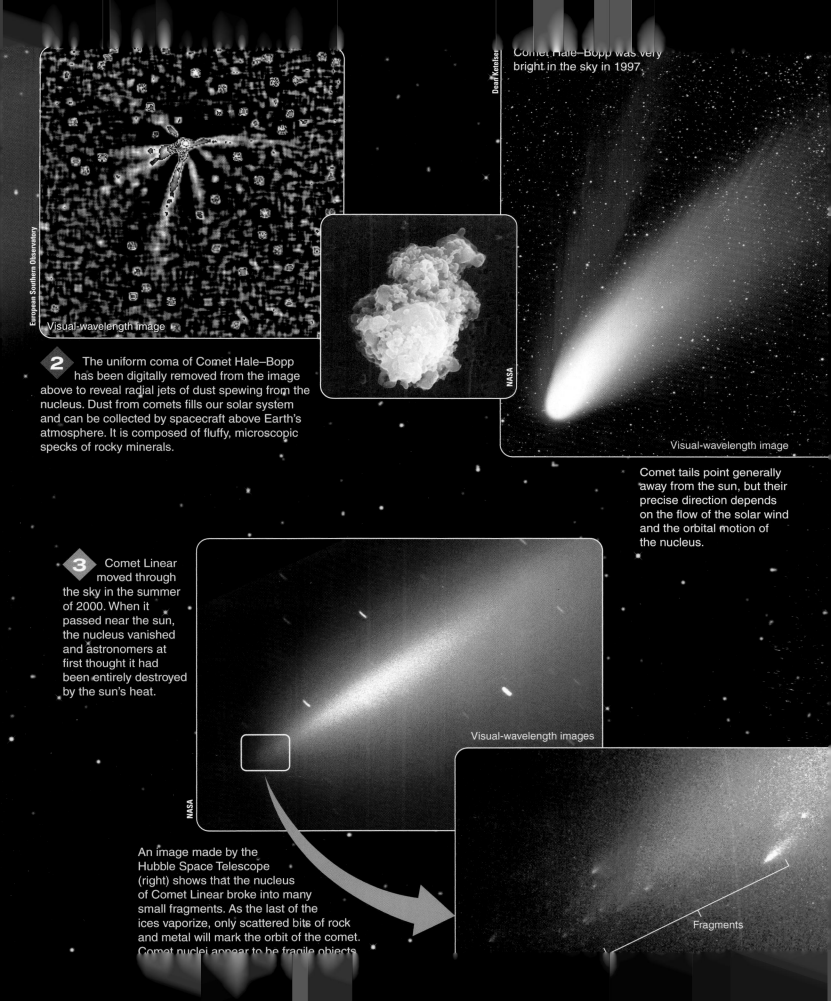

Visual-wavelength image

2 The uniform coma of Comet Hale–Bopp has been digitally removed from the image above to reveal radial jets of dust spewing from the nucleus. Dust from comets fills our solar system and can be collected by spacecraft above Earth's atmosphere. It is composed of fluffy, microscopic specks of rocky minerals.

Comet Hale–Bopp was very bright in the sky in 1997.

Visual-wavelength image

Comet tails point generally away from the sun, but their precise direction depends on the flow of the solar wind and the orbital motion of the nucleus.

3 Comet Linear moved through the sky in the summer of 2000. When it passed near the sun, the nucleus vanished and astronomers at first thought it had been entirely destroyed by the sun's heat.

Visual-wavelength images

An image made by the Hubble Space Telescope (right) shows that the nucleus of Comet Linear broke into many small fragments. As the last of the ices vaporize, only scattered bits of rock and metal will mark the orbit of the comet. Comet nuclei appear to be fragile objects.

Fragments

active region into sunlight, it begins venting gas and dust, and as it rotates into darkness it shuts down. Evidently, the nuclei of comets have a crust of rocky dust that may be material left behind when the ices vaporize. Breaks in that crust can expose ices to sunlight, and vents can occur in those regions. It also seems that some comets have pockets of volatiles buried below the crust. When one of those pockets is exposed and begins to vaporize, the comet can suffer a dramatic outburst.

In 2005, the Deep Impact spacecraft released an instrumented impactor that weighed 816 lb on Earth into the path of comet Tempel 1. Exactly as planned, the nucleus of the comet ran into the impactor at almost 10 km/s (23,000 mph). The impact penetrated the crust of the nucleus and blasted material out into space where the mothership flying past could analyze it (■ Figure 25-15). The burst of vapor and dust was detected from the mothership and also from Earth-based telescopes.

The nuclei of comets are not strong. A nucleus can be ripped apart by the violence of gases bursting through the crust or by tidal forces produced if the comet passes near a massive body. A number of comets have been observed to break into two or more parts while near Jupiter or near the sun. On July 8, 1992, Comet Shoemaker–Levy 9 passed within 1.29 planetary radii of Jupiter's center, well within its Roche limit, and tidal forces ripped the nucleus into at least 21 pieces (■ Figure 25-16a). The fragmented pieces were as large as a few kilometers in diameter, and they released gas and dust and spread into a long string of objects that looped away from Jupiter and then fell back to strike the planet and produce massive impacts over a period of six days in July 1994 (Figure 23-6). The fragmentation of comet nuclei may explain long chains of craters found on the Earth's moon and on some other planetary surfaces in the solar system. A comet breaking into pieces could produce a chain of impact features such as the ones shown in Figure 25-16b and c. The Solar and Heliospheric Observatory (SOHO) was put into space to observe the sun, but it has discovered over 1000 comets called "sun grazers" as they zip in and around the sun. These comets follow long elliptical orbits that bring them 70 times closer to the sun's surface than the planet Mercury (■ Figure 25-17). Most of these comets belong to one of four groups, with the comets in each group having very similar orbits. Like the Hirayama families of asteroids, these comet groups

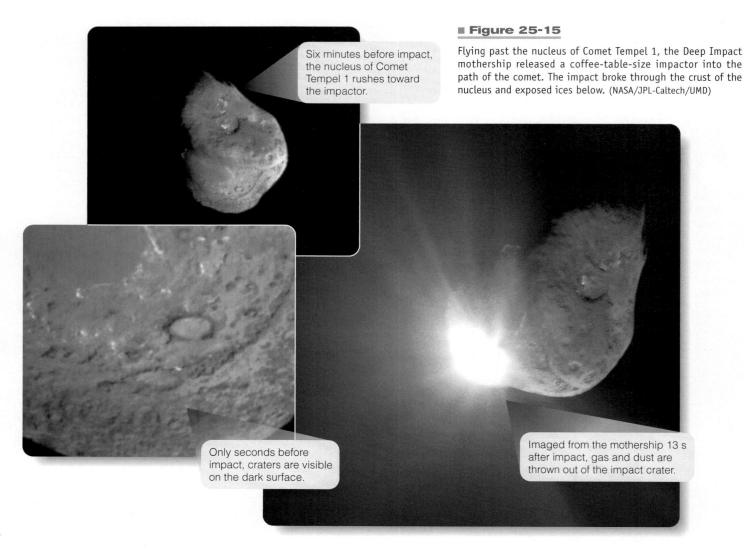

■ **Figure 25-15**

Flying past the nucleus of Comet Tempel 1, the Deep Impact mothership released a coffee-table-size impactor into the path of the comet. The impact broke through the crust of the nucleus and exposed ices below. (NASA/JPL-Caltech/UMD)

Six minutes before impact, the nucleus of Comet Tempel 1 rushes toward the impactor.

Only seconds before impact, craters are visible on the dark surface.

Imaged from the mothership 13 s after impact, gas and dust are thrown out of the impact crater.

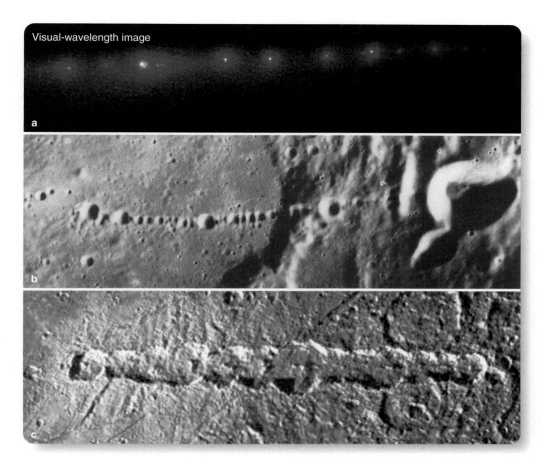

Visual-wavelength image

a

b

c

(a) Tides from Jupiter pulled apart the nucleus of Comet Shoemaker–Levy 9 to form a long strand of icy bodies and dust that fell back to strike Jupiter two years later. (b) A 40-km-long crater chain on the moon and (c) another 140-km-long crater chain on Jupiter's moon Callisto were apparently formed by the impact of fragmented comet nuclei. The impacts that form such chains probably occur within a span of seconds. (NASA)

appear to be made up of fragments of larger bodies. You can jot this down as yet more evidence that comet nuclei are very fragile.

The nuclei of comets range from a few kilometers in diameter up to a few tens of kilometers. Each passage near the sun costs a nucleus many millions of tons of ices, so the nucleus slowly loses its ices until there is nothing left but dust and rock falling along an orbit around the sun. The fate of a comet is clear. The mystery is its origin.

The Origin of Comets

Family relationships among the comets can give you clues to their origin. Most comets have long, elliptical orbits with periods greater than 200 years. These are known as long-period comets. Their orbits are randomly inclined, with comets falling into the inner solar system from all directions. As many circle the sun clockwise as counterclockwise.

In contrast, about 100 or so of the 600 well-studied comets have orbits with periods less than 200 years. These short-period comets follow orbits that lie within 30° of the plane of the solar system, and most revolve around the sun counterclockwise—the same direction the planets orbit. Comet Halley, with a period of 76 years, is a short-period comet.

A comet cannot survive long in an orbit that brings it into the inner solar system. The heat of the sun vaporizes ices and reduces comets to inactive bodies of rock and dust. A comet may last only 100 to 1000 orbits around the sun. Also, encounters with planets, especially Jupiter, can fling a comet into the sun or out of the solar system. Of course, as you have seen, comets hit planets, and that's the end of any such unlucky comets. Even if it didn't vaporize in the sun's heat, a comet couldn't survive more than about half a million years before its career was ended by a planet. The comets visible in our skies can't have survived 4.6 billion years since the formation of the solar system, and that means there must be a continuous supply of new comets. Where do they come from?

In the 1950s, Dutch astronomer Jan Oort proposed that the long-period comets are objects that fall in from the **Oort cloud,** a spherical cloud of icy bodies believed to extend from about 10,000 to 100,000 AU from the sun (■ Figure 25-18). Astronomers estimate that the cloud contains several trillion icy bodies. Far from the sun, they are very cold, lack comae and tails, and are invisible. The gravitational influence of occasional passing stars could perturb a few of these objects to fall into the inner solar system, where the heat of the sun warms their ices and transforms them into comets. Because the Oort cloud is spherical, these long-period comets fall inward from random directions.

It is not outlandish that stars pass close enough to affect the Oort cloud. Data from the Hipparcos satellite show that the star Gliese 710 will pass within 1 ly (about 63,000 AU) of our solar system in about a million years. That may shower the inner solar

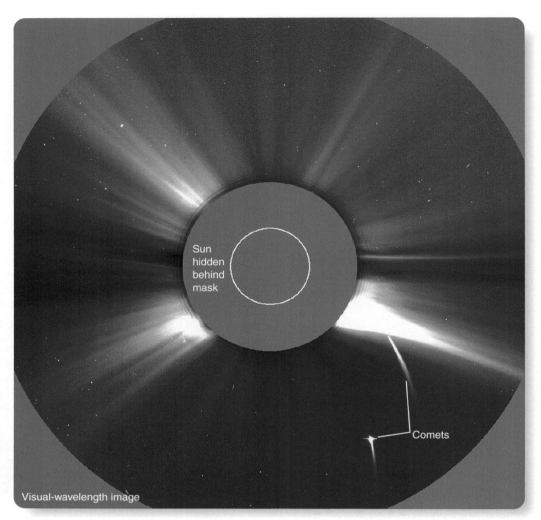

Sun
hidden
behind
mask

Comets

Visual-wavelength image

The SOHO observatory can see comets round-
ing the sun on very tight orbits. Some of
these sun-grazing comets, like the two
shown here, are destroyed by the heat and
are not detected emerging on the other side
of the sun. (SOHO)

system with Oort cloud objects, which, warmed by the sun, will
become comets.

Saying that comets come from the Oort cloud only pushes
the mystery back one step. Where did those icy bodies come from?
Beginning in Chapter 19, you have studied the origin and evolution of our solar system so carefully that the answer leaps out
at you. Those icy bodies are some of the icy planetesimals that
formed in the outer solar nebula.

The bodies in the Oort cloud, however, could not have formed
at their present location. The solar nebula would not have been
dense enough so far from the sun. Also, had they formed from
the solar nebula, you would expect them to be distributed in a
disk and not in a sphere. Rather, astronomers think the future
comets formed as icy planetesimals in the outer solar system near
the present orbits of Uranus and Neptune. As those planets grew
more massive, they swept up many of these planetesimals, but they

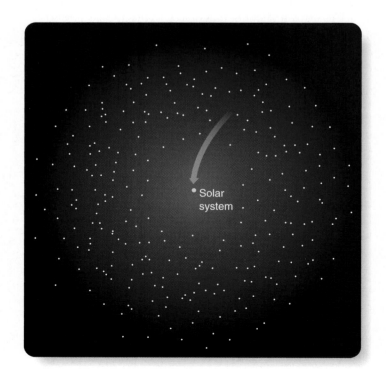

Solar
system

■ **Figure 25-18**

The long-period comets appear to originate in the Oort cloud. Objects that
fall into the solar system from this cloud arrive from all directions.

would have ejected some out of the solar system. Most of those objects vanished into space, but perhaps 10 percent had their orbits modified by the gravity of stars passing nearby and became part of the Oort cloud. These are the objects that form the long-period comets.

Some of the short-period comets, including Comet Halley, appear to have originated in the Oort cloud and had their orbits altered by a close encounter with Jupiter. But many of the short-period comets cannot have begun in the Oort cloud. Interactions with a planet can't capture objects from the Oort cloud into the orbits that many short-period comets occupy. There must be another source of icy bodies in our solar system. To find the answer, you need only look out beyond the orbit of Neptune and study the icy Kuiper Belt objects.

The Kuiper Belt

If you want to have fun, get a bunch of astronomers together, order some pizza, and ask them to decide whether the Kuiper Belt objects are asteroids or comets. The argument will outlast the refreshments. Whatever they are, you first met the Kuiper Belt objects in Chapter 19 when you studied the origin of the solar system. It is important to study them carefully here because they seem to be one of the sources of comets.

In 1951, Dutch-American astronomer Gerard P. Kuiper proposed that the formation of the solar system should have left behind a belt of small, icy planetesimals beyond the Jovian planets and in the plane of the solar system. Such objects were first discovered in 1992 and are now known as Kuiper Belt objects. You should know, however, that in 1943 and 1949, astronomer Kenneth Edgeworth published papers that included a paragraph speculating about objects beyond Pluto. Consequently, you may see the Kuiper Belt referred to as the Edgeworth–Kuiper Belt. Actually, lots of astronomers have speculated about one or more bodies beyond Pluto, so the real controversy isn't in their name but in their origin.

The Kuiper Belt objects are small, icy bodies (■ Figure 25-19) that orbit in the plane of the solar system extending from the orbit of Neptune out to about 50 AU from the sun. Some objects loop out as far as 150 AU, but they seem to have been scattered by gravitational interactions with passing stars. The entire Kuiper Belt, containing as many as 70,000 objects as big as 100 km in diameter, would be hidden in Figure 25-18 behind the yellow dot symbolizing the solar system. Some Kuiper Belt objects are surprisingly large, and one is about 50 percent bigger than Pluto.

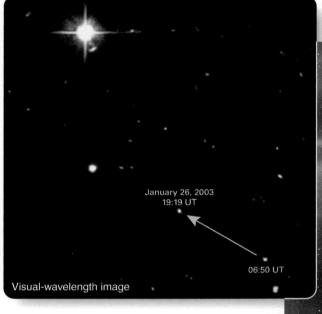

■ **Figure 25-19**

Kuiper Belt objects are small bodies with dark surfaces that are very hard to detect from Earth. In your imagination, you can see them as icy, cratered worlds orbiting far from the sun. The New Horizons spacecraft is planned to fly past at least one Kuiper Belt object after it passes Pluto. Dust, mostly from comets, produces the horizontal glow centered on the sun. (Image: NASA and G. Bernstein; Art: JHUAPL/SwRI)

Presumably these objects are icy planetesimals that formed in the outer solar system beyond Neptune. Planetesimals among the giant planets were ejected, thrown into the sun, or absorbed by the planets, but the KBOs orbiting beyond Neptune are safe from such interactions and could have survived since the formation of the solar system. In fact, they provide clues as to how our solar system formed. In Chapter 24, you learned that some KBOs orbit in resonances with Neptune, and the family called the plutinos provided evidence that Pluto may be one of the largest KBOs. The objects probably became caught in resonances when Neptune migrated outward during or soon after planet formation. The presence of a sharp outer edge to the Kuiper Belt may indicate that the sun formed as a member of a star cluster and that the crowded stars tidally limited the growth of the solar nebula beyond about 50 AU.

Can this belt of ancient, icy worlds generate short-period comets? Mathematical studies show that objects thrown inward from the Kuiper Belt could take up orbits resembling the short-period comets. Collisions and interactions among the KBOs must be a rare but continuing process, and small KBOs or fragments from a collision would resemble the nuclei of comets.

Some of these objects are quite large. One, called Quaoar (pronounced "KWAH-o-wahr"), is about 1270 km in diameter, a bit larger than Pluto's moon Charon. The KBO known as Sedna is even larger. Encounters with such a large body could throw a small object into an extreme orbit. Another object, known as 1998 W31, has its own moon. In fact, binary objects seem quite common in the Kuiper Belt, and that may be evidence that collisions are common.

You can find more evidence that our solar system has a Kuiper Belt by looking at other stars. A number of nearby stars such as Beta Pictoris (Chapter 19) are surrounded by disks of dust believed to be released by objects in Kuiper belts around these stars. Thus the detection of the dust implies the presence of Kuiper belts. It appears that our solar system would be peculiar if it didn't have a Kuiper Belt and plenty of dusty comets.

Building Scientific Arguments

How do comets help explain the formation of the planets?
This scientific argument pulls together a number of ideas. According to the solar nebula hypothesis, the planets formed from planetesimals that accreted in a disk-shaped nebula around the forming sun. In the outer solar nebula, it was cold, and the planetesimals would have contained large amounts of ices. Many of these planetesimals were destroyed when they fell together to make the planets, but some survived. The icy bodies of the Oort cloud and the Kuiper Belt may be the last surviving icy planetesimals in our solar system. When these bodies fall into the inner solar system, they become comets, and the gases they release reveal that they are rich in volatile materials such as water, carbon dioxide, carbon monoxide, methane, and ammonia. These are the ices you would expect to find in the icy planetesimals. Furthermore, comets are rich in dust, and the planetesimals must have included large amounts of dust frozen into the ices when they formed. The nuclei of comets seem to be frozen samples of the ancient solar nebula.

Nearly all of the mass of a comet is in the nucleus, but the light you see comes from the coma and the tail. Build an argument pulling together ideas about cometary nuclei. **What kind of spectra do comets produce, and what does that tell you about their nuclei?**

■　　■　　■

Connections: Your study of comets has revealed that there is a continuous supply of these objects, that they fragment easily when they encounter planets, and that entire comets or fragments of comets can hit planets. Surely that makes you wonder what happens when a comet hits Earth.

25-4 Impacts on Earth

ASTEROIDS AND COMETS FALL through our solar system like runaway trucks on a busy highway. These objects must hit the planets. In fact, it is these collisions that built the planets, and most of the original planetesimals have now been either incorporated into the planets or ejected from the solar system. Nevertheless, planets still get hit, as was dramatically demonstrated in 1994 when Comet Shoemaker–Levy hit Jupiter. Like any other planet, Earth must get hit now and then; where is the evidence?

Impacts and Dinosaurs

Comet nuclei and asteroids may seem small compared to planets, but they hit Earth with tremendous energy. Calculations show that such impacts throw vast quantities of dust into the atmosphere and cause widespread changes in climate. The extinction of entire species, such as the dinosaurs, may have been caused by comet impacts. One of the most important clues is an element called iridium.

In 1980, Luis and Walter Alvarez announced the discovery of an excess of the element iridium in sediments laid down at the end of the Cretaceous period 65 million years ago—just when the dinosaurs and many other species became extinct. Iridium is rare in Earth's crust but common in meteorites, which led the Alvarezes to suggest that a major meteorite impact at the end of the Cretaceous threw vast amounts of iridium-rich dust into the atmosphere. This dust might have plunged Earth into a winter that lasted many years, killing off many species of plants and animals, including the dinosaurs.

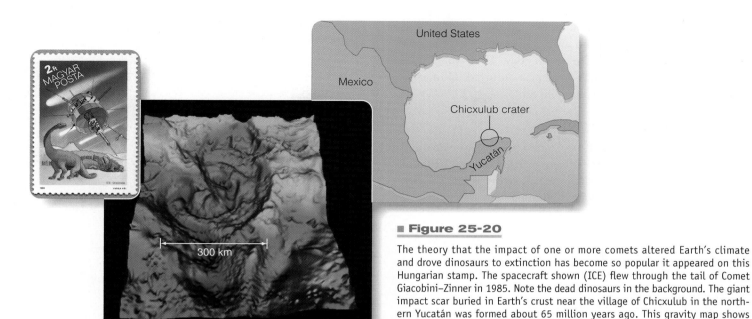

■ Figure 25-20

The theory that the impact of one or more comets altered Earth's climate and drove dinosaurs to extinction has become so popular it appeared on this Hungarian stamp. The spacecraft shown (ICE) flew through the tail of Comet Giacobini–Zinner in 1985. Note the dead dinosaurs in the background. The giant impact scar buried in Earth's crust near the village of Chicxulub in the northern Yucatán was formed about 65 million years ago. This gravity map shows the extent of the crater hidden below limestone deposited since the impact. (Virgil L. Sharpton, Univ. of Alaska, Fairbanks)

This theory was met with skepticism at first, but soon scientists found the iridium anomaly in sediments of the same age all over the world. Others found related elements and mineral forms typical of meteorite impacts. At the same time, theorists studying nuclear weapons predicted that a nuclear war would throw so much dust into the atmosphere that our planet would be plunged into a "nuclear winter" that could last a number of years. A major impact would do the same thing. Within a few years, scientists generally agreed that the Cretaceous extinctions were caused by one or more major impacts.

Mathematical models and observations of the impact of the comet fragments on Jupiter in 1994 have combined to create a plausible scenario of the events following a major impact on Earth. Of course, creatures living near the site of the impact would probably die in the initial shock, and an impact at sea would create tsunamis (tidal waves) many hundreds of meters high that would devastate coastal regions for many kilometers inland even halfway around the world from the impact. But the worst effects would begin after the initial explosion. On land or sea, a major impact would excavate large amounts of pulverized rock and loft it high above the atmosphere. As this material fell back, Earth's atmosphere would be turned into a glowing oven of red-hot rock streaming through the air as a rain of meteors, and the heat would trigger massive forest fires around the world. Soot from such fires has been detected in the layers of clay laid down at the end of the Cretaceous. Once the firestorms cooled, the remaining dust in the atmosphere would block sunlight and produce deep darkness for a year or more. No matter where on Earth an impact occurred, it would almost certainly vaporize large amounts of limestone.

The carbon dioxide released from the limestone would produce intense acid rain. All of these consequences make it surprising that any life could have survived such an impact.

Geologists have located a crater at least 150 km in diameter centered near the village of Chicxulub in the northern Yucatán (■ Figure 25-20). Although the crater is totally covered by sediments, mineral samples show that it contains shocked quartz typical of impact sites and that it is the right age. A gigantic impact formed the crater about 65 million years ago, just when the dinosaurs and many other species died out, and many Earth scientists now believe that this is the scar of the impact that ended the Cretaceous.

At first, these climate-changing impacts were blamed on large meteorites, but the nucleus of a comet would do as much damage. In fact, you have seen in this chapter that the distinction between an asteroid and a comet is not clear-cut.

The theory that an impact caused the extinction at the end of the Cretaceous period was once highly controversial, but the accumulated evidence has made it a widely accepted idea. Now Earth scientists are extending the theory to other extinctions. Chemical evidence from Earth's crust suggests that the extinction at the end of the Permian 250 million years ago was caused by a giant impact. The Permian extinction killed off 95 percent of the species on Earth and set the stage for the rise of the dinosaurs. The dinosaurs may owe their origin as well as their extinction to giant impacts on Earth.

A solar system is a dangerous place to put a planet. Comets, asteroids, and meteoroids constantly rain down on the planets, and Earth gets hit quite often. The Chicxulub Crater isn't the only large impact scar on Earth. For example, a giant crater has been found buried under sediment in Iowa, and another giant impact crater may underlie most of Chesapeake Bay. What would it be like to live on Earth when such an impact occurred? Humanity got a hint in 1908 when something hit Siberia.

The Tunguska Event

On the morning of June 30, 1908, scattered reindeer herders and homesteaders in central Siberia were startled to see a brilliant blue-white fireball brighter than the sun streak across the sky. Still descending, it exploded with a blinding flash and an intense pulse of heat:

> *The whole northern part of the sky appeared to be covered with fire. . . . I felt great heat as if my shirt had caught fire . . . there was a . . . mighty crash. . . . I was thrown on the ground about [7 m] from the porch. . . . A hot wind, as from a cannon, blew past the huts from the north.*

The blast was heard up to 1000 km away, and the resulting pulse of air pressure circled Earth twice. For a number of nights following the blast, European astronomers, who knew nothing of the explosion, observed a glowing reddish haze high in the atmosphere.

Travel in the wilderness of Siberia was difficult early in the 20th century; moreover, World War I, the Bolshevik Revolution, and the Russian Civil War prevented expeditions from reaching the site before 1927. When at last an expedition arrived, it found that the blast had occurred above the Stony Tunguska River valley and had flattened trees in an irregular pattern extending out 30 km (■ Figure 25-21). The trees were knocked down radially away from the center of the blast, and limbs and leaves had been stripped away. The trunks of trees at the very center of the area were still standing, although they had lost all their limbs. No expedition has ever found a crater, so it seems that the explosion, equaling a 12-megaton nuclear weapon, occurred at least a few kilometers above the ground.

All kinds of speculative theories have been proposed in tabloid newspapers and occult books. One popular idea was that a flying saucer with engine trouble tried to make an emergency landing and exploded. More responsible ideas have proposed that Earth was hit by a piece of antimatter or a miniature black hole. The evidence does not necessarily rule out either of these last two ideas, but they violate an important principle in science. They assume more than needed. Science always prefers the simplest adequate explanation for any given phenomenon, and a small asteroid or comet would explain the event.

The comet hypothesis proposes that the head of a small comet struck Earth and exploded in the atmosphere. You might not see such a comet if it was coming from the direction of the sun. Witnesses reported that the fireball in the atmosphere came from the east, which in the morning means from the direction of the sun.

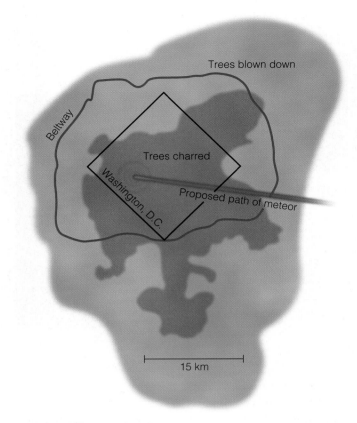

■ **Active Figure 25-21**

The 1908 Tunguska event in Siberia destroyed an area the size of a large city. Here the area of destruction is superimposed on a map of Washington, D.C., and its surrounding beltway. In the central area, trees were burned; and, in the outer area, trees were blown down in a pattern away from the path of the meteor.

Ace Astronomy™ Log into AceAstronomy and select this chapter to see the Active Figure "Impacts on Earth." Learn how often Earth is hit by meteorites of different sizes.

If the object was a comet, whatever tail it had would have been projected straight at Earth and would not have been visible. More likely, say some astronomers, the object was the exhausted core of a comet and had no appreciable tail.

A fragment of a partially exhausted comet should consist of silicate material resembling carbonaceous chondrites, mixed with some remaining ices. Modern studies of the area show that thousands of tons of powdered material resembling carbonaceous chondrites are scattered in the soil. A chunk of dirty ice 50 m in diameter would contain about the right mass and would be totally invisible approaching from the sun. Also, a comet fragment would be so fragile that it would explode in the atmosphere rather than dig a crater, as would a denser, stronger meteorite.

The falling-comet hypothesis has been very popular with astronomers. It does not make any fantastic assumptions, and it explains the observed phenomena, even the reddish glow over

Europe. Some astronomers claim the glow was the faint traces of the comet's scattered dust, trapped high in the atmosphere and illuminated by the summer sun shining over the North Pole.

As compelling as the comet hypothesis may seem, not all astronomers agree. In the early 1980s, a detailed analysis of all the evidence suggested that the object was not cometary. Its speed and direction were not consistent with the motion of a comet. The analysis suggested instead that the object was a small Apollo asteroid, 90 to 190 m in diameter.

In 1993, astronomers studied computer models of objects entering Earth's atmosphere at high speed and concluded that the fragile head of a comet would have exploded much too high in the atmosphere. An iron-rich asteroid, in contrast, would be so strong that it would survive to reach the ground and would form a large crater. The most likely candidate seems to be a stony asteroid about 30 m in diameter. It would explode at just about the right height to produce the Tunguska blast.

One natural question is: Why should a piece of rock explode at all? To understand, you need to recall that a meteor enters Earth's atmosphere at roughly 15 km/s. The air resistance is tremendous; and, as the meteor descends lower and meets denser air, the air resistance increases rapidly. The resistance that the meteor experiences depends on its surface area. When the meteor breaks into smaller pieces, its surface area suddenly increases, so the air resistance suddenly increases, and that causes even more fragmentation. Because air resistance depends on surface area, when the meteor begins to break up, it explodes and vaporizes almost instantly. That is most likely what created the blast that equaled a 12-megaton explosion at Tunguska.

Could such an impact happen again? Earth gets hit by small meteorites every day and by larger objects less often. Impacts by large asteroids may happen many millions of years apart, but they continue to happen. In mid-March 1998, newspaper headlines announced, "Mile-Wide Asteroid to Hit Earth in October 2028." The frightening story was true except that the news media did not mention the uncertainty in the orbit. Within days, astronomers found the asteroid on old photographs, refined the orbit with the new data, and concluded that the asteroid, known as $1997XF_{11}$, will miss Earth by 600,000 miles. There will be no impact by this asteroid in 2028, but there are plenty more asteroids that haven't been discovered yet.

Some people have argued that the danger from asteroid and comet impacts is so great that governments should develop massive nuclear-tipped missiles, ready to blast a meteoroid to pieces long before it can reach Earth. Some astronomers point out that lots of small fragments slamming into Earth may be even worse

than one big impact. Other astronomers argue that the big objects are so rare they can be ignored. The real danger, they say, lies in the smaller, more common meteoroids, which are too small to detect with current telescopes.

Some people may prefer to believe that the Tunguska event was an alien spaceship; that is a titillating idea. But the truth about the Tunguska event is far more exciting. Earth is struck by meteorites every day. Some impacts are large enough to alter the climate, and the future of our civilization on Earth may depend on how much time we have until the next big one.

Ace★Astronomy™ Log into AceAstronomy and select this chapter to see Astronomy Exercise "Cratering." What factors determine the size of a crater?

Building Scientific Arguments

How can a large meteorite impact alter Earth's climate?
This argument is based on theory, but some evidence does exist. Even a meteorite no bigger than a large house strikes Earth with as much energy as a nuclear bomb and can produce a crater a mile in diameter. A large impact could produce a crater 100 km or more in diameter. Rock from Earth's crust would be pulverized and lifted high above the atmosphere. As it fell back in a rain of fiery meteors, it would be so hot it would trigger massive forest fires over much of Earth's surface. Scientists can detect traces of such fires at about the time of the Chicxulub impact, 65 million years ago. Even after the fires cooled, the dust in the atmosphere would prevent sunlight from reaching the ground, and plants dependent on sunlight would be starved of energy. This worldwide darkness might last a year or more before the dust settled out of the atmosphere. By that time, many species would have died out, and the climate could have changed dramatically.

As scary as this is, you need not ask if it will happen. The probability of such an event is 100 percent. Such impacts are common in the solar system. Now extend your argument: **Why does the presence of large amounts of limestone on Earth make a major impact even more devastating to the environment?**

■   ■   ■

Connections: Life on Earth seems fragile and exposed. Are humans just lucky to have survived so long? Does life exist on other worlds? If so, has it, too, survived long enough to develop intelligence? That is the story of the next chapter.

Summary

25-1 | Meteorites

Where do meteors and meteorites come from?
- The vast majority of meteors, including those in meteor showers, appear to be low-density, fragile bits of debris from comets.
- The evidence, including the orbits of meteorites seen to fall, suggests that the meteorites are fragments of asteroids.
- Meteorites can be classified according to their mineral content.
- Iron meteorites are mostly iron and nickel; when sliced open, polished, and etched, they show Widmanstätten patterns. That means the metal cooled from a molten state very slowly.
- Stony meteorites include chondrites, which contain condrules—small, glassy particles that are believed to be very ancient droplets of molten material formed in the solar nebula.
- Stony meteorites that are rich in volatiles and carbon are called carbonaceous chondrites. They are among the least-modified meteorites.
- The achondrites are stony meteorites that contain no chondrules and no volatiles. They appear to have been melted after they formed and, in some cases, resemble solidified lavas.
- Many meteorites appear to have formed as part of larger bodies that melted, differentiated, and cooled very slowly. Later they were broken up, and fragments from the core became iron meteorites, while fragments from the outer layers became stony meteorites.
- Carbonaceous chondrites appear to have formed further from the sun.

25-2 | Asteroids

What are the asteroids?
- Most asteroids lie in a belt between Mars and Jupiter, although some, the Apollo–Amor objects, follow orbits that cross into the inner solar system. If they pass near Earth, they are called Near Earth Objects (NEOs).
- Kirkwood's gaps in the asteroid belt are caused by orbital resonances with Jupiter.
- Two groups of asteroids called the Trojans are caught in the Lagrangian points along Jupiter's orbit 60° ahead and 60° behind the planet.
- Asteroids are irregular in shape and heavily cratered from collisions. Their surfaces are covered by gray, pulverized rock, and some asteroids have such low densities they appear to be fragmented rubble piles.
- C-type asteroids are more common in the outer asteroid belt, where it is cooler. They are darker and may be carbonaceous.
- S-type asteroids are the most common and appear to be the source of the most common kind of meteorites, the chondrites. S-type asteroids are more common in the inner belt.
- M-type asteroids appear to have nickel-iron compositions and may be the cores of broken asteroids.
- A few asteroids, such as Vesta, show evidence of lava flows on their surfaces, but others, such as Ceres, show no evidence of geological activity.
- The asteroids formed as rocky planetesimals between Mars and Jupiter, but Jupiter prevented them from forming a planet. Collisions have broken all but the largest of the asteroids.

25-3 | Comets

Where do comets come from?
- A comet is produced by a lump of rock and ices 10 to 100 km in diameter. In long, elliptical orbits, the icy body stays frozen until it nears the sun. Then some of the ices vaporize and release dust and gas that are blown away to form a tail.
- A type I, or gas, tail is ionized gas carried away by the solar wind.
- A type II, or dust, tail is solid debris released from the nucleus and blown outward by the pressure of sunlight.
- The coma of a comet can be up to a million kilometers in diameter, but the nucleus is rarely more than a few tens of kilometers across.
- Spacecraft flying past comets have revealed that the nucleus of a comet has a very dark, rocky crust and that jets of vapor and dust issue from active regions on the sunlit side.
- The low density of the nuclei show that they are irregular mixtures of ices and silicates, probably containing large voids.
- Comet nuclei are not strong and have been seen to break apart when they pass near planets, especially Jupiter. A comet broke up and hit Jupiter in 1994, and such impacts are probably common in the history of the solar system.
- Comets are believed to have formed as icy planetesimals in the outer solar system. Some were ejected to form the Oort cloud. Comets falling in from the Oort cloud become long-period comets.
- A few short-period comets have been produced when comets passed near Jupiter and had their orbits changed.
- Other icy bodies formed in the outer solar system and now make up the Kuiper Belt beyond Neptune. Objects from the Kuiper Belt that fall into the inner solar system can become short-period comets.

25-4 | Impacts on Earth

What happens when an asteroid or comet hits Earth?
- A major impact on Earth can trigger extinctions because of changes to the climate.
- Material falling back into the atmosphere after a major impact can trigger global forest fires.
- Large amounts of carbon dioxide released into the atmosphere by a major impact can produce intense acid rain.
- A major impact can fill the atmosphere with dust for years and plunge the planet into darkness.
- An impact at Chicxulub in the Yucatán 65 million years ago appears to have triggered the extinction of 75 percent of the species then on Earth, including the dinosaurs. The Great Dying at the end of the Permian may also have been triggered by a major impact.
- A small rocky object entered Earth's atmosphere and exploded over the Stony Tunguska River valley in Siberia in 1908. The explosion devastated forests in an area up to 60 km across.

New Terms

radiant (p. 619)

sporadic meteor (p. 620)

fall (p. 621)

find (p. 621)

iron meteorite (p. 622)

selection effect (p. 622)

Widmanstätten pattern (p. 622)

stony meteorite (p. 622)

chondrite (p. 622)

chondrule (p. 622)

carbonaceous chondrite (p. 623)

CAI (p. 624)

achondrite (p. 624)

stony-iron meteorite (p. 624)

Hirayama family (p. 627)

Kirkwood's gap (p. 630)

Apollo–Amor object (p. 631)

Trojan asteroid (p. 632)

gas (type I) tail (p. 636)

dust (type II) tail (p. 636)

coma (p. 636)

Oort cloud (p. 639)

Review Questions

Ace ⊙ Astronomy™ Assess your understanding of this chapter's topics with additional quizzing and animations at **http://ace .brookscole.com/sf9**

1. If most falls are stony meteorites, why are most finds iron?

2. How do observations of meteor showers reveal one of the sources of meteoroids?

3. How can most meteors be cometary if most, perhaps all, meteorites are asteroidal?

4. Why are meteorites easier to find in Antarctica? Why are stony meteorites better represented among these finds than among finds in the United States?

5. What evidence can you cite that some meteorites have originated inside large bodies?

6. Why couldn't uranium, thorium, and radioactive potassium have melted the planetesimals? What evidence can you cite that some asteroids have had molten interiors?

7. How do Kirkwood's gaps resemble Cassini's division?

8. The first asteroids discovered were much larger than the typical asteroid known today. How does this illustrate a selection effect?

9. Describe three types of asteroids (S, M, and C), and explain a theory to account for their differences.

10. How might the Apollo–Amor objects have originated?

11. What is the difference between a type I tail and a type II tail?

12. If comets are icy planetesimals left over from the formation of the solar system, why haven't they all vaporized by now?

13. What is the difference between the source of long-period comets and that of short-period comets?

14. How did the bodies in the Kuiper Belt and the Oort cloud form? Why are they in different locations in the solar system?

15. What do you see in this image that tells you how big planetesimals were when the solar system was forming?

(Russell Kempton, New England Meteoritical)

16. Discuss the surface of the asteroid Mathilde. What do you see that tells you something about the history of the asteroids?

Visual (NASA)

17. What do you see in this image of the nucleus of Comet Borrelly that tells you how comets produce their comae and tails?

Visual (NASA)

Discussion Questions

1. Futurists suggest that humans may someday mine the asteroids for materials to build and supply space colonies. What kinds of materials could they get from asteroids? (*Hint:* What are S-, M-, and C-type asteroids made of?)

2. If cometary nuclei were heated by internal radioactive decay rather than by solar heat, how would comets differ from what is observed?

3. From what you know now, do you think the government should spend money to locate near-Earth asteroids? How serious is the risk?

Problems

1. Large meteorites are hardly slowed by Earth's atmosphere. Assuming that the atmosphere is 100 km thick and that a large meteorite falls vertically toward the ground, how long does it spend in the atmosphere? (*Hint:* How fast do meteoroids travel?)

2. If a single asteroid 1 km in diameter were fragmented, how many meteoroids 1 m in diameter could it yield? (*Hint:* The volume of a sphere is $\frac{4}{3}\pi R^3$.)

3. What is the orbital velocity of a meteoroid orbiting the sun at the distance of Earth? (*Hint:* See Chapter 5.)

4. If a half-million asteroids, each 1 km in diameter, were assembled into one body, how large would it be? (*Hint:* The volume of a sphere is $\frac{4}{3}\pi R^3$.)

5. The asteroid Vesta has a mass of 2×10^{20} kg and a radius of about 250 km. What is its escape velocity? Could you jump off the asteroid? (*Hint:* See Chapter 5.)

6. The asteroid Pallas has a mass of 2.5×10^{20} kg. What is the orbital velocity of a small satellite orbiting 500 km from the center of Pallas? (*Hint:* See Chapter 5.)

7. What is the maximum angular diameter of Ceres as seen from Earth? Could Earth-based telescopes detect surface features? Could the Hubble Space Telescope (Chapter 6)? (*Hint:* Use the small-angle formula.)

8. At what distances from the sun would you expect to find Kirkwood's gaps where the orbital period of asteroids is one-half of and one-third of the orbital period of Jupiter? Compare your results with Figure 25-10. (*Hint:* Use Kepler's third law.)

9. If the velocity of the solar wind is about 400 km/s and the visible tail of a comet is 100 million km long, how long does it take an atom to travel from the nucleus to the end of the visible tail?

10. If you saw Comet Halley when it was 0.7 AU from Earth and it had a visible tail 5° long, how long was the tail in kilometers? Suppose that the tail was not perpendicular to the line of sight. Is your answer too large or too small? (*Hint:* Use the small-angle formula.)

11. Calculate the orbital velocity of a comet while it is in the Oort cloud. (*Hint:* See Chapter 5.)

12. The mass of a comet's nucleus is about 10^{14} kg. If the Oort cloud contains 2 trillion (2×10^{12}) cometary nuclei, what is the mass of the cloud in Earth masses? (*Hint:* Earth's mass = 6×10^{24} kg.)

Media Cluster

Ace⊗Astronomy™ To access the resources in the Media Cluster, log into AceAstronomy at **http://ace.brookscole.com/sf9** and select Chapter 25.

ACTIVE FIGURES

Build a Comet
This animation shows how radiation from the sun creates the tails of comets. You can compare this to "heat from everywhere," which causes particles to be shed in every direction.

Impacts on Earth
Watch asteroids pummel Earth in this animation and learn about the probability of a "killer asteroid" hitting Earth.

ASTRONOMY EXERCISES

Comets
In this animation you can move a comet to different positions relative to the sun. You can also change the amount of gas and dust in the comet's nucleus and see how this affects the comet's tails.

Cratering
Turn up your speakers for this one, where you can design collisions between a meteorite and a hypothetical planet.

Lab 9: Asteroids and Kuiper Belt Objects
This lab presents an overview of asteroid properties, including an exercise on how to discover asteroids. You also investigate the class of objects known as Kuiper Belt objects.

Critical Inquiries for the Web

1. Some nights are better for looking for meteors than others (see Table 25-1). Meteor showers are associated with comets, but how are these associations made? Several sites on the Internet provide information on meteor showers, including historical data and information on parent comets. Pick a shower whose parent comet is known and summarize how astronomers came to know that the meteors and the comet are related.

2. Chances are very small that you will be killed by an asteroid impact, but if objects are out there that astronomers are not aware of whose orbits intersect Earth, humanity could be in for a surprise one day. Look for information on the LINEAR project and other searches for near-Earth asteroids. How many such objects have been discovered? What is the record for closest known passage of an asteroid to Earth?

3. Search for the IAU Minor Planet Center Web pages and find out what asteroids and what comets have come closest to Earth.

4. Search for information about comets in the sky right now. Are any comets bright enough for you to see? Are any bright comets expected soon?

Exploring *TheSky*

1. Of the five brightest asteroids, which has the most inclined orbit? (*Hint:* Use **Filters** in the **View** menu to turn off everything but the sun, stars, ecliptic, and minor planets. You can use **Tracking Setup** under **Time Skip** in the **Tools** menu to lock onto an object and follow it along the ecliptic as time passes.)

2. Do asteroids go through retrograde motion? (*Hint:* Use **Filters** in the **View** menu to turn off the stars and turn on the Equatorial Grid. See Activity 1 above.)

3. Of the five brightest asteroids, which has the most elliptical orbit? (*Hint:* Use **3D Solar System Mode** in the **View** menu and watch objects orbit the sun.)

4. Of the comets shown, which has the smallest orbit? (*Hint:* Use **3D Solar System Mode** in the **View** menu and watch objects orbit the sun.)

 Go to the Brooks/Cole Astronomy Resource Center (**http://astronomy.brookscole.com**) for critical thinking exercises, articles, and additional readings from InfoTrac College Edition, Brooks/Cole's online student library.

26 | Life on Other Worlds

Did I solicit thee from

Darkness to promote me?

JOHN MILTON, *PARADISE LOST*

As LIVING THINGS, we have been promoted from darkness. We are made of heavy atoms that could not have formed at the beginning of the universe. Successive generations of stars fusing light elements into heavier elements have built the atoms so important to our existence. When a dark cloud of interstellar gas enriched with these heavy atoms fell together to form our sun, a small part of the cloud gave birth to the planet you inhabit. Your atoms were in that cloud and have been part of Earth ever since it condensed from the solar nebula. ▌ If life can originate on other worlds, and if intelligence is a natural result of the evolution of life forms, then you might expect that other intelligent races inhabit other worlds. Visits between worlds seem impossible, but perhaps distant civilizations can be detected by radio. ▌ Alien life forms could

▌ Continued on page 652 ▌

Every life form we know of has evolved to live somewhere on Earth. The Wekiu bug lives with the astronomers at 13,600 feet atop Hawaiian volcano Mauna Kea. It lives in the icy cinders and eats insects carried up by ocean breezes. (Kris Koenig/Coast Learning Systems)

Guidepost

Looking Back

This chapter is either unnecessary or critical, depending on your point of view. If you believe that astronomy is the study of the physical universe above the clouds, then you are done; the last 25 chapters completed your study of astronomy. But if you believe that astronomy is the study of your position in the universe, not just your physical location but also your role as a living being in the evolution of the universe, then everything you have done so far was just preparation for this chapter.

This Chapter

You will begin by tackling a question that has troubled scientists and philosophers for centuries: What is life? You won't get an answer to the question, but just asking it will illuminate the problem and prepare you to search for life on other worlds. In fact, as you read this chapter, you will ask four essential questions:

What is life?

How did life originate on Earth?

Could life begin on other worlds?

Could Earthlings communicate with civilizations on other worlds?

You won't be able to answer these questions yet, but often in science asking a good question is more important than getting an answer.

Looking Ahead

You are different now. You have explored the universe from the phases of the moon to the big bang, from the origin of Earth to the death of the sun. Astronomy is important, not because it is about stars and galaxies but because it is about us. It tells us what we are, and once you know astronomy, you see yourself and your world in a different way. Astronomy changes us. You are different now.

The Nature of Scientific Explanation

Science is a way of understanding the world around you, and the heart of that understanding is the explanations that science provides for natural phenomena. Whether you call these explanations stories, histories, theories, or hypotheses, they are attempts to describe how nature works based on fundamental rules of evidence and intellectual honesty. While you may take these explanations as factual truth, you should understand that they are not the only explanations that satisfy the rules of logic.

A separate class of explanations involves religion, and those explanations can be quite logical. The Old Testament description of the creation of the world, for instance, does not fit scientific observations, but if you accept the existence of an omnipotent being, then the biblical explanation is internally logical and acceptable. Of course, it is not a scientific explanation, but religion is a matter of faith and not subject to the rules of evidence. Religious explanations follow their own logic, and it is wrong to demand that they follow the rules of evidence that govern scientific explanations, just as it is wrong to demand that scientific explanations accept certain religious beliefs on faith. Both kinds of explanations are logical, but the rules are different.

If scientific explanations are not the only logical explanations, then why do people give them such weight? First, you should notice the tremendous success of scientific explanations in producing technological advances in your daily lives. Diseases such as chickenpox are more childhood irritations than life-threatening illnesses thanks to the application of scientific explanations to modern medicine. The power of science to shape your world can lead you to think that its explanations are unique. But, second, the process called science depends on the use of evidence to test and perfect explanations, and the logical rigor of this process gives scientists great confidence in their conclusions.

Scientific explanations have given us tremendous insight into the workings of nature, and consequently both scientists and non-scientists tend to forget that there can be other logical explanations. The so-called conflict between science and religion has been symbolized for centuries by the trial of Galileo. That conflict is easier to understand when you consider the nature of scientific explanations

Galileo's telescope gave him a new way to know about the world.

and the role of evidence in testing scientific understanding.

---

be quite different from Earthlings, but if they are alive, then they must share certain characteristics. Your goal in this chapter is to use your knowledge of science to explain the greatest of mysteries—the origin and evolution of life on Earth and on other worlds (**Window on Science 26-1**).

26-1 The Nature of Life

WHAT IS LIFE? Philosophers have struggled with that question for thousands of years, so it is unlikely that you will find an answer here. But you must have a working model of life before you can speculate on its occurrence on other worlds. To that end, you can identify two important aspects in living things, a physical basis and a unit of controlling information.

The Physical Basis of Life

On Earth, the physical basis of life is the carbon atom (■ Figure 26-1). Because of the way this atom bonds to other atoms, it can form long, complex, stable chains that are capable of extracting, storing, and utilizing energy. Other chemical bases of life may exist. Science fiction stories and movies abound with silicon creatures, living things whose body chemistry is based on silicon rather than carbon. However, silicon forms weaker bonds than carbon does, and it cannot form double bonds as easily. Consequently, it cannot form the long, complex, stable chains that carbon can. Silicon is 135 times more common on Earth than carbon is, yet there are no silicon creatures among us. All Earth life is carbon based. The likelihood that distant planets are inhabited by silicon people seems small, but you should not rule out life based on non-carbon chemistry.

In fact, nonchemical life might be possible. What is required is some mechanism capable of supporting the extraction and utilization of energy that has been identified as life. One could at least imagine life based on electromagnetic fields and ionized gas. No one has ever met such a creature, but science fiction writers conjure up all sorts.

Clearly, you could range far in space and time, theorizing about different bases for alien life, but to make progress you need to discuss what is known best—carbon-based life on Earth. How can a lump of carbon-rich matter live? The answer lies in the information that guides its life processes.

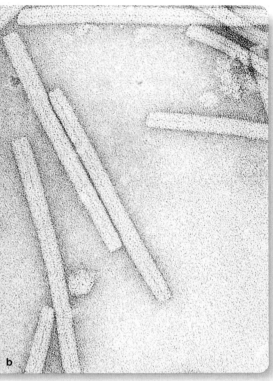

■ Figure 26-1

All living things on Earth are based on carbon chemistry. Even the long molecules that carry genetic information, DNA and RNA, have a framework defined by chains of carbon atoms. (a) Katie, a complex mammal, contains about 30 AU of DNA. (Michael Seeds) (b) Each rod of the tobacco mosaic virus contains a single spiral strand of RNA about 0.01 mm long. (L. D. Simon) All life on Earth stores its genetic information in such carbon-chain molecules.

Information Storage and Duplication

Living cells are tiny chemical factories. They must store all of the recipes for those chemicals in a safe place, use them to fulfill the cell's task, and hand down duplicates of the recipes to offspring. That information is encoded on long molecules.

Study **DNA: The Code of Life** on pages 654–655 and notice three important points:

1 The chemical recipes of life are stored as templates on DNA molecules. The templates automatically guide specific chemical reactions within the cell.

2 The instructions stored in DNA are the genetic information handed down to offspring. When people say, "You have your mother's eyes," they are talking about DNA codes.

3 Notice how the DNA molecule reproduces itself when a cell divides so that each new cell contains a copy of the original information.

Although the DNA molecule must preserve its coded information from damage and make accurate copies, it must also be capable of making mistakes. To see why, you must consider how new DNA recipes are created.

Modifying the Information

You are probably a little surprised to hear that DNA needs to make mistakes, but it is true. If living things are to survive for many generations, then the information stored in their DNA must be able to change as the environment changes. Without change in DNA, a slight warming of the climate, for example, might kill a species of plant, in turn starving the rabbits, deer, and other plant eaters, and leaving the hawks, wolves, and mountain lions with no prey. If the information stored in DNA could never change, then environmental changes would quickly drive life forms to extinction. If life is to survive in a changing world, then the information in DNA must be changeable. Living things must evolve.

Species evolve by **natural selection.** Each time an organism reproduces, its offspring receives copies of the data stored in the DNA, but some variation is possible. For example, most of the rabbits in a litter may be normal, but it is possible for one to get a DNA recipe that gives it stronger teeth. If it has stronger teeth, it may be able to eat something other than the plant the others depend on; and, if that plant is becoming scarce, the rabbit with stronger teeth has a survival advantage. It can eat other plants and so will be healthier than its littermates and have more offspring. Some of these offspring will also have stronger teeth, as the altered DNA recipes are handed down to the new generation. In this way, nature selects and preserves those attributes that contribute to the survival of the species. Those creatures that are unfit die. Natural selection is merciless to the individual, but it gives the species the best possible chance to survive in a changing environment.

The only way nature can obtain new DNA patterns from which to select the best is from DNA molecules that have changed. This can happen through chance mismatching of base pairs—mistakes—in the reproduction of the DNA molecule. Another way

DNA: The Code of Life

1 The key to understanding life is information — the information that guides all of the processes in an organism. In most living things on Earth, that information is stored on a long spiral molecule called **DNA (deoxyribonucleic acid).**

1a The DNA molecule looks like a spiral ladder with rails made of phosphates and sugars. The rungs of the ladder are made of four chemical bases arranged in pairs. The bases always pair the same way. That is, base A always pairs with base T, and base G always pairs with base C.

1b Information is coded on the DNA molecule by the order in which the base pairs occur. To read that code, molecular biologists have to "sequence the DNA." That is, they must determine the order in which the base pairs occur along the DNA ladder.

The Four Bases

A — Adenine

C — Cytosine

G — Guanine

T — Thymine

2 DNA automatically combines raw materials to form important chemical compounds. The building blocks of these compounds are relatively simple **amino acids.** Segments of DNA act as templates that guide the amino acids to join together in the correct order to build specific **proteins,** chemical compounds important to the structure and function of organisms. Some proteins called **enzymes** regulate other processes. In this way, DNA recipes regulate the production of the compounds of life.

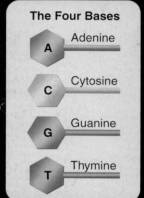

The traits you inherit from your parents, the chemical processes that animate you, and the structure of your body are all encoded in your DNA.

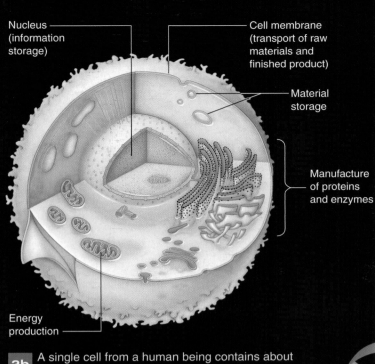

Nucleus
(information
storage)

Cell membrane
(transport of raw
materials and
finished product)

Material
storage

Manufacture
of proteins
and enzymes

Energy
production

2a A cell is a tiny factory that uses the DNA code to manufacture chemicals. Most of the DNA remains safe in the nucleus of a cell, and the code is copied to create a molecule of **RNA (ribonucleic acid).** Like a messenger carrying blueprints, the RNA carries the code out of the nucleus to the work site where the proteins and enzymes are made.

Original DNA

2b A single cell from a human being contains about 1.5 meters of DNA containing about 4.5 billion base pairs — enough to record the entire works of Shakespeare 200 times. A typical human contains a total of about 600 AU of DNA. Yet the DNA in each cell, only 1.5 meters in length, contains all of the information to create a new human. A clone is a new creature created from the DNA code found in a single cell.

Copy DNA

3 DNA, coiled into a tight spiral, makes up the **chromosomes** that are the genetic material in a cell. A **gene** is a segment of a chromosome that controls a certain function. When a cell divides, each of the new cells receives a copy of the chromosomes, as genetic information is handed down to new generations.

3a To divide, a cell must duplicate its DNA. The DNA ladder splits, and new bases match to the exposed bases of the ladder to build two copies of the original DNA code. Because the base pairs almost always match correctly, errors in copying are rare. One set of the DNA code goes to each of the two new cells.

Copy DNA

Cell Reproduction by Division

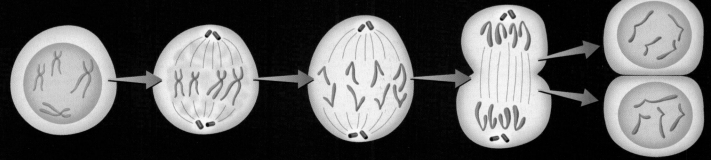

As a cell begins to divide, its DNA duplicates itself.

The duplicated chromosomes move to the middle.

The two sets of chromosomes separate, and . . .

the cell divides to produce . . .

two cells, each containing a full set of the DNA code.

this can occur is through damage to reproductive cells from exposure to radioactivity such as cosmic rays or natural radioactivity in the soil. In any case, an offspring born with altered DNA is called a **mutant.** Most mutations make no difference at all because they change segments of DNA that are not being used. Many mutations are fatal, and the individual dies long before it can have offspring of its own. But in rare cases, a mutation may give a species a new survival advantage. Then natural selection makes it likely that the new DNA message will survive and be handed down, making the species more capable of surviving.

Evolution is not random. Of course, the errors that occur in the DNA code are indeed random, but natural selection is not random. Those changes in the DNA that help a species survive are selected and preserved in future generations. With each passing generation, the species becomes more fit to survive in its environment.

Evolution is natural and automatic, but humans can control it. If you drink milk or eat cheese you are benefiting from the careful breeding of milk cows that has extended over centuries. Do you own a dog or a cat? Controlled evolution has created miniature poodles, huge Doberman pinschers, regal Siamese cats, and tailless Manx cats. But evolution works against us too. Some insecticides don't work anymore because the insects that were susceptible were killed off, and those that survived have a different gene that makes them immune. You have probably heard that some bacteria have built up a resistance to antibiotics. Again, the germs are evolving and handing on the genes—the bits of information—that help them survive.

Building Scientific Arguments

Why can't the information in DNA be permanent?
Sometimes the most valuable scientific arguments are those that challenge common misconceptions. It seems so obvious that DNA codes must not change, but that's a common misconception. The information stored in a creature's DNA provides all of the recipes that make the creature what it is. For example, the DNA in a starfish must contain all the recipes for making the various kinds of proteins needed to consume and digest food. That information must be passed on to offspring starfish, or they will be unable to survive. But the information must be changeable because the environment is changeable. Ice ages come and go, mountains rise, lakes dry up, and ocean currents shift. If the environment changes in some way, one or more of the recipes may no longer work. In this example, a change in the temperature of the ocean water may kill off the specific shellfish the starfish eat. If they can't digest other shellfish, the entire species will become extinct. Natural variation in DNA means that among all the infant starfish in any generation, a few of the recipes are different; if the environment changes, all of the old-style starfish may die, but a few—those with the different DNA—can carry on.

The survival of life depends on this delicate balance between reliable reproduction and the introduction of small variations in DNA information. Now go the next step in your argument. **What are some of the ways these small changes in DNA can arise?**

■   ■   ■

Connections: Life is based not only on information but also on the duplication of information. Today, that process seems so complex that it is hard to imagine how it could have begun.

26-2 The Origin of Life

IF LIFE ON EARTH is based on the storage of information in these long, complex, carbon-chain molecules, how could it have ever gotten started? Obviously, 4.5 billion chemical bases didn't just happen to drift together to form the DNA formula for a human being. The key is evolution. Once a life form begins to reproduce itself, natural selection preserves the most advantageous traits. Over long periods of time spanning thousands, perhaps millions, of generations, the life form becomes more fit to survive. This often means the life form becomes more complex. That means life could have begun as a very simple process that gradually became more sophisticated as it was modified by evolution.

This is a critical point, so let's repeat it. Life is now very complex, but it didn't have to begin by some astonishingly unlikely combination of atoms to make a complicated creature. Life could have begun as a simple organism that automatically made copies of itself. Once reproduction began, natural selection made living things into the complex creatures in the world today. Of course, the key is evidence; what evidence exists to confirm this hypothesis?

The Origin of Life on Earth

The oldest fossils hint that life began in the sea. The first living things on Earth left behind a very poor fossil record. They were at first single-celled creatures; and, even when they became more complex, multicellular creatures, they contained no hard parts such as bones or shells that form good fossils. Yet a few traces of these earliest living things can be found, and they are all ocean creatures.

Although the oldest rocks contain no obvious fossils, microscopes reveal traces of ancient life. Rock from western Australia that is nearly 3.5 billion years old contains microscopic features that may be fossils of living things (■ Figure 26-2), and spectroscopic analysis reveals the presence of organic matter. While the experts debate the details in this controversial field, you can conclude that life was present on Earth as simple organisms at least 3.5 billion years ago. That's roughly one billion years after Earth formed.

A little over a half-billion years ago something happened, perhaps a change in Earth's climate, and life exploded into a wide di-

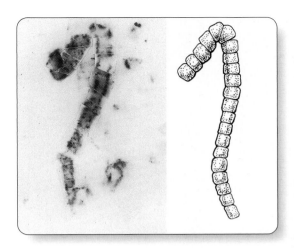

■ Figure 26-2

Among the oldest fossils known, this microscopic filament resembles modern bacterial forms (artist's reconstruction at right). This fossil was found in the 3.5-billion-year-old chert of the Pribara Block in northwestern Australia. (Courtesy J. William Schopf)

■ Figure 26-3

Trilobites made their first appearance in the Cambrian oceans. The smallest were almost microscopic, and the largest were bigger than dinner plates. This example, about the size of a human hand, lived 400 million years ago in an ocean floor that is now a limestone deposit in Pennsylvania. (Grundy Observatory photograph)

versity of complex forms. This marks the beginning of the **Cambrian period** and is sometimes called the *Cambrian explosion.* The best known of the Cambrian creatures may be the trilobites (■ Figure 26-3). Although the Cambrian creatures were diverse, there are no Cambrian fossils of land plants or animals. Evidently, land surfaces were totally devoid of life until only 400 million years ago.

The fossil record shows that life began as simple organisms in the sea soon after Earth formed. Only recently, in geological terms, has life become complex. It isn't difficult to understand how simple creatures evolve to become complex. The problem is trying to understand how nonliving atoms became simple creatures. That's the hard bit right at the beginning.

The key to the origin of this life may lie in an experiment performed by Stanley Miller and Harold Urey in 1952. This **Miller experiment** sought to reproduce the conditions on Earth under which life began. In a closed glass container, the experimenters placed water (to represent the oceans); the gases hydrogen, ammonia, and methane (to represent the primitive atmosphere); and an electric arc (to represent lightning bolts). The apparatus was sterilized, sealed, and set in operation (■ Figure 26-4).

After a week, Miller and Urey stopped the experiment and analyzed the material in the flask. Among the many compounds the experiment produced, they found four amino acids (building blocks of protein), various fatty acids, and urea, a molecule common to many life processes. Evidently, the energy from the electric arc had molded the atmospheric gases into some of the basic components of living matter. Other energy sources, such as hot silica (to simulate hot lava spilling into the sea) and ultraviolet radiation (to simulate sunlight), produce similar results.

More recent studies of the composition of meteorites and models of planet formation suggest that Earth's first atmosphere did not resemble the gases used in the Miller experiment. Earth's first atmosphere was probably composed of carbon dioxide, nitrogen, and water vapor. This finding, however, does not invalidate the Miller experiment. When such gases are processed in a Miller apparatus, some organic molecules appear, although not as many as in the first trials.

The Miller experiment did not create life, of course, nor did it necessarily imitate the exact conditions on the young Earth. That is not the lesson to take from this famous experiment. Rather, it is important because it shows that complex organic molecules form naturally in a wide variety of circumstances. The chemical deck is stacked to deal nature a hand of complex molecules. If you could travel back in time, you would probably find Earth's first oceans filled with a rich mixture of organic compounds in what some have called the **primordial soup.**

The next step on the journey toward life is for the compounds dissolved in the oceans to link up and form larger molecules. Amino acids, for example, can link together to form proteins. This linkage occurs when amino acids join together end-to-end and release a water molecule (■ Figure 26-5). For many years, experts

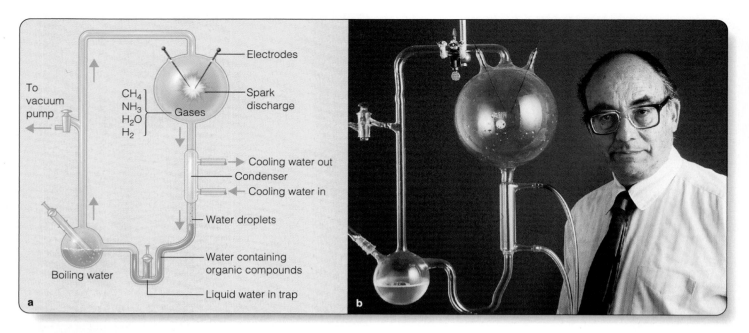

■ **Figure 26-4**

(a) The Miller experiment circulated gases through water in the presence of an electric arc. This simulation of primitive conditions on Earth produced amino acids, the building blocks of proteins. (b) Stanley Miller with a Miller apparatus. (Courtesy Stanley Miller)

have assumed that this process must have happened in sun-warmed tidal pools where evaporation concentrated the broth. But more recent studies suggest that the young Earth was subject to extensive volcanism and large meteorite impacts that periodically modified the climate enough to destroy any life forms exposed on the surface. Rather it seems likely that the early growth of complex molecules took place among the hot springs along the midocean ridges. These complex molecules would not have been photosynthetic—taking energy from sunlight. Rather they would have taken energy from the heat and chemicals emerging

from the hot springs of the midocean ridge. That energy could have powered the growth of long protein chains. Deep in the oceans, they would have been safe from climate changes.

Although these proteins might have contained hundreds of amino acids, they would not have been alive. Not yet. Such molecules would not have reproduced but would have merely linked together and broken apart at random. Because some molecules are more stable than others, however, and because some molecules bond together more readily than others, this automatic **chemical evolution** would have led to the concentration of the varied

■ **Figure 26-5**

Amino acids can link together through the release of a water molecule to form long carbon-chain molecules. The amino acid in this hypothetical example is alanine, one of the simplest.

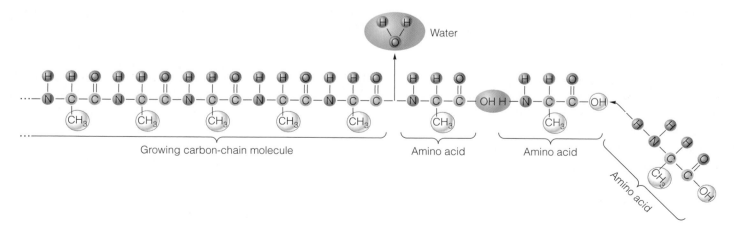

smaller molecules into the most stable larger forms. Eventually, somewhere in the oceans, a molecule took shape that automatically made copies of itself. At that point, the chemical evolution of molecules became the biological evolution of living things.

You might enjoy speculating about an alternative theory that proposes that primitive living things such as reproducing molecules did not originate on Earth but came here in meteorites or comets. Radio astronomers have found a wide variety of organic molecules in the interstellar medium, and some studies have found similar compounds inside meteorites (■ Figure 26-6). Such molecules form so readily that you should be surprised if they were not present in space. A few investigators, however, have speculated that living, reproducing molecules originated in space and came to Earth as a cosmic contamination. If this were true, every planet in the universe would be contaminated with the seeds of life. You may find this theory fun to think about, but it is presently untestable, and an untestable theory is of little use in science.

Whether life originated in the oceans or in space, you still face the problem of how life began—that hard bit at the beginning in which nonliving atoms become reproducing molecules. The exact details are not yet understood, but the important point is that natural processes could have done it. Just because you don't understand something doesn't mean it is impossible. The evidence seems conclusive that life began as reproducing molecules in Earth's oceans.

Which came first, reproducing molecules or the cell? Because you probably think of the cell as the basic unit of life, this question seems to make no sense, but in fact the cell may have originated during chemical evolution. If a dry mixture of amino acids is heated, the acids form long, proteinlike molecules that, when poured into water, collect to form microscopic spheres that function in ways similar to cells (■ Figure 26-7). They have a thin membrane surface, they can absorb material from their surroundings, they grow in size, and they can divide and bud just as cells do. They contain no large molecule that copies itself, however. So it is possible that the structure of the cell originated first and the reproducing molecules later.

An alternative theory proposes that the replicating molecule developed first. Such a molecule would have been exposed to damage if it had been bare, so the first to manufacture or attract a protective coating of protein would have had a significant survival advantage. If this was the case, the protective cell membrane was a later development of biological evolution.

The first living things must have been single-celled organisms much like modern bacteria. Some of the oldest fossils known are **stromatolites,** structures produced by communities of photosynthesizing bacteria that grew in mats and, year by year, deposited layers of minerals that were later fossilized. One of the

■ **Figure 26-6**

A sample of the Murchison meteorite, a carbonaceous chondrite that fell in 1969 near Murchison, Australia. Analysis of the interior of the meteorite revealed evidence of amino acids. Whether the first building blocks of life originated in space is unknown, but the amino acids found in meteorites illustrate how commonly amino acids and other complex molecules occur even in the absence of living things. (Courtesy Chip Clark, National Museum of Natural History)

■ **Figure 26-7**

Single amino acids can be assembled into long proteinlike molecules. When such material cools in water, it can form microspheres, microscopic spheres with double-layered boundaries similar to cell membranes. Microspheres may have been an intermediate stage in the evolution of life between complex molecules and cells holding molecules reproducing genetic information. (Courtesy Sidney Fox and Randall Grubbs)

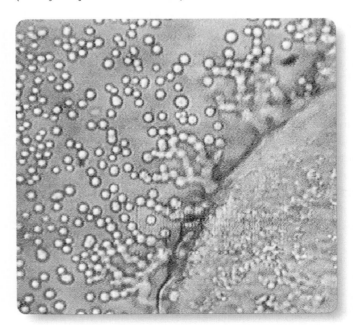

■ **Active Figure 26-8**

A 3.5-billion-year-old fossil stromatolite from western Australia is one of the oldest known fossils (inset). Stromatolites were formed, layer by layer, by mats of bacteria living in shallow water. Such life may have been common in shallow seas when Earth was young. Stromatolites are still being formed today in similar environments. (Mural by Peter Sawyer; photo courtesy Chip Clark, National Museum of Natural History)

Ace◐Astronomy™ Log into AceAstronomy and select this chapter to see the Active Figure "Future of the Sun." Watch Earth's temperature change as the sun ages.

oldest such fossils known is believed to be 3.5 billion years old (■ Figure 26-8). If such bacteria were common when Earth was young, the early atmosphere may have contained a small amount of oxygen produced by the photosynthesis. Atmospheric models suggest that an oxygen abundance of only 0.1 percent would have been sufficient to provide an ozone screen that would protect organisms from the sun's ultraviolet radiation.

How evolution shaped creatures to live in the ancient oceans, to photosynthesize and respire, to become multicellular, and to reproduce sexually is a fascinating story, but this chapter cannot explore those details. You can see that the hard bit in the story is how life began, and there appear to be natural chemical processes that could have led to molecules automatically duplicating themselves. Once some DNA-like molecule formed, evolution was automatic; and, over billions of years, the genetic information stored in living things accumulated those qualities that favored survival, and unsuccessful variations did not get passed down to new generations. As Samuel Butler said, "The chicken is the egg's way of making another egg." In that sense, all living matter on Earth is the physical expression of DNA's automatic tendency to continue its existence.

Perhaps this seems harsh. Human experience goes far beyond mere reproduction. *Homo sapiens* has art, poetry, music, philosophy, religion, science. Perhaps all of the great accomplishments of our intelligence represent more than mere reproduction of DNA. Nevertheless, intelligence, the ability to analyze complex situations and respond with appropriate action, must have begun as a survival mechanism. For example, a fixed escape strategy stored in the DNA is a disadvantage for a creature that frequently moves from one environment to another. A rodent that always escapes from predators by automatically climbing the nearest tree would be in serious jeopardy if it met a hungry fox in a treeless clearing. Even a faint glimmer of intelligence might allow the rodent to analyze the situation and, finding no trees, to choose running over climbing. Intelligence, of which *Homo sapiens* is so proud, may have developed in ancient creatures as a way of making them more versatile without adding tremendous amounts of information to their DNA library.

Any discussion of the evolution of life seems to involve highly improbable coincidences until you consider how many years have passed in the history of Earth. You can read the words—4.6 billion years—so easily, but it is in truth hard to grasp the meaning of such a long period of time.

Geologic Time

Humanity is a very new experiment on planet Earth. You can fit the history of Earth into a single chart such as ■ Figure 26-9. You can take comfort in thinking that creatures like us have walked on Earth for roughly 3 million years, but when you add our history to the chart, you will discover that the entire history of humanity makes up no more than a thin line at the top. In fact, if you tried to represent the entire 4.6-billion-year history of Earth on the chart, the portion describing the rise of complex life after the Cambrian explosion would be an unreadably small segment. It is expanded in Figure 26-9 just to make it legible.

One way to represent the evolution of life is to compress the 4.6-billion-year history of Earth into a one-year-long video program. In such a program, Earth forms as the video begins on January 1; through all of January and February it cools and is cratered, and the first oceans form. Search as you might, you would find no trace of life in these oceans until sometime in March or early April, when the first living things develop. You would need a microscope to see them. The slow development of these simplest of living forms grinds on through the spring and summer of the video. The entire 4-billion-year history of Precambrian evolution lasts until the video reaches mid-November, when the primitive ocean life explodes into the more complex Cambrian organisms such as trilobites and many other specialized creatures (■ Figure 26-10).

While the yearlong video plays on and on, you might amuse yourself by looking at the land instead of the oceans, but you would be disappointed. The land is a lifeless waste with no plants or animals of any kind. Not until November 28 in the video does life appear on the land; but, once it does, it evolves rapidly into a wide range of plants and animals. Dinosaurs, for example, appear about December 12 and vanish by Christmas evening as mammals and birds flourish.

Throughout the one-year run of the video there have been no humans; and even during the last days of the year, as the mammals rise to dominate the landscape, there are no people. In the early evening of December 31, vaguely human forms move through the grasslands, and by late evening they begin making stone tools. The Stone Age lasts till about 11:45 PM, and the first signs of civilization, towns and cities, do not appear until 11:54 PM. The Christian era begins only 14 seconds before the New Year, and the Declaration of Independence is signed with but 1 second to spare.

By imagining the history of Earth as a yearlong video, you can place the rise of life in perspective. Tremendous amounts of time were needed for the first simple living things to evolve in the oceans, and even more time was needed for the evolution of complex creatures that could colonize the land. As life became more complex, it evolved and diversified faster and faster, as if evolution were drawing on a growing library of solutions that had been previously invented with great effort to solve earlier problems. The burst of diversity on land led eventually to the rise of intelligent creatures like us, a process that has taken 4.6 billion years.

If life could originate on Earth and develop into intelligent creatures, perhaps the same thing could have happened on other planets. This raises three questions. First, could life originate on another world if conditions were suitable? No one knows for sure, but you have found natural, chemical processes that could lead to living things, so your answer to this question should probably be yes. The second question is, will life always evolve toward intelligence? Perhaps intelligence arose on Earth under unusual conditions, but you should recall that intelligence appears to be a way to make a species more versatile and consequently more likely to survive. If that is true, then intelligence may develop under a wide range of conditions if enough time is available. But what of the third question: Are suitable conditions so rare that life almost never gets started? The only way to answer that is to search for life on other planets. You can begin in the next section by searching for life on the other planets in our solar system.

Life in Our Solar System

Although science fiction writers imagine life based on something other than carbon chemistry, no one knows of any real examples to illustrate how such life might arise and survive. To make progress in your search for life, you need to limit your discussion to life as it is known on Earth and the conditions it requires. The most important requirement is the presence of liquid water, not only as part of the chemical reactions of life but also as a medium to transport nutrients and wastes within the organism. Also, it seems that life on Earth began in the oceans and developed there for nearly 4 billion years before it was able to emerge onto the land. Certainly, any world where you hope to find life must have liquid water, and that means it must have moderate temperatures.

The liquid water requirement automatically eliminates many worlds in our solar system. The moon is airless, and although some data suggest ice frozen in the soil at its poles, it has never had liquid water on its surface. In the vacuum of the lunar surface, liquid water would boil away rapidly. Mercury too is airless and cannot have had liquid water on its surface for long periods of time. Venus has some traces of water vapor in its atmosphere, but it is much too hot for liquid water to survive. If there were any lakes or oceans of water on its surface when it was young, they must have evaporated quickly. Even if life began there, no traces would be left now.

The inner solar system seems too hot, and the outer solar system seems too cold. The Jovian planets have deep atmospheres; and, at a certain level, they have moderate temperatures where water might condense into liquid droplets. But it seems unlikely that life could begin there. The Jovian planets have no surfaces where oceans could nurture the beginning of life, and currents in

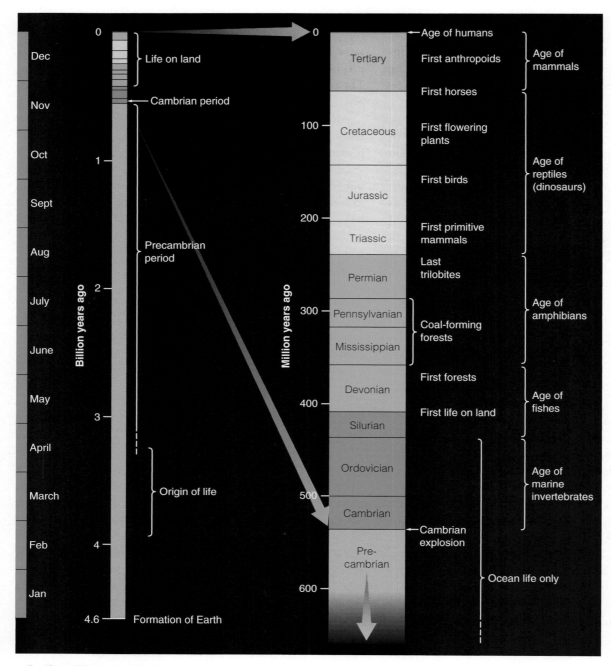

■ Active Figure 26-9

Complex life has developed on Earth only recently. If the entire history of Earth were represented in a time line (left), you would have to magnify the end of the line to see details such as life leaving the oceans and dinosaurs appearing. The age of humans would still be only a thin line at the top of your diagram. If the history of Earth were a yearlong videotape, humans would not appear until the last hours of December 31.

Ace ◐ Astronomy™ Log into AceAstronomy and select this chapter to see the Active Figure "Earth Calendar." Animate this diagram.

the atmosphere seem destined to circulate gas and water droplets from regions of moderate temperature to other levels that are much too hot or too cold for life to survive.

A few of the satellites of the Jovian planets might have suitable conditions for life. Jupiter's moon Europa seems to have a liquid-water ocean below its icy crust (see Figure 23-12), and minerals dissolved in that water would provide a rich broth of possibilities for chemical evolution. Nevertheless, Europa is not a promising site to search for life because conditions may not have remained stable for the billions of years needed for life to evolve beyond the microscopic stage. The subsurface ocean is kept from freezing by tidal heating. If Jupiter's moons interact gravitationally and modify their orbits, Europa may have been frozen solid at some points in history. Such periods of freezing would proba-

■ Figure 26-10

Before the Cambrian explosion, life consisted mostly of single-celled creatures living alone or in organized colonies. During the Cambrian period, life became complex. Anomalocaris (upper left) was about the size of your hand and had specialized organs including eyes, coordinated fins for active swimming, gripping mandibles capable of raking through the seafloor and capturing prey, and a powerful, toothed maw located below its head (upper right). Notice Opabinia at center with its long snout. (D. W. Miller)

Approximate true-color image

■ Figure 26-11

Mars rover Opportunity descended into the crater Endurance and then looked back up the crater wall. In places, Opportunity found layering in the rocks that showed they had formed as layers of sand and silt deposited by flowing water. The rover also found minerals that showed that water was once present at the site. Signs of past life may lie hidden in these rocks. (NASA/JPL/Cornell)

bly prevent life from developing. Drilling through the icy crust of Europa to search for life in its ocean will be a wonderful adventure for future generations, but Europa does not seem to be a good bet to harbor any form of complex life.

Saturn's moon Titan has an atmosphere of nitrogen, argon, and methane and may have lakes or small oceans of liquid methane on its surface. You saw in Chapter 23 how sunlight can convert the methane in the atmosphere to organic smog particles that settle to the surface, and the images radioed back to Earth by the Huygens probe as it parachuted down to the surface of Titan showed that dark material filled drainage channels and lowlands. The chemistry of life that might crawl or swim on such a world is unknown, but life there may be unlikely because of the temperature. The surface of Titan is a deadly $-179°C$ ($-290°F$). Chemical reactions occur slowly or not at all at such low temperatures, so the chemical evolution needed to begin life may never have occurred on Titan.

Mars is the most likely place for life in our solar system. The evidence, however, is not encouraging. In 1976, two robotic spacecraft, Viking 1 and Viking 2, landed at two different places on the Martian surface. The spacecraft scooped up soil samples and subjected them to tests for the presence of living organisms. For example, they gave some soil samples a dose of nutrient-rich water and watched for signs the nutrients were taken up by biological processes. The results seem negative. Although some peculiar chemical processes were detected, no evidence was found that clearly indicated the presence of any living things in the soil.

Rovers on Mars and orbiting spacecraft have found dramatic evidence that liquid water once flowed over the surface (■ Figure 26-11), but that does not mean that life did originate there. Only detailed study of the rocky soil can answer that question, and that means Earth must send a geologist to Mars or bring rocks from Mars back to Earth. Nature has performed the latter task for us.

Meteorite ALH84001 (■ Figure 26-12a) was found on the Antarctic ice in 1984. Years later an analysis of its chemical composition showed that it had originated on Mars. Being a good, skeptical scientist, you are probably wondering how anyone could know that the meteorite came from Mars. The evidence is solid.

■ Figure 26-12

(a) Meteorite ALH84001 is one of many meteorites known to have originated on Mars. It was claimed that the meteorite contained chemical and physical traces of ancient life on Mars, including what appear to be fossils of microscopic organisms (b). The evidence has not been confirmed, and the validity of the claim is highly questionable. (NASA)

The Viking landers measured the abundance of oxygen isotopes on Mars, and the pattern of isotopes was different from Earth and from meteorites. But the isotopes in ALH84001 are a precise match with the Viking results. The meteorite must have come from Mars.

ALH84001 was probably part of debris ejected into space by a large impact on Mars. Some of that debris fell to Earth, and a small number of these meteorites have been found, mostly in Antarctica because that's the easiest place to find meteorites. You might have heard of ALH84001; it was in the news because a team of scientists studied it and announced in 1996 that it contained chemical and physical traces of ancient life on Mars (Figure 26-12b).

The discovery was received with great excitement by the press and the public. It would, indeed, be dramatic evidence if true, because it would show that life could begin on another world. Scientists were excited too, but being professionally skeptical (see Window on Science 19-3), they began testing the results immediately. In many cases, the tests did not confirm the conclusion that life once existed on Mars. Some chemical contamination from water on Earth has occurred, and some chemicals in the meteorite may have originated without the presence of life. The physical features that look like fossil bacteria may be mineral formations in the rock. Although studies of ALH84001 continue, it does not provide unchallenged evidence that life once existed on Mars. Carl Sagan once said, "Extraordinary claims require extraordinary proof," and in this case the evidence is at best inconclusive.

Spacecraft now visiting Mars are revealing the past history of water there and painting a more detailed picture of present conditions. Spacecraft are planned that will eventually land on Mars, collect rocks, and return them to Earth. Nevertheless, conclusive evidence may have to wait until a geologist in a spacesuit can scramble down the dry streambeds of Mars cracking open rocks and searching for fossils.

Your search for life on the other worlds in our solar system has been inconclusive. You didn't find life, but you didn't find evidence to completely rule it out. So far as anyone knows at present, our solar system is bare of life except for Earth. Consequently the search for life in the universe takes you to other planetary systems.

Life in Other Planetary Systems

Might life exist in other solar systems? To consider this question, you need to consider how common planets are and what conditions a planet must fulfill for life to originate and evolve to intelligence. The first question is astronomical; the second is biological. The ability of scientists to discuss the problem of life outside our solar system is severely limited by lack of experience. They know only one planet well—Earth.

In Chapter 19 you learned that planets form as a natural byproduct of star formation and that a number of extrasolar planets have been found circling nearby stars. From this you can conclude that planetary systems are very common.

If a planet is to become a suitable home for life, it must have a stable orbit around its sun. This is simple in a solar system like our own, but in a binary system most planetary orbits are unstable. Most planets in such systems would not last long before they were swallowed up by one of the stars or ejected from the system. Consequently, single stars are the most likely to have planets suitable for life. Because our galaxy contains at least 10^{11} stars, half of which are single, there could be roughly 5×10^{10} planetary systems in which life might exist.

A few million years of suitable conditions does not seem to be enough time to originate life. On our planet it took about a billion years for the first living things to appear and 4.6 billion years for intelligence to evolve. Clearly, conditions on a planet must remain acceptable over a long time. This eliminates massive stars that remain stable on the main sequence for only a few million years. If it takes a few billion years for life to originate and evolve to intelligence, no star hotter than about F5 will do. This is not really a serious restriction, because upper-main-sequence stars are rare anyway.

In previous sections, you saw how life on Earth depends on water. In the past, astronomers have used that to define a **life zone** (or ecosphere) around a star, a region within which planets have temperatures that permit the existence of liquid water. A cool star has a small life zone, and a hot star has a large life zone. The life zone around the sun extends from about the orbit of Venus to the orbit of Mars.

For a number of decades astronomers have used the life zone to eliminate lower-main-sequence stars as candidates for life. The M dwarfs at the bottom of the main sequence are so faint that a planet would have to orbit very close to stay warm. Some astronomers have argued that such a planet would become tidally locked to its star, and one side of the planet would be in perpetual darkness. Water would tend to freeze out on the dark side and end chances for life to develop. Whether the atmosphere could distribute heat to the dark side is unknown. Also, M stars sometimes suffer violent flares that might make nearby planets unsuitable for life. Most astronomers thought the dinky red dwarfs were not good candidates for life. However, extrasolar planets have been discovered orbiting these M stars, and the large numbers of these extrasolar planets have lead some astronomers to suggest that at least some might be stable enough to develop life.

Recent discoveries, however, are making the whole idea of a life zone seem just a little bit of an oversimplification. For one thing, scientists on Earth are finding living things in astonishing places, such as the bottom of icy lakes in Antarctica and far underground in solid rock. Living things have been found in boiling springs where the water is highly acidic. It appears that life is much tougher than some scientists thought. Also, spacecraft visiting other worlds have found conditions that might support life outside the so-called life zone of the sun. The biggest water ocean in our solar system is under the ice on Jupiter's moon Europa. Liquid has flooded parts of Neptune's moon Triton, and Saturn's moon Titan has a complex surface chemistry of carbon-rich organic molecules. It isn't hard to imagine life in these environments, and they all lie outside the conventional life zone of the sun. Perhaps the life zone is most useful as a warning that life is proving to be more varied than many people expected.

One restriction on life is the slow evolution of a planet's star. You learned in Chapter 9 that main-sequence stars gradually grow more luminous as they convert their hydrogen to helium. The sun was only about 70 percent as luminous when it formed as it is today. As a star grows more luminous, it should slowly warm its planets. A planet might have liquid water on its surface long

enough for life to begin but then be sterilized as its star slowly grows more luminous and drives away the planet's atmosphere and oceans. Perhaps a planet must form at just the right distance from its star to remain hospitable to life for billions of years.

The fundamental question is simple. If conditions are right, will life begin? Early in this chapter you found natural chemical processes that seem capable of forming molecules that make copies of themselves. You saw that evolution would automatically shape these molecules to be most likely to survive. Perhaps the fundamental question should be, "What could prevent life from beginning?" Given what you know about life, it seems that it should arise whenever conditions permit, and our galaxy should be filled with planets that are inhabited with living creatures.

Building Scientific Arguments

What evidence can you cite that life is at least possible on other worlds?

A good scientific argument is based on evidence, but it also includes careful analysis. The evidence is limited almost entirely to Earth, but it is promising. Fossils show that life originated in Earth's oceans almost 4 billion years ago, and biologists have proposed relatively simple chemical processes that could have formed these first reproducing molecules. Fossils show that life developed from a very slow beginning into more and more complex creatures that filled the oceans. The pace of evolution quickened dramatically about half a billion years ago, the beginning of the Cambrian period, when life began taking on complex forms; later, when life emerged onto the surface of the land, it evolved rapidly to produce the tremendous diversity of today. Human intelligence has been a very recent development; it is only a few million years old.

If this process occurred naturally on Earth, then it seems reasonable that it could have occurred on other worlds as well. This conclusion hinges mostly on a reasonable belief that the laws of nature apply on other worlds as they do on Earth. Now expand your argument: **What conditions should you expect of other worlds that host life?**

■   ■   ■

Connections: It is both easy and fun to speculate about life on other worlds, but it leads to a simple question: If there really is life on other worlds, why can't we detect it? Why haven't we heard from alien civilizations?

26-3 Communication with Distant Civilizations

IF OTHER CIVILIZATIONS EXIST, perhaps we Earthlings can communicate with them in some way. Sadly, travel between the stars is more difficult in real life than in science fiction—and may

UFOs and Space Aliens

If you discuss life on other worlds, then you might be tempted to use UFO sightings and supposed visits by aliens from outer space as evidence to test your hypotheses. Scientists don't do so for two reasons, both related to the reliability of these observations.

First, the reputation of the sources of UFO sightings and alien encounters does not inspire confidence that these data are reliable. Most people hear of such events via grocery store tabloids, daytime talk shows, or sensational "specials" on viewer-hungry cable networks. You must consider the low reputation of the media that report UFOs and space aliens. Most of these reports are simply made up for the sake of sensation, and you should not try to use them as reliable evidence.

Second, the remaining UFO sightings, those not simply made up, do not survive careful examination. Most are mistakes or unconscious misinterpretations of natural events made by honest people. A number of unbiased studies have found no grounds for believing in UFOs. In short, there is no conclusive evidence that Earth has been visited by aliens from space.

That's too bad. A confirmed visit by intelligent creatures from beyond our solar system would answer many questions. It would be exciting, enlightening, and, like any real adventure, a bit scary. But none of the UFO sightings is dependable, and we are left with no direct evidence of intelligent life on other worlds.

UFOs from space are fun to think about, but there is no evidence that they are real.

in fact be impossible. If physical visits are impossible, perhaps radio communication is the way to say hello. Again, nature places restrictions on such conversations, but the restrictions are not too severe. As you will see, the real problem lies with the life expectancy of civilizations.

Travel between the Stars

Practically speaking, roaming among the stars is tremendously difficult because of three limitations: distance, speed, and fuel. The distances between stars are almost beyond comprehension. It does little good to explain that if you use a golf ball in New York City to represent the sun, the nearest star would be another golf ball in Chicago. It is only slightly better to note that the fastest commercial jet would take about 4 million years to reach the nearest star.

The second limitation is a speed limit—you cannot travel faster than the speed of light. Though science fiction writers invent hyperspace drives so their heroes can zip from star to star, the speed of light is a natural and unavoidable limit that cannot be exceeded. This, combined with the large distances between stars, makes interstellar travel very time consuming.

The third limitation is that you can't even approach the speed of light without using a fantastic amount of fuel. Even if you ignore the problem of escaping from Earth's gravity, you must still use energy stored in fuel to accelerate to high speed and to decelerate to a stop when you reach our destination. To return to Earth, assuming you wish to, you have to repeat the process, and that takes more fuel.

These changes in velocity require a tremendous amount of fuel. If you flew a spaceship as big as a large yacht to a star 5 light-

years (1.5 pc) away and wanted to get there in only 10 years, you would use 40,000 times as much energy as the United States consumes in a year.

Travel for a few individuals might be possible if they accept very long travel times. That would require some form of suspended animation (currently unknown) or colony ships that carry a complete, though small, society in which people are born, live, and die generation after generation as the ship travels through space. Whether the occupants of such a ship would retain the social and emotional characteristics of humans over a long voyage is questionable.

These three limitations not only make it difficult for us to leave our solar system but would also make it difficult for aliens to visit Earth. Reputable scientists have studied "unidentified flying objects" (UFOs) and related phenomena and have never found any evidence that Earth is being visited or has ever been visited by aliens from other worlds (**Window on Science 26-2**). Consequently, humans are unlikely ever to meet an alien face-to-face. The only way to communicate with other civilizations is via radio.

Radio Communication

Nature places two restrictions on radio communication with distant societies. One has to do with simple physics, is well understood, and merely makes the communication difficult. The second has to do with the fate of technological civilizations, is still unresolved, and may severely limit the number of societies detectable by radio.

Radio signals are electromagnetic waves that travel at the speed of light. Because even the nearest civilizations must be a few light-years away, this limits conversations with distant be-

An anticoded message

1	0	1	0	0	1	1	1	1	1
0	0	1	0	1	0	0	1	0	0
0	1	0	1	0	0	1	0	1	0
0	1	0	1	0					

5 rows of 7

1	0	1	0	0	1	1
1	1	1	0	0	1	0
1	0	0	1	0	0	0
1	0	1	0	0	1	0
1	0	0	1	0	1	0

7 rows of 5

1	0	1	0	0
1	1	1	1	1
0	0	1	0	1
0	0	1	0	0
0	1	0	1	0
0	1	0	1	0
0	1	0	1	0

■ **Figure 26-13**

An anticoded message is designed for easy decoding. Here a string of 35 radio pulses, represented as 1s and 0s, can be arranged in only two ways, as 5 rows of 7 or 7 rows of 5. The second way produces a friendly message. Any number of pulses can be used so long as it is the product of two prime numbers. Then the pulses can be arranged in only two ways.

ings. If you ask a question of a creature 10 light-years away, you will have to wait 20 years for a reply. Clearly, the give-and-take of normal conversation will be impossible.

Instead, you could simply broadcast a radio beacon of friendship to announce your presence as an intelligent being. In fact, the human race is already broadcasting a recognizable beacon. Short-wavelength radio signals, such as TV and FM, have been leaking into space for the last 50 years at least. Any civilization within 50 light-years might already have detected Earth.

If you intentionally broadcast a signal, you can anticode it. That is, you could arrange it to make it easy to decode. You could transmit pulses to represent 1s and gaps to represent 0s. A message counting through the first few prime numbers would distinguish your signal from natural sources of radio noise. You could even transmit a picture by sending a string of 1s and 0s that can be arranged in only two ways. One way produces nonsense, but the other way produces a meaningful picture (■ Figure 26-13).

In 1974, at the dedication of the 1000-ft radio telescope at Arecibo, radio astronomers transmitted such a signal toward the globular cluster M13, which is located 26,000 ly from Earth. When the signal finally arrives, any aliens who detect it will be able to arrange its 1679 pulses in only two ways, as 23 rows of 73 or 73 rows of 23. The second arrangement will form a picture that describes life on Earth (■ Figure 26-14).

It took only minutes to transmit the Arecibo message. If more time were taken, a more detailed picture could be sent; and, if you were sure your radio telescope was pointed at a listening civilization, you

could send a long series of pictures. With pictures you could teach aliens your language and tell them all about human life, its difficulties, and its accomplishments.

If you can think of sending such signals, aliens could think of it too. If you pointed your radio telescope in the right direction and listened at the right wavelength, you might hear other intelligent races calling out to one another. This raises two questions: Which stars are the best candidates, and what wavelengths are most likely? You have already answered the first question. Main-sequence G and K stars have the most favorable characteristics,

■ **Active Figure 26-14**

The Arecibo message to M13 begins by counting from 1 to 10 and goes on to describe our solar system and the biochemistry of life on Earth (color added for clarity). Binary numbers give the height of the human figure (1110) and the diameter of the telescope dish (100101111110) in units of the wavelength of the signal, 12.3 cm. (NASA)

Ace◐Astronomy™ Log into AceAstronomy and select this chapter to see the Active Figure "Interstellar Communication." Send messages and probes to nearby stars.

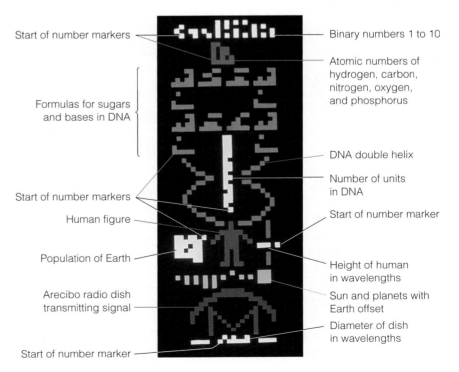

Start of number markers — Binary numbers 1 to 10

Atomic numbers of hydrogen, carbon, nitrogen, oxygen, and phosphorus

Formulas for sugars and bases in DNA

DNA double helix

Number of units in DNA

Start of number markers

Human figure

Start of number marker

Population of Earth

Height of human in wavelengths

Arecibo radio dish transmitting signal

Sun and planets with Earth offset

Diameter of dish in wavelengths

Start of number marker

and maybe M stars are worth a look. But the second question is more complex.

Only certain wavelengths are useful for communication. Wavelengths longer than about 30 cm can't be used because the signal would be lost in the background radio noise from our galaxy. Nor can wavelengths much shorter than 1 cm be used because of absorption within our atmosphere. So only a certain range of wavelengths, a radio window, is open for communication.

This communications window is very wide, so a radio telescope would take a long time to tune over all the wavelengths in the window searching for intelligent signals. Nature may have provided a way to narrow the search, however. Within the communications window lie the 21-cm line of neutral hydrogen and the 18-cm line of OH. The interval between these two lines has been dubbed the **water hole** because the combination of H and OH yields water (H_2O). Water is the fundamental solvent in our life form, so it might seem natural for similar water creatures to call out to each other at wavelengths in the water hole (■ Figure 26-15). But even silicon creatures, if they were competent to build radio telescopes, would be familiar with the 21-cm line of hydrogen and the 18-cm line of OH.

This is not idle speculation. A number of searches for extraterrestrial radio signals have been made, and some major searches are now under way. The field has become known as **SETI,** Search for Extra-Terrestrial Intelligence, and it has generated heated debate. Some scientists and philosophers argue that life on other worlds can't possibly exist, so it is a waste of money to search. Others argue that life on other worlds is common and could be de-

tected with present technology. Congress funded a NASA search for a short time but then ended support in the early 1990s because political leaders feared public reaction. In fact, the annual cost of a major search is only about as much as a single Air Force attack helicopter. The controversy may spring in part from the theological and philosophical controversy that would result from the discovery of intelligent life on another world.

In spite of the controversy, major searches have been made, and others are under way. The NASA SETI project canceled by Congress was completed with private funds as Project Phoenix. The SETI Institute was founded in 1984 and has pursued a number of important searches using radio telescopes at major radio astronomy observatories to listen to the radio emissions of a long list of candidate stars. About two billion radio frequency bands must be examined for each star, so the project depends heavily on computers to search for signs of artificial radio transmissions in the recorded radio emissions of each star.

Astonishing amounts of computer power are needed to search for weak or unusual signals. Consequently, the Berkeley SETI team, with the support of the Planetary Society, has recruited owners of personal computers linked to the Internet to participate in a project called seti@home. Participants download a screen saver that searches data files from the Arecibo radio telescope for meaningful signals whenever the owner is not using the computer. About 4 million people have signed up and are providing 1000 years of computer time per day for the project. For information, locate the seti@home project at http://setiathome.ssl.berkeley.edu/.

META II is a major search for radio signals in the southern sky, and the ingenious search called SERENDIP (Search for Extraterrestrial Radio Emission from Nearby Developed Intelligent Populations) rides piggyback on the 305-m Arecibo Telescope searching for signals wherever the radio astronomers point the telescope. Another major search is scanning the sky for intelligent signals in the form of pulsating light flashes.

The SETI Institute, working with the University of California, Berkeley, is building the Allen Telescope Array, which will eventually include 350 dish antennas (■ Figure 26-16). It will be a powerful radio telescope, but it will also be used to search for radio transmissions from other worlds.

At radio wavelengths, noise is a problem. Poorly designed radio transmitters generate noise at unexpected frequencies, and other electronic devices, such as computers, emit radio frequency signals. Furthermore, society, industry, and governments press to use wider and wider sections of the electromagnetic spectrum. Radio astronomers struggle to hear through the radio babble, and SETI searches suffer because many of the noise signals mimic the patterns expected from other civilizations. It would be ironic if we fail to detect signals from another world because our own world has become too noisy.

Ultimately, the chances of success depend on the number of inhabited worlds in our galaxy, and that number is difficult to estimate.

■ **Active Figure 26-15**

Radio noise from various sources makes it difficult to detect distant signals at wavelengths longer than 30 cm or shorter than 1 cm. In this range, radio emission from H atoms and from OH molecules marks a small wavelength range dubbed the water hole, which may be a likely place for communication.

Ace◐Astronomy™ Log into AceAstronomy and select this chapter to see the Active Figure "Drake Equation." How many civilizations might be ready for communication?

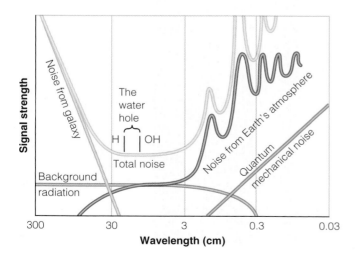

■ Figure 26-16

The Allen Telescope Array now being built in California will eventually grow to include 350 radio dishes, each 6 meters in diameter, in an arrangement precisely designed to maximize resolution. As radio astronomers aim the telescope at galaxies and nebulae of interest, state-of-the-art computer systems will analyze stars in the field of view searching for signals from distant civilizations. (Isaac Gary/SETI Institute)

How Many Inhabited Worlds?

The technology exists, and given enough time the searches will find other inhabited worlds, assuming there are at least a few out there. If intelligence is common, then the signals should be found soon—in the next few decades—but if intelligence is rare in the universe, it may be a very long time before SETI confirms that we are not alone.

Simple arithmetic can provide an estimate of the number of technological civilizations available for communication, N_c. The first proposed formula for N_c is now known as the **Drake equation,** named after radio astronomer Frank Drake, a pioneer in the search for extraterrestrial intelligence. It is better to use a version of the Drake equation modified slightly to make it a bit eas-

ier to understand. The formula gives N_c, the number of communicative civilizations in a galaxy, as

$$N_c = N^* \cdot f_P \cdot n_{LZ} \cdot f_L \cdot f_I \cdot F_S$$

N^* is the number of stars in a galaxy, and f_P represents the fraction of all stars that have planets. If all single stars have planets, f_P is about 0.5. The factor n_{LZ} is the average number of planets in a solar system suitably placed in the life zone, f_L is the fraction of suitable planets on which life begins, and f_I is the fraction of planets on which life forms evolve to intelligence. These factors can be roughly estimated, but the remaining factor is much more uncertain.

F_S is the fraction of a star's life during which the life form is communicative. Here you can assume that a star lives about 10 billion years. If a society survives at a technological level for only 100 years, the chances of communicating with it are small. But a society that stabilizes and remains technological for a long time is much more likely to be in the communicative phase at the proper time to signal to Earth. If you assume that technological societies destroy themselves in about 100 years, F_S is 100 divided by 10 billion, or 10^{-8}. But if societies can remain technological for a million years, then F_S is 10^{-4}. The influence of the factors in the formula is shown in ■ Table 26-1.

If the optimistic estimates are true, there may be a communicative civilization within a few dozen light-years of Earth, and it could be found by searching through only a few thousand stars. On the other hand, if the pessimistic estimates are correct, Earth may be the only planet in our galaxy capable of communication. It all depends on how technological societies function and survive.

Building Scientific Arguments

Why does the number of inhabited worlds that could be detected depend on how long civilizations survive at a technological level?

This scientific argument depends on timing and requires careful analysis. If you turned a radio telescope toward the sky and scanned millions of frequency bands for lots of stars, you

	Table 26-1 I The Number of Technological Civilizations per Galaxy		
		Estimates	
	Variables	**Pessimistic**	**Optimistic**
N^*	Number of stars per galaxy	2×10^{11}	2×10^{11}
f_P	Fraction of stars with planets	0.01	0.5
n_{LZ}	Planets per star in life zone for over 4 billion years	0.01	1
f_L	Fraction of suitable planets on which life begins	0.01	1
f_I	Fraction of life forms that evolve to intelligence	0.01	1
F_S	Fraction of star's life during which a technological society survives	10^{-8}	10^{-4}
N_C	Number of communicative civilizations per galaxy	2×10^{-5}	10×10^6

would be taking a snapshot of the universe at a particular time when you are alive and able to build a radio telescope. To detect other civilizations, they must be in a similar technological stage so they will be broadcasting either intentionally or accidentally. If you searched for decades and detected no signals, it could mean not that life is rare but rather that civilizations do not survive at a technological level for very long. Only a few decades ago, our civilization was threatened by nuclear war, and now we are threatened by the pollution of our environment. If nearly all civilizations in our galaxy are either on the long road up from primitive life forms in oceans or on the long road down from nuclear war or environmental collapse, there may be no one transmitting during the short interval when Earthlings are capable of building radio telescopes with which to listen.

Radio communication between inhabited worlds is limited because it requires very fast computers to search many frequency intervals. Build a new argument to explain: **Why must so many frequencies be searched when the water hole would be a good place to listen?**

■   ■   ■

Connections: Is the human race the only thinking species? If we are, we bear the sole responsibility to understand and admire the universe. Then we are the sole representatives of that state of matter called intelligence. The mere detection of signals from another civilization would demonstrate that we share the universe with others. Although humans may never leave our solar system, such communication would end the self-centered isolation of humanity and stimulate a reevaluation of the meaning of our existence. We may never realize our full potential as humans until we communicate with nonhuman intelligent life.

Study and Review Tools

Summary

26-1 | The Nature of Life

What is life?
- The process of life extracts energy from the surroundings, maintains the organism, and modifies the surroundings to promote the organism's survival.
- Living things have a physical basis—the arrangement of matter and energy that makes life possible. Life on Earth is based on carbon chemistry.
- Living things must also have a controlling unit of information, which is the genetic information passed to each new generation.
- Genetic information for life on Earth is stored in long carbon-chain molecules such as DNA.
- The DNA molecule stores information in the form of chemical bases linked together like the rungs of a ladder. Copied by the RNA molecule, the patterns of bases act as recipes for the manufacture of proteins and enzymes.
- When a cell divides, the DNA molecules split lengthwise and duplicate themselves so that each of the new cells can receive a copy of the genetic information.
- Errors in duplication or damage to the DNA molecule can produce mutants, organisms that contain new DNA information and have new properties.
- Natural selection determines which of these new organisms are best suited to survive, and the species evolves to fit its environment.

26-2 | The Origin of Life

How did life originate on Earth?
- The oldest fossils on Earth are as old as 3.5 billion years. The fossils show that life began in the oceans.
- Life began on Earth as simple organisms like bacteria, and it evolved into more complex creatures.
- The Miller experiment shows that the building blocks of life form naturally under a wide range of circumstances.
- Chemical evolution concentrated simple molecules into larger, stronger molecules, but those molecules did not reproduce copies of themselves.
- Biological evolution begins when molecules begin reproducing.
- Not until about half a billion years ago did life become complex in what is called the Cambrian explosion.
- Life emerged from the oceans only about 0.4 billion years ago, and human intelligence developed over the last 3 million years.

Could life begin on other worlds?
- Life on Earth requires liquid water and moderate temperatures, and that is probably true of life on other worlds.
- No planet in our solar system appears to harbor life at present. Most are too hot or too cold, but life might have begun on Mars before it became too cold and dry.
- Life in the oceans of Europa or Titan seems possible but unlikely.
- Because the origin of life and its evolution into intelligent creatures took so long on Earth, scientists eliminate short-lived stars such as middle- and upper-main-sequence stars as homes for life.

- Main-sequence G, K, and possibly M stars are thought to be likely candidates for searches for life.

- The life zone around a star may be larger than expected, given the wide variety of living things being found in extreme environments on Earth.

26-3 | Communication with Distant Civilizations

Could Earthlings communicate with civilizations on other worlds?

- Because of distance, speed, and fuel, travel between the stars seems almost impossible for humans or for aliens who might visit Earth.

- Radio communication may be possible, but a conversation would be impossible because of very long travel times for radio signals.

- Broadcasting a radio beacon of pulses in mathematical patterns would distinguish the signal from naturally occurring radio emissions and identify a civilization as technological.

- A signal could be anti-coded so it would be easy for another civilization to decode.

- The best place in the radio spectrum for communication is in the water hole, the wavelength range from the 21 cm line of hydrogen to the 18 cm emission line of OH. Even so, millions of radio wavelengths need to be tested to fully survey the water hole for a given target.

- Sophisticated searches are now under way to detect radio transmissions from civilizations on other worlds, but such SETI programs are hampered by limited computer power and radio noise pollution.

- The number of civilizations in our galaxy that are at a technological level and able to communicate may be limited by the lifetime of such civilizations.

New Terms

natural selection (p. 653)

DNA (deoxyribonucleic acid) (p. 654)

amino acid (p. 654)

protein (p. 654)

enzyme (p. 654)

RNA (ribonucleic acid) (p. 655)

chromosome (p. 655)

gene (p. 655)

mutant (p. 656)

Cambrian period (p. 657)

Miller experiment (p. 657)

primordial soup (p. 657)

chemical evolution (p. 658)

stromatolite (p. 659)

life zone (p. 665)

water hole (p. 668)

SETI (p. 668)

Drake equation (p. 669)

Review Questions

Ace⊙Astronomy™ Assess your understanding of this chapter's topics with additional quizzing and animations at **http://ace .brookscole.com/sf9**

1. If life is based on information, what is that information?

2. What would happen to a life form if the information handed down to offspring was always the same? How would that endanger the future of the life form?

3. How does the DNA molecule produce a copy of itself?

4. Give an example of natural selection acting on new DNA patterns to select the most advantageous characteristics.

5. What is the evidence that life on Earth began in the sea?

6. What would make you conclude that liquid water is necessary for the origin of life?

7. What is the difference between chemical evolution and biological evolution?

8. What was the significance of the Miller experiment?

9. How does intelligence make a creature more likely to survive?

10. Why are upper-main-sequence stars unlikely sites for intelligent civilizations?

11. What factors make travel between stars nearly impossible?

12. How does the stability of technological civilizations affect the probability that they can be detected from Earth?

13. What is the water hole, and why would it be a good place to look for other civilizations?

14. The star cluster NGC2264 shown here contains cool red giants plus main-sequence stars ranging from hot blue stars all the way down to red dwarfs. Discuss the likelihood that planets orbiting any of these stars might be home to intelligent life. Don't neglect to estimate the age of the cluster.

Visual

(T. A. Rector, B. A. Wolpa, NRAO/NOAO/ AURA/NSF)

15. If you could search for life in the galaxy shown, would you look among disk stars or halo stars? Discuss the factors that influence your decision.

Visual

(ESO)

Discussion Questions

1. What would you change in the Arecibo message if humanity lived on Mars instead of Earth?

2. What do you think it would mean if decades of careful searches for radio signals for extraterrestrial intelligence turned up nothing? What might that suggest about the future of our own civilization?

3. What philosophical or theological questions might arise if conclusive evidence of intelligent life on other worlds was found?

Problems

1. A single human cell encloses about 1.5 m of DNA, containing 4.5 billion base pairs. What is the spacing between these base pairs in nanometers? That is, how far apart are the rungs on the DNA ladder?

2. If you represented the history of the Earth by a line 1 m long, how long a segment would represent the 400 million years since life moved onto the land? How long a segment would represent the 3-million-year history of human life?

3. If a human generation, the time from birth to childbearing, is 20 years, how many generations have passed in the last million years?

4. If a star must remain on the main sequence for at least 5 billion years for life to evolve to intelligence, how massive could a star be and still harbor intelligent life on one of its planets? (*Hint:* See Chapter 12.)

5. If there are about 1.4×10^{-4} stars like the sun per cubic light-year, how many lie within 100 light-years of Earth? (*Hint:* The volume of a sphere is $\frac{4}{3}\pi R^3$.)

6. Mathematician Karl Gauss suggested planting forests and fields in a gigantic geometric proof to signal to possible Martians that intelligent life exists on Earth. If Martians had telescopes that could resolve details no smaller than 1 second of arc, how large would the smallest element of Gauss's proof have to be? (*Hint:* Use the small-angle formula.)

7. If you detected radio signals with an average wavelength of 20 cm and suspected that they came from a civilization on a distant planet, roughly how much of a change in wavelength should you expect to see because of the orbital motion of the distant planet? (*Hint:* See the Doppler shift in Chapter 7.)

8. Calculate the number of communicative civilizations per galaxy from your own estimates of the factors in Table 26-1.

Media Cluster

Ace◐Astronomy™ To access the resources in the Media Cluster, log into AceAstronomy at http://ace.brookscole .com/sf9 and select Chapter 26.

ACTIVE FIGURES

DNA
Explore DNA in detail and see how individual atoms are linked together to produce each of the four bases. Notice how the molecules "hitch" to the DNA rails and fit with their matching bases.

Future of the Sun
Over billions of years, the sun will gradually mature and die. This animation lets you observe the death of the sun from four different vantage points in our solar system.

Earth Calendar
This animation shows how the formation of Earth and its life forms would look if the entire history of Earth were condensed into one calendar year.

Interstellar Communication
In this animation, you can launch a space probe between distant stars or send radio signals and compare the two methods of communication.

Drake Equation
Explore how different values for factors in the Drake equation affect predictions about how plentiful intelligent life is in a galaxy.

ASTRONOMY EXERCISES

Extrasolar Planets
This simulation presents a hypothetical star system with one planet orbiting a star. Study what variables help astronomers detect extrasolar planets.

Lab 8: Extrasolar Planets
This lab examines how indirect methods such as Doppler shift techniques are used to discover and study extrasolar planets. At the end of the lab you will build your own solar system.

Critical Inquiries for the Web

1. The popular movie *Contact* focused interest on the SETI program by profiling the work of a radio astronomer dedicated to the search for extraterrestrial intelligence. Visit websites that give information about the movie, SETI programs, and radio astronomy, and discuss how realistic the movie was in capturing how such research is done.

2. Where outside the solar system would you look for habitable planets? NASA has increasingly focused its interest on this question. Look for information online about programs dedicated to detecting which planets might support life. What criteria are used to choose targets for the planned searches? What methods will be used to carry out the searches?

 Go to the Brooks/Cole Astronomy Resource Center **(http:// astronomy.brookscole.com)** for critical thinking exercises, articles, and additional readings from InfoTrac College Edition, Brooks/Cole's online student library.

Afterword

The aggregate of all our joys and sufferings, thousands of confident religions,

ideologies and economic doctrines, every hunter and forager, every hero and coward,

every creator and destroyer of civilizations, every king and peasant, every young

couple in love, every hopeful child, every mother and father, every inventor

and explorer, every teacher of morals, every corrupt politician, every superstar,

every supreme leader, every saint and sinner in the history of our species,

lived there on a mote of dust, suspended in a sunbeam.

CARL SAGAN (1934–1996)

*Earth photographed by Voyager 1 from
the edge of the solar system.* (NASA)

Our journey together is over, but before we part company, there is one last thing to discuss—the place of humanity in the universe. Astronomy gives us some comprehension of the workings of stars, galaxies, and planets, but its greatest value lies in what it teaches us about ourselves. Now that you have surveyed astronomical knowledge, you can better understand your own position in nature.

To some, the word *nature* conjures up visions of furry rabbits hopping about in a forest glade. To others, nature is the blue-green ocean depths, and still others think of nature as windswept mountaintops. As diverse as these images are, they are all Earthbound. Having studied astronomy, you see nature as a beautiful mechanism composed of matter and energy interacting according to simple rules to form galaxies, stars, planets, mountaintops, ocean depths, and forest glades.

Perhaps the most important astronomical lesson is that humanity is a small but important part of the universe. Most of the universe is lifeless. The vast reaches between the galaxies appear to be empty of all but the thinnest gas, and the stars are much too hot to preserve the chemical bonds that seem necessary for life to survive and develop. Only on the surfaces of a few planets, where temperatures are moderate, could atoms link together to form living matter.

If life is special, then intelligence is precious. The universe must contain many planets devoid of life, planets where the wind has blown unfelt for billions of years. There may also exist planets where life has developed but has not become complex, planets on which the wind stirs wide plains of grass and rustles dark forests. On some planets, insects, fish, birds, and animals may watch the passing days unaware of their own existence. It is intelligence, human or alien, that gives meaning to the landscape.

Science is the process by which intelligence tries to understand the universe. Science is not the invention of new devices or processes. It does not create home computers, cure the mumps, or manufacture plastic spoons—that is engineering and technology, the adaptation of scientific understanding for practical purposes. Science is understanding nature, and astronomy is understanding on the grandest scale. Astronomy is the science by which the universe, through its intelligent lumps of matter, tries to understand its own existence.

As the primary intelligent species on this planet, we are the custodians of a priceless gift—a planet filled with living things. This is especially true if life is rare in the universe. In fact, if Earth is the only inhabited planet, our responsibility is overwhelming. We are the only creatures who can take action to preserve the existence of life on Earth, and, ironically, our own actions are the most serious hazards.

The future of humanity is not secure. We are trapped on a tiny planet with limited resources and a population growing faster than our ability to produce food. In our efforts to survive, we have already driven some creatures to extinction and now threaten others. If our civilization collapses because of starvation, or if our race destroys itself somehow, the only bright spot is that the rest of the creatures on Earth will be better off for our absence.

But even if we control our population and conserve and recycle our resources, life on Earth is doomed. In 5 billion years, the sun will leave the main sequence and swell into a red giant, incinerating Earth. However, Earth will be lifeless long before that. Within the next few billion years, the growing luminosity of the sun will first alter Earth's climate and then drive off its atmosphere and oceans. Earth, like everything else in the universe, is only temporary.

To survive, humanity must eventually leave Earth and search for other planets. Colonizing the moon and other planets of our solar system will not save us, because they will face the same fate as Earth when the sun dies. But travel to other stars is tremendously difficult and may be impossible with the limited resources we have in our small solar system. We and all of the living things on Earth may be trapped.

This is a depressing prospect, but a few factors are comforting. First, everything in the universe is temporary. Stars die, galaxies die, perhaps the entire universe will someday end. That our distant future is limited only assures us that we are a part of a much larger whole. Second, we have a few billion years to prepare, and a billion years is a very long time. Only a few million years ago, our ancestors were learning to walk erect and communicate.

A billion years ago, our ancestors were microscopic organisms living in the primeval oceans. To suppose that a billion years hence we humans will still be human, that we will still be the dominant species on Earth, or that we will still be the sole intelligence on Earth is the ultimate conceit.

Our responsibility is not to save our race for all eternity but to behave as dependable custodians of our planet, preserving it, admiring it, and trying to understand it. That calls for drastic changes in our behavior toward other living things and a revolution in our attitude toward our planet's resources. Whether we can change our ways is debatable—humanity is far from perfect in its understanding, abilities, or intentions. However, you must not imagine that we and our civilization are less than precious. We have the gift of intelligence, and that is the finest thing this planet has ever produced.

We shall not cease from exploration
And the end of all our exploring
Will be to arrive where we started
And know the place for the first time.

—*T. S. Eliot, "Little Gidding"*

Appendix A

Units and Astronomical Data

Introduction

The metric system is used worldwide as the system of units, not only in science but also in engineering, business, sports, and daily life. Developed in 18th-century France, the metric system has gained acceptance in almost every country in the world because it simplifies computations.

A system of units is based on the three fundamental units for length, mass, and time. Other quantities, such as density and force, are derived from these fundamental units. In the English (or British) system of units (commonly used only in the United States, Myanmar, and Liberia, but not in Great Britain) the fundamental unit of length is the foot, composed of 12 inches. The metric system is based on the decimal system of numbers, and the fundamental unit of length is the meter, composed of 100 centimeters.

Because the metric system is a decimal system, it is easy to express quantities in larger or smaller units as is convenient. You can express distances in centimeters, meters, kilometers, and so on. The prefixes specify the relation of the unit to the meter. Just as a cent is $\frac{1}{100}$ of a dollar, so a centimeter is $\frac{1}{100}$ of a meter. A kilometer is 1000 m, and a kilogram is 1000 g. The meanings of the commonly used prefixes are given in ▪ Table A-1.

The SI Units

Any system of units based on the decimal system would be easy to use, but by international agreement, the preferred set of units, known as the *Système International d'Unités* (SI units) is based on the meter, kilogram, and second. These three fundamental units define the rest of the units, as given in ▪ Table A-2.

The SI unit of force is the newton (N), named after Isaac Newton. It is the force needed to accelerate a 1 kg mass by 1 m/s^2, or the force roughly equivalent to the weight of an apple at Earth's surface. The SI unit of energy is the joule (J), the energy produced by a force of 1 N acting through a distance of 1 m. A joule is roughly the energy in the impact of an apple falling off a table.

▪ Table A-1 ▮ Metric Prefixes

Prefix	Symbol	Factor
Mega	M	10^6
Kilo	k	10^3
Centi	c	10^{-2}
Milli	m	10^{-3}
Micro	μ	10^{-6}
Nano	n	10^{-9}

▪ Table A-2 ▮ SI Metric Units

Quantity	SI Unit	English Unit
Length	Meter (m)	Foot
Mass	Kilogram (kg)	Slug (sl)
Time	Second (s)	Second (s)
Force	Newton (N)	Pound (lb)
Energy	Joule (J)	Foot-pound (fp)

Exceptions

Units can help you in two ways. They make it possible to make calculations, and they can help you to conceive of certain quantities. For calculations, the metric system is far superior, and it is used for calculations throughout this book.

But Americans commonly use the English system of units, so for conceptual purposes this book also expresses quantities in English units. Instead of saying the average person would weigh 133 N on the moon, it might be more helpful to express the weight as 30 lb. Consequently, this text commonly gives quantities in metric form followed by the English form in parentheses: the radius of the moon is 1738 km (1080 miles).

In SI units, density should be expressed as kilograms per cubic meter, but no human hand can enclose a cubic meter, so this unit does not help you grasp the significance of a given density. This book refers to density in grams per cubic centimeter. A gram is roughly the mass of a paperclip, and a cubic centimeter is the size of a small sugar cube, so you can conceive of a density of 1 g/cm³, roughly the density of water. This is not a bothersome departure from SI units because you will not have to make complex calculations using density.

Conversions

To convert from one metric unit to another (from meters to kilometers, for example), you have only to look at the prefix. However, converting from metric to English or English to metric is more complicated. The conversion factors are given in ■ Table A-3.

Example: The radius of the moon is 1738 km. What is this in miles? Table A-3 indicates that 1 mile equals 1.609 km, so

$$1738 \text{ km} \times \frac{1 \text{ mile}}{1.609 \text{ km}} = 1080 \text{ miles}$$

Temperature Scales

In astronomy, as in most other sciences, temperatures are expressed on the Kelvin scale, although the centigrade (or Celsius) scale is also used. The Fahrenheit scale commonly used in the United States is not used in scientific work.

Temperatures on the Kelvin scale are measured from absolute zero, the temperature of an object that contains no extractable heat. In practice, no object can be as cold as absolute zero, although laboratory apparatuses have reached temperatures less than 10^{-6} K. The scale is named after the Scottish mathematical physicist William Thomson, Lord Kelvin (1824–1907).

The centigrade scale refers temperatures to the freezing point of water (0°C) and to the boiling point of water (100°C). One degree centigrade is 1/100th the temperature difference between the freezing and boiling points of water, thus the prefix *centi*. The centigrade scale is also called the Celsius scale after its inventor, the Swedish astronomer Anders Celsius (1701–1744).

The Fahrenheit scale fixes the freezing point of water at 32°F and the boiling point at 212°F. Named after the German physicist Gabriel Daniel Fahrenheit (1686–1736), who made the first successful mercury thermometer in 1720, the Fahrenheit scale is used only in the United States.

It is easy to convert temperatures from one scale to another using the information given in ■ Table A-4.

Powers of 10 Notation

Powers of 10 make writing very large numbers much simpler. For example, the nearest star is about 43,000,000,000,000 km from the sun. Writing this number as 4.3×10^{13} km is much easier.

Very small numbers can also be written with powers of 10. For example, the wavelength of visible light is about 0.0000005 m. In powers of 10 this becomes 5×10^{-7} m.

The powers of 10 used in this notation appear below. The exponent tells you how to move the decimal point. If the exponent is positive, move the decimal point to the right. If the exponent is negative, move the decimal point to the left. For example, 2×10^3 equals 2000.0, and 2×10^{-3} equals 0.002.

$$
\begin{aligned}
&\vdots \\
10^5 &= 100,000 \\
10^4 &= 10,000 \\
10^3 &= 1,000 \\
10^2 &= 100 \\
10^1 &= 10 \\
10^0 &= 1 \\
10^{-1} &= 0.1 \\
10^{-2} &= 0.01 \\
10^{-3} &= 0.001 \\
10^{-4} &= 0.0001 \\
&\vdots
\end{aligned}
$$

■ Table A-3 | Conversion Factors

1 inch = 2.54 centimeters	1 centimeter = 0.394 inch
1 foot = 0.3048 meter	1 meter = 39.36 inches = 3.28 feet
1 mile = 1.6093 kilometers	1 kilometer = 0.6214 mile
1 slug = 14.594 kilograms	1 kilogram = 0.0685 slug
1 pound = 4.4482 newtons	1 newton = 0.2248 pound
1 foot-pound = 1.35582 joules	1 joule = 0.7376 foot-pound
1 horsepower = 745.7 joules/s	1 joule/s = 1 watt

■ Table A-4 | Temperature Scales

	Kelvin (K)	Centigrade (°C)	Fahrenheit (°F)
Absolute zero	0 K	−273°C	−459°F
Freezing point of water	273 K	0°C	32°F
Boiling point of water	373 K	100°C	212°F

Conversions:

$$K = °C + 273$$
$$°C = \tfrac{5}{9}(°F - 32)$$
$$°F = \tfrac{9}{5}°C + 32$$

If you use scientific notation in calculations, be sure you correctly enter numbers into your calculator. Not all calculators accept scientific notation, but those that can have a key labeled EXP, EEX, or perhaps EE that allows you to enter the exponent of ten. To enter a number such as 3×10^8, press the keys 3 EXP 8. To enter a number with a negative exponent, you must use the change-sign key, usually labeled $+/-$ or CHS. To enter the number 5.2×10^{-3}, press the keys 5.2 EXP $+/-$ 3. Try a few examples.

To read a number in scientific notation from a calculator you must read the exponent separately. The number 3.1×10^{25} may appear in a calculator display as 3.1 25 or on some calculators as 3.1 10^{25}. Examine your calculator to determine how such numbers are displayed.

* * *

Astronomy, and science in general, is a way of learning about nature and understanding the universe. To test hypotheses about how nature works, scientists use observations of nature. The tables that follow contain some of the basic observations that support our best understanding of the astronomical universe. Of course, these data are expressed in the form of numbers, not because science reduces all understanding to mere numbers, but because the struggle to understand nature is so demanding that science must use every tool available. Quantitative thinking—reasoning mathematically—is one of the most powerful tools ever invented by the human brain. Thus these tables are not nature reduced to mere numbers but numbers supporting humanity's growing understanding of the natural world around us.

■ Table A-5   I   Constants

Astronomical unit (AU)	$= 1.495979 \times 10^{11}$ m
Parsec (pc)	$= 206{,}265$ AU
	$= 3.085678 \times 10^{16}$ m
	$= 3.261633$ ly
Light-year (ly)	$= 9.46053 \times 10^{15}$ m
Velocity of light (c)	$= 2.997925 \times 10^{8}$ m/s
Gravitational constant (G)	$= 6.67 \times 10^{-11}$ m^3/s^{2}kg
Mass of Earth ($M_\oplus$)	$= 5.976 \times 10^{24}$ kg
Earth equatorial radius ($R_\oplus$)	$= 6378.164$ km
Mass of sun ($M_\odot$)	$= 1.989 \times 10^{30}$ kg
Radius of sun ($R_\odot$)	$= 6.9599 \times 10^{8}$ m
Solar luminosity ($L_\odot$)	$= 3.826 \times 10^{26}$ J/s
Mass of moon	$= 7.350 \times 10^{22}$ kg
Radius of moon	$= 1738$ km
Mass of H atom	$= 1.67352 \times 10^{-27}$ kg

■ Table A-6   I   Units Used in Astronomy

1 angstrom (Å)	$= 10^{-8}$ cm
	$= 10^{-10}$ m
1 astronomical unit (AU)	$= 1.495979 \times 10^{11}$ m
	$= 92.95582 \times 10^{6}$ miles
1 light-year (ly)	$= 6.3240 \times 10^{4}$ AU
	$= 9.46053 \times 10^{15}$ m
	$= 5.9 \times 10^{12}$ miles
1 parsec (pc)	$= 206{,}265$ AU
	$= 3.085678 \times 10^{16}$ m
	$= 3.261633$ ly
1 kiloparsec (kpc)	$= 1000$ pc
1 megaparsec (Mpc)	$= 1{,}000{,}000$ pc

Spectral Type	Absolute Visual Magnitude (M_v)	Luminosity*	Temp. (K)	λ_{max} (nm)	Mass*	Radius*	Average Density (g/cm³)
O5	−5.8	501,000	40,000	72.4	40	17.8	0.01
B0	−4.1	20,000	28,000	100	18	7.4	0.1
B5	−1.1	790	15,000	190	6.4	3.8	0.2
A0	+0.7	79	9900	290	3.2	2.5	0.3
A5	+2.0	20	8500	340	2.1	1.7	0.6
F0	+2.6	6.3	7400	390	1.7	1.4	1.0
F5	+3.4	2.5	6600	440	1.3	1.2	1.1
G0	+4.4	1.3	6000	480	1.1	1.0	1.4
G5	+5.1	0.8	5500	520	0.9	0.9	1.6
K0	+5.9	0.4	4900	590	0.8	0.8	1.8
K5	+7.3	0.2	4100	700	0.7	0.7	2.4
M0	+9.0	0.1	3500	830	0.5	0.6	2.5
M5	+11.8	0.01	2800	1000	0.2	0.3	10.0
M8	+16	0.001	2400	1200	0.1	0.1	63

*Luminosity, mass, and radius are given in terms of the sun's luminosity, mass, and radius.

Table A-8 | The Brightest Stars

Star	Name	Apparent Visual Magnitude (m_v)	Spectral Type	Absolute Visual Magnitude (M_v)	Distance (ly)
α CMa A	Sirius	−1.47	A1	1.4	8.7
α Car	Canopus	−0.72	F0	−3.1	98
α Cen	Rigil Kentaurus	−0.01	G2	4.4	4.3
α Boo	Arcturus	−0.06	K2	−0.3	36
α Lyr	Vega	0.04	A0	0.5	26.5
α Aur	Capella	0.05	G8	−0.6	45
β Ori A	Rigel	0.14	B8	−7.1	900
α CMi A	Procyon	0.37	F5	2.7	11.3
α Ori	Betelgeuse	0.41	M2	−5.6	520
α Eri	Achernar	0.51	B3	−2.3	118
β Cen AB	Hadar	0.63	B1	−5.2	490
α Aql	Altair	0.77	A7	2.2	16.5
α Tau A	Aldebaran	0.86	K5	−0.7	68
α Cru	Acrux	0.90	B2	−3.5	260
α Vir	Spica	0.91	B1	−3.3	220
α Sco A	Antares	0.92	M1	−5.1	520
α PsA	Fomalhaut	1.15	A3	2.0	22.6
β Gem	Pollux	1.16	K0	1.0	35
α Cyg	Deneb	1.26	A2	−7.1	1600
β Cru	Beta Crucis	1.28	B0.5	−4.6	490

Name	Absolute Magnitude (M_v)	Distance (ly)	Spectral Type	Apparent Visual Magnitude (m_v)
Sun	4.83		G2	−26.8
Proxima Cen	15.45	4.28	M5	11.05
α Cen A	4.38	4.3	G2	0.1
α Cen B	5.76	4.3	K5	1.5
Barnard's Star	13.21	5.9	M5	9.5
Wolf 359	16.80	7.6	M6	13.5
Lalande 21185	10.42	8.1	M2	7.5
Sirius A	1.41	8.6	A1	−1.5
Sirius B	11.54	8.6	white dwarf	7.2
Luyten 726-8A	15.27	8.9	M5	12.5
Luyten 726-8B (UV Cet)	15.8	8.9	M6	13.0
Ross 154	13.3	9.4	M5	10.6
Ross 248	14.8	10.3	M6	12.2
ε Eri	6.13	10.7	K2	3.7
Luyten 789-6	14.6	10.8	M7	12.2
Ross 128	13.5	10.8	M5	11.1
61 CYG A	7.58	11.2	K5	5.2
61 CYG B	8.39	11.2	K7	6.0
ε Ind	7.0	11.2	K5	4.7
Procyon A	2.64	11.4	F5	0.3
Procyon B	13.1	11.4	white dwarf	10.8
Σ 2398 A	11.15	11.5	M4	8.9
Σ 2398 B	11.94	11.5	M5	9.7
Groombridge 34 A	10.32	11.6	M1	8.1
Groombridge 34 B	13.29	11.6	M6	11.0
Lacaille 9352	9.59	11.7	M2	7.4
τ Ceti	5.72	11.9	G8	3.5
BD + 5° 1668	11.98	12.2	M5	9.8
L 725-32	15.27	12.4	M5	11.5
Lacaille 8760	8.75	12.5	M0	6.7
Kapteyn's Star	10.85	12.7	M0	8.8
Kruger 60 A	11.87	12.8	M3	9.7
Kruger 60 B	13.3	12.8	M4	11.2

PHYSICAL PROPERTIES (EARTH = ⊕)

Planet	Equatorial Radius (km)	Equatorial Radius (⊕ = 1)	Mass (⊕ = 1)	Average Density (g/cm³)	Surface Gravity (⊕ = 1)	Escape Velocity (km/s)	Sidereal Period of Rotation	Inclination of Equator to Orbit
Mercury	2439	0.382	0.0558	5.44	0.378	4.3	58.646^d	0°
Venus	6052	0.95	0.815	5.24	0.903	10.3	244.3^d	177°
Earth	6378	1.00	1.00	5.497	1.00	11.2	23^{h}56^{m}04.1^s	23°27′
Mars	3396	0.53	0.1075	3.94	0.379	5.0	24^{h}37^{m}22.6^s	25°19′
Jupiter	71,494	11.20	317.83	1.34	2.54	61	9^{h}55^{m}30^s	3°5′
Saturn	60,330	9.42	95.147	0.69	1.16	35.6	10^{h}13^{m}59^s	26°24′
Uranus	25,559	4.01	14.54	1.19	0.919	22	17^{h}14^m	97°55′
Neptune	24,750	3.93	17.23	1.66	1.19	25	16^{h}3^m	28°48′
Pluto	1185	0.18	0.0022	2.0	0.06	1.2	6^{d}9^{h}21^m	122°

ORBITAL PROPERTIES

Planet	Semimajor Axis (AU)	Semimajor Axis (10⁶ km)	Orbital Period (y)	Orbital Period (days)	Average Orbital Velocity (km/s)	Orbital Eccentricity	Inclination to Ecliptic
Mercury	0.3871	57.9	0.24084	87.969	47.89	0.2056	7°0′16″
Venus	0.7233	108.2	0.61515	224.68	35.03	0.0068	3°23′40″
Earth	1	149.6	1	365.26	29.79	0.0167	0°
Mars	1.5237	227.9	1.8808	686.95	24.13	0.0934	1°51′9″
Jupiter	5.2028	778.3	11.867	4334.3	13.06	0.0484	1°18′29″
Saturn	9.5388	1427.0	29.461	10,760	9.64	0.0560	2°29′17″
Uranus	19.18	2869.0	84.013	30,685	6.81	0.0461	0°46′23″
Neptune	30.0611	4497.1	164.793	60,189	5.43	0.0100	1°46′27″
Pluto	39.44	5900	247.7	90,465	4.74	0.2484	17°9′3″

■ **Table A-11 | Principal Satellites of the Solar System**

Planet	Satellite	Radius (km)	Distance from Planet (10^3 km)	Orbital Period (days)	Orbital Eccentricity	Orbital Inclination
Earth	Moon	1738	384.4	27.322	0.055	5°8′43″
Mars	Phobos	14 × 12 × 10	9.38	0.3189	0.018	1.0°
	Deimos	8 × 6 × 5	23.5	1.262	0.002	2.8°
Jupiter	Metis	20	126	0.29	0.0	0.0°
	Adrastea	12 × 8 × 10	128	0.294	0.0	0.0°
	Amalthea	135 × 100 × 78	182	0.4982	0.003	0.45°
	Thebe	50	223	0.674	0.0	1.3°
	Io	1820	422	1.769	0.000	0.3°
	Europa	1565	671	3.551	0.000	0.46°
	Ganymede	2640	1071	7.155	0.002	0.18°
	Callisto	2420	1884	16.689	0.008	0.25°
	Leda	~8	11,110	240	0.146	26.7°
	Himalia	~85	11,470	250.6	0.158	27.6°
	Lysithea	~20	11,710	260	0.12	29°
	Elara	~30	11,740	260.1	0.207	24.8°
	Ananke	15	21,200	631	0.169	147°
	Carme	22	22,350	692	0.207	163°
	Pasiphae	35	23,300	735	0.40	147°
	Sinope	20	23,700	758	0.275	156°
Saturn	Pan	10	133.570	0.574	0.000	0°
	Atlas	20 × 15 × 15	137.7	0.601	0.002	0.3°
	Prometheus	70 × 40 × 50	139.4	0.613	0.003	0.0°
	Pandora	55 × 35 × 50	141.7	0.629	0.004	0.05°
	Epimetheus	70 × 50 × 50	151.42	0.694	0.009	0.34°
	Janus	110 × 80 × 100	151.47	0.695	0.007	0.14°
	Mimas	196	185.54	0.942	0.020	1.5°
	Enceladus	250	238.04	1.370	0.004	0.0°
	Tethys	530	294.67	1.888	0.000	1.1°
	Calypso	17 × 11 × 12	294.67	1.888	0.0	~1°?
	Telesto	12	294.67	1.888	0.0	~1°?
	Dione	560	377	2.737	0.002	0.0°
	Helene	20 × 15 × 15	377	2.74	0.005	0.15°
	Rhea	765	527	4.518	0.001	0.4°
	Titan	2575	1222	15.94	0.029	0.3°
	Hyperion	205 × 130 × 110	1484	21.28	0.104	~0.5°
	Iapetus	720	3562	79.33	0.028	14.72°
	Phoebe	110	12,930	550.4	0.163	150°
Uranus	Cordelia	20	49.8	0.3333	~0	~0°
	Ophelia	15	53.8	0.375	~0	~0°
	Bianca	25	59.1	0.433	~0	~0°
	Cressida	30	61.8	0.462	~0	~0°
	Desdemona	30	62.7	0.475	~0	~0°

(continued)

Planet	Satellite	Radius (km)	Distance from Planet (10^3 km)	Orbital Period (days)	Orbital Eccentricity	Orbital Inclination
Uranus	Juliet	40	64.4	0.492	~0	~0°
(continued)	Portia	55	66.1	0.512	~0	~0°
	Rosalind	30	69.9	0.558	~0	~0°
	Belinda	30	75.2	0.621	~0	~0°
	Puck	85 ± 5	85.9	0.762	~0	~0°
	Miranda	242 ± 5	129.9	1.414	0.017	3.4°
	Ariel	580 ± 5	190.9	2.520	0.003	0°
	Umbriel	595 ± 10	266.0	4.144	0.003	0°
	Titania	805 ± 5	436.3	8.706	0.002	0°
	Oberon	775 ± 10	583.4	13.463	0.001	0°
	Caliban	40	7164	579	0.082	139.2°
	Stephano	~20	7900	676		
	Sycorax	80	12,174	1284	0.509	152.7°
	Prospero	~20	16,100	1950		
	Setebos	~20	17,600	2240	0.539	
Neptune	Naiad	30	48.2	0.296	~0	~0°
	Thalassa	40	50.0	0.312	~0	~0°
	Despina	90	52.5	0.333	~0	~0°
	Galatea	75	62.0	0.396	~0	~0°
	Larissa	95	73.6	0.554	~0	~0°
	Proteus	205	117.6	1.121	~0	~0°
	Triton	1352	354.59	5.875	0.00	160°
	Nereid	170	5588.6	360.125	0.76	27.7°
Pluto	Charon	626	19.7	6.38718	~0	122°

■ Table A-12 I **Meteor Showers**

Shower	Dates	Hourly Rate	Radiant R.A.	Radiant Dec.	Associated Comet
Quadrantids	Jan. 2–4	30	15^h24^m	50°	
Lyrids	April 20–22	8	18^h4^m	33°	1861 I
η Aquarids	May 2–7	10	22^h24^m	0°	Halley?
δ Aquarids	July 26–31	15	22^h36^m	−10°	
Perseids	Aug. 10–14	40	3^h4^m	58°	1982 III
Orionids	Oct. 18–23	15	6^h20^m	15°	Halley?
Taurids	Nov. 1–7	8	3^h40^m	17°	Encke
Leonids	Nov. 14–19	6	10^h12^m	22°	1866 I Temp
Geminids	Dec. 10–13	50	7^h28^m	32°	

■ Table A-13 | Greatest Elongations of Mercury

Evening Sky	Morning Sky
Feb. 24, 2006*	April 8, 2006
June 20, 2006	Aug. 7, 2006
Oct. 16, 2006	Nov. 25, 2006*
Feb. 7, 2007	March 22, 2007
June 2, 2007	July 20, 2007
Sept. 29, 2007	Nov. 8, 2007*
Jan. 22, 2008	March 3, 2008
May 14, 2008	July 1, 2008
Sept. 11, 2008	Oct. 22, 2008*
Jan. 4, 2009	Feb. 13. 2009
April 26, 2009*	June 13, 2009
Aug. 24, 2009	Oct. 6, 2009*
Dec 18, 2009	Jan 27, 2010
Apr. 8, 2010*	May 26, 2010
Aug. 7, 2010	Sep 19, 2010*
Dec. 1, 2010	

Elongation is the angular distance from the sun to a planet.

*Most favorable elongations.

■ Table A-14 | Greatest Elongations of Venus

Evening Sky	Morning Sky
	March 25, 2006
June 9, 2007	Oct. 28, 2007
Jan. 14, 2009	June 5, 2009
Aug. 20, 2010	Jan. 8, 2011
March 27, 2012	Aug. 15, 2012
Nov. 1, 2013	March 22, 2014
June 6, 2015	Oct. 26, 2015
Jan. 12, 2017	June 3, 2017
Aug. 17, 2018	

■ Table A-15 | The Greek Alphabet

A, α	alpha	H, η	eta	N, ν	nu	T, τ	tau
B, β	beta	Θ, θ	theta	Ξ, ξ	xi	Y, υ	upsilon
Γ, γ	gamma	I, ι	iota	O, o	omicron	Φ, ϕ	phi
Δ, δ	delta	K, κ	kappa	Π, π	pi	X, χ	chi
E, ε	epsilon	Λ, λ	lambda	P, ρ	rho	Ψ, ψ	psi
Z, ζ	zeta	M, μ	mu	Σ, σ	sigma	Ω, ω	omega

Group

Atomic masses are based on carbon-12. Numbers in parentheses are mass numbers of most stable or best-known isotopes of radioactive elements.

Atomic number → 11
Symbol → Na
Atomic mass → 22.99

Noble Gases (18)

Transition Elements

IA(1)								VIII						IIIA(13)	IVA(14)	VA(15)	VIA(16)	VIIA(17)	

Period

1
1 H 1.008
IIA(2)
2 He 4.003

2
3 Li 6.941
4 Be 9.012
5 B 10.81
6 C 12.01
7 N 14.01
8 O 16.00
9 F 19.00
10 Ne 20.18

3
11 Na 22.99
12 Mg 24.31
IIIB(3) | IVB(4) | VB(5) | VIB(6) | VIIB(7) | (8) | (9) | (10) | IB(11) | IIB(12)
13 Al 26.98
14 Si 28.09
15 P 30.97
16 S 32.06
17 Cl 35.45
18 Ar 39.95

4
19 K 39.10
20 Ca 40.08
21 Sc 44.96
22 Ti 47.90
23 V 50.94
24 Cr 52.00
25 Mn 54.94
26 Fe 55.85
27 Co 58.93
28 Ni 58.7
29 Cu 63.55
30 Zn 65.38
31 Ga 69.72
32 Ge 72.59
33 As 74.92
34 Se 78.96
35 Br 79.90
36 Kr 83.80

5
37 Rb 85.47
38 Sr 87.62
39 Y 88.91
40 Zr 91.22
41 Nb 92.91
42 Mo 95.94
43 Tc 98.91
44 Ru 101.1
45 Rh 102.9
46 Pd 106.4
47 Ag 107.9
48 Cd 112.4
49 In 114.8
50 Sn 118.7
51 Sb 121.8
52 Te 127.6
53 I 126.9
54 Xe 131.3

6
55 Cs 132.9
56 Ba 137.3
57* La 138.9
72 Hf 178.5
73 Ta 180.9
74 W 183.9
75 Re 186.2
76 Os 190.2
77 Ir 192.2
78 Pt 195.1
79 Au 197.0
80 Hg 200.6
81 Tl 204.4
82 Pb 207.2
83 Bi 209.0
84 Po (210)
85 At (210)
86 Rn (222)

7
87 Fr (223)
88 Ra 226.0
89** Ac (227)
104 Rf (261)
105 Db (262)
106 Sg (263)
107 Bh (262)
108 Hs (265)
109 Mt (266)
110 Ds (269)
111 Uuu (272)
112 Uub (277)
113 Uub (284)
114 Uuq (285)
115 Uub (288)
116 Uuh (289)

Inner Transition Elements

Lanthanide Series 6 *
| 58 Ce 140.1 | 59 Pr 140.9 | 60 Nd 144.2 | 61 Pm (145) | 62 Sm 150.4 | 63 Eu 152.0 | 64 Gd 157.3 | 65 Tb 158.9 | 66 Dy 162.5 | 67 Ho 164.9 | 68 Er 167.3 | 69 Tm 168.9 | 70 Yb 173.0 | 71 Lu 175.0 |

Actinide Series 7 **
| 90 Th 232.0 | 91 Pa 231.0 | 92 U 238.0 | 93 Np 237.0 | 94 Pu (244) | 95 Am (243) | 96 Cm (247) | 97 Bk (247) | 98 Cf (251) | 99 Es (252) | 100 Fm (257) | 101 Md (258) | 102 No (259) | 103 Lr (260) |

The Elements and Their Symbols

Actinium	Ac	Cesium	Cs	Hafnium	Hf	Mercury	Hg	Protactinium	Pa	Tellurium	Te
Aluminum	Al	Chlorine	Cl	Hassium	Hs	Molybdenum	Mo	Radium	Ra	Terbium	Tb
Americium	Am	Chromium	Cr	Helium	He	Neodymium	Nd	Radon	Rn	Thallium	Tl
Antimony	Sb	Cobalt	Co	Holmium	Ho	Neon	Ne	Rhenium	Re	Thorium	Th
Argon	Ar	Copper	Cu	Hydrogen	H	Neptunium	Np	Rhodium	Rh	Thulium	Tm
Arsenic	As	Curium	Cm	Indium	In	Nickel	Ni	Rubidium	Rb	Tin	Sn
Astatine	At	Darmstadtium	Ds	Iodine	I	Niobium	Nb	Ruthenium	Ru	Titanium	Ti
Barium	Ba	Dubnium	Db	Iridium	Ir	Nitrogen	N	Rutherfordium	Rf	Tungsten	W
Berkelium	Bk	Dysprosium	Dy	Iron	Fe	Nobelium	No	Samarium	Sm	Uranium	U
Beryllium	Be	Einsteinium	Es	Krypton	Kr	Osmium	Os	Scandium	Sc	Vanadium	V
Bismuth	Bi	Erbium	Er	Lanthanum	La	Oxygen	O	Seaborgium	Sg	Xenon	Xe
Bohrium	Bh	Europium	Eu	Lawrencium	Lr	Palladium	Pd	Selenium	Se	Ytterbium	Yb
Boron	B	Fermium	Fm	Lead	Pb	Phosphorous	P	Silicon	Si	Yttrium	Y
Bromine	Br	Fluorine	F	Lithium	Li	Platinum	Pt	Silver	Ag	Zinc	Zn
Cadmium	Cd	Francium	Fr	Lutetium	Lu	Plutonium	Pu	Sodium	Na	Zirconium	Zr
Calcium	Ca	Gadolinium	Gd	Magnesium	Mg	Polonium	Po	Strontium	Sr		
Californium	Cf	Gallium	Ga	Manganese	Mn	Potassium	K	Sulfur	S		
Carbon	C	Germanium	Ge	Meitnerium	Mt	Praseodymium	Pr	Tantalum	Ta		
Cerium	Ce	Gold	Au	Mendelevium	Md	Promethium	Pm	Technetium	Tc		

Observing the Sky

Observing the sky with the naked eye is of no more importance to modern astronomy than picking up pretty pebbles is to modern geology. But the sky is a natural wonder unimaginably bigger than the Grand Canyon, the Rocky Mountains, or any other natural wonder that tourists visit every year. To neglect the beauty of the sky is equivalent to geologists neglecting the beauty of the minerals they study. This supplement is meant to act as a tourist's guide to the sky. You analyzed the universe in the regular chapters, but here you will admire it.

The brighter stars in the sky are visible even from the centers of cities with their air and light pollution. But in the countryside, only a few miles beyond the cities, the night sky is a velvety blackness strewn with thousands of glittering stars. From a wilderness location, far from the city's glare, and especially from high mountains, the night sky is spectacular.

Using Star Charts

The constellations are a fascinating cultural heritage of our planet, but they are sometimes a bit difficult to learn because of Earth's motion. The constellations above the horizon change with the time of night and the seasons.

Because Earth rotates eastward, the sky appears to rotate westward around Earth. A constellation visible in the southern sky soon after sunset will appear to move westward, and in a few hours it will disappear below the horizon. Other constellations will rise in the east, so the sky changes gradually through the night.

In addition, Earth's orbital motion makes the sun appear to move eastward among the stars. Each day the sun moves about twice its own diameter, about one degree, eastward along the ecliptic, and consequently each night at sunset, the constellations are about one degree farther toward the west.

Orion, for instance, is visible in the evening sky in January; but, as the days pass, the sun moves closer to Orion. By March, Orion is difficult to see in the western sky soon after sunset. By June, the sun is so close to Orion it sets with the sun and is invisible. Not until late July is the sun far enough past Orion for the constellation to become visible rising in the eastern sky just before dawn.

Because of the rotation and orbital motion of Earth, you need more than one star chart to map the sky. Which chart you select depends on the month and the time of night. The charts given in this appendix show the evening sky for each month.

Two sets of charts are included for two typical locations on Earth. The northern hemisphere charts show the sky as seen from a northern latitude typical of the United States and central Europe. The southern hemisphere star charts are appropriate for readers in Earth's southern hemisphere, including Australia, southern South America, and southern Africa.

To use the charts, select the appropriate chart and hold it overhead as shown in ■ Figure B-1. If you face south, turn the chart until the words *southern horizon* are at the bottom of the chart. If you face other directions, turn the chart appropriately.

■ Figure B-1

To use the star charts in this book, select the appropriate chart for the date and time. Hold it overhead, and turn it until the direction at the bottom of the chart is the same as the direction you are facing.

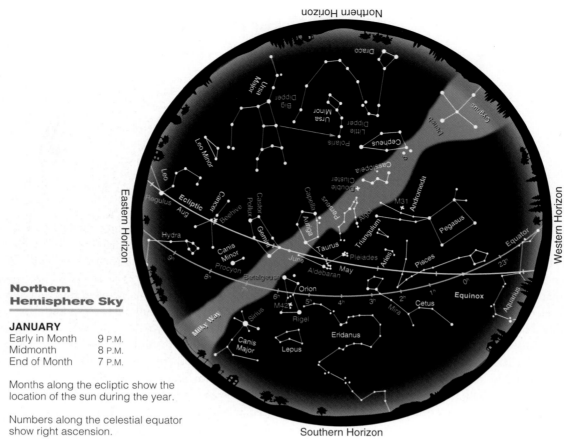

Northern Hemisphere Sky

JANUARY

Early in Month	9 P.M.
Midmonth	8 P.M.
End of Month	7 P.M.

Months along the ecliptic show the location of the sun during the year.

Numbers along the celestial equator show right ascension.

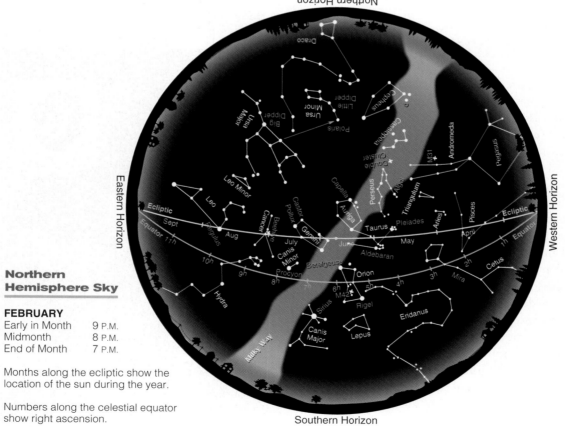

Northern Hemisphere Sky

FEBRUARY

Early in Month	9 P.M.
Midmonth	8 P.M.
End of Month	7 P.M.

Months along the ecliptic show the location of the sun during the year.

Numbers along the celestial equator show right ascension.

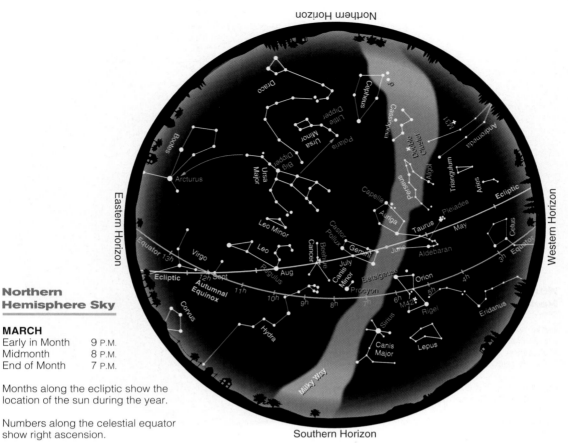

Northern Hemisphere Sky

MARCH

Early in Month	9 P.M.
Midmonth	8 P.M.
End of Month	7 P.M.

Months along the ecliptic show the location of the sun during the year.

Numbers along the celestial equator show right ascension.

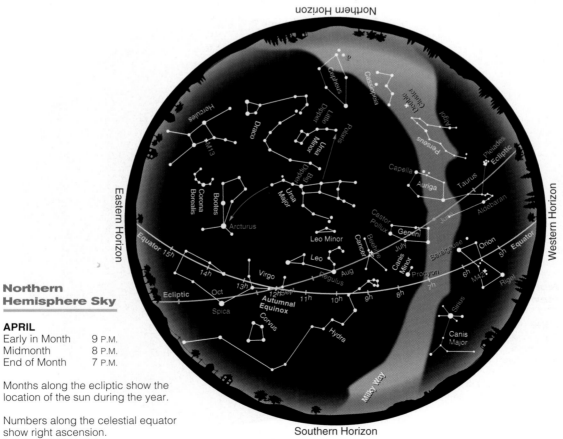

Northern Hemisphere Sky

APRIL

Early in Month	9 P.M.
Midmonth	8 P.M.
End of Month	7 P.M.

Months along the ecliptic show the location of the sun during the year.

Numbers along the celestial equator show right ascension.

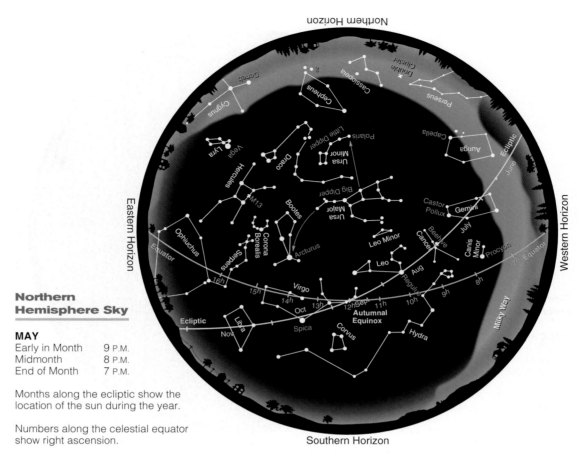

Northern Horizon

Northern Hemisphere Sky

MAY

Early in Month	9 P.M.
Midmonth	8 P.M.
End of Month	7 P.M.

Months along the ecliptic show the location of the sun during the year.

Numbers along the celestial equator show right ascension.

Southern Horizon

Northern Horizon

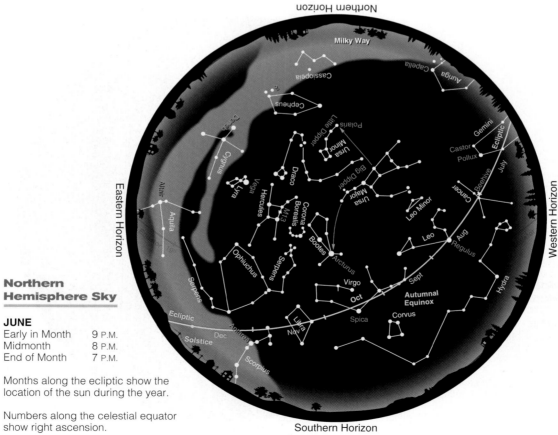

Northern Hemisphere Sky

JUNE

Early in Month	9 P.M.
Midmonth	8 P.M.
End of Month	7 P.M.

Months along the ecliptic show the location of the sun during the year.

Numbers along the celestial equator show right ascension.

Southern Horizon

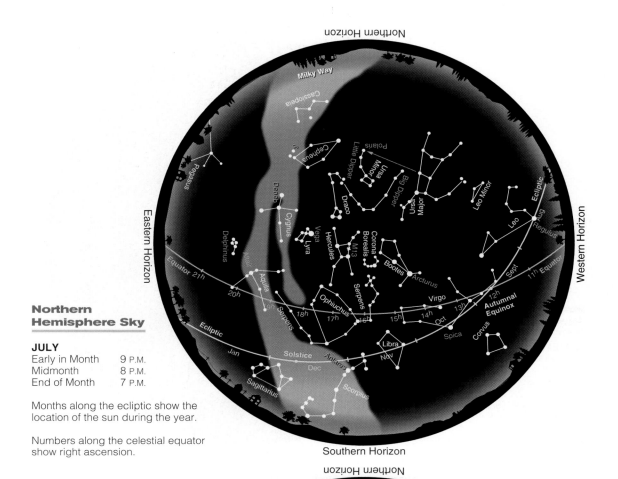

**Northern
Hemisphere Sky**

JULY

Early in Month	9 P.M.
Midmonth	8 P.M.
End of Month	7 P.M.

Months along the ecliptic show the
location of the sun during the year.

Numbers along the celestial equator
show right ascension.

**Northern
Hemisphere Sky**

AUGUST

Early in Month	9 P.M.
Midmonth	8 P.M.
End of Month	7 P.M.

Months along the ecliptic show the
location of the sun during the year.

Numbers along the celestial equator
show right ascension.

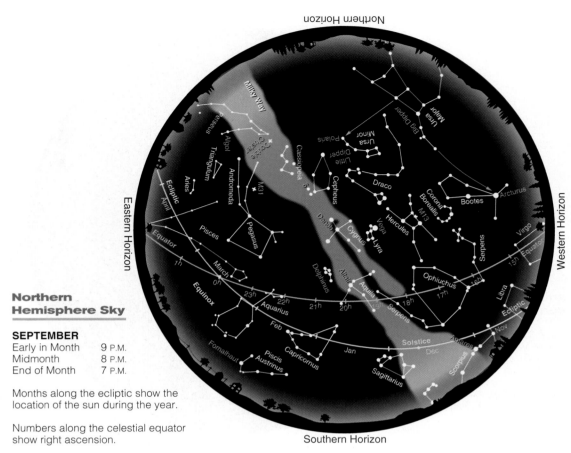

Northern Hemisphere Sky

SEPTEMBER

Early in Month	9 P.M.
Midmonth	8 P.M.
End of Month	7 P.M.

Months along the ecliptic show the location of the sun during the year.

Numbers along the celestial equator show right ascension.

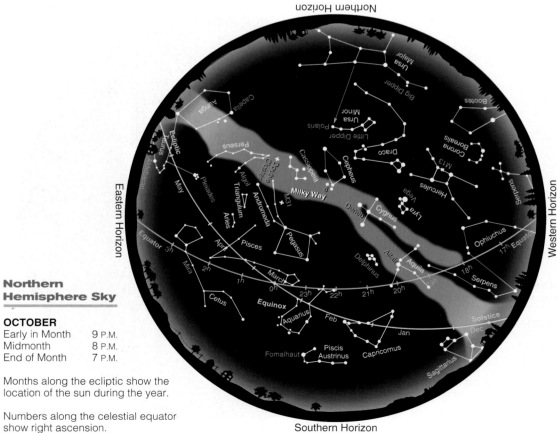

Northern Hemisphere Sky

OCTOBER

Early in Month	9 P.M.
Midmonth	8 P.M.
End of Month	7 P.M.

Months along the ecliptic show the location of the sun during the year.

Numbers along the celestial equator show right ascension.

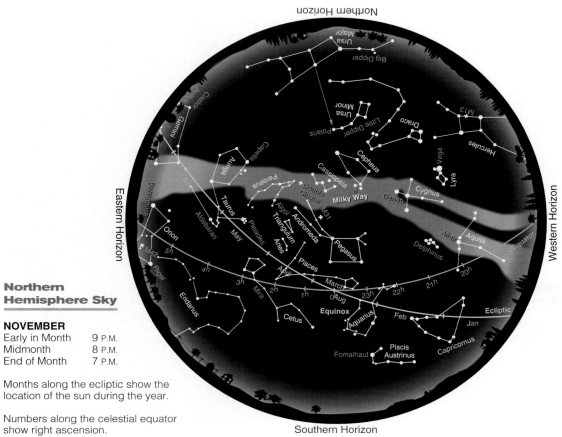

Northern Hemisphere Sky

NOVEMBER

Early in Month	9 P.M.
Midmonth	8 P.M.
End of Month	7 P.M.

Months along the ecliptic show the location of the sun during the year.

Numbers along the celestial equator show right ascension.

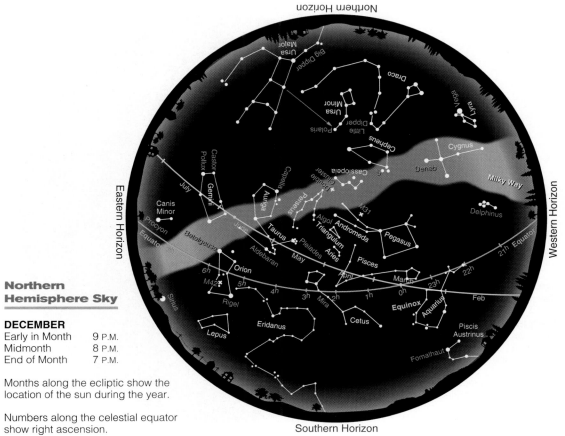

Northern Hemisphere Sky

DECEMBER

Early in Month	9 P.M.
Midmonth	8 P.M.
End of Month	7 P.M.

Months along the ecliptic show the location of the sun during the year.

Numbers along the celestial equator show right ascension.

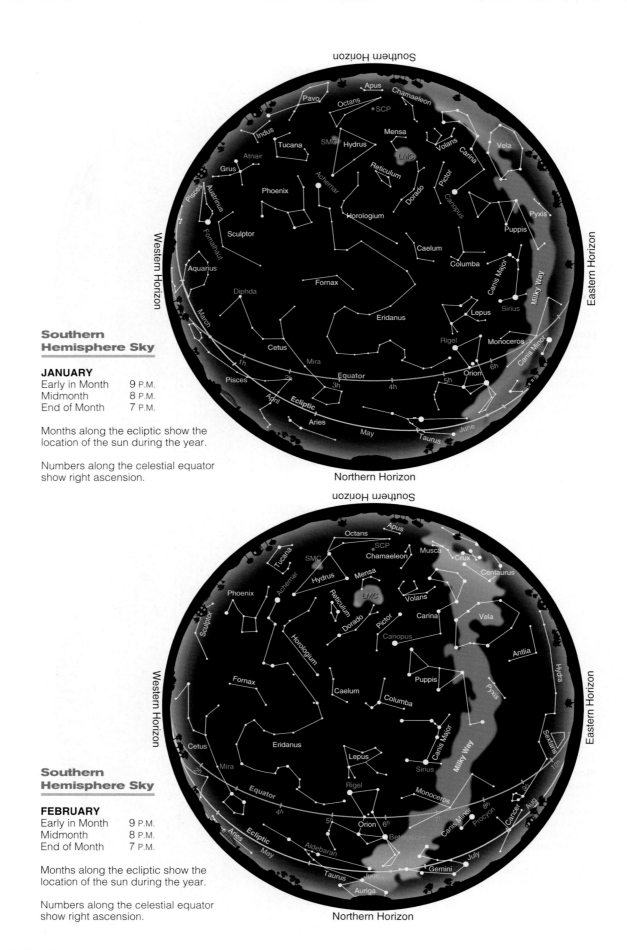

Southern Hemisphere Sky

JANUARY
Early in Month	9 P.M.
Midmonth	8 P.M.
End of Month	7 P.M.

Months along the ecliptic show the location of the sun during the year.

Numbers along the celestial equator show right ascension.

Southern Hemisphere Sky

FEBRUARY
Early in Month	9 P.M.
Midmonth	8 P.M.
End of Month	7 P.M.

Months along the ecliptic show the location of the sun during the year.

Numbers along the celestial equator show right ascension.

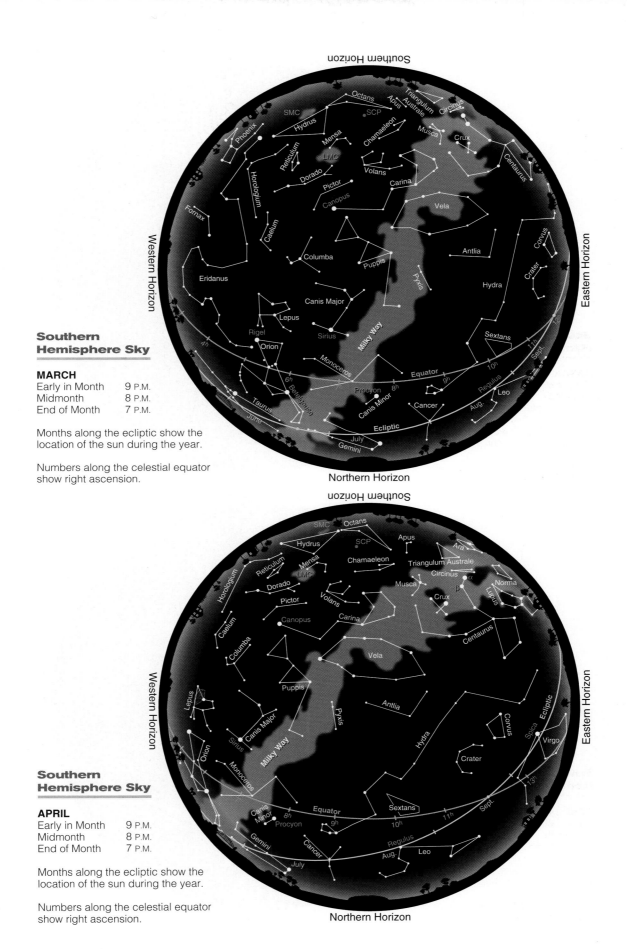

Southern Hemisphere Sky

MARCH

Early in Month   9 P.M.
Midmonth   8 P.M.
End of Month   7 P.M.

Months along the ecliptic show the location of the sun during the year.

Numbers along the celestial equator show right ascension.

Southern Hemisphere Sky

APRIL

Early in Month   9 P.M.
Midmonth   8 P.M.
End of Month   7 P.M.

Months along the ecliptic show the location of the sun during the year.

Numbers along the celestial equator show right ascension.

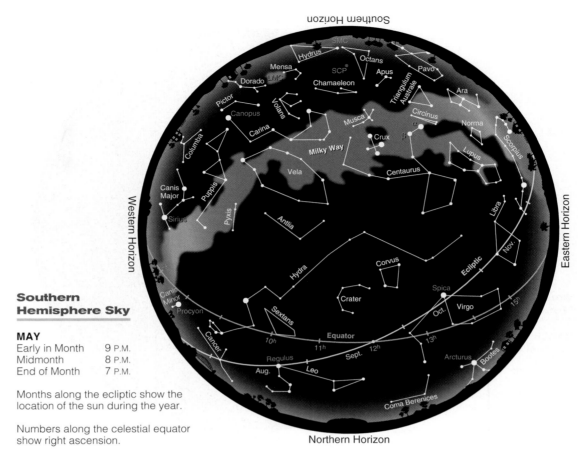

Southern Hemisphere Sky

MAY

Early in Month	9 P.M.
Midmonth	8 P.M.
End of Month	7 P.M.

Months along the ecliptic show the location of the sun during the year.

Numbers along the celestial equator show right ascension.

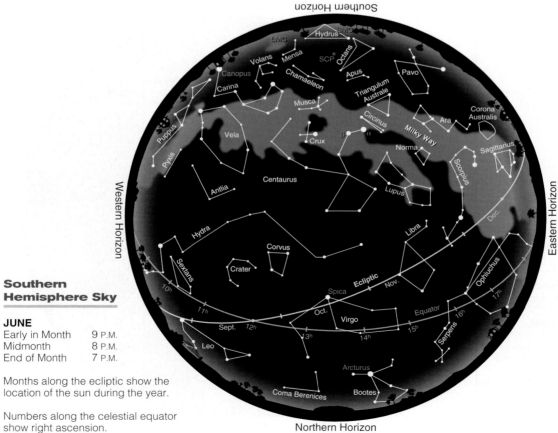

Southern Hemisphere Sky

JUNE

Early in Month	9 P.M.
Midmonth	8 P.M.
End of Month	7 P.M.

Months along the ecliptic show the location of the sun during the year.

Numbers along the celestial equator show right ascension.

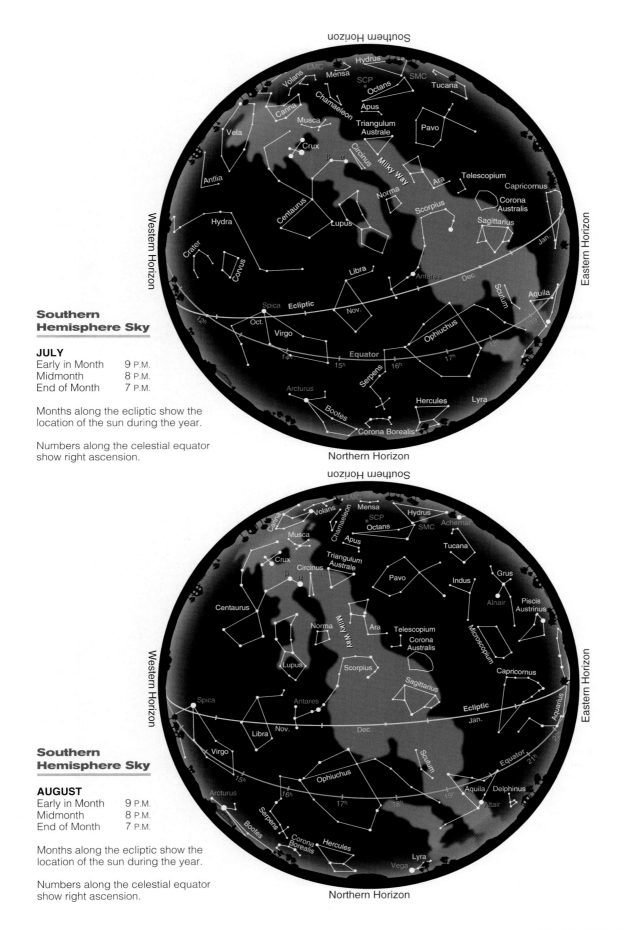

Southern Hemisphere Sky

JULY

Early in Month	9 P.M.
Midmonth	8 P.M.
End of Month	7 P.M.

Months along the ecliptic show the location of the sun during the year.

Numbers along the celestial equator show right ascension.

Southern Hemisphere Sky

AUGUST

Early in Month	9 P.M.
Midmonth	8 P.M.
End of Month	7 P.M.

Months along the ecliptic show the location of the sun during the year.

Numbers along the celestial equator show right ascension.

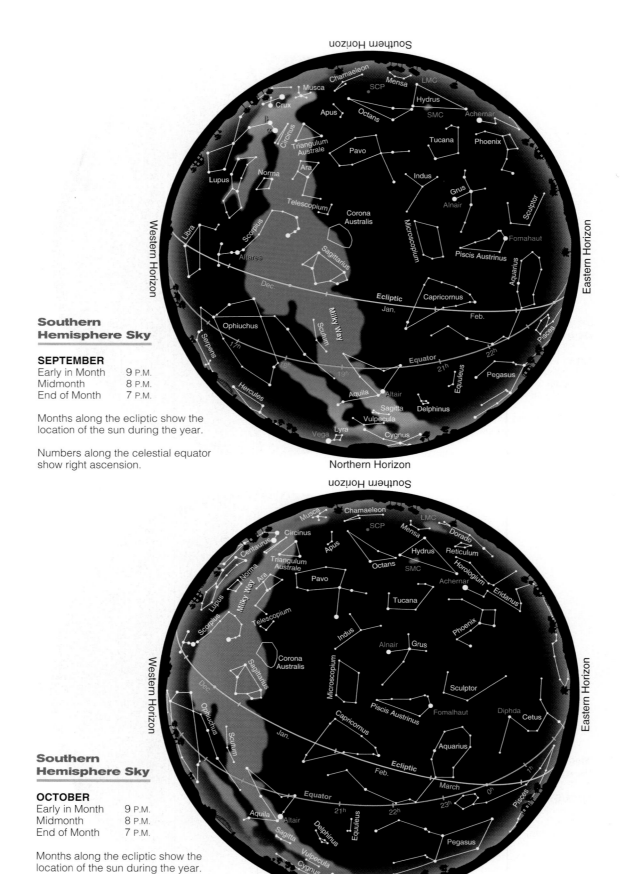

Southern Hemisphere Sky

SEPTEMBER

Early in Month	9 P.M.
Midmonth	8 P.M.
End of Month	7 P.M.

Months along the ecliptic show the location of the sun during the year.

Numbers along the celestial equator show right ascension.

Southern Hemisphere Sky

OCTOBER

Early in Month	9 P.M.
Midmonth	8 P.M.
End of Month	7 P.M.

Months along the ecliptic show the location of the sun during the year.

Numbers along the celestial equator show right ascension.

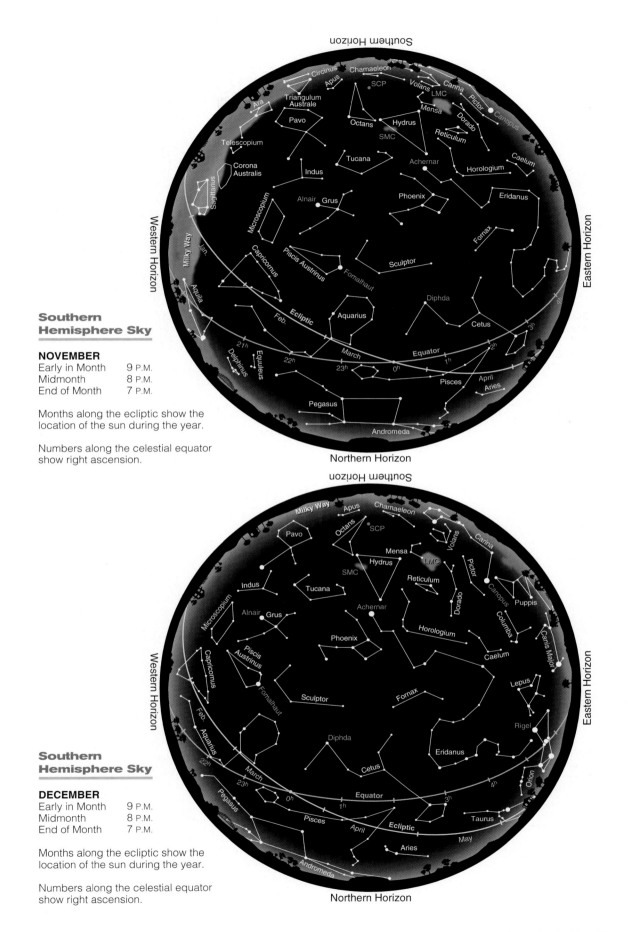

Southern Hemisphere Sky

NOVEMBER

Early in Month	9 P.M.
Midmonth	8 P.M.
End of Month	7 P.M.

Months along the ecliptic show the location of the sun during the year.

Numbers along the celestial equator show right ascension.

Southern Hemisphere Sky

DECEMBER

Early in Month	9 P.M.
Midmonth	8 P.M.
End of Month	7 P.M.

Months along the ecliptic show the location of the sun during the year.

Numbers along the celestial equator show right ascension.

Glossary

absolute age (502) An age determined in years, as from radioactive dating (see also *relative age*).

absolute zero (138) The lowest possible temperature; the temperature at which the particles in a material, atoms or molecules, contain no energy of motion that can be extracted from the body.

absorption line (146) A dark line in a spectrum; produced by the absence of photons absorbed by atoms or molecules.

absorption spectrum (dark-line spectrum) (146) A spectrum that contains absorption lines.

acceleration (89) A change in a velocity; a change in either speed or direction. (See *velocity*.)

acceleration of gravity (87) A measure of the strength of gravity at a planet's surface.

accretion (461) The sticking together of solid particles to produce a larger particle.

achondrite (624) Stony meteorite containing no chondrules or volatiles.

achromatic lens (114) A telescope lens composed of two lenses ground from different kinds of glass and designed to bring two selected colors to the same focus and correct for chromatic aberration.

active optics (121) Optical elements whose position or shape is continuously controlled by computers.

active region (175) An area on the sun where sunspots, prominences, flares, and the like occur.

adaptive optics (121) Computer-controlled telescope mirrors that can at least partially compensate for seeing.

albedo (489) The fraction of the light hitting an object that is reflected.

alt-azimuth mounting (121) A telescope mounting capable of motion parallel to and perpendicular to the horizon.

amino acid (654) One of the carbon-chain molecules that are the building blocks of protein.

Angstrom (Å) (111) A unit of distance; $1 \text{ Å} = 10^{-10}$ m; often used to measure the wavelength of light.

angular diameter (21) A measure of the size of an object in the sky; numerically equal to the angle in degrees between two lines extending from the observer's eye to opposite edges of the object.

angular distance (21) A measure of the separation between two objects in the sky; numerically equal to the angle in degrees between two lines extending from the observer's eye to the two objects.

angular momentum (93) The tendency of a rotating body to continue rotating; mathematically, the product of mass, velocity, and radius.

angular momentum problem (451) An objection to Laplace's nebular hypothesis that cited the slow rotation of the sun.

annular eclipse (48) A solar eclipse in which the solar photosphere appears around the edge of the moon in a bright ring, or annulus. The corona, chromosphere, and prominences cannot be seen.

anorthosite (506) Rock of aluminum and calcium silicates found in the lunar highlands.

aphelion (27) The orbital point of greatest distance from the sun.

apogee (48) The orbital point of greatest distance from Earth.

Apollo–Amor object (631) Asteroid whose orbit crosses that of Earth (Apollo) and Mars (Amor).

apparent visual magnitude (m_V) (16) The brightness of a star as seen by human eyes on Earth.

archaeoastronomy (58) The study of the astronomy of ancient cultures.

asterism (14) A named group of stars not identified as a constellation, e.g., the Big Dipper.

asteroid (456) Small rocky world; most asteroids lie between Mars and Jupiter in the asteroid belt.

astronomical unit (AU) (6) Average distance from Earth to the sun; 1.5×10^8 km, or 93×10^6 miles.

atmospheric window (112) Wavelength regions in which Earth's atmosphere is transparent—at visual, infrared, and radio wavelengths.

aurora (181) The glowing light display that results when a planet's magnetic field guides charged particles toward the north and south magnetic poles, where they strike the upper atmosphere and excite atoms to emit photons.

autumnal equinox (26) The point on the celestial sphere where the sun crosses the celestial equator going southward. Also, the time when the sun reaches this point and autumn begins in the northern hemisphere—about September 22.

Babcock model (177) A model of the sun's magnetic cycle in which the differential rotation of the sun winds up and tangles the solar magnetic field in a 22-year cycle. This is thought to be responsible for the 11-year sunspot cycle.

Balmer series (147) Spectral lines in the visible and near-ultraviolet spectrum of hydrogen produced by transitions whose lowest orbit is the second.

barred spiral galaxy (200) A spiral galaxy with an elongated nucleus resembling a bar from which the arms originate.

basalt (486) Dark, igneous rock characteristic of solidified lava.

belt (561) One of the dark bands of clouds that circle Jupiter parallel to its equator; generally red, brown, or blue-green; believed to be regions of descending gas.

belt–zone circulation (585) The atmospheric circulation typical of Jovian planets. Dark belts and bright zones encircle the planet parallel to its equator.

big bang (203) The theory that the universe began with a violent explosion from which the expanding universe of galaxies eventually formed.

binary star (200) One of a pair of stars that orbit around their common center of mass.

binding energy (143) The energy needed to pull an electron away from its atom.

black body radiation (139) Radiation emitted by a hypothetical perfect radiator; the spectrum is continuous, and the wavelength of maximum emission depends only on the body's temperature.

blueshift (153) The shortening of the wavelengths of light observed when the source and observer are approaching each other.

bow shock (484) The boundary between the undisturbed solar wind and the region being deflected around a planet or comet.

breccia (506) A rock composed of fragments of earlier rocks bonded together.

bright-line spectrum (146) See *emission spectrum*.

butterfly diagram (174) See *Maunder butterfly diagram*.

CAI (624) Calcium–aluminum-rich inclusions found in some meteorites; believed to be very old.

Cambrian period (657) A geological period 0.6 to 0.5 billion years ago during which life on Earth became diverse and complex. Cambrian rocks contain the oldest easily identifiable fossils.

capture hypothesis (510) The theory that the moon formed elsewhere in the solar system and was later captured by Earth.

carbonaceous chondrite (623) Stony meteorite that contains both chondrules and volatiles. These may be the least altered remains of the solar nebula still present in the solar system.

Cassegrain focus (120) The optical design of a reflecting telescope in which the secondary mirror reflects light back down the tube through a hole in the center of the objective mirror.

catastrophic hypothesis (451) Explanation for natural processes that depends on dramatic and unlikely events, such as the collision of two stars to produce our solar system.

celestial equator (20) The imaginary line around the sky directly above Earth's equator.

celestial sphere (20) An imaginary sphere of very large radius surrounding Earth and to which the planets, stars, sun, and moon seem to be attached.

center of mass (95) The balance point of a body or system of bodies.

charge-coupled device (CCD) (123) An electronic device consisting of a large array of light-sensitive elements used to record very faint images.

chemical evolution (658) The chemical process that led to the growth of complex molecules on the primitive Earth. This did not involve the reproduction of molecules.

chondrite (622) A stony meteorite that contains chondrules.

chondrule (622) Round, glassy body in some stony meteorites; believed to have solidified very quickly from molten drops of silicate material.

chromatic aberration (114) A distortion found in refracting telescopes because lenses focus different colors at slightly different distances. Images are consequently surrounded by color fringes.

chromosome (655) One of the bodies in a cell that contains the DNA carrying genetic information.

chromosphere (48) Bright gases just above the photosphere of the sun.

circular velocity (94) The velocity required to remain in a circular orbit about a body.

circumpolar constellation (21) Any of the constellations so close to the celestial pole that they never set (or never rise) as seen from a given latitude.

closed orbit (95) An orbit that returns to its starting point; a circular or elliptical orbit. (See *open orbit*.)

collisional broadening (155) The smearing out of a spectral line because of collisions among the atoms of the gas.

coma (636) The glowing head of a comet.

comet (456) One of the small, icy bodies that orbit the sun and produce tails of gas and dust when they near the sun.

comparative planetology (476) The study of planets by comparing the characteristics of different examples.

comparison spectrum (124) A spectrum of known spectral lines used to identify unknown wavelengths in an object's spectrum.

composite volcano (528) A volcano built up of layers of lava flows and ash falls. These are steep sided and typically associated with subduction zones.

condensation (461) The growth of a particle by addition of material from surrounding gas, one atom or molecule at a time.

condensation hypothesis (510) The hypothesis that the moon and Earth formed by condensing together from the solar nebula.

condensation sequence (461) The sequence in which different materials condense from the solar nebula at increasing distances from the sun.

constellation (14) One of the stellar patterns identified by name, usually of mythological gods, people, animals, or objects; also, the region of the sky containing that star pattern.

continuous spectrum (146) A spectrum in which there are no absorption or emission lines.

convection (164) Circulation in a fluid driven by heat; hot material rises, and cool material sinks.

convective zone (171) The region inside a star where energy is carried outward as rising hot gas and sinking cool gas.

corona (54, 530) The faint outer atmosphere of the sun; composed of low-density, very hot, ionized gas. On Venus, round networks of fractures and ridges up to 1000 km in diameter.

coronagraph (165) A telescope designed to photograph the inner corona of the sun.

coronal hole (181) An area of the solar surface that is dark at X-ray wavelengths; thought to be associated with divergent magnetic fields and the source of the solar wind.

coronal mass ejection (CME) (181) Gas trapped in the sun's magnetic field.

cosmic microwave background radiation (203) Radiation from the hot clouds of the big bang explosion. Because of its large redshift, it appears to come from a body whose temperature is only 2.7 K.

cosmic ray (132) A subatomic particle traveling at tremendous velocity that strikes Earth's atmosphere from space.

Coulomb barrier (169) The electrostatic force of repulsion between bodies of like charge; commonly applied to atomic nuclei.

Coulomb force (142) The repulsive force between particles with like electrostatic charge.

critical point (558) The temperature and pressure at which the vapor and liquid phases of a material have the same density.

dark-line spectrum (146) See *absorption spectrum*.

debris disk (468) A disk of dust found by infrared observations around some stars. The dust is debris from collisions among asteroids, comets, and Kuiper belt objects.

decameter radiation (557) Radio signals from Jupiter with wavelengths of about 10 m.

decimeter radiation (557) Radio signals from Jupiter with wavelengths of about 0.1 m.

deferent (65) In the Ptolemaic theory, the large circle around Earth along which the center of the epicycle moved.

density (156) The amount of matter per unit volume in a material; measured in grams per cubic centimeter, for example.

deuterium (170) An isotope of hydrogen in which the nucleus contains a proton and a neutron.

diamond ring effect (48) A momentary phenomenon seen during some total solar eclipses when the ring of the corona and a bright spot of photosphere resemble a large diamond set in a silvery ring.

differential rotation (176) The rotation of a body in which different parts of the body have different periods of rotation; this is true of the sun, the Jovian planets, and the disk of the galaxy.

differentiation (462) The separation of planetary material according to density.

diffraction fringe (115) Blurred fringe surrounding any image caused by the wave properties of light. Because of this, no image detail smaller than the fringe can be seen.

DNA (deoxyribonucleic acid) (654) The long carbon-chain molecule that records information to govern the biological activity of the organism. DNA carries the genetic data passed to offspring.

Doppler broadening (155) The smearing of spectral lines because of the motion of the atoms in the gas.

Doppler effect (153) The change in the wavelength of radiation due to relative radial motion of source and observer.

Drake equation (669) A formula for the number of communicative civilizations in our galaxy.

dust (type II) tail (636) The tail of a comet formed of dust blown outward by the pressure of sunlight. (See *gas tail*.)

dynamo effect (177) The process by which a rotating, convecting body of conducting matter, such as Earth's core, can generate a magnetic field.

east point (20) The point on the eastern horizon exactly halfway between the north point and the south point; exactly east.

eccentric (62) In astronomy, an off-center circular path.

eccentricity, *e* (76) A measure of the flattening of an ellipse. An ellipse of $e = 0$ is circular. The closer to 1 *e* becomes, the more flattened the ellipse.

eclipse season (50) That period when the sun is near a node of the moon's orbit and eclipses are possible.

eclipse year (51) The time the sun takes to circle the sky and return to a node of the moon's orbit; 346.62 days.

ecliptic (24) The apparent path of the sun around the sky.

ejecta (518) Pulverized rock scattered by meteorite impacts on a planetary surface.

electromagnetic radiation (110) Changing electric and magnetic fields that travel through space and transfer energy from one place to another—for example, light, radio waves, and the like.

electron (138) Low-mass atomic particle carrying a negative charge.

ellipse (76) A closed curve enclosing two points (foci) such that the total distance from one focus to any point on the curve back to the other focus equals a constant.

elliptical galaxy (200) A galaxy that is round or elliptical in outline; it contains little gas and dust, no disk or spiral arms, and few hot, bright stars.

emission line (146) A bright line in a spectrum caused by the emission of photons from atoms.

emission spectrum (bright-line spectrum) (146) A spectrum containing emission lines.

energy (96) The capacity of a natural system to perform work—for example, thermal energy.

energy level (144) One of a number of states an electron may occupy in an atom, depending on its binding energy.

enzyme (654) Special protein that controls processes in an organism.

epicycle (65) The small circle followed by a planet in the Ptolemaic theory. The center of the epicycle follows a larger circle (deferent) around Earth.

equant (66) The point off-center in the deferent from which the center of the epicycle appears to move uniformly.

equatorial mounting (121) A telescope mounting that allows motion parallel to and perpendicular to the celestial equator.

escape velocity (95) The initial velocity an object needs to escape from the surface of a celestial body.

evening star (28) Any planet visible in the sky just after sunset.

evolutionary hypothesis (451) Explanation for natural events that involves gradual changes as opposed to sudden catastrophic changes—for example, the formation of the planets in the gas cloud around the forming sun.

excited atom (144) An atom in which an electron has moved from a lower to a higher orbit.

extrasolar planet (469) A planet orbiting a star other than the sun.

eyepiece (114) A short-focal-length lens used to enlarge the image in a telescope; the lens nearest the eye.

fall (621) A meteorite seen to fall. (See *find*.)

false-color image (123) A representation of graphical data in which the colors are altered or added to reveal details.

field (92) A way of explaining action at a distance; a particle produces a field of influence (gravitational, electric, or magnetic) to which another particle in the field responds.

filament (165) On the sun, a prominence seen silhouetted against the solar surface.

filtergram (165) An image (usually of the sun) taken in the light of a specific region of the spectrum—e.g., an H-alpha filtergram.

find (621) A meteorite that is found but was not seen to fall. (See *fall*.)

fission hypothesis (510) The theory that the moon formed by breaking away from Earth.

flare (181) A violent eruption on the sun's surface.

focal length (114) The distance from a lens to the point where it focuses parallel rays of light.

folded mountain range (486) A long range of mountains formed by the compression of a planet's crust—for example, the Andes on Earth.

forward scattering (562) The optical property of finely divided particles to preferentially direct light in the original direction of the light's travel.

frequency (111) The number of times a given event occurs in a given time; for a wave, the number of cycles that pass the observer in 1 second.

galaxy (8) A very large collection of gas, dust, and stars orbiting a common center of mass. The sun and Earth are located in the Milky Way Galaxy.

Galilean satellites (455) The four largest satellites of Jupiter, named after their discoverer, Galileo.

gas (type I) tail (636) The tail of a comet produced by gas blown outward by the solar wind. (See *dust tail*.)

gene (655) A unit of DNA containing genetic information that influences a particular inherited trait.

general theory of relativity (102) Einstein's more sophisticated theory of space and time, which describes gravity as a curvature of space-time.

geocentric universe (61) A model universe with Earth at the center, such as the Ptolemaic universe.

geosynchronous satellite (94) An Earth satellite in an eastward orbit whose period is 24 hours. A satellite in such an orbit remains above the same spot on Earth's surface.

giant (192) Large, cool, highly luminous star in the upper right of the H–R diagram; typically 10 to 100 times the diameter of the sun.

global warming (490) The gradual increase in the surface temperature of Earth caused by human modifications to Earth's atmosphere.

globular cluster (198) A star cluster containing 50,000 to 1 million stars in a sphere about 75 ly in diameter; generally old, metal-poor, and found in the spherical component of the galaxy.

gossamer ring (563) The dimmest part of Jupiter's ring produced by dust particles orbiting near small moons.

granulation (163) The fine structure visible on the solar surface caused by rising currents of hot gas and sinking currents of cool gas below the surface.

grating (124) A piece of material in which numerous microscopic parallel lines are scribed; light encountering a grating is dispersed to form a spectrum.

gravitational collapse (460) The stage in the formation of a massive planet when it grows massive enough to begin capturing gas directly from the nebula around it.

greenhouse effect (490) The process by which a carbon dioxide atmosphere traps heat and raises the temperature of a planetary surface.

grooved terrain (566) Region of the surface of Ganymede consisting of parallel grooves; believed to have formed by repeated fracture of the icy crust.

ground state (145) The lowest permitted electron orbit in an atom.

half-life (457) The time required for half of the atoms in a radioactive sample to decay.

heat of formation (463) In planetology, the heat released by the infall of matter during the formation of a planetary body.

heavy bombardment (466) The period of intense meteorite impacts early in the formation of the planets when the solar system was filled with debris.

heliocentric universe (63) A model of the universe with the sun at the center, such as the Copernican universe.

helioseismology (167) The study of the interior of the sun by the analysis of its modes of vibration.

Hirayama family (627) Family of asteroids with orbits of similar size, shape, and orientation; believed to be fragments of larger bodies.

horizon (20) The line that marks the apparent intersection of Earth and the sky.

horoscope (28) A chart showing the positions of the sun, moon, planets, and constellations at the time of a person's birth; used in astrology to attempt to read character or foretell the future.

hypothesis (77) A conjecture, subject to further tests, that accounts for a set of facts.

infrared radiation (111) Electromagnetic radiation with wavelengths intermediate between visible light and radio waves.

intercrater plain (515) The relatively smooth terrain on Mercury.

interferometry (122) The observing technique in which separated telescopes combine to produce a virtual telescope with the resolution of a much-larger-diameter telescope.

interstellar medium (188) The gas and dust distributed between the stars.

inverse square law (90) The rule that the strength of an effect (such as gravity) decreases in proportion as the distance squared increases.

Io flux tube (558) A tube of magnetic lines and electric currents connecting Io and Jupiter.

ion (142) An atom that has lost or gained one or more electrons.

ionization (142) The process in which atoms lose or gain electrons.

Io plasma torus (558) The doughnut-shaped cloud of ionized gas that encloses the orbit of Jupiter's moon Io.

iron meteorite (622) A meteorite composed mainly of iron-nickel alloy.

irregular galaxy (201) A galaxy with a chaotic appearance, large clouds of gas and dust, and both population I and population II stars, but without spiral arms.

isotopes (142) Atoms that have the same number of protons but a different number of neutrons.

joule (J) (96) A unit of energy equivalent to a force of 1 newton acting over a distance of 1 meter; 1 joule per second equals 1 watt of power.

Jovian planet (454) Jupiterlike planet with large diameter and low density.

jumbled terrain (509) Strangely disturbed regions of the moon opposite the locations of the Imbrium Basin and Mare Orientale.

Kelvin temperature scale (138) The temperature, in Celsius (centigrade) degrees, measured above absolute zero.

kinetic energy (96) Energy of motion; depends on mass and velocity of a moving body.

Kirchhoff's laws (146) A set of laws that describe the absorption and emission of light by matter.

Kirkwood's gap (630) Region in the asteroid belt in which there are very few asteroids; caused by resonances with Jupiter.

Kuiper belt (456) The collection of icy planetesimals believed to orbit in a region from just beyond Neptune out to about 50 AU.

large-impact hypothesis (511) The theory that the moon formed from debris ejected during a collision between Earth and a large planetesimal.

L dwarf (150) A type of star that is even cooler than the M stars.

life zone (665) A region around a star within which a planet can have temperatures that permit the existence of liquid water.

light-gathering power (115) The ability of a telescope to collect light; proportional to the area of the telescope objective lens or mirror.

light pollution (117) The illumination of the night sky by waste light from cities and outdoor lighting, which prevents the observation of faint objects.

light-year (ly) (7) The distance light travels in 1 year.

limb (164, 498) The edge of the apparent disk of a body, as in "the limb of the moon."

limb darkening (164) The decrease in brightness of the sun or other body from its center to its limb.

line of nodes (50) The line across an orbit connecting the nodes; commonly applied to the orbit of the moon.

liquid metallic hydrogen (557) A form of hydrogen under high pressure that is a good electrical conductor.

lobate scarp (514) A curved cliff such as those found on Mercury.

lunar eclipse (42) The darkening of the moon when it moves through Earth's shadow.

Lyman series (147) Spectral lines in the ultraviolet spectrum of hydrogen produced by transitions whose lowest orbit is the ground state.

Magellanic Clouds (201) Small, irregular galaxies that are companions to the Milky Way; visible in the southern sky.

magnetic carpet (166) The widely distributed, low-level magnetic field extending up through the sun's visible surface.

magnetosphere (484) The volume of space around a planet within which the motion of charged particles is dominated by the planetary magnetic field rather than the solar wind.

magnifying power (116) The ability of a telescope to make an image larger.

magnitude scale (16) The astronomical brightness scale; the larger the number, the fainter the star.

main sequence (192) The region of the H–R diagram running from upper left to lower right, which includes roughly 90 percent of all stars.

mantle (483) The layer of dense rock and metal oxides that lies between the molten core and Earth's surface; also, similar layers in other planets.

mare (498) One of the lunar lowlands filled by successive flows of dark lava; from the Latin word for *sea*.

mass (90) A measure of the amount of matter making up an object.

Maunder butterfly diagram (174) A graph showing the latitude of sunspots versus time; first plotted by W. W. Maunder in 1904.

Maunder minimum (175) A period of less numerous sunspots and other solar activity from 1645 to 1715.

meteor (457) A small bit of matter heated by friction to incandescent vapor as it falls into Earth's atmosphere.

meteorite (457) A meteor that has survived its passage through the atmosphere and strikes the ground.

meteoroid (457) A meteor in space before it enters Earth's atmosphere.

micrometeorite (501) Meteorite of microscopic size.

midocean rise (486) One of the undersea mountain ranges that push up from the seafloor in the center of the oceans.

Milankovitch hypothesis (30) The hypothesis that small changes in Earth's orbital and rotational motions cause the ice ages.

Milky Way (9, 196) The hazy band of light that circles the sky, produced by the combined light of billions of stars in our Milky Way Galaxy.

Milky Way Galaxy (9, 197) The spiral galaxy containing the sun; visible at night as the Milky Way.

Miller experiment (657) An experiment that reproduced the conditions under which life began on Earth and manufactured amino acids and other organic compounds.

minute of arc (21) An angular measure; each degree is divided into 60 minutes of arc.

molecule (142) Two or more atoms bonded together.

momentum (89) The tendency of a moving object to continue moving; mathematically, the product of mass and velocity.

morning star (28) Any planet visible in the sky just before sunrise.

multiringed basin (501) Very large impact basin in which there are concentric rings of mountains.

mutant (656) Offspring born with altered DNA.

nadir (20) The point on the bottom of the sky directly under your feet.

nanometer (nm) (111) A unit of length equal to 10^{-9} m.

natural law (77) A conjecture about how nature works in which scientists have overwhelming confidence.

natural motion (87) In Aristotelian physics, the motion of objects toward their natural places—fire and air upward and earth and water downward.

natural selection (653) The process by which the best traits are passed on, allowing the most able to survive.

neap tide (98) Ocean tide of low amplitude occurring at first- and third-quarter moon.

nebula (188) A cloud of gas and dust in space.

nebular hypothesis (451) The proposal that the solar system formed from a rotating cloud of gas.

neutrino (170) A neutral, massless atomic particle that travels at or nearly at the speed of light.

neutron (141) An atomic particle with no charge and about the same mass as a proton.

Newtonian focus (120) The focal arrangement of a reflecting telescope in which a diagonal mirror reflects light out the side of the telescope tube for easier access.

node (50) A point where an object's orbit passes through the plane of Earth's orbit.

north celestial pole (20) The point on the celestial sphere directly above Earth's North Pole.

north point (20) The point on the horizon directly below the north celestial pole; exactly north.

nuclear fission (169) Reaction that splits nuclei into less massive fragments.

nuclear fusion (169) Reaction that joins the nuclei of atoms to form more massive nuclei.

nucleus (of an atom) (141) The central core of an atom containing protons and neutrons; carries a net positive charge.

objective lens or mirror (114) The main optical element in an astronomical telescope. The large lens at the top of the telescope or large mirror at the bottom.

oblateness (556) The flattening of a spherical body; usually caused by rotation.

occultation (597) The passage of a larger body in front of a smaller body.

Oort cloud (639) The cloud of icy bodies—extending from the outer part of our solar system out to roughly 100,000 AU from the sun—that acts as the source of most comets.

open orbit (95) An orbit that does not return to its starting point; an escape orbit. (See *closed orbit*.)

outflow channel (540) Geological feature on Mars that appears to have been caused by sudden flooding.

outgassing (463) The release of gases from a planet's interior.

ovoid (600) Geological feature on Uranus's moon Miranda thought to be produced by circulation in the solid icy mantle and crust.

ozone layer (489) In Earth's atmosphere, a layer of oxygen ions (O_3) lying 15 to 30 km high that protects the surface by absorbing ultraviolet radiation.

paradigm (68) A commonly accepted set of scientific ideas and assumptions.

parallax (64) The apparent change in the position of an object due to a change in the location of the observer. Astronomical parallax is measured in seconds of arc.

partial eclipse (44, 45) A lunar eclipse in which the moon does not completely enter Earth's shadow; a solar eclipse in which the moon does not completely cover the sun.

Paschen series (147) Spectral lines in the infrared spectrum of hydrogen produced by transitions whose lowest orbit is the third.

passing star hypothesis (450) The proposal that our solar system formed when two stars passed near each other and material was pulled out of one to form the planets.

path of totality (46) The track of the moon's umbral shadow over Earth's surface. The sun is totally eclipsed as seen from within this path.

penumbra (42) The portion of a shadow that is only partially shaded.

penumbral eclipse (54) A lunar eclipse in which the moon enters the penumbra of Earth's shadow but does not reach the umbra.

perigee (48) The orbital point of closest approach to Earth.

perihelion (27) The orbital point of closest approach to the sun.

permitted orbit (143) One of the energy levels in an atom that an electron may occupy.

photon (111) A quantum of electromagnetic energy; carries an amount of energy that depends inversely on its wavelength.

photosphere (47) The bright visible surface of the sun.

planet (6) A nonluminous object, larger than a comet or asteroid, that orbits a star.

planetary nebula (192) An expanding shell of gas ejected from a star during the latter stages of its evolution.

planetesimal (461) One of the small bodies that formed from the solar nebula and eventually grew into protoplanets.

plastic (483) A material with the properties of a solid but capable of flowing under pressure.

plate tectonics (486) The constant destruction and renewal of Earth's surface by the motion of sections of crust.

plutino (612) One of the icy Kuiper belt objects that, like Pluto, are caught in a 3:2 orbital resonance with Neptune.

polar axis (121) The axis around which a celestial body rotates.

positron (170) The anti-particle of the electron.

potential energy (96) The energy a body has by virtue of its position. A weight on a high shelf has more potential energy than a weight on a low shelf.

precession (22) The slow change in the direction of Earth's axis of rotation; one cycle takes nearly 26,000 years.

pressure (P) wave (482) In geophysics, a mechanical wave of compression and rarefaction that travels through Earth's interior.

primary lens or mirror (114) The main optical element in an astronomical telescope. The large lens at the top of the telescope tube or the large mirror at the bottom.

prime focus (120) The point at which the objective mirror forms an image in a reflecting telescope.

primeval atmosphere (488) Earth's first air, composed of gases from the solar nebula.

primordial soup (657) The rich solution of organic molecules in Earth's first oceans.

prominence (48, 180) Eruption on the solar surface; visible during total solar eclipses.

protein (654) Complex molecule composed of amino acid units.

proton (141) A positively charged atomic particle contained in the nucleus of atoms; the nucleus of a hydrogen atom.

proton–proton chain (170) A series of three nuclear reactions that build a helium atom by adding together protons; the main energy source in the sun.

protoplanet (462) Massive object resulting from the coalescence of planetesimals in the solar nebula and destined to become a planet.

protostar (188) A collapsing cloud of gas and dust destined to become a star.

quantum mechanics (143) The study of the behavior of atoms and atomic particles.

radial velocity (V_r) (153) That component of an object's velocity directed away from or toward Earth.

radiant (619) The point in the sky from which meteors in a shower seem to come.

radiation pressure (466) The force exerted on the surface of a body by its absorption of light. Small particles floating in the solar system can be blown outward by the pressure of the sunlight.

radiative zone (170) The region inside a star where energy is carried outward as photons.

radio interferometer (127) Two or more radio telescopes that combine their signals to achieve the resolving power of a larger telescope.

ray (500) Ejecta from a meteorite impact, forming white streamers radiating from some lunar craters.

reconnection (181) The process in the sun's atmosphere by which opposing magnetic fields combine and release energy to power solar flares.

red dwarf (192) Cool, low-mass star on the lower main sequence.

redshift (153) The lengthening of the wavelengths of light seen when the source and observer are receding from each other.

reflecting telescope (114) A telescope that uses a concave mirror to focus light into an image.

refracting telescope (114) A telescope that forms images by bending (refracting) light with a lens.

regolith (506) A soil made up of crushed rock fragments.

relative age (502) The age of a geological feature referred to other features. For example, relative ages reveal that the lunar maria are younger than the highlands.

resolving power (115) The ability of a telescope to reveal fine detail; depends on the diameter of the telescope objective.

resonance (512) The coincidental agreement between two periodic phenomena; commonly applied to agreements between orbital periods, which can make orbits more or less stable.

retrograde motion (64) The apparent backward (westward) motion of planets as seen against the background of stars.

revolution (23) The motion of an object in a closed path about a point outside its volume; Earth revolves around the sun.

rift valley (486) A long, straight, deep valley produced by the separation of crustal plates.

RNA (ribonucleic acid) (655) A long carbon-chain molecule that uses the information stored in DNA to manufacture complex molecules necessary to the organism.

Roche limit (562) The minimum distance between a planet and a satellite that holds itself together by its own gravity. If a satellite's orbit brings it within its planet's Roche limit, tidal forces will pull the satellite apart.

rotation (23) The turning of a body about an axis that passes through its volume; Earth rotates on its axis.

saros cycle (52) An 18-year $11\frac{1}{3}$-day period after which the pattern of lunar and solar eclipses repeats.

Schmidt-Cassegrain focus (120) The optical design of a reflecting telescope in which a thin correcting lens is placed at the top of a Cassegrain telescope.

scientific model (19) An intellectual concept designed to help you think about a natural process without necessarily being a conjecture of truth.

scientific notation (6) The system of recording very large or very small numbers by using powers of 10.

secondary atmosphere (488) The gases outgassed from a planet's interior; rich in carbon dioxide.

secondary crater (500) A crater formed by the impact of debris ejected from a larger crater.

secondary mirror (120) In a reflecting telescope, the mirror that reflects the light to a point of easy observation.

second of arc (21) An angular measure; each minute of arc is divided into 60 seconds of arc.

seeing (115) Atmospheric conditions on a given night. When the atmosphere is unsteady, producing blurred images, the seeing is said to be poor.

seismic wave (482) A mechanical vibration that travels through Earth; usually caused by an earthquake.

seismograph (482) An instrument that records seismic waves.

selection effect (622) An influence on the probability that certain phenomena will be detected or selected, which can alter the outcome of a survey.

semimajor axis, *a* (76) Half of the longest axis of an ellipse.

SETI (668) Search for Extra-Terrestrial Intelligence.

shear (S) wave (482) A mechanical wave that travels through Earth's interior by the vibration of particles perpendicular to the direction of wave travel.

shepherd satellite (577) A satellite that, by its gravitational field, confines particles to a planetary ring.

shield volcano (528) Wide, low-profile volcanic cone produced by highly liquid lava.

sidereal drive (121) The motor and gears on a telescope that turn it westward to keep it pointed at a star.

sidereal period (41) The period of rotation or revolution of an astronomical body relative to the stars.

sinuous rille (499) A narrow, winding valley on the moon caused by ancient lava flows along narrow channels.

small-angle formula (54) The mathematical formula that relates an object's linear diameter and distance to its angular diameter.

smooth plain (515) Apparently young plain on Mercury formed by lava flows at or soon after the formation of the Caloris Basin.

solar constant (182) A measure of the energy output of the sun; the total solar energy striking 1 m^2 just above Earth's atmosphere in 1 second.

solar eclipse (45) The event that occurs when the moon passes directly between Earth and the sun, blocking your view of the sun.

solar nebula hypothesis (452) The theory that the planets formed from the same cloud of gas and dust that formed the sun.

solar system (6) The sun and the nonluminous objects that orbit it, including the planets, comets, and asteroids.

solar wind (166) Rapidly moving atoms and ions that escape from the solar corona and blow outward through the solar system.

south celestial pole (20) The point of the celestial sphere directly above Earth's South Pole.

south point (20) The point on the horizon directly above the south celestial pole; exactly south.

special relativity (101) The first of Einstein's theories of relativity, which dealt with uniform motion.

spectral class or type (149) A star's position in the temperature classification system O, B, A F, G, K, and M. Based on the appearance of the star's spectrum.

spectral sequence (149) The arrangement of spectral classes (O, B, A, F, G, K, M) ranging from hot to cool.

spectrograph (123) A device that separates light by wavelength to produce a spectrum.

spicule (165) Small, flamelike projection in the chromosphere of the sun.

spiral arm (9) Long, spiral pattern of bright stars, star clusters, gas, and dust that extends from the center to the edge of the disk of spiral galaxies.

spiral galaxy (200) A galaxy with an obvious disk component containing gas; dust; hot, bright stars; and spiral arms.

sporadic meteor (620) A meteor not part of a meteor shower.

spring tide (98) Ocean tide of high amplitude that occurs at full and new moon.

star (6) A celestial object composed of gas held together by its own gravity and supported by nuclear fusion occurring in its interior.

stony-iron meteorite (624) A meteorite that is a mixture of stone and iron.

stony meteorite (622) A meteorite composed of silicate (rocky) material.

stromatolite (659) A layered fossil formation caused by ancient mats of algae or bacteria that build up mineral deposits season after season.

strong force (169) One of the four forces of nature; the strong force binds protons and neutrons together in atomic nuclei.

subduction zone (486) A region of a planetary crust where a tectonic plate slides downward.

subsolar point (524) The point on a planet that is directly below the sun.

summer solstice (26) The point on the celestial sphere where the sun is at its most northerly point; also, the time when the sun passes this point, about June 22, and summer begins in the northern hemisphere.

sunspot (162) Relatively dark spot on the sun that contains intense magnetic fields.

supergiant (193) Exceptionally luminous star 10 to 1000 times the sun's diameter.

supergranule (164) A large granule on the sun's surface including many smaller granules.

synodic period (41) The period of rotation or revolution of a celestial body with respect to the sun.

T dwarf (151) A very-low-mass star at the bottom end of the main sequence with a cool surface and a low luminosity.

temperature (138) A measure of the velocity of random motions among the atoms or molecules in a material.

terminator (498) The dividing line between daylight and darkness on a planet or moon.

terrestrial planet (454) Earthlike planet—small, dense, rocky.

theory (77) A system of assumptions and principles applicable to a wide range of phenomena that have been repeatedly verified.

thermal energy (139) The energy stored in an object as agitation among its atoms and molecules.

tidal coupling (498) The locking of the rotation of a body to its revolution around another body.

tidal heating (567) The heating of a planet or satellite because of friction caused by tides.

total eclipse (42, 45) A solar eclipse in which the moon completely covers the bright surface of the sun; a lunar eclipse in which the moon completely enters Earth's dark shadow.

transition (147) The movement of an electron from one atomic orbit to another.

transition region (164) The layer in the solar atmosphere between the chromosphere and the corona.

Trojan asteroid (632) Small, rocky body caught in Jupiter's orbit at the Lagrangian points, 60° ahead of and behind the planet.

ultraviolet radiation (112) Electromagnetic radiation with wavelengths shorter than visible light but longer than X rays.

umbra (42) The region of a shadow that is totally shaded.

uncompressed density (460) The density a planet would have if its gravity did not compress it.

uniform circular motion (60) The classical belief that the perfect heavens could move only by the combination of constant motion along circular orbits.

valley networks (540) Dry drainage channels resembling streambeds found on Mars.

Van Allen belt (485) One of the radiation belts of high-energy particles trapped in Earth's magnetosphere.

velocity (89) A rate of travel that specifies both speed and direction.

vernal equinox (26) The place on the celestial sphere where the sun crosses the celestial equator moving northward; also, the time of year when the sun crosses this point, about March 21, and spring begins in the northern hemisphere.

vesicular basalt (505) A porous rock formed by solidified lava with trapped bubbles.

violent motion (87) In Aristotelian physics, motion other than natural motion. (See *natural motion.*)

water hole (668) The interval of the radio spectrum between the 21-cm hydrogen radiation and the 18-cm OH radiation; likely wavelengths to use in the search for extraterrestrial life.

wavelength (110) The distance between successive peaks or troughs of a wave; usually represented by λ.

wavelength of maximum intensity (λ_{max}) (139) The wavelength at which a perfect radiator emits the maximum amount of energy; depends only on the object's temperature.

weak force (169) One of the four forces of nature; the weak force is responsible for some forms of radioactive decay.

west point (20) The point on the western horizon exactly halfway between the north point and the south point; exactly west.

white dwarf (192) The remains of a dying star that has collapsed to the size of Earth and is slowly cooling off; at the lower left of the H–R diagram.

Widmanstätten pattern (622) Bands in iron meteorites due to large crystals of nickel–iron alloys.

winter solstice (26) The point on the celestial sphere where the sun is farthest south; also, the time of year when the sun passes this point, about December 22, and winter begins in the northern hemisphere.

Zeeman effect (175) The splitting of spectral lines into multiple components when the atoms are in a magnetic field.

zenith (20) The point on the sky directly overhead.

zodiac (28) The band around the sky centered on the ecliptic within which the planets move.

zone (561) One of the yellow-white regions that circle Jupiter parallel to its equator; believed to be areas of rising gas.

Answers to
Even-Numbered Problems

Chapter 1:
2. 3475 km; **4.** 1.05×10^8 km; **6.** about 1.2 seconds; **8.** 75,000 years;
10. about 27

Chapter 2:
2. 4; **4.** 2800; **6.** A is brighter than B by a factor of 170; **8.** 66.5°;
113.5°

Chapter 3:
2. a) full; b) first quarter; c) waxing gibbous; d) waxing crescent;
4. 29.5 days later on about March 30; 27.3 days later on about March 24;
6. 6850 arc seconds or about 1.9°; **8.** a) The moon won't be full until
Oct. 17; b) The moon will no longer be near the node of its orbit;
10. August 12, 2026 [July 10, 1972 + 3 × $(6585\frac{1}{3}$ days)]. In order
to get Aug. 12 instead of Aug. 11, you must take into account the number
of leap days in the interval.

Chapter 4:
2. Retrograde motion: Jupiter, Saturn, Uranus, Neptune, and Pluto; Never
seen as crescents: Jupiter, Saturn, Uranus, Neptune, and Pluto; **4.** Mars,
about 18 seconds of arc; the maximum angular diameter of Jupiter is 50 seconds of arc; **6.** $\sqrt{27} = 5.2$ years

Chapter 5:
2. The force of gravity on the moon is about $\frac{1}{6}$ the force of gravity on
Earth; **4.** 7350 m/s; **6.** 5070 s (1 hr and 25 min); **8.** The cannonball
would move in an elliptical orbit with Earth's center at one focus of the
ellipse; **10.** 6320 s (1 hr and 45 min)

Chapter 6:
2. 3m; **4.** Either Keck telescope has a light-gathering power that is
1.56 million times greater than the human eye; **6.** No, his resolving power
should have been about 5.8 seconds of arc at best; **8.** 0.013 m (1.3 cm
or about 0.5 inches); **10.** about 50 cm (From 400 km above, a human is
about 0.25 seconds of arc from shoulder to shoulder.)

Chapter 7:
2. 150 nm; **4.** by a factor of 16; **6.** 250 nm; **8.** a) B; b) F; c) M; d) K;
10. about 0.58 nm

Chapter 8:
2. 730 km; **4.** 9×10^{16} J; **6.** 0.222 kg; **8.** about 3.5 times;
10. 400,000 years

Chapter 19:
2. It will look $206,265^2 = 4.3 \times 10^{10}$ times fainter, which is about
26.6 magnitudes fainter; 22.6 mag; **4.** about 3.3 times the half-life,
or 4.3 billion years; **6.** large amounts of methane and water ices;
8. about 1300

Chapter 20:
2. about 17 percent; **4.** 81×10^6 yr; they have been subducted;
6. 0.22 percent

Chapter 21:
2. The rate at which an object radiates energy is proportional to its surface
area, which is proportional to its radius squared (r^2). However, the energy
an object has stored as heat is proportional to its mass and hence to its volume, and that is proportional to its radius cubed (r^3). So the cooling time
will be proportional to the amount of heat divided by the rate of cooling,
which is the same as the radius cubed divided by the radius squared (r^3/r^2).
That shows that the cooling time is proportional to the radius (r); and
that means that the bigger an object is, the longer it takes to cool; **4.** No.
Their angular diameter would be only 0.5 second of arc. They would be
visible in photos taken from orbit around the moon; **6.** 0.5 second of arc
(assuming an astronaut seen from above is 0.5 meter in diameter); no;
8. 10.0016 cm; **10.** Mercury, $V_e = 4250$ m/s; moon, $V_e = 2380$ m/s;
Earth, $V_e = 11,200$ m/s

Chapter 22:
2. 33,400 km (39,500 km from the center of Venus); **4.** 61 seconds of
arc; **6.** 260 km; **8.** 82 seconds of arc

Chapter 23:
2. 35 Earth days; **4.** 4.4°; **6.** about 0.056 nm; **8.** 5.2 m/s

Chapter 24:
2. 42 seconds of arc; **4.** 256 m/s; **6.** 8.8 s; yes; **8.** 12.3 km/s;
10. 1.04×10^{26} kg (17.2 Earth masses)

Chapter 25:
2. one billion; **4.** 79 km; **6.** 0.18 km/s; **8.** 3.28 AU and 2.5 AU;
10. 9 million km (0.06 AU); too small; **12.** 33 Earth masses

Chapter 26:
2. 8.9 cm; 0.67 mm; **4.** about 1.3 solar masses; **6.** 380 km;
8. pessimistic, 2×10^{-5}; optimistic, 10^7

Credits

This page constitutes an extension of the copyright page. We have made every effort to trace the ownership of all copyrighted material and to secure permission from copyright holders. In the event of any question arising as to the use of any material, we will be pleased to make the necessary corrections in future printings. Thanks are due to the following authors, publishers, and agents for permission to use the material indicated.

Chapter 1. 2–3: T. Rector, University of Alaska, and WIYN/NOAO/AURA/NSF **4, left:** M. Seeds **4, right:** USGS **5, left:** NASA infrared photograph **5, right:** NASA **6: NOAO 6:** NASA **8:** NOAO **9, bottom:** Anglo-Australian Telescope Board **9, top:** Detail of galaxy map from M. Seldner, B. L. Siebers, E. J. Groth, and P. J. E. Peebles, *Astronomical Journal 82* (1977) **10:** Bill Schoening/NOAO/AURA/NSF

Chapter 2. 12–13: Courtesy of Kris Koenig **33:** USPS **15, top:** From Duncan Bradford, *Wonders of the Heavens,* Boston: John B. Russell, 1837 **16:** William Hartman **18:** Roger Ressmeyer/Corbis **20:** AURA/NOAO/NSF **23:** Royalty-Free/Corbis **27:** NASA

Chapter 3. 36–37: NSO/AURA/NSF **38, top:** Yerkes Observatory **38, bottom:** T. Scott Smith **39:** UCO/Lick Observatory **43:** © 1982 by Dr. Jack B. Marling **44:** Celestron International **47:** Daniel Good **48, left:** Daniel Good **48, right:** NOAO **49:** Daniel Good **49, left:** UCO/Lick Observatory **54, top:** Laurence Marschall

Chapter 4. 59, top: Jamie Backman **59, bottom:** Benelux Press/Index Stock Imagery **60:** Courtesy NPS Chaco Culture National Historical Park **64:** Yerkes Observatory **64:** From *Cosmographica* by Peter Apian (1539) **68:** NOAO and Nigel Sharp **69:** Grundy Observatory **71:** Yerkes Observatory **74:** The Granger Collection, New York **77:** From the collection of John Coolidge III

Chapter 5. 86: NASA/JSC **88:** NASA **91:** ESA/STScI and NASA **95:** NASA **99:** Nigel Sharp/NOAO/AURA/NSF **106:** Larry Mulvehill/The Image Works **107:** NASA/JSC

Chapter 6. 108: Robert Frost, "The Star-Splitter," *New Hampshire: A Poem with Notes and Grace Notes* (New York: Henry Holt and Co., 1923), pp. 27–30. **108–109:** NOAO/AURA/NSF **110:** NOAO/AURA/NSF **116:** NASA **117, top:** Courtesy of William Keel **117, center:** NOAA **117, bottom:** NOAO/AURA/NSF **119, left:** NOAO/AURA/NSF **119, right:** ESO **120:** AURA/NOAO/NSF **121, left:** W. M. Keck Observatory **121, right:** Paul Kalas **123, left and bottom:** C. Hawk, B. Savage, N. A. Sharp, NOAO/WIYN/NSF **123, right:** NOAO/WIYN/NSF **124:** NASA/CXC/Rutgers/J. Hughes **126, bottom:** NRAO **126, top:** Courtesy Seth Shostak/SETI Institute **127:** NRAO/AUI/NSF **127, bottom:** NRAO/AUI **127, center:** David Parker/SPL/Photo Researchers, Inc. **129, left:** William Keel/IRAF **129, center:** NASA **129, bottom:** Kris Koenig/Coast Learning Systems **130, 131:** NASA **134, left:** ESO **134, right:** NASA/CXC/PSU/S. Park

Chapter 7. 136, 137: ESO **138:** NASA Hubble Heritage Team/STScI/AURA **147:** AURA/NOAO/NSF **150:** AURA/NOAO/NSF **152:** Adapted from Thomas R. Geballe, Gemini Observatory, from a graph that originally appeared in *Sky and Telescope Magazine,* February 2005, p. 37. **158:** T. Rector, University of Alaska and WIYN/NURO/AURA/NSF

Chapter 8. 160–161: NASA/SOHO **163:** Daniel Good **164:** P. N. Brant, G. Scharmer, G. W. Simpson, Swedish Vacuum Solar Telescope, La Palma **165, left:** NOAA/SEL/USAF **165, right:** NOAO/NSO **166:** SOHO/ESA/NASA **167:** Stanford-Lockheed Institute for Space Research, Palo Alto, CA and NASA GSFC **172, left:** Brookhaven National Laboratory **172, right:** Courtesy SNO **174, background:** Royal Swedish Academy of Science **174, right, and 175, top left:** NASA **175, top right:** J. Harvey/NSO and HAO/NCAR **175, center:** M. Seeds **175, bottom left:** SOHO/EIT, ESA and NASA **175, bottom right:** SOHO/EIT, ESA and NASA **179:** K. Strassmeier, Vienna, AURA/NOAO /NSF **180, left:** TRACE/NASA **180, right:** Sacramento Peak Observatory **180, bottom:** SOHO, EIT, ESA and NASA **181, top left:** NASA **181, center left:** SOHO/MDI, ESA and NASA **181, background:** © Jan Curtis **181, bottom left:** SOHO/LASCO, ESA, and NASA **181, bottom right:** Yohkoh/ISAS/NASA **184, top :** NOAO and Daniel Good **184, bottom:** NASA/SOHO

Perspectives. 186–187: T. A. Rector and B. A. Wolpa, NRAO/NOAO/AUI/NSF **188, left:** ESO **188, right:** NOAO and Nigel Sharp **189, left:** NASA, ESA, The Heritage Hubble Team **189, right:** ESO **190, center:** NASA **190, bottom:** NASA/CXC/SAO **191, top right:** Anglo-Australian Telescope Board **191, center right:** Dan Gezari, Dana Backman, and Mike Werner **191, left:** ESO **191, bottom :** NASA **192:** M. Seeds **193:** NASA/ESA/HEIC/Hubble Heritage Team **194:** Hubble Heritage Team, STScI/AURA/NASA **195:** NASA **196:** NASA, ESA, J. Hester and A. Loll, Arizona State University **197:** AURA/NOAO/NSF **198:** NASA/Heritage Hubble Team; STScI/AURA **199:** R. Williams and the Hubble Deep Field Team, STScI, NASA **200, top right:** AURA/NOAO/NSF **200, center and bottom:** Anglo-Australian Telescope Board **201, top right:** NASA/Hubble Heritage Team **201, top left:** NASA/Hubble Heritage Team **201, center left:** Rudolph E. Schild **201, center right:** G. J. Jacoby, M. J. Pierce/ AURA/NOAO/NSF **201, bottom left:** AURA/NOAO/NSF **201, bottom right:** © R. J. Dufour, Rice University **202, top:** Brad Whitmore, STScI, and NASA **202, bottom:** Caltech **203:** NASA/WMAP Science Team **204:** SOHO/MDI **205:** K. Lanzetta, SUNY, A. Schaller for STScI, and NASA

Chapter 19. 448–449: NASA **451:** Janet Seeds **454:** John Hopkins University, Applied Physics Laboratory, NASA **455:** Celestron International **456, 457:** UCO/Lick Observatory **457, bottom right:** Grundy Observatory **458, left:** Daniel Good **458, right:** R. Kempton, New England Meteorical Services **462:** NASA **464:** JPL/NASA **467:** NASA **468, center:** C. R. O'Dell, Rice, NASA **468, top right:** M. McCaughran, Max Plank Inst. für Astronomie, C. R. O'Dell, Rice, NASA **468, bottom left:** J. Bally, H. Throop, C. R. O'Dell, NASA **469:** D. Padgett, IPAC/Caltech, W. Brandner IPAC, K. Stapelfeldt, JPL/NASA **470, top:** C. Burrows and J. Krist, STScI and NASA **470, center left:** A. Weinberger, E. Becklin, UCLA, G. Schneider, U. of Arizona, and NASA **470 center right:** Joint Astronomy Center **470, bottom left:** NASA **470, bottom right:** NASA, ESA, Kalas, Graham, and Clampin **473:** ESO **475:** NASA

Chapter 20. 476–477: Jean-Bernard Carillet/Getty Images **478:** NASA **478, top right:** UCO/Lick Observatory **484:** From *Virus Ultrastructure: Electron Micrograph Images.* Copyright 1995 by Linda M. Stannard. Reprinted with permission from Linda M. Stannard. Found at: http://web.uct.ac .za/depts/mmi/stannard/adeno.html **485:** Jimmy Westlake **486, left:** William K. Hartmann **486 top center:** Janet Seeds **486, top right and top left;** National Geophysical Data Center **487, bottom:** M. Seeds **491:** Ozone Processing Team/NASA/Goddard Space Flight Center **493, top:** William K. Hartmann **493, bottom:** USGS

Chapter 21. 496–497: JSC/NASA **498:** UCO/Lick Observatory **499:** NASA **500, 501:** NASA **502, left:** NASA **502 center and right:** UCO/Lick Observatory **503:** Phyllis Leber **504:** NASA **505, left:** UCO/Lick Observatory **505, right and bottom, and 506:** NASA **507 :** M. Seeds **508:** NASA/Clementine **509, right:** NASA **509, left:** Don Davis **512, 513, 514, 515:** NASA **516:** Martin Slade/JPL/NASA **518, 520, 521:** NASA

Chapter 22. 522: © Calvin J. Hamilton **523 and 525, top and bottom:** NASA **526, top:** M. Seeds **526, top:** USGS **527:** NASA **528, bottom left:** USGS **529:** NASA **530:** PhotoDisc/Getty Images **531, left:** NASA **531, right:** © 1992, David P. Anderson, Southern Methodist University **532, left:** Courtesy of Lowell Observatory **534, right:** USGS **535:** NASA **535:** Philip James, University of Toledo, AURA/NOAO/NSF **201, bottom left:** AURA/NOAO/NSF **201, bottom right:** Steven Lee, University of CO, Boulder, and NASA **536, 537:** NASA **538:** NASA/JPL/Cornell **539:** NASA **540:** NASA/JPL/Malin Space Science Systems **541, right:** NASA/USGS **541, left:** © Calvin J. Hamilton **542:** Malin Space Science Systems/NASA **543:** NASA/JPL/Arizona State University **544:** NASA/JPL/Cornell/USGS **545, top:** NASA **545, bottom left:** Malin Space Science Systems/NASA **545, bottom right:** NASA **546:** NASA/JPL/Malin Space Science Systems **547, left:** Damon Simonelli and Joseph Ververka, Cornell University/NASA **547, right, and 550:** NASA

Chapter 23. 552–553: NASA/JPL **555:** NASA/JPL/Space Science Institute/University of Arizona **557:** NASA/JPL/University of Arizona **559, top:** John P. Clarke, University of Michigan, NASA **559, bottom and 560, top left:** NASA **560, top right:** NASA/JPL/University of Arizona **560, left:** Hughes Aircraft Company **560, bottom:** NASA/JPL **561, bottom:** NASA/JPL/Caltech **562:** NASA/JPL/Space Science Institute **563, bottom, and 564, left:** NASA **564, center:** Mike Skrutskie **564, right:** University of Hawaii **565-569:** NASA **570, top:** NASA/JPL/Bruce A. Goldberg **570, center:** NASA/JPL/John Hopkins University Applied Physics Lab **570, bottom:** NASA/JPL **572:** NASA, ESA, J. Clarke, Boston University and X. Levay, STScI **573:** NASA/STScI **574:** NASA and E. Karkoschka **575, 576, 577:** NASA **577:** NASA/JPL/Science Institute **579, 580** NASA/ESA/JPL/University of Arizona **581, 582:** NASA/JPL/Space Science Institute **585:** NASA/JPL

Chapter 24. : 588–589: NASA/JHUAPL/SwRI **590, 592, 593:** NASA **595:** Lawrence Sromovsky, UW-Madison Space Science and Engineering Center **596:** Courtesy Floyd Herbert, LPL **597:** ESO **599, 601:** NASA **601, center and bottom right; 602, bottom left:** USGS **602, bottom right:** P. Bridges **604:** NASA, L. Sromovsky, and P. Fry, UW-Madison **605:** NASA **605, left:** NASA and Erich Karkoschka **605, right:** U.C. Berkeley/W. M. Keck Observatory **606:** NASA **608:** Courtesy of Lowell Observatory **609:** NASA **610, top:** NASA/McGill/V. Kaspi et al. **610, bottom left:** NASA **610, bottom right:** R. Albrecht, ESA/ESO Space Telescope European Coordinating Facility, NASA **614:** NASA and Heidi Hammel

Chapter 25. 616–617: NASA/JPL **618:** Courtesy Tod Lauer **620:** M. Seeds **621:** NASA/JPL-Caltech/M. Kelley, University of Minnesota **622:** USGS **622, 623:** M. Seeds **623:** Courtesy Dr. Monica Kress **624:** Courtesy of Russell Kemton, New England Meteoritical **625, top:** Hubble Heritage Team/STScI/AURA/NASA **625, bottom; 628, 629:** NASA **632:** Reprinted with permission of MIT Lincoln Laboratory, Lexington, Massachusetts **633:** Courtesy of Gareth

Index

fission, nuclear, 169
fission hypothesis, **510**
FitzGerald, Edward, 136
flares, **181**
Fleming, Alexander, 610
flooding stage, 481
floppy mirrors, 121
focal length, **114**
foci, 76
folded mountain ranges, **486**
*Forerunner of Dissertations on the Universe,
    Containing the Mystery of the Universe,
    The* (Kepler), 75
forward scattering, **562**
fossils, 656, 657, 660, 665
Franklin, Benjamin, 590
Fraunhofer, Joseph von, 136
frequency, **111**
Frost, Robert, 108
full moon, 40, 41
fusion. *See* nuclear fusion

G

galaxies, **8**
    clusters of, 8–9
    intelligent life in, 669
    See also Milky Way Galaxy
Galilean satellites, **455**, 565–572
    Callisto, 566
    comparative data on, 571
    Europa, 567–569
    Ganymede, 566–567
    history of, 571–572
    Io, 555–556, 558, 559, 569–571, 572
Galileo Galilei, 56, 68–72, 87–89, 572,
    573, 603
Galileo spacecraft, 554, 556, 557, 560
Galle, Johann, 603
gamma rays, 112
    solar energy and, 170–171
gamma-ray telescopes, 130
Ganymede, 566–567
Gaposchkin, Sergei, 152
gas giant planets, 594
gas (type I) tails, **636**
Gaspra asteroid, 628
Gemini telescopes, 110, 119
general theory of relativity, **102–105**
genes, **655**
geocentric universe, **61, 64**

geologic time, 661
geosynchronous satellites, **94**
gibbous moon, 40, 41
Global Oscillation Network Group
    (GONG), 168
global warming, **490–491**
gossamer rings, **563**
Grand Maximum, 182
granulation, **163**
grating, **124**
gravitational collapse, **460**
gravity
    acceleration of, **87**
    Einstein's theory of, 102–105
    Galileo's study of, 87–89
    general relativity theory and, 102–105
    mutual gravitation and, 90–92
    Newton's law of, 86, 90–92, 94, 97,
        99–100
    orbital motion and, 92–100
    tidal forces and, 97–99
Great Dark Spot, 604
Great Nebula of Orion. *See* Orion Nebula
Great Red Spot, 560
Great Square of Pegasus, 14
Greek alphabet, 15, 685
Greek astronomy, 60–63
greenhouse effect, **490**
    on Earth, 490–491
    on Venus, 524, 531–532
grooved terrain, **566**
ground state, **145**

H

half-life, **457–458**
*Harmonice Mundi (The Harmony of the
    World)* (Kepler), 76
Hawaiian Islands, 529, 539
heat
    infrared radiation and, 565
    solar flow of, 162
    temperature and, 138, 139
    thermal energy and, **139**
    tidal, **567**, 568, 569, 571, 572
heat of formation, **463**, 465
heavy bombardment, **466**, 502
heliocentric universe, **63**, 66–67
helioseismology, **167–168**
helium, spectral analysis of, 156
Herschel, Caroline, 591

Herschel, William, 590–591, 603
Hipparchus, 16, 62, 63
Hipparcos satellite, 639
Hirayama, Kiyotsugu, 627
Hirayama families, **627**
history of astronomy, 56–83
Hodges, E. Hulitt, 618
Homo sapiens, 660
Hoover, Herbert, 588
horizon, **20**
horoscope, **28**
Hubble, Edwin P., 130
Hubble Space Telescope (HST), 130, 132,
    468, 469
Huygens, Christian, 573
Huygens probe, 579, 580, 663
hydrogen
    liquid metallic, **557**
    nuclear fusion of, 169–170
    spectral analysis of, 148
hydrogen atom, 142, 144
hypothesis, 77, 503

I

Iapetus, 582
ice ages, 30, 175, 182
ice giant planets, 594
ICE spacecraft, 636, 643
Ida asteroid, 629
imagination, scientific, 484
imaging systems, 123
impact cratering, 500–501
    See also craters
infrared astronomy, 128–129, 132
Infrared Astronomy Satellite (IRAS), 630
infrared radiation, **111**, 128–129, 132
    heat detected from, 565
    human emission of, 141
Infrared Telescope Facility (IRAF), 129
infrared telescopes, 128–129, 131, 132
Inquisition, 71, 72
intelligent life, 676
    development of, 660
    estimates of, 669
intensity, 17
intercrater plains, **515**
interference, 128
interferometry, **122**
International Astronomical Union, 14,
    527

International Ultraviolet Explorer (IUE), 129, 132
inverse square law, **90**
Io, 555–556, 558, 559, 569–571, 572
Io flux tube, **558**
Io plasma torus, **558**
ionization, **142**
ions, **142**
iridium, 642
iron meteorites, 621, **622**, 624
Ishtar Terra, 530
isotopes, **142**

J

James Webb Space Telescope, 131, 132
John Paul II, Pope, 71, 72
joule (J), **96**, 140, 677
Jovian planets, **454**–455, 464–465, 554–555
   See also specific planets
jumbled terrain, **509**
Jupiter, 555–572
   atmosphere of, 455, 554, 558–559, 560–561
   comet impact on, 563–564, 638, 639
   composition of, 556–557
   history of, 564–565
   interior temperature of, 565
   magnetic field of, 557–558, 559, 596
   moons of, 69–70, 72, 555–556, 565–572, 662–663, 683
   ring of, 559, 562–563
   rotation of, 556
   size and mass of, 555–556
   spacecraft exploration of, 552, 554, 556, 557, 559
   statistical data about, 557

K

K stars, 667
Keck Foundation, 575
Keck telescopes, 121, 122
Kelvin temperature scale, **138**, 678
Kennedy, John, 502
Kepler, Johannes, 74–79, 86, 93, 96–97, 102
kiloparsecs, 679
kinetic energy, **96**
Kirchhoff, Gustav, 146

Kirchhoff's laws, 145, **146**
Kirkwood, Daniel, 573, 630
Kirkwood's gaps, **630**, 631
Kitt Peak National Observatory, 119, 120, 618
Kuhn, Thomas, 68
Kuiper, Gerard P., 456, 641
Kuiper belt, **456**, 612, 641–642

L

L dwarfs, **150**–151, 152
Lagrangian points, 631–632
Lakshmi Planum, 530
Laplace, Pierre-Simon de, 451
Large Binocular Telescope (LBT), 119, 122
large-impact hypothesis, **511**
lava flows. *See* volcanic activity
laws
   of gravity, 86, 90–92
   of motion, 76–78, 89–90, 91
   See also names of specific laws
lenses, telescope, 114, 124
Leonid meteor shower, 619, 621
Leverrier, Urbain Jean, 603
Levy, David H., 563
life
   DNA as code of, 653, 654–655, 656
   evolution of, 653, 656, 660–661, 662
   extraterrestrial communication with, 665–670
   intelligent, 660, 669–670, 676
   number of worlds with, 669–670
   origin of, 656–661
   physical basis of, 652, 653
   in the solar system, 661–664
   in stellar planetary systems, 664–665
life zone, **665**
light
   characteristics of, 111
   forward scattering of, **562**
   interaction of matter and, 144–145
   space-time curvature and, 103–104, 105
   spectrographic analysis of, 123–124
   speed or velocity of, 101–102
   starlight, 138–141
   visible spectrum of, 111, 112
light buckets, 122
light-gathering power, **115**, 128
light pollution, **117**
light-year (ly), 7, 679

limb, **164**, 498
limb darkening, **164**
linear diameter, 45
LINEAR telescope, 631, 632
line of nodes, **50**–51
liquid metallic hydrogen, **557**
Little Ice Age, 175, 182
lobate scarps, **514**, 515, 516, 517
LONEOS (Lowell Observatory Near Earth Object Search), 631
long-period comets, 639, 640
Lowell, Percival, 534, 608, 609
lunar eclipses, **42**–44
   dates of forthcoming, 44
   partial, **44**
   penumbral, **44**
   predicting, 49–53
   total, **42**–43, 44, 53
lunatic, 36
Luther, Martin, 67, 80
Lyman series, **147**

M

M stars, 665, 668
Magellan spacecraft, 525
magnetic carpet, **166**, 167
magnetic fields
   of the Earth, 484–485, 486
   of Ganymede, 566
   of Jupiter, 557–558, 559, 596
   of Mercury, 516
   of Neptune, 596, 605
   of Saturn, 572–573, 596
   of stars, 177–178
   of the sun, 175, 176–177, 178–181
   of Uranus, 594–596
magnetosphere, **484**
magnifying power, **116**
magnitude
   apparent visual magnitude, **16**
   classifying stars by, 16–17
   intensity and, 17
magnitude scale, **16**, 18
main-sequence stars
   life zone around, 665
   properties of, 680
mantle, **483**
mare (maria), **498**, 502, 507, 508–509
Mare Imbrium, 502, 505, 509
Mare Tranquillitatis, 505

National Science Foundation, 575
natural law, 77
natural motion, **87**
natural selection, **653**
nature, 676
neap tides, **98**
Near Earth Asteroid Tracking (NEAT)
    system, 634
near-Earth objects (NEOs), 631
NEAR spacecraft, 456, 628
nebular hypothesis, **451**
Neptune, 603–608
    atmosphere of, 554, 604–605,
        607–608
    discovery of, 603–604
    history of, 607
    interior of, 605
    magnetic field of, 596, 605
    moons of, 606–607, 684
    rings of, 599, 605–606
    rotation of, 604–605
    spacecraft exploration of, 590, 606
    statistical data about, 605
Nereid, 606
neutrinos, **170**, 171–172
neutrons, **141**
Newgrange, 58, 59
New Horizons spacecraft, 641
new moon, 40, 41
newton (N), 677
Newton, Isaac, 76, 84, 86–100, 123, 677
Newtonian focus, **120**
nodes, **50**, 51
north celestial pole, **20**
northern hemisphere star charts,
        688–693
north point, **20**
nuclear fission, **169**
nuclear fusion, **169**
    hydrogen, 169–170
    solar, 168–172
    temperature and, 172
nucleus
    of atoms, **141**
    of comets, 457, 635, 636, 638–639

O

Oberon, 600, 601
objective lens, **114**
objective mirror, **114**
oblateness, **556**

observatories
    location of, 117–118, 122
    See also names of specific observatories
observed densities, 461
occultations, **597**
ocean tides, 97–99
Olympus Mons volcano, 529, 539, 541
Oort, Jan, 639
Oort cloud, **639**–641
open orbit, **95**
optical telescopes, 113–122, 128
orbital motion, 92–100
orbital velocity, 92–93
orbits
    asteroid, 630, 631–632
    closed, **95**
    comet, 456–457, 639
    Earth, 6, 24, 25, 30, 32
    Kepler's laws of, 76–78
    lunar, 38–39
    Newton's laws of, 92, 97, 99–100
    open, **95**
    permitted, **143**
    planetary, 25, 28, 72–79, 453, 454, 682
    space-time curvature and, 102–103, 104
organic molecules, 578, 657, 659
Orion constellation, 14, 16, 22, 687
Orionids meteor shower, 620
Orion Nebula, disks surrounding stars in,
        467–468
outflow channels, **540**, 542
outgassing, **463**
ovoids, **600**
oxygen, 489, 492
ozone layer, **489**, 491–492

P

P (pressure) waves, **482**–483
paradigms, **68**
parallax, **64**, 73
    stellar, 591
parsec (pc), **679**
partial lunar eclipse, **44**
partial solar eclipse, **45**, 46
Paschen series, **147**
passing star hypothesis, **450**–451
path of totality, **46**
Payne-Gaposchkin, Cecilia, 151–152
Pegasus, 14, 15
penumbra, **42**
    lunar eclipse, 42

solar eclipse, 46
sunspot, 174
penumbral eclipse, **44**
perigee, **48**
perihelion, **27**
periodic table of the elements, 686
permitted orbits, **143**
Perseids meteor shower, 619
phases of the moon, 39, 40–41
Philolaus, 60
Phobos, 547–548
Phoebe, 579–580, 581
photons, **111**
    atomic absorption of, 144, 145
    big bang story and, 425–427, 428
    production of, 139
    wavelength of, 139–140
photosphere, **47**, 138, 162–164
photosynthesis, 489
Piazzi, Giuseppe, 630
Pioneer spacecraft, 524, 525, 552, 557
planetary motion, 64, 72–79
    Kepler's laws of, 76–78
Planetary Society, 668
planetesimals, **461**–462, 465
    asteroids and, 633, 634
    comets and, 642
    meteorites and, 625–627
planets, **6**
    atmospheres of, 455
    cataclysmic impacts and, 612
    comet impacts with, 563–564, 638, 639,
        642
    comparative study of, 476, 478–480,
        481
    craters on, 454, 466, 467
    densities of, 455, 480
    developmental stages of, 480–481
    extrasolar, 469–472
    formation of, 452–453, 459–467
    Jovian, 454–455, 464–465, 554–555
    life on, 661–664
    orbits of, 25, 28, 72–79, 453, 454, 682
    properties of, 682
    rings surrounding, 465–466
    rotations of, 453, 465
    terrestrial, 454–455, 465, 479–480
    types of, 453–456, 465, 466–467
    See also specific planets
plastic, **483**
plate tectonics, **486**–487, 533
Plato, 60, 64